Lecture Notes in Electrical Engineering

Volume 925

The book series *Lecture Notes in Electrical Engineering* (LNEE) publishes the latest developments in Electrical Engineering—quickly, informally and in high quality. While original research reported in proceedings and monographs has traditionally formed the core of LNEE, we also encourage authors to submit books devoted to supporting student education and professional training in the various fields and applications areas of electrical engineering. The series cover classical and emerging topics concerning:

- Communication Engineering, Information Theory and Networks
- Electronics Engineering and Microelectronics
- Signal, Image and Speech Processing
- Wireless and Mobile Communication
- Circuits and Systems
- Energy Systems, Power Electronics and Electrical Machines
- Electro-optical Engineering
- Instrumentation Engineering
- Avionics Engineering
- Control Systems
- Internet-of-Things and Cybersecurity
- Biomedical Devices, MEMS and NEMS

For general information about this book series, comments or suggestions, please contact leontina.dicecco@springer.com.

To submit a proposal or request further information, please contact the Publishing Editor in your country:

China

Jasmine Dou, Editor (jasmine.dou@springer.com)

India, Japan, Rest of Asia

Swati Meherishi, Editorial Director (Swati.Meherishi@springer.com)

Southeast Asia, Australia, New Zealand

Ramesh Nath Premnath, Editor (ramesh.premnath@springernature.com)

USA, Canada

Michael Luby, Senior Editor (michael.luby@springer.com)

All other Countries

Leontina Di Cecco, Senior Editor (leontina.dicecco@springer.com)

** This series is indexed by EI Compendex and Scopus databases. **

Bhuvan Unhelker · Hari Mohan Pandey ·
Gaurav Raj
Editors

Applications of Artificial Intelligence and Machine Learning

Select Proceedings of ICAAAIML 2021

 Springer

Editors
Bhuvan Unhelker
University of South Florida
Sarasota–Manatee
Sarasota, FL, USA

Hari Mohan Pandey
Bournemouth University
Poole, UK

Gaurav Raj
School of Engineering and Technology
Sharda University
Greater Noida, India

ISSN 1876-1100 ISSN 1876-1119 (electronic)
Lecture Notes in Electrical Engineering
ISBN 978-981-19-4833-6 ISBN 978-981-19-4831-2 (eBook)
https://doi.org/10.1007/978-981-19-4831-2

This Springer imprint is published by the registered company Springer Nature Singapore Pte Ltd.
The registered company address is: 152 Beach Road, #21-01/04 Gateway East, Singapore 189721,
Singapore

Contents

About the Editors

Bhuvan Unhelker is an accomplished IT professional and Professor of IT at the University of South Florida, USA. He is also Founding Consultant at Method Science and Co-Founder and Director at PlatiFi. His research areas are business analysis & requirements modeling, software engineering, Big Data strategies, agile processes, mobile business, and green IT. His domain experience is banking, financial, insurance, government, and telecommunications. He has designed, developed, and customized a suite of industrial courses, which are regularly delivered to business executives and IT professionals globally, including those in Australia, the USA, the UK, China, India, Sri Lanka, New Zealand, Singapore, and Malaysia. His thought leadership is reflected through multiple Cutter Executive Reports, various journals, and 21 books. He is a winner of the Computerworld Object Developer Award (1995), Consensus IT Professional Award (2006), and IT Writer Award (2010). He has a Doctorate in the area of "Object Orientation" from the University of Technology, Sydney, in 1997. He is Fellow of the Australian Computer Society, Senior Member of IEEE, Professional Scrum Master, and Lifetime Member of the Computer Society of India.

Hari Mohan Pandey is Associate Professor in the Department of Computing and Informatics at Bournemouth University, the UK. He is specialized in computer science and engineering. His research area includes artificial intelligence, soft computing techniques, natural language processing, language acquisition, and machine learning algorithms. Hari is Author of various books in computer science engineering. He has published over 100 scientific papers in reputed journals and conferences. He is serving on the editorial board of reputed journals as Action Editor, Associate Editor, and Guest Editor. He is Reviewer of top international conferences such as GECCO, CEC, IJCNN, BMVC, and AAAI. He has delivered expert talks as Keynote and Invited Speaker. He has also given lectures in the international summer/winter schools. He has been given the prestigious award "The Global Award for the Best Computer Science Faculty of the Year 2015," an award for INDO-US project "GENTLE," and a Certificate of Exceptionalism from the Prime Minister of India. Previously, he worked as Sr. Lecturer in the Computer Science Department at Edge Hill University, the UK, and Research Fellow in Machine Learning at the School of Technology at Middlesex University London, London.

Gaurav Raj is currently working at Sharda University, Greater Noida, India, and has done B.Tech. in CSE from UPTU, M. Tech. in SE from NIT, Allahabad, and Ph.D. in CSE from Punjab Technical University, India. His areas of research are artificial intelligence, cloud computing, software engineering, and web services and security. He has more than 11 years of teaching experience. He has several research papers to his credit published in reputed journals and conferences. He also has several patents published to his credit. He has vast experience in organizing faculty development programs, workshops, seminars, and conferences.

Firefly Algorithm and Deep Neural Network Approach for Intrusion Detection

Miodrag Zivkovic⬤, Nebojsa Bacanin⬤, Jelena Arandjelovic⬤, Ivana Strumberger⬤, and K. Venkatachalam⬤

1 Introduction

Machine learning (ML) refers to computer systems and programs that adapt and improve their performance by learning from a given task. ML methods enable computers to learn from data inputs and then use statistical analysis to generate outputs within a specific range. As a result, ML facilitates decision making process because models can be created from sample data, and decision-making process can be automated using data inputs. Deep learning (DL) has recently become a key machine learning area that can be successfully applied for many real-life challenges. Deep learning may utilize labeled datasets to educate the system (supervised learning), although it does not always require a datasets with labels to train itself (unsupervised learning).

The DL attempts to replicate the way the human brain combines light and sound into vision and hearing, and then extracts or transforms data features (or representations) via a layer of nonlinear processing units in a cascade. There are two types of methods depending on whether they are used for classification or to analyze patterns—supervised and unsupervised. Deep learning consists of a network of several "layers" of core processing units linked together. Deep Learning has been proved to absorb the most info and exceed people in a variety of cognitive tests and

M. Zivkovic · N. Bacanin (✉) · J. Arandjelovic · I. Strumberger
Singidunum University, Danijelova 32, 11000 Belgrade, Serbia
e-mail: nbacanin@singidunum.ac.rs

M. Zivkovic
e-mail: mzivkovic@singidunum.ac.rs

J. Arandjelovic
e-mail: jelena.arandjelovic.16@singimail.rs

I. Strumberger
e-mail: istrumberger@singidunum.ac.rs

K. Venkatachalam
Department of Computer Science and Engineering, CHRIST (Deemed to be University),
Bangalore, India
e-mail: venkatachalam.k@christuniversity.in

© The Author(s), under exclusive license to Springer Nature Singapore Pte Ltd. 2022
B. Unhelker et al. (eds.), *Applications of Artificial Intelligence and Machine Learning*,
Lecture Notes in Electrical Engineering 925,
https://doi.org/10.1007/978-981-19-4831-2_1

as a result of these traits, has emerged as a promising method in the AI field [17]. As already noted above, DL models can be applied for classification tasks in many domains, among other things in the area of computer security. In this very important area, DL may represent foundation for creating efficient intrusion detection system (IDS). The IDS is a part of hardware and software that identifies and mitigates threats and assaults, collects and analyzes data on hostile actions before reporting it to the system administrator.

The IDS use two methods to detect potential security breaches on computer systems: first one is signature-based detection which correlates data activity with a signature or pattern stored in the signature database. The second one identifies any irregularity and issues alarm and it's called behavior-based or statistical anomaly-based detection. Because it learns what normal system behavior is, it is referred to as an expert system. Moreover, the IDS systems are often divided into five categories [13]: network IDS (NIDS), host IDS (HIDS), application protocol IDS (AIPDS), protocol-founded IDS (PIDS) and hybrid IDS (HIDS). However, above mentioned traditional IDS systems suffer from some critical drawbacks. For example, in most cases detection rate is very low and in some cases, these systems trigger false alarms by generating false positive events. These issues arise from the fact that these systems do not perform well on classifying between real and false threats. At the other hand, it is known that the ML approaches do very good job in classification and that is why many most recent IDS systems are based on ML algorithms [10].

Also, the ability of ML methods to perform accurate classification to a large extension depends on the data quality. The datasets which are used by IDS systems are mostly high-dimensional with many redundant and irrelevant features along with many samples. That is why the data preprocessing phase is very important in ML approaches. The nature-inspired metaheuristics, such is swarm intelligence, can be applied to such high-dimensional data to select those features that have the most important influence to overall classification and to discard features which are irrelevant.

1.1 Research Goals and Contributions

The basic goal of proposed research is to develop better classifier for IDS by using hybrid deep neural network (DNN) and enhanced firefly algorithm (FA) approach. The DNN is used for classification and the FA swarm intelligence algorithm is utilized in data preprocessing phase for feature selection. In overall, by using this approach, the process of training DNN is shortened and the system is able to differentiate between false and real threats with a greater accuracy.

For validation purposes, well-known KDD Cup 99 and NSL-KDD public benchmark datasets from Kaggle and UCL repositories were utilized.

The contribution of the presented research can be summed in the following:

- developing framework based on enhanced FA approach and DNN with a goal to improve performance for the IDS systems by establishing better classification accuracy;
- reducing the time which is needed for training DNN used for IDS classification tasks and
- implementing an enhanced firefly algorithm (FA) metaheuristics that will specifically target the known deficiencies of the basic FA approach.

1.2 Structure of the Paper

This manuscript has been assembled in the following order. Section 2 introduces fundamental theoretical background related to DNN and swarm intelligence metaheuristics along with relevant literature sources. Section 3 exhibits the basic FA metaheuristics, and afterwards it presents the proposed enhanced method. Section 4 shows empirical results, evaluation and discussion. Lastly, Sect. 5 delivers the conclusion and proposes future studies in this area.

2 Theoretical Background and Literature Review

Feed-forward artificial neural network (ANN) formed by many hidden layers is known as the DNN [2]. At minimum, the DNN is formed by exactly one input, one output, and three hidden layers, as shown in Fig. 1. Besides DNN, in the ML

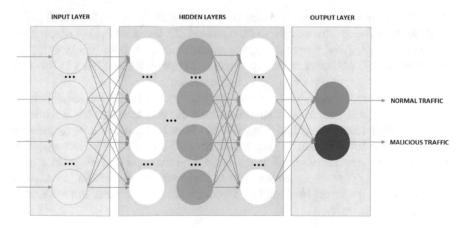

Fig. 1 Typical DNN architecture

subdomain which is known as deep learning (DL), many other models exist, such are deep recurrent neural network (DRNN), deep restricted Boltzmann machine (DE-RBM), generative adversarial network (GAN), etc.

Swarm intelligence (SI) belong to artificial intelligence (AI) methods that entails a group study of how individuals in a population interact with one another on a local level. Foundation of SI algorithms is based on the following five principles: proximate, stability, diverse response, adaptability and quality. The search process of metaheuristics algorithms, including SI is conducted by mechanisms of exploration and exploitation. Through Exploration (or diversification) algorithms try to cover as much of the search space as possible. The algorithm is "exploring" the search area, therefore increasing the chances to find the optimal solution. Contrariwise, exploitation (or intensification) describes the process of using already known information and directing the search to promising solutions.

The SI approaches were used in solving different problems from the various application domains, including cloud computing [3, 7, 9], wireless sensor networks [4, 24, 26, 28], predicting number of COVID-19 cases [25, 27], machine learning [6, 11], classification of brain tumor MRI images [8], feature selection [5] and global optimization problems [20].

Based on the literature survey, many SI-based approaches have been adapted for improving classification performance of ML method used in IDS systems [14, 16]. For example, in [22] an efficient IDS was developed by combining the PSO algorithm and deep belief network (DBN). Similarly, in [15], an IDS based on the hybridization between the ABC metaheuristics and random neural network (RNN) is proposed. PSO-XGBoost hybrid algorithm was proposed in [12]. Besides mentioned ensamble methods based on SI and ML for IDS can also be found in the literature [1].

3   Proposed Method

Firefly algorithm (FA) represents among the most promising and up-to-date SI metaheuristics for solving difficult optimisation issues [23]. Despite of being adapted for various engineering and numerical optimization challenges, both benchmark [18] and practical [19, 24], there are still problems, especially those related to ML that can be successfully tackled by applying this approach.

As it is the case in every other nature-inspired algorithm, for modeling natural systems, the FA incorporates some simplification rules. Solution's quality is modeled by the brightness intensity of individual (firefly) from the population. Brightness of a firefly defines its attraction, which is then translated into the encoded objective function for the purpose of simplicity, a firefly's brilliance at a given point x may be $I(vecx) sim 1/f(vecx)$ in the simplest example of a minimization issue. As a consequence, if the target function has a higher value in this case, the associated firefly will be less brilliant.

The changes in illumination strength and attractiveness are monotonously decreasing functions since the range from the origin rises as the illumination intensity and attractiveness drops, and vice versa. It may be expressed like [23]:

$$I(r) = \frac{I_0}{1 + \gamma r^2} ,$$ (1)

where $I(r)$ stands for the light intensity, r represents the distance between two fireflies, and I_0 denotes the power of light at the origin. The air soak up some of the light, making it appear gloomy and the light absorption coefficient γ is used to indicate air absorption.

Most of the time, the integrated effect of the inverse square law and absorption may be expressed in the shape also known as a Gaussian:

$$I(r) = I_0 e^{-\gamma r^2}$$ (2)

The attractiveness of the individual firefly β is equivalent to its light intensity, as indicated:

$$\beta(r) = \beta_0 e^{-\gamma r^2} ,$$ (3)

where β_0 represents the appealingness at $r = 0$. Estimation of an exponential function requires a lot of computing resources, for that reason the preceding statement could be calculated like this:

$$\beta(r) = \frac{\beta_0}{1 + \gamma r^2}$$ (4)

The following parameters influence a firefly's migration on the way the brighter, and hence more attractive firefly j [23]:

$$x_i(t) = x_i(t) + \beta_0 r^{-\gamma r_{i,j}^2}(x_j - x_i) + \alpha(rand - 0, 5),$$ (5)

In equation above, $rand$ represents uniformly distributed random number in range of 0 and 1, β_0 is attraction at $r = 0$, $r_{i,j}$ represents range between i and j, the randomization parameter is denoted by the symbol α.

To compute the distance between fireflies, we were using The Cartesian distance form bellow:

$$r_{i,j} = ||x_i - x_j|| = \sqrt{\sum_{k=1}^{D}(x_{i,k} - x_{j,k})^2} ,$$ (6)

The algorithm's convergence speed is heavily influenced by the parameter γ. Theoretically, this metric measures attraction variance and has a value of $[0, +\infty)$. In the equation above, D represents a number of parameters in the problem formulation. For the most challenges settings $\beta_0 = 0$ and $\alpha \in [0, 1]$ are suitable settings.

There are two distinct circumstances in the FA, both of which are linked to the value of γ:

- if $\gamma = 0$, then $\beta = \beta_0$. It means the air surrounding firefly is absolutely free of contaminants. In this situation, β is always the largest it can be, and fireflies take the largest possible steps towards other fireflies. Because exploitation is maximal while exploration is minor, the exploration-exploitation balance is out of whack.
- if $\gamma = \infty$, then $\beta = 0$. There was a heavy fog surrounding the fireflies in this situation, and they couldn't observe other fireflies. In this case, the fireflies move randomly, and the exploration is intensified, with almost no exploitation.

3.1 Enhanced FA Metaheuristics

Some flaws of the basic FA were noticed by conducting practical simulations on bound-constrained and constrained benchmarks [18]. Basically, drawbacks of the FA can be summed as follows: insufficient exploration in early iterations and poor balance betwixt exploitation and exploration.

During the preliminary stages of algorithm's execution, the fireflies do not posses a comprehensive awareness of their surroundings, and in order to execute more efficiently, they need to study more about its structure. However, in the original FA implementation, exploration, that is essential for better understanding a problem solution space, is too weak and in some runs, the algorithm gets stuck in the local optimum domain. This issue can also be viewed from the perspective of exploitation-exploration balance. In early iteration, this trade-off is moved more towards exploitation disabling algorithm's ability to converge to optimum portion of the search area, leading to the poorer values of mean and standard deviation.

There are many ways to address those issues. In this approach, an enhanced FA (eFA) method is devised that establishes better exploitation-exploration balance and overcomes low exploration power in early iterations by utilizing a principle of opposition-based learning (OBL) [21]. The OBL proved to be able to substantially improve metaheuristics search process.

The idea in eFA is relatively simple, yet very efficient. In each iteration, for the best solution in population x^*, opposite solution $x^{*,o}$ is generated. Let x_j^* represent the j-the parameter of best individual x^*, while the $x_j^{*,o}$ marks its opposite number, that can be obtained in the following manner:

$$x_j^{*,o} = lb_j + ub_j - x_j \tag{7}$$

where $x_j \in [lb_j, ub_j]$ and $lb_j, ub_j \in R, \forall j \in 1, 2, 3, ...D$. Values lb_j and ub_j denote the lower and upper boundary of j-th parameter, while D stands for the number of individual's dimensions (parameters).

When opposite solution $x^{*,o}$ of the current best x^* is generated at the end of each iteration, by applying simple greedy selection principle, better solution is retained, while worse is discarded from population. In this way, if the search did not converge to optimum domain in early iterations, current best opposite solution may be in the correct portion of the search area, i.e. exploration is enhanced. Moreover, in later iterations exploitation process may be improved. Taking all into consideration, pseudo-code of the eFA method is shown in Algorithm 1.

Algorithm 1. Pseudo-code of the eFA

Create the starting population of fireflies x_i, $(i = 1, 2, 3, ..., SN)$
Light intensity I_i at position x_i is determined by $f(x)$
Specify the light absorption coefficient γ
Specify the number of rounds IN
while (**do**$t < IN$)
  for (**do**$i = 1$ to SN)
    for (**do**$j - 1$ to i)
      if (**then**$I_j < I_i$)
        Advance the firefly j on the way to firefly i in d dimension (Eq (5))
        Attractiveness decreases with range r as $\exp[-\gamma r]$
        Verify the new individual, replace the worst firefly with a better one, and refresh the
light intensity
      end if
    end for
  end for
  Rank all individuals and obtain the currently best x^*
  Generate $x^{*,o}$ with Eq. (7)
  Perform greedy selection between $x^{*,o}$ and x^*
end while

In the provided pseudo-code, the total amount of fireflies in the population is SN, the current iteration is t, the total amount of algorithm iterations is IN. Proposed eFA utilizes only one more fitness function evaluation than the original FA and based on conducted practical simulations with unconstrained benchmarks obtains significantly better performance.

4 The eFA and DNN Framework for IDS Classification Experiments

The eFA-DNN framework proposed in this research is similar as one shown in [13]. The eFA is given the task to perform the feature selection, while the DNN is used for classifying task. For experimental purposes, two IDS datasets were used. The NSL-KDD is a Kaggle dataset that is freely obtainable to the public. The second dataset utilized in the experiments is the KDD Cup 99 dataset, that belongs to the standard UCI benchmarks. Both datasets were cleaned by utilizing the min-max normalization approach, after passing through the one-hot encoding process for categorical features, to achieve homogenity [13].

The following equation is used to normalize the dimension values of the data in the range [0, 1] using min-max normalization:

$$t = \frac{v - min_d}{max_d - min_d}(train\_max_d - train\_min_d) + train\_min_d \qquad (8)$$

where t is the data value's converted value. The original minimum value is denoted by v in the dimension d and max_d refers to the original maximum value of dimension d. Also, $train\_max_d$ refers to the transformed maximum value and $train\_min_d$ refers to the transformed minimum value of the dimension d.

Since both used datasets consist of a relatively high number of features, the eFA is used for feature selection. Each eFA individual is encoded as $l - bit$ string, where l represents the number of features in the dataset. If the value of i-th position is 0, the i-th feature is not included for training the DNN classifier. However, if its value 1, the feature is included.

Each eFA individual is evaluated by its fitness, that is proportional to the DNN classification accuracy for the testing set and reverse-proportional to the number of selection features. This implies that for each fitness evaluation, the DNN classifier is trained on the training set and evaluated on the test set. To make a fair comparative analysis with other approaches, the same fitness function formulation is used as in [13] for i-th solution:

$$fit_i = \alpha \cdot (1 - acc_i) + (1 - \alpha) \cdot (SF/TF), \qquad (9)$$

fit_i and acc_i are fitness and accuracy of i-th solution, respectively, SF is the number of selected features from the dataset, while the TF denote the total number of features in the dataset. The α is parameter that controls the relative influence of the accuracy and the number of selected features to the fitness function. As in [13], for simulations in this paper α is set to 0.9, which means that the accuracy has higher influence on fitness than the number of selected features.

In this research, the DNN classifier with the following characteristics is used: one input, one output layer, three hidden layers with 32, 64 and 128 nodes, each layer except the output uses rectified linear unit (ReLU) activation function, Adam optimizer with the learning rate of 0.001 is used for training and the output layer uses sigmoid activation, since there are only two target classes. The framework is developed in Python and Python core libraries along with the numpy, pandas and scikitlearn were utilized. Proposed framework flowchart is given in Fig. 2.

Similarly as in [13], both datasets were split into the training and test sets by applying $train_test_split()$ method with 70% of data utilized in training phase and 30% in testing phase. The DNN classifier is trained in 100 epochs with 10-fold cross-validation. Simulations with each dataset are executed in 30 independent runs and obtained mean values were reported.

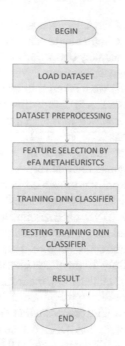

Fig. 2 Proposed eFA-DNN framework for IDS classification

Table 1 Performance validation of eFA-DNN for NSL-KDD dataset

Model	SMO-DNN	PCA-DNN	DNN	FA-DNN	eFA-DNN
Accuracy	0,994	0,938	0,914	0,982	**0,996**
Precision	0,995	0,934	0,891	0,991	**0,996**
Recall	0,995	0,918	0,882	0,989	**0,996**
F-score	0,996	0,937	0,905	0,993	**0,997**
Sensitivity	0,996	0,938	0,908	0,992	**0,997**
Specificity	0,996	0,926	0,898	0,991	**0,997**

Table 2 Performance validation of eFA-DNN for KDD Cup 99 dataset

Model	SMO-DNN	PCA-DNN	DNN	FA-DNN	eFA-DNN
Accuracy	0,928	0,898	0,909	0,925	**0,931**
Precision	**0,927**	0,884	0,896	0,921	**0,927**
Recall	0,928	0,898	0,909	0,925	**0,931**
F-score	0,927	0,882	0,894	0,922	**0,930**
Sensitivity	0,928	0,898	0,909	0,925	**0,930**
Specificity	0,930	0,885	0,882	0,919	**0,931**

To validate enhancements of the suggested eFA over the basic implementation, the basic FA (FA-DNN) was also implemented and included in comparative analysis. Besides original FA, the spider monkey optimization DNN framework (SMO-DNN) proposed in [13], principal component analysis DNN (PCA-DNN) and standard DNN without feature selection (DNN) were taken into consideration for comparisons. It is noted that the all frameworks taken in comparative analysis were tested under the same conditions as approach proposed in this manuscript and that all results were retrieved from [13]. Comparisons between eFA-DNN and other techniques for NSL-KDD and KDD Cup 99 datasets are reported in Tables 1 and 2, respectively. Best results are marked bold and average metrics values are reported.

As an overall conclusion from the presented comparisons, the proposed eFA-DNN framework manages to establish best metrics' values for both datasets. It is important to note that the eFA-DNN obtains substantially better results than the framework based on the original FA implementation (FA-DNN).

Moreover, the eFA-DNN outscores state-of-the-art nature-inspired framework SMO-DNN proposed in [13]. Only for precision metrics in the KDD Cup 99 dataset, the eFA-DNN and SMO-DNN establish the same results. The power behind the eFA-DNN lies in the fact that it allows fast convergence of the attributes, with optimum set of parameters for coherent classifying process while reducing the dimensionality. As a result, superior performance of the eFA-DNN can be summarized with the achieved accuracy of 0,996 on the NSL-KDD dataset, and 0.931 on the KDD Cup 99.

Visualization of accuracy metrics by using cat and whiskers diagrams for eFA-DNN and FA-DNN comparison for NSL-KDD and KDD Cup 99 datasets is shown in Figs. 3a and 3b, respectively.

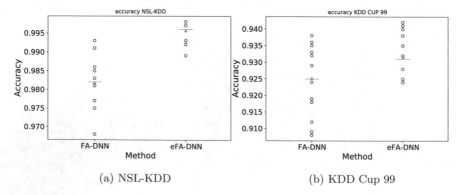

(a) NSL-KDD   (b) KDD Cup 99

Fig. 3 eFA-DNN vs. FA-DNN accuracy comparison for NSL-KDD dataset **a** and KDD Cup 99 dataset **b**

5   Conclusion

This paper proposes hybrid eFA-DNN framework for classification challenge in IDS domain, that has been tested by using the standard IDS benchmark datasets—NSL-KDD and KDD Cup 99. The eFA-DNN first employs the eFA metaheuristics to decrease the number of features from the datasets and then on the reduced dimensions datasets, the DNN classification is performed.

The proposed eFA-DNN has been evaluated by comparing it to the already established DNN models, including the hybridized PCA-DNN and SMO-DNN approaches along with the basic DNN method. The eFA-DNN framework obtained the best performances for all observed performance indicators. Also, the eFA-DNN was compared with the FA-DNN, which is based on the standard FA metaheuristics and managed to completely outscore the basic version.

As one of the constraints of the suggested approach it can be noted that it has been utilized for binary classification only. In future, we plan to test the performances of the suggested method for the multiclass classification problems. Moreover, since the proposed eFA showed good performance, we plan to adapt it further and utilize it for other optimization problems from various application domains, including cloud systems, sensor networks, and other machine learning challenges.

References

1. Aburomman AA, Reaz MBI (2016) A novel SVM-KNN-PSO ensemble method for intrusion detection system. Appl Soft Comput 38:360–372
2. Ahmad J, Farman H, Jan Z (2019) Deep learning methods and applications. In: Deep learning: convergence to big data analytics. Springer, pp 31–42
3. Bacanin N, Bezdan T, Tuba E, Strumberger I, Tuba M, Zivkovic M (2019) Task scheduling in cloud computing environment by grey wolf optimizer. In: 2019 27th telecommunications forum (TELFOR). IEEE, pp 1–4
4. Bacanin N, Tuba E, Zivkovic M, Strumberger I, Tuba M (2019) Whale optimization algorithm with exploratory move for wireless sensor networks localization. In: International conference on hybrid intelligent systems. Springer, pp 328–338
5. Bezdan T, Cvetnic D, Gajic L, Zivkovic M, Strumberger I, Bacanin N (2021) Feature selection by firefly algorithm with improved initialization strategy. In: 7th conference on the engineering of computer based systems, pp 1–8
6. Bezdan T, Stoean C, Naamany AA, Bacanin N, Rashid TA, Zivkovic M, Venkatachalam K (2021) Hybrid fruit-fly optimization algorithm with k-means for text document clustering. Mathematics 9(16):1929
7. Bezdan T, Zivkovic M, Antonijevic M, Zivkovic T, Bacanin N (2020) Enhanced flower pollination algorithm for task scheduling in cloud computing environment. In: Machine learning for predictive analysis, pp 163–171. Springer
8. Bezdan T, Zivkovic M, Tuba E, Strumberger I, Bacanin N, Tuba M (2020) Glioma brain tumor grade classification from MRI using convolutional neural networks designed by modified FA. In: International conference on intelligent and fuzzy systems. Springer, pp 955–963
9. Bezdan T, Zivkovic M, Tuba E, Strumberger I, Bacanin N, Tuba M (2020) Multi-objective task scheduling in cloud computing environment by hybridized bat algorithm. In: International conference on intelligent and fuzzy systems. Springer, pp 718–725

10. Buczak AL, Guven E (2015) A survey of data mining and machine learning methods for cyber security intrusion detection. IEEE Commun Surv Tutor 18(2):1153–1176

11. Gajic L, Cvetnic D, Zivkovic M, Bezdan T, Bacanin N, Milosevic S (2021) Multi-layer perceptron training using hybridized bat algorithm. In: Computational vision and bio-inspired computing. Springer, pp 689–705

12. Jiang H, He Z, Ye G, Zhang H (2020) Network intrusion detection based on PSO-Xgboost model. IEEE Access 8:58392–58401

13. Khare N, Devan P, Chodhary, Lal C, Bhattacharya S, Singh G, Singh S, Yoon B (2020) SMO-DNO: spider monkey optimization and deep neural network hybrid classifier model for intrusion detection. Electronics 16–18 (2020)

14. Mishra S, Sagban R, Yakoob A, Gandhi N (2021) Swarm intelligence in anomaly detection systems: an overview. Int J Comput Appl 43(2):109–118

15. Qureshi AUH, Larijani H, Mtetwa N, Javed A, Ahmad J et al (2019) RNN-ABC: a new swarm optimization based technique for anomaly detection. Computers 8(3):59

16. Ravindranath V, Ramasamy S, Somula R, Sahoo KS, Gandomi AH (2020) Swarm intelligence based feature selection for intrusion and detection system in cloud infrastructure. In: 2020 IEEE congress on evolutionary computation (CEC). IEEE, pp 1–6

17. Smys S, Chen JIZ, Shakya S (2020) Survey on neural network architectures with deep learning. J Soft Comput Paradigm (JSCP) 2(03):186–194

18. Strumberger I, Bacanin N, Tuba M (2017) Enhanced firefly algorithm for constrained numerical optimization. In: 2017 IEEE congress on evolutionary computation (CEC). IEEE, pp 2120–2127

19. Strumberger I, Tuba E, Bacanin N, Zivkovic M, Beko M, Tuba M (2019) Designing convolutional neural network architecture by the firefly algorithm. In: 2019 international young engineers forum (YEF-ECE). IEEE, pp 59–65

20. Strumberger I, Tuba E, Zivkovic M, Bacanin N, Beko M, Tuba M (2019) Dynamic search tree growth algorithm for global optimization. In: Doctoral conference on computing, electrical and industrial systems. Springer, pp 143–153

21. Tizhoosh HR (2005) Opposition-based learning: a new scheme for machine intelligence. In: International conference on computational intelligence for modelling, control and automation and international conference on intelligent agents, web technologies and internet commerce (CIMCA-IAWTIC 2006), vol 1, pp 695–701

22. Wei P, Li Y, Zhang Z, Hu T, Li Z, Liu D (2019) An optimization method for intrusion detection classification model based on deep belief network. IEEE Access 7:87593–87605. https://doi.org/10.1109/ACCESS.2019.2925828

23. Yang XS (2009) Firefly algorithms for multimodal optimization. In: International symposium on stochastic algorithms. Springer, pp 169–178

24. Zivkovic M, Bacanin N, Tuba E, Strumberger I, Bezdan T, Tuba M (2020) Wireless sensor networks life time optimization based on the improved firefly algorithm. In: 2020 international wireless communications and mobile computing (IWCMC). IEEE, pp 1176–1181

25. Zivkovic M, Bacanin N, Venkatachalam K, Nayyar A, Djordjevic A, Strumberger I, Al-Turjman F (2021) Covid-19 cases prediction by using hybrid machine learning and beetle antennae search approach. Sustain Urban Areas 66:102669

26. Zivkovic M, Bacanin N, Zivkovic T, Strumberger I, Tuba E, Tuba M (2020) Enhanced grey wolf algorithm for energy efficient wireless sensor networks. In: 2020 zooming innovation in consumer technologies conference (ZINC). IEEE, pp 87–92

27. Zivkovic M, Venkatachalam K, Bacanin N, Djordjevic A, Antonijevic M, Strumberger I, Rashid TA (2021) Hybrid genetic algorithm and machine learning method for Covid-19 cases prediction. In: Proceedings of international conference on sustainable expert systems: ICSES 2020, vol 176. Springer, p 169

28. Zivkovic M, Zivkovic T, Venkatachalam K, Bacanin N (2021) Enhanced dragonfly algorithm adapted for wireless sensor network lifetime optimization. In: Data intelligence and cognitive informatics. Springer, pp 803–817

Dimensionality Reduction Method for Early Detection of Dementia

Ambili Areekara Vasudevan, A. V. Senthil Kumar, and Sivaram Rajeyyagari

1 Introduction

Dementia affects millions of people worldwide, and its prevalence make a noteworthy effect on forbearing, lives, their families' lives, their well-being. It is a neurodegenerative brain disorder. On the report of World Health Organization (WHO), dementia affects nearly 47 million people around the world. The expected growth in this figure is huge and likely to be 82 million by 2030 and 150 million by 2050 [1]. This disorder has become a growing mortality factor worldwide because of the dramatic rise in dementia cases [2]. The diagnosis of dementia is an arduous task. Early diagnosis of dementia has a hand in identifying the proper doctoring, precludes or slackening cognitive dissonance, and tries to obtain better treatment and future planning [3].

AD, dementia with Lewy bodies, Frontotemporal Dementia, and Vascular Dementia (VD) are the most prevalent dementia forms. In about 75% of general cases, Alzheimer's disease is liable. Mixed dementia is a mixture of both Alzheimer's and VD [4]. Some tests exist for detecting dementia such as Mini Mental State Examination (MMSE), Abbreviated Mental Test Score (AMTS), Cognitive Abilities Screening Instrument (CASI), etc. These tests help to detect the early stage of dementia. Neuro images even struggle with the curse of dimensionality. High dimensionality will lead to over fitting and also affect the efficiency of the classification [5]. Dimensionality approaches will help to solve this problem. The attributes

A. Areekara Vasudevan (✉) · A. V. S. Kumar
PG Research and Computer Application, Hindustan College of Arts and Science, Coimbatore, India
e-mail: ambili9009@gmail.com

A. V. S. Kumar
e-mail: avsenthilkumar@yahoo.com

S. Rajeyyagari
Department of Computer Science College of Computing and Information Technology, Shaqra University, Shaqra, Saudi Arabia
e-mail: dr.sivaram@su.edu.sa

© The Author(s), under exclusive license to Springer Nature Singapore Pte Ltd. 2022
B. Unhelker et al. (eds.), *Applications of Artificial Intelligence and Machine Learning*,
Lecture Notes in Electrical Engineering 925,
https://doi.org/10.1007/978-981-19-4831-2_2

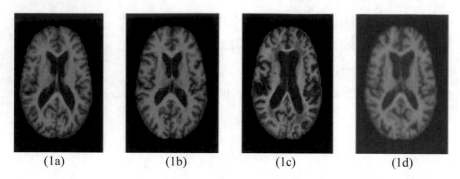

(1a) (1b) (1c) (1d)

Fig. 1 Various brain images of dementia; **a** normal brain; **b** very mild dementia; **c** mild dementia; **d** severe dementia

derived from the data have been used in modern dementia detection studies. But high dimensionality is still a gap in this field.

Figure 1 shows different states of dementia. The ultimate aim is the diagnosis of dementia stages with low dimensional features with help of machine learning techniques. Early detection of dementia using image processing and machine learning techniques are useful for medical field as well as the common peoples.

We put forward a dimensionality reduction method by using a combination of LDA and firefly swarm intelligence algorithm named LDA-FA along with a convolutional neural network classifier. The lower-dimensional data set also subsumes the details derived from a classification model with higher dimensional characteristics, leading to a minimum loss of required features. Dimensional reduction tactics can allow the classifier to select the indispensable attributes, and this would also prevent the attribute from harming the classifier's efficiency. This novel method provides an optimized dimensionality method for detecting dementia.

This research extends recent findings in [16] by examining the use of feature reduction approaches in the categorization of dementia To demonstrate the robustness of the suggested approach, a large database of MR images with over 1000 patients was used, as well as different subdivisions of training-test sets. In this work cortical and subcortical features of mri with reduced dimensionality are used for the classification of the disease. The structure of this document is as follows:

2 Related Works

Dementia is a brain disease that affects cognitive functions including memory, rationality, and thinking. It is brought about by old age or a harrowing injury, with AD accounting for 60–70% of cases [6]. Various studies on dementia detection do not focus on dimensionality reduction. In [7] described dementia detection using the deep learning method which discriminates demented and non-demented brain from the samples.

Table 1 Accuracy of different dimensionality method

Dimensional reduction Method	Accuracy (%)
SAE [13]	76.2
PSO [15]	91.89
PCA[21]	76.9
PCA + Wavelet [23]	95
RELM with PCA [22]	77.30
DTCWT[24]	92.65

Most researchers use neuro image data for the study of brain disorders. There are many redundant and irrelevant features of high-dimensional neuro imaging data. This increases the need for dimensionality reduction to get more accurate and appropriate features for classification. Principal Component Analysis (PCA), linear discrimination analysis, and independent component analysis are the most commonly used methods [8]. Davatzikos et al. [9] and [10] proposed the PCA to minimize the dimensionality of extracted characteristics from MRI brain scans for stratification between demented and non-demented MRI. [11] structural MRI is used LDA for dimensionality reduction and classification of the stages of dementia.

Seo, Kangwon et al. [12] come up with a new multiple-based dimension reduction technique on preprocessed images termed Locally Linear Embedding (LLE) to adequately reduce the feature dimensionality without forfeiting the efficiency of AD exploration. [13] offers different stages of dementia detection method with high accuracy using Sparse Auto Encoder to reduce the dimensionality of features which are extracted from MRI brain scan. Several swarm intelligence algorithms are used for data classification and optimization along with deep learning. These swarm intelligence algorithms help the classifiers to prevent falling into local minimum [14]. [15] suggested Alzheimer's dementia detection by decision tree classifier with dimensionality reduction method. The dimensionality of features extracted by PCA is minimized by particle swarm optimization (PSO). The work in [16] suggests two algorithms by reducing the dimensionality of the data to overcome the small sample size problem to diagnose neurodegenerative disorders using a deep neural network.

Existing dementia detection research has focused on the help of optimized neural network techniques. Selecting suitable characteristics is important as optimum level accuracy. To improve the system consistency, we put forward LDA and firefly to annihilate the extraneous features which reduce the accuracy of the classifier. Table 1 shows the accuracy of several classifications with dimensionality reduction methods.

3 Methodology

This section illustrates the detailed methodology that is used to classify Dementia. Figure 2 shows the flow diagram of the proposed method. It has the following steps:

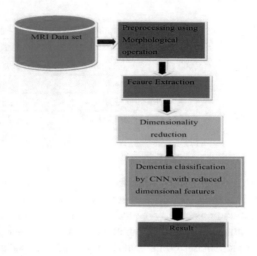

Fig. 2 Flow diagram of proposed system

Dataset collection and description of the dataset, preprocessing the images, classification of dementia using CNN and performance measures for the system. Each step of the proposed system is outlined as follows:

3.1 Materials and Subjects

The algorithm of the suggested technique is based on three steps. The first stage involves data preprocessing, the second stage involves feature extraction and dimensionality reduction from input photos, and the third stage is dementia classification. For the classification of dementia stages, we devised a CNN-based technique. The Open Access Series of Imaging Studies (OASIS) database was used to collect the necessary data for this article. In this study, we examine at MRI brain scans to detect dementia. The acquired dataset is divided into four categories: Normal, very mild dementia, mild dementia, and severe dementia. For the suggested model, a dataset of 50 images is employed, with an equal number of images from each category. Patients with the condition ranged in age from 25 to 90 years.

3.2 Pre-processing

MRI images were preprocessed in the manner described by (17). Data from MRIs contains noisy data that must be reduced before computer processing. Data preprocessing is a crucial phase in the data mining process that involves modifying or removing data before it is utilized to ensure or improve performance.

3.3 Feature Reduction

There are a variety of functions that are obsolete and unwanted after they are extracted. The overall system's stability and efficiency will be affected as a result of storing and using all of the features. For feature reduction, a variety of approaches have been proposed. In our project, we will propose a new feature reduction approach that combines LDA and the firefly algorithm, and it demonstrates the system performance when the reduced function is fed into the classifier. **Linear Discriminant Analysis (LDA).** LDA is a supervised dimensionality reduction approach that, as compared to other methods, allows for the greatest separation of features belonging to different groups. This approach maximizes the ratio of between class scatters matrix to with-in-class scatter matrix in any given data collection, ensuring maximum separation. We can define within-class scatter matrix.

$$S_w = \sum_{j=1}^{c} \sum_{i=1}^{n_j} \left(x_i^j - \mu_j \right) \left(x_i^j - \mu_j \right)^T \tag{1}$$

where x_i^j represent i^{th} sample of class j, μ_j indicates mean of class j, c stands for number of classes and μ_j is the number of samples in class j.

Between classes scatter matrix is

$$S_b = \sum_{j=1}^{c} (\mu_i - \mu)(\mu_j - \mu) \tag{2}$$

where, μ is the mean of classes.

Firefly algorithm. Firefly Algorithm (FA) was expanded by Xin-She Yang in 2008. FA algorithm is gleaned from spasmodic firefly patterns and activities [18]. This method is primarily adapted to solve complicated problems dependent on firefly characteristics [19]. In FA, firefly due to their unisex is attracted by another firefly based on the brightness. The attraction of any two fireflies increases when they are close together and decreases when they are far apart due to the light of the fireflies. The light of a firefly's elegance is therefore directly proportional to the interval between two fireflies [20]. A firefly's brightness is determined by the environment of the objective function f(x) in which x = x_1, x_2, x_3,......x_n.

In FA the brightness of every firefly can be gauged with the help of an objective function. Thus the initial population of the firefly is derived to deciding the brightness of every firefly for the generated population. Dependent on the light intensity (x_i, x_j) of two fireflies, the interval between two fireflies is determined. When the distance between two fireflies (x_i, x_j) is less adjacent fireflies are clustered and assess the attraction between them.

Fireflies' attraction is connected to the light intensity of the neighboring fireflies, so the variance of attractiveness β with the distance r can now be defined as,

$$\beta = \beta_0 e^{-\gamma r^2} \tag{3}$$

where β_0 is the attraction at $r = 0$.

When a firefly I is attracted to a more attractive firefly j, its movement is determined by

$$x_i(t + 1) = x_i(t) + B_0 e^{-\gamma r_{ij}^2}(x_j - x_i) + \alpha\left(rand - \tfrac{1}{2}\right) \tag{4}$$

where α is the randomization parameter, rand is a random number generator uniformly distributed in [0, 1].

4 Result and Discussion

The results are assessed based on the classifier's output using the reduced features from the LDA-FA method. The classifier's performance was assessed using accuracy, sensitivity, and specificity. The following is a summary of the proposed model:

Input: MRI Dataset

Output: Classification of Dementia

Preprocessing: Normalize and enhance the image with morphological operations

Feature Extraction: Shape and texture features are extracted from the MRI.

Dimensional Reduction: Apply the extracted features to the LDA-FA for dimensional reduction.

Classification: Input the reduced features to CNN for classification of Dementia.

The efficiency of the classifier is evaluated by providing the inputs selected by the LDA-FA. Using the parameters of True positive (TP), False negative (FN), True negative (TN), and False positive (FP), the classification results deliver specificity (Spe), accuracy (Acc), and sensitivity (Sen) of an image. Table 1 demonstrate different methods. We apply the dimensional reduction method to reduce the dataset and provide the most suitable features to the neural network to improve the classification accuracy. Table 2 shows the accuracy, sensitivity, and specificity of existing dimensional reduction methods. Figure 3 depicts the performance of various dimensionality reduction methods.

$$Acc = \frac{TN + TP}{TN + TP + FN + FP}$$

$$Sen = \frac{TP}{TP + FN}$$

$$Spe = \frac{TN}{FP + TN}$$

Table 2 Performance table

Method	Accuracy (%)	Sensitivity (%)	Specificity (%)
CNNPCA [20]	81	75	76
CSLBT [17]	90.1	89	77
Proposed	95.5	97	82.5

Fig. 3 Parameter analysis of various dimensionality reduction method

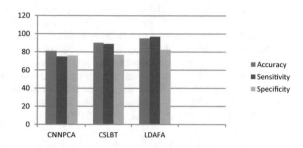

5 Conclusion

For the classification of dementia datasets, a hybrid Linear Discriminant analysis (LDA)—firefly dependent neural network model is used in this research. The dataset used in the analysis was obtained from a freely accessible UCI machine learning library, which contained redundant and obsolete attributes in its natural form. This work included preprocessing for enriching the nature of the input image, feature extraction method to extract shape and texture features and the extracted features are then subjected to an LDA-firefly hybrid algorithm for dimensionality reduction. The proposed method strengthened the classification performance by selecting the most redundant and optimized features.

References

1. Ahmed MR, Zhang Y, Feng Z, Lo B, Inan OT, Liao H (2019) Neuroimaging and machine learning for dementia diagnosis: recent advancements and future prospects. IEEE Rev Biomed Eng 12:19–33. https://doi.org/10.1109/RBME.2018.2886237
2. Zheng C, Xia Y, Pan Y, Chen J (2016) Automated identification of dementia using medical imaging: a survey from a pattern classification perspective. Brain Inform 3(1):17–27. https://doi.org/10.1007/s40708-015-0027-x
3. Raeper R, Lisowska A, Rekik I (2018) Cooperative correlational and discriminative ensemble classifier learning for early dementia diagnosis using morphological brain multiplexes. IEEE Access 6:43830–43839
4. Bansala D, Khannaa K, Chhikaraa R, Duab RK, Malhotrab R (2019) Analysis of classification and feature selection techniques for detecting dementia. In: International conference on sustainable computing in science, technology and management (SUSCOM-2019)

5. Mwangi B, Tian TS, Soares JC (2014) A review of feature reduction techniques in neuroimaging. Neuroinformatics 12(2):229–244. https://doi.org/10.1007/s12021-013-9204-3
6. Aram S, Hooshyar D, Park KW, Lim HS (2017) Early diagnosis of dementia from clinical data by machine learning techniques. Appl Sci 7(651):1–17
7. Ucuzal H, Arslan AK, Çolak C (2019) Deep learning based-classification of dementia in magnetic resonance imaging scans. In: 2019 international artificial intelligence and data processing symposium (IDAP), Malatya, Turkey, pp 1–6. https://doi.org/10.1109/IDAP.2019.8875961
8. Raghavendra U, Acharya UR, Adeli H (2019) Artificial intelligence techniques for automated diagnosis of neurological disorders. Eur Neurol 82:41–64. https://doi.org/10.1159/000504292
9. Davatzikos D, Resnick S, Wu X, Parmpi P, Clark C (2008) Individual patient diagnosis of AD and FTD via High-dimensional pattern classification of MRI. J Neuroimage 1220–1227
10. Mahmood R, Ghimire B (2013) Automatic detection and classification of Alzheimer's disease from MRI scans using principal component analysis and artificial neural networks. In: 2013 20th international conference on systems, signals and image processing (IWSSIP), Bucharest, Romania, pp 133–137. https://doi.org/10.1109/IWSSIP.2013.6623471
11. Vemuri P et al (2011) Antemortem differential diagnosis of dementia pathology using structural MRI: differential-STAND. NeuroImage 55(2):522–531. https://doi.org/10.1016/j.neuroimage.2010.12.073
12. Seo K et al (2019) Visualizing Alzheimer's disease progression in low dimensional manifolds. Heliyon 5(8):e02216. https://doi.org/10.1016/j.heliyon.2019.e02216
13. Alkabawi EM, Hilal AR, Basir OA (2017) Feature abstraction for early detection of multi-type of dementia with sparse auto-encoder. In: 2017 IEEE international conference on systems, man, and cybernetics (SMC), Banff, pp 3471–3476. https://doi.org/10.1109/SMC.2017.8123168
14. Bhattacharya S et al (2020) A novel PCA-firefly based XGBoost classification model for intrusion detection in networks using GPU. Electronics 9(2):219. https://doi.org/10.3390/electronics9020219
15. Sweety M, Evanchalin, Jiji W (2014) Detection of Alzheimer disease in brain images using PSO and decision tree approach, pp 1305–1309. https://doi.org/10.1109/ICACCCT.2014.7019310
16. Segovia F, Górriz JM, Ramírez J, Martinez-Murcia FJ, García-Pérez M (2018) Using deep neural networks along with dimensionality reduction techniques to assist the diagnosis of neurodegenerative disorders. Logic J IGPL 26(6):618–628
17. Yang XS (2010) Nature-inspired metaheuristic algorithms. Luniver Press, New York
18. Ambili AV, Senthil Kumar AV, El Emary IMM (2021) CNN approach for dementia detection using convolutional SLBT feature extraction method. In: Smys S, Tavares JMRS, Bestak R, Shi F (eds) Computational Vision and Bio-Inspired Computing. AISC, vol 1318. Springer, Singapore. https://doi.org/10.1007/978-981-33-6862-0_29
19. Moazenzadeh R, Mohammadi B, Shamshirband S, Chau KW (2018) Coupling a firefly algorithm with support vector regression to predict evaporation in northern Iran. Eng Appl Comput Fluid Mech 12:584–597
20. Yang XS (2009) Firefly algorithms for multimodal optimization. In: Watanabe O, Zeugmann T (eds) Stochastic algorithms: foundations and applications. SAGA 2009. LNCS, vol 5792. Springer, Heidelberg. https://doi.org/10.1007/978-3-642-04944-6_14
21. Miah ASM et al (2021). Alzheimer's disease detection using CNN based on effective dimensionality reduction approach. In: Vasant P, Zelinka I, Weber GW (eds) Intelligent Computing and Optimization. ICO 2020. AISC, vol 1324. Springer, Cham. https://doi.org/10.1007/978-3-030-68154-8_69
22. Lama RK, Gwak J, Park JS, Lee SW (2017) Diagnosis of Alzheimer's disease based on structural MRI images using a regularized extreme learning machine and PCA features. J Healthc Eng 2017:5485080. https://doi.org/10.1155/2017/5485080. Epub 18 June 2017. PMID: 29065619, PMCID: PMC5494120

23. Sivapriya TR, Kamal ARNB (2013) Hybrid feature reduction and selection for enhanced classification of high dimensional medical data. In: 2013 IEEE international conference on computational intelligence and computing research, Enathi, India, pp 1–4. https://doi.org/10.1109/ICCIC.2013.6724237

24. Cao P et al (2017) Nonlinearity-aware based dimensionality reduction and over-sampling for AD/MCI classification from MRI measures. Comput Biol Med 1(91):21–37. https://doi.org/10.1016/j.compbiomed.2017.10.002 Epub 2017 Oct 6 PMID: 29031664

25. Alam S, Kwon GR, Kim JI, Park CS (2017) Twin SVM-based classification of Alzheimer's disease using complex dual-tree wavelet principal coefficients and LDA. J Healthc Eng 2017:8750506. https://doi.org/10.1155/2017/8750506. Epub 16 August 2017. PMID: 29065660, PMCID: PMC5576415

Prognostication in Retail World: Analysing Using Opinion Mining

Neelam Thapa and Anil Kumar Sagar

1 Introduction

In this advancement and development of the technological era, we cannot deny how important and major role data plays in it. But the increase of generation of data is hard to handle with the old data management systems. That's where Big Data came to focus. Big Data is nothing but a huge multitude of data that enlarges rapidly concerning time. This day retailers have easy access to data or we can say big data. And using this data, retailers will have a clear vision in decision making in which their performance will improve. Big Data also helped the retail industry through its analytics method by providing better decisions for sales and improving customer experience. It empowers to comprehend with regards to when, where, and why clients purchase. Investigation of this information helps us in expecting and interfacing with the client's retail needs. The well-known business this day is the retail business. Some of them are Amazon, Netflix, Starbuck. Amazon is a well-known online store. They have an effortless approach to all their customer's data including reviews, customer name, browsing records, methods of pay and even location with all this data amazon aims for good sale of products. Netflix has all the data like viewing choice and behaviour of the viewers from every part of the world. They can also suggest a movie or series of a particular client's taste. Starbucks uses data analytics technologies and methodologies; they mainly focus on consumer behaviour of a particular location even before launching their store. As it is easy to gain the clients in a retail world, it is equally easy for a retailer to lose them if the client experiences a certain situation where they might feel a lack of customer's service, then they might not want to come back to the same retailer again and they can just opt for another one. So, in that kind

N. Thapa (✉) · A. K. Sagar
Department of Computer Science and Engineering, Sharda University, Greater Noida, India
e-mail: 2020461403.neelam@pg.sharda.ac.in

A. K. Sagar
e-mail: anil.sagar@sharda.ac.in

© The Author(s), under exclusive license to Springer Nature Singapore Pte Ltd. 2022
B. Unhelker et al. (eds.), *Applications of Artificial Intelligence and Machine Learning*,
Lecture Notes in Electrical Engineering 925,
https://doi.org/10.1007/978-981-19-4831-2_3

of situation the sales might drop drastically for the product. To avoid those problems, we performed opinion mining or sentiment analysis of the customers' reviews. Here, in this paper we extracted the dataset from Kaggle and the dataset is about amazon customer review. Opinion mining makes use of NLP to find a variety of viewpoints on a stated issue in a straight-pool of text. Sentiment analysis technology is used to mark the judgements on a degree of positive to negative, facilitating to spot the direction in behaviour and frame of mind. And also, a proposed model will be generated by using classification model.

2 Literature Survey

On this viewpoint, a couple of studies have been described. Some researchers put forward a point of view for data mining and use association rule and RFM analysis [1]. There was also utilisation of SBM and ANN techniques and got accuracy of 99.3 and 86.22% respectively [2]. Some studied this field and made a comparison between combine classifier and random classifier and it got results of 19.4 and 5.5 respectively [3]. There was study based on the mail survey of total 1951 people using the SERVQUAL variable model [4]. Some other researcher did comparison analysis between Logistic regression light FM and multilayer perceptron and due to which he got results as 79.40, 81.6 and 92.4% respectively [5]. Three classification model that is NB, J48 and J48 had a higher security of 89.40% were also done [6]. Regression Tree model random forest model least absolute string cage and selection operation and Logistic regression models and also multilayer perceptron and did a comparison of all [7]. Some utilised RNN which is dependent on RFM variables and the results showed RNN brings out the correct prediction with the help of RFM values [8]. Some used opinion mining and also classifiers like Naive Bayes, Decision tree and got 94.3% for Naive bayes than decision tree [9]. The use Internet of things (IOT) and focused on prediction analysis and also used real-time analysis was also done [10].

Some utilise CRM data mining Framework, Naive bayes and neural network and showed that neural network classifiers worked better [11]. Support vector machines, baseline classifiers and got at 8.6, 7.5% respectively while comparing the models both [12]. The churn prediction, data mining for the enhancement of customer services [13]. The utilisation of predictive models that is SVM and regression model were also done and result to which regression model worked better [14]. Some used prediction analysis, linear regression, coefficient of determination, Akaike information criterion on e-commerce customers comments [15].

Big data analytics is been used many retail companies, and it follows certain steps while analysing the data.

Risk Management. Use Case: Banco de Oro, a Philippine banking organization, utilizes Big Data investigation to recognize false exercises and disparities. The association uses it to limit a rundown of suspects or main drivers of issues. Product development and innovation. Use Case: Rolls-Royce, perhaps the biggest producer of fly motors for carriers and military across the globe, utilizes Big Data examination to dissect how effective the motor plans are and if there is any requirement for upgrades. Best and quick decision making within organisation. Use Case: Starbucks utilizes Big Data examination to settle on essential choices. For instance, the organization uses it to choose if a specific area would be reasonable for another outlet or not. They will investigate a few unique components, like populace, socioeconomics, availability of the area, and then some. Improve customer experience. Use Case: Starbucks utilizes Big Data examination to settle on essential choices. For instance, the organization uses it to choose if a specific area would be reasonable for another outlet or not. They will investigate a few unique components, like populace, socioeconomics, availability of the area, and then some.

3 Proposed Model

Here, in this proposed model will be extracting the data set of Amazon customer review and with the help of opinion mining this review will be mined their opinion mining on emotional AI using sentimental analysis technique. And with the help of classification model the accuracy of opinion mining will be obtained.

3.1 Sentimental Analysis

Sentiment analysis can categorize the mostly used word, good review and the bad review into the different categories. Sentimental analysis thoroughly includes data mining, machine learning and artificial intelligence. Further it has various types.

Finely-Grained Sentiment Analysis. Finely-grained sentiment analysis, it breaks down the text according to its contrariety. Solely depending upon the comparison to a five-star rating system. Example: If the rating system has five-star rating which implies that the rating is in good scale. So, this analysis will understand that all the text lies in good review category. And if it has one-star then it lies on bad review.

Emotion Detection. Emotion detection instead of detecting positive and negative emotions in the text specified. As ML technique been used this analysis can differentiate a human language level whether it's a cheerfulness, annoyance, fury, surprise or even a sarcastic comment. Example: If we write a review like "This clothing line is just nailing it", it will understand the sarcasm of the text that it is a good review.

Intention-Based Analysis. Intention-based analysis distinguishes between act and opinion. Example: If a client is expressing its disappointment on certain product, then it will focus on that specified issue.

Expectation-Based Analysis. Expectation-based analysis just collects only the precise part of review which has been referred to as good or bad. Example: If a client complains about the camera quality of the cell, then the analysis will give back negative review only on camera quality instead of the whole product.

3.2 Classification Model

The initial stage in categorization is always to comprehend the issues and discover possible characteristics and identification. There are two stages in classification training and testing. In the training stage, the models are being trained on available data sets and in the testing stage, the models are being tested. Numerous variables like correctness, inaccuracy, call-back define the completion of the estimation.

Below is the flowchart of how a classification model works with huge datasets (Fig. 1).

Naïve Bayes. The statistically categorized approach that depends on Bayes theorem is known as naive Bayes. And it's the most basic supervised learning algorithm available. This classification model is quick and correct and we can say the most trustworthy method can handle enormous amounts of data with a sufficient amount of correctness and pace.

$$P(h|d) = p(d|h)p(h)\big/ p(d) \tag{1}$$

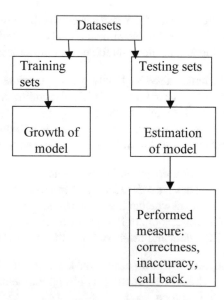

Fig. 1 Flowchart of classification of model

p(h) = denotes the likelihood that the hypothesis 'h' is correct

p(d) = denotes the data probabilities

p(h|d) = stated the data 'd', p(h|d) is the probability of hypothesis 'h'

p(d|h) = likelihood of data 'd' if hypothesis 'h' is true 't'

Naive Bayes is a wonderful method of how the easiest answer sometimes can be supreme. Despite recent improvements in ML, it proved to be not just easy, quick, correct but also dependable. It is used in various applications, nonetheless, it excels in NLP i.e., Natural language processing issues. The text which is provided is calculated according to the statistical probabilities and analyse all the text then give doubt the mightiest probabilities. The below mentioned flowcharts shows how the huge dataset is being fed, and then move on to another stage where all the gigantic data have to go through training and testing. Then it goes for ML modelling, where everything is put in determined classes (Fig. 2).

XGBoost. This classifier uses knowledge of gradient boost base which is an evolved decision tree, ml algorithm. ANN Surpasses every other algorithm or architecture in prediction issues involving unorganised information (pictures, texts etc.). It is free and easy to access i.e., open-source application. Further it can be described by boosting and gradient. One of the assembled approaches is boosting. Here, if the current model makes any miscalculation, then the new model will come to be used by adding the new models and clear all the mistakes making it to be more precise.

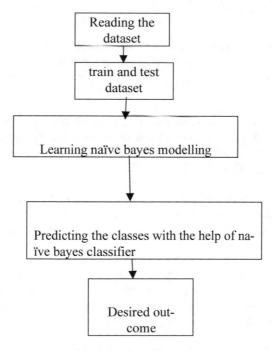

Fig. 2 Workflow of Naïve Bayes classification model

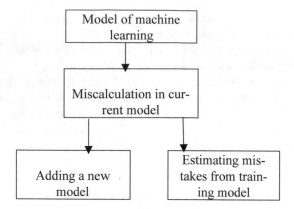

Fig. 3 Workflow of the gradient boosting

In the gradient boosting the brand-new model is added on to check and walk on the prediction of any available mistakes, after which the result was put together to arrive at the predicted outcome (Fig. 3).

3.3 Techniques and Tools

There are many techniques and tools available to modernize the big data world. These techniques and tools help us in getting a desirable outcome, some of them we have discussed in this paper.

Some of the techniques are:

Data Mining. It alludes to strategies utilized to extricate designs from information, such principal learning, group investigation, arrangement furthermore, relapse, which can be accustomed to deciding for instance the qualities of fruitful workers or even decide client buy conduct.

Neural Network. This technique follows how a human neural brain works. We can say it was inspired by it from the massive data on the useful ones and will proceed further, as it can detect the aim of a hidden pattern. Neural nets refer to naturally or artificially networks of cells.

Machine Learning. It is a man-made consciousness method which permits PCs to adjust conduct dependent on exact information in order to settle on astute choices dependent on data. Machine learning have many algorithms which helps us to refine and get the desire outcome from the gigantic datasets.

Predictive Modelling. It utilizes cluster of models toward anticipate likelihood used out from occasion happening, and can also appeal to models for anticipating all the possible outcomes which make the client be able to strategically pitch a spare item. It is able to predict all the required outcome.

Natural Language Processing. It is a professional skill of programming, man-made awareness also historical background, all this data assessment instrument uses computations to explore human (trademark) languages. It actually follows how a human language works, and try to find miscalculation is available.

Decision Tree. As the name suggests, a decision tree is a structure with a set of data domains in small scale subtending a connected free at the same time. It can give us the desirable outcome more precisely.

Some of the tools used are:

Excel. It can handle a large quantity of data mostly with the help of extra plugins, and also offers a number of effective instructions. The data which does not come up to the specific information margins then excel is one of the very handy tools.

Tableau. As the digital marketing sector is increasing, we need a better visualisation application and for that tableau is one of the sophisticated as well as rapidly expanding applications. The original information is simplified into a very comprehensible style where the particular individual is from a technical field or not. It is created in such a way so that it can be understood.

Python. A python is very easily understandable programming language. This programming language can handle huge amount of data. The provided data in python can be cleaned so that all the missing value, duplicate vales can be removed. We can use any machine learning models easy in this programming language. Currently, there are rapid increase in usage of this language in market.

Hadoop. It is a free open-source application that measure enormous information in uniform amounts on a dispersed structure. It allows piling up, lay aside as well as interrogating all the large data uniformly.

4   Data

Datafiniti's by-product data collection has compiled a collection of some 34,000 customer reviews of Amazon goods such as the E-book, Firestick, and far more. For each product, the dataset includes basic product information, a ranking, comment content, and much more (Fig. 4).

5   Evaluation Outcome

The above pie chart shows the distribution between two sentiments i.e., positive and negative. Here 57.84% of used text from the amazon client's review falls under

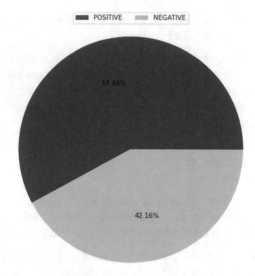

Fig. 4 Distribution of sentiments

positive sentiment, which means the clients have written good reviews on a certain product. And 42.16% are negative sentiment. The client was not satisfied and put up some negative reviews on the disliked product (Figs. 5 and 6).

The below picture shows the most frequently used words from a client while writing a review on particular product that they bought. Here, it shows the words that fall under good reviews. This picture only shows the reviewed words when the client was satisfied (Fig. 7).

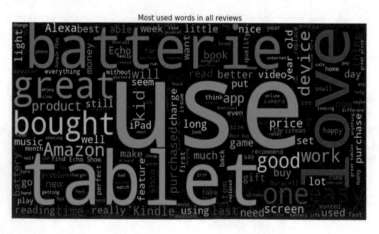

Fig. 5 Mostly used words in all the reviews

Fig. 6 Mostly used words in positive review

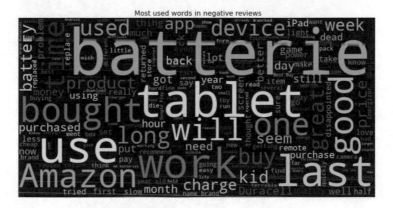

Fig. 7 Mostly used words in negative review

The reviewed words that the client gave to a certain product, which was all negative as they were not satisfied with the product or might be with customer service, were all listed here with the help of word cloud (Fig. 8).

A confusion matrix was created to show the accuracy of the used model, in this case its naïve bayes classifier model. It contains various shades of same colour; the darker shade of the colour shows the value in highest and the lighter shade shows the lesser value. Here, in the x-axis represent predicted values and y-axis represent actual value. As we move the x–y axis alongside we can compare and see the difference between actual and predicted accuracy of the model. This confusion matrix with heatmap made it easy for us to understand this data visualization (Fig. 9).

In this confusion matrix we have used XGBoost model in comparison to naïve bayes model. If we compare the both model's confusion matrix, we can see that this model has reach a certain accuracy more than naïve bayes.

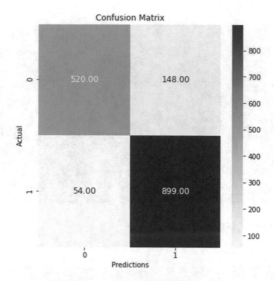

Fig. 8 Confusion matrix with Naïve Bayes model

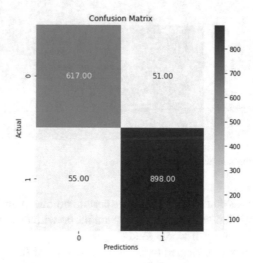

Fig. 9 Confusion matrix with XGBoost model

6 Conclusion

Big data plays a vital role in analysing the issues or we can say enormous amounts of information. The retailer has easy access to all these enormous information and they also strategized all its sales and client attracting technique according to the data they are being fed. Whether it's a market-led or a real-time retailer, the business has been increasing with the help of big data. The retailer has to go through the challenges like they should make sure all the information that they are collecting is precise and recognisable. Presently all the retailer whether it's small or big have application, so all the data might come from different places and different organisation, they should also make sure that all the data are safe and also update its technologies according to the day-to-day time. And winning over their client is very important. As we studied further, we found that there are many tools and techniques in big data analysis, which cleans all the required data and can also manage data which are lost. Here, in the present study we are showing how big data is useful in the world of retail business, how everything is just in the doorstep of all the clients and also provide them a sense of security. And there was a used of machine learning models i.e., Naïve bayes and XGBoost classification model for the provided dataset of amazon client review, and got an accuracy of 88 and 93% respectively.

References

1. Chen M-C, Chiu A-L, Chang H H (2005) Mining changes in customer behavior in retail marketing
2. Dolatabadi SH, Keynia F (2017) Designing of customer and employee churn prediction model based on data mining method and neural predictor
3. Eichinger F, Nauck DD, Klawonn F Sequence Mining for Customer Behaviour Predictions in Telecommunications
4. Baumann C, Burton S, Elliott G, Kehr HM (2007) Prediction of attitude and behavioural intentions in retail banking
5. Kulkarni N (2020) Customer behaviour prediction
6. Raju SS, Dhandayudam P (2018) Prediction of customer behaviour analysis using classification algorithms
7. Subroto, Christianis M (2021) Rating prediction of peer-to-peer accommodation through attributes and topics from customer review
8. Salehinejad H, Rahnamayan S (2016) Customer shopping pattern prediction: a recurrent neural network approach
9. Songpan W (2017) The analysis and prediction of customer review rating using opinion mining
10. ten Bok BGJ (2016) Innovating the retail industry; an IoT approach
11. Femina Bahari T, Sudheep Elayidom M (2015) An efficient CRM-data mining framework for the prediction of customer behaviour
12. Ravnik R, Solina F, Zabkar V (2014) Modelling in-store consumer behaviour using machine learning and digital signage audience measurement data
13. Ahmed AA, Maheswari Linen D (2017) A review and analysis of churn prediction methods for customer retention in telecom industries
14. Arif M, Qamar U, Khan FH, Bashir S (2018) A survey of customer review helpfulness prediction techniques
15. Wu S-H, Hsieh Y-H, Chen L-P, Yang P-C, Fanghuizhu L (2019) Temporal model of the online customer review helpfulness prediction with regression methods
16. Cadsawan L-A et al (2015) A financial and operational analysis of Banco De Oro
17. Smith-Gillespie A, Muñoz A, Morwood D, Aries T (2020) Rolls-Royce
18. Shirdastian H, Laroche M, Richard M-O (2017) Using big data analytics to study brand authenticity sentiments: The case of Starbucks on Twitter
19. Balaji MS, Roy SK (2016) Value co-creation with Internet of things technology in the retail industry

20. Belarbi H, Tajmouati A, Bennis H, Tirari MEH (2014) Predictive analysis of big data in retail industry
21. (Marcel) van Eupen MGH (2014) Big data opportunities for the retail sector a model proposal
22. Seetharaman A, Niranjan I, Tandon V, Saravanan AS (2016) Impact of big data on the retail industry
23. Aktas E, Meng Y (2017) An exploration of big data practices in retail sector
24. Santoro G, Fiano F, Bertoldi B, Ciampi F (2019) Big data for business management in the retail industry

Impact of Resolution Techniques on Chlorophyll Fluorescence Wheat Images Using Classifier Models to Detect Nitrogen Deficiency

Parul Datta, Bhisham Sharma, and Sushil Narang

1 Introduction

Wheat is a universal staple food of the world; just like water, there is no possibility that any person on earth can live without this food made of wheat [1, 2]. India accounts for the most incredible success story of the Green Revolution leading to increase in production of wheat. Due to this, it has become the second-largest producer of wheat in the world. However, it appears that the yield has stabilized, and there is a great sense of uncertainty due to climatic and environmental vectors. The breeders need to remain vigilant for producing better seeds so that food security remains within bounds. Documented statistics show that more than 80% of all seeds in India are saved by farmers. Hence, all these farmers need to equip themselves with better knowledge and modern instrumentation for assessing the determinants that impact their crop growth at any stage. Chlorophyll fluorescence imaging is an emerging technique that is a non-invasive technique for determining Photo System II (PSII) activity [3]. The best application of this is its use in understanding plant physiology, especially when some stress needs to be tracked. Plants suffer with two types of stresses during breeding: Biotic stress that occurs due to living organisms such as weed, parasites, etc. Abiotic stress occurs due to environmental conditions such as deficient water, nutrient deficiency, etc.

Due to the sensitivity of PSII activity to abiotic and biotic factors, it has become a necessary procedure for understanding photosynthetic mechanisms and assessing

P. Datta (✉) · B. Sharma · S. Narang
Chitkara University School of Engineering and Technology, Chitkara University, Baddi, Himachal Pradesh, India
e-mail: parul.datta@chitkarauniversity.edu.in

B. Sharma
e-mail: bhisham.sharma@chitkarauniversity.edu.in

S. Narang
e-mail: sushilk.narang@chitkara.edu.in

© The Author(s), under exclusive license to Springer Nature Singapore Pte Ltd. 2022
B. Unhelker et al. (eds.), *Applications of Artificial Intelligence and Machine Learning*,
Lecture Notes in Electrical Engineering 925,
https://doi.org/10.1007/978-981-19-4831-2_4

how plants respond to environmental changes [4]. With this technique's help and concern for food security, the research has become essential for further progress. The reason is that this technology enables the investigation of spatial-temporal hetero-geneities in the fluorescence emission pattern within cells, leaves, and entire plants [5]. It has been primarily used in horticultural research to diagnose biotic or abiotic stresses in both pre and postharvest conditions, as well as to detect biotic or abiotic stresses in plants or plant products.

Numerous studies have demonstrated that chlorophyll fluorescence is an excel-lent indicator of nutrient deficiency [6, 7]. As a result, using this technique in the precision agriculture and horticulture industry can help avoid over-fertilization while still ensuring optimal productivity. Majority of studies [8–10] have concentrated on Nitrogen (N) because it is the macronutrient that plants require in the greatest quan-tities for development. About 50–80% of nitrogen requirement is in leaves as they perform photosynthesis for producing food for the plant. The fluorescence imaging system can be used to quantify differences in plant nitrogen absorption and chloro-phyll content. This system is being proposed as a viable alternative to Ultraviolet A (UVA wavelength 320 and 400 nm) laser illumination technique. At the same time plants with an abundant supply of nitrogen can be distinguished from those with an insufficient supply. The changes in the ratios were attributed to a decrease in fluo-rescence emission at 690 and 740 nm caused by a decrease in chloride content in the presence of a depleted N supply [11].

In this research work, an attempt will be made to find the impact of improving the resolution of Chlorophyll fluorescence images on classification models that detect nitrogen deficiency stress in the wheat. The field of Chlorophyll fluorescence imaging is based on the computations of the pixels that represent the photosynthetic activity of the plant. If the images are suffering from noise or have low resolution quality due to entrance of ambient light and other factors the accuracy of the classification models is bound to get impacted either in a positive or in a negative manner [12–14].

The fundamental information that is carried by captured image is almost entirely dependent upon the spatial resolution factor of the camera sensors. The spatial reso-lution helps to capture the various patterns of the objects embodied in the image. The captured patterns can reveal a lot of information that may be useful for space explo-ration, military, climatic or agricultural reasons. Bad choice of special resolution and other environmental factors while capturing can lead to low quality representation of the object's details. It is observed whenever a finer resolution is selected there will be always better representation of details of the objects in the images. In the context of our research work, it is observed that the analysis of wheat canopy morphology can be made more accurate by improving the quality of the image. This can be done with the help of super resolution methods such as bilinear, cubic etc.

The paper is divided as follows: Sect. 2 describes the related work. Section 3 includes the experimental setup of the study. Section 4 consists of the results and discussions of the paper. Section 5 concludes the paper.

2   Related Work

The extensive literature study shows that chlorophyll fluorescence imaging is increasingly popular in wheat research, and this section examines the rise of current concerns and problems associated with its application. This part of the article also argues that it is necessary to move away from the conventional manual techniques of tracking abiotic and biotic stress in order to be more effective.

The photosynthetic process of plants can result in one of two outcomes: Either the light energy is lost through internal conversion (as heat) or the light energy is released as light (mostly fluorescence). The focus of this research is on the latter outcome, which is chlorophyll fluorescence. When the plant emits fluorescence, it can be caught using a camera and other equipment processes in order to better understand its internal responses to a given stress or stimulus, which is then analyzed. While chlorophyll fluorescence emission from whole leaf systems is too dim to be detected with the naked eye, it can be observed in lit chlorophyll solution extracts, which are a type of solution extract [14]. The fluorescence of chlorophyll reaches its maximum intensity in the red area of the spectrum (685 nm), and it can be seen all the way up to the infrared region at around 800 nm. Each of these activities is in direct competition with the other for a limited pool of absorbed energy, and any change in the energy consumption of one process leads to a commensurate change in the energy consumption of the others. As a result of this quality, chlorophyll fluorescence can be used as a non-invasive, quick, and reliable probe of photochemistry without the need for additional equipment. Specifically, a Chlorophyll Fluorometer is a device that is designed to measure the fluorescence emission of chlorophyll [15]. Chlorophyll Fluorometer devices are classified into two types: The first captures fluorescence in the continuous excitation state (also known as Kautsky Fluorescence Induction), and the second catches fluorescence in the discrete excitation state (also known as discrete excitation state) [16].

Due to the fact fluorescence is a signal; its measurement may be disturbed by interference by the ambient light, as well as by some amount of background noise [17]. During such situations, pulse-modulated Chlorophyll Fluorometer devices are utilized to separate the required signal from interference from the surrounding environment [18]. If fluorescence emissions are analyzed in this context, the fundamental premise of the process is that peak fluorescence emission occurs in the red area of the spectrum (685 nm) and extends into the infrared region to approximately 800 nm at healthy temperatures [16]. Both of these strategies are used quite frequently in contemporary literature [19–21], as seen by the number of examples that can be found. It is possible to utilize both of these techniques to detect and identify many factors that influence the photochemistry of a plant. A variety of plant-related issues can be investigated using these instruments, such as water scarcity and shortage in nutrients, salt stress, chill stress, yield estimation, pathogen attack, and a variety of other issues. Recent developments in this area demonstrate that the excitation signals from Chlorophyll fluorescence are processed as signals and then captured as an image. A large number of researchers can take advantage of image processing techniques while also benefiting from advancements in machine and deep learning algorithms as a result.

Nutritional deficiencies in plants have a negative impact on the function of the photosynthetic system. From recent research, it can also be concluded that there is a substantial relationship between potential leaf photosynthesis and the highest possible crop yield. An examination at the level of the plant canopy might also give information about the plant's overall health. As opposed to analysis and sampling of individual leaves of the plant, the advantage of doing analysis at the plant canopy level is the simplicity with which it can be accomplished. When studying the responses of a crop such as wheat, the examination of the wheat canopy is a better alternative for understanding the responses of the wheat plant to various stimuli or stresses [25]. This is due to the fact that the acreage of this agricultural plantation is more than that of other crops throughout the world. In addition to doing analyses at the leaf and canopy levels, researchers are conducting analyses on virtually every part of the wheat plant, including the roots, ears, and kernels. However, from the perspective of Chlorophyll fluorescence, it is often done at the level of individual leaf samples or at the level of the entire canopy of the plant. Researchers in the hyper-spectral imaging sector are also undertaking study at the canopy level to analyses agricultural yields and damage, which can be seen as a positive development. There is evidence of studies that have used thermal imaging for analyzing wheat crop production and other features, as well as for other purposes. However, there are more citations regarding the use of visible light spectrum for building systems for detecting nitrogen stress in the plants [26]. A python library named as Plants and computer vision (plant CV) [35] is an extensive resource for building stress detection systems in crops and plants. The library supports segmentation of plant parts through multiple image processing methods such as Otsu, Triangle etc. The plant CV methods also allow developers to extract shape features and texture features of the plants and at the same time allow algorithms that can track the changes in the different parts of the plants.

In terms of technology implementation, the fundamental advantage of adopting imaging modalities is that most of these methods are non-invasive and offer a variety of advantages over other approaches [27–29]. In general, it can be observed that statistical analysis of imaging data is carried out extensively for the purposes of correlation and prediction of wheat crop growth, as well as for describing and describing the various aspects of wheat crop growth, particularly different types of traits in relation to specific environmental conditions. Analysis of Variance (ANOVA), correlation, and factor analysis are some of the most widely used statistical approaches [24]. Time-series analysis, data smoothing methods such as sliding window averaging, and a variety of other techniques are available. When these statistical methods are applied to wheat research, they help to advance the field by estimating, correlating, forecasting, and modeling a variety of elements that influence the growth of the wheat plant. Recent improvements, as seen by recent advances in modern literature, demonstrate that machine learning and deep learning algorithms are having a significant impact on wheat research right now. Deeper readings show that the most frequently used algorithm in context of wheat research is neural networks [30] and most of the work is directed towards identifying and classifying various types of pathogens and diseases that occur in wheat crops such as red rust. Due to availability issues related to public chlorophyll fluorescence datasets in this context and limited annotated dataset many researchers have resorted to the use of deep learning models for building wheat crop related models such as Wheatnet [31–33]. In this research work, an attempt has been made to understand the impact of the pre-process methods that improve the quality of the images.

3   Experimental Setup

In this section, various steps taken to construct a suitable experimental step for analyzing the impact of resolution on the classification model constructed for detecting nitrogen deficiency in the wheat canopy plants are discussed. The dataset was obtained from a publicly accessible repository [34]. It contains chlorophyll fluorescence images of the wheat variety PBW550. There were 1200 images in the dataset. The images capture the plants' canopy in vegetative state, and all the images have three channels (RGB). The block diagram in Fig. 1 can be referred to:

From Fig. 1, it can be observed that the images must undergo the process of improvement using resolution methods. Chlorofluorescence wheat images dataset was used to perform the experiment. Cubic interpolation resolution technique was applied on the dataset and images were scaled to $2\times$ and $3\times$. Then the segmentation

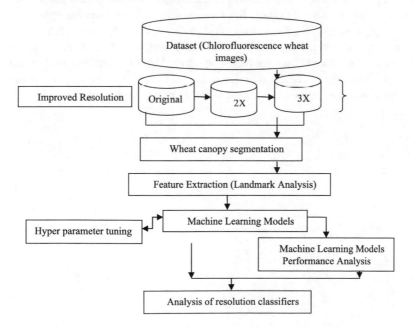

Fig. 1 Block diagram of the proposed method

was performed on the wheat canopy. Next landmark analysis was done in which the key features were extracted. In the next step, machine learning models were applied where hyper parameters were tuned for the learning algorithm. Lastly, the performance of the classifiers was analyzed.

The following steps accounts for the details involved in achieving the proposed work.

Step 1: Image improvement through resolution techniques

Resolution methods were applied on chlorofluorescence images to increase their crispness and pixel information. The corresponding images were recreated using resolution methods. Table 1 shows the results of resolution on wheat images.

Step 2: Segmentation results

Table 2 gives the output of the images after application of the canopy segmentation algorithm. The segmentation of the wheat canopy has been used using python plant CV library.

With the wheat canopy segmentation, one can distinguish between the backdrop and foreground of the images visually easily. The segmentation technique aids in the identification of different parts of the wheat canopy that are undergoing photosynthetic activity. According to a preliminary analysis of the information contained inside the photos, each image contains bimodal data, with the variance between the foreground and the background being far too tiny. Low variance between the foreground and background make the segmentation technically hard. It has been determined that the initial excitation (F_o) of the photosynthetic activity is characterized by low contrast and saturation in the images, and that as the photosynthetic activity becomes more vigorous, the number of pixels and the intensity of pixel increase (F_m). As a result, algorithms for segmenting the foreground (wheat canopy) and background (pot and experiment room background), such as Otsu, can be used to extract the wheat canopy from the backdrop image. Experimentation has revealed that alternative methods of segmentation, when used in conjunction with wheat canopy segmentation, do not produce accurate wheat canopy segmentation.

Table 1 Resolution output

Wheat Canopy Original	Resolution Results (Scale= 2)	Resolution Results (Scale= 3)

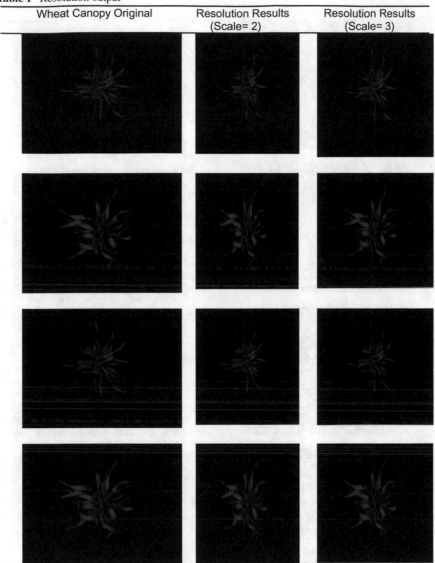

Step 3: Wheat Canopy Landmark Analysis

In conjunction with computer vision techniques, images data collection systems can be used to increase the temporal resolution at which plant phenotypes can be recorded non-destructively, allowing for more accurate measurements of plant morphology. The forms of plant leaves and the corresponding plant canopy

P. Datta et al.

Table 2 Segmentation output

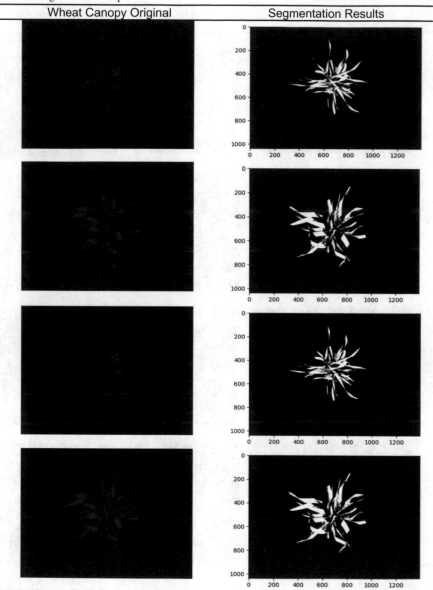

Table 3 Nitrogen deficient and control images

Nitrogen Deficient Images	Control Images

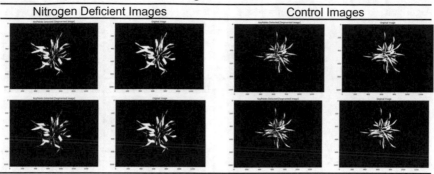

formed are important characteristics for biologists because they can aid in the identification of plant species, the measurement of their health, the analysis of their growth patterns, and the understanding of the relationships between different kinds of plant's responses. Landmarks are often geometric points that are placed on a shape's contours and represent biological aspects for which investigation has been carried out. The shape, tips of leaves, as well as the pedicel and branch angles, are all potential landmarks for understanding the changes that occur when a plant responds to a particular stress. In the context of our research work, we have drawn landmarks using cv2 python library (Table 3).

It can be observed visually that there is a significant difference in the way the landmarks get drawn on the wheat canopy shape. Clearly, it depends on the shape of the wheat canopy generated as per control or nitrogen condition. In case of the nitrogen deficient canopy there wheat canopy get fragmented and the leaves have lesser area. Hence, the number of landmarks increase and are more dispersed. The dataset of these landmarks points will be subjected to machine learning algorithms so that the process of identifying Nitrogen deficiency can be automated.

4 Results and Discussions

This section gives details on the process of evaluation of the system at each step and discusses technically the performance of all the algorithms involved in this research work. Table 4 gives the final values of the performance metrics of all the algorithms used for construction of the Nitrogen deficiency detection binary classifier. In total, four machine learning models were evaluated and after the due cross validation using 10 k-fold the classifier was appropriated for the above mentioned goals of the study. An extensive regime for hyper-parameter tuning was used to finally arrive at the best trade between the performance metrics. The important hyper -parameters values are also mentioned in the algorithm.

Table 4 Impact of resolution on accuracy and prediction time

Algorithm	Base performance		2 × Resolution scale (Cubic)		3 × Resolution scale (Cubic)	
	Average accuracy	Average prediction time (ms)	Average accuracy	Average prediction time (ms)	Average accuracy	Average prediction time (ms)
Support Vector Machine (SVM) (c = 1, gamma = 1/n_features, kernel = poly)	0.84	14.94	0.87	18.94	0.89	20.14
DecisionTreeClassifier (depth = 6 weights = balanced max_features = sqrt alpha = 0.1)	0.95	2.46	0.96	3.21	0.98	3.99
Naive Bayes (var_smoothing = 1e–09 sample_weight = individual)	1	2.73	1	3.13	1	3.73
KNeighborsClassifier (neighbors = 6 weights = distance algorithm = kd_tree leaf_size = 30 distance = minkowski)	0.91	5.86	0.93	6.18	0.95	6.99

In order to determine the consistency of the best algorithm's performance, a 10 k-fold methodology was employed. This method allows for an in-depth evaluation of the consistency of performance. The Interquartile Range (IQR) method is used to calculate the averages of the data. Avoiding bias in performance calculations is possible by computing averages with IQR. With regard to the resolution, the classifier models' performance has improved. This verification supports the fact that the increase in resolution enhances the effectiveness of increased computational information in increasing image correctness. An increase in the number of landmarks points and accompanying score values was seen when a cubic-based interpolation method was utilized to improve the resolution.

Increased resolution also increases the amount of time necessary to train the classifier, which is a negative effect. It's due to the fact that the quantity of data points utilized for training the 2× and 3× feature set is increasing as time marches by. While the SVM method performs poorly, the Naive Bayes algorithm is the best performer on the scale. As an additional point to note, the performance of Naive Bayes remains constant at both the 2× and 3× scales. The decision tree technique is the second-best algorithm in terms of accuracy and forecasting time.

5 Conclusions

In this research work, we have constructed a system where nitrogen deficiency is detected using Naïve Bayes algorithm. Extensive experimentation and fine-tuning of the supervised machine Learning Models were undertaken, and empirically, it was found that there was no need to either improvise or shift to deep learning algorithms for higher levels of accuracy as Naïve Bayes algorithm was yield to 100% accuracy and at the same time on prediction time was also good. To the best of our knowledge, little work has been done in the context of assessing chlorophyll fluorescence imaging quality in terms of resolution and using them for detecting nitrogen stress in wheat plants. The current research work has been done on a single variety of the wheat crop. For further research work additional wheat varieties may be used, and also the plant material may be imaged on a plant growth chamber that can control the environment (such as temperature, humidity, light intensity and other factors), which could be an added advantage.

References

1. Sharma I, India BR, Prasad P, Bhardwaj SC (2017) Recent molecular technologies for tackling wheat diseases. Achieving Sustain Cultivation Wheat 1:385–416
2. Ghosh M, Swain DK, Jha MK, Tewari VK (2020) Chlorophyll meter-based nitrogen management in a rice–wheat cropping system in Eastern India. Int J Plant Prod 14(2):355–371
3. Moustakas M, Calatayud Á, Guidi L (2021) Editorial: chlorophyll fluorescence imaging analysis in biotic and abiotic stress. Front Plant Sci 12:658500
4. Moustakas M, Calatayud A, Guidi L (2021) Chlorophyll fluorescence imaging analysis in biotic and abiotic stress
5. Singh B, Jasrotia P (2020) Impact of integrated pest management (IPM) module on major insect-pests of wheat and their natural enemies in North-western plains of India. J Cereal Res 12(2):100185
6. Feng W, Li X, Wang Y-H, Wang C-Y, Guo T-C (2013) Difference of chlorophyll fluorescence parameters in leaves at different positions and its relationship with nitrogen content in winter wheat plant. Acta Agron Sin 38(4):657–664
7. Bhusal N, Sharma P, Sareen S, Sarial AK (2018) Mapping QTLs for chlorophyll content and chlorophyll fluorescence in wheat under heat stress. Biol Plant 62(4):721–731
8. Yang C et al (2021) [Method for estimating relative chlorophyll content in wheat leaves based on chlorophyll fluorescence parameters]. Ying yong sheng tai xue bao bian ji wei yuan hui=J Appl Ecol 32(1):175–181
9. Plaza-Bonilla D, Lampurlanés J, Fernández FG, Cantero-Martínez C (2021) Nitrogen fertilization strategies for improved Mediterranean rainfed wheat and barley performance and water and nitrogen use efficiency. Eur J Agron 124:126328
10. Sharma S, Singh P, Choudhary OP, Neemisha (2021) Nitrogen and rice straw incorporation impact nitrogen use efficiency, soil nitrogen pools and enzyme activity in rice-wheat system in north-western India. Field Crops Res 266:108131
11. Spyroglou I, Rybka K, Rodriguez RM, Stefański P, Valasevich NM (2021) Quantitative estimation of water status in field-grown wheat using beta mixed regression modelling based on fast chlorophyll fluorescence transients: a method for drought tolerance estimation. J Agron Crop Sci 207(4):589–605

12. Ni Z, Lu Q, Huo H, Zhang H (2019) Estimation of chlorophyll fluorescence at different scales: a review. Sensors 19(13):3000
13. Jia M et al (2021) Estimation of leaf nitrogen content and photosynthetic nitrogen use efficiency in wheat using sun-induced chlorophyll fluorescence at the leaf and canopy scales. Eur J Agron 122:126192
14. Lawson T, Vialet-Chabrand S (2018) Chlorophyll fluorescence imaging. Methods Mol Biol (Clifton, N.J.) 1770:121–140
15. Baker NR, Rosenqvist E (2004) Applications of chlorophyll fluorescence can improve crop production strategies: an examination of future possibilities. J Exp Bot 55(403):1607–1621
16. Wu W-M, Chen H-J, Li J-C, Wei F-Z, Wang S-J, Zhou X-H (2013) Effects of nitrogen fertilization on chlorophyll fluorescence parameters of flag leaf and grain filling in winter wheat suffered waterlogging at booting stage. Acta Agron Sin 38(6):1088–1096
17. Thakur V, Pandey GC (2020) Effect of water scarcity and high temperature on wheat productivity, pp 251–275
18. Lysenko V (2011) Fluorescence kinetic parameters and cyclic electron transport in guard cell chloroplasts of chlorophyll-deficient leaf tissues from variegated weeping fig (Ficus benjamina L.). Planta 235(5):1023–1033
19. Lamb JJ, Eaton-Rye JJ, Hohmann-Marriott MF (2012) An LED-based fluorometer for chlorophyll quantification in the laboratory and in the field. Photosynth Res 114(1):59–68. https://doi.org/10.1007/s11120-012-9777-y
20. Takeuchi A, Yoshida H, Shibata M (2009) Development of simplified PAM chlorophyll fluorometer for vegetation condition monitoring
21. Lichtenthaler HK, Buschmann C, Knapp M (2005) How to correctly determine the different chlorophyll fluorescence parameters and the chlorophyll fluorescence decrease ratio RFd of leaves with the PAM fluorometer. Photosynthetica 43(3):379–393
22. Mishra AN (2018) Chlorophyll fluorescence: a practical approach to study ecophysiology of green plants, pp 77–97
23. Takayama K (2014) Chlorophyll fluorescence imaging for plant health monitoring, pp 207–228
24. Upadhyay K (2020) Correlation and path coefficient analysis among yield and yield attributing traits of wheat (Triticum aestivum L.) genotypes. Arch Agric Environ Sci 5(2):196–199
25. Hupp S, Rosenkranz M, Bonfig K, Pandey C, Roitsch T (2019) Noninvasive phenotyping of plant-pathogen interaction: consecutive in situ imaging of fluorescent pseudomonas syringae, plant phenolic fluorescence, and chlorophyll fluorescence in Arabidopsis leaves. Front Plant Sci 10:1239
26. Sarkar U, Banerjee G, Ghosh I (2021) A machine learning based fertilizer recommendation system for paddy and wheat in West Bengal, pp 163–174
27. Cao J et al (2021) Wheat yield predictions at a county and field scale with deep learning, machine learning, and google earth engine. Eur J Agron 123:126204
28. Detecting and Distinguishing Wheat Diseases using Image Processing and Machine Learning Algorithms 2020
29. Gómez D, Salvador P, Sanz J, Casanova JL (2021) Modelling wheat yield with antecedent information, satellite and climate data using machine learning methods in Mexico. Agric Forest Meteorol 300:108317
30. Fan Y, Ma S, Wu T (2020) Individual wheat kernels vigor assessment based on NIR spectroscopy coupled with machine learning methodologies. Infrared Phys Technol 105:103213
31. Watt M, Fiorani F, Usadel B, Rascher U, Muller O, Schurr U (2020) Phenotyping: new windows into the plant for breeders. Ann Rev Plant Biol Ann Rev 71:689–712
32. Lee T et al (2017) WheatNet: a genome-scale functional network for Hexaploid bread wheat, Triticum aestivum. Mol Plant 10(8):1133–1136
33. Genaev M, Ekaterina S, Afonnikov D (2020) Application of neural networks to image recognition of wheat rust diseases
34. https://data.mendeley.com/research-data/?type=DATASET&search=wheat%20canopy%20Sukhjit%20sandhu
35. https://pypi.org/project/plantcv/

Exploring Practical Deep Learning Approaches for English-to-Hindi Image Caption Translation Using Transformers and Object Detectors

Paritosh Bisht⑩ and **Arun Solanki**⑩

1 Background

In the past few years, significant progress has been made using deep learning techniques in the fields of computer vision and natural language. This has mainly been due to the availability of massive datasets that are being created using data sourced from the internet. Another major reason has been the substantial increase in compute available for training and deployment of large models. This has also resulted in an increased interest in multi-modal machine learning techniques and their applications for handling multi-modal data. The aim of multi-modal machine learning is draw inferences from multi-modal data using a single machine learning system for prediction. This also involves bridging the gap between different types of data which are digitally represented in completely different ways like text and images. Theoretically the field deals with challenges in representing, translating, mapping, aligning, fusing and learning the different modalities or datatypes [1]. On the other hand, a lot of real-world applications deal with multi-modal data. One such application is automatic image captioning which involves creating text annotations in natural language to describe images [2–5].

Recent work has shown that Image captioning requires high quality data due to the difficulties faced in representing and mapping the two modalities [6]. It has also been found that it is very difficult to evaluate captions. This has resulted in the creation of specialized tasks and datasets like Visual Question Answering and Visual Dialog [7, 8]. Nonetheless, problems with handling multimodality remains a major challenge in the field [9–11]. It has hence been thought that the deep learning architectures proposed for image captioning are made practically viable mainly by being trained

P. Bisht (✉) · A. Solanki
Gautam Buddha University, Greater Noida 201308, India
e-mail: bishtparitosh@gmail.com

A. Solanki
e-mail: asolanki@gbu.ac.in

© The Author(s), under exclusive license to Springer Nature Singapore Pte Ltd. 2022   47
B. Unhelker et al. (eds.), *Applications of Artificial Intelligence and Machine Learning*,
Lecture Notes in Electrical Engineering 925,
https://doi.org/10.1007/978-981-19-4831-2_5

on large well annotated datasets like MSCOCO captions and Conceptual Captions [12, 13]. With the rise of transformer models and the recent trends in computer vision and natural language, the current successful approaches are mainly focused on creating large models trained on even more massive datasets [14]. Unfortunately, these approaches are not viable for low resource applications as most of the data is in English and there are few such datasets available for Image captioning in other languages like Hindi. The task of Multimodal caption translation has hence been proposed to enable captioning in other languages [15, 16]. It involves utilizing bilingual caption pairs (or multilingual caption sets) that can be used for training specialized models for automatic translation of captions. Captions generated in English can then be translated into other languages and image features can also be used to improve the quality of translation [17, 18].

Previous work on image caption translation using deep learning for Hindi has mainly focused on extending Machine translation systems by introducing multimodality inspired by Image captioning systems [19–22]. Many of these involve encoding and fusing features extracted from the images to the intermediary outputs of machine translation systems. While these approaches show some improvement over text only baselines, they also increase model complexity, interpretability and computational requirements. On the other side, large text only transformer models pre-trained on massive parallel corpora for bilingual and multilingual translation can also be fine-tuned for downstream tasks like caption translation without using any image information. It is not clear which approach is better for real world use and the subsequent work aims to answer this question [23, 24].

2   Dataset, Methodology and Experimental Setup

Proposed work aims to explore various deep learning models that can be trained for automatic translation of captions from English to Hindi. It focuses on models based on sequence to sequence and transformer architectures which are extensively used for text translation [25–27]. Variants of these models with multi-modal capabilities have also been taken.

2.1   Dataset

Hindi Visual Genome (HVG) dataset has been selected for training and evaluation [28, 29]. The dataset consists of 28 k English and Hindi caption pairs aligned to a region of interest in a reference image for each instance of the dataset. It contains 4 disjoint subsets: 'train', 'dev', 'test' and 'challenge test' sets. While the first three sets have been made from the same distribution, the later has been made by taking the most ambiguous caption pairs where the ambiguity can be resolved by visual understanding of the associated image region. The models selected are trained on

image_id	X	Y	width	height	eng	hi
13	41	368	244	329.0	225	a white microwave oven	एक सफेद माइक्रोवेव ओवन
23	86	389	37	146.0	333	a man reading a book	एक आदमी एक किताब पढ़ रहा है
33	108	610	321	170.0	73	a black car with a yellow door	पीले दरवाजे के साथ एक काली कार
53	179	91	198	64.0	88	a gray laundry basket	एक ग्रे कपड़े धोने की टोकरी
73	227	25	456	171.0	104	white ambulance on the street	सड़क पर सफेद एम्बुलेंस
243	939	92	96	127.0	162	a seagull is standing	एक सीगल खड़ा है

Fig. 1 A few data samples from the 'train' set of Hindi visual genome

the 'train' set and utilize 'dev' set for evaluation while training. For prediction and final evaluation, the 'test' and 'challenge test' sets have been used (Fig. 1).

2.2 Experimental Setup

All the models have been trained on a system with the following system configuration: Ubuntu 20.04 laptop with Intel i7-8750 h, 16 GB 2400 MHz DDR4 RAM, Nvidia GTX 1070 Mobile with 8 GB VRAM and 512 GB NVMe PCIe-SSD. The models have been implemented in python 3.7 with PyTorch 1.7.

2.3 Pre-processing Dataset

All captions in the dataset have been pre-processed by removing all special characters including punctuation for both languages. As majority of the captions consist of short phrases instead of grammatically complete sentences, there is no significant loss of information. Roman characters have also been removed from the target Hindi captions retaining only Devanagari characters. All characters for input English captions are also lower-cased to ensure homogeneity. For the images in the dataset, regions of interests have been extracted from the images and stored as separate images referred to as 'box images'. The associated full images have also been stored as 'full images'. Further processing on both the set of images is to be done during run-time.

2.4 Run-Time Data Handling

For all of the experiments, a standard data handling pipeline has been created. It consists of a custom torch dataset for loading text and image data from the file system and a custom data-loader for creating batches of the data samples from the dataset with associated samplers and collator. Distinct word vocabularies for Hindi and English data have been created from the words in the 'train' and 'dev' sets. These have also been used to tokenize and numericalize the captions while sampling. English captions have been tokenized using NLTK toolkit and Hindi captions using the Indic-NLP library. Tokens <SOS> and <EOS> have been added to each caption at the start and end of the sequence. The text data samples in the batch have then been sorted in decreasing order of length. These have finally been padded to the maximum length caption from the batch by adding <PAD> tokens. While training, a few random perturbations have been added to the process for better generalization.

Images when used have been loaded while sampling, resized according to the model requirements and normalized according to Torchvision defaults. While training, random square crop of maximum size has been taken from the resized images while center crop has been taken for evaluation and prediction.

2.5 Training and Evaluation

For training and evaluation of all sequence to sequence and core transformer models, PyTorch Lightning framework has been used. All the model hyperparameters were manually tuned by grid search along with trial and error. The best performing models for each category were selected. These have then been trained for a maximum of 40 epochs with check-pointing and early stopping on the average training loss defined after each epoch. The final checkpoint saved has been selected for evaluation. Training and validation logs have also been saved using Tensorboard.

All the models have been validated on several natural language automatic evaluation metrics to judge overall the performance of the models. Word (WD-ACC) and character (CH-ACC) accuracy have been used for verifying the general correspondence between the target and generated captions. These are determined by taking the uni-gram precision of the set of words and characters respectively in the output captions to the target captions. The output Hindi captions have also been evaluated against target Hindi captions on BLEU-4 and METEOR algorithms for verifying n-gram precision [30, 31]. WER, TER and chrF metrics have also been calculated for evaluating word substitution, word edit rates and character n-gram match scores respectively [32–34]. Finally, RIBES has also been calculated as an alternative evaluation metric that is not based on word boundaries [35]. These have all been evaluated using the Vizseq toolkit. The output captions from the models have also been validated subjectively by native speakers for grammatical correctness, legibility and quality.

For the final pre-trained transformer implementation, Huggingface Transformers library with the Seq2SeqTrainer has been used for training and the Translation-Pipeline has been used for evaluation. METEOR and BLEU-4 scores have also been evaluated during training and used for optimal model selection. Training and validation logs have been saved using weights and biases.

To ensure reproducibility, all random number generator seeds have been set to 0. Models have been trained multiple times to ensure that there was no significant difference between multiple runs with the final trained instance taken for evaluation (Fig. 2).

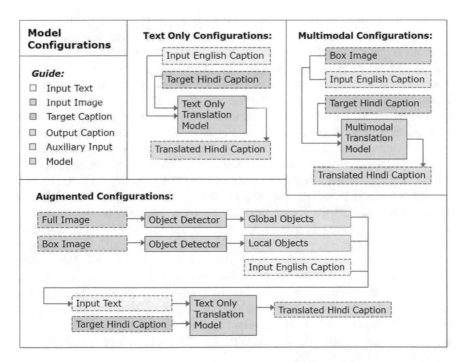

Fig. 2 Model data-flow diagrams while training for text only (top middle), multi-modal (top right) and augmented text (bottom) configurations

3 Model Architectures and Configurations Tested

The model architectures selected for caption translation are divided into three types of configurations based on the input to the model. In the text only configurations, the model directly takes the source Hindi and target English captions as input. In the multi-modal configurations, along with the text captions, box images are also processed as input. In the augmented text configurations, object classes corresponding to the objects detected are taken from the box and/or full images by a distinct object detector model. These are then added to the source English captions and processed as input. This has been done with the expectation that with the information from both the different text and image modalities expressed in a singular form, the model would generalize for both the input data and correct inconsistent information learned from the caption data by validating with the appropriate class data. Final output are the translated Hindi captions from all the configurations.

3.1 Class A (Seq2Seq Caption Translation)

These set of models are based on sequence to sequence (seq2seq) architecture for machine translation which use recurrent neural network layers to store information about the sequential relationships in the data [25, 26]. The architecture consists of a distinct encoder and decoder. The encoder works by generating an intermediate representation of the input sample which is then propagated to the decoder. The decoder uses this information and the learned relationships between in the output data to generate the output sequence one token at a time. These relations are usually represented in a high dimensional embedding space for each token [36]. The model learns alignment between the two sequences automatically as it sees more data.

In the proposed implementation for caption translation, the encoder and decoder consist of 512-dimension word embeddings with 20% dropout each for English and Hindi tokenized captions respectively. The decoder slowly learns to generate the next token and teacher forced learning has been used to optimize the process with 50% of the generated tokens. Cross-entropy loss has then been calculated for each generated output and gradients accumulated for the entire batch and propagated accordingly [37]. RELU activation function has been used throughout the architecture [38]. During training, ADAM optimizer has been used with learning rate of 1e–3 and default decay weights for optimizing the gradient descent [39]. While during evaluation and prediction, Greedy search has been used for selecting the best token generated by the decoder to generate the Hindi translated caption. The maximum length of the generated caption has been fixed to 24 words with early truncation on encountering the <EOS> token.

In "Config A1" (Text Only Seq2Seq without Attention), two LSTM layers of 1024 nodes have been used in both encoder and decoder for modeling sequences [40]. LSTM outputs along with the hidden and cell states from the encoder have been passed to the decoder which have been used to initialize the decoder LSTM layers. LSTM outputs from the decoder have then been fed to a linear layer to generate probabilistic predictions for each token of the Hindi vocabulary which have then been used to generate the output captions word by word. Both of these configurations have been trained with 512 batch size for 40 epochs. In "Config A2" (Text Only Seq2Seq with Attention), a bidirectional LSTM of (1024 × 2) nodes has been used in the encoder for modeling English data, and a single layer of 2048 nodes has been used for modeling Hindi data. LSTM hidden and cell states from both layers have each been concatenated separately and propagated to the decoder to initialize the decoder LSTM. LSTM outputs from the encoder have also been send to the decoder where self-attention has been applied to create an intermediate context vector. This vector has then been fed to the decoder LSTM to generate the output token sequence at each step. In "Config A3" (Multimodal Seq2Seq with Attention), an additional image encoder which consists of a feature extractor and an image attention module has also been added to the network. The feature extractor consists of Torchvision VGG-19 model pre-trained on ImageNet [41, 42]. The last two layers have been replaced by a two layer dense network to generate features corresponding to the English vocabulary size. The image attention module then applies multiplicative attention to the extracted image features for generating a context vector of encoder LSTM size. This has then been concatenated with the outputs from the Decoder LSTM to generate the output tokens. This model variant has been trained with 64 batch size and training early stopped at 37 epochs.

3.2   Class B (Transformer Caption Translation)

These set of models are based on the vanilla transformer architecture for machine translation which use feed-forward and self-attention layers to store relationships between the tokens of sequential data [27, 43]. Similar to the seq2seq architecture, the design consists of a distinct encoder and decoder for input and output sequential data respectively. Both of the modules consist of sequences of transformer blocks. These blocks each consist of a multi-head attention layer for learning different types of relations in the data, and a feed forward layer for storing the relationships. The multi-head attention layer has multiple probing heads each with a key, query and value relation vector for scaled dot product attention. Data is input to the two layers through adding and normalizing residual connections with the older input for better generalization. Output from the last encoder transformer block is fed to the second decoder block multi-head attention key and query relations along with the data from the first decoder block. The first decoder block consists of an extra masked multi-head self-attention layer for modeling the output sequence data to generalize for the words not in the output vocabulary. Input to the both the modules is through word embeddings similar to sequence-to-sequence networks but with added position embedding for sequence localization during learning.

In the proposed implementation, 768-dimension word and linear position embeddings of the English and Hindi caption token sequences have been fed as input to the transformer encoder and decoder respectively. The number of encoder and decoder blocks have been set to 4, with 16 heads of multi-head attention. Feed-forward layer dimension has been set to 4 times the embedding size, with 10% dropout and GELU activation function used across the entire architecture [44]. ADAM optimizer with learning rate of 1e-4 has been used with default decay weights. During evaluation and prediction, Greedy search has been utilized for token selection during evaluation and the maximum caption length has been fixed to 48 words with early truncation on encountering the <EOS> token similar to the seq2seq architecture.

In "Config B1" (Text Only Transformer), the transformer architecture as described above has been directly used with English and Hindi captions from the preprocessed dataset and no information from the image data has been taken. In "Config B2" (Augmented Transformer with Box Image Objects Appended), object classes detected on the box images with 60% or more accuracy by the torchvision Mask-RCNN object detector pre-trained on the MSCOCO dataset have been added to the input English captions with a special <BBOX> token to separate the two and then tokenized together [45, 46]. In "Config B3" (Augmented Transformer with Box and Full Image Objects Appended), object classes detected on first the box images have been appended with the special <BBOX> separator token and then the object classes from the full images have been appended with a special <FULL> separator token. In "Config B4" (Uni-modal Transformer with Box Image Objects Prepended), the object classes detected have been appended before the captions and in the "Config B5" (Augmented Transformer with Box and Full Image Objects Prepended), both image sets detected have been appended before the captions along with the <BBOX> and <FULL> separator token. All of these model variants have been trained with 196 batch size for 40 epochs. In the final "Config B6" (Multi-modal Transformer), a separate module for creating patch and position embeddings from input box images has been added to the network. These embeddings are fed to the transformer architecture along with the text data for machine translation. This configuration is inspired by previous work on Vision transformer and Multimodal transformer architectures [47–50]. The box images taken have been resized to ($112 \times 112 \times 3$) and patch embeddings have been created for each image using a convolutional layer with padding and stride equal to the patch size. A uniform patch size of (16×16) has been taken, and hence patch embeddings of size 64 have been created for each input image. Corresponding position embeddings have been also created and concatenated to the patch embeddings. Finally, the encoder word and position embeddings corresponding to English source captions have been concatenated together and the resultant tensor has been fed to the transformer encoder. This model variant has been trained with batch size of 64 for 40 epochs.

3.3 Class C (Pre-trained Transformer Caption Translation)

As a final comparison, the MarianMT pre-trained transformer was taken which was created by Jörg Tiedemann from University of Helsinki using the Marian C++ library [51, 52]. It is a highly efficient transformer model optimized from bilingual translation task. The "Config C1" (Text Only Finetuned MarianMT) uses this model which has been loaded using the Huggingface transformer library with the pre-trained weights from OPUS-MT dataset available on the model hub. Marian tokenizer based on sentence-piece tokenization has been used for model inputs and outputs [53]. The weights have been fine-tuned by training the model with batch size of 32 on the training set for 15 epochs, check-pointing on each epoch and selecting the checkpoint with the best metric scores.

4 Results

Results from the evaluation metrics show that most of the model configurations chosen are able to translate the English captions to Hindi with reasonably high precision (Table 1, 2).

The Class A (seq2seq) configurations are able to achieve around 68% word and 82% character precision on the 'Test' set, and around 51% word and 70% character precision on the 'Challenge Test' set. The big drop in precision implies that the seq2seq models are not able to generalize across more ambiguous captions. This is also highlighted by the n-gram based metric scores where the configurations are able to achieve respectable scores of around 36 BLEU-4, 0.59 METEOR and 0.58 RIBES on the 'Dev' and 'Test' sets but are only able to achieve around 20 BLEU-4, 0.42 METEOR and 0.41 RIBES on the 'Challenge Test' set. The error rates (Table 3) also show a similar trend with 35–43% increase in word and translation error rates, and around 30% decrease in ChrF.

Table 1 Evaluation metrics for all the configurations on the 'Test' set

Models	WD-ACC	CH-ACC	BLEU-4	METEOR	RIBES	WER	TER	ChrF
Config A1	0.684	0.819	35.635	0.600	0.583	45.784	0.432	0.544
Config A2	0.674	0.808	34.289	0.584	0.570	47.018	0.444	0.529
Config A3	0.682	0.816	35.365	0.598	0.581	46.144	0.438	0.540
Config B1	0.727	0.863	41.815	0.653	0.630	41.684	0.391	0.618
Config B2	0.735	0.867	41.250	0.662	0.627	41.272	0.388	0.620
Config B3	0.727	0.861	42.637	0.654	0.625	41.015	0.383	0.617
Config B4	0.724	0.861	40.228	0.648	0.608	42.378	0.399	0.611
Config B5	0.721	0.862	39.853	0.645	0.615	42.776	0.405	0.608
Config B6	0.717	0.857	36.232	0.630	0.599	48.483	0.496	0.588
Config C1	0.753	0.884	45.074	0.682	0.639	38.560	0.369	0.656

Table 2 Evaluation metrics for all the configurations on the 'Challenge Test' set

Models	WD-ACC	CH-ACC	BLEU-4	METEOR	RIBES	WER	TER	ChrF
Config A1	0.511	0.706	20.191	0.420	0.411	62.410	0.589	0.375
Config A2	0.508	0.695	20.301	0.418	0.411	61.691	0.583	0.374
Config A3	0.516	0.697	20.366	0.427	0.422	61.877	0.585	0.377
Config B1	0.631	0.807	33.508	0.564	0.522	50.557	0.461	0.530
Config B2	0.621	0.799	32.913	0.556	0.518	50.800	0.468	0.519
Config B3	0.621	0.805	32.189	0.553	0.513	51.160	0.465	0.522
Config B4	0.616	0.797	31.569	0.548	0.518	52.366	0.477	0.514
Config B5	0.611	0.796	31.278	0.541	0.506	53.584	0.491	0.508
Config B6	0.589	0.782	28.206	0.511	0.479	60.206	0.579	0.471
Config C1	0.718	0.869	46.522	0.666	0.612	41.069	0.366	0.645

Table 3 Select configurations compared with existing work

Models	'Test Set' scores		'Ch-Test Set' scores	
	BLEU-4	RIBES	BLEU-4	RIBES
Proposed (Config A1)	35.64	0.583	20.19	0.411
Proposed (Config B1)	41.82	0.630	33.51	0.522
Proposed (Config C1)	**45.07**	0.639	**46.52**	**0.612**
Meetei et al. (NIT Silchar 2019)	28.45	0.630	12.58	0.480
Laskar et al. (NIT Silchar 2020)	40.51	**0.803**	33.57	**0.754**
Parida et al. (Idiap NMT 2019)	41.32	—	30.94	—
Parida et al. (ODIANLP 2020)	40.85	—	38.50	—
Kaur et al. (Model 7)	**47.52**	0.693	**42.52**	0.507

The Class B (transformer) configurations also show a drop in precision across the evaluation sets. They are able to achieve around 73% word and 87% character precision on 'Test' set, but only around 63% word and 80% character precision on the 'Challenge Test' set. Similarly, the configurations are able to achieve higher scores of around 43 BLEU-4, 0.66 METEOR and 0.62 RIBES on the 'Dev' and 'Test' sets but only around 32 BLEU-4, 0.55 METEOR and 0.51 RIBES on the 'Challenge Test' set. The error rates also show the same trend with 27–35% increase in word and translation error rates, and around 20% decrease in ChrF. It can be seen that the transformer-based models outperformed the seq2seq models on all metrics by a significant margin. The drop in metric scores across evaluation sets is also lower which implies better generalization to more ambiguous captions. On the other hand, the multi-modal transformer tested is negatively impacted by adding image modality. This may be due to the relatively low number of training data images which resulted in the patch embedding not being very effective. No difference was observed with different patch or embedding sizes. Similarly, augmented text transformer configurations also show negligible difference from the text only configurations on all the metrics. But on the other hand, they also show no significant deterioration in the translated caption quality and a few captions even had minor improvements on human evaluation. The models can also be used with no object classes attached with almost identical results to the text only approaches. Subjectively, this is found to be a better method as it is modular, requires significantly less compute for training and evaluation and is overall also more data efficient than the multi-modal approach. From the different configurations, the box only configurations seem to perform better than when both box and full images are used, though the difference is small. There is also negligible difference between the append and prepend configurations probably due to the short length of the captions.

Finally, it can be seen that the Class C (pre-trained transformer) configuration of fine-tuned MarianMT outperforms all the other models by a significant amount on all the metrics. This configuration also did not show the drop in metric scores observed in the previous models which implies that it is able to model even the ambiguous captions accurately. This can be seen as it achieves 75% word and 88% character precision on the 'Test' set, and 72% word and 87% character precision on the 'Challenge Test' set. It also manages to achieve comparably higher scores of 46 BLEU-4, 0.68 METEOR and 0.64 RIBES on the 'Test' set and 46 BLEU-4, 0.67 METEOR and 0.61 RIBES on the 'Challenge Test' set than the other models. Similarly, this is only a difference of 6.5% in WER, under 1% in TER and 4.5% decrease in ChrF. The reason behind this difference compared to other models is probably due to relationships the model had previously learned from the data it was pre-trained on which helped it resolve ambiguous words accurately.

5 Comparison with Existing Work

When comparing the results obtained in this study with the existing work (Table 3), it can be seen that the models proposed have been able to compare favorably with the best performing models. While the multimodal model presented by Kaur et al. [19] is able to achieve the highest BLEU-4 score on the 'Test' set, the pre-trained transformer configuration presented in this study stomps all the systems on the 'Challenge Test' set. It has also been able to achieve the second score in the 'Test' set outperforming rest of the systems. The pre-trained transformer configuration also scored second on both 'Test' and Challenge Test' set in the RIBES metric behind the scores achieved by Laskar et al. [20] which are surprisingly higher. On the other hand, the vanilla text only transformer configuration has also been able to achieve respectable scores that are comparable against all the other models. While the text only seq2seq configuration has only been able to beat the system from Meetei et al. [22] on BLEU-4 score, it still shows respectable performance for a simple baseline.

6 Conclusion

Today, in a large number of computer vision and natural language applications, deep neural networks are being used to solve specialized problems [54–56]. Image caption translation is one such application where several deep learning based multi-modal approaches have been proposed in the past. Results obtained in this study contradict some of the previous results by finding that simpler text only approaches show better performance and efficiency over multimodal approaches. This may be due to the information that can be gained from image modality not being significant enough compared to the text modality when the amount of data is low and sparse as is in our case with the Hindi Visual Genome dataset. On the other hand, it can also be seen that the text modality is sufficient enough to get most of the information. This is highlighted by the unimodal text only approaches which show better results over the tested multimodal systems on even the 'Challenge Test' sets. With the availability of efficient text only transformer-based models pre-trained on massive amounts of Hindi-English bilingual data like the MarainMT tested that has been able to show amazing results, this conclusion is even more imperative. This may also be true for some other low resource multi-modal machine learning applications where text is the primary modality and the other modalities are auxiliary. For applications with large amounts of data available, like the other language pairs used in past image caption translation studies, multimodal approaches may still be better as highlighted by past studies. On the other hand, systems that use intermediate text modality (augmented text models) give a nice alternative to multimodal approaches if significant information is present in other modalities.

In the future several more studies are required in the fields of Hindi multimodal machine learning and subsequently for English-Hindi image caption translation task. There is a dire need for new massive and dense multimodal datasets to fuel further advances in this area. There is also a need for novel massive multilingual multi-modal transformer models and semi-supervised methods for training such models. New techniques and advances are also required in the areas of multi-modal representation, fusion and alignment. Until then, variants of text only transformers similar to ones proposed in the study are adequate for the task of Hindi image caption translation.

References

1. Baltrušaitis T, Ahuja C, Morency L (2019) Multimodal machine learning: a survey and taxonomy. IEEE Trans Pattern Anal Mach Intell 41:423–443
2. Anderson P (2018) Bottom-up and top-down attention for image captioning and visual question answering. In: 2018 IEEE/CVF conference on computer vision and pattern recognition, pp 6077–6086
3. Xu K et al (2015) Show, attend and tell: neural image caption generation with visual attention. ICML
4. Bernardi R et al (2016) Automatic description generation from images: a survey of models, datasets, and evaluation measures. J Artif Intell Res 55:409–442
5. Tanti M, Gatt A, Camilleri K (2018) Where to put the image in an image caption generator. Nat Lang Eng 24:467–489
6. Shekhar R, Takmaz E, Fernandez R, Bernardi R (2019) Evaluating the representational hub of language and vision models. arXiv abs/1904.06038
7. Agrawal A et al (2015) VQA: Visual Question Answering. Int J Comput Vis 123:4–31

8. Das A et al (2017) Visual dialog. In: 2017 IEEE conference on computer vision and pattern recognition (CVPR), pp 1080–1089
9. Agarwal S, Bui T, Lee JY, Konstas I, Reiser V (2020) History for visual dialog: do we really need it? ACL
10. Goyal Y, Khot T, Summers-Stay D, Batra D, Parikh D (2017) Making the V in VQA matter: elevating the role of image understanding in visual question answering. In: 2017 IEEE conference on computer vision and pattern recognition (CVPR), pp 6325–6334
11. Cao J, Gan Z, Cheng Y, Yu L, Chen YC, Liu J (2020) Behind the scene: revealing the secrets of pre-trained vision-and-language models. In: Vedaldi A, Bischof H, Brox T, Frahm JM (eds) Computer Vision – ECCV 2020. ECCV 2020. LNCS, vol 12351. Springer, Cham. https://doi.org/10.1007/978-3-030-58539-6_34
12. Chen X (2015) Microsoft COCO captions: data collection and evaluation server. arXiv abs/1504.00325
13. Sharma P, Ding N, Goodman S, Soricut R (2018) Conceptual captions: a cleaned, hypernymed, image alt-text dataset for automatic image captioning. ACL
14. Radford A (2021) Learning transferable visual models from natural language supervision. arXiv abs/2103.00020
15. Caglayan O (2019) Multimodal Machine Translation. Doctoral dissertation, Université du Maine
16. Barrault L, Bougares F, Specia L, Lala C, Elliott D, Frank S (2018) Findings of the third shared task on multimodal machine translation. WMT
17. Guasch SR, Costa-Jussà M (2016) WMT 2016 multimodal translation system description based on bidirectional recurrent neural networks with double-embeddings. WMT
18. Huang P, Liu F, Shiang S, Oh J, Dyer C (2016) Attention-based multimodal neural machine translation. WMT
19. Kaur J, Josan G (2020) English to Hindi multi modal image caption translation. J Sci Res 64:274–281
20. Laskar SR, Singh RP, Pakray P, Bandyopadhyay S (2019) English to Hindi multi-modal neural machine translation and Hindi image captioning. WAT@EMNLP-IJCNLP
21. Bojar O (2019) Idiap NMT system for WAT 2019 multimodal translation task. WAT@EMNLP-IJCNLP
22. Meetei LS, Singh TD, Bandyopadhyay S (2019) WAT2019: English-Hindi translation on Hindi visual genome dataset. WAT@EMNLP-IJCNLP
23. Caglayan O (2016) Does multimodality help human and machine for translation and image captioning? WMT
24. Caglayan O, Madhyastha P, Specia L, Barrault L (2019) Probing the need for visual context in multimodal machine translation. NAACL-HLT
25. Sutskever I, Vinyals O, Le QV (2014) Sequence to sequence learning with neural networks. NIPS
26. Bahdanau D, Cho K, Bengio Y (2015) Neural machine translation by jointly learning to align and translate. CoRR abs/1409.0473
27. Vaswani A (2017) Attention is all you need. arXiv abs/1706.03762
28. Parida S, Bojar O, Dash S (2019) Hindi visual genome: a dataset for multimodal English-to-Hindi machine translation. Computación y Sistemas 23
29. Nakazawa T (2020) Overview of the 7th workshop on Asian translation. WAT@AAC/IJCNLPL
30. Papineni K, Roukos S, Ward T, Zhu W (2002) Bleu: a method for automatic evaluation of machine translation. ACL
31. Banerjee S, Lavie A (2005) METEOR: an automatic metric for MT evaluation with improved correlation with human judgments. IEEvaluation@ACL
32. Klakow D, Peters J (2002) Testing the correlation of word error rate and perplexity. Speech Commun 38:19–28
33. Snover MG, Dorr B, Schwartz R, Micciulla L (2006) A study of translation edit rate with targeted human annotation

34. Popovic M (2015) ChrF: character n-gram F-score for automatic MT evaluation. WMT@EMNLP
35. Isozaki H, Hirao T, Duh K, Sudoh K, Tsukada H (2010) Automatic evaluation of translation quality for distant language pairs. EMNLP
36. Mikolov T, Sutskever I, Chen K, Corrado G, Dean J (2013) Distributed representations of words and phrases and their compositionality. NIPS
37. Murphy K (2012) Machine learning - a probabilistic perspective. Adaptive computation and machine learning series
38. Glorot X, Bordes A, Bengio Y (2011) Deep sparse rectifier neural networks. AISTATS
39. Kingma DP, Ba J (2015) Adam: a method for stochastic optimization. CoRR abs/1412.6980
40. Hochreiter S, Schmidhuber J (1997) Long short-term memory. Neural Comput 9:1735–1780
41. Simonyan K, Zisserman A (2015) Very deep convolutional networks for large-scale image recognition. CoRR abs/1409.1556
42. Deng J, Dong W, Socher R, Li L, Li K, Fei-Fei L (2009) ImageNet: a large-scale hierarchical image database. In: 2009 IEEE conference on computer vision and pattern recognition, pp 248–255
43. Narang S (2021) Do transformer modifications transfer across implementations and applications? arXiv abs/2102.11972
44. Hendrycks D, Gimpel K (2016) Gaussian error linear units (GELUs). arXiv: Learning
45. He K, Gkioxari G, Dollár P, Girshick RB (2020) Mask R-CNN. IEEE Trans Pattern Anal Mach Intell 42:386–397
46. Lin TY et al (2014) Microsoft COCO: common objects in context. In: Fleet D, Pajdla T, Schiele B, Tuytelaars T. (eds) Computer Vision – ECCV 2014. ECCV 2014. LNCS, vol 8693. Springer, Cham. https://doi.org/10.1007/978-3-319-10602-1_48
47. Tan HH, Bansal M (2019) LXMERT: learning cross-modality encoder representations from transformers. EMNLP
48. Li LH, Yatskar M, Yin D, Hsieh C, Chang K (2019) VisualBERT: a simple and performant baseline for vision and language. arXiv abs/1908.03557
49. Lu J, Batra D, Parikh D, Lee S (2019) ViLBERT: pretraining task-agnostic visiolinguistic representations for vision-and-language tasks. NeurIPS
50. Yao S, Wan X (2020) Multimodal transformer for multimodal machine translation. ACL
51. Junczys-Dowmunt M et al (2018) Marian: fast neural machine translation in C++. arXiv abs/1804.00344
52. Tiedemann J, Thottingal S (2020) OPUS-MT – building open translation services for the world. EAMT
53. Kudo T, Richardson J (2018) SentencePiece: a simple and language independent subword tokenizer and detokenizer for neural text processing. EMNLP
54. Solanki A, Pandey S (2019) Music instrument recognition using deep convolutional neural networks. Int J Inf Technol 1–10
55. Rawat A, Solanki A (2020) Sequence imputation using machine learning with early stopping mechanism. In: 2020 international conference on computational performance evaluation (ComPE), pp 859–863
56. Tayal A, Gupta J, Solanki A, Bisht K, Nayyar A, Masud M (2021) DL-CNN-based approach with image processing techniques for diagnosis of retinal diseases. In: Multimedia systems

Saving Patterns and Investment Preferences: Prediction Through Machine Learning Approaches

Sachin Rohatgi⊙, P. C. Kavidayal, and Krishna Kumar Singh⊙

1 Introduction

Uttarakhand has done a lot on the economic front since its inception. It is one of the fastest-growing states in India, thanks to the massive growth in capital investment arising from conducive industrial policy and generous tax benefits. The economic development of the state will contribute to the economic development of the country. However, the support of the investors in capital formation and economic development can't be overlooked. The savings and investments made by the individual investors are beneficial for the economic development of the state and the country. At the same time, individual investors are becoming choosier in selecting the different investment instruments available all thanks to the recent financial crisis that emerged and financial bubbles busted. The financial service providers are finding it difficult to provide a tailor-made solution based on their needs to the individual investors. Many finance theories proposed by researchers are based on the individual investor's behavior in the financial market. In the behavioral finance theory, the researchers suggest that the individual investor behavior is influenced by many demographic factors like age, gender, occupation, qualification, etc. A literature review for the current study has mainly focused on identifying the factors affecting the investment decisions or study the impact of these factors on the investment behavior. However, there is seldom any research on identifying and predicting the factors affecting saving and investment patterns of the individual investors from the Uttarakhand region. To address the identified gaps authors have taken a two-steps approach:

S. Rohatgi (✉) · P. C. Kavidayal
Kumaun University, Nainital, Uttarakhand, India
e-mail: mr.rohatgi@gmail.com

K. K. Singh
Symbiosis Centre for Information Technology, Pune, India
e-mail: krishnakumar@scit.edu

1) In the first step, the machine learning tools (KNN, Tree, SVM, Random Forest, Neural Network. Linear Regression, AdaBoost) are used to predict the saving patterns and investment preferences and after comparing the result based on R-value, the linear regression seems to be the best method for the prediction.
2) In the second step, two regression equations were constructed, one taking saving patterns as the dependent variable and the second one taking investment preferences as the dependent variable.

2 Review of Literature and Hypothesis Development

The presence of competitiveness in today's business environment ensures that the business should understand consumer behavior in a better way so that they can grow vertically. Many scholars have earlier defined the meaning of consumer behavior. It is the process by which consumers are satisfying their needs by selecting, consuming, and disposing of the products and services. According to Solomon the companies and the business need to understand the mechanism which helps in satisfying consumer needs [15]. By doing so they can provide the right product and services to the consumers. After an in-depth understanding of consumer buying behavior, the companies can predict the future buying behavior of the consumers and make the marketing strategies accordingly. In financial sectors, the investors are the consumers. It is therefore imperative for the financial service providers to gauge the financial needs of the investors. For this, they can develop the financial products and strategies in such a way that the investors' needs can be satisfied to a large extent. The financial needs of the investors can be explored and studied through the behavioral finance approach. This approach provides a way to access investor behavior and helps in understanding the reasons for market anomalies. Many scholars have tried to establish a relationship between the demographic and societal factors with the saving patterns and investment preferences. Kansal and Singh have studied the impact of 11 parameters on the financial decisions taken by women in today's world. They conclude that decisions taken by women are not different from the decision taken by men [4]. K.K. Singh et al. discussed a database model for the stock market and further they propose a green database model for the Indian stock market [12]. Mehta and Sharma in their descriptive study tried to know the behavior of the individual investor in a specific region in India. They conclude that the primary motive of the investment for a common investor is tax saving. Further, they found that middle-age group investors are risk-takers and the basic idea of investing in the equity market is to get high returns [7]. K.K. Singh and Dimri have focused on the theory of behavioral finance in the stock market. They conclude that the green database design will help the efficiency of the database. They have also proposed a model as score based financial forecasting method [13]. Chen et al. conclude that the investor's personality affects the short term and long-term investment preferences. Further, they have used the machine learning algorithm to establish that the investors with the openness and extraversion characteristics tend to earn more profits in the long run [1]. Sah in her study found that women are more concerned about meeting their current expenses like medical expenses therefore, they are inclined towards short-term investments. Although women are now these days more educated but still, they depend on their family members, friends, and relatives for investment-related information and for making investment decisions. Women do not want to take much risk and therefore wanted to invest in safer options like bank deposits and gold [11]. R. Singh and Sailo in their study conclude that investment was independent of the gender, marital status, and income of the investors however, the family size has an impact on the investment [14]. Mak and Ip concluded that demographic, psychological, and sociological variables affect the investment behavior and found that age, income level, education, gender, investment experience, and marital status are the main variables affecting investment behavior [6]. Lad in his study found that there exists a significant relationship between the education level of the investors and the choice of investment avenue. However, the majority of the investors prefer conventional investment sources and are unaware of the contemporary ones [5]. Ezekiel and Prince Oshoke stated that the education level, occupation, and marital status are the main decisive factors of the individual investor behavior. They conclude that the people in professional practice had more access to information and are willing to take more risks while deciding their portfolio [2]. According to Tripathy and Patjoshi there is no difference between the awareness level of the rural investor's gender and educational qualifications. Further, they said that bank deposits, gold, and real estate are the main investment preferences for the investors [16]. Prasad et al. in their study have tried to establish a relationship between the type of investment (low, medium, high) and the demographic variables (age, gender, annual income, education, marital status). They concluded that the annual income has the greatest impact [9]. Rohatgi et al. in their study used the machine learning tool in predicting the stock market. They have used Generalized Linear Model, Deep learning, Decision Tree, Random Forest, Gradient Boosted Trees, and Support Vector Machines on the BSE index. They conclude that the Gradient Boosted Trees is the efficient one [10]. The literature discussed above shows that the main factors that affect investment behavior are age, gender, income level, education, marital status, geographical area, and no of dependents in the family. However, little research has been done on the behavior of individual investors. In order, to analyze the individual behavior of the investors, the following hypothesis is submitted:

H_1: The saving patterns of the small investors can be predicted by the eight variables—age, gender, qualification, occupation, annual income, marital status, dependents in the family, and geographical area.

H_2: The investment preferences of the small investors can be predicted by the eight variables—age, gender, qualification, occupation, annual income, marital status, dependents in the family, and geographical area.

3 Data Analysis

Establishing a relationship of saving patterns with age, gender, qualification, occupation, marital status, annual income, dependents in the family, geographical area through machine learning.

In this study, authors have used anaconda orange to see machine aspects of factors and their predictions. Figure 1 shows the model creation in which authors took data of saving patterns as a dependent variable and variables like age, gender, occupation, education, annual income, marital status, dependents in the family, and geographical area as independent variables and used a sampler to segregate data. 70% of data is being used for training machines and the remaining data is used for testing.

The machine got trained on KNN, Calibrated Learner, Random Forest, AdaBoost, Linear Regression, Tree, Neural Network, and SVM algorithms. Table 1 shows comparative results and performance of the different algorithms on the given dependent and independent variables.

Table 2 shows the prediction results of all these algorithms and their performance on saving patterns (dependent variable). R square of all algorithms are negative however the best value is coming for linear regression (–0.071) which shows the best possible result in prediction but because of negative values, relationships are not very stable. Further authors are trying to establish a relationship of saving

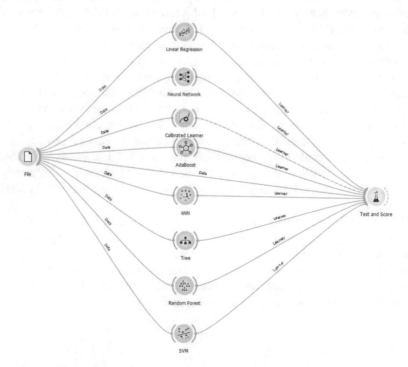

Fig. 1 The machine learning model of saving patterns

Table 1 Test and score results of saving patterns on the machine learning process

Evaluation Results				
Model	MSE	RMSE	MAE	R2
KNN	0.183	0.428	0.264	0.722
Tree	0.048	0.218	0.030	0.928
SVM	0.145	0.381	0.229	0.780
Random Forest	0.126	0.355	0.221	0.809
Neural Network	0.710	0.842	0.419	-0.076
Linear Regression	0.120	0.347	0.719	0.818
AdaBoost	0.049	0.222	0.022	0.926

Model Comparison by MSE							
	KNN	Tree	SVM	Random Forest	Neural Network	Linear Regression	AdaBoost
KNN		1.000	0.892	0.992	0.051	0.946	1.000
Tree	0.000		0.011	0.010	0.026	0.029	0.472
SVM	0.108	0.989		0.830	0.040	0.742	0.995
Random Forest	0.008	0.990	0.170		0.036	0.582	0.997
Neural Network	0.949	0.974	0.960	0.964		0.960	0.973
Linear Regression	0.054	0.971	0.258	0.418	0.040		0.991
AdaBoost	0.000	0.528	0.005	0.003	0.027	0.009	

patterns and investment preferences with age, gender, qualification, occupation, marital status, annual income, dependents in the family, geographical area through machine learning.

For the analysis of the relationship between investment preferences as a dependent variable and age, gender, occupation, education, annual income, marital status, dependents in the family and geographical area as independent variables authors have used anaconda orange to see machine aspects of factors and its predictions. Figure 2 shows the model creation in which authors took data of investment preferences as a dependent variable and age, gender, occupation, education, annual income, marital status, dependents in the family, and geographical area as independent variables and used a sampler to segregate data. 70% of data is being used for training machines and the remaining data is used for testing. The machine got trained on KNN, Calibrated Leaner, Random Forest, AdaBoost, Tree, Neural Network, and SVM algorithms. Table 3 shows comparative results and performance of the different algorithms on the given dependent and independent variable. Table 4 shows the prediction results of all these algorithms and their performance on Investment Preferences "Dependent variable". R square of all algorithms are in the negative and the best value is coming for KNN (–0.034) and the second-best value is coming for linear regression (–0.103) which shows the best possible result in prediction but because of negative values, relationships is unstable. As the results are shown by all the algorithms are negative for both the dependent variables i.e., Saving Patterns and Investment Preferences, still linear regression seems to be the best method for prediction. Therefore, in the next step, two regression equations were constructed.

Table 2 Prediction results of saving patterns through algorithms in machine learning

SP3	KNN	Random Forest	AdaBoost	Linear Regression	Tree	Neural Network	SVM	Fold	Age	Gender	Qualification	Occupation
1	2	2.25	3	1.99814	3	2.61066	1.52314	1	2	0	3	2
1	2.2	1.40381	1	2.29229	1	2.19311	1.45325	1	3	1	3	3
2	2.2	2.40667	3	2.22709	3	2.10967	1.70843	1	1	0	3	0
1	2.4	1.99714	3	1.57299	2.66667	1.97069	1.36683	1	4	1	2	1
3	2	3.06429	4	2.48545	4	2.75957	2.13785	1	4	1	4	2
1	1	2.21667	1	2.03804	1.66667	2.04272	1.65028	1	5	0	3	2
3	2.2	1.75714	2	2.03268	1	1.39211	1.29763	1	4	0	3	1
3	1.6	2.505	1	1.93108	1	1.64707	1.49561	1	4	0	4	1
1	2.6	3.04917	3.2381	1.49331	3.5	2.43601	1.63377	1	1	1	2	0
1	1	1.2	1	1.54845	1	0.693705	0.871066	1	3	1	3	1
4	1.4	1.31333	1	1.85372	1	1.62756	0.914478	1	1	0	3	0
2	1.4	1.60667	3	2.24033	1	1.92886	1.32897	1	2	0	3	1
3	1.6	1.64667	1	2.17841	1	2.08634	1.47774	1	3	1	3	3
1	1.4	1.105	1	1.62253	2	1.68199	1.36846	1	6	1	3	1
3	2	1.325	1	2.00926	1.33333	1.71515	1.2294	2	2	0	3	1
1	2.8	2.76476	3	2.01934	3.5	2.24141	2.35796	2	4	0	3	2
1	1.6	1.02	1	1.89272	1	1.63627	1.33411	2	3	1	4	1
1	1.8	2.23452	1	1.44763	1.66667	1.23197	1.03815	2	5	0	2	2
1	2.4	2.27698	3	1.42546	1	1.58064	1.38013	2	4	0	2	1
1	2.8	2.41667	2.33333	2.30325	2	2.51123	1.82738	2	2	0	4	1
3	1.6	1.50714	2.5	1.51732	2	1.67916	1.09921	2	1	0	2	0

(continued)

Table 2 (continued)

SP3	KNN	Random Forest	AdaBoost	Linear Regression	Tree	Neural Network	SVM	Fold	Age	Gender	Qualification	Occupation
1	2.4	1.99	1	1.78216	1	1.38737	1.11855	2	4	0	3	1
1	2.8	2.68095	1	2.31913	1	2.03739	1.57438	2	3	1	2	3
1	2.2	3.10429	3	2.22089	3.25	1.95199	1.46644	2	3	0	2	3
1	2.8	1.54405	2.2	1.99915	2.75	2.3041	1.90021	2	2	1	3	1
4	2.2	1.54048	1	2.19001	1.33333	1.94687	1.93681	2	3	0	4	1
1	2.2	1.635	1	2.25657	3.25	1.63165	1.6344	2	5	1	3	2
1	2.6	3.27417	3	2.53597	4	2.58112	2.55106	2	3	1	4	1
2	2.8	2.4825	2.33333	2.15338	1.75	2.93547	1.85738	3	6	0	3	5
1	1.6	1.525	2	2.13082	1	1.68668	1.50681	3	2	0	3	1
4	2.6	2.0325	3	2.24853	1.75	2.5375	1.86958	3	6	1	3	5
1	1	2.105	2.2	1.9118	1	1.40305	1.09422	3	2	1	3	1
1	1	1.44	1	2.1735	1	1.72112	1.26467	3	2	0	2	2
4	1.6	1.79286	1	2.19921	1	1.44917	1.09597	3	2	1	4	2
1	1.8	2.53492	1	1.07678	2	1.2222	1.24648	3	6	0	2	1

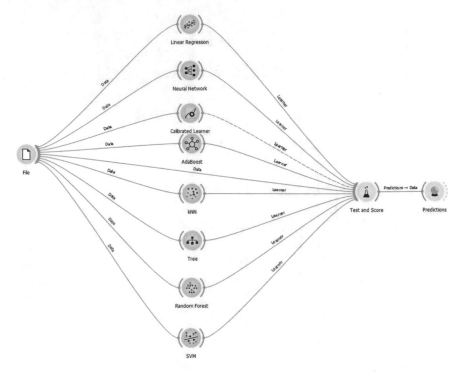

Fig. 2 The machine learning model of investment patterns

Table 3 Test and score results of Investment patterns on the machine learning process

Evaluation Results				
Model	**MSE**	**RMSE**	**MAE**	**R2**
KNN	0.186	0.903	0.787	-0.034
Tree	1.108	1.053	0.846	-0.404
SVM	1.028	1.041	0.855	-0.303
Random Forest	0.895	0.946	0.837	-0.135
Neural Network	0.960	0.900	0.841	-0.217
Linear Regression	0.870	0.933	0.843	-0.103
AdaBoost	1.244	1.115	0.860	-0.576

Model Comparison by MSE							
	KNN	**Tree**	**SVM**	**Random Forest**	**Neural Network**	**Linear Regression**	**AdaBoost**
KNN		0.042	0.045	0.167	0.084	0.247	0.005
Tree	0.958		0.656	0.932	0.810	0.929	0.302
SVM	0.955	0.344		0.825	0.732	0.167	0.118
Random Forest	0.833	0.068	0.175		0.258	0.604	0.039
Neural Network	0.916	0.190	0.226	0.742		0.745	0.068
Linear Regression	0.753	0.071	0.143	0.396	0.255		0.054
AdaBoost	0.995	0.656	0.852	0.961	0.932	0.585	

Table 4 Prediction results of investment preferences through algorithms in machine learning

SF1	KNN	AdaBoost	Linear Regression	Neural Network	Tree	Random Forest	SVM	Fold	Age	Gender	Qualification	Occupation
4	4.2	4	4.26281	5.69075	4	4.40056	4.28335	1	3	1	3	0
3	3	3	3.09435	3.18694	3	3.59444	3.11409	1	5	0	2	2
4	3.6	4	3.82547	3.35097	4	3.93333	4.0684	1	4	1	3	1
4	4	4	3.85664	3.54316	4	4.18778	4.23124	1	2	0	3	1
4	4.2	4	4.07481	3.29252	4	4.1625	4.44665	1	1	1	2	0
5	5	5	5.0038	4.88372	5	4.72	4.7043	1	6	1	4	2
4	4	4	4.12597	5.13283	4	4.05833	4.20322	1	1	1	3	0
3	4	3	3.04266	3.00169	3	3.56667	3.65463	1	2	0	2	0
3	4	3	3.19473	4.08375	3	4.525	4.14713	1	2	0	3	3
5	4.4	5	5.12308	4.66546	5	4.08667	4.52585	1	2	0	2	0
3	3.4	3	3.05584	0.932659	3	3.48611	4.08274	1	1	0	2	0
5	4.6	5	5.04275	4.68348	5	4.35	4.5336	1	2	0	2	2
4	3.8	4	3.73694	3.04103	4	4.05429	4.12638	1	1	1	3	0
3	3.4	3	3.27134	2.68667	3	3.665	3.88391	1	1	1	3	0
3	1.8	1	1.23115	0.881067	1	1.73429	1.45418	1	6	0	3	1
5	4.4	5	4.74338	4.9112	5	4.775	4.75227	1	6	0	3	5
4	4	4	3.95614	4.18979	4	3.925	4.04626	1	6	1	3	1
4	4	4	4.09539	4.11785	4	4.06667	4.1369	1	4	1	4	1
4	4.6	4	3.91035	3.79249	4	4.05714	3.84033	1	2	1	3	1
5	5	5	4.9618	5.11759	5	4.85952	4.91356	1	2	1	3	1
4	4	4	4.3087	4.20707	4	3.9375	4.02954	1	4	0	3	1

(continued)

Table 4 (continued)

SF1	KNN	AdaBoost	Linear Regression	Neural Network	Tree	Random Forest	SVM	Fold	Age	Gender	Qualification	Occupation
4	3.4	4	4.23512	4.28714	4	3.95	4.11949	1	1	1	1	0
5	5	5	4.85878	5.02603	5	4.95	4.73274	1	2	1	3	0
4	4	4	3.95614	4.18979	4	3.925	4.04626	1	6	1	3	1
5	5	5	5.20799	5.57172	5	5	5.10514	1	4	0	4	1
5	5	5	4.14266	4.29691	5	4.76369	4.19923	1	2	0	3	1
1	2.2	1	1.20471	1.20074	1.66667	1.67095	1.43243	1	2	0	3	0
3	3	3	3.06129	2.65574	3	3.11833	2.84154	1	3	0	4	1
5	5	5	5.00708	5.08423	5	4.775	5.08397	1	6	1	4	2
4	3.6	4	4.10091	4.15081	4	3.9506	3.94903	1	2	1	2	0
4	3.4	4	4.23512	4.28714	4	3.95	4.11949	1	1	1	1	0
5	4.4	5	5.01903	4.81454	5	4.73095	4.82482	1	1	0	3	0
4	3.4	4	3.71213	3.83252	4	3.85095	3.91531	1	1	1	3	0
5	5	5	4.86228	4.67327	5	4.80571	4.83307	1	6	0	3	3
5	5	5	4.52602	4.83348	5	5	4.75706	1	5	0	4	1

3.1 Regression Equation and Data

Regression Equation

To explore the relationship between the saving and investment patterns and the impact of demographic variables on the saving and investment patterns of small investors from Uttarakhand a thorough literature review was conducted. Based on the literature review a hypothesis was framed and two regression equations were constructed

$$Saving\ frequency_t = \alpha + \beta_1 age_t + \beta_2 gender_t + \beta_3 marital\ status_t + \beta_4 geographical\ area_t$$
$$+\beta_5 annual\ income_t + \beta_6 dependents\ in\ the\ family_t + \beta_7 occupation_t$$

$$(1)$$

where saving frequency (saving frequency $_t$) is a function of age (age $_t$), gender (gender $_t$), marital status (marital status $_t$), geographical area (geographical are $_t$), annual income (annual income $_t$), dependents in the family (dependents in the family $_t$) and occupation (Occupation $_t$). Where α is a regression constant and β (i = 1,2,3,4.6) represents the coefficient of each independent variable.

$$Investment\ preferences_t = \alpha + \beta_1 age_t + \beta_2 gender_t + \beta_3 marital\ status_t + \beta_4 geographical\ area_t$$
$$+\beta_5 annual\ income_t + \beta_6 dependents\ in\ the\ family_t + \beta_7 occupation_t$$

$$(2)$$

where Investment preferences (Investment preferences $_t$) is a function of age (age $_t$), gender (gender $_t$), marital status (marital status $_t$), geographical area (geographical area $_t$), annual income (annual income $_t$), dependents in the family (dependents in the family $_t$) and occupation (occupation $_t$). Where α is a regression constant and β (i = 1,2,3,4.6) represents the coefficient of each independent variable.

The first regression Eq. (1) discusses whether and how the six variables identified, affect the saving frequency of the small investors in Uttarakhand. The regression Eq. (2) examines the investment preferences of the small investors and how these preferences are affected by the variables like age, gender, marital status, etc.

Data

An organized survey was regulated to gather information from the respondents. The small investors being the target population from a province of Uttarakhand are approached for giving the responses. In the state, there are two divisions, the Garhwal division, and the Kumaun division. The Garhwal division has seven districts and the Kumaun division has six districts. Two districts from every division have been chosen as the sampling region for the collection of data. Areas were considered dependent on the similar portrayal of hills and plains region. Areas of Dehradun and Pauri have been taken from the Garhwal division, and from the Kumaun division, districts of Nainital and Udham Singh Nagar have been picked as the sample for information assortment. The choice of respondents was done dependent on arbitrary examining. According to the objectives of the study, information is gathered from the population through a poll regulated on 540 people. After an assortment of information, the information cleaning was done to recognize certain missing qualities and eliminate them from the informational index. After the data cleaning, 448 respondents were available. Table 5 gives the descriptive statistics of the small investor's characteristics.

Premise Analysis

The validity of regression models is based on the fulfillment of the underlying assumptions be it linearity or multicollinearity. For the effective use of regression models, these underlying assumptions must be fulfilled. According to Poole and Farrell the regression models are valid only when these conditions are tested and satisfied [8]. Before regression analysis, linearity and multicollinearity must be checked.

Linearity Analysis

Table 6 shows the significance of the linear relationship for Eq. 1. In Table 6, the p-value for all the given variables is less than (<) 0.0001, indicating that there exists a significant relationship between all independent variables and dependent variables. Therefore, the underlying assumption of linearity for the regression is accomplished. Table 7. shows the linearity for Eq. 2. In this, the p values of all the variables except age, and gender are less than (<) 0.01.

Multicollinearity

According to Hart and Sailor if tolerance (T) is below 0.20, multicollinearity is a serious problem [3]. Further, the variables are correlated if the value of variance inflation factors (VIF) is between 1 and 10. These values indicate the low multicollinearity and therefore the assumption of multicollinearity is fulfilled. Table 8 shows the values of the coefficient of determination (R^2) and the variance inflation factors (VIF) for the regression Eq. 1.

As all the basic assumptions are in line, the multiple regression Eq. 1 and 2 are assumed to be valid and further regression analysis can be conducted.

4   Result Analysis and Discussion

In this section, the regression analysis is done to check the relationship between the eight independent variables and the dependent variable saving pattern and investment preferences of a small investor. Tables 9 and 10. depicts the result of the regression analysis.

4.1   Effect of Eight Independent Variables on the Saving Patterns

Table 9 depicting the regression analysis of eight independent variables on the saving patterns. It is clearly shown in the table that these variables have a significant effect

Table 5 Descriptive statistics

Characteristics		Frequency	Percentage
Age	18–22	75	16.7
	23–27	87	19.4
	28–32	87	19.4
	33–37	95	21.2
	38–42	47	10.5
	Above 42	57	12.7
	Total	448	100.0
Gender	Male	256	57.1
	Female	192	42.9
	Total	448	100.0
Qualification	Up to 12th	3	0.7
	Graduation	153	34.2
	Post-graduation	182	40.6
	Professional	110	24.6
	Total	448	100.0
Occupation	Student	83	18.5
	Salaried	172	38.4
	Business	111	24.8
	Professional	49	10.9
	House-Hold	11	2.5
	Retired	22	4.9
	Total	448	100.0
Annual income	Up to Rs. 2,50,000	100	22.3
	Rs 2,50,001–Rs 5,00,000	216	48.2
	Rs 5,00,001–Rs 10,00,000	110	24.6
	Above Rs 10,00,000	22	4.9
	Total	448	100.0
Marital status	Unmarried	167	37.3
	Married	281	62.7
	Total	448	100.0
Dependents in the family	0	122	27.2
	1–3	204	45.5

(continued)

Table 5 (continued)

Characteristics		Frequency	Percentage
	4–6	115	25.7
	Above 6	7	1.6
	Total	448	100.0
Geographical area	Nainital district	138	30.8
	Udham Singh nagar district	145	32.4
	Dehradun district	118	26.3
	Pauri district	47	10.5
	Total	448	100.0

Table 6 Multiple correlation coefficient for regression Eq. 1

Co-efficient[a]

Equation 1	Unstandardized coefficients		Standardized coefficients	t	Sig.
	B	Std. Error	Beta		
(Constant)	1.594	0.303		5.264	0
Age	– 0.1	0.055	– 0.13	– 1.808	0
Gender	– 0.166	0.118	– 0.067	– 1.414	0
Qualification	0.158	0.08	0.1	1.983	0
Occupation	0.078	0.061	0.08	1.274	0
Annual income	– 0.18	0.083	– 0.119	– 2.174	0
Marital status	0.119	0.155	0.047	0.766	0
Dependents in the family	0.184	0.087	0.116	2.109	0
Geographical area	0.147	0.06	0.117	2.444	0

a: Dependent variable: How frequently do you save

on the saving patterns of small investors. Thus age, gender, qualification, occupation, annual income, marital status, dependents in the family, and geographical area can be determined as important predictors of saving patterns of the small investors in the state of Uttarakhand.

The regression coefficient as per Eq. 1. For the age is –0.13, gender is –0.067, qualification is 0.1. The occupation has a regression coefficient of 0.08, annual income has a co-efficient of –0.119. The marital status has a co-efficient of 0.047, dependents in the family have a regression coefficient of 0.116, whereas the geographical area has a co-efficient of 0.117. The p values of all these variables are less than ($<$) 0.01. R-squared is a statistical measure to show how close the data is to the fitted regression line. For this model, the value of R^2 0.64 indicates the proportion of the variance of saving patterns that are explained by all the eight variables i.e., age,

Table 7 Multiple correlation coefficient for regression Eq. 2

Co-efficient[a]

Equation 2	Unstandardized coefficients		Standardized coefficients	T	Sig.
	B	Std. Error	Beta		
(Constant)	1.677	0.223		7.536	0
Age	0.034	0.041	0.061	0.826	0.227
Gender	0.055	0.086	0.031	0.639	0.145
Qualification	0.033	0.059	0.029	0.569	0
Occupation	− 0.074	0.045	− 0.105	− 1.644	0
Annual income	0.018	0.061	0.017	0.3	0
Marital status	0.054	0.114	0.03	0.478	0
Dependents in the family	0.01	0.064	0.009	0.161	0
Geographical area	− 0.023	0.044	− 0.025	− 0.511	0

a: Dependent Variable. What is your current investment preferences

Table 8 The measure of tolerance and VIF

Co-efficient[a]

Equation 1	Unstandardized coefficients		Standardized coefficients	T	Sig.	Collinearity statistics	
	B	Std. Error	Beta			Tolerance	VIF
(Constant)	1.594	0.303		5.264	0		
Age	− 0.1	0.055	− 0.13	− 1.808	0	0.417	2.396
Gender	− 0.166	0.118	− 0.067	− 1.414	0	0.964	1.037
Qualification	0.158	0.08	0.1	1.983	0	0.851	1.175
Occupation	0.078	0.061	0.08	1.274	0	0.549	1.821
Annual income	− 0.18	0.083	− 0.119	− 2.174	0	0.73	1.37
Marital status	0.119	0.155	0.047	0.766	0	0.584	1.711
Dependents in the family	0.184	0.087	0.116	2.109	0	0.723	1.384
Geographical area	0.147	0.06	0.117	2.444	0	0.942	1.062

a. Dependent Variable: How frequently do you save

gender, qualification, occupation, annual income, marital status, dependents in the family, and geographical area. The R^2 value is ensuring that 64% of the observed variations are explained by the regression model. Therefore, it can be elucidated that the null hypothesis is going to be accepted and statistically it can be inferred that the saving patterns of the small investors can be predicted by the eight variables—age,

Table 9 Result of regression analysis for Eq. 1

Age	-0.13
Gender	-0.067
Qualification	0.1
Occupation	0.08
Annual income	-0.119
Marital status	0.047
Dependents in the family	0.116
Geographical area	0.117
R^2	0.64

Table 10 Result of regression analysis for Eq. 2

Age	0.061
Gender	0.031
Qualification	0.029
Occupation	-0.105
Annual income	0.017
Marital status	0.03
Dependents in the family	0.009
Geographical area	-0.025
R^2	0.74

gender, qualification, occupation, annual income, marital status, dependents in the family and geographical area.

4.2 Effect of Eight Independent Variables on the Investment Preferences

Table 10 depicting the regression analysis of eight independent variables on the investment preferences. It is demonstrated in the table that these variables have a significant effect on the investment preferences of small investors. Thus age, gender, qualification, occupation, annual income, marital status, dependents in the family, and geographical area can be determined as important predictors of investment preferences of the small investors in the state of Uttarakhand.

The regression coefficient as per Eq. 2 for age is 0.061, gender it is 0.031, for qualification it is 0.029, for the occupation, it is -0.105, annual income has a regression coefficient of 0.017, marital status has a coefficient of 0.03, whereas dependents in the family have a regression coefficient of 0.009 and the geographical area has -0.025 as a regression coefficient. For this model, the value of R^2 0.74 indicates the proportion of the variance of investment preferences that are explained by all the eight variables i.e., age, gender, qualification, occupation, annual income, marital status, dependents in the family, and geographical area. The R^2 value is ensuring that 74% of the observed variations are explained by the regression model. From the results of the regression model, 2 six out of eight variables have a significant effect on the investment preferences of the small investors. The two variables identified as age, and gender are excluded as the age (p-value = 0.227 > 0.1) and gender (p value = 0.145 > 0.1). Thus, only qualification, occupation, annual income, marital status, dependents in the family, and geographical area are statistically significant predictors for investment preferences. By leveraging the result shown in Table 10, it is concluded that the investment preferences are related to the qualification, occupation, annual income, marital status, and dependents in the family. They can be used as a predictor to forecast investment preferences.

4.3 Discussion

Concerning the results shown in Tables 9 and 10. it is understood that there is no significant difference between the variables. Like occupation is affecting the saving patterns and investment preferences but in the opposite direction. Occupation has a positive effect on the saving patterns and a negative effect on investment preferences. On the same lines, the annual income is affecting the saving patterns and investment preferences but in the opposite direction. It harms the saving patterns and has a positive effect on investment preferences. Whereas qualification is affecting the dependent variable in the same direction. Marital status also affects the saving patterns and investment preferences but in a positive direction. Further independent variable dependents in the family also affect the saving patterns and the investment preferences in the positive directions. Lastly, the geographical area has a positive effect on the saving patterns but a negative impact on the investment preferences. Furthermore, it is assessed that broadly three attributes namely occupation, annual income, and geographical area affect the saving patterns and investment preferences in opposite directions.

5 Conclusion

The financial industry plays a crucial role in the economic development of a country. Therefore, the companies coming with the new financial products must keep a check on the small investors saving patterns and investment preferences. Since the financial crisis of 2008, the investors are becoming more circumspect, especially against the new financial products. They feel hesitant in investing in these financial products. Therefore, companies should plan and bring those types of financial products which are designed as per the needs of these small investors. However, with the iverse population and geographical region, it is difficult for the financial service providers to understand the small investors saving patterns and investment preferences. This study has first used the machine learning tool to predict the value of the dependent variables and after assessing the results given by the six algorithms it was found that linear regression can be the nearest to predict the dependent variables. In the second step, the two regression models were constructed based on two dependent variables i.e., Saving patterns and Investment Preferences. Further, the study has tried to address the real-world challenges and research gap and this study has:

1) Identified the variables affecting the saving patterns of the small investors.
2) Identified the variables affecting the investment preferences of the small investors.

This study helps the financial service providers to understand the behavior of the small investors better by analyzing their characteristics. The service providers can gauge the relationship between occupation, annual income, and geographical area. The 1st variable, the occupation has a positive effect on the saving patterns which means every category of a respondent to be a salaried, professional, student, retired, household or even a businessman everyone saves, however, these categories of small investors have a negative relation with investment preferences, means everyone is not investing. Similarly, the second variable, annual income has a negative relationship with the saving patterns, as the annual income increases the spending increases so saving decreases. However, it has a positive relationship with investment preferences. Lastly the third variable, the geographical area also has a positive impact on saving patterns as people from all areas invest, however as far as investment preferences are concerned it has a negative relation, which means the people from the selected regions don't have the investment opportunities. They don't find viable investment options. So financial services providers can seize on these gaps and try to reach these investors. For future research, it is suggested that the regression results can be used and implemented in developing a data mining model that can further help the companies in understanding the investment behavior better. Further apart from the variables taken in the study some new variables can be tested for the regression analysis.

References

1. Chen TH, Ho RJ, Liu YW (2019) Investor personality predicts investment performance? A statistics and machine learning model investigation. Comput Hum Behav 101:409–416
2. Ezekiel A, Oshoke Prince (2020) The influence of demographic factors on investment behaviour of individual investors: a case study of Edo State, Nigeria. Economy 7(1):69–77
3. Hart MA, Sailor DJ (2009) Quantifying the influence of land-use and surface characteristics on spatial variability in the urban heat island. Theoret Appl Climatol 95:3–4
4. Kansal P, Singh S (2013) Investment behaviour of Indian Investors: Gender Biasness. In: Seventh national conference on Indian capital market: emerging issues
5. Lad S (2018) A study of investment pattern and awareness of rural investors. Sankalan 3(1):59–65

6. Mak MKY, Ip WH (2017) An exploratory study of investment behaviour of investors. Int J Eng Bus Manage 9:1–12
7. Mehta K, Sharma R (2015) Individual investors' behavior : in demographical backdrop. SCMS J Indian Manage 12(3)
8. Poole MA, O'Farrell PN (1971) The assumptions of the linear regression model. Trans Inst Brit Geogr 52:145
9. Prasad S, Kiran R, Sharma RK (2021) Examining saving habits and discriminating on the basis of demographic factors: a descriptive study of retail investors. Int J Finan Econ 26(2):2859–2870
10. Rohatgi S, Kumar Singh K, Jasuja D (2021) Comparative analysis of machine learning algorithm to forecast Indian stock market. In: 2021 international conference on advance computing and innovative technologies in engineering. IEEE, Greater Noida, India
11. Sah VP (2017) A study on investment behavioural patterns of women investors. CVR J Sci Technol 13:107–110
12. Singh KK, Dimri P, Rawat M (2014) Green database model for stock market: a case study of Indian stock market. In: Proceedings of the 5th international conference-the next generation information technology summit, IEEE, Noida, India
13. Singh KK, Dimri P (2016) Score based financial forecasting method by incorporating different sources of information flow into integrative river model. In: Proceedings of the 6th international conference - cloud system and big data engineering, confluence. IEEE, Noida, India
14. Singh R, Sailo S (2017) Impact of demographic factors on saving and investment patterns of bank employees in Aizawl. Asian J Manage 8(4).1304–1310
15. Solomon MR (2012) Consumer Behavior: Buying, Having, Being, 10th edn. Prentice Hall India Learning Private Limited, Delhi
16. Tripathy P, Patjoshi PK (2020) Investment alternatives and preferences of rural investors: a case study of Barang block, Odisha, India. J Criti Rev 7(19):5565–5571

A Machine Learning Based Approach for Detection of Distributed Denial of Service Attacks

Raghavender Kotla Venkata

1 Introduction

Intrusion detection is very important for information systems. An Intrusion Detection System (IDS) can detect and prevent different kinds of intrusions. Denial of Service (DoS) is one of the common attacks that cause limited damage. However, DDoS causes much damage to information systems by denying services of server machines with large volumes of fake requests (flooding). This problem arises in the Wide Area Network (WAN). History revealed sever DDoS attacks on major IT companies like Yahoo. Different approaches came into existence to overcome the problem of DDoS attacks.

The SDN technology is employed in [1] and [5] to separate the data layer from the configurations layer in order to predict DDoS traffic. Such solutions need the support from SDN platforms. There are many detection models found based on machine learning approaches. The machine learning based approaches are found in [4, 6, 8] and [10]. The concept of autoencoders is used in [2] for prediction of DDoS attacks. Anomaly detection based procedure is employed in [3] while HTTP flood kind of DDoS attack is explored in [10].

Many feature selection algorithms are also found in the literature in order to improve the performance of prediction models. Information entropy based feature selection is employed in [11]. Information gain is another method used in [12, 14, 15] and [16]. Correlation based approach is employed in [14] along with information gain and SVM for feature selection. From the literature it is found that there is need for improved and efficient feature selection model for leveraging DDoS prediction performance. Our contributions in this paper are as follows.

R. Kotla Venkata (✉)
G.Narayanamma Institute of Technology and Science (For women) (Autonomous) Shaikpet, Hyderabad, Telangana, India
e-mail: drkvraghavender@gmail.com

An algorithm named Hybrid Feature Selection (HFS) is proposed and implemented to overcome the problem of curse of dimensionality.

A prototype application is built to demonstrate proof of the concept.

The algorithm is evaluated and the results are compared with the state of the art DDoS prediction models that are based on machine learning.

The remainder of the paper is structured as follows. Section 2 provides review of literature. Section 3 presents the proposed system in detail. Section 4 presents experimental results while Sect. 5 concludes the paper besides giving directions for future work.

2 Related Work

DDoS detection models are developed using machine learning techniques. However, there is need for feature selection in order to have more accurate results. In this section, the existing literature is reviewed on both DDoS prediction models and the feature selection methods. Myint et al. [1] used the technology known as SDN for predicting DDoS attacks. This approach is decoupled from the data layer of the network. Ali and Li [2] focused on the neural network based models known as autoencoders to predict DDoS attacks. Trejo et al. [3] focused on the machine learning approaches with anomaly detection to achieve DDoS prediction models. Doshi et al. [4] used machine approaches to detect DDoS attacks. Their work was related to the Internet of Things (IoT) use cases. Rahman et al. [5] also employed SDN approach along with machine learning to achieve better prediction performance.

Hou et al. [6] focused on NetFlow analysis along with machine learning to have improved prediction performance. Gu et al. [7] employed a hybrid feature selection approach and K-means algorithm with semi-supervised approach to detect DDoS attacks. Different machine learning models are used in [8] for prediction of DDoS attacks. However, they found that there is need for efficient feature selection. Intrusion detection research is carried out in [9] and [10] using machine learning techniques. The focus of the [10] is HTTP food attacks in the application layer using machine learning.

Biesiada et al. [11] used entropy method in order to achieve feature ranking. Then the feature ranking was used to find features that are more useful. Alhajet al. [12] on the other hand focused on information gain approach in order to find good features. The work done in [11] and [12] used different measure in order to find suitable features. However, the hybrid of these two is made in the proposed HFS in this paper. Ladha [13] made a good review of many feature selection methods. Roobaertet al. [14] used correlation and information gain along with SVM classifier to identify good features for prediction performance. Sui [15] also focused on the information gain measure to have a feature selection algorithm as done in [12]. Information gain is the measure used by Shaltout et al. [16] also for prediction performance. Saini et al. [17] focused on different ML algorithms for DDoS attack detection. Pande et al. [18] used both machine learning and deep learning methods to detect such attacks. Pande et al. [19] also contributed toward detection of DDoS attacks using different ML approaches. From the literature, we found that there are many feature selection methods available. However, the combination of two methods to have a hybrid approach is still desired.

3 Proposed Framework for DDoS Attack Detection

A framework is proposed to detect DDoS attacks. The design of the framework is to detect DDoS attacks using machine learning approach. The approach is based on supervised learning. There are many supervised learning models that exist already. However, their performance depends on the quality of training given. The proposed framework incorporates a feature selection algorithm which improves performance of DDoS attack prediction.

3.1 Problem Definition

Given a network traffic data as input, it is challenging to know whether the traffic is genuine or related to malicious attacks such as DDoS attacks that flood fake requests towards the intended target. This is the challenging problem considered.

3.2 The Framework

The proposed framework is presented in Fig. 1. The framework takes the NSL-KDD dataset which is widely used for the research associated with DDoS and other attacks. The given dataset is subjected to pre-processing which divides the dataset into training set and testing set. The training set is then subjected to feature selection method. The proposed feature selection method is known as Hybrid Feature Selection (HFS) method as provided in Sect. 3.3.

Once feature selection is completed, the selected features are provided to the ML algorithm. The ML algorithm gets trained with the selected features. The dimensionality reduction helps the underlying DDoS detection model to perform well in terms of accuracy and performance. The resultant DDoS detection model is able to classify intrusions associated with DDoS attacks. The proposed algorithm and the performance metrics are provided in the following sub sections.

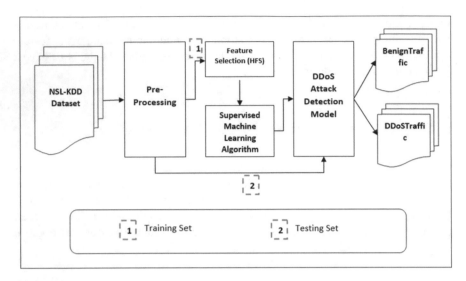

Fig. 1 Proposed framework for DDoS attack detection

3.3 Proposed Algorithm

The algorithm is based on the two metrics that can be used in combination to determine useful features. Without feature selection, the DDoS detection models based on ML techniques may fail to perform or deteriorate in the performance. Entropy computes uncertainty while the gain computes reduction in entropy.

Algorithm: Weight Based Feature Subset algorithm
Inputs: NSL-KDD Dataset **D**, threshold for gain t1, threshold for entropy t2
Outputs: Chosen Features F'
01   Initialize features vector F
02   F = GetFeatures(D)
03   For each feature f in F
04   Find gain g
05   Find entropy e
06   IF $g>=t1$ and $e>=t2$ THEN
07   Add f to F'
08   END IF
09   End For
10   Return F'

Algorithm 1: Hybrid feature selection algorithm

As shown in Algorithm 1, the NSL-KDD dataset is used as input. It is denoted as D. The aim of the algorithm is to find features that are capable of contributing to class label or detection of DDoS attacks. Towards this end entropy and gain are the two statistical measures used. The entropy measure is computed using Eqs. 1 and 2 while Eq. 3 computes gain value.

$$H(X) = - \sum_{x \in X} p(x) \log \log p(x) \tag{1}$$

$$H(Y) = - \sum_{y \in Y} p(y) \log \log p(y) \tag{2}$$

$$\text{Information gain} = H(y) - H(y/x) \tag{3}$$

With these metrics, the proposed feature selection algorithm is capable of finding features that are more useful rather than using all features. Thus it is able to reduce time and space complexity besides improving performance of the DDoS detection models that are based on machine learning approaches.

3.4 Metrics for Performance Evaluation

The confusion matrix shown in Fig. 2 is used to compute performance of the proposed detection models. The models are evaluated with metrics known as precision, recall, accuracy and F-Measure (F1-Score).

As presented in Fig. 2, based on the true positive, false positive, false negative and true negative values obtained by the detection models, different metrics are derived as shown in Eqs. 4, 5, 6 and 7.

$$Precision = \frac{TP}{TP + FP} \tag{4}$$

$$Recall = \frac{TP}{TP + FN} \tag{5}$$

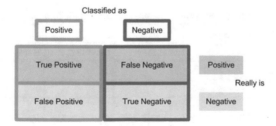

Fig. 2 Shows confusion matrix

$$F_1 = 2.\frac{\text{precision} \cdot \text{recall}}{(\text{precision}) + \text{recall}} \tag{6}$$

$$Accuracy = \frac{TP + TN}{TP + TN + FP + FN} \tag{7}$$

These performance metrics are used for evaluation of the DDoS attack models with and without feature selection. The results are also evaluated using the execution time that is computed as end time—start time where the former is found before algorithm starts while the latter is found after algorithm completes.

4   Experimental Results

Experiments are made with the DDoS attack detection models made up of ML algorithms such as K-NN, NB, DT, RF and SVM with and without feature selection algorithm known as HFS. The results are also evaluated with execution time. In spite of challenges like lack of adequate training samples, the proposed system could be evaluated with available data and the results are compared with existing ones.

As shown in Table 1, the results of ML based DDoS detection models without feature selection are provided in terms of accuracy, precision, recall and F-Score.

As presented in Fig. 3, DDoS detection methods are provided in horizontal axis and the vertical axis provides performance percentage. It is understood that different models exhibited different performance. However, the performance of SVM showed highest accuracy and K-NN exhibited least accuracy. Nevertheless, the performance of these models is increased when the proposed feature section is employed in the training phase. It is evidenced in the results shown in Fig. 4.

As presented in Table 2, the accuracy, precision, recall and F-score are provided for different detection models.

As presented in Fig. 4, the proposed feature selection algorithm named HFS is used along with different classifiers. The usage of HFS has its influence on the results. The empirical results showed that there is increase in the performance of all models when HFS is used. For instance, K-NN showed 77% accuracy without feature selection

Table 1 Performance comparison without feature selection algorithm employed

ML based DDoS detection model	Performance (%)			
	Accuracy	Precision	Recall	F-Score
K-NN	77	75	74	74.49664
Naïve Bayes	84	88	79	83.25749
Decision Tree	75	77	69	72.78082
Random Forest	84	71	95	81.26506
SVM	88	90	80	84.70588

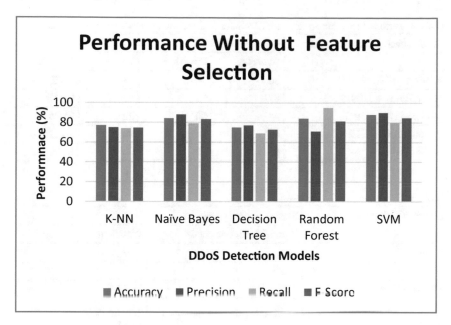

Fig. 3 Performance of DDoS detection models without feature selection

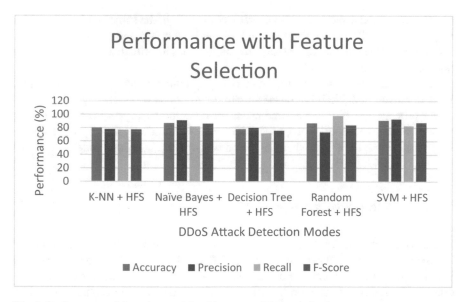

Fig. 4 Performance of detection models with proposed feature selection

Table 2 Performance comparison with DDoS attack detection models with feature selection

DDoS attack detection model	Performance (%)			
	Accuracy	Precision	Recall	F-Score
K-NN + HFS	80	78	77	77.49677
Naïve Bayes + HFS	87	91	82	86.2659
Decision Tree + HFS	78	80	72	75.78947
Random Forest + HFS	87	74	98	84.32558
SVM + HFS	91	93	83	87.71591

Table 3 Execution time comparison

DDoS detection model	Execution time (seconds)	
	Without feature selection	With feature selection
K-NN	28.43	23.72
Naïve Bayes	33.24	31.13
Decision Tree	20.38	18.87
Random Forest	14.21	12.32
SVM	14.19	12.34

and the same model showed 80% accuracy with HFS. In the same fashion, the SVM model showed 88% accuracy while the SVM + HFS model showed 91% accuracy. This has revealed that the feature selection method has its influence on the improving quality of prediction.

As presented in Table 3, the time taken by the prediction models with and without HFS employed in the training phase are provided.

As presented in Fig. 5, the time taken for each model is provided. The results show with and without the proposed HFS method. The time taken by the SVM + HFS and RF + HFS is less when compared with other prediction models.

5 Conclusion and Future Work

In this paper machine learning based DDoS attack detection models are investigated. There are many existing classifiers such as RF, DT, LG and SVM. These classifiers when used as the prediction models suffer from performance degradation due to quality of training set. This problem is known as curse of dimensionality which is indicated by irrelevant and redundant features. Therefore, to overcome this drawback, a feature selection algorithm known as HFS is defined in this paper. The dataset is divided into training and testing set. The training process includes selection of features as well. Thus the time complexity is reduced besides improving performance of the prediction models. With the HFS, the aforementioned classifiers proved to have

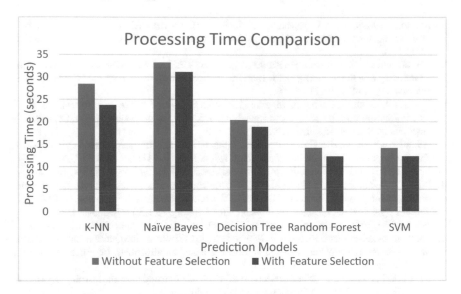

Fig. 6 Shows performance difference in terms of execution time

better performance. The empirical results showed that there is increased performance with feature selection. Thus the DDoS prediction models are better used with good feature selection. In future, we intend to define hybrid feature selection model that optimizes the prediction process further.

References

1. Oo MM, Kamolphiwong S, Kamolphiwong T (2017) The design of SDN based detection for distributed denial of service (DDoS) attack. In: International computer science and engineering conference, pp 258–263
2. Ali S, Li Y (2019) Learning multilevel auto-encoders for DDoS attack detection in smart grid network. IEEE Access 4:1–13
3. Trejo LA, Ferman V, Medina-Pérez MA, Fernando (2019) DNS-ADVP: a machine learning anomaly detection and visual platform to protect top-level domain name servers against DDoS attacks. IEEE Trans J 7:1–13
4. Doshi R, Apthorpe N, Feamster N (2018) Machine learning DDoS detection for consumer Internet of Things devices. In: IEEE symposium on security and privacy workshops, pp 29–35
5. Rahman O, Quraishi MAG, Lung C-H (2019) DDoS attacks detection and mitigation in SDN using machine learning. In: 2019 IEEE world congress on services (SERVICES), pp 184–189
6. Hou J, Fu P, Cao Z, Xu A (2018) Machine learning based DDos detection through NetFlow analysis. In: Milcom 2018 track 3 - cyber security and trusted computing, pp 565–570
7. Gu Y, Li K, Guo Z, Wang Y (2019) Semi-supervised K-means DDoS detection method using hybrid feature selection algorithm. IEEE Transl Content Min Permit Acad Res 7:64351–64365
8. Khuphiran P, Leelaprute P, Uthayopas P, Ichikawa K, Watanakeesuntorn W (2018) Performance comparison of machine learning models for DDoS attacks detection. IEEE, pp 1–4

9. Mishra P, Varadharajan V, Tupakula U, Pilli ES (2018) A detailed investigation and analysis of using machine learning techniques for intrusion detection. IEEE Commun Surv Tutorials 21(1):1–46

10. S. Indraneel and Venkata Praveen Kumar Vuppala. HTTP Flood attack Detection in Application Layer using Machine learning metrics and Bio inspired Bat algorithm. *Applied Computing and Informatics*, p1–12. (2017).

11. Biesiada J, Duch W, Kachel A, Maczka K Pałucha S (2005) Feature ranking methods based on information entropy with Parzen windows. In: International conference on research in electrotechnology and applied informatics, pp 1–9

12. Alhaj TA, Siraj MM, Zainal A, Elshoush HT, Elhaj F (2016) Feature selection using information gain for improved structural-based alert correlation. Plos One 11(11):1–18

13. Ladha L (2011) Feature selection methods and algorithms. Int J Comput Sci Eng 3(5):1–11

14. Roobaert D, Karakoulas G, Chawla NV (2006) Information gain, correlation and support vector machines. In: Guyon I, Nikravesh M, Gunn S, Zadeh LA (eds) Feature Extraction. STUDFUZZ, vol 207, pp 463–470. Springer, Heidelberg. https://doi.org/10.1007/978-3-540-35488-8_23

15. Sui B (2013) Information gain feature selection based on feature interactions, pp 1–69

16. Shaltout NA, El-Hefnawi M, Rafea A, Moustafa A (2014) Information gain as a feature selection method for the efficient classification of influenza based on viral hosts. In: Proceedings of the world congress on engineering, vol 1, pp 1–8

17. Saini P, Behal S, Bhatia S (2020) Detection of DDoS attacks using machine learning algorithms. https://doi.org/10.23919/INDIACom49435.2020.9083716

18. Pande S, Khamparia A (2019) A review on detection of DDoS attack using machine learning and deep learning techniques. https://doi.org/10.13140/RG.2.2.33777.63848

19. Pande S, Khamparia A, Gupta D, Thanh DNH (2021) DDOS detection using machine learning technique. In: Khanna A, Singh AK, Swaroop A (eds) Recent Studies on Computational Intelligence. SCI, vol 921. Springer, Singapore. https://doi.org/10.1007/978-981-15-8469-5_5

Convolutional Neural Network Based Automatic Speech Recognition for Tamil Language

S. Girirajan and A. Pandian

1 Introduction

ASR is a process of converting raw speech signal into corresponding text transcription. It was really challenging task due to variation such as speaker modulation, speaker attributes, background noise etc. The task of ASR system is categorized into Three phase. First stage is recognizing the phones from a raw speech signal and also it involves features selection or dimensionality reduction. It will extract the useful features from the speech signal based on task specific knowledge. Such a feature selection is performed by using Mel Frequency Cepstral Coefficient (MFCC) [1]. In the second stage, word is estimated based on the likelihood of the phones, it is called as lexicon model and final stage is to frame the sentence by considering the grammatical sequence of the particular language, it is called as Language model. Phonemes are generated directly from frames of the speech signal by using frame-level classification.

Speech signals are in variable-length that needs to be mapped with variable-length phonetic symbols or words. Due to this reason traditional ASR system uses Hidden Markov Model (HMM) and achieved high accuracy rate. HMM handles the variable length sequence in successful way similarly In HMM we have states with probability distribution that are used to analyze the temporal behavior of the speech signals. Another powerful model that estimates the probability distribution of speech signal is Gaussian Mixture Model (GMM). Later by using Expectation Maximization algorithm these two models are combined to achieve high accuracy rate for ASR system [2–4]. Due to wide usage of neural network models in various fields,

S. Girirajan (✉) · A. Pandian
Department of Computer Science and Engineering, SRM Institute of Science and Technology, Kattankulathur, Tamil Nadu, India
e-mail: girirajans.cse@gmail.com

A. Pandian
e-mail: pandiana@srmist.edu.in

© The Author(s), under exclusive license to Springer Nature Singapore Pte Ltd. 2022   91
B. Unhelker et al. (eds.), *Applications of Artificial Intelligence and Machine Learning*,
Lecture Notes in Electrical Engineering 925,
https://doi.org/10.1007/978-981-19-4831-2_8

researchers decided to use Artificial Neural Network model along with HMM to improve the accuracy as well as performance insist of GMM [5–7]. At the beginning deep learning models were used to design an ASR for mono-phone HMM model. Later it was used to recognize the large vocabulary continuous speech with tri-phone HMM model [8–10]. CNN is an premier, state of art, ANN design architecture that are used in image based classification and it shows high performance as well as accuracy in image recognition task [11, 12]. Due to the success in image related task, researchers decided to implement CNN for ASR system. Palaz et al. [13–15] proposed CNN-based acoustic model. In this work researcher designed a model that works in two stages, feature learning and classifier stage. Both feature learning and classifier works together by reducing the cost function based on relative entropy. Feature learning consist of several convolutional layers, filters that are placed in first convolutional layer are used to extract the information and modeled between first and second convolutional layer. After extracting the features, it will be classified by using fully connected layer and softmax layer. Due to its complex structure CNN are widely used in image classification for better performance. In order to use CNN in speech related task, speech signal need to be converted into a spectrogram, it is a plot 2-D representation of frequency over time [16, 17]. The 2-D representation of amplitude over time can also be plotted to input the CNN model. For CNN based speech recognition input can be either Amplitude over time or frequency over time (spectrogram).

This paper is organized as follows: Sect. 2 discusses some relevant literature, Sect. 3 CNN overview, Sect. 4 Experimental setup, Sect. 5 Result and Discussion and Sect. 6 Conclusion and Future work.

2 Literature Review

Abdel-Hamid et al. [18] proposed CNN model for ASR system and conducted two set of experiments by using Full weight sharing (FWS) and Limited weight sharing (LWS), they found that LMS gives better accuracy relatively 6–10 percentage of error rate is reduced when compared with FWS. Since LWS learns feature patterns of different frequency bands. In the study [19] researcher described CNN model and its applications over radiology related task. It discussed various issues that occurred in radiology and how to overcome those issues by using CNN layer together with pooling and fully connected layers. In this paper [20], features are extracted from the audio signals by using CNN architecture that consists of 3 convolution layers and achieved 97.46% accuracy in TIDIGIT corpus. In [21], deep CNN is applied to recognize text transcript from the noisy speech signal. This deep CNN consist of filters that reduce the noise from the given speech signal, pooling and input feature maps to get high accuracy. This model is evaluated by using Aurora4task and AMI dataset and word error rate (WER) is reduced significantly. In this paper [22], CNN model is designed to recognize the set of predefined words from the given speech signal. Model is implemented by using 1-D CNN that takes raw speech signal as input and dataset used for the experiment consist of total 30 words from that 10 isolated words are considered. This model achieved 97.4 and 88.7% accuracy for training and testing phase respectively. The high variations in accuracy level of training and testing phase is due to the noises that are present in speech signal. In [23], researcher proposed a novel approach called NovoGrad by using 1-D CNN, introduced a new layer-wise optimizer. In this paper [24], researcher developed an ASR system by using CNN for Telugu language one of the Indian regional languages. The model implemented by using dataset that is collected from airport inquiry system and achieved higher accuracy when compared with previous state of art methodologies. In this study [25], they used deep CNN to design a ASR system for recognizing the digits in Pashto language. To implement the model data is collected from 50 different speakers (male: 25, female: 25).Each speaker utters a digits from 0 to 9. After extracting the features from the speech signal by using MFCC. Preprocessed data is given as the input for CNN model that is designed with 4 hidden layers and achieved 84.17% accuracy. In this paper [26], researcher used CNN with convolutional and fully-connected layer to recognize the Arabic characters that produce similar sound. Raw speech single preprocessed by using MFSC and then preprocessed data given as the input for the CNN model. This proposed model achieves nearly 80% accuracy. In this study [27], designed a ASR for recognizing the digits, features from the speech signal is extraction by using MFCC. Extracted features are given as the input for back-propagation Neural Network. The two important stages in CNN model is filtering and pooling. The CNN model takes the input in the form of spectrogram image. Due to complexity CNN will not consider entire pixels for matching the speech signal insist of that it will consider the part of image pixels. This process of matching the pixels is called filtering [28]. Pooling is the process of dimensionality reduction in the feature maps. The features that are learned by filtering process are arranged in feature map in specific order.

3   Convolution Neural Network

CNN is widely used in image processing and computer vision related task. Input for the CNN model will be in 3-dimensional or 2-dimensional. So speech signal need to be converted into a spectrogram, it is a plot 2-D representation of frequency over time. The general architecture of the CNN consists of Input, Hidden and output Layer (Fully connected layer). Pixels of the image fed into the input layer in the form of array. Mathematical operations are performed over hidden layers to extract the features from the input image. Hidden layer consist of various other layers like Convolution Layer, Activation using linear unit (ReLU) and Pooling layer that perform feature extraction from the image. In convolutional layer features are extracted from the input images. Filters are used on the convolutional layer to create the feature map that indicates presence of detected feature in the input. Feature maps are generated by sliding the filter across the input image and multiplication operation is performed between the values of the filter and part of the input image that are covered by filter. Feature maps that are generated to give the information such as corners and edges of the image. Feature map of the image further rectified by applying ReLU activation function over convolution layer [29]. Many techniques are available to carry out pooling but widely used approaches are max and average pooling. Maximum activation is extracted by using max pooling and average of all activation will be considered by average pooling. In full connected (FC) layer classification is carried out by considering the feature that is learned in previous layers.

Feature map (fm) can be computed from the given input spectrogram image S by using Eq. (1) shown below.

$$fm^{(n)} = f\left\{ \sum_{i=0}^{k} \sum_{i=0}^{k} W_{i,j}^n S_{x+i,y+j}^{n-1} + b^n \right\} \tag{1}$$

$S_{x+i,y+j}^{n-1}$ denotes the upper layer of the feature map, feature weights can be denoted by $W_{i,j}^n$, b denotes the bias value. Activation function can be denoted as f, similarly n and m denotes the current layer and previous layer feature in 2-dimension. FC layer performs the classification by using soft-max function as given in Eq. (2) shown below.

$$FC_i^n = fm_i^{n-1} W^n + b^n \tag{2}$$

FC_i^n denotes the output of fully connected layer n, fm_i^{n-1} denotes the feature map of $n-1$ layer and b^n denotes the offset term. The output layer consists of probabilities of classes for each image input. This probability of class is calculated by using Eq. (3) shown below.

$$P(\hat{x}) = max(P(x_i))0 < i < n \tag{3}$$

P denotes the probability of classes, n denotes the number of classes. If $\hat{x} = x_i$ then the result of the prediction will be consider as correct.

CNN gives better accuracy due to features such as weight sharing, no limitation in usage of layers, pooling and local connections. A value of filters remains same since it is used to identify the similar patterns in different part of the image. Due to this memory required for processing is reduce without affecting the runtime.

CNN takes the image as input. In this case speech signal will be converted into spectrogram by representing as frequency vs time plot. Figure 1 shows the CNN architecture used in ASR.

Due to different vocal tract length each speaker will have some variations in speech signal. Pattern in Speech signal of same unit of speech will have slight variation in frequency based on the shape of vocal tract. This variation can be stabilized by applying convolutional and pooling layer over the frequency axis that increases the performance of CNN.

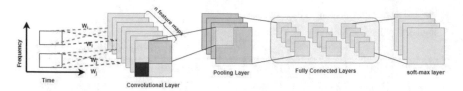

Fig. 1 CNN architecture

4 Experimental Setup

ASR can be implemented in three stages, Data preprocessing, Feature Extraction and recognition. The Proposed ASR architecture using CNN is shown in Fig. 2. The effectiveness of the suggested variants was evaluated by using Open sourced high-quality multi-speaker speech data set. Recorded excellent quality of speech signals for Tamil, Telugu, and Malayalam sentences were available in this dataset. Volunteers are used to prepare audio speech signals. Along with the wave file this data set also contains the text transcript for corresponding audio speech file. The data set consists of 153 h of male and 7440 min of female training data set together with text transcription for Tamil, Telugu and Malayalam languages. At first stage preprocessing is done over the input data. In second stage from the preprocessed data required features are extracted. At the last stage by using the extracted features, CNN will reduce the dimensionality of the input vector and predicts the word class. System configuration to implement the proposed work is windows 10 operating system with NVIDIA GTX 1650 GPU is used. The entire work is implemented in python 3.7 by using Librosa, Tenserflow and Keras.

Table 1 shown below describes the various properties of the dataset. Both male and female volunteers are used to create a corpus. Source to generate the dataset is collected from Wikipedia. Volunteers are in the age limit of 21 to 35. The Open sourced high-quality multi-speaker speech data set contains speech corpus for various languages like Tamil, Telugu, Malayalam, Gujarati. In this proposed work we used only Tamil Language alone for training and testing purpose.

From the dataset we have take some short speech to train and test the model. For the experiment purpose we have taken 10 classes of data samples. Each class takes 100 samples of utterance and length of each utterance is less than 3 s. Equal number of utterance is taken from both male and female speaker. Table 2 shows the list of utterance considered for testing and training the proposed model.

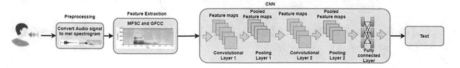

Fig. 2 ASR architecture using CNN

Table 1 Dataset details along with properties

Language	Gender	No.of Sentences	Words		Syllables		Phonemes	
			Total	Unique	Total	Unique	Total	Unique
Tamil	Male	1956	13,545	6159	48,049	1642	107,570	37
	Female	2335	15,880	6620	56,607	1696	126,659	37

Table 2 Short speech data taken from dataset

1	2	3	4	5
முந்தையது	மெதுவாக	வீடு	சகோதரி	நிறுத்து
6	7	8	9	10
பள்ளி	விளையாட்டு	மாணவர்	பயணம்	விலங்குகள்

4.1 Feature Extraction

Speech signal consist of multiple parameters such as various speakers, Source of data, acoustic variance, speaking rate. These parameters need to be considered for the performance of ASR. Essential features are extracted from the speech signal to reduce the uncertainty. In the proposed work, feature extraction techniques like MFSC and GFCC are used to achieve better accuracy rate.

Mel Frequency Spectral Coefficients (MFSC)
To extract more closely related set of features, short-time Fourier transform (STFT) dimensionality is reduced further by using MFSC [18]. To generate the mel spectrogram from the given audio signal sequence of operation need to be followed as shown in Fig. 3. Initially audio signal is divided into block of frames with fixed length. This avoids the loss of information between the adjacent frames by overlapping it. Before applying STFT, spectral characteristics of signal are improved by smoothing windows. It reduces the discontinuities and spectral leakage. Window size is fixed as 2048 and number of samples is set to 512. Next these frames are normalized by using STFT, signal with high frequency are normalized to low frequency. For a given time both temporal and frequency resolution are provided by STFT. After normalization, signals are passed to the set of filters that resample the signal to get better resolution by applying mel-filter banks. Finally logarithm serves to transform a multiplication into an addition. It is part of the computation of the MFSC.

$$MFSC_n = log\left(\sum_{l=0}^{l} M_n(L) + |F(L)|^2\right), n = 1 \ldots \ldots N \qquad (4)$$

$M_n(L)$ denotes the filter banks, n denotes the number of filter banks. Energy spectrum is denoted by $|F(L)|^2$.

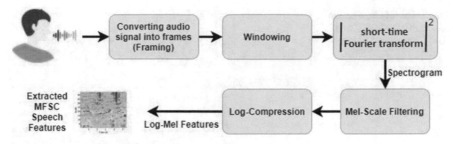

Fig. 3 Process involved in MFSC feature extraction

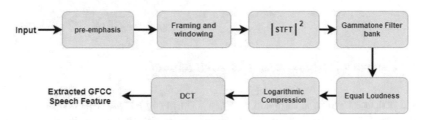

Fig. 4 Process involved in GFCC feature extraction

Gammatone-Frequency Cepstral Coefficient (GFCC)

The set of procedures need to be followed in extracting features by using GFCC are shown below in Fig. 4. To simulate the auditory process of human ear GammaTone Filter banks (GTFB) were applied in speech recognition as Eq. (5).

$$g(t) = xt^{n-1}e^{-2\pi bt}\cos(2\pi c_f t + \varphi). \tag{5}$$

x denotes the constant by default it is set to 1, filter order is denoted by n, φ denotes the phase shift between the filters, b denotes the bandwidth and c_f denotes the center frequency.

In GFCC input frames are windowed for 32 and 16 ms is set for overlapping frames. STFT is applied to get the spectrum. Later this spectrum is passed to GTFB, based on the center frequency of each frames equal loudness is applied. Then Logarithm is calculated on output of each filter. Finally to convert spectral to cepstral domain discrete cosine transform (DFT) is applied.

Combined MFSC and GFCC

In this proposed work MFSC and GFCC are combined together to extract the features from the given speech signal that exploit advantages of both MFSC and GFCC as shown in Fig. 5. This approach finds the most efficient feature from the input to improve the performance and accuracy of the ASR system. In this proposed work, 24 static and 24 derived parameters are used in equal share between MFSC and GFCC to improve the speed of the proposed ASR system.

Fig. 5 Process involved in
concatenating MFSC and
GFCC feature extraction

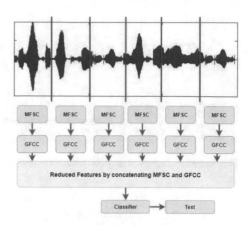

4.2   Convolutional Neural Network Model

In the proposed work, along with input and output layer we have taken 14 hidden layers for CNN. This hidden layer includes convolution layer, pooling, dense, dropout, normalization and flattened layers. In order to learn the features, convolution and pooling layers are used. Later these features were classified by using fully connected and softmax layer. Activation functions like ReLU and softmax were used to normalize the different range of values. ReLU will consider the output value as zero when input is less than zero otherwise input value will be taken as output. Based on the above constrain ReLU is used on convolution and pooling layer alone. Softmax will normalize the output in the range of 0 to 1. Since output layer is connected to fully connected layer. The resultant of fully connected layer should be either 0 or 1. By considering this softmax activation function is used in fully connected layer. Adam optimizer is used to train the model with cross-entropy loss function. For the loss function batch size is fixed as 50 with 1000 epoch. Layers used in CNN along with parameters are listed in Table 3.

Table 3 Layers used in CNN along with parameters

	No. of Layers	No. of Filters	Size of the filters	No. of Nodes	Activation
Convolution layer	5	32	(3,3)		
Max pooling	5		(2,2)		
Activation	5				ReLU
Normalization	1				
Flatten	1				
Dense	1			128	
Dropout	1				
Output				20	Soft-max

5 Result and Discussion

In speech signal, dimension of each vector is normalized to zero mean and unit variance. Dimension of the output is maintained by using padding. In the proposed work we have taken speech signal as input, so padding is done based on the silence to fit the length of the speech signal. Later features are extracted from this padded speech signal. CNN takes the input in the form feature coefficient vector. Sample rate is set to 1600 to extract features, STFT length is set to 512 and hop size is set to 160 samples. For GFCC number of cepstral coefficient are set to 13. First and second order derivation of GFCC is used to extract features from the speech signal. Similarly 128 bands log mel spectrogram are used for MFSC feature extraction. Size of the vector generated by MFSC and GFCC is (38, 185) and (124, 185). Convolutional layer that present first in CNN will compute the input feature vector by using 32 ReLU kernel with 3×3 filter size. The feature maps that are generated by using ReLU will be down sampled by using max pooling layer with 2×2 filter size. Later normalization is carried out to reduce the covariant shift on each activation layer. After this remaining convolution and pooling layers that are present in CNN will be computed in same manner. Dense and flatten layer consist of 128 nodes. The final layer is softmax that consist of 20 fully connected neuron for final classification. In dense layer retention probability is set to 0.25. The experiment is carried out separately for MFSC and GFCC and then results obtained by these feature extraction technique is compared with combined MFSC and GFCC feature extraction technique from that we found that high accuracy is obtained in GFCC feature extraction. The proposed model compared with previous start of art techniques and found that proposed model increase the accuracy and performance. Accuracy is measured based on correct and incorrect prediction by considering the precision, recall and F1 score.

$$recall(R) = \frac{T_p}{T_p + F_N} \tag{6}$$

$$precision(P) = \frac{T_p}{T_p + F_p} \tag{7}$$

$$F1\ Score = \frac{P \times R}{P + R} \tag{8}$$

Proposed model accuracy based on different feature extraction technique is shown in Table 4.

Table 4 Performance based on feature extraction techniques

	Training	Testing
MFSC	74.12	71.36
GFCC	79.86	75.43
MFSC + GFCC	90.34	86.02

Fig. 6 Epoch vs accuracy

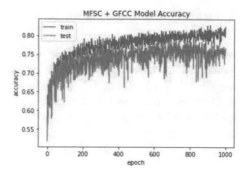

Fig. 7 Epoch vs loss

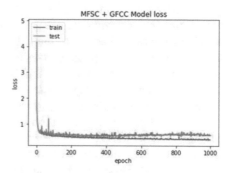

Figure 6 and 7 shows the loss and accuracy in each frame based on the epoch for the combined MFSC and GFCC feature extraction technique. For the experimental purpose we have used 1000 epochs.

Confusion matrix for Combine MFSC and GFCC model is shown in Table 5. From the confusion matrix we found that words like பள்ளி, விளையாட்டு, மாணவர், விலங்குகள் are classified well and words such as சகோதரி, நிறுத்து, பயணம் are not classified properly. Precision and recall of the proposed model classification is shown Table 6.

Table 5 Confusion matrix for proposed model

	1	2	3	4	5	6	7	8	9	10
1	8	0	0	0	0	0	0	0	1	1
2	1	8	0	1	0	0	0	0	0	0
3	0	1	8	0	0	1	0	0	0	0
4	0	0	1	9	0	0	0	0	0	0
5	0	1	3	2	1	0	2	2	1	0
6	0	0	0	0	0	10	0	0	0	0
7	0	0	0	0	0	0	10	0	0	0
8	0	0	0	0	0	0	0	10	0	0
9	1	1	1	0	2	0	4	0	2	0
10	0	0	0	0	0	0	0	0	0	10

Table 6 Accuracy of proposed model

	Precision	Recall
முந்தையது	0.885	0.721
மெதுவாக	0.812	0.814
வீடு	0.864	0.802
சகோதரி	0.612	0.608
நிறுத்து	0.745	0.756
பள்ளி	0.945	1.00
விளையாட்டு	0.962	1.00
மாணவர்	0.932	1.00
பயணம்	0.762	0.712
விலங்குகள்	0.981	1.0

6 Conclusion

The proposed work is carried out with limited set of words that are frequently used, to check the accuracy level of CNN. In the proposed work multiple experiments were conducted based on the different feature extraction techniques like MFSC, GFCC and combine MFSC + GFCC feature extraction. Later the input is processed to CNN for classification with limited number layers. The proposed model performed well when compare with previous state of art methodology. It achieves accuracy around 90.63% training accuracy and 81.25% testing. Experiments were done with single syllable data. Further we decided to work with large dataset with multiple syllable words.

References

1. Davis SB, Mermelstein P (1990) Comparison of parametric representations for monosyllabic word recognition in continuously spoken sentences. In: Readings in speech recognition. Elsevier, pp 65–74
2. Jiang H (2010) Discriminative training for automatic speech recognition: a survey. Comput Speech Lang 24(4):589–608
3. He X, Deng L, Chou W (2008) Discriminative learning in sequential pattern recognition—A unifying review for optimization-oriented speech recognition. IEEE Sig Process Mag 25(5):14–36
4. Deng L, Li X (2013) Machine learning paradigms for speech recognition: an overview. IEEE Trans Audio Speech Lang Process 21(5):1060–1089
5. Dahl GE, Ranzato M, Mohamed A, Hinton GE (2010) Phone recognition with the mean-covariance restricted Boltzmann machine. Adv Neural Inf Process Syst 23
6. Yu D, Deng L, Dahl G (2010) Roles of pre-training and fine-tuning in context-dependent DBN-HMMs for real-world speech recognition. In: Proceedings NIPS workshop deep learning, unsupervised feature learning
7. Seide F, Li G, Chen X, Yu D (2011) Feature engineering in context-dependent deep neural networks for conversational speech transcription. In: Proceedings IEEE workshop automation speech recognition understanding (ASRU), pp 24–29
8. Seide F, Li G, Yu D (2011) Conversational speech transcription using context-dependent deep neural networks. In: Proceedings Interspeech, pp 437–440
9. Sainath TN, Kingsbury B, Ramabhadran B, Fousek P, Novak P, Mohamed A (2011) Making deep belief networks effective for large vocabulary continuous speech recognition. In: IEEE Workshop Automation Speech Recognition Understanding (ASRU), pp 30–35
10. Pan J, Liu C, Wang Z, Hu Y, Jiang H (2012) Investigation of deep neural networks (DNN) for large vocabulary continuous speech recognition: why DNN surpasses GMMs in acoustic modeling. In: Proceedings ISCSLP
11. Eickenberg M, Gramfort A, Varoquaux G, Thirion B (2017) Seeing it all: convolutional network layers map the function of the human visual system. Neuroimage 152:184–194
12. Sultana F, Sufian A, Dutta P (2018) Advancements in image classification using convolutional neural network. In: Fourth international conference on research in computational intelligence and communication networks (ICRCICN), pp 122–129, Kolkata, India, November 2018
13. Palaz D, Collobert R, Doss MM (2013) End-to-end phoneme sequence recognition using convolutional neural networks. arXiv preprint arXiv:13122137
14. Palaz D, Doss MM, Collobert R (2015) Convolutional neural networks-based continuous speech recognition using raw speech signal. In: 2015 IEEE international conference on acoustics, speech and signal processing (ICASSP). IEEE
15. Palaz D, Collobert R (2015) Analysis of CNN-based speech recognition system using raw speech as input. In: Proceeding of Interspeech 2015 (No. EPFL-Conf-210029)
16. Qian Y, Woodland PC (2017) Very deep convolutional neural networks for robust speech recognition. In: 2016 IEEE workshop spoken language technology SLT 2016 - Proceedings, vol 1, no 16, pp 481–488. https://doi.org/10.1109/SLT.2016.7846307
17. Gouda SK, Kanetkar S, Harrison D, Warmuth MK (2018) Speech recognition: keyword spotting through image recognition
18. Abdel-Hamid O, Mohamed A, Jiang H, Deng L, Penn G, Yu D (2014) Convolutional neural networks for speech recognition. IEEE/ACM Trans Audio Speech Lang Process 22(10):1533–1545
19. Yamashita R, Nishio M, Do RKG, Togashi K (2018) Convolutional neural networks: an overview and application in radiology. Insights Imag 9(4):611–629
20. Haque MA, Verma A, Alex JSR, Venkatesan N (2020) Experimental evaluation of CNN architecture for speech recognition. In: Luhach A, Kosa J, Poonia R, Gao XZ, Singh D (eds) First International Conference on Sustainable Technologies for Computational Intelligence. AISC, vol 1045. Springer, Singapore. https://doi.org/10.1007/978-981-15-0029-9_40

21. Qian Y, Bi M, Tan T, Yu K (2016) Very deep convolutional neural networks for noise robust speech recognition. IEEE/ACM Trans Audio Speech Lang Process 24(12):2263–2276
22. Jansson P (2018) Single-word speech recognition with convolutional neural networks on raw waveforms
23. Li J et al (2019) Jasper: an end-to-end convolutional neural acoustic model. In: INTERSPEECH, April 2019
24. Nagajyothi D, Siddaiah P (2018) Speech recognition using convolutional neural networks. Int J Eng Technol 7(4.6):133–137
25. Zada B, Ullah R (2020) Pashto isolated digits recognition using deep convolutional neural network. Heliyon 6(2)
26. Rajagede RA, Dewa CK, Afiahayati (2017) Recognizing Arabic letter utterance using convolutional neural network. In: Proceedings - 18th IEEE/ACIS international conference on software engineering, artificial intelligence, networking and parallel/distributed computing (SNPD), pp 181–186, Kanazawa, Japan
27. Ali MA, Hossain M, Bhuiyan MN (2013) Automatic speech recognition technique for Bangla words
28. Seide F, Li G, Chen X, Yu D (2011) Feature engineering in context-dependent deep neural networks for conversational speech transcription. In: Proceedings IEEE workshop automatic speech recognition understanding (ASRU), pp 24–29
29. Morgan N (2012) Deep and wide: multiple layers in automatic speech recognition. IEEE Trans Audio Speech Lang Process 20(1):7–13

Identification of Wheat and Foreign Matter Using Artificial Neural Network and Genetic Algorithm

Neeraj Julka and A. P. Singh

1 Introduction

Wheat cereals are considered to be the primary source of nutrition consumed by humans in India and abroad after rice. Moreover, protein content in heat in much larger in comparison to other grains [1]. Also, the statistics indicate that wheat crop ranks top among other crops grown in India [2]. The first step after the harvest of agricultural produce is the grading of harvested crops [2] including wheat. In general, the grading of wheat grains is determined manually by visual inspectors. This method is very much time consuming and suffers from the drawback of inconsistency in the results. However, in the recent past, machine vision-based grading equipment has played an increasingly important role such tasks. Machine Vision is generally used to extract features to determine crop quality parameters without adversely affecting food crops [1]. Many research studies have been attempted to determine the ability of morphological characteristics to classify various types of grains and foreign matter using artificial intelligence techniques [3–5]. A few studies also used colour characteristics to identify grains [6, 7]. In this study, the slight variations in the results obtained with different illumination sources were found to be the primary discussion. Moreover, little research has been carried out on shape and size features in the classification of durum wheat [8, 9].Consequently, it shows machine learning algorithms, provide better results in the classification of agricultural produce using colour, morphological and textural characteristics. Several studies have also been executed to classify agricultural produce by machine vision systems. The work presented by [10] on classification of wheat and dockage using a support vector machine and neural network. In this work, images of wheat kernels and impurities acquired by a camera were used for the classification. Different morphological characteristics such

N. Julka (✉) · A. P. Singh
Department of Electronics and Communication Engineering, Sant Longowal Institute of Engineering and Technology, Longowal, Sangrur, Punjab 148106, India
e-mail: neerajjulkasliet@gmail.com

as volume, area, etc. were extracted from each component of the grain and impurity. The results obtained using machine learning algorithms showed that the ANN is more efficient than the SVM in terms of classification accuracy, 94.5 and 86.8% respectively. [11] introduced a approach that is based on offline machine vision system in which fifty-two features were extracted used to determine the accuracy for the classification of wheat kernels of different qualities. The results compared the hybrid approach for classification in terms of accuracy. [12], suggested a system that would classify three types of wheat. In this work, neural network with multilayer perceptron network (MLP) was trained to classify wheat grains. The various geometric features were obtained and considered as input to ANN like length of major and minor axes, diameter, perimeter and entropy information. The system showed an accuracy of more than 95% to classify the wheat grains using these features. [13] proposed a system that classifies three different verities of wheat grains using ANN. Seven inputs were fed as input to ANN with single hidden layer and a single output. The morphological features were extracted from grains such as area, length, compactness and perimeter. In the test phase, 99.78% ANN accuracy was achieved. Another study was presented based on morphological characteristics for the classification of four varieties of wheat of Iran [17]. Using image processing, ten geometric characteristics have been extracted from each grain. Out of these ten only nine, features were selected and fed as input of ANN's. The accuracy obtained with the proposed network is 85.72% for different wheat varieties. The present work focuses primarily on the grading of wheat kernel using Hybrid GA-ANN approach. The authors of the present work earlier developed a similar type of machine vision system for detecting foreign matter in wheat kernels [18] using shape and size features. The primary goal of this work is to study the ability of the (ANN) machine learning algorithm and the (GA) optimization technique to analyze the features extracted from the images. In this work, MATLAB software 2020a was opted to perform image processing and extraction of features by the hybrid GA-ANN algorithms on foreign matter and wheat grains. Consequently, illumination systems are setup in a controlled environment and wheat kernels and foreign matter are placed. Current research has highlighted the detection of wheat grains and foreign matter, making it more practical. The results of the proposed work are quite successful in detecting foreign matter in wheat kernels. Hence, overall grading accuracy has increased with the proposed GA-ANN model in comparison to ANN models reported earlier so far.

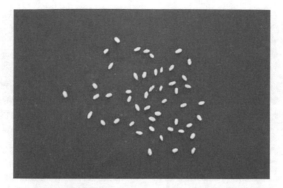

Fig. 1 Identification of frame captured taken from wheat grains

2 Proposed Methodology

2.1 Preparation of Samples

The wheat variety used in this research is the PBW 550 Unat that was taken from the Akal Seeds Academy, Mastuana Sahib, District Sangrur, Punjab, India as shown in Fig. 1. In general, harvested wheat kernels usually contain foreign material such as stones, straw and chaff. After preparation of the image samples, these were pre-processed in the Machine vision and Motion Control Lab at SLIET, Longowal and consequently the features were extracted for further development as detailed below:

2.2 Image Segmentation

Segmentation consists of dividing the distinct parts of an image that do not have any relationship to each other. Hence, the main task is to extract the features in real-time environment from the segmented parts. In order to extract morphological, color and texture features, a complete function was implemented in the MATLAB environment and considered in this work. Within this function, a limit is set to predefine values for objects for the thresholding process used for segmentation. Also, the noise was selected for the pre-processing are removed using morphological operations. The next step is to treat objects that are not threshold limits as noise. The noise was selected for the pre-processing are removed using morphological operations as shown in Fig. 2. In the final step, the features of the threshold objects of wheat grains and foreign matter in the image are extracted and are labeled and stored as the input data of the ANN.

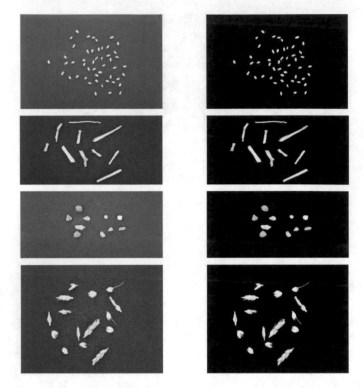

Fig. 2 Pre-processing and segmentation

2.3 Feature Extraction

In the next step, a total of sixty one features (fifteen morphological, eighteen color and twenty eight texture features) were extracted from wheat and foreign matter and the same are tabulated in Table 1. Total 518 samples of wheat kernels and foreign matter were used in the present work. A total of 362 samples were used for training and 156 samples for testing. Optimal number of features was selected from the available features to improve the overall detection accuracy.

2.4 Detection using ANN

MATLAB image processing and optimization toolboxes were used to extract the sixty one features of each wheat kernel and foreign matter respectively. The primary function of ANN can learn from extracted features of dataset. The neural network is composed of different layers such as input layer, hidden layer and output layer. Input layer contains the number of features for a specific problem and hidden layer

Table 1 Features extracted from each grain and foreign matter from each object

Morphological features

1	Area	9	Rectangular Aspect Ratio
2	Major Axis Length	10	Area Ratio
3	Minor Axis Length	11	Maximum Radius
4	Perimeter	12	Minimum Radius
5	Length	13	Radius Ratio
6	Width	14	Standard Deviation of all Radii
7	Thinness Ratio	15	Haralick Ratio
8	Aspect Ratio		

Color features

1	Red Mean	10	Red Variance
2	Green Mean	11	Green Variance
3	Blue Mean	12	Blue Variance
4	Red Range	13	Hue Variance
5	Green Range	14	Saturation Variance
6	Blue Range	15	Variance
7	Saturation Range	16	Hue
8	Hue Range	17	Saturation
9	Variance Range	18	Variance

Texture features

1	Short Run Emphasis	15	Energy:matlab
2	Long Run Emphasis	16	Entropy
3	Gray Level Non-Uniformity	17	Homogeneity:matlab
4	Run Percentage	18	Homogeneity
5	Run Length Non-Uniformity	19	Maximum Probability
6	Low Gray Level Run Emphasis	20	Sum of Squares: variance
7	High Gray Level Run Emphasis	21	Sum Average
8	Autocorrelation	22	Sum Variance
9	Contrast	23	Sum Entropy
10	Correlation:matlab	24	Difference Variance
11	Correlation	25	Information measure of Correlation1
12	Cluster Prominence	26	Information measure of Correlation 2
13	Cluster Shade	27	Inverse Difference Normalized
14	Dissimilarity	28	Inverse Difference Moment Normalized

acts a channel between input and output layer. Output layer shows the number of classification categories in which the neural network categories in its respective category. The weight and biases of the input is capable to obtain the correct output [14] The trial and error method adopted the number of neurons in the hidden layers and type of activation function [14, 15]. In this work, neural network was used and tested with different algorithms. It has been concluded that Levenberg Marquardt is the best method for grading of agricultural produce among different neural algorithms. The numbers of neurons in the hidden layer was tested by random trials. The highest accuracy of the classification was recorded and the outcome was stored as a confusion matrix.

2.5 Selection of Optimal Features Using GA

Genetic Algorithm (GA) gives the best possible solutions of the specific problem for similar applications as proposed by [16]. Such a set of feasible solutions is referred to as a population. The initial generation of the population is randomly generated. The solutions are assessed using a specific fitness function. Chromosomes are represented as strings. Crossover and Mutation are the techniques used to generate new populations from the old. A new population is generated from the old, referred to as generation or iteration. The process of generating a new population terminates when the predefined condition is met. The number of iterations assumes as the max limit as shown in Fig. 3.To select robust features a large number of extracting features from images for a specific problem. After selection of optimal features, classification algorithm was applied. Selection of optimal features reduces the computational time and complexity. Moreover, it improves overall the accuracy of ANN classifier. The main contribution of this work is to select an optimal number of features from subsets of morphological, color and texture features from subsets of features using genetic algorithm (GA).

3 Experimental Results and Discussions

The proposed GA parameters for experimental results such as population, number of generations, crossover and mutation probabilities. Table 2 shows the results of various algorithms with neural network to select the best in term of accuracy and opted for this proposed work. The results obtained in the form of confusion matrix of ANN and hybrid GA-ANN are shown in Tables 3 and 4 respectively. As a result, GA algorithm gives the best ability to select an optimal number of features in distinguishing wheat grains and foreign matters. In testing phase, the ANN and GA-ANN gives an detection accuracy was 97.4 and 98.7% respectively. The performance graphs shows of ANN and GA-ANN in terms of MSE 0.0036052 at epoch 7 and 0.0033746 at epoch 12 are shown in Fig. 4. The Learning coefficient (Mu) of the GA algorithm reaches the predetermined value and execution of the program is terminated and the value of MSE is obtained. The minimum value of the MSE directly concerns to a minimum misclassification. The state diagrams shows the changes the gradient, learning coefficient and validation values of ANN and GA-ANN are shown in Fig. 5 (a–b). As observed in Fig. 5, the gradually decreasing the value of gradient in initially epochs and stops

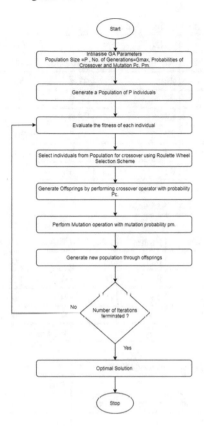

Fig. 3 GA flowchart

in epoch 12 at 2.5631 e–08 and 2.1336 e–08 in epoch 15 for ANN and hybrid algorithm. The value of Learning coefficient (Mu) had a gradually decreasing pattern and reached in epoch 12 to 1e–13 for ANN and epoch 15 to 1e–14 for the GA-ANN. In addition, Fig. 6 shows the regression plots of input data. The results showed that the use of ANN and hybrid GA-ANN structures with 10 hidden layers provided the 100% classification accuracy for the training. The structures 82-10-4 and 30-10-4 provided the 97.4 and 98.7% classification accuracy for testing conditions of ANN and hybrid GA-ANN respectively. As is evident, the highest classification accuracy was achieved with hybrid GA-ANN.

The value of Learning coefficient (Mu) had a gradually decreasing pattern and reached in epoch 12 to 1e–13 for ANN and epoch 15 to 1e–14 for the GA-ANN.

Table 2 Comparison of different algorithms

		ANN		GA-ANN	
No.	Algorithm function	No. of features	Accuracy	No. of features	Accuracy
1	**LM**	**82**	**97.4**	**30**	**98.7**
2	BFG	82	96.9	30	97.2
3	RP	82	98.0	30	98.5
4	SCG	82	99.0	30	97.2
5	CGB	82	97.1	30	98.1
6	CGF	82	97.8	30	97.9

Table 3 Confusion matrix of ANN classifier

	Train				Test			
	Wheat	Straw	Stones	Chaff	Wheat	Straw	Stones	Chaff
Wheat	153	0	0	0	68	0	0	0
Straw	0	59	0	0	0	20	0	0
Stones	0	0	88	0	0	0	40	0
Chaff	0	0	0	62	0	4	0	24
Accuracy	100%				97.4%			

Table 4 Confusion matrix of hybrid model GA-ANN

	Train				Test			
	Wheat	Straw	Stones	Chaff	Wheat	Straw	Stones	Chaff
Wheat	153	0	0	0	68	0	0	0
Straw	0	59	0	0	0	20	0	0
Stones	0	0	88	0	0	0	40	0
Chaff	0	0	0	62	0	2	0	26
Accuracy	100%				98.7%			

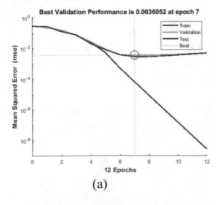

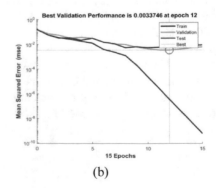

(a)   (b)

Fig. 4 Performance graphs **a** ANN **b** GA-ANN

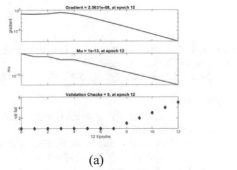

(a)   (b)

Fig. 5 State diagram of ANN and GA-ANN

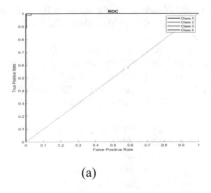

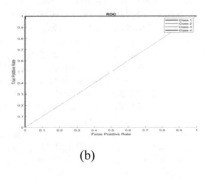

(b)

(a)

Fig. 6 Regression diagrams of ANN and GA-ANN

4 Conclusion

In this work, the detection of foreign materials in wheat kernels using Hybrid GA-ANN algorithm was discussed and proposed. This study was based on image processing and machine learning algorithms. Based on machine learning, ANN and hybrid GA-ANN models were implemented and proposed to achieve convincing results. Therefore, the use of GA with ANN in hybrid mode has the ability to classify the wheat and non-wheat components and can be used for similar practical applications in agriculture sector. The extracted features proposed in the present work provide excellent results for detection of foreign matter in wheat kernels. The GA algorithm was used to select an optimal number of features for achieving the highest classification accuracy using Hybrid-ANN model.

References

1. Saini M, Singh J, Prakash NR (2012) Analysis of wheat grain varieties using image processing: a review. Int J Sci Res ISSN (Online Impact Factor) 3(6):2319–7064
2. Yaman K et al (2001) Dinamik Çizelgeleme Için Görüntü Işleme Ve Arima Modelleri Yardimiyla Veri Hazirlama, March 2001
3. Majumdar S, Jayas DS (2000) i. m 43(6):1669–1675
4. Majumdar S, Jayas DS (1999) Classification of bulk samples of cereal grains using machine vision. J Agric Eng Res 73(1):35–47. https://doi.org/10.1006/jaer.1998.0388
5. Utku H, Koksel H (1998) Use of statistical filters in the classification of wheats by image analysis. J Food Eng 36(4):385–394. https://doi.org/10.1016/S0260-8774(98)00072-7
6. Luo X, Jayas DS, Symons SJ (1999) Identification of damaged kernels in wheat using a colour machine vision system. J Cereal Sci 30(1):49–59. https://doi.org/10.1006/jcrs.1998.0240
7. Majumdar S, Jayas DS (2000) Classification of cereal grains using machine vision: Ii. colormodels. Trans ASAE 43(6):1677–1680
8. Majumdar S, Jayas DS (2000) Classification of cereal grains using machine vision: III. Texture models. Trans Am Soc Agric Eng 43(6):1681–1687. https://doi.org/10.13031/2013.3068
9. Majumdar S, Jayas DS (2000) iv. c 43(6):1689–1694
10. Punn M, Bhalla N (2013) Classification of wheat grains using machine algorithms
11. Ebrahimi E, Mollazade K, Babaei S (2014) Toward an automatic wheat purity measuring device: a machine vision-based neural networks-assisted imperialist competitive algorithm approach. Meas J Int Meas Confed 55:196–205. https://doi.org/10.1016/j.measurement.2014.05.003
12. Abdullah NA, Quteishat AM (2015) Wheat seeds classification using multi-layer perceptron artificial neural network
13. Yasar A, Kaya E, Saritas I (2016) Classification of wheat types by artificial neural network. Int J Intell Syst Appl Eng 4(1):12. https://doi.org/10.18201/ijisae.64198
14. Ardabili S et al (2020) Modelling temperature variation of mushroom growing hall using artificial neural networks. In: Várkonyi-Kóczy A (eds) Engineering for sustainable future. INTER-ACADEMIA 2019. LNNS, vol 101, pp 33–45. Springer, Cham. https://doi.org/10.1007/978-3-030-36841-8_3
15. Sabanci K, Aydin C (2019) Determination of classification parameters of barley seeds mixed with wheat seeds by using ANN. Inf Secur Educ J 6(1):21. https://doi.org/10.6025/isej/2019/6/1/21-25
16. Khehra BS, Pharwaha APS (2017) Comparison of genetic algorithm, particle swarm optimization and biogeography-based optimization for feature selection to classify clusters of microcalcifications. J Inst Eng Ser B 98(2):189–202. https://doi.org/10.1007/s40031-016-0226-8
17. Khoshroo A, Arefi A, Masoumiasl A, Jowkar GH (2014) Classification of wheat cultivars using image processing and artificial neural networks. Agric Commun 2(1):17–22
18. Julka N, Singh AP (2019) Machine vision based detection of foreign material in wheat kernels using shape and size descriptors. Int J Adv Sci Technol 28(20):736–749

Efficient Classification of Heart Disease Forecasting by Using Hyperparameter Tuning

Divya Lalita Sri Jalligampala, R. V. S. Lalitha, T. K. Ramakrishnarao, Kalyan Ram Mylavarapu, and K. Kavitha

1 Introduction

Heart is the essential part of human body through which blood will be pumped to other parts to function properly. The functionality of heart is affected with conditions called as heart disease. One of the serious diseases in the world is heart disease. The mortality rate due to this disease is increasing gradually. Recent reports from WHO says that 17.9 million people died in the year 2019, due to cardiovascular diseases which occupies 32% of all deaths globally. Out of these 85% deaths caused due to heart stroke and attack. The risk factors for heart disease are like Diabetes, High BP, Alcohol consumption, Poor Diet, No physical exercise, High Cholesterol etc. In the current scenario, early prediction and diagnosis of heart disease is crucial which in turn needy for medical practitioners, doctors, and others. Machine Learning is currently an emerging field in healthcare industry for identification of diseases, easy diagnosis, classification of medical images etc. For effective disease prediction system, different Machine Learning methods were developed like SVM, Random Forest, KNN, Logistic Regression, Naive Bayes etc. and these are applied on the dataset for the prediction [15]. But, performance of a model is the key point in Data science. So, later Deep Learning (or) Deep Neural Network came into existence which is a subbranch of Artificial Intelligence. A lot of applications are there for Deep Learning like Image processing, Self-Driving cars, Healthcare, Speech Recognition, Language Translation, Virtual Assistants, Natural Language processing and others. Deep learning gained much prominence because of its mastery performance in terms of accuracy in case of massive amount of data. It does not require knowledge on

D. L. S. Jalligampala · R. V. S. Lalitha (✉) · T. K. Ramakrishnarao · K. R. Mylavarapu
CSE, Aditya College of Engineering & Technology, East Godavari, Andhra Pradesh Surampalem, India
e-mail: lalitha517@gmail.com

K. Kavitha
CSE, Gokaraju Rangaraju Institute of Engineering and Technology, Hyderabad, Telangana, India

© The Author(s), under exclusive license to Springer Nature Singapore Pte Ltd. 2022   115
B. Unhelker et al. (eds.), *Applications of Artificial Intelligence and Machine Learning*,
Lecture Notes in Electrical Engineering 925,
https://doi.org/10.1007/978-981-19-4831-2_10

domains, handles complex applications, no feature extraction separately, usage of end-to-end approach and a lot. All these features automatically affect performance of a model. Deep Learning creates an Artificial Neural Network that can learn and make decision on its own. Coming to health sector there are different models in deep learning to develop a disease prediction system [4]. Still people expectations are high especially in terms of health. So, to further optimize predictive performance of a Neural Network model, a technique used is Hyper parameter tuning which can find good set of values to parameters [8]. By using these tuned values, a model will be trained and tested, and then performance is evaluated and compared. Section 2 discussed about Related Works, Sect. 3 is about the Methodology, Sect. 4 gives Results and Discussion and Sect. 5 is regarding the conclusion of this research work.

2 Related Works

Sajja performed comparison between efficiency of different machine learning algorithms like SVM, Logistic Regression, KNN, Naive Bayes, and deep learning models CNN for predicting heart disease where CNN model bagged 94% accuracy [1]. Syed Nawaz Pasha first experimented different machine learning algorithms like SVM, KNN, DT on heart disease data set for early prediction of disease. Their performance is not that much accurate so deep learning technique Artificial Neural Network is used to further improve the performance of previously used algorithms [2]. V. Sharma used neural networks as classifier for predicting diagnosis of heart disease. Different algorithms are used, and its performance measured in terms of evaluation metrics, finally results are compared [3]. Mehmood experimented one of the deep learning algorithms Convolutional Neural Networks for the early prediction of heart disease. Performance is measured in terms of evaluation metrics against base learners and concluded using CNN for heart disease prediction achieved highest accuracy [4]. S. P. Rajamohana experimented different machine learning and deep learning models for the classification of heart disease prediction to find which methods are accurate than others [5]. Shankar presented a methodology to predict heart disease by using neural network algorithm and achieved accuracy of a method in between the range of 85% to 88%. And further improved the accuracy by modifying attributes [6]. Jae Kwon Kim developed Neural Network based heart disease prediction system by using Feature Correlation analysis on Korean Dataset. And the Performance of model evaluated using metric ROC curve [7]. Dangare investigated Multilayer Perceptron Neural Network model to develop heart disease prediction system with 13 attributes and 15 attributes and its performance is evaluated and compared [8]. Javid used different machine learning algorithms like SVM, Random Forest and K-NN and deep learning models like short term memory and gated recurrent unit neural networks on heart disease dataset taken from UCI database. To further improve the performance, voting based ensemble method is used and results shows that accuracy is improved by 2.1% [9]. Mantovani proposed an approach to use hyper parameter tuning as the optimization techniques on one of the machine learning algorithms J48

Decision trees to improve the performance the classifier [10]. Mahesh experimented a machine learning algorithm Random Forest on heart disease dataset taken from UCI database. This process starts with data preprocessing by using some statistical methods to find correlation between attributes in the dataset and then normalization is used for transformation. To improve the accuracy, parameters of random forest algorithm are tuned by using grid search method and PCA is sued to reduce dimensions. Finally, performance is measured by using metrics like Accuracy, Precision-Recall curve and Receiver Operating curve [11]. R. Soares de Andrades used three machine learning algorithms Random Forest, SVM, Gradient Boosting with hyper parameter tuning on ECG and other clinical reports to find best set of parameters for the above algorithms. Performance is evaluated and observed that this model is able to perform best accuracy and sensitivity [12]. Ambesange used different logistic regression algorithms to effectively predict heart disease. For this first pre-processing is performed like transformation, outlier removal, feature selection. Then Hyper parameter tuning techniques grid search and random search are performed to tune the parameters of these algorithms. Performance is evaluated using different metrics and achieved better accuracy [13]. Mate proposed a methodology to develop an efficient heart disease prediction system by using gradient boosting approach and then hyper parameter tuning technique is used to further improve the accuracy of a model [14]. Asvinth analyses different machine learning algorithms like Decision tree, Logistic Regression and Naive Bayes on heart disease dataset to develop prediction system. To improve accuracy of these algorithms, gridsearchcv parameter tuning method is used. From the results it is observed that Logistic Regression produced highest accuracy with 93% [15].

3 Methodology

Deep learning neural network model is applied on dataset to develop heart disease prediction system and its performance is calculated in terms of metrics. Later to further enrich the performance of a model, Hyper parameter tuning is performed where it finds best set of values for different parameters then model is trained by using these tuned values and its performance is measured. Finally, results are compared to find the model with best accuracy.

In this methodology, Heart Disease Dataset is taken as input to the model. The dataset is accessible from different databases namely UCI, Cleveland, Long Beach, and Hungary. In this, the dataset from UCI Databases is considered, which includes health related data concerning risk factors for getting heart disease. This repository consisting of 303 records and 14 attributes. First 13 attributes represent information of different symptoms like Chest pain, High BP, Cholesterol, Obesity, Physical Inactivity, Poor Diet etc. and last attribute acting as outcome of model that is class label for the task. The following table shows sample dataset (Table 1).

On this dataset data pre-processing is performed to find any missing values and then to better understand data, histograms are constructed with respect to attributes.

Table 1 Sample data records with attributes

	Age	Sex	Cp	Trestbps	Chol	Fbs	Restecg	Thalach	Exang	Oldpeak	Slope	Ca	Thal	Target
0	63	1	3	145	233	1	0	150	0	2.3	0	0	1	1
1	37	1	2	130	250	0	1	187	0	3.5	0	0	2	1
2	41	0	1	130	204	0	0	172	0	1.4	2	0	2	1
3	56	1	1	120	236	0	1	178	0	0.8	2	0	2	1
4	57	0	0	120	354	00	1	163	1	0.6	2	0	2	1

To find relationship between attributes, correlation matrix with heat map is used which gives information about how strong attributes are related each other.

Neural Network Model for Heart Disease Prediction Using Deep Learning
Deep Learning is a sub branch of Artificial Intelligence. Its structure is similar to our human brain with number of neurons formed as artificial neural network, which makes computers to perform tasks like human beings without their intervention. Deep Learning has a lot of applications, especially in medical sector. It plays a major role for analysis and early prediction of disease by using different types of models. In keras, the simplest way to construct a model is Sequential. So, in this methodology, a sequential neural network model is used for classifying heart disease prediction. It is best suitable when we are having stack of layers like input layer and output layer. To define model, first Dense is used as the layer type and layers are added to this by using add method one at a time. Sequential constructor then takes some parameters like input in terms of number of attributes from the dataset, activation to specify the activation function used, name of the layer, kernel initializer, learning rate and so on. Next compile model by using two parameters: Loss and Optimizer. Then fit () function is used to train model with corresponding values of training data, validation mechanism, target data, epochs and batch_size and then test the test data. Finally, performance is evaluated by using function: evaluate () and then make prediction on new data by using function: predict ().

3.1 Classification of Heart Disease Prediction Using Neural Network Model with Hyper Parameter Tuning

Hyperparameter Tuning
Deep Learning models are carrying several parameters which are influencing the performance of the model like Learning Rate, Activation Function, Kernel Initializer, epochs, batch_size and so on. So, to get best results from our model to maximize the performance, there is a need to set appropriate values of these parameters. To make this happen, Hyper parameter tuning is used where it finds a set of optimal parameters for our model. So that our model can achieve more accuracy significantly by reducing bias and variance. The following steps describe how classification of

heart disease prediction can be performed by using Hyper parameter tuning with selected model.

Step 1: Take the dataset and divided it into training dataset and test dataset.
Step 2: Neural Network model is applied on the training data set.
Step 3: Perform Hyperparameter Tuning to tune best parameters to perform the task.
Step 4: Train the dataset with tuned parameters and then test the test dataset.
Step 5: Finally, evaluation metrics measures the performance.

The following figure represents the architecture of a neural network model with hyperparameter tuning (Fig. 1).

Manual Hyper Parameter Tuning
In this, we must check each possible value for all parameters to the model manually. First, we select some values for parameters then model will be trained based on those values and performance is measured. This process will be repeated with different values for parameters until to get required accuracy. For instance, there are six parameters considered like kernel_initializer, activation_function, dropout, learning rate, epochs, and batch_size. For these parameters, 720 combinations of parameter values will be possible (Table 2).

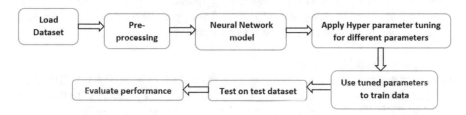

Fig. 1 Block diagram of neural network model with hyper parameter tuning

Table 2 Accuracy of sequential model with random parameter values

S. No	Model with corresponding parameter values	Accuracy
Set I	initializer = normal, activation = relu, dropout = 0.1, lr = 0.1, epochs = 50, batch_size = 10	68.85
Set II	initializer = normal, activation = linear, dropout = 0.1, lr = 0.01, epochs = 50, batch_size = 10	63.93
Set III	initializer = uniform, activation = relu, dropout = 0.2, lr = 0.001, epochs = 40, batch_size = 20	72.13
Set IV	initializer = uniform, activation = tanh, dropout = 0.1, lr = 0.1, epochs = 30, batch_size = 20	73.77
Set V	initializer = normal, activation = tanh, dropout = 0.3, lr = 0.001, epochs = 30, batch_size = 20	68.85

But this process will take more amount of time, minimizes the performance because human selected random values will not produce accurate output and practically it becomes complex if number of parameters are high.

3.2 Automatic Hyper Parameter Tuning

A Hyper parameter tuning that can produce better accuracy by reducing loss automatically. GridSearchCV method is used generate best set of values of all parameters to build a model and then performance of model is evaluated.

This method is available in scikit-learn package. The process includes the following steps: Define the model, provide different range of values for all hyperparameters, define a method for finding hyperparameter values and use some evaluative criteria to judge/the model and use a cross validation method.

Finding Optimal Parameters by Using GridSearchCV

GridSearchCV is hyperparameter tuning method which determines optimal values for our model. This method will take two parameters: Estimator which represents the model we are using, and param_grid represents the dictionary to iterate through parameters which we want to check. Cross validation is used to run on every set of parameters. And it is given to fit () function. Finally, we can get the best parameter values with its score by using best_score () and best_params methods. The following parameters are used with corresponding set of values.

Learning Rate and Dropout
The set of possible values for learning_rate from 0 to 1 like [0.0001, 0.001, 0.01, 0.1] and for dropout [0.1, 0.2, 0.3, 0.4, 0.5] are taken and passed to Dictionary of GridSearchCV. The method finds dropout_rate = 0.1 and learning_rate = 0.1 as optimal values.

Activation Function and Kernel Initialization
The set of possible values for activation_function is taken as ['softmax', 'relu', 'tanh', 'linear'] and kernel_initialization is taken as ['uniform', 'normal', 'zero'] and passed to GridSearchCV. The method returns best activation_fucntion is linear and kernel_initialization is linear as the following.

Epochs and Batch_size
The set of possible values taken for epochs is [10, 50, 100] and batch_size is [10, 20, 40] and passed to GridSearchCV. The method produced best epochs are100 and batch_size is 10.

All the generated optimal Hyper parameter values are now used as input values to model and then performance is evaluated in terms of evaluation metrics.

Table 3 Representation of confusion matrix

Actual/Predicted class label	Positive (C1)	Negative (C2)
Positive (C1)	True positives	False Positives
Negative (C2)	False Negatives	True Negatives

3.3 Evaluation Metrics

The model performance is measured with help of Confusion matrix. Where it represents information about correctly and incorrectly labelled records classified by model. It can be represented as (Table 3):

Where True positives therefore says how many tuples correctly classified as C1, False positives—the number of tuples with actual class label C1 is incorrectly classified as C2, False negatives—the number of tuples with actual class label C2 is incorrectly classified as C1, True negatives—the tuples correctly classified as C2. Now by using these values, evaluation metrics are calculated as below:

Accuracy. The proportion of correctly classified values from whole observations.

$$Accuracy = (TP + FP)\big/(TP + FP + TN + FN) \qquad (1)$$

Precision. The proportion of correctly classified positive predictions from the whole positive predictions.

$$Precision = (TP)\big/(TP + FP) \qquad (2)$$

Recall. The proportion of correctly classified positive values and total positive values.

$$Recall = (TP)\big/(TP + FN) \qquad (3)$$

F1-Score. The mean of recall and precision.

$$F1-score = (2 * precision * recall)\big/(precision + recall) \qquad (4)$$

4 Results and Discussion

4.1 Outcome of Data Pre-processing

On input dataset, pre-processing is performed but no missing values found and then histogram is constructed to all attributes of the dataset for better understanding of attributes behavior and patterns (Fig. 2).

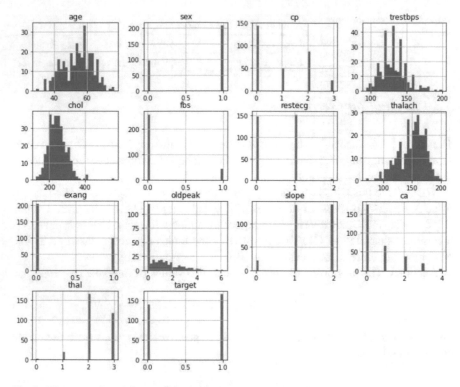

Fig. 2 Histogram for attributes of dataset

Next correlation matrix with heatmap is constructed to find correlation between different attributes in the dataset means which attributes are strongly related each other and which are not. So that we can select strongly related attributes for the further process to get effective results (Fig. 3).

4.2 Performance Evaluation

For sequential model without hyper parameter tuning, performance is calculated in terms of accuracy, precision, recall and f1-score and is shown (Table 4).

Next, the performance of a sequential model with hyper parameter tuning with gridsearchcv () measured in terms of accuracy, precision, recall and f1-score for heart disease data which is given below (Tables 5 and 6).

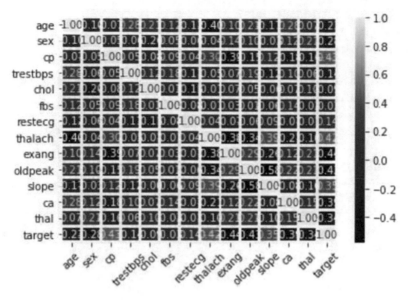

Fig. 3 Correlation matrix with heat map for attributes of dataset

Table 4 Performance of a sequential model

Model	Accuracy	Precision	Recall	F1-score
Neural Network-Sequential model	78.68	83	77	77

Table 5 Performance of a sequential model with hyper parameter tuning

Model	Accuracy	Precision	Recall	F1-score
Sequential model with Hyperparameter tuning	83.60	87	82	83

Table 6 Performance comparison of a model without and with hyper parameter tuning

Model	Accuracy	Precision	Recall	F1-score
Neural Network-Sequential	78.68	83	77	77
Sequential model with Hyperparameter tuning	83.60	87	82	83

4.3 Comparison of Model Performance with and without Hyper Parameter Tuning

The performance comparison of a sequential model with and without hyper parameter tuning as visually represented as (Fig 4).

This comparison shows that a model with Hyperparameter tuning produces better result than a model without Hyper parameter tuning.

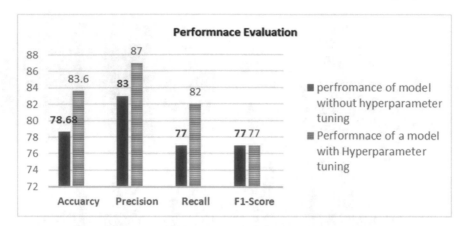

Fig. 4 Visual representation of performance comparison of a sequential model with and without hyper parameter tuning

5 Conclusion

Heart disease is one of the diseases which is having highest mortality count. For the early prediction of this disease deep Learning is used which is an emerging mechanism in medical field. To classify the heart disease prediction, a sequential model is used on dataset with random values for different parameters. Then the model performance is evaluated with respect to metrics like accuracy, recall, precision, and F1-score. As parameters are critical for the performance of a model, so their values must be optimal. For this Hyper parameter tuning technique is used which obtains best set of values for parameters. Based on number of parameters taken in the model, different combinations will be formed to create a model and then their performance will be evaluated manually. But these human judged parameter values will not be effective and even the process becomes complex if size of parameters is high. So gridsearchcv () method is used to find optimal values of all parameters. And by using these tuned values, a model will be trained and its performance is evaluated. From the comparisons, heart disease prediction can be classified using a sequential model with gridsearchcv () generated highest accuracy than a model without Hyper parameter tuning. In future, hybrid models will be developed with different feature selection techniques, image-based analysis to increase accuracy of a model.

References

1. Sajja TK, Kalluri HK (2020) A deep learning method for prediction of cardiovascular disease using convolutional neural network. Revue d'Intelligence Artificielle 34(5):601–606. https://doi.org/10.18280/ria.340510
2. Pasha SN et al (2020) Cardiovascular disease prediction using deep learning techniques. IOP Conf Ser Mater Sci Eng 981:022006

3. Sharma V, Rasool A, Hajela G (2020) Prediction of heart disease using DNN. In: 2020 second international conference on inventive research in computing applications (ICIRCA), pp 554–562. https://doi.org/10.1109/ICIRCA48905.2020.9182991

4. Mehmood A, Iqbal M, Mehmood Z et al (2021) Prediction of heart disease using deep convolutional neural networks. Arab J Sci Eng 46:3409–3422. https://doi.org/10.1007/s13369-020-05105-1

5. Rajamhoana SP, Devi CA, Umamaheswari K, Kiruba R, Karunya K, Deepika R (2018) Analysis of neural networks based heart disease prediction system. In: 2018 11th international conference on human system interaction (HSI), pp 233–239. https://doi.org/10.1109/HSI.2018.8431153

6. Shankar V, Kumar V, Devagade U et al (2020) Heart disease prediction using CNN algorithm. SN Comput Sci 1:170. https://doi.org/10.1007/s42979-020-0097-6

7. Kim JK, Kang S (2017) Neural network-based coronary heart disease risk prediction using feature correlation analysis. J Healthcare Eng 13, Article ID 2780501. https://doi.org/10.1155/2017/2780501

8. Dangare C, Apte S (2012) A data mining approach for prediction of heart disease using neural networks. Int J Comput Eng Technol 3(3)

9. Javid I, Zager A, Ghazali R (2020) Enhanced accuracy of heart disease prediction using machine learning and recurrent neural networks ensemble majority voting method. Int J Adv Comput Sci Appl 11(3):110369. https://doi.org/10.14569/IJACSA.2020.0110369

10. Mantovani RG, Horváth T, Cerri R, Vanschoren J, de Carvalho ACPLF (2017) Hyper-parameter tuning of a decision tree induction algorithm. In: 5th Brazilian conference on intelligent systems, BRACIS 2016, Recife, Pernambuco, Brazil, pp 37–42, 9 October 2016–12 October 2016. Institute of Electrical and Electronics Engineers, Piscataway

11. Sonth MV, Ambesange S, Sreekanth D, Tulluri S (2020) Optimization of random forest algorithm with ensemble and hyper parameter tuning techniques for multiple heart diseases, 27 November 2020. https://doi.org/10.13140/RG.2.2.12451.68649

12. Soares de Andrades R, Grellert M, Beck Fonseca M (2019) Hyperparameter tuning and its effects on cardiac arrhythmia prediction. In: 2019 8th Brazilian conference on intelligent systems (BRACIS), pp 562–567. https://doi.org/10.1109/BRACIS.2019.00104

13. Ambesange S, Vijayalaxmi A, Sridevi S, Venkateswaran, Yashoda BS (2020) Multiple heart diseases prediction using logistic regression with ensemble and hyper parameter tuning techniques. In: 2020 fourth world conference on smart trends in systems, security and sustainability (WorldS4), pp 27–832. https://doi.org/10.1109/WorldS450073.2020.9210404

14. Priya RL, Jinny SV, Mate YV (2021) Early prediction model for coronary heart disease using genetic algorithms, hyper-parameter optimization and machine learning techniques. Health Technol 11:63–73. https://doi.org/10.1007/s12553-020-00508-4

15. Asvinth A, Hiremath M (2020) A computational model for prediction of heart disease based on logistic regression with GridSearchCV. Int J Sci Technol Res 9(03). ISSN 2277-8616

16. Gupta S, Sedamkar RR (2021) Genetic algorithm for feature selection and parameter optimization to enhance learning on Framingham heart disease dataset. In: Balas VE, Semwal VB, Khandare A, Patil M (eds) Intelligent Computing and Networking. LNNS, vol 146. Springer, Singapore. https://doi.org/10.1007/978-981-15-7421-4_2

17. Sharma S, Parmar M (2020) Heart diseases prediction using deep learning neural network model. Int J Innov Technol Exploring Eng (IJITEE) 9(3). ISSN 2278-3075

18. Kayiram K, Laxman Kumar S, Pravallika P, Sruthi K, Lalitha RVS, Krishna Rao NV (2020) Fashion compatibility, recommendation system, convolutional neural networks, sentiment analysis. In: International conference, ACCES 2020, GRIET, Hyderabad, 18th and 19th September 2020

19. Lalitha RVS, Divya Lalitha Sri J, Kavitha K, Rayudu Srinivas RRT, Sujana C (2021) Prediction and analysis of corona virus disease (COVID-19) using Cubist and OneR. IOP Conf Ser Mater Sci Eng 1074:012022. https://doi.org/10.1088/1757-899X/1074/1/012022

20. Nawaz MS, Shoaib B, Ashraf MA (2021) Intelligent cardiovascular disease prediction empowered with gradient descent optimization. Heliyon 7(5):e06948. https://doi.org/10.1016/j.heliyon.2021.e06948. PMID: 34013084, PMCID: PMC8113842

LS-Net: An Improved Deep Generative Adversarial Network for Retinal Lesion Segmentation in Fundus Image

A. Mary Dayana and W. R. Sam Emmanuel

1 Introduction

Diabetic Retinopathy (DR) is the leading cause of vision loss in most working-age adults, especially in people with Diabetes Mellitus (DM). The International Diabetes Federation (IDF) 2019 [1] reported that 463 million adults worldwide have critical diabetic complications and its incidence is likely to increase 700 million by 2045. In this context, regular screening and timely treatment of the eye can prevent blindness in diabetic patients. However, fundus images obtained through the screening process require a professional ophthalmic expert to identify the typical pathological signs of the retina as they are dispersed over the eye characterizing low contrast and irregular shapes. As shown in Fig. 1, the lesions allied with DR are Microaneurysms (MAs), Hemorrhages (HEMs), Hard Exudates (HEs), and Soft Exudates (SEs). The dark red lesions such as MAs, and HEMs are the early indicators of DR. Exudates are bright yellow colored lipid deposits within the retina formed due to the leakage of deteriorated blood capillaries. In this scenario, an automated computer- assisted diagnosis (CAD) would help ophthalmologists identify the ocular disease more precisely and accurately.

Although recent research on retinal lesion segmentation has demonstrated considerable improvement in segmentation accuracy, these are still dependent on massive data to produce satisfactory results during training.

Generative Adversarial Networks (GANs) [2] is now becoming popular, and their outstanding capabilities in image processing has gained the attention of researchers

A. Mary Dayana (✉) · W. R. Sam Emmanuel
Department of Computer Science, Nesamony Memorial Christian College, Marthandam,
Affiliated to Manonmaniam Sundaranar University, Tirunelveli, India
e-mail: mary_dayana_csa@nmcc.ac.in

W. R. Sam Emmanuel
e-mail: sam_emmanuel@nmcc.ac.in

© The Author(s), under exclusive license to Springer Nature Singapore Pte Ltd. 2022   127
B. Unhelker et al. (eds.), *Applications of Artificial Intelligence and Machine Learning*,
Lecture Notes in Electrical Engineering 925,
https://doi.org/10.1007/978-981-19-4831-2_11

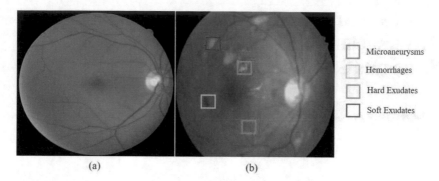

Microaneurysms

Hemorrhages

Hard Exudates

Soft Exudates

(a)   (b)

Fig. 1 Sample fundus images **a** Healthy Retina **b** Unhealthy Retina

worldwide [3]. The applications of GANs in medical imaging include image segmentation [4], image classification, image synthesis, image augmentation, image style transfer, and super-resolution tasks. Deep learning-based generative methods outperform the existing neural network models in the medical image segmentation tasks. Conditional GANs (cGANS) [5], a variation of GAN has an innate ability to produce reliable image features in a controlled manner by combining deep neural networks with adversarial learning concepts. The objective of the proposed work is to develop an improved deep generative model (LS-Net) based on conditional GAN for retinal multi-lesion segmentation. The following is how the rest of the paper is organized: The review of research related to GAN and cGAN-based techniques for lesion segmentation is presented in Sect. 2. In Sect. 3, the method used for the proposed study is outlined. In Sects. 4 and 5, the experimental data and the results are discussed. A brief conclusion is summarized in Sect. 6.

2 Related Work

In recent years, deep learning algorithms have become popular in segmenting medical images. Researchers [6–8] applied deep learning techniques based on U-Net [9] to segment retinal lesions. The study in [10, 11] developed a patch-based method using a deep Convolutional Neural Network (CNN) to segment multiple lesions in fundus images. Most of the reviewed methods [12, 13] employed deep CNNs as their base models to detect lesions in fundus images. At present, Generative Adversarial Networks (GANs) initially proposed by [2] have achieved tremendous potential in the processing of medical images. GAN has been used to detect brain tumors, segment the skin lesions [14], detect anomalies, generate electronic health records, generate high-resolution realistic images, enhance training data and synthesize retinal images from vessel trees. GANs fundamental idea is adversarial training that jointly optimizes the network with a Generator and a Discriminator. GANs can enhance the contrast and visual details of the lesion borders. Subsequently, the performance of segmentation is improved.

Conditional GAN (cGAN) proposed by [5] is an extended form of GAN conditioned on some extra information to generate images. Zhou et al. [15] developed an architecture called DR-GAN based on Conditional GAN for fine-grained lesion synthesis in retinal images to achieve robust segmentation. Ahn et al. [16] proposed an image synthesis framework for DR using conditional GAN to observe disease progression based on an adversarial learning mechanism. Singh et al. [17] formulated a network model based on cGAN to segment retinal optic disc. The study in [18] proposed a Convolutional Neural Network to generate lesion images using cGAN. Sarp et al. [19] introduced an algorithm using cGAN for wound border segmentation and tissue classification. Rammy et al. [20] proposed a patch-based conditional GAN for retinal blood vascular segmentation and achieved state-of-art results. A lightweight conditional GAN was developed by [21] for the semantic segmentation of road surface areas. Similarly, the author in [22] utilized a pixel2pixel cGAN for the segmentation of spectral images. Xiao et al. [4] formulated a technique to increase the performance of DR lesion segmentation by integrating HEDNet [23] and a Conditional Generative Adversarial Network (cGAN). The lesion segmentation performance is attained by adding the adversarial loss to the segmentation loss. A deep ensemble CNN based on U-Net [9] was developed by [7] for exudate detection. Furthermore, Conditional GAN was utilized to alleviate the class imbalance problem and to enhance the dataset. Though numerous approaches have been developed based on deep learning with GAN, most existing methods [20, 24] segment only the blood vessel structures. However, segmentation of subtle retinal lesions is crucial for DR diagnosis in clinical practice. Therefore, a deep generative model is developed to segment the DR lesions by applying a conditional setting to the Generative Adversarial Network.

3 Methodology

The purpose of the proposed framework (LS-Net) is to develop an adversarial learning framework for pixel-level segmentation of lesions in the retinal fundus images with controllable lesion information. The model is a paired image-to-image translation network trained with original fundus lesion images and corresponding ground truths. The methodology includes three steps: Generator model building, discriminator model building, and network training with spectral normalization and a modified loss.

3.1 Conditional GANs

Conditional GAN (cGAN) is a new variation of GAN initially proposed by [5]. Unlike GAN, cGANs learn to map an input image m and a random noise vector b to an output image n, whereas GANs learn to map a random noise vector b to an output image n. The conditional GAN has two competing networks: A generator and a discriminator. The generator G is trained in such a way to produce realistic segmented outputs that are indistinguishable from the actual input. The discriminator D attempts to differentiate the real input and the generated output. Adding conditions to the model with further lesion information can help direct the data generation process. Conditional GANs learn to map an input image to the output image and are trained adversarially. In addition, it also learns a loss function to train this mapping. The objective function is written as in Eq. (1).

$$L_{cGAN}(G, D) = E_{m,n}[log\,log\,D(m, n)] + E_{m,b}[log\,log(1 - D(m, G(m, b)))]$$

(1)

The optimal solution is attained using the minimax game-theoretic approach in which the generator G minimizes the objective function and the discriminator D attempts to maximize it.

3.2 Image-To-Image Translation Network

The baseline of the proposed Image-to-Image Translation Network was adapted from [25]. The two key features of the proposed LS-Net deep generative lesion segmentation model were Encoder-Decoder based U-net [9] generator and a patch-based Markovian discriminator (Patch-GAN) [25].

Generator Architecture. U-net proposed by [9] is a supervised method designed to produce accurate segmentation with limited training images. The proposed encoder-decoder based deep generative LS-Net with U-Net baseline has a contracting and an expansive path with skip connections. As in [20], the hidden details of the input image are extracted using the encoder network and the size is reduced with down-sampling. At each stage, the decoder network reconstructs the features with up-sampling to predict the final segmentation map. Skip connections pass the input features from the same level of an encoder to the same decoder level with conditional information to preserve the spatial information during down-sampling. Introducing skip connections in the encoder-decoder architecture enables feature reusability as well as stabilize training and convergence.

Discriminator Architecture. The patch-based discriminator architecture Patch-GAN [25] is proposed to detect the high-frequency components in the fundus image. It is trained to discriminate between the actual and generated output. It works on over-lapping patches and penalizes each patch focusing on portions of the image rather

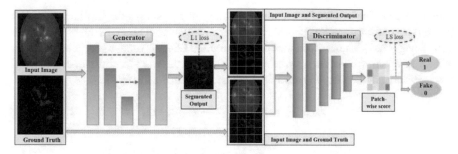

Fig. 2 Structure of the proposed LS-Net method

than over the entire image [26]. In this approach, each rectangular patch is treated as a single image and generates a probability map on each patch that can be averaged to give a single score. Assuming independence among all patches, the discriminator processes the input as a Markov random field. The reduced patch size permits fast network convergence and produces segmentation maps with high resolution [20]. The architecture of the proposed deep generative lesion segmentation LS-Net model is demonstrated in Fig. 2.

3.3 Spectral Normalization

Spectral Normalization (SN) suggested by [27] is a weight normalization strategy that helps stabilize the training of discriminator networks. Spectral normalization replaces every weight W in the network with W/σW, and σW is described in Eq. (2).

$$\sigma(W) = \|W_q\| = p^T W_q \tag{2}$$

Here, p and q are the random vectors in the domain and co-domain matrix. With Spectral Normalization, it is possible to renormalize the weight whenever it is updated. As in [27], the power iteration method is used to estimate the spectral norm of each layer. In addition, it controls the Lipshitz constant of the discriminator to mitigate the mode collapse and exploding gradient problem. As a result, spectral normalization proves to be more efficient and stabilizes the training process for segmentation.

3.4 Loss Function

A loss function is chosen to improve the accuracy and training stability. The network is trained with the loss function and the weights are updated to minimalize the loss function [28]. The loss function also determines the measure of the similarity between the input and the output image. GAN with usual adversarial loss will not produce realistic images [25]. Therefore, a modified loss called LS_{loss} is used, a grouping of conditional adversarial loss and additional L1 loss terms. The Conditional-Adversarial Loss is defined as in Eq. (1), and the L1 loss is defined as in Eq. (3).

$$L_{L1}(G) = E_{m,n,b}\big[\|n - G(m, b)\|_1\big] \tag{3}$$

$$LS_{loss} = arg_{G}^{minmax}\,_{D}\,L_{cGAN}(G, D) + \lambda L_{L1}(G) \tag{4}$$

The integration of the L1 loss and the adversarial loss generates a modified loss as described in Eq. (4), controlled by a hyperparameter lambda (λ).

4 Experimental Setup

4.1 Dataset

The Indian Diabetic Retinopathy Image Dataset (IDRiD) [29] is a publicly available dataset that is used for pixel-level annotation of the four distinct types of DR lesions: MAs, HEMs, HEs, and SEs. The IDRiD dataset comprises 54 images for training and 27 images for testing. Among the 81 pathological images, all the 81 images have MAs and HEs, 80 images have HEMs and only 40 images have SEs. All the images and their corresponding ground truths have a resolution of 4288×2848.

4.2 Evaluation Metrics

The performance of the proposed deep generative adversarial learning model is evaluated using two metrics: F1 Score and Area Under the Precision-Recall curve (PR – AUC). The F1 Score is an evaluation metric used to balance precision and recall, and it is defined as in Eq. (5). The Precision and Recall are computed using Eqs. (6) and (7).

$$F1 = 2 \cdot \frac{Precision \cdot Recall}{Precision + Recall} \tag{5}$$

$$Precision = \frac{TP}{TP + FP} \tag{6}$$

$$Recall = \frac{TP}{TP + FN} \tag{7}$$

where TP, FP and FN denotes the True Positive, False Positive and False Negative values.

4.3   Network Training

An Encoder-Decoder-based U-net with skip connections is employed as a Generator and a 70 × 70 Markovian PatchGAN is used as a discriminator. The input to the network was the retinal image and its ground truth taken from the IDRiD dataset. The Generator generates the segmented output while the discriminator compares the input image and its ground truth pair with the input image and its segmented output pair to guess how realistic they are. The PatchGAN discriminator splits the input into small rectangular patches for further processing and estimates the probability of a given sample. The Generator is trained so that the discriminator cannot differentiate the original image from the generated image. Alternatively, the discriminator is trained adversarially to detect the generated image as real or fake. While training, the discriminator is trained first, and then the generator is trained. Subsequently, the combined model is trained by setting the loss parameters and weights looping through several epochs. During adversarial training, spectral normalization [27] was utilized to increase the model stability and a modified loss LS_{loss} was applied to refine the results obtained through the final segmentation process.

4.4   Implementation Details

The proposed framework is implemented in Python using Keras. Leaky ReLU [30] activations are used in the encoder and regular ReLUs are used in the decoder of the generator. The PatchGAN discriminator is employed using Leaky ReLU with a slope value 0.2 [26]. A patch size of 128 is used for the Soft Exudates, Hard Exudates, and Hemorrhages, and a patch size of 64 is used for the Microaneurysm as in [4]. The weight of the modified LS_{loss} hyperparameter λ value is 0.5. The learning rate was set as 0.0001. The network was trained to run for 300 epochs, and the learning rate decayed after 200 epochs. The adam optimizer [31] is used to optimize the objective function, with a momentum factor of 0.9. Until the model converges, the network is trained with fundus images that have distinct types of lesions.

5　Results and Discussion

5.1　Results

The model's performance is reported in terms of the Precision-Recall (PR) curve and F1 Score metrics. The comparative evaluation of the proposed method with other representative methods in terms of PR-AUC and F1-score is depicted in Table 1. The original image from the IDRiD test set with its ground truth and the resultant segmented outputs are illustrated in Fig. 3.

Table 1 Comparative analysis with existing GAN based methods on IDRiD test dataset

Author	Methodology	PR-AUC				F1-score			
		MA	HEM	HE	SE	MA	HEM	HE	SE
Gullon [28]	UNet + GAN	0.31	0.43	0.71	0.41	0.46	0.47	0.70	0.40
Xiao et al. [4]	HEDNet + cGAN	0.43	0.48	0.84	0.48	0.42	0.45	0.69	0.43
Proposed Method	LS-Net + cGAN	0.54	0.57	0.84	0.65	0.53	0.55	0.78	0.61

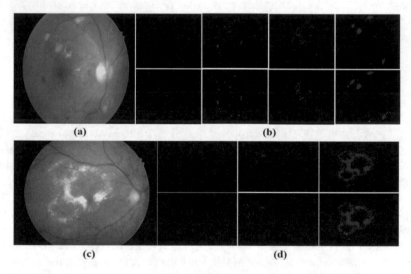

(a)　　　　　　　　　　　(b)

(c)　　　　　　　　　　　(d)

Fig. 3　a Original retinal fundus image from the IDRiD test set representing all four types of lesions **b** Top row—ground truth of MA, HEM, HE, and SE. Bottom row—segmented output **c** Fundus image representing three types of lesions **d** Top row—ground truth of MA, HEM, and HE. The bottom row represents the segmented output

5.2 Discussion

Conditional GANs is an adversarial learning network that can learn the scale-invariant features of the input image and can segment the lesion region precisely. The proposed encoder-decoder-based deep generative model employed with spectral normalization improves the network stability, robustness, and generalization properties. Different loss functions have experimented with the proposed model based on adversarial training to assess the pixel-level multi-lesion segmentation performance. Using L1 loss or cGAN loss alone produces blurred results and, at the same time, presents visual artefacts. The combination of L1 loss and cGAN loss yields clear and consistent results. The proposed LS-Net model achieves 0.54, 0.57, 0.84, and 0.65 for AUC and 0.53, 0.55, 0.78, and 0.61 for F1-Score and outperforms all the compared techniques. However, [4] reported the same performance for Hard Exudate segmentation in terms of AUC by attaining 0.84 but obtains lesser scores in other measures. The comparative performance results of the LS-Net model with existing GAN based methods are depicted in Figs. 4 and 5. The obtained results indicate that achieving the best performance in all classes is not possible because of the inter-variability features of lesions. Even though the segmentation performance of HE is good enough, there are still a few false positives in the segmentation of MAs. The reason is due to the inconspicuous size of the MA lesion. From the experimental outcomes, we observe that the proposed LS-Net architecture trained with modified LS_{loss} function achieves superior performance, increasing the training stability while preserving the detailed features of the generated image in pixel-level lesion segmentation.

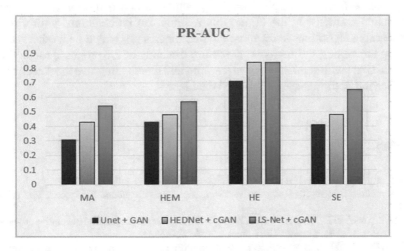

Fig. 4 Performance of the proposed method in terms of PR-AUC

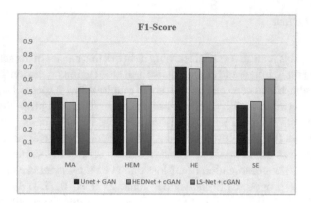

Fig. 5 Performance of the proposed method in terms of F1-Score

6 Conclusion

A conditional GAN based deep adversarial learning framework has significantly improved the performance of lesion segmentation over other traditional deep learning algorithms. The proposed architecture mainly focused on enhancing DR diagnosis via lesion segmentation in fundus images with deep adversarial learning. The advantage of the proposed work relies on segmenting the lesions with fewer false positives besides improving generalizability and the stability of segmentation using spectral normalization. Network training requires only a small dataset; however, it renders a superior performance in pixel-level segmentation and outperforms the existing GAN-based methods. In the future, this method can be combined with different deep learning algorithms to diagnose the pathologies associated with other medical imaging modalities. Furthermore, the future direction of the proposed method is to integrate GAN with semi-supervised learning and thereby improve the detection and classification performance with limited data labels.

References

1. International Diabetes Federation: IDF Diabetes Atlas Ninth edition 2019
2. Goodfellow IJ et al (2014) Generative adversarial nets. Adv Neural Inf Process Syst 3:2672–2680
3. Pan Z, Yu W, Yi X, Khan A, Yuan F, Zheng Y (2019) Recent progress on generative adversarial networks (GANs): a survey. IEEE Access 7:36322–36333
4. Xiao Q et al (2019) Improving lesion segmentation for diabetic retinopathy using adversarial learning. In: Karray F, Campilho A, Yu A (eds) Image analysis and recognition. ICIAR 2019. LNCS, vol 11663, pp 333–344. Springer, Cham. https://doi.org/10.1007/978-3-030-27272-2_29
5. Mirza M, Osindero S (2014) Conditional generative adversarial nets
6. Playout C, Duval R, Cheriet F (2019) A novel weakly supervised multitask architecture for retinal lesions segmentation on fundus images. IEEE Trans Med Imag 38(10):2434–2444

7. Zheng R et al (2018) Detection of exudates in fundus photographs with imbalanced learning using conditional generative adversarial network. Biomed Opt Express 9(10):4863–4878
8. Sambyal N, Saini P, Syal R, Gupta V (2020) Modified U-Net architecture for semantic segmentation of diabetic retinopathy images. Biocybern Biomed Eng 40(3):1094–1109
9. Ronneberger O, Fischer P, Brox T (2015) U-Net: convolutional networks for biomedical image segmentation. In: Navab N, Hornegger J, Wells W, Frangi A (eds) Medical image computing and computer-assisted intervention – MICCAI 2015. MICCAI 2015. LNCS, vol 9351, pp 234–241. Springer, Cham. https://doi.org/10.1007/978-3-319-24574-4_28
10. Lam C, Yu C, Huang L, Rubin D (2018) Retinal lesion detection with deep learning using image patches. Invest Ophthalmol Vis Sci 59:590–596
11. Mary Dayana A, Sam Emmanuel WR (2020) A patch - based analysis for retinal lesion segmentation with deep neural networks. In: Pandian A, Palanisamy R, Ntalianis K (eds) Proceeding of the international conference on computer networks, big data and IoT (ICCBI - 2019). ICCBI 2019. LNDECT, vol 49. Springer, Cham. https://doi.org/10.1007/978-3-030-43192-1_75
12. Guo S, Li T, Kang H, Li N, Zhang Y, Wang K (2019) L-Seg: an end-to-end unified framework for multi-lesion segmentation of fundus images. Neurocomputing 349:52–63
13. Gondal WM, Kohler JM, Grzeszick R, Fink GA, Hirsch M (2018) Weakly-supervised localization of diabetic retinopathy lesions in retinal fundus images. In: Proceedings - International conference image processing ICIP, pp 2069–2073, September 2018
14. Lei B et al (2020) Skin lesion segmentation via generative adversarial networks with dual discriminators. Med Image Anal 64:101716
15. Zhou Y, Wang B, He X, Cui S, Shao L (2020) DR-GAN: conditional generative adversarial network for fine-grained lesion synthesis on diabetic retinopathy images. IEEE J Biomed Heal Inform
16. Ahn S, Pham QTM, Shin J, Song SJ (2021) Future image synthesis for diabetic retinopathy based on the lesion occurrence probability. Electron 10(6):1–12
17. Singh VK et al (2018) Retinal optic disc segmentation using conditional generative adversarial network. Front Artif Intell Appl 308:373–380
18. Ikeda Y, Doma K, Mekada Y, Nawano S (2021) Lesion image generation using conditional GAN for metastatic liver cancer detection. J Image Graph 9(1):27–30
19. Sarp S, Kuzlu M, Pipattanasomporn M, Guler O (2021) Simultaneous wound border segmentation and tissue classification using a conditional generative adversarial network. J Eng 2021(3):125–134
20. Rammy SA, Abbas W, Hassan NU, Raza A, Zhang W (2020) CPGAN: conditional patch-based generative adversarial network for retinal vessel segmentation. IET Image Process 14(6):1081–1090
21. Cira CI, Manso-Callejo MÁ, Alcarria R, Pareja TF, Sánchez BB, Serradilla F (2021) Generative learning for postprocessing semantic segmentation predictions: a lightweight conditional generative adversarial network based on pix2pix to improve the extraction of road surface areas. Land 10(1):1–15
22. Mishra P, Herrmann I (2021) GAN meets chemometrics: segmenting spectral images with pixel2pixel image translation with conditional generative adversarial networks. Chemom Intell Lab Syst 215:104362
23. Xie S, Tu Z (2015) Holistically-nested edge detection. In: Proceedings of the IEEE international conference on computer vision, pp 1395–1403
24. He J, Jiang D (2020) Fundus image segmentation based on improved generative adversarial network for retinal vessel analysis. In: 2020 3rd international conference on artificial intelligence and big data (ICAIBD), pp 231–236. IEEE
25. Isola P, Zhu JY, Zhou T, Efros AA (2017) Image-to-image translation with conditional adversarial networks. In: Proceedings - 30th IEEE conference on computer vision and pattern recognition, CVPR 2017, pp 5967–5976
26. Mahmood F et al (2019) Deep adversarial training for multi-organ nuclei segmentation in histopathology images. IEEE Trans Med Imag 99:3257–3267

27. Miyato T, Kataoka T, Koyama M, Yoshida Y (2018) Spectral normalization. In: International conference on learning representations - ICLR 2018
28. Gullón N (2019) Retinal lesions segmentation using CNNs and adversarial training
29. Porwal P et al (2019) IDRiD: diabetic retinopathy - segmentation and grading challenge. Med Image Anal
30. Xu B, Wang N, Chen T, Li M (2015) Empirical evaluation of rectified activations in convolutional network
31. Kingma DP, Ba JL (2015) Adam: A method for stochastic optimization. In: 3rd international conference on learning representations, ICLR 2015, pp 1–15

A Novel Approach for Analysis of Air Quality Index Before and After Covid-19 Using Machine Learning

Rajesh Kumar Tiwari, Ajay Kumar Pathak, and Tapan Kumar Dey

1 Introduction

These days pure air has turned out to be a precious resource. Unwanted substances like dust, smoke, chemical elements and their compounds are polluting the air every now and then. Obnoxious air quality has already given rise to a number of health problems. There are several factors which decide the air quality like, chemical compounds, atmospheric parameters, and emissions from natural sources [1]. Air Quality Index (AQI) level is useful in understanding the suitability of air for respiration, and the related consequences that may arise. The AQI level is inversely proportional to the air quality, which implies that places with poorer air quality will have higher AQI levels and vice-versa. The rise and fall of several pollutants, month-wise and year-wise was evaluated. It was seen that the level of most of the pollutants decreased considerably in 2020 and 2021. The month-wise levels of most of the pollutants show a valley-like structure in the monsoon seasons, from July to September [2]. The EPA (United States Environmental Protection Agency) has devised the following pollutants as the determinants of AQI: Ground-level ozone, particle pollution (also known as particulate matter, including PM2.5 and PM10), Carbon Monoxide, Sulphur Dioxide and Nitrogen Dioxide [3]. For developing Air Quality Index (AQI), sub-indices are created for each pollutant, where after they are aggregated [1]. Each pollutant has its own source and unwelcoming effects. The AQI indices have further been categorized into several categories as given below [4].

R. K. Tiwari (✉)
RVS College of Engineering and Technology, Jamshedpur, India
e-mail: rajeshkrtiwari@yahoo.com

A. K. Pathak
YBN University, Ranchi, India

T. K. Dey
NIT, Jamshedpur, India

© The Author(s), under exclusive license to Springer Nature Singapore Pte Ltd. 2022   139
B. Unhelker et al. (eds.), *Applications of Artificial Intelligence and Machine Learning*,
Lecture Notes in Electrical Engineering 925,
https://doi.org/10.1007/978-981-19-4831-2_12

Machine learning (ML) is a subset of artificial intelligence (AI). Based on its ability of learning, ML provides experience without being explicitly programmed. There is different approach of learning in ML which is also known as Models of Machine Learning. Based on the learning approaches, ML techniques can be defined as: guided, unguided, semi-guided and reinforcement [5]. Guided ML approach: In guided approach, we use the concept of labeled data. Using the labelled data, the machine learns. Once the model is intelligent enough, the model can easily predict the output. It also helps in prediction of future results for unseen data. Unguided ML approach: In unguided approach, the data are unknown or unlabelled. Due to unlabelled data, the machine itself learn and predict for output. Semi-Guided ML approach: It is hybrid model where both labelled and unlabelled data helps the machine to learn. Once the learning process is over, the machine can predict the output. Reinforcement ML approach: In this approach, based on the reward and punishment, machine learn and predict the output. Reinforcement ML is also called as reward-punishment ML approach. Manufacturing of robots and related field are the best example of reinforcement ML.

Deep Learning (DL) is another sub-part of Artificial Intelligence and also subset of Machine Learning, in other words we can say it is deeper study or learning of ML.

2 Background and Motivation

Air is the most vital natural resource for all forms of life to exist. The process of addition of unwanted particles and dangerous substances to the air, which tamper its quality is known as Air Pollution. Today, pollution has become an indispensable problem for all the countries across the globe. As pollution increases, the air quality lowers and after a certain threshold, the Air becomes unfit for inhalation. Air Quality Index (AQI) is a vital metric to help in assessing the quality of air. The level of several pollutants, viz. Ozone (O_3), Particulate Matter-2.5 and 10 (PM2.5, PM10), Carbon Monoxide (CO), Sulphur Dioxide (SO_2) and Nitrogen Dioxide (NO_2) are taken into account for determining the AQI level at a particular place.

Due to the statistical properties of ARIMA (which is also known as autoregressive integrated moving average model), it can be used for time series forecasting [10, 11]. U. Brunelli et al. [12, 13], have proposed a forecasting model for partial dynamics pollutants. Sharda and Patil [15] have made a comparison between ARMA and ANN. G. Bontempi et al. [14] have proposed a Jupyter notebook on Kaggle, which aimed at spotting the decrease in AQI levels during lockdown. The available Jupyter notebook served as a strong reference during the earlier part of the Exploratory Data Analysis. It is a common practice to impute missing values via different means. For example, by replacing all the missing values with the mean value or the median value or the modal value in the column. However, missing values for the given could not be gauged using such methods [16, 17]. The reference notebook suggested reasons such as, the stations did not have devices to capture each pollutant and issues in meter reading as the potential causes for missing values [2]. The Corona Virus Disease or Covid-19 is

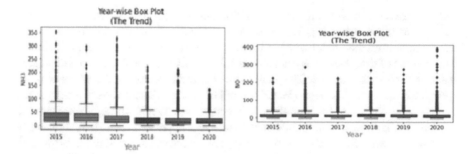

Fig. 1 Year-Wise major pollutants

a pandemic that has claimed millions of lives till date. To contain its spread, countries are imposing nationwide lockdowns. With the imposition of lockdown, life came to a still. Almost everything flourished online pertaining to which, the air quality started getting better and better. Intelligent usage of visual aids such as graphs and charts were used in the reference to bring forth the rejuvenation of air quality in the event of lockdown, as evident through the decrease in AQI levels [18]. The improvement in the air quality was well observed via the decrease in the AQI levels. Year-wise in major pollutants (NO, NO2, NOx, NH3, CO, SO2, O3, BTX, Particulate Matter) graph for last five years in India is shown below. Year-wise in major pollutants (NO, NO2, NOx, NH3, CO, SO2, O3, BTX, Particulate Matter) in India is shown below (Fig. 1).

3 Proposed Work

In our proposed work, we want to predict the air quality index of a particular region and for that we require all the gases details like the pollutant concentration, their percentage, the maximum level, minimum level of the concentration etc. In this proposed work we have taken datasets of different cites of India. We have used multiple linear regression and polynomial regression analysis techniques to calculate the AQI level of different regions. For outlier, we have used the Box-Plot technique. The present work imbibed the concepts of Machine Learning. The dataset of cities of India were split into 67% training data and 33% testing data. Models on Multiple Linear Regression and Polynomial Regression were trained and tested. In Multiple linear regression, we try to fil two or more explanatory variable into the response variable and observed the resultant data [6]. The columns O3 (Ozone), PM2.5 (Particulate Matter 2.5), PM10 (Particulate Matter 10), CO (Carbon Monoxide), SO2 (Sulphur Dioxide) and NO2 (Nitrogen Dioxide) were used as the independent columns and the column for AQI level as the target column or the dependent column. The model was fitted with training data and tested over the test dataset, yielding an r^2 score of 0.9142. In machine learning Polynomial regression is consider a special kind of linear regression in which the target variable and independent variables are having curvilinear relationship [7]. The same independent and dependent columns were used to first train the model with the training data, and then test it with the test data. The r^2 score of the model was evaluated as 0.9287 with degree 2, 0.9315 with degree 3 and 0.7136 with degree 4. The geographical coordinates of each city under study were retrieved using the Geocode API offered by Google. The details of a city named Brajrajnagar could not be obtained and hence, it was eventually removed from the dataset. The latitude and longitude of each city was appended to the data frame. The average AQI was computed for every city, and color codes assigned according to the severity of the AQI level. All this information was used to map the cities on the map of India, color-coded according to their AQI levels. Whenever a certain city was clicked on, a card indicating its mean AQI and its impact popped-up. Big cities like Ahmedabad had reported AQI levels as high as 2049 in the month of February, 2018. However, when life came to a standstill during lockdown, the fact that most of the industries went non-functional, automobiles stopped running, construction activities came to a halt, so on and so forth actually enriched the air quality.

4 Multiple Linear Regression

Multiple Linear Regression is a statistical tool used to predict the outcome of a dependent variable on the basis of the values taken by two or more independent variables, which may be continuous or categorical. It is an extension of the Ordinary Linear Squares (OLS) Regression Model that worked with just one independent and one dependent variable. The following formula is used to facilitate the working of Multiple Linear Regression [8].

$$Pi = Z0 + Z1xi1 + Z2xi2 + \ldots + Zpxip + \epsilon \tag{1}$$

Pi = dependent variable, xi = explanatory variables, Z0 = y-intercept (constant term).

Zp = slope coefficients for each explanatory variable,

ϵ = the model's error term (also known as the residuals), As Multiple Linear Regression takes 'multiple' independent variables into account, and calculates the line of best fit to establish the relationship between the independent and dependent variables, hence its name [8]. To make way for Multiple Linear Regression on the given dataset, the independent variables, namely, Ozone, PM2.5, PM10, CO, SO2 and NO2 were extracted to the X variable and the AQI level as the Y variable. After the entire dataset is split into training and testing data, in the ratio of 0.67:0.33. The model is trained with the training data and thereafter run on the test dataset to predict the independent variable. The r2 score was evaluated for the model as 0.9142. Therefore, the coefficient of determination resulted in 91.42%.

5 Polynomial Regression

Linear Regression is good if we have a linear or nearly linear dataset. On the other hand, we may have values which do not distribute linearly. These values may distribute quadratic ally or cubically or close to a function of some nth-degree polynomial. Under such circumstances, a model based on linear regression (or multiple linear regression) cannot promise to align with the dataset. Under such cases, Polynomial Regression, which is a special case of Linear Regression comes into picture. With Polynomial Regression, one has the liberty to choose the degree of the polynomial which would accord with the distribution of the variables. In Polynomial Regression, we model the relationship between dependent (M) and independent variables (n). The equation can be written as Sample Head

$$M = b0 + b1n1 + b2n1^2 + b3n1^3 + \cdots + bnn1^n \qquad (2)$$

In Polynomial regression, the original features are converted into Polynomial features of required degree (2, 3, ..., n) and then modelled using a linear model [9]. Overfitting and underfitting are common issues that arise out of Polynomial Regression. Underfitting refers to the condition when the model performs poorly on the training data, because it doesn't successfully capture the relationship between the independent and the dependent variables. On the other hand, overfitting is the condition when a model does not fare well with testing data, as the model is too generalized over the training data [10]. To ensure that the chosen model remains cognizant of all such issues, the r^2 score is calculated, which helps in determining the suitable degree for the Polynomial Regression equation. The r^2 score gives an estimation of the model's alignment to the relationship of the independent and the dependent variables. After splitting the entire dataset into 67% training data and 33% testing data, the Polynomial Regression model was trained with the training data, with degrees varying from 2 to 4. The models were tested for test data and respective r^2 scores were evaluated. The model with degree 3 gave the highest r^2 score of 0.9315, and hence, was preferred over others.

6 Experimental Result

This section gives an inclusive evaluation of the results obtained from the present work. The experiment had been implemented over the cloud on Google CoLab with Python 3.7 as the programming language. However, the entire access to the environment was done via Google Chrome V89 web browser running on a system with Intel Core i3 processor, 4 GB RAM and 64-bit Windows 10 Operating System [19]. The coefficient of determination for the Linear Regression Model was 91.42%. On the other hand, the coefficient of determination for the Polynomial Regression model with degree 2 was evaluated as 92.87%, for degree 3 was 93.15% while for degree 4 was 71.36%.

6.1 Dataset

Here is a glimpse at the top 5 rows of the dataset (Fig. 2).

Columns with object data type. The alphanumeric and date kind of values are stored in object data type. The following attributes have object as their data type.

City. The attribute 'City' holds the names of different cities for which the observations were collected over the time range. There are 24 cities in total. The maximum observations, 2009 in number were recorded in Lucknow, Ahmedabad, Chennai, Delhi, Mumbai and Bengaluru while Aizawl contributed to just 113 observations. There were 26,219 number of not-null entries in the City column.

Date. The next attribute for Date recorded the date of each observation. The observations started on 1st of January, 2015 and continued till 1st of July, 2020. There are 26,219 number of not-null entries in the column.

AQI Bucket. AQI Bucket refers to the categorization as Good, Moderate, Satisfactory, Poor, Very poor and Severe on the basis of the AQI level of a city and the established norms.

Columns with float64 as the data type.: The non-integral numeric values, or floating-point numbers have float64 as their data type. The following attributes are of float64 type.

PM2.5. PM stands for Particulate Matter. PM2.5 are fine inhalable particles, with diameters that are generally 2.5 μm and smaller [11]. The PM2.5 column recorded the level of PM2.5 contamination in each city on a particular date. This row has 21,930 not-null entries with 0.04 as the smallest observation and 949.99 as the largest.

PM10. PM10 are inhalable particles, with diameters that are generally 10 μm and smaller [11]. The PM10 column has 15,453 records of the levels of PM10 concentrations in different cities. The lowest recorded concentration was 0.01 whereas the highest was 1000.

	City	Date	PM2.5	PM10	NO	NO2	NOx	NH3	CO	SO2	O3	Benzene	Toluene	Xylene	AQI	AQI_Bucket
0	Ahmedabad	2015-01-01	NaN	NaN	0.92	18.22	17.15	NaN	0.92	27.64	133.36	0.00	0.02	0.00	NaN	NaN
1	Ahmedabad	2015-01-02	NaN	NaN	0.97	15.69	16.46	NaN	0.97	24.55	34.06	3.68	5.50	3.77	NaN	NaN
2	Ahmedabad	2015-01-03	NaN	NaN	17.40	19.30	29.70	NaN	17.40	29.07	30.70	6.80	16.40	2.25	NaN	NaN
3	Ahmedabad	2015-01-04	NaN	NaN	1.70	18.48	17.97	NaN	1.70	18.59	36.08	4.43	10.14	1.00	NaN	NaN
4	Ahmedabad	2015-01-05	NaN	NaN	22.10	21.42	37.76	NaN	22.10	39.33	39.31	7.01	18.89	2.78	NaN	NaN

Fig. 2 Top five row of our dataset

NO. The column NO indicates the concentration of Nitric Oxide in the air. This column has 22,986 not-null values. 390.68 was the highest recorded value and 0.02 was the least.

NO2. The column NO2 signifies the concentration of Nitrogen Dioxide in the city under observation on a particular day. There are 23,002 not-null values present in the column. The highest value being 362.21 while the lowest being 0.01.

NOx. NOx is a collective term used to indicate a variety of Nitrogen Oxides, viz. Nitrous Oxide (N2O), Nitric Oxide (NO), Dinitrogen dioxide (N2O2), Dinitrogen Trioxide (N2O3), Nitrogen Dioxide (NO2), Dinitrogen Tetroxide (N2O4) and Nitrogen Pentoxide (N2O5) [12]. The column NOx was used to assess the concentration of NOx in the air. Out of the 22,176 not-null values, the highest value was recorded as 467.63 surprisingly, the lowest record was 0.NH3. This column stores the level of Ammonia concentration recorded every day in different cities. There are 16,372 not-null entries wherein the highest value is 352.89 and 0.01 being the lowest.

CO. CO is the chemical formula for Carbon Monoxide. Therefore, this row stores the Carbon Monoxide concentration for each record in the data frame. There are 24,258 not-null entries and the highest concentration was found to be 175.81 while the lowest was surprisingly, 0.

SO2. This column holds the concentration of Sulphur Dioxide in different cities over the period. There are 22,675 not-null entries wherein the highest value was recorded as 193.86 while 0.01 was the lowest value.

O3. This column holds the ozone concentration across different cities. There are 22,559 not-null entries present in the data frame. The highest value is 257.73 and the lowest value is 0.01.

Benzene. The Benzene column holds 20,932 not-null values indicating the Benzene concentration. The highest value for the column is 455.03 while 0 was the lowest value for this column too.

Toluene. There are 18,664 not-null values indicating the concentration of Toluene. The highest value for the column is 454.85 while the lowest being 0 again.

Xylene. This column has just 9,412 not-null values which indicate the concentration of Xylene across cities. The highest value was recorded as 170.37. This column too had the lowest value as 0.

AQI. This column reflects the AQI (Air Quality Index) level of a particular city. The column has 21,937 values with 2049 as the highest value while 13 as the lowest value.

The original dataset was modified to add few more columns as shown (Fig. 3).

The newly added columns have been discussed below.

Avg_AQI_Index. The categorization of AQI levels into different classes on the basis of the guidelines established by the EPA was achieved via the given column. The classes are Good, Moderate, Unhealthy for Sensitive Groups, Unhealthy, Very Unhealthy and Hazardous.

AQI_Color. The EPA has also assigned certain colours to different AQI bands to distinguish them visually. The colours added were green, beige, orange, red, purple and dark red.

latitude. This column holds the latitudinal coordinates of each city.

	City	AQI	Avg_AQI_Index	AQI_Color	latitude	longitude
0	Ahmedabad	452.122939	Hazardous	darkred	23.021624	72.579707
1	Aizawl	34.765766	Good	green	23.743524	92.738291
2	Amaravati	95.299643	Moderate	beige	16.509668	80.518454
3	Amritsar	119.920959	Unhealthy for Sensitive Groups	orange	31.634308	74.873679
4	Bengaluru	94.318325	Moderate	beige	12.979120	77.591300

Fig. 3 Modified dataset after adding new features

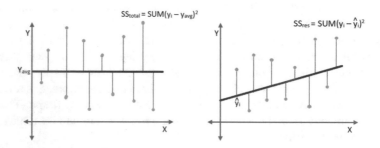

Fig. 4 Residual sum of square errors

longitude. This column holds the longitudinal coordinates of each city.

6.2 Evaluation Matrix

The r^2 score was used as the evaluation metric. R-squared or r^2 score is a statistical measure that represents the goodness of fit of a regression model. The ideal value for r-square is 1. The closer the value of r-square to 1, the better is the model fitted [13].

The formula for calculating r^2 score is given below. $R^2 = 1 - (SSres/SStot)$ where, SSres stands for residual sum of squares, and SStot is the total sum of squares. Residual sum of squares is calculated by the summation of squares of perpendicular distance between data points and the best fitted line. The graphs given below indicate the calculation of SStotal and SSres [13] (Fig. 4).

6.3 Result and Discussion

The Multiple Linear Regression model gave quite acceptable predictions of the AQI levels in cities. The following graph is an indicator of the same (Fig. 5).

The r^2 score was evaluated as 0.9142., Hence, the coefficient of determination was evaluated as 91.42%., On the other hand, the Polynomial Regression model on

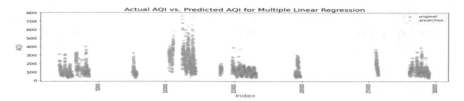

Fig. 5 Actual AQI Vs. predicted AQI for multiple linear regression

degree 3 performed slightly better than the Multiple Linear Regression model, as was evident from the higher r^2 score. The r2 score was evaluated as 0.9315. Thus, the coefficient of determination was evaluated as 93.15%. When the average AQI levels of different cities were computed, Ahmedabad reported the highest mean AQI level as 452.12 while Aizawl had the lowest AQI level as

A careful study of the given dataset and applied techniques reveal significant insights about the purity of air in different cities. However, there are a lot of missing values in the dataset [20]. An intelligent choice was not to impute missing values on account of several reasons as discussed in Sect. 2. Before the models were trained and tested, it was attempted to remove all the missing values. Better data collection facilities would definitely avoid any such inappropriateness. Integration of data collection methods with Real-time APIs would further smoothen the whole process and enhance the overall effectiveness. The data collected may be grouped station-wise in a city, to further understand the air quality across different areas in a city.

7 Conclusion

To preserve the environment is our first and foremost duty. The present work can help in understanding how certain pollutants affect the quality of air we breathe. Carefully pondering over the dataset can help in understanding how terribly the AQI has risen over the years. Those smoggy days are an invitation to a dreadful future.

The visual appeal of graphical aids can help in grasping the touch points at a glance. The strong decline of AQI levels in the wake of lockdown helped in restoring the air quality to a great extent. The predictions extracted via Machine Learning models could play a major role in adopting preventive measures well suited to the need of the situation. The Governmental standards of different pollutants may be revisited and amended or even strictly enforced. Health workers can develop a better understanding of the concentration of pollutants in a city, and accordingly suggest suitable place changes to their patients. A glance at the map of the nation can deliver the intent of assessing the severity of air pollution in a city almost instantly. Lockdown was just a period. But the positive effect that it brought about in the environment needs to prevail. With the help of the project, a quantitative estimation of the air quality may be achieved. The project can help in undertaking pollutant-specific action. With insights backed with scientific proof, the project was modelled with a view to be of utility in combating the nuisance of air pollution for a cleaner and greener environment.

References

1. Ghorani-Azam A, Riahi-Zanjani B, Balali-Mood M (2016) Effects of air pollution on human health and practical measures for prevention in Iran. J Res Med Sci 21:1–12
2. Tiwari RK (2020) Human age estimation using machine learning techniques. Int J Electron Eng Appl 8(1):01–09. https://doi.org/10.30696/IJEEA.VIII.I.2020.01-09
3. Liu X, Liu H (2016) Effects of air pollution on human health and practical measures for prevention in Iran. J Res Med Sci 21:1–12
4. Ghorani-Azam A, Riahi-Zanjani B, Balali-Mood M: Data publication based on differential privacy in V2G network, vol 9, no 2, pp 34–44. https://doi.org/10.30696/IJEEA.IX.I.2021. 45-53
5. Raimondo G, Montuori A, Moniaci W, Pasero E, Almkvist E (2007) A machine learning tool to forecast PM10 level. In: Science, San Antonio, TX, USA, pp 1–9, 14–18 January 2007
6. Garcia JM, Teodoro F, Cerdeira R, Coelho RM, Kumar P, Carvalho MG (2016) Developing a methodology to predict PM10 concentrations in urban areas using generalized linear models. Environ Technol 37:2316–2325. [CrossRef] [PubMed]
7. Bhanarkar AD et al (2005) Assessment of contribution of SO2 and NO2 from different sources in Jamshedpur region, India. Atmos Environ 39(40):7745
8. Sahoo C (2020) Cloud computing and its security measures. Int J Electron Eng Appl 8(I):10–19. https://doi.org/10.30696/IJEEA.VIII.I.2020.10-19
9. Wang J, Christopher SA (2003) Intercomparison between satellite derived aerosol optical thickness and PM2. 5 Mass: Impliances for air quality studies. Geophys Res Lett 30(21)
10. Siew LY, Chin LY, Mah P, Wee J (2008) Arima and integrated arfima models for forecasting air pollution index in Shah Alam, Selangor. Malays J Anal Sci 12(1):257–263
11. Zhu J (2015) Comparison of ARIMA model and exponential smoothing model on 2014 air quality index in Yanqing county Beijing, China. Appl Comput Math 4(6):456
12. Mitchell TM (2009) Machine learning. In: Proceedings of the IJCAI International Joint Conference on Artificial Intelligence, Pasadena, CA, USA, July 2009
13. Gayathri M, Poorviga A, Vasantha Raja SS (2021) Prediction of breast cancer stages using machine learning, vol 7, no 1, pp 36–42. https://doi.org/10.30696/IJEEA.IX.I.2021.36-42
14. Ai M, Liu H (2021) Privacy-preserving of electricity data based on group signature and homomorphic encryption. Int J Electron Eng Appl 9(1):08–18. https://doi.org/10.30696/IJEEA.IX. I.2021.08-18
15. Pradeep M, Ragul K, Varalakshmi K: Voice and gesture based home automation system. Int J Electron Eng Appl 9(2):11–20. https://doi.org/10.30696/IJEEA.IX.I.2021.11-20
16. Nallakaruppan MK, Senthil Kumaran U (2018) Quick fix for obstacles emerging in management recruitment measure using IOT based candidate selection. Serv Oriented Comput Appl 12(3–4): 275–284
17. Nallakaruppan MK, Ilango HS (2017) Location aware climate sensing and real time data analysis. In: Computing and Communication Technologies (WCCCT), 2017 World Congress on IEEE (2017)
18. Poobrasert O, Luxsameevanich S, Chompoobutr S, Satsutthi N, Phaykrew S, Meekanon P (2020) Heuristic-based usability evaluation on mobile application for reading disability. Int J Electron Eng Appl 8(II):11–21. https://doi.org/10.30696/IJEEA.VIII.II.2020.11-21

19. Yang X (2021) Power grid fault prediction method based on feature selection and classification algorithm. Int J Electron Eng Appl 9(2):34–44. https://doi.org/10.30696/IJEEA.IX.I.2021.34-44

20. Daga D, Saikia H, Bhattacharjee S, Saha B (2021) Privacy-preserving of electricity data based on group signature and homomorphic encryption: a conceptual design approach for women safety through better communication design. Int J Electron Eng Appl 9(3):01–11. https://doi.org/10.30696/IJEEA.IX.III.2021.01-11

Embedding of Q-Learning in Sine Co-Sine Algorithm for Optimal Multi Robot Path Planning

H. K. Paikray, P. K. Das, and S. Panda

1 Introduction

Mobile robots are being used everywhere in today's date and application of mobile robot is increasing rapidly. Robots are increasingly being used in every sector, such as home maintenance, Education, Cocking, security, defense, automated transportation, and health sector. Due to so many applications research and more and more emphasis are giving in multi robot field. A multi robot can do a lot of things if operated with the right techniques [1, 6]. More and more applications are focusing on motion planning. Choosing right motion planning technique is a demanding and vital task [7, 11]. When it performs the task it directly affects the performance of the robot in the workspace. The objective must be taken in mind before choosing any technique are smoothening of robot trajectory, obstacle avoidance, most significantly, the length of shortest path [1, 2]. Based on prior experimental information, that is, the initial and goal positions, and the obstacles positions, etc., are provided to a multiple robot, control techniques are based on either a approach of sensor-based or model-based [4, 8–10].

H. K. Paikray (✉) · S. Panda
Department of Computer Application, VSSUT, Burla, Odisha, India
e-mail: hemantakpaikray@gmail.com

S. Panda
e-mail: suchetapanda_mca@vssut.ac.in

P. K. Das
Department of Information Technology, VSSUT, Burla, Odisha, India
e-mail: pradiptadas_it@vssut.ac.in

© The Author(s), under exclusive license to Springer Nature Singapore Pte Ltd. 2022   151
B. Unhelker et al. (eds.), *Applications of Artificial Intelligence and Machine Learning*,
Lecture Notes in Electrical Engineering 925,
https://doi.org/10.1007/978-981-19-4831-2_13

2 Formulation of the Problem

Consider that a mobile robot is moving in environment with obstacle an initial point (M_s, N_s) and goal (M_g, N_g), and the position of obstacle is at (M_o, N_o). The surrounding consists of robots and obstacles. Another important point is there are multiple robot as well as multiple obstacles. Obstacles are both static and dynamic in nature. As we know that robots are roaming in the workspace, a robot can also have a collision with another robot, which means robotacts as obstacles dynamically to each other. The main objective is to find the shortest path without collision with any static obstacle and dynamic obstacles (also with other mobile robot). Apart from this, the robot should make the shortest and safest route and also ensures there is no collision occurs during the execution. Keeping these criteria in mind, the objective function for navigation in this section has been designed. The motion of other robots as well impedes each other's movement as there possibly will be aninter-collision with each other. Therefore, in the formulation problem, the motion of the robot is also considered in order to understand and keep itself away from collision.

2.1 Objective Function Creation for Optimized Navigation

To avoid obstacles and maintains collision free path the objective function is designed, importantly the path with shortest length from initial pointto the goal point. These three purposes must be enclosed to perform the tasks with greater performance and minimal computational cost.

(a) **Length of the route with minimal distance**
   The main purpose in trajectory planning of the robot involves the computation of a shortest path. The robot must pursue the route between the initial position and goal position that should be minimal in length. The shortest possible path can be generated using the Euclidean distance. On each approach, the location of the robotmust be updated accurately updated to obtain the path with shortest travel length in between its position and the goal position. The objective function is always dependent upon the best position and the goal position of the mobile robot. Mathematically it is written as:

$$f_1(M, N) = d[(M_{mr}(i), N_{mr}(i)), (M_t, N_t)] \tag{1}$$

   where, $(M_{mr}(i), N_{mr}(i))$ is the coordinate position of a mobile robot at i^{th} position. There exist n numeral points in the interval of the initial and the goal position and should produce an outcome with minimal value. The total path lengthis equal to the sum of lengths among all these different locations of the mobile robot which is illustrated as:

$$E_{mpl} = \sum_{i=1}^{n} d[(M_{mr}(i), N_{mr}(i)), (M_t, N_t)]$$

$$= \sum_{i=1}^{n} mpl \tag{2}$$

$$mpl$$
$$= \sqrt{(N_{mr}(i+1) - N_{mr}(i))^2 + (M_{mr}(i+1) - M_{mr}(i))^2}$$
$$+ \sqrt{(N_{mr}(n) - N_t)^2 + (M_{mr}(n) - M_t)^2} \tag{3}$$

where, $(M_{mr}(n), N_{mr}(n))$, is the co-ordinate of the mobile robot at n^{th} position (previous location before goal).

(b) **Avoid obstacle for safe navigation**

For smooth and safe navigation, obstacles avoidance should be considered for a path with a minimum length of travel. The function depends upon the obstacle and robot location that is, location of obstacle is j and the location of robot is i. The function is symbolized as:

$$f_2(M, N) - [(M_o(j), N_o(j)), (M_{mr}(i), N_{mr}(i))] \tag{4}$$

There must be safe and shortest path between the robot at i^{th} location and obstacle at j^{th} location. This means the robot must move to the safest, closest position to the obstacle. The path length in total between them is symbolized as:

$$E_{oa} = \sum_{j=1}^{m} \sum_{i=1}^{n} d[(M_o(j), N_o(j)), (M_{mr}(i), N_{mr}(i))]$$

$$= \sum_{j=1}^{m} \sum_{i-1}^{n} oa \tag{5}$$

where, number of obstacles is equals to m m.

$$oa = \sqrt{(N_{mr}(i) - N_o(j))^2 + (M_{mr}(i) - M_o(j))^2} \tag{6}$$

(c) **Smoothness function of trajectory**

This approach deals with creating a smooth route from source to destination while avoiding self collision between robot and other obstructions. The minimization of angle variation is explained here from the Euclidean path i.e. ith location of the robot to its goal. It is symbolized as:

$$f_3(M, N) = |\alpha[A(i), A(i+1)], \alpha[A(i), B]| \tag{7}$$

where, $\alpha[A(i), A(i+1)]$ Corresponds to the angle between the paths of robot i^{th} *and* $(i+1)^{th}$ *location and* $\alpha[A(i), B]$ is the angle between the path of the robot at i^{th} *location* and goal.

$$E_{ts} = \sum_{i=1}^{n} |\alpha[A(i), A(i+1)], \alpha[A(i), B]|$$

$$and \ \alpha[A(i), B] = \tan^{-1}\left[\frac{N_t - N_{mr}(i)}{M_t - M_{mr}(i)}\right] \tag{8}$$

In total, the multi-objective function is achieved as:

$$f(mof) = w_1 \sum_{i=1}^{n} E_{mpl}(i) + w_2 \sum_{i=1}^{n} \sum_{j=1}^{m} E_{oa}(i)(j) + w_3 \sum_{i=1}^{n} E_{ts}(i) \tag{9}$$

where, the weight function w_1, w_2 and w_3, illustrates the degree of outcome of individual objectives of the smoothness function. They must comply with the subsequent limitation:

$$w_1 + w_2 + w_3 = 1 \tag{10}$$

Symbolizes the ultimate fitness of the multi-objective functions is.

$$f_{fitness} = \frac{1}{f(mof) + \varepsilon} \tag{11}$$

where, ε is a small value that is added to make it safe. The function is a multi-objective one and is assumed as aminimization problem as all the elements are needed to be minimized in order to achieve the best possible solution.

2.2 Movement of Mobile Robot

The Objective function is designed for guiding the robot to transit from the initial location to the goal. It is considered as a multi-objective function that involves finding a shortest route through a smooth angle during obstacles avoidance. Apart from this, it also necessary to take the movement of the mobile robot as one robot can obstruct the navigation of another robot in multi-mobile robot navigation. In this situation, one robot is set as the dynamic barrier to other robots. Hence, it is rightly said that in order to do the job efficiently the obstacle movements must be considered. The position of the obstacles changes at each time due to the dynamic nature. The movement is assumed to be a linear ora circular trajectory.

(a) **Production of linear trajectory**
    The mobile robot (itself treated as a dynamic obstacle) that moves with velocity V_{mr} in a straight path by forming a slope of the angle θ_{mr}. The interconnection between the new position $(M_{mr}(t+1), N_{mr}(t+1)$ and present position $(M_{mr}(t), N_{mr}(t))$ is symbolized as:

$$M_{mr}(t+1) = M_{mr}(t) + V_{mr} \times \cos\theta_{mr} \tag{12}$$

$$N_{mr}(t+1) = N_{mr}(t) + V_{mr} \times \sin\theta_{mr} \tag{13}$$

(b) **Production of circular trajectory**

Movement of a mobile robot can also made when making the circular trajectory having center coordinates (M_c, N_c) with the radius value of r_c units. The updated robot's position is symbolized as:

$$M_{mr}(t+1) = M_c + r_c \times \cos\beta \tag{14}$$

$$N_{mr}(t+1) = N_c + r_c \times \sin\beta \tag{15}$$

where, β value varies between $0\,to\,2\pi$.

3 Projected Optimization Methods

The architecture as well as the optimization techniques for navigation of the mobile robots is furnished in this section. In addition to this, a controller has been explained here for dealing with any conflict situation during the navigation of multiple robots in the same terrain. Finally, an explanation of the hybridization mechanism performed with the proposed approach is described.

3.1 Q-Learning Algorithm

Q-learning is an efficient and productive reinforcement learning technique in the field of machine learning [3, 5]. It works on the strategy of penalty and reward which allow the agents to learn from the environment and perform an action by transition of states to acquire a reward or penalty based on the feedback received from the environment or by exploring the unstructured environment.

The experience is acquired by the agent in Q-learning through the process of intensification and diversification. In Q-learning, intensification and diversification is well-adjusted through definite approach such as Boltzmann approach, $\varepsilon-$ greedy policy. In Boltzmann approach, the agent adopts an exponential probability distribution through the Q-value of the current state due to all actions. Whereas, $\varepsilon-$ greedy policy choose an action from set of actions through the probability ε, which is necessary to be adjusted manually. Consider there are set of states $S = \{s_1, s_2, \ldots s_n\}$ in the environment and each states have set of actions $SA = \{a_1, a_2, \ldots a_m\}$. An agent selects an action $a_t \in SA$ at instant t in the state $s_t \in S$ to transit to the subsequent state $s_{t+1} \in S$ through the transition process and acquire an immediate reward r_{t+1}

from the environment. Consider the agent in the state s_t and is expected to decide the subsequent state by executing an action a_t. Then Q-value is calculated through the following Equation:

$$Q(S_t, a_t) = re(S_t, a_t)$$
$$+ \gamma \, \underset{a'}{Max} \, Q\big(\delta(S_t, a_t), a'\big) \qquad (16)$$

where, $\delta(S_t, a_t)$ represents the next state due to selection of action a_t at state S_t. $re(S_t, a_t)$ is the reward achieved by exercising an action a_t at state S_t. γ denotes the learning factor, whose value varies in between 0 and 1. St $+$ 1 be the next state selected by execution of the action at state S_t. Then Q(δ (St, at), a') = Q(St $+$ 1, a'). Therefore, selection of a' that maximizing Q (St, a_t) is an interesting problem. The pseudo code of conventional Q-learning algorithm is presented in algorithm1.

The pitfalls of the conventional Q-learning algorithm are précised as follows (1).It may execute the action randomly and obtain an optimal Q-value; (2) Every movement of the agent, it is necessary to acquire the appropriate action through the optimal Q-value of that state by fetching the entry of the Q-table from the storage; (3) the learning rate and convergence speed becomes very slow. Traditional Q-learning requires enormous computation to evaluate Q-value at a particular state through all set of possible actions and necessary to store Q-value in huge space for r all possible actions at a particular state which leads to slower the convergence rate. In traditional Q-learning with 'n' number of states each with 'm' number of actions, the required dimension of the Q-table is ($m \times n$). Due to the above flaws of the traditional Q-learning, a modified Q-learning has been proposed to overwhelm the weaknesses of that traditional Q-learning. In modified Q-learning the earlier actions can be affected by the feedback of the successive states. If the action taken by the current state is false, then the preceding actions should be penalty otherwise the preceding actions should be rewarded. In the modified form, the classical Q-learning used to store the Q-value of the best state-action and thus saves the storage area. In the modified Q-learning, two storage spaces are to store the Q-value and corresponding lock variable for a particular state. So, it is necessary ($2 \times n$) dimension for Q-table to store Q-value for n number of states. In this way, the modified Q-learning save mn $-$ 2n $=$ n (m $-$ 2) amount of the memory as compared to the traditional approach. The traditional Q-value presented in Eq. 16 has been modified as follows: where α is the learning rate and its value is adopted as follows:

$$Q(s_t, a_t) = (1 - \alpha)Q(s_t, a_t) + \alpha\left[re_t + \gamma \, \underset{a_{t+1}}{max} \, Q(s_{t+1}, a_{t+1})\right] \qquad (17)$$

$$\alpha = 1/1 + \text{total of time svisited to state } s_t \qquad (18)$$

$$Q_{t+1}(s_t, a_t)$$

$$= \begin{cases} Q_t(s_t, a_t) + \alpha \Delta Q_t(s_t, a_t) \; if \; s = s_t \; and \; a = a_t \\ Q_t(s_t, a_t) \;\; otherwise \end{cases} \tag{19}$$

$$\Delta Q_t(s_t, a_t) = \left\{ re_t + \alpha \max_{a'_t} \left[Q_t(\delta(s_t, a_t), a'_t) \right] \right\} - Q_t(s_t, a_t) \tag{20}$$

where δ is the transition function. The possible action in robot path planning are the movement of robots in all possible direction in the frontal side and there is no movement of back direction is considered in this problem. Reward is calculated as a cost function and cost function is calculated in terms of reaching at the goal position, robot collides with obstacles or teammates and smooth movement of robots. The objective of the reward function is expressed to control a collection of robots to realize this. The term re_t^i represents the reward established by the robot i at instant t and it is formulated as follows:

$$re_t^i = \left(re^{goal} \right)_t^i + \left(re^{collision} \right)_t^i + \left(re^{rotvel} \right)_t^i \tag{21}$$

The term $\left(re^{goal} \right)_t^i$ represents the reward awarded, when the robot i reached at goal position at time t and it is calculated as follows:

$$\left(re^{goal} \right)_t^i = \begin{cases} re_{arrv} \; if \; \| P_i^t - G_i \| < 0.01 \\ \omega_g \left(\| P_i^{t-1} - G_i \| - \| P_i^t - G_i \| \right) otherwise \end{cases} \tag{22}$$

The term $\left(re^{collision} \right)_t^i$ is the penalty, when the robot crashes with its teammates or obstacles present in the workspace.

$$\left(re^{collision} \right)_t^i = \begin{cases} r_{collision} \; if \; \left\| P_i^t - P_j^t \right\| < 2 \, Ror \, \| P_i^t - B_k \| < R \\ 0 \; otherwise \end{cases} \tag{23}$$

To boost the robot to proceed in smoothness path, a small penalty $\left(re^{rotvel} \right)_t^i$ is familiarized to penalize the large rotational velocities.

$$\left(re^{rotvel} \right)_t^i = \omega_\omega |\omega_i^t| \; if \; |\omega_i^t| > 0.7 \tag{24}$$

Here, we set $re_{arrv} = 15$, $r_{collision} = -15$ and $\omega_\omega = -0.1$ in the learning process. The pseudo code of modified Q-learning algorithm is presented as follow:

Algorithm 1: Modified Q-learning algorithm MQL(X_i)

Input: Information of environment G, Initial state s_1 and goal state s_g,$\alpha, \gamma, \varepsilon, \xi$,$max\_iter$and represent the position vector (X_i) into states$_1$using $10x_i + y_i$

Output : Optimal Q-table

1. Initialization
2. Set the $Q_{n \times m} = \{0\}$;
3. $s = s_i$;
4. $iter = 1$
5. Repeat
6. {
7.   choose an random action $a_i$from $A = \{a_1, a_2, \ldots, a_m\}$and execute $a_i$and move to next state
8.   Calculate the learning factor αusing equation 3.
9.   Evaluate the reward re_tusing equation 6.
10.   Calculate the error signal $\Delta Q_t(s_t, a_t)$ using equation 5.
11.   Update Q-table through $Q_{t+1}(s_t, a_t)$ using equation 4
12. $iter = iter + 1$
13.   } until ($\|Q_t - Q_{t-1}\| > \xi$&&$iter < max\_iter$)
14. $max = Q(s_1, a_1)$
15. For $i = 1$to$n$
16.   if($Q(s_i, a_i) > max$)
17. $max = Q(s_i, a_i)$
18. $s = s_i$
19.   Return s;

3.2 Sine–Cosine Algorithm (SCA)

It is a meta-heuristic algorithm based on the population size to solve the optimization problems. It produces a number of random initial solutions and compels them to fluctuate towards or outwards the best potential area through a sine cosine based mathematical model. Despite the consequences, two processes such as exploration and exploitation are two varying population based optimization algorithms are emphasized. In the exploration phase, an optimization algorithm combines the arbitrary solutions from the solution set with high rate of arbitrariness for finding the promising area in the problem space. However, in the exploitation phase, progressive modification has been done with the arbitrariness and the arbitrary variations obtained here is considerably less as compared to the exploration. In the present context of SCA, each time the position is updated through the subsequent equation.

$$x_i^d(t + 1) = x_i^d(t) + \alpha_1 \times \sin(\alpha_2)$$
$$\times \left| \alpha_3 pbest_i^d(t) - x_i^d(t) \right| \tag{25}$$

$$x_i^d(t + 1) = x_i^d(t) + \alpha_1 \times \cos(\alpha_2)$$
$$\times \left| \alpha_3 pbest_i^d(t) - x_i^d(t) \right| \tag{26}$$

where $x_i^d(t)$ is the location of the ith element at d^{th} dimension and t^{th} iteration; α_1, α_2 and α_3 are the random numbers, $pbest_i^d(t)$ is the personal best location of the ith element in d^{th} dimension and $|\,|$ is the absolute value of the expression. The Eq. 25 and Eq. 26 are merged and selected based on greed α_4 y strategy as illustrates:

$$x_i^d(t+1) = \begin{cases} x_i^d(t) + \alpha_1 \times \sin(\alpha_2) \times \left|\alpha_3\, pbest_i^d(t) - x_i^d(t)\right|, \alpha_4 \prec d_i \\ x_i^d(t) + \alpha_1 \times \cos(\alpha_2) \times \left|\alpha_3\, pbest_i^d(t) - x_i^d(t)\right|, \alpha_4 \geq d_i \end{cases} \tag{27}$$

The four main parameters $\alpha_1, \alpha_2, \alpha_3\ and\ \alpha_4$ require further elaboration. Where, dynamically varies between $0 \leq \alpha_4 \leq 1$. The parameter α_3 denotes the effects of the target during the iterations which is a random weight. The amplitude movement in the problem space is indicated by α_2. The parameter α_1 determines the movement's towards the next solution that may exist either in between the target and the solution or outside the potential area. For selecting the sine or cosine function a greedy selection parameter used and denoted as α_4. It is required to maintain the stability between two processes such as exploration and exploitation. The exploration potential of the algorithm is benefited by these properties if the possible location revise mechanism is remodelod in a proficient way to evade the local optima in the solution space. The global best position is used in the SCA function for avoiding the local optima and increases the unexplored paths in the problem space. Hence, the formula for updating the proposed position is mathematically formulated as below.

$$x_i^d(t+1)$$
$$= \begin{cases} gbest + \alpha_1 \times \sin(\alpha_2) \times \left|\alpha_3 gbest - x_i^d(t)\right|, \alpha_4 \prec d_i \\ gbest + \alpha_1 \times \cos(\alpha_2) \times \left|\alpha_3 gbest - x_i^d(t)\right|, \alpha_4 \geq d_i \end{cases} \tag{28}$$

$$d_i = 0.5 - 0.5 \frac{iter}{itermax} \tag{29}$$

$$\alpha_1 = 2 - \frac{iter}{max\_iter} \tag{30}$$

where g_{best} is the best position of the population, and α_4 are the random numbers lies in between [0, 1].

4   Necessity of Hybridization and Proposed Algorithm for Multi-robot Path Planning

SCA suffers overflow diversity like other population based algorithm. In SCA, the solution is updated based on the current state solution and coefficient α_1 and support the exploration of the search space by reallocating the solution far from the current state. The coefficient α_3 is also support the exploration during the search process. Hence, in each iteration, the solution losses its own feature and always reallocate to its new position and get trapped at local optima. The above mention drawback unable to generate the true solution. Therefore, it is necessary to embedded the improved Q-leaning into SCA to overcome above flaws and improve the convergence rate.

Algorithm 2: QSCA ($\overrightarrow{x}^i$)

Initialize: $\overrightarrow{x}^j$ randomly for total population. N is the population size
1.   **While** (termination conation not reached)
2.   **For** $t = 1 \, to \, max\_i \, ter$
3.   **For** $i = 1 to N$
4.   **For** $d = 1 to D$
5.   **S=Call MQL(x_i)**
6.   Convert State S to particle qp_i using $(\boldsymbol{qp_i})_x = \boldsymbol{S/10}, (\boldsymbol{qp_i})_y = \boldsymbol{S\%10}$
7. $\boldsymbol{Q_{gbest}} = qp_i$
8. $pbest_i{=}x_i$
9. $SCA_{best} = min \, i \, mum(f(x_i))$
10.   If $(Q_{gbest} < SCA_{best})$ then
11. $gbest = Q_{gbest}$
12.   Else
13. $gbest = SCA_{best}$
14.   Evaluate α_1 and d_i using Eq. 30 and Eq. 29
15.   Update position using Eq. 28
16.   Update best solution $gbest$
17.   **End for**
18.   **For End**
19.   **For End**
20.   **Return the fittest solution** $gbest$
21.   **End while**

Algorithm of path planning

Input: v_i^p, (M_s, N_s) and (M_g, N_g) are the velocity, initial position and goal position for n robots respectively, $1 \le i \le n$ and the threshold value ε
Output: Optimal Trajectory of path OTP_i is generated for the robot R_i from (M_s, N_s) to (M_g, N_g)
1.   **Begin**

2.   **For** $i = 1 \, to \, n$
3. $x^{curr_i} \leftarrow M_s; y^{curr_i} \leftarrow N_s;$
4.   **End for**
5.   **For every** robot $i = 1$ to n
6.   **While** $(Curr_i \ne G_i)$ // $curr_i = (M_s, N_s), G_i = (M_g, N_g)$ //
7.   Call QSCA(x^i);
                 // $x^i \leftarrow (x^{curr_i}, y^{curr_i})$ //
8.   Move to (x^{curr_i}, y^{curr_i});
9.   **End for**
10.   **End for**
11. **End**

(a) Preliminary stage of simulation with 10 robots and 14 obstacles.

(b) Intermediate stage of simulation

(c) Final stage of simulation (all robotsreached at theircorresponding location after 30 steps)

Fig. 1 Simulation result for 10 robot and 14 obstacles

5 Computer Simulation

The simulation of the path planning problem of multiple mobile robots has been carried out through QSCA using programming in C. The experiment has been performed with two different scenarios that is, 10 numbers of circular-shaped robots with the radius of 6 pixels with 14 obstacles of different shapes, and 15 robots with 20 number of obstacles. Figure 1(a) shows the initial environment for simulation (Figs. 2 and 3, Table 1).

  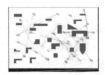

(a) Preliminary stage of simulation with 15 robots and 20 obstacles.

(b) Intermediate stage of simulation

(c)Final simulation (all robots reached at their corresponding location after 30 steps)

Fig. 2 Simulation result for 15 robot and 120 obstacles

  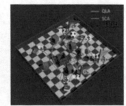

(a) Initial situation with Three robots and seven obstacles

(b) Intermediate movement of robots

(c)All robots reach their destination

Fig. 3 Experiment through real robot

Table 1 Robot specification (GCTronic' e-puck)

SL NO	Types of parameter	Values
1	Rotational motor	2
2	LED	10
3	Distance Sensor	8
4	Position Sensor	2
5	Light Sensor	8
6	Minimum and maximum detectable distance	3 cm/6 cm
7	Range of sensor(6 to 7 cm)	0–1075
8	IR Sensor	8
9	Processor	60 MHz
10	Programmable flash memory	14 Bits
11	Robot diameter	7.4 cm
12	Weight	150 g
13	Switch	16 position rotating switch

6 Experiment on E-puck Robot

7 Performance Analysis

The overall performance of the multi-robot path planning has been analyzed and evaluated in terms of two metrics such as average total trajectory path deviation (ATTPD) and the average untraveled total trajectory path distance (AUTPD). The average total trajectory path deviation (ATTPD) is denoted as $\sum_{i=1}^{n} (T P_{k-real} - \sum_{r=1}^{n} T P_{ir} | j)$

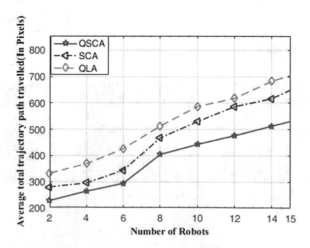

Fig. 4 Total travel length with different no. of steps in QSCA, SCA, QLA

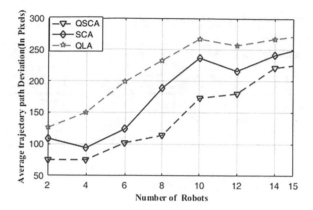

Fig. 5 Path deviation with different no. of robot

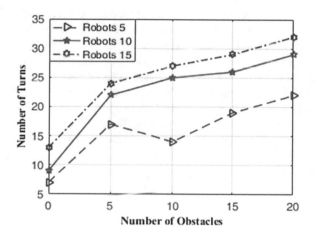

Fig. 6 Energy utilization with different no of turn

where TP_{k-real} is the actual path from an starting position S_k to goal position G_k for robot R_k in j^{th} run, $\sum_{r=1}^{j} TP_{ir}|j$ average path travelled in j^{th} run by robot R_k. Similarly, the average untraveled total trajectory path distance (AUTPD) in j^{th} run is $\sum_{i=1}^{n} \|G_k - S_k\|$ (Figs. 4, 5 and 6, Table 2).

Table 2 Comparison of simulation and experimental results

Algorithm	Parameters					
	Travel length (in pixels)		Path deviation		Energy utilization throw no of turn	
	Simulation	Experiment	Simulation	Experiment	Simulation	Experiment
QLA	725	745	275	315	32	39
SCA	682	698	257	285	29	35
QSCA	549	601	229	245	22	26

8   Conclusion and Future Works

The path planning problem of multi-robot has been carried out using QLA-SCA algorithm. The issue of premature convergence as well as the intent by the local optima has been resolved during the enhancement of projected algorithm. In the multi-robot problem of path planning, the fitness function is developed through the consideration of different constraint. Further, optimization has been done through the proposed algorithm. The algorithm has also been evaluated for subsequent locations of every robot present in the workspace. The projected algorithm generates smooth and collision-free trajectory path since source position to goal position. The performance calculation and evaluation of the proposed algorithm has been carried out to justify the effectiveness of the algorithm. The result achieved from simulation environment and experimental environment that QSCA (Q-learning sine cosine algorithm) outperforms QLA and SCA. The future direction of the proposed algorithm will focus on in vision based path planning [12].

References

1. Kashyap AK, Parhi DR, Pandey A (2021) Multi-objective optimization technique for trajectory planning of multi-humanoid robots in cluttered terrain. ISA Transactions
2. Kashyap AK et al (2020) A hybrid technique for path planning of humanoid robot NAO in static and dynamic terrains. Appl Soft Comput 96:106581
3. Das PK, Behera HS, Panigrahi BK (2016) A hybridization of an improved particle swarm optimization and gravitational search algorithm for multi-robot path planning. Swarm Evolut Comput 28:14–28
4. Paikray HK, Das PK, Panda S (2021) Optimal multi-robot path planning using particle swarm optimization algorithm improved by sine and cosine algorithms. Arab J Sci Eng 46(4):3357–3381
5. Das PK, Behera HS, Panigrahi BK (2016) Intelligent-based multi-robot path planning inspired by improved classical Q-learning and improved particle swarm optimization with perturbed velocity. Eng Sci Technol Int J 19(1):651–669
6. Mohanta JC, Keshari A (2019) A knowledge based fuzzy-probabilistic roadmap method for mobile robot navigation. Appl Soft Comput 79:391–409
7. Ali AA et al (2016) An algorithm for multi-robot collision-free navigation based on shortest distance. Robot Auton Syst 75:119–128

8. Das PK et al (2016) A hybrid improved PSO-DV algorithm for multi-robot path planning in a clutter environment. Neurocomputing 207:735–753
9. Orozco-Rosas U, Montiel O, Sepúlveda R (2019) Mobile robot path planning using membrane evolutionary artificial potential field. Appl Soft Comput 77:236–251
10. Pandey A, Parhi DR (2016) Multiple mobile robots navigation and obstacle avoidance using minimum rule based ANFIS network controller in the cluttered environment. Int J Adv Robot Autom 1(1):1–11
11. Tian S et al (2021) Multi-robot path planning in wireless sensor networks based on jump mechanism PSO and safety gap obstacle avoidance. Future Gener Comput Syst 118:37–47
12. Wen S et al (2021) A multi-robot path-planning algorithm for autonomous navigation using meta-reinforcement learning based on transfer learning. Appl Soft Comput 110:107605

Image-Based Number Sign Recognition for Ethiopian Sign Language Using Support Vector Machine

Ayodeji Olalekan Salau◉, Nigus Kefyalew Tamiru, and Deepak Arun◉

1 Introduction

Sign language is one of the means of communication used by hearing and speech impaired people to communicate, share their feelings, attitudes, and interact with normal people and their peers [1, 2]. According to the World Health Organization (WHO) [3], there are 360 million people worldwide who are deaf, of which 32 million of them are children, which represent approximately 5% of the worlds population. According to the central statistics agency report in 2007, it's estimated that more than 1.5 million deaf people live in Ethiopia [4]. Hearing challenged persons can communicate with each other and normal people by using speech, reading and writing, and sign language. Communication through sign language is preferable for normal and hearing impaired people. However, there is a communication gap between the normal and hearing impaired people, since normal people have no or little knowledge of sign language. Therefore, to minimize the communication gap between normal and hearing impaired people in Ethiopia, this study presents a method for Ethiopian sign language (ETHSL) recognition based on local numbers.

A. O. Salau (✉)
Department of Electrical/Electronics and Computer Engineering, Afe Babalola University, Ado-Ekiti, Nigeria
e-mail: ayodejisalau98@gmail.com

N. K. Tamiru
Department of Electrical and Computer Engineering, Debre Markos University, Debre Markos, Ethiopia

D. Arun
Saveetha School of Engineering, Saveetha Institute of Medical and Technical Sciences, Chennai, India

© The Author(s), under exclusive license to Springer Nature Singapore Pte Ltd. 2022　　167
B. Unhelker et al. (eds.), *Applications of Artificial Intelligence and Machine Learning*,
Lecture Notes in Electrical Engineering 925,
https://doi.org/10.1007/978-981-19-4831-2_14

The rest of this paper is structured as follows. Section 2 presents a review of related works, while Sect. 3 presents the proposed SVM method. The experimental results and discussion are presented in Sect. 4, and Sect. 5 concludes the paper.

2 Related Works

Video-based finger spelling recognition system was proposed for Ethiopian Sign Language (ETHSL) in [4] to recognize Amharic alphabet signs using image pre-processing techniques, global thresholding, center of mass, and finite state of automata. The method achieved 91% recognition when tested using 238 ETHSL finger spellings.

Indian sign language (ISL) recognition system using Principal Component Analysis (PCA) features was proposed in [5]. Support Vector Machine (SVM) classifier was employed as a knowledge base for sign language recognition. The study yielded a recognition accuracy of 96% for 200 input images. However, the feature extraction stage was improved for better recognition. Authors in [6] designed a system for the recognition of ISL using a SVM classifier. The accuracy of the proposed system is 95.52% using PCA features and SVM classifier. A Myanmar sign language recognition system was presented in [7] to recognize human hand shapes using PCA and SVM algorithms. The proposed system achieved a recognition efficiency of 89%.

Three very simple but efficient classifiers were proposed in [8] to test the performance of different classifiers (KNN, Artificial Neural Network (ANN), and SVM) for HCI using hand gestures. These classifiers achieved an accuracy of 77.5, 82.5 and 91% for KNN, ANN and SVM respectively. The classifiers used were successful to recognize the correct matches. A method for image based Arabic sign language recognition system was proposed in [9]. The system was based on Histogram of oriented gradients (HOG) descriptors and one-versus-all SVM. The proposed system's recognition rate was 63.5%.

Authors in [10] presented a hand gesture (HG) detection and identification method for ETHSL. Gabor filter with PCA was used to extract features from HG digital images, and ANN was used to recognize ETHSL using the extracted features. A recognition accuracy of 98.53% was attained for ETHSL translation. Despite the fact that the authors obtained a 98.53% accuracy and divided their dataset into three sets: testing, training, and validation, the authors did not give a comparison of results to existing state-of-the-art algorithms that worked on ETHSL. Furthermore, the authors were unable to translate basic and specialized Amharic alphabet signs into their own alphabets.

The authors in [11] employed Hidden Markov Model (HMM) to train features taken from frames of films of Amharic word signs. A dataset was created using three signers, each of whom performed each sign twenty times. The authors used fifteen of the twenty movies for training purposes and the remaining five for testing. The proposed method was put to test using videos that had been collected for educational reasons. Using eight features (area, centroid, bounding box, major axis, minor axis, eccentricity, orientation, and perimeter) learned by HMMs, the authors achieved an overall recognition accuracy of 86.9%, whereas using only three basic features (centroid, area, and orientation), they achieved an overall recognition accuracy of 83.5%.

According to the authors of [12], there are just a few literatures that have accomplished ETHSL recognition. In their paper, Amharic text was transformed to ETHSL. Macromedia Flash 8.0 and ActionScript 2.0 were used to model and construct a 2-dimensional avatar. Although the authors claim a recognition accuracy of 66.6%, no comparative analysis was provided to compare the suggested system's performance to that of existing systems.

In [13], the authors developed an ETHSL classification approach for converting Amharic alphabets to English equivalents. The authors used ANN with PCA driven features and Haar-like features to achieve their results. The developed system can only recognize ten simple Amharic letter signs from a collection of images. The work of [14] later improved on the research work in [13]. For continuous ETHSL, the author attempted to create an offline candidate hand gesture selection and trajectory determination method. The video sequence was used to extract candidate Ethiopian Manual Alphabet (ETHMA) frames for this recognition scheme. Hand movement trajectories were calculated by the device, which achieved a recognition accuracy of 71.88%. The proposed system is constructed employing PCA features and SVM classification technique, which have not been used in prior works for ETHSL recognition.

This paper presents a SVM method for an automatic Amharic sign language (AMSL) recognition system that analyzes local number sign images of signers performing AMSL and converts the detected AMSL signs to text. SVM classifier was used to learn three distinct features of the AMSL signs: shape, motion, and color. In addition, an AMSL dataset with 480 samples of ten different indicators was used to evaluate the proposed SVM algorithm.

3 Ethiopian Sign Language Number System

3.1 Amharic Sign Language System

Ethiopian Sign Language (ETHSL) was introduced to assist hearing impaired people in Ethiopia to communicate effectively. However, recognition of Ethiopia AMSL is still a major issue. In this study, we have only considered the first ten local number signs of ETHSL which involve the use of a single hand shape. The local numbers greater than ten are dynamic and were not considered in this study. A total of 10 local number signs were used as shown in Fig. 1.

In Fig. 1, the representation of numerical signs of ETHSL are presented in which the numbers such as ፩(one), ፪(two), ፫(three), ፬(four), ፭(five), ፮(six), ፯(seven), ፰(eight), ፱(nine), and ፲(ten) are static single hand shapes. The remaining numbers

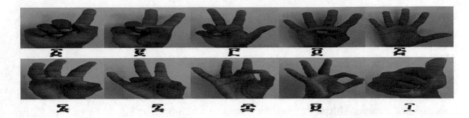

Fig. 1 Ethiopian local numbers fingerspelling

also make use of dynamic single hand sign notations. There is no zero (0) number in Ethiopian local numbers. So, zero (0) doesn't have a corresponding sign.

3.2 Proposed System

The proposed systems architecture comprises of three main stages: Image processing, feature extraction, and SVM modeling. The systems architecture is shown in Fig. 2.

3.2.1 Local Number Sign Image Acquisition

Local number sign (LNS) images were obtained using a Smart Mobile Phone. To obtain the LNS images, the right hand of three male and two female sign language students were employed. Sixteen sample images were captured for ten ETHSL local numbers from each student totaling up to 160 local number sign images. All the collected images have black background using Image J Application Software. Images in JPG format were used for training, validation, and testing. A sample of the original image employed is shown in Fig. 3(a) and the preprocessed nine (፱)number sign

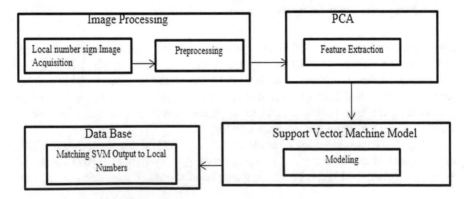

Fig. 2 Architecture of the proposed sign language recognition system

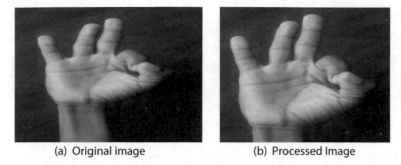

| (a) Original image | (b) Processed Image |

Fig. 3 Signs for the local number '፱'

image is shown in Fig. 3(b). Therefore, the total dataset used in this work is 800 samples which has an original pixel size of 640 by 480.

3.2.2 Image Preprocessing

Image pre-processing was performed to make ETHSL local numbers recognition process simpler. The image pre-processing steps applied are shown in Fig. 4.

(i) **Size Normalization**

In order to normalize the size of the captured images, the images were resized to the same size. To achieve this, the sign image size is cropped to an image size of 128 × 96 using the algorithm presented in Table 1.

Table 1 shows that the original sample images have a size of 640 by 480 which is large. This size requires more effort and a longer time for RGB color conversion, image contrast adjustment, image segmentation, smoothing, noise removal and feature extraction. Therefore, the original images are converted (normalized) to standard size (128, 96) using image size normalization algorithm.

(ii) **RGB Color Conversion**

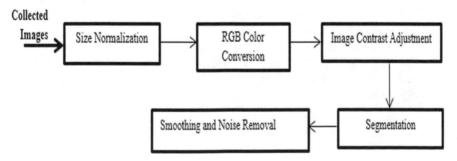

Fig. 4 Image preprocessing procedure

Table 1 Image size normalization algorithm

Algorithm for image size normalization
Input: Sample of video file: V
Input: Sample of image I
Output: An array of resized image (128, 96)
Begin
Image = Original image I (640,480)
Image = resize (Image, 128, 96*)*
Image = image_to_array (Image)
Return Image

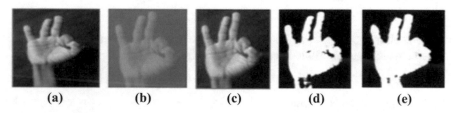

(a)   **(b)**   **(c)**   **(d)**   **(e)**

Fig. 5 Image preprocessing results of sign ' g̱'

Processing the acquired images in RGB color is computationally heavy and takes more processing time when compared with grayscale or binary form [10]. Consequently, the RGB sign images were converted into grayscale using Python built-in functions (RGB to gray converter) and further converted to a binary image as shown in Fig. 5a–e.

(iii) **Image Contrast Adjustment**

After the RGB color images are converted to grayscale, there is a need to enhance the quality and adjust the contrast of the grayscale images in order not to affect the performance of the system. The result of this enhancement can be seen in Fig. 5(c). The quality of the image is seen to improve and easier to process.

(iv) **Segmentation**

Segmentation is used to discriminate the hand shape of the local number sign from the background as shown in Figs. 5d,e.

(v) **Smoothing and noise Removal**

Removal of noise was performed by using nonlinear median filter to improve the image quality. The nonlinear median filter was observed to simultaneously reduce the image noise and preserve the sharp edges. This is depicted in Fig. 5(e).

3.2.3 Feature Extraction

After image segmentation, smoothing and noise removal, the processed data is transformed into a reduced representation of feature vectors using a feature extraction technique [15, 16]. To extract features from the images, we used a statistical method called principal component analysis (PCA) for local number sign image extraction.

3.2.4 Principal Component Analysis

Some of the techniques used for feature extraction are Linear Discriminant Analysis (LDA), Independent Component Analysis, and PCA [4, 5, 17]. Among these techniques, PCA is a powerful tool for image formation, data patterns, and pattern detection. The other main purpose of PCA is to reduce the large dimensionality of a data space by avoiding irrelevant information. The mathematical formulation of PCA is given in Eqs. (1) to (4).

1. Mean and Variance

$$Mean(X') = \frac{1}{n} \sum_{i=1}^{n} Xi \qquad (1)$$

2. Standard Deviation

$$SD = \sqrt{\frac{1}{n} \left(\sum_{i=1}^{n} (Xi - X') \right) 2} \qquad (2)$$

3. Covariance

$$Cov(X, Y) = \frac{\sum_{i=1}^{n} (Xi - X')(Yi - Y')}{n - 1} \qquad (3)$$

4. Eigen Values and Eigen Vectors

$$[A - \lambda I]X \qquad (4)$$

3.2.5 Training, Testing, and Validation

The acquired data is divided into three parts for training, validation, and testing. This is the most common approach because splitting the dataset into three sets such as training set, testing set, and validation set helps to improve the data generalization.

3.3 Support Vector Machine Modeling

After dividing the data into the training, testing, and validation set, the next step performed was to train and then model the SVM using binary classification. However, multiclass SVM is used for some problems. Usually, it is achieved by binarizing one class versus all other classes [18, 19]. Multiclass SVM was used to develop a more efficient model. The proposed multiclass SVM model is shown in Fig. 6.

Kernel functions like Polynomial, Gaussian radial basis, and Sigmoid were employed to divide data that couldn't be separated by a linear function. Gaussian radial basis kernel function was used for training and testing to give a high prediction accuracy. The mathematical formula of the radial basis kernel function used is given by Eq. (5).

$$K(x, y) = exp[\frac{-\|x - y\|2}{2\sigma 2}]$$ (5)

where: $\|x\|^2$ is the squared Euclidean distance between the two feature vectors and σ is the free parameter.

For any ETHSL number recognition problem which is a multiple classification problem, recognition is achieved through the combination of binary problems. SVM modeling was achieved by training the multi-class SVM (one versus all method). Once the local number sign image was recognized, it is translated to its corresponding local numbers.

4   Results and Discussion

In this section, the proposed method was subjected to a thorough experimental examination. The experimental setup was presented first, followed by a description of the baseline technique which uses local number sign images.

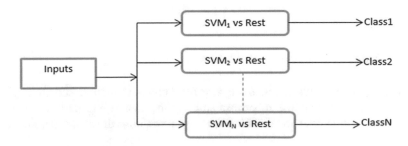

Fig. 6 Multi-class support vector machine model

Table 2 List of numbers and occurrence per class

Arabic Numbers	Ethiopian numbers	Number of Occurrence per Class	Arabic Numbers	Ethiopian numbers	Number of Occurrence per Class
1	፩	80	6	፮	80
2	፪	80	7	፯	80
3	፫	80	8	፰	80
4	፬	80	9	፱	80
5	፭	80	10	፲	80
Total	**800**				

4.1   Dataset

At present, there is no standard dataset available online for Ethiopian gesture-based communication. As a result, the dataset for this study was generated by collecting gesture images from five people who are Debre Markos college special needs graduates. The images were gathered in the day time in a hall. This was done to diminish the clamors and lighting impact of the images. Table 2 presents the number of occurrences per class and the total number of captured images.

4.2   Performance Evaluation

To calculate the performance of a SVM model, accuracy is one of the most widely used measurement method. Earlier, different models were also evaluated using this matrix [19]. The mathematical formula of Accuracy is:

$$Accuracy = \frac{C}{S} \tag{6}$$

where: C is number of samples that are correctly classified and S is the total number of samples.

4.3   Test Result

The proposed ETHSL system was trained, validated and tested using the SVM model classifier. To validate the model, the training dataset was divided into a batch size of 32. In the case of our study, training dataset is 480; the selected batch size is 32 and 50 epochs was used to train the proposed SVM model. Therefore, the dataset (480) was divided into 15 steps each with 32 samples. The SVM model weights was updated after each batch of thirty-two samples with 50 epochs, the model runs through the whole dataset 50 times. Therefore, there are a total of 750 batches during the entire training process. The SVM model trained fast due to the selected parameters and the amount of data that was used.

The training loss achieved is 2.7614 in the first epoch and the last is 0.0062. This indicates that the training loss decreases as the number of epoch's increases. The trained SVM model's accuracy is 99.33% with a test loss accuracy of 9.7%. The comparison of training and testing accuracy of the SVM model, training accuracy, training loss, validation accuracy and validation loss is clearly shown in Fig. 7.

Figure 7 shows the training loss, validation loss, training accuracy and validation accuracy using blue color, yellow color, green color and red color respectively. The results show that as the number of epoch increases, the training accuracy also increases but the testing loss decreases as it approaches the end of the epoch.

Additionally, a confusion matrix is used for the visualization of the SVM model's recognition accuracy using the local number sign class. This gives a clear idea about the local number signs that are misclassified. Diagonal elements in the confusion matrix shows the true positive rate for unknown samples using the multi-class SVM

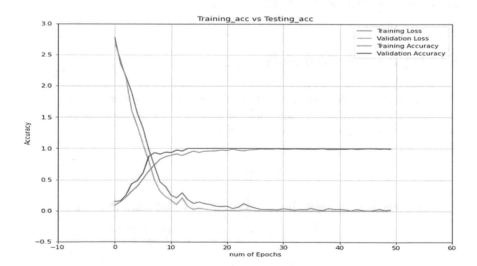

Fig. 7   Training and testing accuracy for local number signs

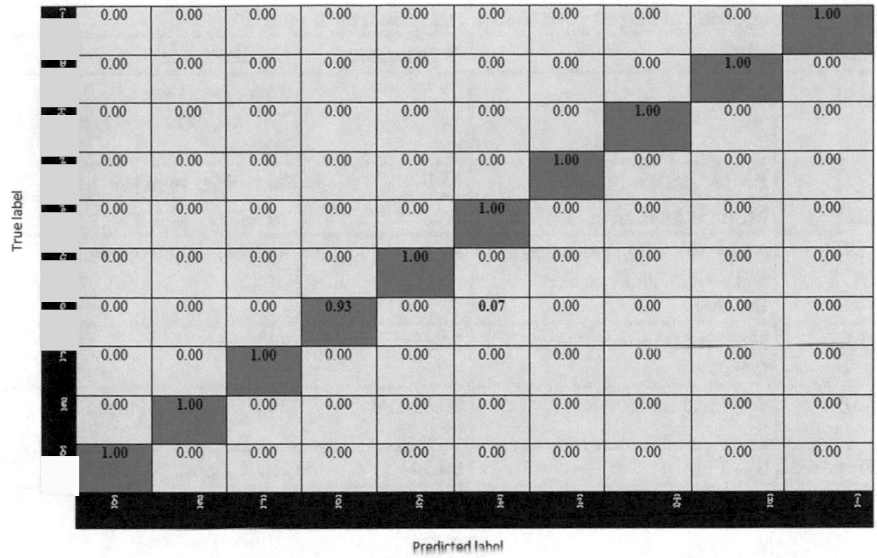

0.00	0.00	0.00	0.00	0.00	0.00	0.00	0.00	0.00	1.00
0.00	0.00	0.00	0.00	0.00	0.00	0.00	0.00	1.00	0.00
0.00	0.00	0.00	0.00	0.00	0.00	0.00	1.00	0.00	0.00
0.00	0.00	0.00	0.00	0.00	0.00	1.00	0.00	0.00	0.00
0.00	0.00	0.00	0.00	0.00	1.00	0.00	0.00	0.00	0.00
0.00	0.00	0.00	0.00	1.00	0.00	0.00	0.00	0.00	0.00
0.00	0.00	0.00	0.93	0.00	0.07	0.00	0.00	0.00	0.00
0.00	0.00	1.00	0.00	0.00	0.00	0.00	0.00	0.00	0.00
0.00	1.00	0.00	0.00	0.00	0.00	0.00	0.00	0.00	0.00
1.00	0.00	0.00	0.00	0.00	0.00	0.00	0.00	0.00	0.00

True label

Predicted label

Fig. 8 Confusion matrix of the proposed model

approach. The classification result is good because of the use of more robust features as well as the efficient classifier. The recognition accuracy for each class of local number sign is shown in the confusion matrix in Fig. 8.

The results of the confusion matrix shown in Fig. 8 shows that numbers, one (፩), two (፪), three (፫), five (፭), six (፮), seven (፯), eight (፰), nine (፱)and ten (፲)were correctly recognized since the local number signs have different hand shapes but number four (፬)which has quite a similar hand shape with number six was not recognized correctly. Thus, the SVM model gives inaccurate results for the local number signs, four and six. Generally, most classes of local number signs have high recognition accuracy of almost 100%, while only one class (local sign number four) has low efficiency of 7%. A comparison of results with works on ETHSL is presented in Table 3. The results show that the proposed SVM method outperforms existing methods for ETHSL.

Table 3 Comparison of the proposed method with existing methods

Author	Method	% Accuracy	Collected data
[10]	ANN	98.53	There are a total of 170 images in 5 samples of 34 letters
[11]	HMM	86.9	20 videos, 34 letters
[12]	Machine translation architecture	66.6	34 letters
[13]	ANN with PCA driven features and ANN with Haar-like features	88.08 and 96.22	34 letters, 1000 videos
[14]	Modified Housdorff Distance (MHD)	71.88	34 letters
[20]	ANN and SVM	80.82 and 98.06	34 letters, 1710 videos
Proposed	SVM	99.33	160 samples, 34 letters

5 Conclusion

In this paper, a SVM model was successfully developed to recognize Ethiopian local number signs with their corresponding numbers. The SVM model achieves a better recognition rate as compared to other previous works related to Ethiopian sign language recognition. However, there are some aspects that were not addressed in this paper which require further study. These include: This study focused on recognizing only ten static local numbers. This can be extended to include the other numbers in future research works. Also in this study, we used a static image dataset to recognize local number signs but in the future a video dataset can be used.

References

1. Rivera-Acosta M, Ortega-Cisneros S, Rivera J, Sandoval-Ibarra F (2017) American sign language alphabet recognition using a neuromorphic sensor and an artificial neural network. Sensors 17:2176
2. Abeje BT, Salau AO, Mengistu AD et al (2022) Ethiopian sign language recognition using deep convolutional neural network. Multimed Tools Appl. https://doi.org/10.1007/s11042-022-127 68-5
3. Neumann K, Chadha S, Tavartkiladze G, Bu X, White KR (2019) Newborn and infant hearing screening facing globally growing numbers of people suffering from disabling hearing loss. Int J Neonatal Screen 5:7
4. Gebretinsae E (2017) Video-based finger spelling recognition for Ethiopian sign language using center of mass and finite state of automata. Contemp Issues Summit Harv Boston USA 13(1):46–60
5. Girme M, Marathe P (2017) Indian sign language recognition system using pca features. Int J Res Appl Sci Eng Technol (IJRASET) 1192–1197. https://doi.org/10.22214/ijraset.2017.10171

6. Deshpande PD, Kanade SS (2018) Recognition of Indian sign language using SVM classifier. Int J Trend Sci Res Dev 2(3):1053–1058. ijtsrd11104
7. Tun M, Lwin T (2019) Real-time Myanmar sign language recognition system using PCA and SVM. Int J Trend Sci Res Dev 3:2361–2366. https://doi.org/10.5281/zenodo.3591477.svg
8. Anand SAU (2012) Performance comparison of three different classifiers for hci using hand gestures. Int J Mod Eng Res (IJMER) 2:2857–2861
9. Alzohairi RAR, Alshehri W, Aloqeely S, Alzaidan M, Bchir O (2018) Image based Arabic sign language recognition system. Int J Adv Comput Sci Appl 9(3):185–194. https://doi.org/10.14569/IJACSA.2018.090327
10. Admasu YF, Raimond K (2010) Ethiopian sign language recognition using artificial neural network. In: 10th IEEE International Conference on Intelligent Systems Design and Applications, pp 995–1000. https://doi.org/10.1109/isda.2010.5687057
11. Gimbi T (2014) Recognition of isolated signs in Ethiopian sign language. Master's Thesis, Addis Ababa University, p 137
12. Tesfaye M (2010) Machine translation approach to translate Amharic text to Ethiopian sign language. In: 2nd Proceedings of IEEE International Conference, pp 252–273.
13. Zerubabel L (2008) Ethiopian finger spelling classification: a study to automate Ethiopian sign language. Master's Thesis, Addis Ababa University, Addis Ababa, Ethiopia, p 96
14. Tsegaye A (2011) Offline candidate hand gesture selection and trajectory determination for continuous Ethiopian sign language. Master's Thesis, Addis Ababa University, Addis Ababa, Ethiopia, p 117
15. Jiang X, Ahmad W (2019) Hand gesture detection based real-time american sign language letters recognition using support vector machine. In: 2019 IEEE International Conference on Dependable, Autonomic and Secure Computing, International Conference on Pervasive Intelligence and Computing, International Conference on Cloud and Big Data Computing, International Conference on Cyber Science and Technology Congress (DASC/PiCom/CBDCom/CyberSciTech), pp 380–385
16. Salau AO, Jain S (2019) Feature extraction: a survey of the types, techniques, and applications. In: 5th IEEE International Conference on Signal Processing and Communication (ICSC), Noida, India, pp 158–164. https://doi.org/10.1109/ICSC45622.2019.8938371
17. Tabassum T, Mahmud I, Uddin MdP, Nitu A (2020) Enhancement of single-handed bengali sign language recognition based on hog features. J Theor Appl Inf Technol 98(5):743–756
18. Ekbote J, Joshi M (2017) Indian sign language recognition using ANN and SVM classifiers. In: International Conference on Innovations in Information, Embedded and Communication Systems (ICIIECS), pp 1–5
19. Kalam MA, Mondal MNI, Ahmed B (2019) Rotation independent digit recognition in sign language. In: International Conference on Electrical, Computer and Communication Engineering (ECCE), pp 1–5
20. Tamiru NK, Tekeba M, Salau AO (2022) Recognition of Amharic sign language with Amharic alphabet signs using ANN and SVM. Vis Comput. 38:1703–1718.https://doi.org/10.1007/s00371-021-02099-1

BIC Algorithm for Exercise Behavior at Customers' Fitness Center in Ho Chi Minh City, Vietnam

Nguyen Thi Ngan⬚ and Bui Huy Khoi⬚

1 Introduction

The 2019 Coronavirus (COVID-19) pandemic has impacted many aspects of people's lives worldwide, such as food, food, medicine, even the gym market. Around the world, the gym market is on a downward trend because of the pandemic caused by the Covid-19 virus. The ban on mass gatherings was announced in many countries and territories. People are restricted from going out and gyms are closed. However, in the Vietnamese gym market, the big guys are having a tough competition on this fertile, fertile, and safe land. The pandemic has been well controlled in Vietnam. The fight against the pandemic has achieved many positive signals. People are health conscious and create a movement to take part in the exercise. Regular exercise is emerging as one of the best protective factors - an important part of the overall strategy to promote and keep us healthy during our ongoing struggle with the COVID-19 pandemic. Not exercising is a risk factor for many chronic diseases that contribute to the severity of COVID-19 (if you are unlucky enough to catch the virus), such as cardiovascular disease, type 2 diabetes, obesity, and chronic lung disease. According to The Centers for Disease Control and Prevention report, patients with underlying health conditions are six times more likely to be hospitalized with COVID-19. They also had a 12-fold higher mortality rate than those without the underlying medical condition. Therefore, health care services (such as exercise services in fitness centers) are increasingly interested in consumers. In recent years, there have been many studies in the world on the practice of sports, such as the study of Zhang and Li [1], Anwong et al. [2]. In the world, the fitness industry has developed strongly. In Vietnam, this industry has also developed more than in previous years. Currently, the number of gyms/fitness

N. Thi Ngan · B. Huy Khoi (✉)
Industrial University of Ho Chi Minh City, Ho Chi Minh City, Vietnam
e-mail: buihuykhoi@iuh.edu.vn

N. Thi Ngan
e-mail: nguyenthingan@iuh.edu.vn

© The Author(s), under exclusive license to Springer Nature Singapore Pte Ltd. 2022
B. Unhelker et al. (eds.), *Applications of Artificial Intelligence and Machine Learning*,
Lecture Notes in Electrical Engineering 925,
https://doi.org/10.1007/978-981-19-4831-2_15

centers in Ho Chi Minh City is increasing rapidly, densely arranged in inner and outer districts. These gyms/centers are very diverse, price and additional services to attract customers to their centers. Some well-known centers can be mentioned, such as California, Fit24, Citigym, Unifit, Nowfit, Showing the development of Vietnam's fitness industry is developing strongly. Realizing the need and potential in this field is not small, so this study identifies the needs, attitudes, and behaviors of consumers, besides collecting market information, and related information to serve as a basis for analyzing and evaluating the model of factors influencing consumer behavior, making comments and assessments as the basis for proposing a business model for the fitness industry to develop. Many of the customers attend fitness centers looking to improve their health [3]. Therefore, the study uses the BIC algorithm for exercise behavior at customers' fitness centers in Ho Chi Minh City, Vietnam.

2 Literature Review

2.1 Exercise Behavior (EB)

In a study by Biddle [4], Exercise is necessary for life, but not everyone is motivated enough, qualified enough for the exercise to have benefited their health. The use of services at the fitness center so that they have a training environment and motivation to practice more often effects of exercise to improve or maintain good health. Despite many recommendations to maintain exercise, about 30% of adults do not exercise. Fitness behavior depends on a multitude of barriers and factors such as exercise ability, weather, environment, personality, self-efficacy… Therefore, fitness centers as a place to develop exercise [5].

2.2 Usefulness (US)

The usefulness of Anwong et al.'s study [2] proves that "Exercise not only brings health but also helps to have a fit body and reduce stress. Höglund and Normen [6] showed that exercise and sports help control weight so that customers have a beautiful body. Since then, the need for health care has been paid more attention and fitness centers with full equipment, training instructors to achieve the best results. The studies by Anwong et al. [2], Riseth et al. [5], and Zhang and Li [1] all found helpfulness to have a positive effect on exercise behavior. Most users decide to exercise because exercise brings health and fitness benefits to them. From the problems mentioned above, we give hypothesis 1:

H1: Utility has a positive effect on exercise behavior in fitness centers.

2.3 Ease of Use (EU)

Ease of use is researched by Zhang and Li [1] proving that: "Ease of use in the location of the fitness center is convenient for practitioners to move, besides ease of use of the center's machinery". In addition, Ease of use was studied The long-term use of the fitness center by members, Riseth et al. [5] suggested that for members to stay with the center for a long time, the exercise equipment must be easy to use easy to access. Besides, research by Anwong et al. [2] shows that advertising on the media also attracts customers to the center more. Since then, ease of use is a very important factor for the fitness behavior of customers in the fitness center. From the problems mentioned above, we give hypothesis 2:

> H2: Ease of use has a positive effect on exercise behavior in fitness centers.

2.4 Barrier (BR)

The study of Aghenta [7] showed that "barriers are the reasons that prevent clients from practicing at the center. Throughout the process before, during, and after the decision to exercise, there are barriers to the client's training. Some typical barriers are: not knowing how to practice, fear of doing wrong exercises, no motivation, and lack of training goals. Besides, Research by Anwong et al. [2], Riseth et al. [5] also mentions that Coach support, exercise programs, and economics affect training. Barriers affect the decision to exercise at a fitness center or not [8]. Therefore, from the problems proposed, we give hypothesis 3:

> H3: Barriers have a negative effect on exercise behavior in fitness centers.

2.5 Facilities (FAC)

The physical facilities of the sports center must be covered with an area of 60 m^2 or more, the distance between the training equipment is 1 m with light from 150 lx or more. The sound system is in good working condition. There is a toilet, changing area, a place to store belongings for practitioners, and a first-aid bag according to the regulations of the Ministry of Health. The rules include the following main contents: training time, participants in the exercise, subjects not allowed to take part in training, measures to ensure safety when practicing [9]. Facility researched by Anwong et al. [2], factors affecting the decision to use a fitness center in Bangsean Chonburi show the influence of facilities in the fitness center sports. The fitness center has full, quality, modern gym equipment, a clean and airy space, and comes with many attached services to attract and keep customers to stay with the center more [10]. From the problems mentioned above, the author gives the theory:

Therefore, hypothesis 4 (H4): Facilities have a positive effect on exercise behavior in fitness centers.

2.6 Price and Promotion (PP)

Prices and promotions are factors that reflect the decision to sign up for a workout and attract customers to come and stay with the center [11]. This factor is studied by Anwong et al. [2] and reflects the influence on customers' exercise behavior. There are many promotions for new and old members and the price of the center. It also tells the level of the center, which classes the center, is allocated to customers. It can be concluded that price and promotion are important factors for customers to decide to exercise at a fitness center. Therefore, hypothesis 5 is given as:

H5: Price and promotion have a positive effect on exercise behavior in the fitness center.

2.7 Service Quality (SQ)

Service quality is based on the satisfaction of additional services that increase revenue for the center [12] such as Aerobic, Yoga, Zumba, and Personal trainer (if customers have demand). Provide services that not only focus on serving but also focus on experience so that customers are loyal to the center [13]. Anwong et al. [2] found that service quality is customer satisfaction, determined by the comparison between perceived quality and expected quality. Then, before going to practice, customers put their trust in the quality of service expressed through the factors to be satisfied with what they expected. It can be concluded that the belief that service quality is important for customers to exercise at a fitness center is one factor promoting the development of the fitness industry. Therefore, hypothesis 6 is suggested:

H6: Service quality has a positive effect on exercise behavior in fitness centers.

3 Methodology

3.1 Sample

The statistical results described in the chart above show that the number of females accounts for a higher proportion with 56.1%, while male-only accounts for 43.9%. The survey sample group under the age of 18–30 years old accounted for the highest proportion with 45.5%, the second-highest was the sample group aged 31 to 45 years old with 27.8%, followed by the group, below 18 years old with a rate of 17.7%. And

Table 1 Statistics of Sample

Characteristics		Amount	Percent (%)
Sex and Age	Male	87	43.9
	Female	111	56.1
	Below 18	35	17.7
	18–30	90	45.5
	31–45	55	27.8
	Above 45	18	9.1
Customer	Student	33	16.7
	Freelance	68	34.3
	Skilled labor	64	32.3
	Stay-at-home parent	22	11.1
	Other	11	5.6
Income/Month	Below 5 million VND	30	15.2
	5.1–10 million VND	92	46.5
	10.1–20 million VND	69	34.8
	Over 20 million VND	7	3.5

the last group is the group over 45 years old with the rate of 9.1%. Table 1 describes statistics of sample characteristics as the study of Hair et al. [14].

The group of customers who are Freelance accounts for the highest proportion at 34.3%, followed by the group of Skilled labor customers with 32.3%, the student group of customers ranked third with the rate of 16.7%, the customer group is stay-at-home parent ranked fourth with 11.1% and the group with the lowest rate is Other customers with 5.6%.

The survey sample group with income from 5.1 to 10 million accounted for the highest proportion with 46.5%, the second was the sample group with income from 10.1 to 20 million, accounting for the second-highest rate with 34.8%, followed to the sample group Under 5 million with the rate of 15.2%. And the last group is the group of over 20 million with a rate of 3.5%.

3.2 Reliability Test and BIC Algorithm

Cronbach's Alpha coefficient of 0.6 or more is acceptable [15–17] in case the concept being studied is new or new to the subject with respondents in the research context. However, according to Nunnally et al. [18], Cronbach's Alpha (α) does not show which variables should be discarded and which should be kept. Therefore, besides Cronbach's Alpha coefficient, one also uses Corrected item-total Correlation (CITC) and those variables with Corrected item-total Correlation greater than 0.3 will be kept.

BIC (Bayesian Information Criteria) was used to choose the best model for R software. BIC has been used in the theoretical context for model selection. As a regression model, BIC can be applied, estimating one or more dependent variables from one or more independent variables [19]. An essential and useful measurement for deciding a complete and straightforward model is the BIC. Based on the BIC information standard, a model with a lower BIC is selected. The best model will stop when the minimum BIC value [19–21].

4 Results

4.1 Reliability Test

Factors and items are in Table 2, Cronbach's Alpha coefficient of greater than 0.6 and Corrected Item - Total Correlation (CITC) is higher than 0.3 reliable enough to carry out further analysis and lower than 0.3 it is not reliable as BR4. There are some new items in Table 2. The mean of items is from 3.3535 to 4.0556 is suitable for research data.

4.2 BIC Algorithm

R report shows every step of searching for the optimal model. BIC selects the best 5 models as Table 3.

There are six independent and one dependent variable in the model. Usefulness (US) influences Exercise Behavior (EB) with a Probability is 100% and Ease of use (EU), Barrier (BR), Facilities (FAC), Price and promotion (PP), and Service Quality (SQ) influence Exercise Behavior (EB) with probability is 97.1, 8.7, 29.3, 4.3, and 53.5%.

4.3 Model Evaluation

According to the results from Table 4, BIC shows model 1 is the optimal selection because BIC −1.335) is minimum. Usefulness (US), Ease of use (EU), and Service Quality (SQ) impact Exercise Behavior (EB) is 53% in Table 4. BIC finds model 1 is the optimal choice and three variables have a Probability of 32.8%. The above analysis shows the regression Eq. 1 below is statistically significant.

Table 2 Reliability

Factor	α	CITC	Item	Code	Mean
US	0.830	0.626	Exercise to improve health	US1	3.7020
		0.675	Exercise is very helpful to control	US2	3.8384
		0.743	Exercise to have a beautiful body	US3	3.7424
		0.595	Workout at the fitness center helps	US4	3.8081
EU	0.826	0.609	Easily find information about the service in the media	EU1	3.9040
		0.629	The location of the gym is easy to get around	EU2	3.8131
		0.738	Hours of operation of the spiritual center actively	EU3	3.8535
		0.658	Easy to learn and practice with the central machine	EU4	3.8939
BR	0.628	0.470	Don't know how to practice, afraid to practice wrong	BR1	3.7273
		0.573	Want to be effective but not economical enough for using a coach service	BR2	4.0556
		0.590	No motivation, lack of training goals	BR3	3.8030
		0.114	Without a coach to guide, design the exercise program should give up	BR4	3.6717
FAC	0.892	0.762	Fully equipped, quality, modern gym equipment	FAC1	3.6667
		0.756	The gym space is cool, beautiful, and clean	FAC2	3.7323
		0.760	The center has parking space, changing room, sauna…	FAC3	3.8535
		0.767	Many additional services to increase choice such as personal trainer, boxing class, Yoga…	FAC4	3.6970
PP	0.792	0.606	The price is commensurate with the facilities of the sports center	PP1	3.7323
		0.630	Price commensurate with the training program with the Trainer	PP2	4.0606
		0.664	There are many promotions for new and old members	PP3	3.8030
SQ	0.841	0.794	Center for many classes such as Yoga, Dance,	SQ1	3.6818
		0.550	Aerobic, Dance to change the exercise	SQ2	3.5859
		0.700	The training in the center has the expected results	SQ3	3.6515
		0.675	Attentive and friendly staff	SQ4	3.8939
EB	0.843	0.723	Keep practicing for a long time	EB1	3.4697
		0.727	Start exercising now	EB2	3.3535
		0.674	Introduce friends and family to practice	EB3	3.8131

Table 3 BIC model selection

EB	Probability (%)	SD	Model 1	Model 2	Model 3	Model 4	Model 5
Intercept	100.0	0.29880	0.1130	0.3693	0.2070	−0.0001	0.2734
US	100.0	0.06790	0.6020	0.6251	0.5632	0.5510	0.6072
EU	97.1	0.07040	0.1976	0.2115	0.2115	0.1989	0.1996
BR	8.7	0.02117					−0.0044
FAC	29.3	0.05328			0.1060	0.0091	
PP	4.3	0.01294					
SQ	53.5	0.06188	0.1073			0.0096	0.1029

Table 4 Model test

Model	nVar	R^2	BIC	Post prob
Model 1	3	0.530	−1.335	0.328
Model 2	2	0.515	−1.329	0.240
Model 3	3	0.526	−1.319	0.152
Model 4	4	0.537	−1.315	0.120
Model 5	4	0.531	−1.290	0.034

$$\mathbf{EB = 0.1130 + 0.6020US + 0.1976EU + 0.1073SQ} \tag{1}$$

5   Conclusions

BIC algorithm determined the factors affecting the exercise behavior in the fitness center of customers in Ho Chi Minh City, Vietnam. We have provided the theoretical bases to build the proposed research model for the topic. The study was carried out according to qualitative and quantitative research methods, including 6 scales. After testing Cronbach's Alpha, the scales have good measurability with 22 observed variables. The BIC algorithm shows the model to be consistent with the actual data. According to the analysis results from the BIC algorithm that tested the suitability with the proposed hypotheses, the practice behavior of customers in the fitness center in Ho Chi Minh City is determined by 3 factors, there is 53% variation of factors affecting the Exercise Behavior (EB). The research results show that the factors that have the most impact on exercise behavior are Usefulness with a Beta coefficient equal to 0.6020, Ease of use with a Beta coefficient of 0.1976, and Service Quality has a Beta coefficient of 0.1073.

Implications for Usefulness

We can see that the factor US2 has the highest rating (Mean = 3.8384), while US1 has the lowest rating (3.7020). The average value of the factors in the Useful factor is greater than 3, with the range from 3.7020 to 3.8384. The research results show customers are interested in the effects of exercise because of the benefits that exercise brings to customers, so that is the reason. The most powerful way for customers to overcome all barriers is to exercise and have the healthiest and most beautiful body.

Some of the author's recommendations are: (1) the center must invest in body machines so that customers can always see the value of exercise, which is to control weight, muscle, and body fat. (2) The center needs a variety of exercises for customers to change the atmosphere and exercises to reduce stress after a working day such as boxing, dance, dance, yoga.… (3) The center always creates weight loss challenges, body contests so that customers are more motivated and more interesting when exercising. (4) The center must use quality standard equipment so that customers can exercise without affecting their health.

Implications for Ease of use

The factor EU1 has the highest rating (Mean = 3.9040), while EU2 has the lowest rating (Mean = 3.8131). The mean value of the factors in the perceived ease of use factor is greater than 3, with the range from 3.8131 to 3.9040. The factor that affects the exercise behavior of customers in the fitness center is the ease of use and Beta = 0.1976, which is the second most important concern of customers. Managers need to promote their strengths to help customers choose to use them, but besides that, the author has some recommendations: (1) Expanding branches and centers to cover the area of that neighborhood. (2) The trainer gives dedicated instructions so that new customers who do not know how to practice with machines can practice on their own when they do not have enough money to practice with the trainer. (3) Run ads around the center to attract customers close to home.

Implications for Service Quality

SQ4 factor has the highest rating (Mean = 3.8939), while SQ2 has the lowest rating (Mean = 3.5859). The mean value of the factors in the perceived ease of use factor is greater than 3, with the range from 3.5859 to 3.8939. One factor that affects the exercise behavior of customers in the fitness center is Service quality and the coefficient Beta = 0.1073, which is the fourth concern of customers. Managers need to promote their strengths to help customers choose to use them, but besides that, the author has some recommendations: (1) Strengthen customer care so that customers feel Beloved, friendly staff of the center, well-trained staff. (2) Improve the Coach's qualifications so that the Trainer helps the client to achieve the expected effectiveness in training. (3) If the coach plans appropriate exercise, nutrition and helps the client change their body shape, that customer will be loyal and always be a source of introduction to the center. (4) There are many accompanying services such as boxing, dance, and yoga teachers so that customers can see the variety and quality when practicing at the center.

6 Limitations and Future Scope

BIC Algorithm for Exercise Behavior only stops at factors affecting Exercise Behavior at Customers' Fitness Center in Ho Chi Minh City, Vietnam, represented as Usefulness (US), Ease of use (EU), and Service Quality (SQ), the above factors are not completed for Exercise Behavior. With the situation of the Covid-19 epidemic in Vietnam, we could not survey 203 people. Therefore, it is not possible to fully represent the existing population in Ho Chi Minh City. Future studies can be carried out with a larger sample, with a larger population, not stopping at Ho Chi Minh City but can also expand to other large cities or provinces in Vietnam such as Hanoi, Can Tho, Da Nang as popular places to practice sports in sports centers, from which it is possible to research and gather more knowledge and have Suggestions can be made to help managers overcome the limitations.

Acknowledgements This research is funded by the Industrial University of Ho Chi Minh City, Vietnam.

References

1. Zhang W, Li Y (2014) A study on consumer behavior of commercial health and fitness club—a case of consumers in Liverpool. Am J Ind Bus Manag 4(1):58–69
2. Anwong P, Lertbuddhalak S, Khamloy W (2020) The Factors affecting the decision to use the fitness center in Bangsean Chonburi. KKBS J Bus Adm Account 4(3):57–76
3. Fernández-Martínez A, Murillo-Lorente V, Sarmiento A, Álvarez-Medina J, Nuviala A (2021) Exercise addiction and satisfaction of fitness center users as precursors to the intention of continuing to engage in physical activity. Sustainability 13(1):129
4. Biddle SJ, Hagger MS, Chatzisarantis NL, Lippke S (2007) Theoretical frameworks in exercise psychology
5. Riseth L, Nøst TH, Nilsen TI, Steinsbekk A (2019) Long-term members' use of fitness centers: a qualitative study. BMC Sports Sci Med Rehabil 11(1):1–9
6. Höglund K, Normen L (2002) A high exercise load is linked to pathological weight control behavior and eating disorders in female fitness instructors. Scand J Med Sci Sports 12(5):261–275
7. https://digitalcommons.wku.edu/cgi/viewcontent.cgi?article=2367&context=theses
8. Parfitt G, Hughes S (2009) The exercise intensity–affect relationship: evidence and implications for exercise behavior. J Exerc Sci Fit 7(2):S34–S41
9. Olson TL, Terbizan DJ (2017) Resting metabolic rate and body composition change in women following different exercise training programs. J Bones Muscle Study 1(01):1–7
10. Ha Nam Khanh G (2015) Study of the factors affecting customers loyalty for Gym service at KIM center, Vietnam. Int J Sci Eng Res (IJSER) 2347–3878. ISSN (Online)
11. Azizah S (2018) Alternative development designs of a culinary center in Surabaya that attract more customers. IPTEK J Proc Ser (6):110–114

12. Xu H, Zhuang Y, Gu H, Xu X, Zhang Y (2019) Pattern-based service composition for user satisfaction and service revenue. Int J Serv Technol Manag 25(5–6):585–600
13. Turaga NK, Patibandla JK, Turaga M, Bandaru SR (2020) Factors influencing to complain by first time customers and loyal customers in DTH services. J Crit Rev 7(12):1076–1080
14. Hairetal JF (2009) Multivariate Data Analysis: A Global Perspective. Prentice Hall, Hoboken
15. Nunnally JC (1978) Psychometric Theory, 2d edn. McGraw-Hill, New York
16. Peterson RA (1994) A meta-analysis of Cronbach's coefficient alpha. J Consum Res 21(2):381–391
17. Slater SF (1995) Issues in conducting marketing strategy research. J Strateg Mark 3(4):257–270
18. Nunnally JC (1994) Psychometric Theory 3E'. Tata McGraw-hill education, New York
19. Raftery AE, Madigan D, Hoeting JA (1997) Bayesian model averaging for linear regression models. J Am Stat Assoc 92(437):179–191
20. Kaplan D (2021) On the quantification of model uncertainty: a Bayesian perspective. Psychometrika 86(1):215–238
21. Raftery AE (1995) Bayesian model selection in social research. Sociol Methodol 25:111–163

Medicine Supply Chain Using Ethereum Blockchain

Amrita Jyoti, Gopal Gupta, Rashmi Mishra, and Ankit Jaiswal

1 Introduction

The FBI and IACC (International Anti Counterfeiting Coalition) revealed that counterfeiting is one of the most critical criminal activities of the twenty-first century, and it is thriving gradually by bringing in new faux drug company makers into the market. The increased access to medications via online pharmacies and unauthorized distribution channels makes it difficult to ensure product safety in the supply chain [1]. Counterfeit medicine is a huge challenge for the pharmaceutical trade worldwide. Because of technology's emergence, the distribution of those fake medicines is additionally enhanced day after day. Furthermore, Due to the functioning of more nodes from the failure threshold, the network failure probability is very low, along with robust fault tolerance. That is why to stop this issue, a secure drug supply-chain management system is required to trace the drug delivery process at each [5]. Every stage starts from the supplier's raw material then moves on to the following supply-chain process: manufacturing product, distribution stage, pharmacy, clinics, and consumers, respectively, to protect consumers from forged drugs. Moreover,

A. Jyoti (✉)
Ajay Kumar Garg Engineering College, Ghaziabad, India
e-mail: amritajyoti2013@gmail.com

G. Gupta
ABESIT, Ghaziabad, India
e-mail: gopalmiet@gmail.com

R. Mishra
Krishna Engineering College, Mohan Nagar, Ghaziabad, India
e-mail: rashmimishra1987@gmail.com

A. Jaiswal
Infosys, Bangalore, India
e-mail: ankit.12r23@gmail.com

© The Author(s), under exclusive license to Springer Nature Singapore Pte Ltd. 2022   193
B. Unhelker et al. (eds.), *Applications of Artificial Intelligence and Machine Learning*,
Lecture Notes in Electrical Engineering 925,
https://doi.org/10.1007/978-981-19-4831-2_16

within computer science, the most recent invention that can overcome such problems is Blockchain Technology. Blockchain will not only provide security to the provision-chain process but will also oversee the delivery with efficiency. So, this paper aims to find out the issues in the drug supply chain management system and solve the problem by using Blockchain's Decentralized applications. In this paper, we have implemented a DSCM that stands for drug supply chain management system built on Ethereum Blockchain and has the ability to continuously observe and keep track of the drug delivery process in the smart pharma industry due to its varied features. The reason behind the fact that why most of the industry requires to shift to Blockchain technology is because it provides a distributed decentralized ledger electronically where all the pair nodes within the network will see and authenticate the transactions connected info. An added factor is its agreement algorithmic program that empowers the network to store the sole valid info within the repository and rule out the matter of duplicate dealing. Our paper will analyze the journey of drugs using Ethereum Blockchain and other objectives of this paper are to gather practical knowledge of supply chain management, improve operational & stream-lining operations, maintain all the critical databases. It will lower the workforce and manual paperwork used to manage the records offline and maintain the efficiency, integrity, and uniformity of the data.

2 Related Previous Work

Nowadays, Blockchain applications do not seem to be solely restricted to cryptocurrency; however, they are also being utilized in other fields such as agriculture, healthcare, finance, education, transportation and supply chain. Blockchain solutions for supply chain management and logistics have recently received increasing acceptance that they provide an immutable and transparent recording of transactions between a distrustful of stakeholders [2].

In 2018, an agriculture supply chain management system named AgriBlockIoT was developed on Blockchain. This system manages the supply chain of food products and traces the original source from where it comes [3]. AgriBlockIoT system was built with Hyperledger fabrics and Ethereum, which are two different Blockchain network platforms [4]. The application of AgriBlockIoT uses different sensors to ensure that the data generated gets stored in a secure, immutable, and transparent manner.

Electronic medical records (EMR) systems are used for securing the health care trade and are built on a Blockchain network [7]. The characteristics of Blockchain are in line with the requirements of EMR.

A Dutch company Guard time, designed and developed a Blockchain-assisted framework with the goal of patient's identity verification and also for their data security. According to this framework, all nation individuals will get cards of their own, which utilizes Blockchain to correspond Electronic medical records (EMR) information [8]. The hash code utilized in the consensus algorithm is relegated with an update whenever a transaction occurs within the network. Due to the Blockchain audit records inside the EMR are tamper-proof and are beyond modification [11].

Tamper-resistant logs likely could be utilized to store data status in the existing healthcare database. Any updates in the database (such as registered records) will be allotted a timestamp and password while composing the transaction in a block. Thus, the Blockchain-based system guarantees the integrity and the security of medical data.

3 Proposed Method

3.1 Overview of Drug SCM Procedure

The privateness and safety of the drug delivery chain of the pharmaceutical enterprise have protracted through the decentralized distributed nature of blockchain technology [9]. The entire drug supply chain management (DSCM) mechanism is conferred with contributors and the blockchain community, as shown in Fig. 1. Participants will control and update the entire supply. The product designed is web-based, primarily to showcase the transfer of medications on the Blockchain. It is a smart contract that sites the problems of saving crucial data necessary at distinct levels of the supply chain and making it verifiable through all stakeholders within the supply chain. It traverses all actions from the starting point to the end. This smart contract is designed to run on any system with a meta mask installed within the browser. The suppliers, manufacturers, wholesalers, distributors, hospitals & doctors, and pharmacies are stakeholders of the network whose data are stored within the blockchain-based system. The product provides exact information anywhere and anytime across the entire chain. All handovers are visible within the supply chain when problems are detected, immediate access to alerts and updates, all materials supply can be traced back, all parties have seamless collaboration between them, reduce paperwork and speed up process. The resources of the DSCM structure are raw materials, drugs, record repositories, and order. Client application-based front-end is provided to each user of the system wherein they can communicate with the blockchain network. Without any problems, they can carry our transaction. During the entire DSCM process, all the client application users can track the drug delivery status. On the other side, a separate data library is an information storage pool, also referred to as the stored-off Blockchain. System users can see the whole data of medicine, raw materials, and helpful info like price, manufacture date, expiry date, etc. There are peer nodes in our system to run the consensus algorithm to achieve distributed ledger consistency purposes [6]. Within our system firstly, the raw material was sent by the supplier to the drug company. For login and to perform the transactions, each user has a web application portal. Suppose a manufacturer places orders for raw material, any peer node of the network can validate this transaction and after that, the order is received by the supplier. Afterward, the supplier performs the confirm order event when delivery of the raw material order is ready. On the other hand, using our proposed blockchain system by following the scenario mentioned above, hospitals/doctor can place the order to the medicine company if they want.

3.2 Three-Layer Architecture

The implementation of proposed approach is conducted using a three-layer architecture, comprising of data storage layer, blockchain layer, and provenance database layer, as in Fig. 2.

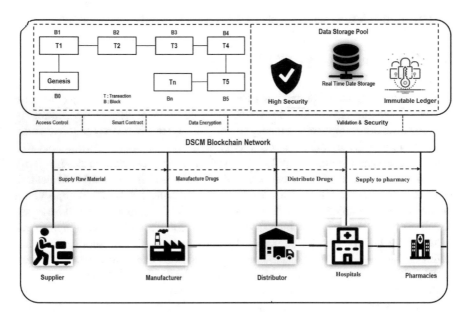

Fig. 1 Drug supply chain management

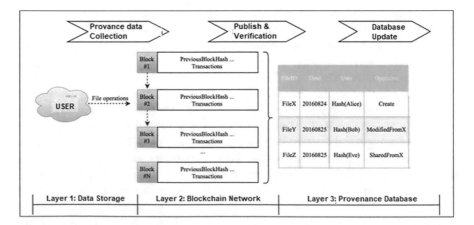

Fig. 2 Three-layer architecture for proposed approach

3.3   Detail Architecture of Proposed DSCM System

When it comes to stopping counterfeit drugs within the drug delivery chain, a tech-nology that could monitor every step in the supply chain at the individual drug level and verify an immutable chain of transaction ledger is blockchain technology. The crucial and primary aim of the blockchain network is to save data in a distributed manner where multiple transactions are contained in each block. For safety reasons, encryption and hashed methods are used to save transections. The proposed appli-cation provides the smart contract and distributed ledger functionalities as a service that may be a service-oriented framework for the users. By using the front-end web application of our system, end-user (suppliers, manufacturer, wholesaler, distributors, doctors & hospitals, and pharmacies) can perform the transaction like raw material supply, medicine orders, update orders, update the drug's data, update the records, data sharing, deliver drugs, entire drug delivery tracking, customers management, drug management, etc. [10]. The main objective of this smart contract is to provide a safe SCM application to the stakeholders and stop counterfeit drugs. Due to integrity management features and the security of Blockchain, our proposed method is safe. By using this DSCM system, every user can track drug delivery. Our DSCM appli-cation is developed on the permissioned blockchain system i.e., an application can be used by only specific participants that are another unique feature of our applica-tion. For enrollment, users are provided with certificates and confidential credentials, which are then validated by the administrator in this network and can use this safe blockchain network. A consensus algorithm is responsible for the user's enrollment with the private network to manage transaction orders and perform transactions. To save the data and for transaction execution, this application has a distributed ledger in simulated environments wherein each peer node features a smart contract in this blockchain network. After the execution, these nodes validate the transaction and the transaction block is written into the ledger. Avoid invalid transactions from saving in history and to keep transaction history as one is the primary objective of the consensus algorithm. To provide a unique hash code and a digital signature for each transaction, the consistency of the ledger is kept by using the consensus protocol. All the events, logs of each transaction, transaction records, and actions performed by every user are kept within the proposed system i.e., disturbed ledger.

3.4   Smart Contract of DSCM

Providing the users, a conflict-free and transparent way to exchange money, shares, property, or anything where a broker or third-party agent is not involved is known as a smart contract. The blockchain platform has this as its best feature. On technical ground, multiple lines of programming code that enforce the agreement between two parties without paying anything to a broker or third-party agent is a smart contract. A predefined set of terms on which both parties have agreed is part of this code. When the particular condition is the same as database events, then this contract is triggered automatically. Users can manage their assets among different parties and access rights using smart contracts. These smart contracts are entirely safe and protected from tampering and deletion as they are managed and stored in a distributed ledger of a blockchain platform. The blockchain network's consensus algorithm automatically triggers and enforces a smart contract, which is a safe way to implement terms and business conditions in some lines of programming code. Some crucial problems occur while writing the smart contracts; for writing the smart contract, a new programming language is used i.e., Solidity, due to this, it is hard to maintain and the learning rate is very low. The transactions are carried out among every peer node of the network in sequence or periodically and due to this, it takes more time to complete as the execution time of transactions is low. We overcome this issue by deploying the smart contracts not for all nodes but only for particular nodes in the network. Thus, the transactions can be validated by a defined set of nodes, and the performance of the application also increases remarkably.

3.5   Transactions Execution Procedure in DSCM

The transactional process of the drug supply chain management system is explained in this section: Fig. 3 shows how a transaction should be accomplished in the blockchain network. Firstly, users connect to the blockchain system through a given front-end of a client application using their registered credentials and complete their transaction requests. Users may be able to perform the transaction only after when administrator registered them to a blockchain network. The user has to login using the registered credentials on the client application, in order to submit a transaction proposal in the blockchain network. And then, that proposal is sent to each and every peer node. Either committers or endorsers are the two categories in which peer nodes are divided. The endorser peers will extract the read and write data while executing the transaction in the simulated environment known as the RW set. The client again submits the signed transaction with every RW sets to the consensus manager. After that consensus manager ordered the data into a block and deliver the transaction into the committer's nodes. Then, the committers nodes validate the transaction with the current world state, and then transaction information is saved into the ledger. At Last, as per the written data, the ledger gets updated. Then, the notification for transaction status, either submitted or not, is sent to the client by committer peers. Infura and REST API helps to establish the connection of client applications with the blockchain network.

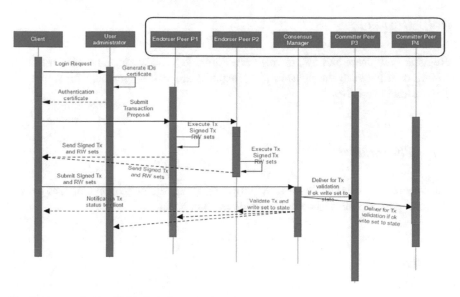

Fig. 3 Transaction in the blockchain network

4 Stakeholders

4.1 Supplier

Supplier creates a new batch with details of the farm and supplies raw materials to the manufacturer.

4.2 Manufacturer

Manufacturer updates information of raw materials details (like batch ID and consumption units) that are used to manufacture new batch medicine and quantity.

4.3 Distributor

The distributor is responsible for distributing the medicine to Hospitals and do verification on quality and condition.

4.4 Hospitals

Register new users and assign roles according to their work and verify medicine before distributed to the pharmacies. Hospitals are also responsible for producing new medicine batches.

4.5 Pharmacies

Pharmacies are responsible for providing the right medicine to consumers as per the doctors' prescription and complete medicine batch status.

5 Scope

The Blockchain is a ledger, a database that keeps everything in one place like a log that cannot be modified without going away with a mark within the ledger in the act of information changing. Blockchain makes the recording of both analysis and treatment of information magnificent in addition to medical histories. Presently, info is siloed and since individuals recognize that tons of data enclosed in medical records are fabricated, there is an absence of assurance that causes a delay in diagnosing and treating the patients in a hospital. In a Blockchain-enabled world, these patients will be able to own their whole non-fabricated medical histories in a wallet — from birth to their current existence. These EHRs will consist of all medical records, procedures and every little thing. When a doctor interviews a patient, the medic has access to the patient's complete records, creating identification of cancer, as an example, much more accessible. However, Blockchain means a lot of to health care than providing help for the purpose of care. Blockchain firmly shares health information, standardizes data formatting along with enhancing health care transactions overall. Its impact will affect each leading participant within the care cycle, from patient and supplier to money handler, drug company, and even researchers and federal regulators. It will be fair to point out the fact that Blockchain is far and wide in aid. Its application in healthcare has not reached its full potential yet; however, the outcomes are profound because it comes ancient, in step with trade insiders.

6 Result and Evaluation

We have explained the operation of the DSCM, the system works when you interact with a blockchain network. As we have already discussed, each participant in the system is assigned a user interface of the client application, where you can start your business immediately after you confirm your identity. The client application and interacts with a blockchain network using the REST server, the composer of each of the requests will be sent to the REST server and saves the start-up of the operation of the blockchain network. The client is a web portal for the program, which is designed to deal with drugs, where the manufacturers are able to add, update, and delete information about the medicinal product on the blockchain network. The administration Portal for the Pharmaceutical allows you to create, read, update, and delete operations (CRUD) with the help of the software's user interface. The users can also update, drug information, by sending an update request via a user interface of the blockchain network. After the server responds, and the REST is displayed in JSON format, which contains the meta-data. Our website indicates that all of the operations that are performed via the world wide web, the application of the blockchain network, it has to be a surgery, and a few are active participants in the system. The attributes of the portal and of the dates, time and date, the location of the transaction, operation, type of operation, and the systems of the participants, who are involved in this effort, the start-up. Our site shows a log of all the operations that are carried out on the blockchain network.

6.1 Deployment Costs

The gas limit is referred as the maximum number of gas which we willing to spend on a particular transaction. A higher limit of gas referred to do more computational work for the execution of the smart contract. The standard ETH transfer involves a gas limit of 21,000 units of gas. In Ethereum, the gas limit is measured in terms of the unit of gas. The gas limit is calculated as follows:

$$gas\,Limit = G_{transaction} + G_{txda\,\tan onzero} \times data\,ByteLength \qquad (1)$$

where,

$G_{transaction}$ is 21000 gas.

$G_{txda\,\tan onzero}$ is 68 gas.

dataByteLength is the data size in bytes.

6.2 Price of the Gas Used in deploy the Smart Contract

The gas price is multiplied by the limit of gas is referred as the total cost of the transaction which is similar to pay the gas at the station of the gas such as $3.50 (gas price) per gallon (unit). The ten units referred to a transaction price of $35. When the transaction is sent by the user, the price of the gas is specified in Gwei/Gas (1 Gwei equals 0.000000001 ETH) and the total commission pays equal to the product of the

Table 1 Gas used and cost of deploying Smart Contract in DSCM

Smart contract name	Used gas	Total gas for deploying
DSCMSupplyChainStorage	6,200,513	0.10400746 ETH
DSCMSupplyChain	5,899,063	0.09978223 ETH
DSCMSupplyChainUser	1,595,106	0.02980103 ETH

gas price and gas used. A low price of gas is preferred by most of the consumers since the inexpensive transactions is allowed on the Ethereum blockchain.

The actual deployment costs in gas for the Supply ChainStorage.js, Supply-Chain.js and SupplyChainUser.js contracts are shown in Table 1.

7 Conclusion and Future Work

Blockchain technology has changed the traditional method of supply chain management in order to be reliable, automated, reliable, current, audio, and transparent. This is to ensure that the entire process of a solid and healthy, our supply chain has been affected by the problem of fake medicines, the system completely. The main objective and novelty of our proposed system is the use of a blockchain-based pharmaceutical company in the supply system. I have performed several experiments in order to test the performance of the system with the help of some fertility metrics such as end-to-end path, the operating system response time, and latency. This system helps pharmaceutical companies to eliminate the problem of counterfeit drugs, and greatly enhance their business. In future work, we will be increasing the size of the network and implement it in real-time, in order to make it possible for pharmaceutical companies to control the operation of and the validity of the system. In this section, we briefly describe the most important challenges for the implementation of blockchain-based provenance tracking and monitoring solutions in the pharmaceutical industry.

References

1. Metcalf DS, Bass J, Hooper M et al (2019) Blockchain in Healthcare: Innovations that Empower Patients, Connect Professionals and Improve Care. CRC Press, Taylor & Francis Group, Boca Raton, FL
2. Pandey P, Litoriya R (2020) Securing E-health networks from counterfeit medicine penetration using Blockchain. Wirel Pers Commun 117(1)
3. Dubovitskaya A, Xu Z, Ryu S, Schumacher,M, Wang F (2017) Secure and trustable electronic medical records sharing using Blockchain. In: AMIA Annual Symposium Proceedings, vol. 2017. Am. Med. Inform. Assoc. 2017, 650
4. Hang L, Choi E, Kim DH (2019) A novel EMR integrity management based on a medical blockchain platform in hospital. Electronics 8:467

5. Clauson KA, Breeden EA, Davidson C, Mackey TK (2018) Leveraging blockchain technology to enhance supply chain management in healthcare. Blockchain Healthc. Today
6. Ali MS, Vecchio M, Putra GD, Kanhere SS, Antonelli F (2020) A decentralized peer-to-peer remote health monitoring system. Sensors 20:1656
7. Yu C, Xu X, Yu S, Sang Z, Yang C, Jiang X (2020) Shared manufacturing in the sharing economy: concept, definition and service operations. Comput Ind Eng 146:106602
8. Lalic B, Majstorovic V, Marjanovic U, von Cieminski G, Romero G (Eds.) (2020) Advances in Production Management Systems. The Path to Digital Transformation and Innovation of Production Management Systems. Springer International Publishing, Heidelberg, pp 259–266. https://doi.org/10.1007/978-3-030-57993-7_30
9. Wu X, Lin Y (2019) Blockchain recall management in pharmaceutical industry. Procedia CIRP 83:590–595
10. Zhang A, Lin X (2018) Towards secure and privacy-preserving data sharing in e-health systems via consortium blockchain. J Med Syst 42(8):140
11. Blockchain Threat Report-Mcafee.com (2018). www.mcafee.com/enterprise/en-us/assets/reports/rp-blockchain-security-risks.pdf. Accessed 21 Jan 2020

Human Activity Recognition Using Single Frame CNN

V. Aruna, S. Aruna Deepthi, and R. Leelavathi

1 Introduction

The main advantage of CNN is it can extract the features from the images without any human supervision. It can be implemented fast and is easy to understand. The accuracy of image prediction is very high among all other algorithms. The limitations of ANN are it requires a processor that has the same processing power of their structure. It is difficult for complex data. CNN's follow a model which is hierarchical that works on building a network.

Overfitting indicates the model cannot solve the problem as it is too complex. It occurs when the model tries to predict data that is too noisy. It can be prevented by training with more data, reducing the parameters, by changing the network structure. If there is less parameter there may be a chance of less over fit, so it improves the performance of the model.

Underfitting occurs when the model is unable to classify the data that was trained on. Human activity recognition in short HAR [1] is a field of study that is concerned with detecting the moment or action of a person based on video data or sensor data. It is useful in many applications like surveillance, health care which includes computer–human interaction. There are many approaches with good accuracy. In this model, an accuracy of 95% is achieved. For more accuracy, LSTM (long short term memory) method, early fusion and late fusion methods are used. There are two methods of human activity recognition: one is vision-based and the other is sensor-based. Vision based HAR is used in this paper. In vision -based HAR it requires a camera to track everything in the environment based on that video it detects [2] the activity. It needs a higher amount of processing power. It provides reliable data.

V. Aruna · S. A. Deepthi (✉) · R. Leelavathi
Vasavi College of Engineering, Hyderabad, India
e-mail: sadeepthi@staff.vce.ac.in

V. Aruna
e-mail: v.aruna@vce.ac.in

© The Author(s), under exclusive license to Springer Nature Singapore Pte Ltd. 2022   205
B. Unhelker et al. (eds.), *Applications of Artificial Intelligence and Machine Learning*,
Lecture Notes in Electrical Engineering 925,
https://doi.org/10.1007/978-981-19-4831-2_17

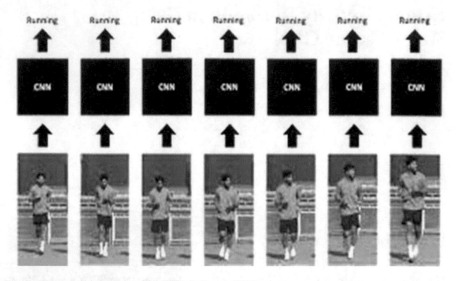

Fig. 1 Human activity recognition (Single frame CNN)

It is used in many applications like surveillance, recognizing military activities and health care. It is most suitable for a [4] security system. Human activity recognition is a common field in research areas of computer vision.

The practical definition of Deep Learning is "'It's a sub-domain of ML (machine learning) algorithms in the type of a neural network that applies a cascade of layers of working units to derive features and make perceptive approximations about new data". Deep learning [5] can also be called a deep neural network (Fig. 1).

2   Literature Survey

Human activity recognition has been studied for years and researchers have stated a variety of solutions to approach this problem. Existing [6] approaches typically use inertial sensors, vision sensors, and a mixture of both. Machine Learning and limit-based algorithms are regularly applied. The approaches that consolidate both vision and inertial sensors have likewise been proposed. Another fundamental piece of this algorithm is information handling. The [8] nature of the info highlights enormously affects the exhibition. Some past works are centered on creating the most valuable features from the time series data collection. The normal methodology is to break down the sign in both time and frequency domain.

The basic goal is to analyze all the frames in the video to identify the activities that are taking place within the video. A video incorporates a spatial perspective. A few activities [9] (example, skateboarding, swimming, running, etc.) can likely be distinguished by utilizing a single outline but for more complex actions (example, strolling vs. running, twisting vs. falling) might require more than 1 frame's data to distinguish it accurately. The most Advanced Activity recognition we know today is used in Tesla for their Level 3 autonomous driving in their vehicles. In the early fusion approach we merge the individual networks by a fusion layer to get the output. This approach helps the model to learn both temporal and spatial information of the data. The model [10] predicts for each frame and the probabilities are merged using the fusion layer. First download the data and then extract it.

3 Proposed Methods and Results

Import Required Libraries like OpenCV, Pafy, and Numpy and configure Numpy, Python and Tensor flow seeds for consistent results. The random number generator is initialized by the seed() method. To generate a random number, this random number generator requires a starting value (seed value) equal to 23. The dataset UCF50 is an Action recognition dataset that includes 50 action categories which consist of real YouTube videos. Select a few videos from each class of the dataset randomly that is used to display them on the top of the frame. This gives a quick overview of the dataset appearance. Read and Pre-process the Dataset. It uses classification architecture in order to train on a video classification dataset, so pre-processing of the dataset should be done first. Create a function called Create dataset (), which uses Frame extraction () function to create the final pre-processed dataset. A list that contains its associated [7] labels. Call the create dataset method which returns features and labels. Then convert the class label to one hot encoded vector. To categorical method is [11] used for converting the video labels into one-hot-encoded vectors. The two Jumpy arrays, in which one contains all the labels in one hot encoded format and the other contains all images. Next split the dataset into training and a testing set. This is used for model validation accuracy (Fig. 2).

3.1 CNN Classification Model

Now create a CNN Classification model with two CNN layers (Fig. 3).

First write a function that designs the model create model (). It uses Sequential models for model creation.

In the CNN model, the input image is first subjected to the convolution layer. This convolution layer extracts information from the input image and produces the output by convolving the input image with a filter or kernel. Once the convolution [3] operation is done, apply the RELU activation function for the maximum output. Next, the output [12] is taken to batch normalization which standardizes the inputs. Then it is given to the global average pooling layer which generates one feature map for each category of the classification tasks. Finally, the dense layer feeds the outputs

Fig. 2 Data with its label

to all its neurons; each neuron produces one output to the next layer. The stacked CNN model is created now (Fig. 4).

Use the **plot model** function to verify the final model structure. This function [13] helps to create a network which is complex, and also to make sure that the built network is correct.

3.2 Compile and Train the Model

Keras supports a callback called Early Stopping that allows stopping the training process before completion. This callback provides the performance measure for monitoring the trigger, and it terminates the training process if it is triggered. Add Early [14] Stopping Callback, optimizer, metrics values and loss to the model and then start training. Then using the feature's test set and label's test set, evaluate the trained model on the and save the trained model (Fig. 5).

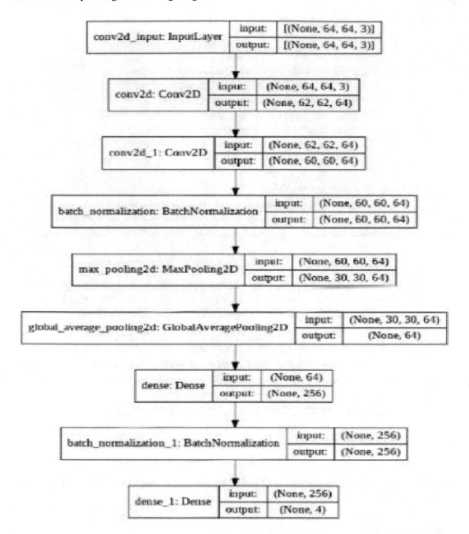

Fig. 3 Flowchart of CNN model

3.3 Plot Accuracy Curves and Model's Loss

Define a function plot metric that gives metric values using metric names as identifiers and then create a range object which is used as a time and now plot the graph and add title to the plot. Now creating and training of the model is done, next is to test its performance on some test videos. Create a function that uses the Pay library to

```
Model: "sequential"

Layer (type)                  Output Shape         Param #
=================================================================
conv2d (Conv2D)               (None, 62, 62, 64)   1792

conv2d_1 (Conv2D)             (None, 60, 60, 64)   36928

batch_normalization (BatchNo  (None, 60, 60, 64)   256

max_pooling2d (MaxPooling2D)  (None, 30, 30, 64)   0

global_average_pooling2d (Gl  (None, 64)           0

dense (Dense)                 (None, 256)          16640

batch_normalization_1 (Batch  (None, 256)          1024

dense_1 (Dense)               (None, 4)            1028
=================================================================
Total params: 57,668
Trainable params: 57,028
Non-trainable params: 640

Model Created Successfully!
```

Fig. 4 Model creation

```
Epoch 1/50
3840/3840 [==============================] - 312s 81ms/step - loss: 0.7191 - accuracy: 0.6829 - val_loss: 1.9478
Epoch 2/50
3840/3840 [==============================] - 313s 82ms/step - loss: 0.3746 - accuracy: 0.8613 - val_loss: 0.0967
Epoch 3/50
3840/3840 [==============================] - 311s 81ms/step - loss: 0.2712 - accuracy: 0.9046 - val_loss: 0.1064
Epoch 4/50
3840/3840 [==============================] - 311s 81ms/step - loss: 0.2181 - accuracy: 0.9218 - val_loss: 0.2642
Epoch 5/50
3840/3840 [==============================] - 310s 81ms/step - loss: 0.1930 - accuracy: 0.9331 - val_loss: 0.0105
Epoch 6/50
3840/3840 [==============================] - 311s 81ms/step - loss: 0.1659 - accuracy: 0.9441 - val_loss: 0.0253
Epoch 7/50
3840/3840 [==============================] - 313s 81ms/step - loss: 0.1476 - accuracy: 0.9518 - val_loss: 0.0298
Epoch 8/50
3588/3840 [=========================>..] - ETA: 19s - loss: 0.1453 - accuracy: 0.9518
```

Fig. 5 Output of the trained model

download any YouTube video and finally by just passing the URL, it returns the video title.

Calling the Create Dataset Method
The above data with labels is taken and some of the outputs are shown. Data is extracted for images pushups, Horse riding, playing guitar and skate boarding images from the above Dataset (Fig. 6).

```
[9] features, labels = create_dataset()

    Extracting Data of Class: PushUps
    Extracting Data of Class: HorseRiding
    Extracting Data of Class: PlayingGuitar
    Extracting Data of Class: SkateBoarding
```

Fig. 6 Output of extracting features

Fig. 7 Skate boarding using single frame CNN

Using Single-Frame CNN Method

Now define a function that gives a single prediction from the whole video. This function makes predictions by taking 'n' frames from the entire video. Finally, it averages the predictions of those n frames to get the activity class name finally from that entire video. Calculating the average probability of the trained activity's maximum is the output of the activity present in the video (Fig. 7).

To Predict on Live Videos using with and without Moving Average function.
Create a function that uses moving average to perform predictions on live videos. This function behaves like a simple classifier used to predict video frames, if window size hyper parameter is set to 1.

- Create a fixed-size Deque Object that implements moving/rolling average functionality.
- Read the video file by using the Video Capture object.
- Get the width and height of the video
- Write the over layer video files by using Video Writer object
- Resize the Frame to fixed Dimensions.
- Dividing the normalized frame by 255, so that each pixel's value is in the range of 0–1.
- Send the normalized frame of the image to the model to get the Predicted Probabilities.
- Overlay the class name text on top of the Frame.
- Close the Video Capture and Video Writer objects and release all the resources it contains.

Without Using Moving Average:
Without using moving average means window size is set to 1(the window size used in the moving average process). This method is used for testing the model on the YouTube videos containing multiple activities. It takes each frame from the video

Fig. 8 Playing Guitar using moving average

and predictions are made according to the activities in the trained model. As the predictions are made for each frame, the flickering of the activity labels will be more. A simple, elegant solution to avoid this flickering is by [19] using a moving average. Set the window size to 1. Create the output YouTube Video Path. Call the predict_on_live_video method to start the Prediction and finally play the Video file in the Jupyter Notebook.

Using Moving Average:
Moving average is the average of all 'n' predictions. This method takes 20 frames from the video and predictions are made according to the activities in the trained model. Then average the predictions and finally the output is labeled with the resultant average prediction. As it considers the averaging prediction, the flickering is less. Set the window_size to 20. Create the Output YouTube Video Path. Call the predict_on_live_video method to start the prediction and play the Video file in Jupyter Notebook. This approach is far better than the previous method of predicting each frame independently (Fig. 8).

Calculation of moving average technique.

$$P_f = \frac{\sum_{i=-n+1}^{0} P_i}{n} \tag{1}$$

n = No. of frames to be averaged.
Pf = Final predicted probability.
P = Probability predicted for the current frame.
P_{-1} = Probability predicted for the last frame.
P_{-2} = Probability predicted for the penultimate frame.

P_{-n+1} = Probability predicted for the (n-1) [the] last frame is the mathematical formula by which moving/rolling average can be calculated in the real time scenario. By using the above formula flickering can be avoided. In the below graph X axis

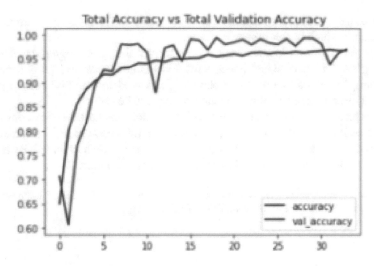

Fig. 9 Accuracy curve

represents the number of epoch's and Y axis represents the total accuracy. With the model 95% of accuracy has been achieved (Fig. 9).

4 Conclusion

In this paper, Human Activity Recognition is implemented using the single frame CNN, a video classifier is designed that extracts the features from the frames. It has 3 steps, first is pre-processing the dataset which requires to avoid unnecessary computation, the next step is training and compiling the model and the last step is testing the trained model. The training has been done for different activities which are divided into two sets. The YouTube URL (that contains the activity) has been passed, to function and it calculates and returns the average probability of all the trained activities highest among is the activity present in the video. In a single frame CNN it runs an image classification model on each frame and returns the average of all individual probabilities to get a final probability vector.

5 Future Scope

In the LSTM method, it extracts features from the person's pose and body in the video for each frame and use those extracted features to the LSTM network. LSTM is a part of RNN which can handle long video sequences of data. It supports multiple sequences parallel to the input data. The advantage of LSTM is it can learn from the data directly, which is raw time series data.

The future research is encouraged to perform training on large datasets which are complex. An accurate and better result for a large dataset (which is formed by combining 4 huge datasets) can be obtained. The model should be able to tell the behaviour exactly. For that the large datasets have to be trained.

In this model, the temporal nature of videos are not considered, but by considering the time series data for more accurate predictions this model can be improved. To consider the time series data RNN and LSTM approaches can be adapted. By training for more activities, it can be used in many fields of applications.

Acknowledgements We are thankful to the authorities of Vasavi College of Engineering for allowing us to use resources for the completion of this paper.

References

1. Poppe R (2010) A survey on vision-based human action recognition. J Healthc Eng Image Vis Comput 28:976–990
2. Moeslund TB, Hilton A, Krüger V (2006) A survey of advances in vision-based human motion capture and analysis. Comput Vis Image Underst 104:90–126
3. Fan L, Huang W, Gan C, Ermon S, Gong B, Huang J (2018) End-to-end learning of motion representation for video understanding. In: Proceedings of the IEEE Conference on Computer Vision and Pattern Recognition, pp 6016–6025. vv
4. Dang LM, Min K, Wang H, Piran MJ, Lee CH, Moon H (2020) Sensor-based and vision-based human activity recognition: a comprehensive survey. Pattern Recognit. Article 107561
5. Redmon J, Farhadi A (2017) YOLO9000: better, faster, stronger. In: Proceedings of the IEEE Conference on Computer Vision and Pattern Recognition, pp 7263–7271
6. Shabani AH, Clausi D, Zelek JS (2011) Improved spatio-temporal salient feature detection for action recognition. In: Proceedings of the British Machin fe Vision Conference (Dundee), pp 1–12
7. He K, Zhang X, Ren S, Sun J (2016) Deep residual learning for image recognition. In: 2016 IEEE conference on computer vision and pattern recognition (CVPR), pp 770–778
8. Thomee B, Shamma DA, Friedland G, Elizalde B, Ni K, Poland D et al (2016) Yfcc100 m: the new data in multimedia research. Commun ACM 59(2):64–73
9. Ahmad J, Muhammad K, Lloret J, Baik SW (July 2018) Efficient conversion of deep features to compact binary codes using fourier decomposition for multimedia big data. IEEE Trans Ind Inform 14(7)
10. Girshick RB (2015) Fast R-CNN, CoRR, vol abs/1504.08083
11. Shikha MS, Kumar R, Aggarwal A, Jain S (2020) Human activity recognition. Int J Innov Technol Explor Eng (IJITEE) 9(7). ISSN: 2278-3075
12. Singh D et al (2017) Human activity recognition using recurrent neural networks. In: Holzinger A, Kieseberg P, Tjoa A, Weippl E (eds) Machine Learning and Knowledge Extraction. CD-MAKE 2017. LNCS, vol 10410, pp 267–274. Springer, Cham. https://doi.org/10.1007/978-3-319-66808-6_18
13. Bulbul E, Çetin A, Doğru IA (2018) Human activity recognition using smartphone. In: 2018 2nd International Symposium on Multidisciplinary Studies and Innovative Technologies
14. Porwal K, Gupta R, Naik TG, Vijayarajan V (2020) Recognition of human activities in a controlled environment using CNN. In: 2020 International conference on smart electronics and communication (ICOSEC)

Monitoring Pedestrian Social Distance System for COVID-19

S. Prasanth Vaidya◉ and Marni Srinu

1 Introduction

Social distance has proved a very useful step to slow the circulation of the disease in the fight against COVID-19 [5]. People are encouraged to minimise their meetings and reduce the likelihood of spreading the illness. A social distance detecting device to assure social distance protocol in public areas and workplaces to monitor the fact that people track the camera in real-time video feeds over an extremely distant distance is designed [19]. For instance, employees at work can combine the device with their CC-television camera systems and watch whether or not people retain a distance from one other [7].

According to a recent study, social distance is a crucial containment tool and is necessary to avoid SARS-CoV-2 infection, because people with moderate or no symptoms can carry corona virus and transmit others [8].

Figure 1 shows that the best way to reduce infectious physical contact is through proper social distancing, thus reducing the infection rate. This lower peak may be compatible with existing hospital infrastructure, allowing patients fighting the coronavirus pandemic to receive better care [12].

Social distancing is a way of controlling infectious disease spread. As the name suggests, social distancing means physical distancing, reducing close connections and reducing the spread of an infectious disease like coronavirus. The most effective non-pharmaceutical way to suppress the growth of a disease is social distancing. If people are not closer together, it cannot spread germs [9]. The guidelines for coronavirus is given in Fig. 2 where the minimum distance between pedestrians is 2 m or 6 ft and no group of pedestrians is allowed [3].

S. Prasanth Vaidya (✉) · M. Srinu
Aditya Engineering College, Surampalem, AP, India
e-mail: vaidya269@gmail.com

M. Srinu
e-mail: srinu.marni@aec.edu.in

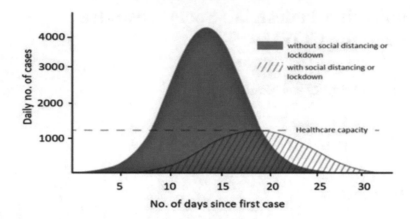

Fig. 1 Consequence of social distancing with number of cases blending with accessible healthcare capacity

Fig. 2 Physical distance guidelines for coronavirus

During the pandemic, a quantitative framework is constructed to investigate how individuals trade off the utility benefit of social participation against the internal and external health hazards that come with social connections [18]. In this aspect, our main goal is to estimate the distance between the individuals exposed to the public places. It is classified into three modules i.e., for the detection of object, to detect the type of object, and to calculate and give the distance between the objects [14, 17].

With this, the object is said to maintain the minimum and estimated distance so as to reduce the pandemic effects.

The literature survey is given in Sect. 2 followed by methods used in Sect. 3 and Sect. 4 provides proposed method, Sect. 5 provides result analysis and finally conclusion in Sect. 6.

2 Literature Survey

Punn et al. [12] presented a Deep Learning (DL) schema for computerized social distance supervising by means of monitoring images. The framework proposed utilizes YOLO v3 model for human detection to monitor individuals identified using bounding boxes and assigned identifications. Ahmad et al. [1] targets on recognising human in ROI using object tracking model and iOpenCV library. Distance is calculated among the people by comparing the values of fixed pixels. In segmented tracking area, the distance between central points and the boundary between individuals is measured. In order to keep the distance safe, unsafe distances between the persons can be detected. Nadikattu et al. [10] proposes an innovative method of tracking the people position in a sensor-based outdoor environment. They developed the concept for novel device in these COVID-19 environments. If someone is within six feet of the person, the device will give alert. The method is precise and can help to keep social distance. The detection tool has been developed to alert people by analyzing a video feed to keep a safe distance from others. For pedestrian detection, the camera video frame was used as a source and a pre-trained YOLOv3 based open-source object recognition model was used. citenaveen2021social. Saponara et al. [15] implemented new deep learning technique using YOLOv2 approach to detect and track human in different scenarios is developed. It is also utilized to measure and classify the distance among pedestrians, as well as to verify automatically that rules of social distancing are followed or not. As a result, the goal of this study is to see if and how people follow rules of social distancing to reduce the growth of virus. Hou et al. [6] presented a paper for detecting pedestrian using deep learning model. This model presents an alert system for pedestrians about social distancing. Shao et al. [16] presented a paper in pedestrian detection from UAV images.

3 Technologies Used

3.1 Python

Python is a programming language that is general, dynamic, high-level and interpreted. It supports object-based programming approach to application development. The python library Numpy, TKinter is used.

NumPy. NumPy is the Numerical Python for array work. It also works in the field of linear algebra and matrices. We can use it freely, and it is an open source project. NumPy aims to supply an array object up to 50 times faster than the Python lists [11]. Numpy is an array processing package for general purposes. It provides a multidimensional object for high performance arrays and the tools required for these arrays. Numpy can also be used as an efficient multidimensional container for generic data in addition to its obvious scientific applications [4].

3.2 OpenCV

OpenCV is the software library for Open Source and Machine Learning. In the object detection function, people are detected by using an object detector in video streams. The implementation is kept smooth and orderly by manipulating frames in this configuration [2].

3.3 YOLOv3

YOLOv3 is a regression algorithm that predicts the classes and bounding boxes for the input frame for one round of the algorithm, instead of selecting the interesting area of the image [13]. YOLOv3 is the most recent version of the YOLO—You Only Look Once popular object detection technique. The model published detects 80 distinct subjects, but above all, it is extremely quick and almost as accurate as Single Shot MultiBox (SSD). YOLO algorithm works using Residual blocks and Bounding box regression.

Residual Blocks. The first step is to split the image into different grids. The size of each grid is $S \times S$. The Fig. 3 shows the division into grids of an entry sample image with red color.

Bounding Box Regression. A bounding box is a contour that shows an object in an image. The following characteristics contain each bounding box in the image: Height (bh), Width (bw), Class (for example, person, car, traffic light and so on), Center of Bounding Box (bx, by).

4 Propounded Monitoring Pedestrian Scheme

In this section, the propounded system implementation in terms of modules is described. The proposed Architecture is provided in Fig. 4. The detail steps are provided in the following subsections.

Fig. 3 Forming residual blocks using YOLOv3 for sample image

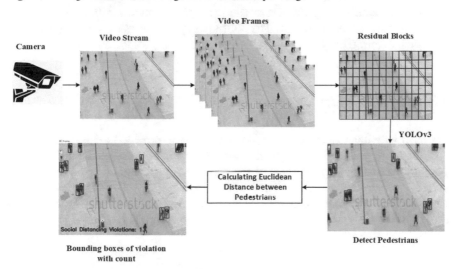

Fig. 4 Monitoring pedestrian social distance system for Covid-19

4.1 Camera Perspective Transformation

The first step is to convert the input video from an arbitrary viewpoint to the view required. The most straightforward way of transformation, since the input frames are monocular (shot from one camera), consists of choosing four places in the perspective view, in which the user monitors and maps social distance at the corners. These sites in the real world should also create parallel lines. This assumes that everyone is on the same flat floor.

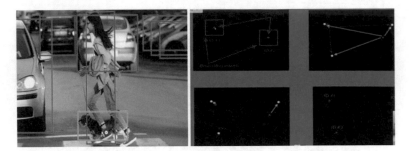

Fig. 5 Pedestrian detection and tracking

4.2 Pedestrian Detection and Tracking

In pedestrian detection, YOLO mechanism is used, which is trained on the coco dataset that identifies the person class. The next step is to track and draw the boundary box around the observed object. The YOLOv3 system is based on the pedestrian detection system. The pedestrian detection and tracking is shown in Fig. 5.

4.3 Distance Calculation

For each individual in the frame, a bounding box is provided to approximate the position of the person. i.e., as a human location, taking the centroid boundary box. Then by estimating (x, y), in view of the bottom of the bounding box of each individual, the view is determined. The last step is to calculate the distance and scale the distance between each couple of pedestrians. The Euclidean distance is given in Eq. 1.

$$d(p, q) = \sqrt{\sum_{i=1}^{n} (q_i - p_i)^2} \tag{1}$$

For example two points (1, 2) and (4, 6) are taken, then the Euclidean distance between those two points are 5.

4.4 Distance Violation with Count

In this final step, after calculating pedestrian detection with distance between then using Euclidean distance, the violation count is calculated from the distance metrics between pedestrians. The distance violation is shown in red boxes and who are following the distance are shown with green bounding boxes. The number of red boxes count is also provided for analysis who are violating the count for further measures in overcoming the social distance violations.

5   Testing and Analysis

In this section, the testing and output analysis of the proposed method is discussed. Here the sample videos of size $640 \times 480 \times 3$ are considered as shown in Fig. 6.

The output of the sample videos with pedestrian detection following social distance are provided with green bounding box where as the pedestrian who are violating the social distance are given with red bounding box with number of violating in each and every frame. The output frames of the sample videos with bounding boxes of red and green with count is show in Fig. 7. For the sample videos, time of the video, number of violations in the video and accuracy is provided in the Table 1.

For further analysis sample video (a) is considered where thirteen random frames with detection of pedestrian violations count is shown in Fig. 8.

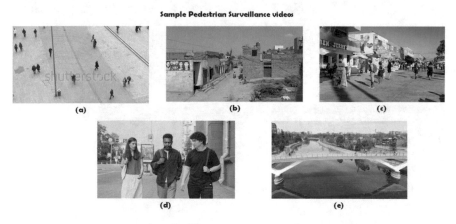

Sample Pedestrian Surveillance videos

(a)   (b)   (c)

(d)   (e)

Fig. 6 Sample pedestrian surveillance videos

Table 1 Videos data with time, violations count and accuracy

Videos	Time (Seconds)	Violations count	Accuracy (%)
a	29	11	100
b	9	0	100
c	14	11	90
d	9	3	100
e	6	2	100

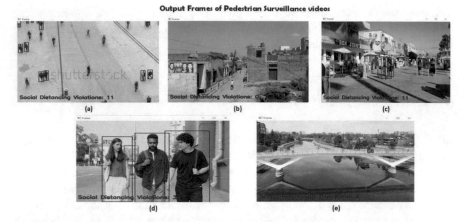

Fig. 7 Output frames of pedestrian surveillance videos

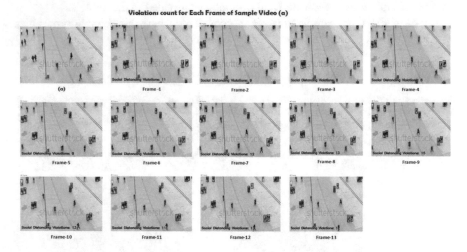

Fig. 8 Violations count for random frames of sample video **a**

6 Conclusion

For pedestrian detection, pre-trained YOLOv3 paradigm is used. The pair way centre distances between identified boundary boxes are measured using the Euclidean distance. The resemblance of gross length is employed to control social distance violations among people. Experimental results have shown that the schema determines pedestrians who are too confined and breach social separation efficiently. It is an effective real-time framework in which the social separation monitoring procedure via object detection and tracking methodologies is automated, where each participant is identified using bounding boxes in real time. The boundary boxes generated help to locate the clusters or groups of persons that satisfy the proximity property computed using pairs.

The future scope of the paper is to analyse the number of pedestrians who are having the chances of getting affected with covid-19 based on predictions and also to calculate the temperature of the pedestrian.

References

1. Ahamad AH, Zaini N, Latip MFA (2020) Person detection for social distancing and safety violation alert based on segmented ROI. In: 2020 10th IEEE international conference on control system, computing and engineering (ICCSCE). IEEE, pp 113–118
2. Bradski G, Kaehler A (2000) OpenCV. Dr. Dobb's J Softw Tools 3
3. Dreamstime: social distance poster (2018). https://www.dreamstime.com/social-distancing-en-image184286069
4. Harris CR, Millman KJ, van der Walt SJ, Gommers R, Virtanen P, Cournapeau D, Wieser E, Taylor J, Berg S, Smith NJ et al (2020) Array programming with NumPy. Nature 585(7825):357–362
5. Harris R (2020) Face Covid. How to respond effectively to the Corona crisis. https://drive.google.com/filc/d/1MZJybtT9KmiE9Dw9EKvPJsd9Ow7gXaMe/view. Accessed 20
6. Hou YC, Baharuddin MZ, Yussof S, Dzulkifly S (2020) Social distancing detection with deep learning model. In: 2020 8th international conference on information technology and multimedia (ICIMU). IEEE, pp 334–338
7. Kishore D, Rao CS (2021) Quaternion polar complex exponential transform and local binary pattern-based fusion features for content-based image retrieval. In: Microelectronics, electromagnetics and telecommunications. Springer, pp 769–776
8. Li Y, Guo F, Cao Y, Li L, Guo Y (2020) Insight into Covid-2019 for pediatricians. Pediatr Pulmonol 55(5):E1–E4
9. Liu W, Liao S, Ren W, Hu W, Yu Y (2019) High-level semantic feature detection: a new perspective for pedestrian detection. In: Proceedings of the IEEE/CVF conference on computer vision and pattern recognition, pp 5187–5196
10. Nadikattu RR, Mohammad SM, Whig D, et al (2020) Novel economical social distancing smart device for Covid19. Int J Electr Eng Technol 11(4)
11. Oliphant TE (2006) A guide to NumPy, vol 1. Trelgol Publishing USA
12. Punn NS, Sonbhadra SK, Agarwal S, Rai G (2020) Monitoring Covid-19 social distancing with person detection and tracking via fine-tuned yolo v3 and DeepSort techniques. arXiv preprint arXiv:2005.01385
13. Redmon J, Farhadi A (2018) YOLOv3: an incremental improvement. arXiv preprint arXiv:1804.02767
14. Sanivarapu PV (2021) Multi-face recognition using CNN for attendance system. In: Machine learning for predictive analysis. Springer, pp 313–320
15. Saponara S, Elhanashi A, Gagliardi A (2021) Implementing a real-time, AI-based, people detection and social distancing measuring system for Covid-19. J Real-Time Image Process 1–11
16. Shao Z, Cheng G, Ma J, Wang Z, Wang J, Li D (2021) Real-time and accurate UAV pedestrian detection for social distancing monitoring in Covid-19 pandemic. IEEE Trans Multimedia
17. Singh A, Vaidya SP (2019) Automated parking management system for identifying vehicle number plate. Indones J Electr Eng Comput Sci 13(1):77–84

18. Soundrapandiyan R, Mouli PC (2015) Adaptive pedestrian detection in infrared images using background subtraction and local thresholding. Procedia Comput Sci 58:706–713
19. Vaidya SP, Mouli PC (2020) A robust and blind watermarking for color videos using redundant wavelet domain and SVD. In: Smart computing paradigms: new progresses and challenges. Springer, pp 11–17

A Study and Comparative Analysis on Different Techniques Used for Predicting Type 2 Diabetes Mellitus

Middha Karuna and Agrawal Shilpy

1 Introduction

Diabetes is a long-lasting disease instigated due to the presence of the high glucose level in the blood. Millions of people all over world are affected by this lifestyle related health issue. It is caused due to Sedentary lifestyle and high BMI. This chronic disease can be identified by conducting several chemical and physical tests. As per IDF, ~425 million people are currently living with diabetes, the number will reach ~629 million by year 2045 [1]. Death Rate to rise by 25% in next decade.

Frequent urination, hunger and urge to drink more water are some of the symptoms indicating increasing levels of glucose. Serious issues involving diabetes have been on a constant rise and it may lead to irreparable loss in the body. This may lead to lifetime diseases such as cardiovascular ailment, eye complications, kidney related issues, and ulcers in the foot etc. [2].

1.1 Classification of Diabetes Mellitus: There Are Primarily Three Types of Known Diabetes Mellitus

Type-1 Diabetes Mellitus (T1DM), a cause often recognized as insulin deficiency happens when the body is not able to generate enough insulin. Mistaken destruction of the insulin making cells often termed as beta cells by the immune system, thinking them as intruders may lead to rise in blood sugar levels.

Type-2 Diabetes Mellitus (T2DM), popularly known as T2DM is a condition which is little different from Type 1. In this condition there is no shortage of insulin generation in the body, rather the ability of the body to use it efficiently goes down

M. Karuna (✉) · A. Shilpy
School of Engineering and Sciences, G D Goenka University, Gurugram, Haryana 122103, India
e-mail: karuna114@gmail.com

© The Author(s), under exclusive license to Springer Nature Singapore Pte Ltd. 2022   225
B. Unhelker et al. (eds.), *Applications of Artificial Intelligence and Machine Learning*,
Lecture Notes in Electrical Engineering 925,
https://doi.org/10.1007/978-981-19-4831-2_19

and this leads to increase in glucose levels. This increase, over a period of time takes a shape of abnormal functioning of the pancreas.

Gestational Diabetes - A short term condition at the time of pregnancy, that usually doesn't last long after delivery but still can take the shape of a life-threatening Type 2 Diabetes, is usually known as Gestational Diabetes. A possible and timely control can be adopted with the help of medications, regular workout and a healthy and balanced diet.

1.2   Possible Reasons for Diabetes

Genetics and external environmental factors are key contributors causing diabetes.

1.   Obesity is one the most critical factor that can cause diabetes.
2.   Passing of traits to the offspring's is another popular cause behind diabetes.
3.   Consistent high blood sugar levels increase the risk of diabetes.
4.   With age the body becomes more susceptible to the risk of diabetes.

1.3   Possible Problems Due to Diabetes:

a.   Heart diseases: Diabetes vividly increases the risk of various cardiovascular problems.
b.   Eyes or vision related complications (Retinopathy).
c.   Hearing loss or disability.
d.   Kidney related complications (Nephrotic syndrome).
e.   Skin disorders.
f.   Nerves related impairments.

The remaining paper is further divided into 5 sections. Section 2 covers the existing predictive analysis techniques. Section 3 includes tabular comparison of various predictive analysis techniques used in diabetes detection. Section 4 lists the advantages and disadvantages of various techniques employed in diabetes detection and prediction. Section 5 concludes the paper and touches upon the future scope. References are listed in the end.

2 Existing Predictive Analysis Techniques

2.1 Machine Learning (ML) Algorithms can be broadly classified as

a. **Supervised Learning**

The term supervised means that the learning process is being supervised or directed by the datasets. The datasets are labelled in terms of inputs and desired output. So, the machine is learning from the fact that the labelled inputs would yield a labelled output. The volume and quality of data used to train and test is very critical in this supervised learning process. Techniques like regression and various classifications are majorly used in Supervised learning. e.g., Logistic Regression, Support Vector Machine and K Nearest Neighbors algorithm etc.

Some of the commonly used supervised machine learning techniques in diabetes prediction are described below.

- **Support Vector Machine (SVM)** – It is a technique that represents number of attributes in a dataset as vectors in a hyperplane or set of hyperplanes. The classification objective of the technique is achieved by the choosing the hyperplane that has the largest distance to the trained data point of any class. The lower the margin the minimal the generalization error [4].
- **Logistic Regression (LR)** - It is a statistics-based technique, which utilizes a function that converts log-odds to probability to model a binary dependent variable. This model represents a dependent variable value with true or false, generally labeled "0" and "1". This technique is used when we must figure out the correlation between an independent variable and contrary dependent variable. This is contrasting to linear regression, where the dependent variable is more continuous in nature [5].
- **K-Nearest Neighbors (KNN)** - In KNN, the goal is to find the distance between query and other data points in a sample. The (K) closest examples to the query has been picked in case of classification to pick the most frequent label. In case of regression the average of the labels is generally picked. This technique is also popularly known as lazy learning.
- **Random Forest (RF)** - A classification technique which is most widely used in reducing the dimensionality of the dataset. It is a classifier that is generally comprised of a forest (few decision trees). It is one of the most contributing technique and is often used as part of the collection in order solve problems related to classification and regression. It efficiently grades the variable as per the order of importance [4].
- **Decision Tree- (ID3)** - The technique is simple to implement and starts with construction of tree like structure based on the input attributes. The traversal from root to the leaf is done as a decision-making step and the it can only be done until the conditions are satisfied [4]. The technique doesn't demand any

prior domain knowledge. This technique is vulnerable to large datasets due to the increase in complexity.

- **Naïve Bayes (NB)** - It is a probability driven classification technique, means the core basis of prediction is a feasibility of an outcome. Solves multi class problems with ease, and the implementation is biased towards assumptions. If the assumptions fall into the right space than the performance significantly improves and is immense compared to other models. Mostly suitable for categorical data, it disappoints when dealing with numerical data.

b. **Unsupervised Learning**

Datasets involved in the learning process are not labelled and thereby the logic or pattern is being inferred by the machine. The hidden patterns are unfolded using techniques like clustering, attribute derivation, finding anomalies, reducing dimensionalities.

2.2 Deep Learning (DL) Algorithm

These algorithms are strong-growing and it works quite very much like a human mind. Deep learning algorithms are based on functioning of neural network. Research [3] shows that different types of problems may require multilevel information. As per the proven research work, Deep learning techniques not just improves the overall outcome but also helps in reducing the rate of classification error. The issues associated with computation and bad data (noise) has also been effectively handled by deep learning techniques. Techniques associated with DL are Artificial Neural Network, Deep Neural Network, Recurrent Neural Network, Probabilistic Neural Network.

Some of the commonly used deep learning techniques in diabetes prediction are described below.

- **Artificial Neural Network (ANN)** - ANN is a system inspired from the human brain functioning. It mimics the functioning of biological neural network. It is comprised of three layers. The first layer or the input layer is an entry point for the data and is responsible for passing the data to the network. The hidden layer generally performs some computations. The output layer represents the unique classes/outcome [4].
- **Deep Neural Network (DNN)** - DNN is basically an extension of ANN structure and its existence is primarily attributed to the availability of infrastructure and compute power. It is a feed forward style of network which always works in a forward direction. It can handle complex and huge dimensions of data set, for example it can handle feature extraction and representation millions of attributes [2].
- **Recurrent Neural Network (RNN)** – RNN is a specialized type of ANN that only work with sequential data vectors. It is the ability to preserve the information that makes it suitable for handling complex solutions like voice assistants. It encounters

vanishing gradients problem which leads to issues with longer sequences of data processing [6].

- **Probabilistic Neural Network (PNN)** - It is a feed forward type of ANN, which is purely a derivation of statistical algorithm and Bayesian Network, mainly used for classification and pattern recognition problems. It is slower than typical ANN setup and usually need more memory space for model storage which makes it a tad slower as well.
- **Genetic Algorithm (GA)** – It is an adaptive search driven optimization technique which is inspired from genome based evolutionary concepts. It is generally used to compute true solutions to problems around optimization and search.
- **Long short -term memory (LSTM)** - LSTM solves most of the issues encountered by RNN but at a computational time cost. The memory blocks availability and processing generally take more time and that is the reason why it is becomes a hard choice to train the data. It is popularly used in the field of natural language processing (NLP) [6].

2.3   Nature-Inspired Algorithm

Human beings, birds, animals and all living creatures evolve in order to overcome a problem and to survive. Nature has always been a great tutor for human beings along with all other living creatures. There are many such nature-based phenomenon that became a baseline for solving various issues. The algorithms mainly inspired from such nature-based phenomenon are primarily used to solve search and optimization related problems. It is hard to classify the nature-based algorithms into different types but still at a high level we can categorize it as Evolution-based algorithms and Swarm Intelligence based algorithms. Genetic Algorithm is one of the examples of evolution-based algorithm and Ant Colony Algorithm can be a good example of swarm intelligence-based algorithm.

3   Performance Analysis of Various Predictive Analysis Techniques

Various independent techniques and hybrid models are used in prediction of Type 2 Diabetes Mellitus.

Table 1 briefly covers the Machine learning techniques employed in the past and the corresponding accuracy yielded using these techniques in diabetes prediction. It lists down the Machine learning techniques applied over the last decade and concludes at high level that Random Forest algorithm works with significantly higher rate of accuracy compared to other Machine learning techniques.

Table 2 briefly covers the Deep learning techniques employed in the past and the corresponding accuracy yielded using these techniques in diabetes prediction. It lists

Table 1 Machine learning techniques

Author/Reference	Year	Methods	Accuracy
Kumari VA and Chitra R [7]	2013	Support Vector Machine (SVM)	78%
Kandhasamy JP and Balamurali S [8]	2015	J48 K-Nearest Neighbors (KNN) Random Forest SVM	73.82% 70.18% (k* = 1), 72.65% (k* = 3), 73.17% (k* = 5) 71.74% 73.34%
Iyer A et al. [9]	2015	J48 Naïve Bayes	76.9% 79.6%
Yuvaraj N and SriPreethaa KR [10]	2017	Random Forest Decision Tree- (ID3) Naïve Bayes	94.0% 88.0% 91.0%
Sisodia D and Sisodia DS [11]	2018	Decision Tree Naïve Bayes SVM	73.82% 76.30% 65.10%
Tigga NP and Garg S. [12]	2019	Logistic Regression SVM Decision Tree Naïve Bayes Random Forest K-NN	74.4% 74.4% 69.7% 68.9% 75.0% 70.8%
Reddy DJ et al. [13]	2020	LR SVM KNN RF NB GB	80.64% 79.15% 87.61% 98.48% 77.34% 87.31%

k*, refers to cross-validation value

down the Deep learning techniques applied over the last decade and at high level concludes that Deep Neural Network method works with significantly higher rate of accuracy compared to corresponding Deep learning techniques.

Table 3 briefly covers the hybrid models employed in the past and the corresponding accuracy yielded using these models in diabetes prediction.

4   Comparison of Techniques Employed for Diabetes Prediction

This section covers the advantages and disadvantages associated with various techniques employed for diabetes prediction. There are over 100 plus Machine learning and Deep learning techniques but this section briefly touches upon only the popular ones employed for diabetes prediction (Table 4).

Table 2 Deep learning techniques

Author/Reference	Year	Methods	Accuracy
Ebenezer OO and, Khashman A [14]	2014	Multilayer ANN	82%
Soltani Z and Jafarian A [15]	2016	Probabilistic Neural Network (PNN)	Training 89.56% Testing 81.49%
Ashiquzzaman A et al. [16]	2017	DNN, with dropout	88.41%
Vijayashree J and Jayashree J [17]	2017	PCA, DNN and ANN	82.67% (DNN with RFE) 76.77% (DNN with PCA) 78.62% (ANN with RFE) 70% (ANN with PCA)
Ayon SI and Islam M. [3]	2019	Deep Neural Network	98.35% k* = 5 97.1% k* = 10
Spänig S et al. [18]	2019	DMLP	84%
Ryu KS et al. [19]	2020	2 hidden layer DMLP with dropout	80.11%

Table 3 Hybrid models

Author/Reference	Year	Methods	Accuracy
Aslam MW et al. [20]	2013	Genetics + KNN (GPKNN), Genetics + SVM (GPSVM)	80.5% 87.0%
Varma KVSRP et al. [21]	2014	Decision tree + Gaussian	75%
Choubey DK and Paul S [22]	2015	Genetic Algorithm (GA), J48graft Decision Tree (J48graft DT)	74.78%
Gill NS and Mittal P [23]	2016	Support Vector Machine + Neural Network	96.09%
Choubey DK and Paul S. [24]	2017	RBF NN GA_RBFNN	76.08% 77.39%
Wu H et al. [25]	2018	KNN + LR	95.42%
Swapna G et al. [26]	2018	Deep Learning (CNN-LSTM with SVM)	95.7%
Ghosh SK and Ghosh A. rc4.5 [27]	2019	Clustering based analysis -Intuitionistic Fuzzy Set and Multigranulation Rough Set Model	90.9%
Kannadasan K et al. [28]	2019	DNN framework using stacked autoencoders and SoftMax layer	86.26%
Rc5.8 Saleh Albahli [29]	2020	K means clustering, Random Forest, XG Boost and Logistic Regression	97.5% (k* = 10)

k*, refers to cross-validation value

Table 4 Advantages and disadvantages of various techniques

Techniques	Advantages	Disadvantages
K-Nearest Neighbors (k-NN)	• Easy implementation model • Faster and efficient training process	• Disappointing with huge date as time is not a liberty in all scenarios • Humongous storage space required • Representation of knowledge underneath is poor
Random forest	• Good accurate classifier • Works well with larger dimensions (X * Y)	• Trains fast but predicts slow • Slow and sluggish to evaluate • Hard to interpret outcome
Decision tree	• Easy and understandable presentation • Effective with numerical and categorical data • Minimum preprocessing effort and assumptions required	• Predictions can change immensely even with the slightest of change in data • Vulnerable to large datasets due to the increase in complexity • Unbalanced classifier may yield to unqualified output
Support Vector Machine (SVM)	• Equally efficient with semi-structured and unstructured datasets. Mainly covers pictures and text • Accomplish precise and robust output • Resolves overfitting issues with ease	• Longer training duration with large datasets • Kernel selection crucial function and can disrupt the classification decisions • Variable weight interpretation most difficult aspect
Naïve Bayes approach	• Handles missing values efficiently • Categorical in nature • Works well even with small sample size	• Works well with independent attributes only • Drop in accuracy, based on dependencies amongst attributes • Strong data distribution assumptions can be often very costly
Artificial neural network	• Efficient extraction of features from irrelevant data • Effective with nonlinear and complex associations	• Hidden layers lead to overfitting issues • Weight initialization strategies not always plausible • Only applicable to • Numerical data
Logistic regression	• Easy implementation alongside efficient training • Handles non linearity effectively	• Continuous outcome prediction ineffective • Vulnerable to overfitting issues and inaccurate classification rates • Stable results with large sample size

(continued)

Table 4 (continued)

Techniques	Advantages	Disadvantages
CNN	• Automatic extraction of attributes • Less processing time	• Very high compute cost • Need large amount of data for effective outcome
MLP	• Solve classification and regression problems effectively • Works well on different types of datasets like text and images	• Inadequate in modern and advanced fields • May need very high dimensions of data
RNN	• Effective memorizing capabilities used for generating high accuracy rates • Handles complex problems effectively	• Encounters diminishing gradients problem • Weights and bias calculation a challenging task and may impact the accuracy rate
LTSM	• Solves vanishing gradient problem • Heavily applicable in Natural Language Processing applications	• Demand very high compute duration as the operations is storage intensive • Hard to train data
Neural network	• High fault tolerance alongside high processing speed	• High processing time is not always a justified decision • Only applicable in selective scenarios
Genetic Algorithm (GA)	• Parallel processing • Applicable in adaptive and evolutionary approach	• No assurance of global optima • Response time can change invariably

5 Conclusion and Future Scope

Diabetes mellitus, one of the deadliest and irreversible disease around the world. The severity of this disease makes the phenomenon of prevention and early detection very essential. Around 80% of the severe complications can be averted or significantly reduced with early detection and identification. Clinical efficiency of the detection process is vital and has been attempted to be addressed in the research work done so far.

This paper is an attempt to represent an inclusive study around research conducted so far using various machine learning techniques like Logistic Regression, Decision Tree, Random Forest etc. The scope of most of the research work has been attributed around a well-known publicly available dataset (**PIMA Indian Diabetic dataset**). The metrics emphasized in this comparison paper is mainly the rate of Accuracy achieved in various related studies. The popular classification methods along with their pros and cons have listed in a tabular manner in order to give researcher comprehensive view of the techniques utilized so far. The separate comparison table has also been included for advance deep learning classification techniques adopted so far.

Identification of different types of diabetes using a single classifier is something that can be looked up as future work as this has not been touched upon yet. This paper also aims to help the researchers to adopt an ensembled technique with an objective of achieving a better rate of accuracy for predicting Type 2 Diabetes.

The study mainly comprised of findings related to the medical conditions similar to diabetes and can be essentially enhanced to cover other medical conditions or diseases as well. Additionally, the focus is primarily around the misclassification but there is a huge potential towards accuracy and compute speed which needs to be tapped. Also, the approach to include different datasets in terms of sample size and dimensions would be a key to generalize the study and have a deeper impact. In the existing study a handful set of classifiers has been utilized to predict outcomes. The hybridization of techniques will also help to further stabilize the results and make the application more robust and flexible.

References

1. Lu Y, Li Y, Li G, Lu H (2020) Identification of potential markers for type 2 diabetes mellitus via bioinformatics analysis. Mol Med Rep 22(3):1868–1882. https://doi.org/10.3892/mmr.2020. 11281
2. Zhu T, Li K, Herrero P, Georgiou P (2020) Deep learning for diabetes: a systematic review. IEEE J Biomed Heal Inform 2194(c):1–14. https://doi.org/10.1109/JBHI.2020.3040225
3. Islam Ayon S, Milon Islam M (2019) Diabetes prediction: a deep learning approach. Int J Inf Eng Electron Bus 11(2):21–27. https://doi.org/10.5815/ijieeb.2019.02.03
4. Choudhury A, Gupta D (2019) A survey on medical diagnosis of diabetes using machine learning techniques. In: Kalita J, Balas V, Borah S, Pradhan R (eds.) Recent Developments in Machine Learning and Data Analytics. Advances in Intelligent Systems and Computing, vol 740. Springer, Singapore. https://doi.org/10.1007/978-981-13-1280-9_6
5. Kumar YJN, Kameswari Shalini N, Abhilash PK, Sandeep K, Indira D (2019) Prediction of diabetes using machine learning. Int J Innov Technol Explor Eng 8(7):2547–2551. https://doi. org/10.35940/ijrte.e6290.018520
6. Larabi-Marie-Sainte S, Aburahmah L, Almohaini R, Saba T (2019) Current techniques for diabetes prediction: review and case study. Appl Sci 9(21). https://doi.org/10.3390/app921 4604
7. Chitra K (2018) Classification of diabetes disease using support vector machine, vol 3, no 2, pp 1797–1801. https://www.researchgate.net/publication/320395340
8. Kandhasamy JP, Balamurali S (2015) Performance analysis of classifier models to predict diabetes mellitus. Procedia Comput Sci 47(C):45–51. https://doi.org/10.1016/j.procs.2015. 03.182
9. Iyer A, Jeyalatha S, Sumbaly R (2015) Diagnosis of diabetes using classification mining techniques. Int J Data Min Knowl Manag Process 5(1):01–14. https://doi.org/10.5121/ijdkp.2015. 5101
10. Yuvaraj N, SriPreethaa KR (2019) Diabetes prediction in healthcare systems using machine learning algorithms on Hadoop cluster. Cluster Comput 22. https://doi.org/10.1007/s10586-017-1532-x
11. Sisodia D, Sisodia DS (2018) Prediction of diabetes using classification algorithms. Procedia Comput Sci 132(Iccids):1578–1585. https://doi.org/10.1016/j.procs.2018.05.122
12. Tigga NP, Garg S (2020) Prediction of type 2 diabetes using machine learning classification methods. Procedia Comput Sci 167(2019):706–716. https://doi.org/10.1016/j.procs.2020. 03.336

13. Jashwanth Reddy D et al (2020) Predictive machine learning model for early detection and analysis of diabetes. Materials Today: Proceedings, no xxxx. https://doi.org/10.1016/j.matpr. 2020.09.522

14. Olaniyi EO, Adnan K (2014) Onset diabetes diagnosis using artificial neural network. Int J Sci Eng Res 5(10):754–759. http://www.ijser.org

15. Soltani Z, Jafarian A (2016) A new artificial neural networks approach for diagnosing diabetes disease type II. Int J Adv Comput Sci Appl 7(6):89–94. https://doi.org/10.14569/ijacsa.2016. 070611

16. Ashiquzzaman A et al (2018) Reduction of overfitting in diabetes prediction using deep learning neural network. In: Kim K, Kim H, Baek N (eds.) IT Convergence and Security 2017. LNEE, vol 449, pp 35–43. Springer, Singapore. https://doi.org/10.1007/978-981-10-6451-7_543

17. Vijayashree J, Jayashree J (2017) An expert system for the diagnosis of diabetic patients using deep neural networks and recursive feature elimination. Int J Civ Eng Technol 8(12):633–641

18. Spänig S, Emberger-Klein A, Sowa JP, Canbay A, Menrad K, Heider D (2019) The virtual doctor: an interactive clinical-decision-support system based on deep learning for non-invasive prediction of diabetes. Artif Intell Med 100(February):101706. https://doi.org/10.1016/j.art med.2019.101706

19. Ryu KS, Lee SW, Batbaatar E, Lee JW, Choi KS, Cha HS (2020) A deep learning model for estimation of patients with undiagnosed diabetes. Appl Sci 10(1). https://doi.org/10.3390/app 10010421

20. Aslam MW, Zhu Z, Nandi AK (2013) Feature generation using genetic programming with comparative partner selection for diabetes classification. Expert Syst Appl 40(13):5402–5412. https://doi.org/10.1016/j.eswa.2013.04.003

21. Varma KV, Rao AA, Lakshmi TS, Rao PN (2014) A computational intelligence approach for a better diagnosis of diabetic patients. Comput Electr Eng 40(5):1758–1765. https://doi.org/10. 1016/j.compeleceng.2013.07.003

22. Choubey DK, Paul S (2015) Ga_J48graft Dt: a hybrid intelligent system for diabetes disease diagnosis. Int J Bio-Sci Bio-Technol 7(5):135–150. https://doi.org/10.14257/ijbsbt.2015.7.5.13

23. Gill NS, Mittal P (2016) A computational hybrid model with two level classification using SVM and neural network for predicting the diabetes disease. J Theor Appl Inf Technol 87(1):1–10

24. Paul S, Choubey DK (2017) GA_RBF NN: a classification system for diabetes. Int J Biomed Eng Technol 23(1):71. https://doi.org/10.1504/ijbet.2017.10003045

25. Wu H, Yang S, Huang Z, He J, Wang X (2018) Type 2 diabetes mellitus prediction model based on data mining. Inform Med Unlocked 10(August 2017):100–107. https://doi.org/10.1016/j. imu.2017.12.006

26. Swapna G, Vinayakumar R, Soman KP (2018) Diabetes detection using deep learning algorithms. ICT Express 4(4):243–246. https://doi.org/10.1016/j.icte.2018.10.005

27. Ghosh SK, Ghosh A (2020) A novel clustering-based gene expression pattern analysis for human diabetes patients using intuitionistic fuzzy set and multigranulation rough set model, vol 1154

28. Kannadasan K, Edla DR, Kuppili V (2019) Type 2 diabetes data classification using stacked autoencoders in deep neural networks. Clin Epidemiol Glob Heal 7(4):530–535. https://doi. org/10.1016/j.cegh.2018.12.004

29. Albahli S (2020) Type 2 machine learning: an effective hybrid prediction model for early type 2 diabetes detection. J Med Imaging Heal Inform 10(5):1069–1075. https://doi.org/10.1166/ jmihi.2020.3000

RGB Based Secure Share Creation in Steganography with ECC and DNN

S. Ahmad⬛ and M. R. Abidi

1 Introduction

Sharing sensitive information over a public network, such as the Internet, may be risky because privacy and confidentiality aren't guaranteed. The work presented in this paper is mainly concerned with secret communication, specifically steganography and cryptography. Steganographic methods (approach by modification) eventually change the statistical characteristics of images during the embedding process. Such non-natural distortions can be observed and can lead to the indication of the presence of a secret image. Steganalysis refers to the techniques for detecting the presence of non-natural distortions in order to distinguish source data (cover) from media with a hidden message (known as stego). During steganalysis, it is impossible to discern the single secret hidden in the cover image. However, if the cover image contains multiple secret images, the generated cover image or stego image becomes vulnerable to steganalysis. The work presented here has used the cryptographic technique known as Elliptic Curve Cryptography (ECC) to vanquish the steganalysis attack in multiple hidden secrets. The algorithm flow of Elliptic Curve Cryptography is shown in Fig. 1.

2 Steganographic Techniques

Steganography is the process of hiding data within other file types. Steganographic embedders fall into three categories: frequency, spatial, and deep learning. Statistical techniques are used by frequency domain steganographers to hide information in an image's frequency coefficients [1]. Data is hidden in the raw pixel bits of an image using deep learning architectures like GANs [2, 3]. Deep learning steganographic

S. Ahmad (✉) · M. R. Abidi
Department of Electronics Engineering, Z.H.C.E.T., AMU Aligarh, Aligarh 202002, India
e-mail: sahmad19@gmail.com

© The Author(s), under exclusive license to Springer Nature Singapore Pte Ltd. 2022   237
B. Unhelker et al. (eds.), *Applications of Artificial Intelligence and Machine Learning*,
Lecture Notes in Electrical Engineering 925,
https://doi.org/10.1007/978-981-19-4831-2_20

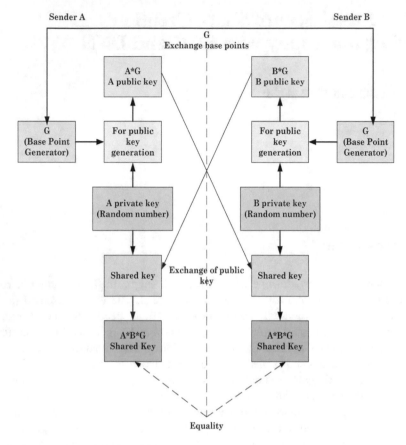

Fig. 1 A cryptosystem based on ECC

embedders [4] have proven to be very effective at concealing large amounts of data while avoiding detection by current steganalyzers. Many different network designs have been proposed, including SteganoGAN, HiDDeN, and BNet.

3 Elliptic Curve Cryptography [ECC]

In steganography, the hidden data is only known by the sender and recipient. It is impossible to discover the single secret hidden in the cover image during a steganalysis attack. To encrypt the image, Ali Soleymani [5, 6] presented an Elliptic Curve Cryptography-based cryptosystem. I. Yasser in [7] convert pixels to point (Xn, Yn) using Eq. (1):

$$X_n = nk + i$$

$$Y_n = \sqrt{X_n^3 + aX_n + b}, \qquad i = 0, 1, 2, 3 \ldots \tag{1}$$

If i does not solve the Eq. (1) for each pixel, the process will proceed until the first i does. This procedure, known as the Koblitz Method, is fully explained and implemented in [8]. F. Amounas in [9] suggested a new mapping technique based on matrix characteristics.

3.1 Proposed Model

The proposed method involves combining the principles of Kreuk [10] and Baluja [11]. It is intended for multi-image steganography, which involves concealing three or more images inside a single cover image.

1. **Preparation and Hiding network:** The secret image to be concealed is prepared by the preparation network. The container image is generated by the hidden network, which takes the output of the preparation and hidden networks as inputs. The input to this network is a $N \times N$ pixel region with the depth concatenated RGB channels from the cover picture and the converted channels of the hidden image.
2. **Revealed Network:** It acts as a decoder and is used by the receiver, the receiver only gets the container image neither the cover nor the secret image. To show the hidden image, the decoder network removes the cover image.
3. **Error Propagation:** The system is trained by lowering the following error where C reflects the cover, C' is container or stego image, S is original secret image S' is decoded hidden secrets respectively, and β is how to weigh their reconstruction errors as:

$$L(C, C', S, S') = \left\| C - C' \right\|^2 + \beta \left\| S - S' \right\|^2 \tag{2}$$

3.2 Implementation Specifications

The model makes use of the ADAM-Adaptive Moment Estimation optimizer [12], which determines a different learning rate for each parameter. It is computationally efficient and needs very little memory, making it ideal for the proposed model. The model was trained on the Linnaeus 5 dataset with images measuring 128×128 pixels. For the full model, the loss is expressed as:

$$LOSS = \beta_C \left\| C - C' \right\|^2 + \beta_S \left\| S_1 - S_1' \right\|^2 + \beta_S \left\| S_2 - S_2' \right\|^2 + \beta_S \left\| S_3 - S_3' \right\|^2$$

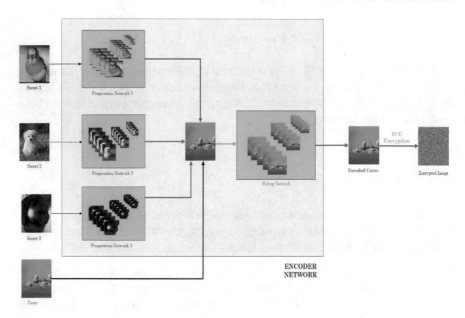

Fig. 2 The hiding network and ECC encryption

3.3 Modelling of Architecture

The architecture model is shown in Figs. 2 and 3.

3.4 Secure Share Creation Using RGB

The original RGB image's pixel values are retrieved and interpreted as a matrix $P \times Q$. Multiple shadows also called shares such as $shadow1$, $shadow2$,...$shadown$ are constructed and reconstructed using the extracted pixel values [13]. Figure 4 shows the block diagram of the RGB Based Secure Share Creation.

$$\mathbf{R} = \begin{bmatrix} 227 & 227 & 228 & 225 \\ 226 & 225 & 227 & 225 \\ 226 & 226 & 226 & 227 \\ 227 & 226 & 227 & 227 \end{bmatrix}$$

The R_{b1} and R_{b2} matrices are used to perform the following procedure before shadow construction.

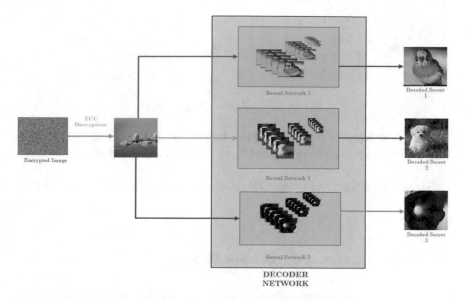

Fig. 3 The reveal network and ECC decryption

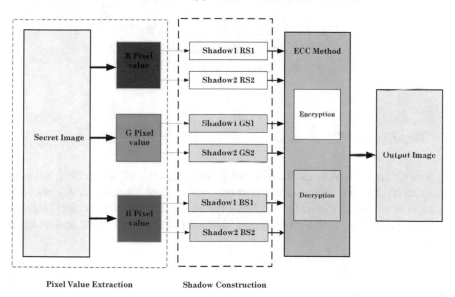

Fig. 4 Block diagram of RGB based secure share creation

$$R_{b1} = \begin{bmatrix} 113 & 113 & 114 & 112 \\ 113 & 112 & 113 & 112 \\ 113 & 113 & 113 & 113 \\ 113 & 113 & 113 & 113 \end{bmatrix}, \quad R_{b2} = \begin{bmatrix} 114 & 114 & 114 & 113 \\ 113 & 113 & 114 & 113 \\ 113 & 113 & 113 & 114 \\ 114 & 113 & 114 & 114 \end{bmatrix}$$

$$B_{R1} = 128 - R_{b1}, \quad B_{R2} = R_{b2}$$

The red component shadows are created by combining the basic and key matrices K_m (obtained by permutation of either R, G, B) with the XOR operation.

$$R_{S1} = B_{R1} \oplus K_m, \quad R_{S2} = B_{R1} \oplus B_{R2}, \quad R_{S3} = B_{R2} \oplus R_{S1}, \quad R_{S4} = R_{S1} \oplus R$$

3.5 Reconstruction of the Shadow Images

Multiple shadows are restored using simple XOR operations to obtain the original image in the shadow reconstruction process. That is to say,

$$R = K_m \oplus R_{S4} \oplus R_{S3} \oplus R_{S2} \oplus R_{S1}$$
$$G = K_m \oplus G_{S4} \oplus G_{S3} \oplus G_{S2} \oplus G_{S1}$$
$$B = K_m \oplus B_{S4} \oplus B_{S3} \oplus B_{S2} \oplus B_{S1}$$

4 Results and Discussion

In this section, we will see how the pre-processing network and hidden network function of the deep convolutional network works. How we generate the encoded cover shares and implement the ECC encryption and decryption operation on them. We use the Structural Similarity (SSIM) index of the original cover and encoded cover.

4.1 Results for Multiple Hidden Secrets

Secret images are passed through different pre-processing networks so that enough padding is added to each Conv2D layer so as to keep the output image in the same dimension as of cover image. The output of the hiding network called the encoded cover image or container image or stego image is shown in Fig. 6 (Fig. 5).

| (a) Secret-1 | (b) Secret-2 | (c) Secret-3 | (d) Cover |

Fig. 5 Multiple secret and cover images

Fig. 6 Encoded cover image

4.2 Splitting of Channels

The ECC encryption of an encoded cover image is accomplished by splitting it into its individual channels, i.e. red, green, and blue as described in Fig. 7.

Fig. 7 Red, Green and Blue channels of encoded cover image

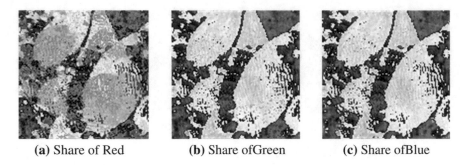

(a) Share of Red   (b) Share ofGreen   (c) Share ofBlue

Fig. 8 Shares of the channels

4.3 Creating Shares or Shadows of the Corresponding Channels

The total number of shares generated in this work is four for each channel, and only few shares are being demonstrated in Fig. 8.

4.4 Encrypting the Generated Shares

The encryption of shares shown in Fig. 9 is carried out using elliptic curve cryptography, with $a = 1$, $b = 3$, and $p = 257$ as the prime number. We can pick an affine point or generator point G from either of the 264 points.

$$y^2 = x^3 + x + 3 \, mod \, 257 \tag{3}$$

If we take the value of random integer $k = 10$, Bob's private key $y = 5$ and generator point $G = (2, 28)$ then Bob's public key is given by:

$$P_B = yG$$

Fig. 9 Encrypted shares

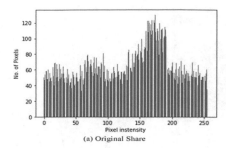
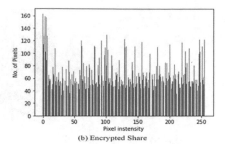

(a) Original Share   (b) Encrypted Share

Fig. 10 Histogram comparison of original and encrypted shares

$$P_B = 5 \times (2, 28) \tag{4}$$

A pair of points shown in Eq. (5) is referred to as P_C:

$$P_C = [(kG), (P_M + kK_B)] \tag{5}$$

4.5 Comparison of Histogram Between the Original Shares and the Encrypted Shares

Figure 10 shows the histograms of the original shares and encrypted shares, as we can see that their histograms are entirely different from each other. Thus we conclude that there is no significant relationship between them and the original shares and encrypted shares are different images.

4.6 Decryption and Stacking of Shares

PC is sent as a cipher message from Alice to Bob. As Bob receives the encrypted message PC and using his private key, y, he multiplies it by kG and adds the second point in the encrypted message to compute M, which corresponds to the plaintext message M as described by the Eq. (6).

$$P_M = (P_M + kK_B) - [y((kG))] \tag{6}$$

We decrypt all the shares of red, green, and blue channels using ECC decryption and stacked them together to get the encoded cover image shown in Fig. 11.

Fig. 11 Encoded cover image

| **(a)** Secret-1 | **(b)** Secret-2 | **(c)** Secret-3 |

Fig. 12 Decoded secret images

Fig. 13 Full model result of multiple hidden secrets

4.7 Revealed Network

After passing the encoded cover image through the revealed network, the embedded secret images are extracted from the encoded cover image. The results are shown in Fig. 12. Fig. 13 shows the full model result of the multiple hidden secrets (Table 1).

4.8 SSIM of Cover, Container, Secret and Decoded Secret Images

Figure 14 shows the similarity index of the cover image and container or encoded image and also the similarity index of the original secret and the decoded secret.

Table 1 The PSNR and SSIM value of encoded images and decoded secret images

	Encoded (SSIM)	Encoded (PSNR)	Decoded Secret (SSIM)	Decoded Secret (PSNR)
Baluja's model	0.92	28.41	0.92	28.06
HIGAN	0.94	30.95	0.94	29.67
Daun's model	0.95	36.71	0.96	36.97
Proposed model	0.99	32.74	0.99	34.52

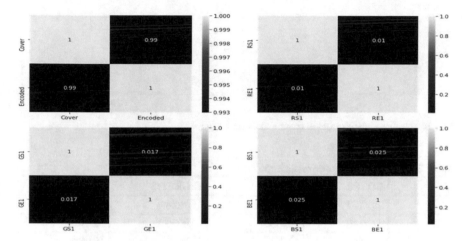

Fig. 14 SSIM of multiple hidden secrets

The distribution of error in each pixel of both the cover and secret images after the encoding and decoding procedure is shown in Fig. 15 for the most appropriate value of β, i.e. 1.

Table 2 shows the pixel-wise average errors on a scale of 256 for different values of reconstruction error β. The errors in cover and secret images is minimum for β = 1.

4.9 Complexity of Computation

The RGB Based Secure Share Creation's computational complexity for the encryption and decryption of the secret image is analysed and compared to existing approaches. The amount of shares involved has a direct relationship with the computational complexity. If there are n shares concealing the secret of the input original color image, then our proposed scheme's complexity will be $O(n)$. Table 3 shows a comparison of the computational complexity of certain existing state-of-the-art methodologies with the proposed technique.

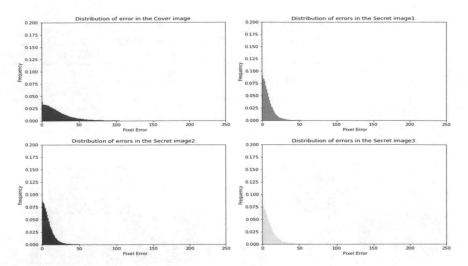

Fig. 15 Distribution of errors in cover and secret images

Table 2 Pixel-wise average errors in a 256 scale

β	COVER	SECRET 1	SECRET 2	SECRET 3
1	28.7727	11.6339	11.2826	14.0695
0.75	32.7146	12.3792	17.8990	16.4168
0.25	41.1467	13.2736	11.8664	19.9658

Table 3 Performance of previous schemes in comparison

Schemes	Computational complexity
Lin and Tsai (2004)	$O(n log^2 n)$
Yang (2004)	$O(1)$
Yang et al. (2007)	$O(n log^2 n)$
Shyu (2009)	$O(1)$
Wu and Sun (2013)	$O(n)$
Gu et al. (2014)	$O(1)$
Proposed scheme	$O(n)$

5 Conclusion

The work presented here is an innovative deep learning approach for image steganography that shows how existing image steganography algorithms are frequently poor in terms of payload ability. Multiple hidden secrets is a proposed technique that produces an end-to-end mapping from the cover image, hidden image to embedded image, then to the encrypted image, and then from the encrypted image to the decoded image. It outperforms conventional approaches while remaining highly resilient. Watermarking or a more secure central storage technique for passwords or key operations are just two of the many reasons to employ this method of data concealment. Regardless, the technology is simple to use and undetectable. The more you understand its features and functionality, the more versatile you will be. And if we talk about the future work if lossless neural networks are achieved in the future, we will be able to implement them with modern cryptographic techniques. Building cryptographic algorithms that can work with neural networks is also a promising potential area to examine. Furthermore, combining neural networks with image steganography opens up a wide range of possibilities in this area.

References

1. Li B, He J, Huang J, Shi YQ (2011) A survey on image steganography and steganalysis. J Inf Hiding Multimed Signal Process 2(2):142–172. 10.1.1.648.4107
2. Mishra R, Bhanodiya P (2015) A review on steganography and cryptography. In: 2015 International Conference on Advances in Computer Engineering and Applications, pp 119–122. https://doi.org/10.1109/ICACEA.2015.7164679
3. Zhang KA, Cuesta-Infante A, Xu L, Veeramachaneni K (2019) SteganoGAN: high capacity image steganography with GANs. arXiv:1901.03892 [cs, stat]
4. Heaton J (2018) Ian goodfellow, yoshua bengio, and aaron courville: deep learning. Genet Program Evol Mach 19:305–307. https://doi.org/10.1007/s10710-017-9314-z
5. Soleymani A, Nordin MJ, Ali ZM (2013) A novel public key image encryption based on elliptic curves over prime group field. JOIG 1(1):43–49. https://doi.org/10.12720/joig.1.1.43-49
6. Singh LD, Singh KM (2015) Image encryption using elliptic curve cryptography. Elsevier Procedia Comput Sci 54:472–481. https://doi.org/10.1016/j.procs.2015.06.054
7. Yasser I, Mohamed MA, Samra AS, Khalifa F (2020) A chaotic-based encryption/decryption framework for secure multimedia communications. Entropy 22:1253–1276. https://doi.org/10.3390/e22111253
8. Bh P, Chandravathi D, Roja PP (2010) Encoding and decoding of a message in the implementation of elliptic curve cryptography using Koblitz's method. (IJCSE) Int J Comput Sci Eng 02(05):1904–1907
9. Amounas F, Kinani EHE (2012) Security enhancement of image encryption based on matrix approach using elliptic curve. Int J Inf Netw Secur 1(2):54–59
10. Kreuk F, Adi Y, Raj B, Singh R, Keshet J (2020) Hide and speak: towards deep neural networks for speech steganography. arXiv:1902.03083[cs, eess, stat]
11. Baluja S (2020) Hiding images within images. IEEE Trans Pattern Anal Mach Intell 42:1685–1697. https://doi.org/10.1109/TPAMI.2019.2901877
12. Zeiler MD (2012) ADADELTA: an adaptive learning rate method. arXiv:1212.5701[cs]
13. Geetha P, Jayanthi VS, Jayanthi AN (2018) Optimal visual cryptographic scheme with multiple share creation for multimedia applications. Comput Secur 78:301–320. https://doi.org/10.1016/j.cose.2018.07.009
14. Duan X, Guo D, Liu N, Li B, Gou M, Qin C (2020) A new high capacity image steganography method combined with image elliptic curve cryptography and deep neural network. IEEE Access 8:25777–25788. https://doi.org/10.1109/ACCESS.2020.2971528
15. Fu Z, Wang F, Cheng X (2020) The secure steganography for hiding images via GAN. J Image Video Proc 2020:46. https://doi.org/10.1186/s13640-020-00534-2

16. Hodeish ME, Bukauskas L, Humbe VT (2016) An optimal (k, n) visual secret sharing scheme for information security. Procedia Comput Sci 93:760–767. https://doi.org/10.1016/j.procs.2016.07.288
17. Hu D, Wang L, Jiang W, Zheng S, Li B (2018) A novel image steganography method via deep convolutional generative adversarial networks. IEEE Access 6:38303–38314. https://doi.org/10.1109/ACCESS.2018.2852771
18. Nagaraj S, Raju GSVP, Rao KK (2015) Image encryption using elliptic curve cryptograhy and matrix. Procedia Comput Sci 48:276–281. https://doi.org/10.1016/j.procs.2015.04.182
19. Xiang Z, Sang J, Zhang Q, Cai B, Xia X, Wu W (2020) A New Convolutional neural network-based steganalysis method for content-adaptive image steganography in the spatial domain. IEEE Access 8:47013–47020. https://doi.org/10.1109/ACCESS.2020.2978110

Model to Detect and Correct the Grammatical Error in a Sentence Using Pre-trained BERT

R. Vijaya Prakash⊙, **M. Sai Teja, G. Deepthi, C. Namratha, D. Nikhil Sai,**
and P. Manish Raj

1 Introduction

Linguistic correction is especially important in today's technology when we rely on phrase framing and construction more than ever. Technology has become so ingrained in our lives that we require it to travel through each stage of our lives to preserve our capacity to recognize correct and incorrect language. The phrase "Natural Language Processing" refers to the automated computer processing of human languages [1]. Some of the most prevalent NLP applications include sentiment classification, chatbots, voice recognition, machine translation, spell checking, term searching, and information retrieval. Natural Language Understanding and Natural Language Generation are two components of NLP.

The act of turning supplied information into meaningful representations and assessing those parts of the language is known as natural language understanding. The technique of creating meaningful phrases and sentences in natural language from an internal representation is known as natural language generation (NLG). NLP has several phases. Tokenization is the transformation of strings into tokens, which are small structures or units that may be utilized for tokenization. Normalize words into their base or root form by stemming them, for example, Affectations, Affects, Affections, Affected, Affection, Affecting, and so forth. On the other hand, lemmatization considers the morphological analysis of the void to accomplish so. The Lemma combines many forms of a word, maps numerous words into one common root, and the output is a valid term. POS Tags are grammatical terms that refer to parts of speech. Named entity recognition is a technique for recognizing named entities including a person's name, a company's name, numbers, or a location. Chunking is

R. Vijaya Prakash (✉)
SR University, Warangal, India
e-mail: r.vijayaprakash@sru.edu.in

M. Sai Teja · G. Deepthi · C. Namratha · D. Nikhil Sai · P. Manish Raj
SR Engineering College, Warangal, India

the process of taking little bits of information and combining them together to make larger chunks [2].

The Transformers [3] are a unique NLP concept that aims to tackle pattern problems and long connections. The aim of transformer is to manage input and output relationships with complete attention and repetition. We may use BERT [3] as well as Universal Language Model Fine-Tuning (ULMFit) [4] for this system. It aims to pre-train deep bidirectional representations from unlabeled text using both left and right context conditioning. As a consequence, with just one extra output layer, the pre-trained BERT model may be fine-tuned to create efficient models for a variety of NLP applications.

We choose Bert over ULMFit because BERT gives more accurate results than ULMFit. Models were created using BERT, a method of pretraining language representations. These algorithms may be used to text data to extract high-quality linguistic features. They can also be fine-tuned with your own data to create cutting-edge predictions for tasks such as categorization, entity identification, and question answering. Quicker development, less data, and better results are all advantages of fine-tuning.

2   Related Work

According to early work in Grammatical error correction (GEC) and Grammatical error Detection (GED), several good rules were extensively used to discover and fix flaws that appeared in texts on a regular basis in the early 1990s. If any of the rules are violated, Macdonald et al. [5] presented a computational grammar checker that will identify an error. Although machine learning algorithms are utilized in most modern techniques, certain rules-based approaches still exist, such as [6]. Many learning-based techniques involve two phases, the most essential of which is solid feature engineering before constructing a typical machine learning model. To find mistakes, researchers utilized several methods, including Maximum Entropy-Based Classifiers (MEC) [7] and LFG-Based Features [8]. Tetreault et al. [9], used a MEC to identify preposition mistakes, are one of several classic studies that have been modified to cope with error categories. This type of model was employed by N Han et al. [10] to detect article use errors, whereas Convolution Neural Networks (CNN) were used in [11]. To detect verb form issues, the fundamental template matching approach was changed [12]. Several GEC efforts have investigated the use of synthetic ungrammatical data and huge unlabeled correct texts for GED. The utility of synthetic ungrammatical data for GED has been investigated using a variety of approaches. Foster et al. recommended that some rules be used to produce mistakes, with the goal of improving error detection performance through data augmentation [13]. In addition, [14] used a language model that was trained on a vast and diverse domain dataset to create features. In recent years, CNN-based error detection techniques, such as those used in the GEC task [15], have garnered a lot of attention. Bidirectional LSTM was used by Reiet al. [17] to create a neural sequence labelling model for erroneous token annotation based on the entire text representation [16]. They increase the sequence labelling model's capabilities by developing better word embedding utilizing character embedding [18]. The model's second job is to persuade it to learn increasingly precise and broad representations of each word and sentence.

3 Methodology

BERT is a software that pre-trains profound bidirectional assertions from unlabelled text using both left and right framework conditioning. The pre-trained BERT model may be ideal for building cutting-edge NLP models for a range of applications since it has an extra output layer. BERT architecture [19] is divided into two types: $BERT_{BASE}$ and $BERT_{LARGE}$. The Encoder stack of $BERT_{BASE}$ contains 12 layers, but the Encoder stack of $BERT_{LARGE}$ has 24 levels. BERT designs (BASE and LARGE) contain larger feedforward networks (with 768 and 1024 hidden units, respectively), as well as more attention heads (12 and 16).

Figure 1 shows how our BERT model for sentence categorization for the CoLA dataset works. BERT is a large neural network design with many parameters ranging from 100 million to over 300 million. When a BERT model is trained from scratch on a small dataset, overfitting might occur. As a result, it is preferable to start with a pre-trained BERT model that has been trained on a large dataset. Model fine-tuning refers to the process of further training the model using our comparatively limited sample. Figure 2 depicts BERT's fine-tuning for identifying whether a statement is grammatically accurate. At the conclusion of each phrase, we must use the special <SEP> token. This token is a by-product of two-sentence challenges, in which BERT is given two sentences and instructed to figure out what they mean. We must include the special <CLS> token to the beginning of every phrase for classification jobs. This token has special significance. The BERT model is built up of 12 Transformer layers, as illustrated in Fig. 2. Each transformer takes in a list of token embeddings and returns the same number of embeddings.

The flowchart in Fig. 3 shows how the system operates. When we begin the process, the first step is to provide the data (sentence input) to the system, which will then break the sentence into words (tokenization), which we will use to feed the data (input) to the model. Then, for sequence classification, we use Bert's pre-trained model, fine-tune it, and start developing the model using training and validation data. Then there is the confirmation of the outcomes and the loss of training. The scores are converted to probabilities by the SoftMax layer as specified in Eq. 1. The cell with the highest probability is selected for this time step, and the word associated with it is output.

$$\sigma(\vec{Z})i = \frac{e^{z_i}}{\sum_{j=1}^{k} e^{z_i}} \tag{1}$$

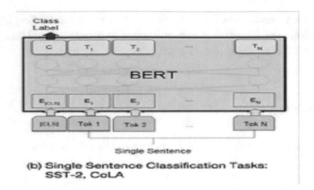

Fig. 1 BERT architecture

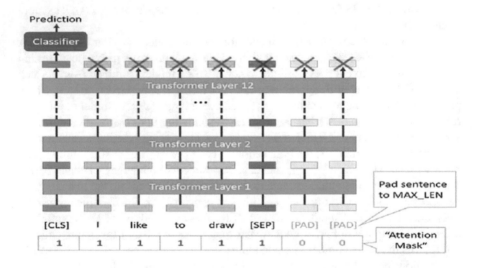

Fig. 2 Bert 12 layer transformer architecture

e^{z_i} and e^{z_j} are standard exponential function for input vector output vector respectively for k classes in a multi-class classifier.

MCC (Mathew's Correlation Coefficient) is used to evaluate the training model's performance on the test set (CoLA Data train set). Using the method in Eq. 2, the MCC may be computed directly from the confusion matrix. The MCC is a correlation coefficient that compares observed and anticipated binary classifications. It returns a value between -1 and $+1$. A coefficient of $+1$ denotes a perfect forecast, a value of 0 denotes a random prediction, and a coefficient of -1 denotes complete disagreement between prediction and observation.

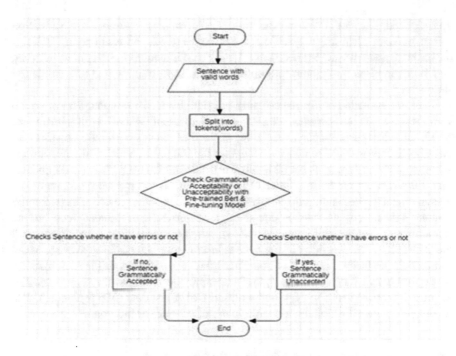

Fig. 3 Flowchart depicts the process of Grammatical errors in the proposed work

$$MCC = \frac{TN \times TP - FP \times FN}{\sqrt{(TN + FN)(FP + TP)(TN + FP)(FN + TP)}} \tag{2}$$

Finally, the model evaluates if a sentence is grammatically accurate or not, allowing language learners to determine whether a written sentence contains mistakes on their own.

4 Results

We used the CoLA dataset, which comprises hundreds of grammatically correct and grammatically flawed statements, to carry out our work. The primary purpose of this dataset has been identified to underline the fact that it may be utilized to do grammatical correction. The Corpus of Linguistic Acceptability (CoLA) contains 10,657 phrases from 23 linguistics papers that have been professionally annotated by their original authors for acceptability (grammaticality). We will parse the data with pandas and examine a few characteristics and data points. That number of training phrases was discovered to be 8,551. We are interested in two characteristics in data: phrases and labels, which indicate if something is grammatically acceptable or not (0 = unacceptable, 1 = acceptable). Then we'll split our training set in half, using

90% for training and 10% for validation (7,695 training samples, 856 validation samples). With a learning rate of 2e-5, a batch size of 32, and four epochs, BERT's pre-trained model was fine-tuned. Table 1 and Fig. 4 illustrates the accuracy, training, and validation loss throughout four epochs using 7,695 training and 856 validation samples.

Then we evaluated performance using test data from 516 samples, 354 of which were positive. On test samples, we also utilized an attention mask with a batch size of 32. To check performance (MCC), we utilized Mathew's correlation coefficient. Each batch comprises 32 sentences, except for the last batch, which contains four test sentences (516 test samples per 32 batch size). The Batch0 MCC value is 0 indicates that it is a random prediction, Batch1 the MCC value is -ve indicates that there is a disagreement between prediction and observations in the sentences. Similarly, Batch8 sentences the proposed model produced the perfect forecast of the sentences. For the remaining batches the MCC value is positive ranging from 0.27 to 1.0, indicates that there is a proper forecasting of sentences taken in these batches. The average MCC score of all the batches is 0.571. Figure 5 displays the MCC score for each batch.

Figures 6 and 7 illustrate the model's comparison findings; the time it took to acquire the results was 5 to 10 s, and we used CoLA test data (516 sentences). The model determines if a statement is grammatically correct or incorrect.

Table 1 Accuracy, training and validation results for different epoch's

Training loss	Validation loss	Validation accuracy	Training time	Validation time	Epoch
0.51	0.38	0.84	0:00:50	0:00:02	1
0.32	0.41	0.85	0:00:50	0:00:02	2
0.20	0.44	0.86	0:00:50	0:00:02	3
0.14	0.52	0.85	0:00:50	0:00:02	4

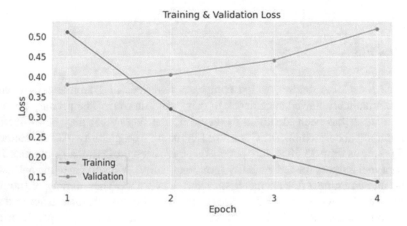

Fig. 4 Training and validation loss

MCC Score per Batch
MCC Score (-1 to +1)
1.0
0.8
0.6
0.4
0.2
0.0
0 1 2 3 4 5 6 7 8 9 10 11 12 13 14 15 16
Batch #

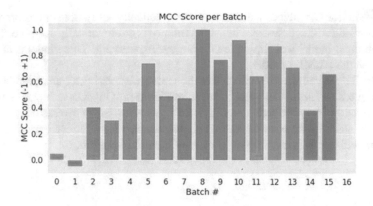

Fig. 5 MCC score per batch

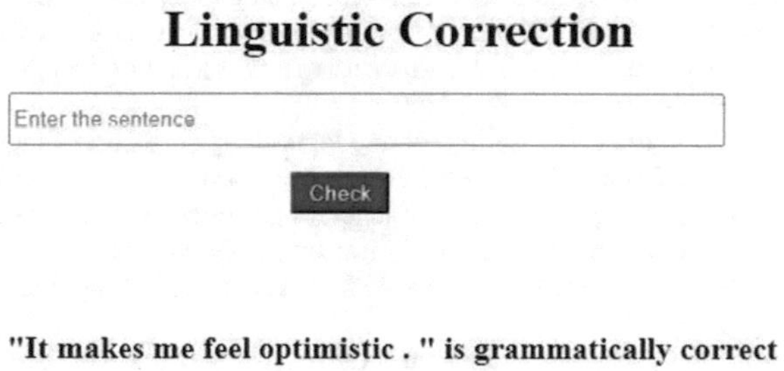

Linguistic Correction
Enter the sentence
Check
"It makes me feel optimistic . " is grammatically correct

Fig. 6 Proposed model for a correct sentence

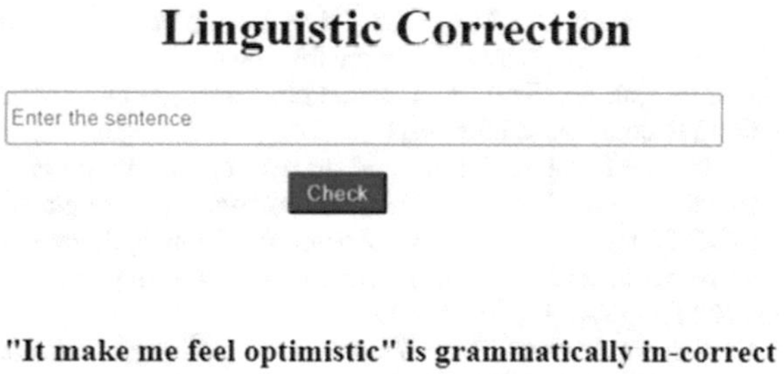

Linguistic Correction
Enter the sentence
Check
"It make me feel optimistic" is grammatically in-correct

Fig. 7 Proposed model for incorrect sentence

5 Conclusion

The potential of utilizing a pretrained BERT model and fine-tuning the pretrained BERT model was investigated in this study. The attention masking of words, learning rate of 2e-5, Batch size of32, and epochs 4 were also investigated in this research for effective prediction of whether sentences are grammatically acceptable or unsuitable. The suggested algorithm assesses whether a sentence is grammatically accurate. The BERT Pretrained model was fine-tuned during our experiment. This model is more accurate than the ULMFit model.

While BERT may be able to predict all of the sentence's missing tokens, predicting the correct words might easily lead to redundant editing. Our research shows that simply rephrasing the entire sentence with BERT yields too many diverse outcomes. Instead, previous mistake span detection may be required for effective grammatical error repair, and this is something we'll be looking into in the future.

References

1. Young CW, Eastman CM, Oakman RL (1991) An analysis of illformed input in natural language queries to document retrieval systems. Inf Process Manag 27(6):615–622
2. Li Y, Anastasopoulos A, Black AW (2020) Towards Minimal supervision BERT-based grammar error correction (Student Abstract). In: Proceedings of the AAAI Conference on Artificial Intelligence, vol 34, no 10, pp 13859–13860
3. Church KW, Gale WA (1991) Probability scoring for spelling correction. Stat Comput 1:93–103
4. Kim M, Choi S, Jin J, Kwon H (2015) Adaptive context-sensitive spelling error correction techniques for the extremely unpredictable error generating language environments. In: 2015 IEEE International Conference on Computer and Information Technology; Ubiquitous Computing and Communications; Dependable, Autonomic and Secure Computing; Pervasive Intelligence and Computing, pp 927–930
5. Macdonald N, Frase L, Gingrich P, Keenan S (1982) The writer's workbench: computer aids for text analysis. IEEE Trans Commun 30(1):105–110
6. Foster J, Vogel C (2004) Parsing Ill-formed text using an error grammar. Artif Intell Rev 21:269–291
7. Chodorow M, Tetreault J, Han NR (2007) Detection of grammatical errors involving prepositions. In: Proceedings of the Fourth ACL-SIGSEM Workshop on Prepositions (SigSem 2007). pp 25–30. Association for Computational Linguistics, USA
8. Przepiórkowski A, Patejuk A (2020) From lexical functional grammar to enhanced universal dependencies. Lang Resour Eval 54:185–221
9. Tetreault JR, Chodorow M (2008) The ups and downs of preposition error detection in ESL writing. In: Proceedings of the 22nd International Conference on Computational Linguistics–Volume 1 (COLING 2008), pp 865–872. Association for Computational Linguistics, USA
10. Han NR, Chodorow M, Leacock C (2006) Detecting errors in English article usage by non-native speakers. Nat Lang Eng 12(2):115–129
11. Krizhevsky A, Sutskever I, Hinton GE (2017) ImageNet classification with deep convolutional neural networks. Commun ACM 60(6):84–90
12. Lee J, Seneff S (2008) Correcting misuse of verb forms. In: ACL Proceedings of the 46th Annual Meeting of the Association for Computational Linguistics, pp 15–20, Columbus, Ohio, USA
13. Foster J Andersen E (2009) Generate: generating errors for use in grammatical error detection. In: Proceedings of the Fourth Workshop on Innovative Use of NLP for Building Educational Applications, pp 82–90. Association for Computational Linguistics, Boulder, CO, USA
14. Gamon M (2010) Using mostly native data to correct errors in learners writing. In: Human Language Technologies: Conference of the North American Chapter of the Association of Computational Linguistics, Proceedings, pp 163–171. Association for Computational Linguistics, USA

15. Wang Q, Tan Y Automatic grammatical error correction based on edit operations information. In: Cheng L, Leung A, Ozawa S (eds.) Neural Information Processing. ICONIP 2018. LNCS, vol 11305, pp 494–505. Springer, Cham (2018). https://doi.org/10.1007/978-3-030-04221-9_44
16. Rei M, Yannakoudakis H (2016) Compositional sequence labeling models for error detection in learner writing. In: 54th Annual Meeting of the Association for Computational Linguistics, ACL, pp 7–12, 2016. The Association for Computer Linguistics Berlin, Germany
17. Hochreiter S, Schmidhuber J (1997) Long short-term memory. Neural Comput 9(8):1735–1780
18. Rei M, Crichton GK, Pyysalo S (2016) Attending to characters in neural sequence labeling models. In: 26th International Conference on Computational Linguistics, Proceedings of the Conference: Technical Papers, pp 309–318. ACL
19. Vaswani A et al (2017) Attention is all you need. In: Advances in Neural Information Processing Systems, pp 6000–6010

Crop Recommendation System for Precision Agriculture Using Fuzzy Clustering Based Ant Colony Optimization

T. P. Ezhilarasi and K. Sashi Rekha

1 Introduction

For millennia, India has practiced natural farming, which has evolved to include contemporary methods due to globalization. In India, this has resulted in plant diseases. Depending on the land utilized for agriculture, various approaches have advocated farming methods & fertilizer usage. As a result, new strategies & methods, such as PAG (Precision Algorithm), have been developed for combating crop disease. PAG is a recommender system & "site-specific" agricultural method. PAG provides various benefits in terms of agricultural production & plant decisions, but it also has various drawbacks. PAG planting suggestions are determined by a variety of factors. PAG's goal is to find site-specific factors able to solve crop selection problems. Even though the site-specific approaches have increased productivity, their outcomes must be monitored because not all PAGs offer correct results. Precision is required in agricultural production, as errors result in economic and technical damages. Agricultural predictions have been studied to improve their efficiency & reliability [1]. As a result, the goal of this research is to evaluate food production using datasets that include critical aspects including historical information on weather, moisture, precipitation, & previous agricultural production. The following is a list of the contributions made by this task: Firstly, a large amount of historical agricultural production & temperature records is obtained, & data pre-processing work is performed. Then, regarding plant recommendations, ACO-Fuzzy, a model for the prediction based on an Improved Deep Convolutional Neural Network with ACO is used. For every collection of datasets, ACO is used to select the best architecture for Fuzzy sub-models. Optimization is the process of making the best or most effective use of situations or resources to acquire an optimized solution for a given problem. The techniques involved will result in productive outcomes with a high level

T. P. Ezhilarasi (✉) · K. Sashi Rekha
Department of Computer Science, Saveetha School of Engineering, Chennai, Tamil Nadu, India
e-mail: ezhil.9944@gmail.com

© The Author(s), under exclusive license to Springer Nature Singapore Pte Ltd. 2022   261
B. Unhelker et al. (eds.), *Applications of Artificial Intelligence and Machine Learning*,
Lecture Notes in Electrical Engineering 925,
https://doi.org/10.1007/978-981-19-4831-2_22

of accuracy. Thus, the aim is to provide an optimal solution as well as focus on giving a robust solution that is needed in organizations. The main objective of every optimization algorithm is to yield an optimum solution. So, the processes are iteratively continued and compare all the possible solutions until the best or most satisfactory solution is found. Optimization technique plays a major role in this new digital era. In today's world, all engineering design and industries were now adopted this optimization technique [2]. Another technique namely data mining also widely used to analyze and extract meaningful information from huge amounts of data. Optimization techniques provide well support to implement various data mining algorithms. Most of the techniques like clustering, feature extraction, and classification are well hybridized with optimization techniques for effective implementation.

Support Vector Machine and kernel methods support optimization algorithms to implement the data mining process effectively [3].

Algorithms like Ant Colony Optimization (ACO), Genetic Algorithm (GA) method, Artificial immune systems, Bees Algorithm, Honey-Bees Mating Optimization (HBMO), Algorithm, Shuffled Frog Leaping Algorithm (SFLA), Cuckoo Search (CS), Firefly Algorithm, Eagle Strategy, Particle swarm optimization are some of the nature-inspired algorithms [4]. PSO, ACO, and Cuckoo search algorithms are famous among recent bio-inspired algorithms because they are widely used in multidisciplinary domains. Thus, we review the applications of these algorithms in the social media and agriculture domain. It will help researchers to know about the implementation process of ACO and PSO algorithms in the social and agriculture domain. This review aims to create awareness about the future scope of those two algorithms in the two fields. Ant colony optimization algorithm serves as a population-based metaheuristic technique for finding the best path through graphs [5].

2 Related Work

The benefits of another vigorous system (FSL) over a classic method (ACO) in both regular & crisis settings were investigated in this research utilizing Fluffy areas, Activity, State, Reward, Activity (SARSA) Learning (FSL) & underground ACO approaches. These approaches' numerical models were constructed. In the East Aghili channel in Iran, three water shortages of 10, 20, and 30% were explored for the reproduction interaction [6]. The presentation of the produced models was evaluated using water depth and conveyance indicators. The outcomes uncovered that the FSL and ACO strategies offered practically a similar execution for the ordinary activity condition with high and worthy pointers. Be that as it may, the FSL technique beat the ACO strategy as far as execution in three thoughts about crisis activities [7]. The practice of evaluating & collecting valuable data from enormous volumes of data is known as data mining. Banking, retail, medical, & farming seem to be just a few of the industries that use data mining. Data gathering is used in agriculture to investigate a variety of biotic and abiotic factors. Agricultural production has been utilized to help producers solve their problems. Precision farming is indeed a new agricultural strategy that employs research data on soil properties, soil characteristics, and plant production information to recommend the good plant to producers depending on their area factors. This decreases the number of times a crop is chosen incorrectly and increases productivity. This challenge is tackled in this research by providing recommender systems for site-specific characteristics utilizing an ensemble method using qualified majority methodology utilizing Random tree, CHAID, K-Nearest Neighbor, and Naive Bayes as a learner to suggest crops with excellent precision. As a result, farmers may plant the appropriate crop, enhancing their yield as well as the nation's productiveness. Their future study will focus on improving an existing data set with such a high number of variables and incorporating yield prediction.

Soil CEC is an important feature that determines the state of soil richness. Even though it is difficult to measure, soil physicochemical parameters that can be easily evaluated tend to predict it. Scientists have used a variety of soil characteristics to predict soil CEC and infer peso-move capabilities. We proposed a mixture calculation to determine factors that affect soil CEC: a development of underground ACO in combination with a versatile organization-based fluffy derivation framework. The ACO-ANFIS development calculation's potential force in establishing a framework for differentiating the most determinant borders of rural soils [9]. This paper [10] presents a coordinated displaying strategy for multi-rules land-use appropriateness evaluation (LSA) utilizing grouping rule disclosure (CRD) by subterranean ant settlement improvement (ACO) in ArcGIS.

Different connected data parameters, such as silt fixing, river depth, velocity, dregs size, Flow rate, extract percentage, amount of passage & sub-passages, & river profoundness upstream of the residual ejection, have been used to create the proposed methodology. Using a few genuine analyzed data, the evaluation limitation of the constructed half-breed simulations is investigated [12]. The instrument was utilized to deal with land use appropriateness arrangement in the examination region for watered agribusiness. The resultant guide was then contrasted with present watered land to show spatial dissemination of flooded land appropriateness and to uncover future capability of land use improvement around here. In this examination, an Artificial Neural Networks (ANN) model is created to explore the connection between bioethanol creation and the working boundaries of enzymatic hydrolysis and maturation measures. The working boundaries of the hydrolysis interaction that impact the decreasing sugar fixation are the substrate stacking, α-amylase focus, amyl glucosidase fixation, and strokes speed. The working boundaries of the maturation interaction that impact the ethanol focus are the yeast fixation, response temperature, and unsettling speed. The allure capacity of the model is coordinated with subterranean ant province advancement (ACO) to decide the ideal working boundaries which will augment lessening sugar and ethanol fixations [13–16]. The rest of this paper was laid out as follows. Section 2 discusses similar research on crop production recommendations. The development framework for yield estimation is described in full in Sect. 3. The suggested model's results are discussed in Sect. 4 & compared with the actual version. Lastly, in Sect. 5 of the paper, consider future research.

3 Proposed Methodology

Agriculture was among the most important areas that affect a country's economic development. In a nation like India, agriculture is the main source of income for the majority of people. Several technological advances are being deployed in farming, such as Machine Learning algorithms, to make it much easier for producers to cultivate & enhance their production. Present a homepage with the following cases: plant recommendations, fertilizer recommendations, & crop diseases predictions, all of which have been applied in this project. The user can provide soil data to the agricultural recommendations program and the program would suggest which plant the user would produce. This paper is a recommendation system for the myanimelist.net platform, a website where users can rate animes and build profiles where they write animes they have seen, animes they want to see in the future, animes they dropped, and so on. We developed two systems: the first uses a Collaborative Filtering technique, and the second is a hybrid of Collaborative Filtering and Content-Based techniques that perform dimensional reduction through the Fuzzy Clustering algorithm (Fuzzy C-Means).

CNNs can adjust to the crop feature extraction process utilizing their hierarchical representational structure, whereas NNs (Neural Networks) find key predictors. Furthermore, using CNNs necessitates prior knowledge & experience, limiting its generalization possibilities. As a result, CNNs with ACO has been presented as a method for analyzing agricultural yield estimates. This method collects historic plant & weather information, which is then preprocessed & used by the suggested plan to make crop recommendations. The role of ACO in the system is to determine the best architecture for ACO sub-models built from sets of data. FCACO's hyper-parameters have been fine-tuned to refine its cellular structure. Figure 1 illustrates FCACO's proposed technique.

Groundwater, soil sampling, composting, plant dung, rooftop garden media, organic wastes, & greenhouse effect media analysis were among the experiments carried out. Certain specialist talks were offered by individual articles on particular concerns, & data on farming was supplied through a website that gives statics &

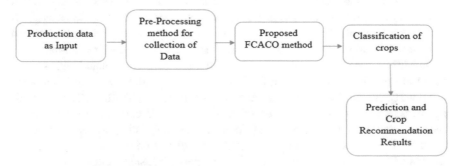

Fig. 1 FCACO based crop recommendation

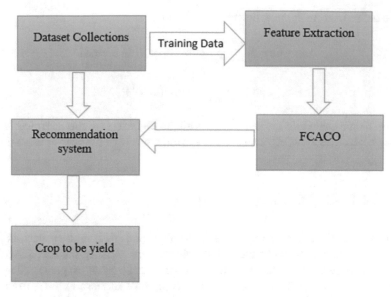

Fig. 2 Proposed methodology

actual information relating to agricultural economic entities. The goal of the agricultural site is to deliver information to everybody at any time and from any location, enabling farmers to adopt new technologies. Flume is being used to download Web sites for data agriculture websites, allowing them to be saved on the Hadoop distributed database. This structure is utilized to meet the demands of farm officers & agriculture experts by assisting them in evaluating & developing crop development suggestions based on historical data. To speculate on the influence of rising temperatures on crops and the steps that need to be done to address the problem in a certain district. The practice of picking characteristics & essentials presented in the form of a subset is referred to as feature extraction. The filter method occurs during the data preprocessing stage of model development. As shown in Fig. 2, feature selection is the process of normalizing data to remove technological variability by substituting incomplete data, as these missing values distract the decision-making processes.

3.1 Collaborative Filtering

The principal idea for the collaborative filtering technique, especially for users, is to automatically predict (filter) what a customer's interests are by collecting (collaborating) preferences or taste data from more customers, who are his nearest neighbors, that is, those users, whose amines were similar and who gave similar rates to them. So, a first naive approach is based on applying the k-nearest neighbor on the user-item matrix, and then using the ratings from those neighbors to calculate predictions for the active user, with the following Eqs. 1 and 2:

$$r_{u,i} = \overline{r_u} + k \sum_{u^1 \epsilon U} \text{simil}(u,\ u^1)\left(r_{u^1,i} - \overline{r_{u^1}}\right) \tag{1}$$

$$k = 1/ \sum_{u^1 \epsilon U} |\text{simil}(u,\ u^1)| \tag{2}$$

The recommendations are the amines with the highest predictions. However, this approach becomes slow, especially if the system has a high number of users.

3.2 Ant Colony Optimization (ACO)

Ants can smell their food in the best suitable shortest path from their nest. In ants, they have a special type of chemical called pheromone. Initially, the ants move from their colony randomly to find the food from their nest. The shortest path is computed based on the ant which brings more food in less time. When the ants move from their nest in search of food they disperse pheromones from all parts of their body in their path so that it is helpful for other ants to sense and direct them to their food. The pheromone is dispersed when ants reached their colony. If ants face any hurdles to finding their food along the way, they separate into two paths and the shortest path is the path that contains a high density of pheromone. The pheromone with less density in other paths will get evaporated after some time.

Pseudocode for ACO Algorithm
Begin
Start the parameters and pheromone track
Generate the initial population (ants)
Calculate the fitness of each individual
Find the optimal way for each ant
Determine the global optimal ant
Update the p trade of pheromone
Check whether termination = True
End.

3.3 Fuzzy Systems

The design of fuzzy clustering is briefly explained in this section. The Fuzzy clustering according to the nth rule which is denoted as Rn is represented in Eq. 3

$$R_n : \text{ if } a_1(k) \text{ is } x_{i1} \text{ And} \ldots \text{ And } a_n(k) \text{ is } x_{in} \tag{3}$$

whereas the input variables $a1(k) = an(k)$, the control output variable is $b(k)$ and the fuzzy set is denoted as xij then the action recommended is denoted as ui(t) and it is a fuzzy singleton. For the fuzzy set A and the current membership in Eq. 4 a Gaussian member function is utilized as,

$$M(x) = \exp\{-\frac{(x-m)^2}{\sigma^2}\} \tag{4}$$

whereas center and is fuzzy set A. In the fuzzy theory, the AND operation of fuzzy is implemented in inference engine with the help of algebraic product is computed by Eq. 5

$$\emptyset_i\left(\underset{x}{\rightarrow}\right) = \prod_{j=1}^{n} M_{ij}(xj) = \exp\left\{-\sum_{j=1}^{n}\left(\frac{x_j-m_{ij}}{\sigma_{ij}}\right)^2\right\} \tag{5}$$

3.4 Fuzzy Clustering

The fuzzy sets and the fuzzy rules are introduced by this section in each of the input variables that are generated by the fuzzy clustering which is proposed already. When there are a lot of inputs in fuzzy clustering, the antecedent part splitting takes a long time, and many of the fuzzy rules created are redundant. The development of fuzzy rules via fuzzy clustering, which is done automatically, eliminates the labor of designing and generating superfluous rules.

Equation 6 is used to construct a new fuzzy rule for the very first incoming data kx, as well as the breadth and center of the Gaussian fuzzy set of xj.

$$m1j = aj(0) \text{ and } \sigma 1j = \sigma \text{fixed, for } b = 1, \ldots, n \tag{6}$$

whereas sfixed determines the fuzzy sets width. For the incoming data which is succeeding kn, find in Eq. 7.

$$I = \arg \max_{1\leq i\leq r(k)} \emptyset_i\left(\underset{x}{\rightarrow}(k)\right) \tag{7}$$

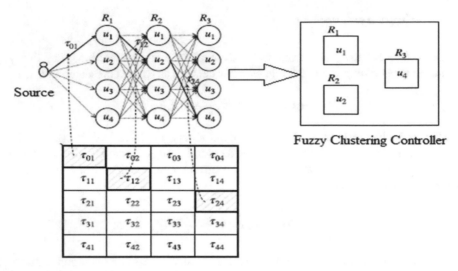

Fig. 3 Relationship between ACO and fuzzy clustering controller

whereas the count of existing rules is denoted as r(k) at time k (Fig. 3).

The Fuzzy Clustering is Summarized as

Asr(n) at time n
The FC is summarized as
IF the first incoming data is xn THEN do
{A new, fluctuating rule that assigns the constant breadth and center of each fluid trend is built with equations. - 5.}
Else for every new incoming data xn do
{Manipulate equation –6 IF $\phi 1 < \phi k$
THEN
{r(n+1) = r(n) + 1
For b=1,2,3,..n
{Manipulate equation -7
IF M1, b(aj) >p
THEN {the Aij fuzzy set is used in the r(k+1) rule of the antecedent part}
ELSE{A new fuzzy set is generated,
  hj(k+1) = hj(k)+1,set width & center of the new fuzzy set with the help of equation – 8}}}}

So, one of the biggest problems with recommendation systems based on CF is the Curse of Dimensionality: these systems are highly sensitive to the increasing number of users in the system. Classical ways to alleviate this issue are to perform Dimensional Reduction like, for example, Singular Value Decomposition (SVD). We use a technique that, at the same time, reduces dimensionality and combines Content-based information with the CF system:

- Given the items matrix (M × F), where M is the number of items and F is the number of binary features, perform Fuzzy Clustering, using the C-Means algorithm, on this binary matrix, to obtain a soft-assignment with probability membership of an item to a cluster, obtaining an item-cluster matrix (M × C), with C number of clusters.
- We combine the item-cluster matrix with the user-item matrix (or Utility matrix) with dimension (N × M), N number of users, obtaining through weighted mean the user-clusters matrix of dimension (N × C), where each element represent the probability of a user belonging to a cluster.
- We perform, like in the CF system, K-Nearest Neighbors on the user-cluster matrix, and get prediction using the same approach.

(M × F) --> Fuzzy Clustering --> (M × C) --> |(N × C) --> K Nearest Neighbors --> Recommendations.

where N = number of users, M = number of anime, F = number of anime features, C = number of clusters, K = number of neighbors.

3.5 Fuzzy Clustering ACO

Algorithm FCACO

Step 0: (Initialization) set the cluster number $r = 0$ and iteration counter $t = 0$.
Step 1: Set the ant counter $g = 0$, iteration $t = t + 1$.
Step 2: According to Eq. 7, select the fuzzy controllers Na.
Step 3: Set the evaluation counter of the controller $q = q + 1$.
Step 4: for $k = 1$ to the member of the pre-defined time step {update n}.
Step 5: Compute the value F quality.
Step 6: If $(q < Ka)$ then go to step 3.
Step 7: with the help of Eq. 8, update the pheromone trail Tij (t).
Step 8: At the end.
If($i <$ predefined number).
then {go to step 1} Else {stop}.

4 Results and Discussion

4.1 Dataset

The data for this research came from two main factors. Crop yields information was collected from faostat3.fao.org (https://data.world/thatzprem/agriculture-india), while weather information was gotten from the Indian water portal.

Org (https://www.timeanddate.com/weather/india/new-delhi/historic). Crop yields historic (Monthly/Yearly) information was in a CSV (Comma Separated Values) style, while the weather information included specifics on record rainfall for particular regions. In this investigation, 6 years of historical information were used.

4.2 Data Pre-processing

Because less important data is a part of the climatic dataset characteristics, pre-processing is a key step in this research. As part of the pre-processing procedure, irrelevant data is eliminated and only vital data are considered. As a result, a variety of historical types of data are pre-processed & integrated. Each month's Wind Speed, Maximum/Minimum/Average Temperatures, Dew, Moisture, & Wind Load have been included in the weather data. The mean temperature in a given location over a month is known as the average temperature, whereas the world's highest temperature in a particular region is known as the maximum temperature, & the lowest temperature value in the region is known as the minimum temperature. The dataset was divided into sixty-four pairings, with 60% of the information being used for learning and 40% for assessment.

4.3 Evaluation Metrics and Tuning Step

To evaluate the quality of our systems, we used:
   The root-mean-square error (RMSE) measure, defined as in Eq. 8:

$$\text{RMS Errors} = \sqrt{\frac{\sum_{i=1}^{n}\left(\widehat{y}_i - y_i\right)^2}{n}} \tag{8}$$

where n is the number of recommendations, y^t is the predicted rate, and yt is the real rate the user gave to the anime t. the mean absolute error (MAE) measure, defined as in Eq. 9:

$$\text{MAE} = \frac{1}{n}\sum_{i=1}^{n}\left|f_i - y_i\right| = \frac{1}{n}\sum_{i=1}^{n}|e_i| \tag{9}$$

where fi is the predicted rate for anime i, and yi is the real rate the user gave to it.
   These two measures were used during the training phase to decide whether parameters within the system perform better. For the recommender system based on the collaborative filtering technique, we performed training on the number of neighbors used to compute recommendations, while for the one based on fuzzy clustering, we tuned the number of clusters too.

4.4 Graph and Tables

Collaborative Filtering vs Fuzzy Clustering—Tuning the number of neighbors with the following tests have been made using a fixed number of clusters (60) for the recommender system based on fuzzy clustering as shown in Figs. 4 and 5.

Fuzzy Clustering—Tuning Number of Clusters

The following tests have been made using a fixed number of neighbors (12) for the recommender system based on fuzzy clustering as shown in Figs. 6 and 7.

Collaborative Filtering vs Fuzzy Clustering—Average computation time The following tests have been made using a machine with the following characteristics as shown in Fig. 8:

- Operating System: Ubuntu 16.04 64 bit
- CPU: Intel i5-4570
- RAM: 8 GB

The graphs reported above show us four important facts:

- The Recommender System based on Collaborative Filtering is slightly more accurate on average than the one based on Fuzzy Clustering.

Fig. 4 Comparison of collaborative filtering with fuzzy clustering

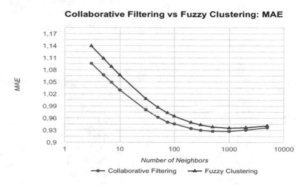

Fig. 5 Collaborative filtering vs. fuzzy clustering ACO comparison

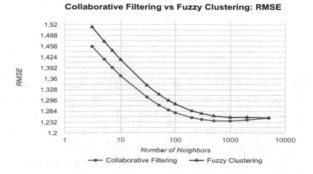

Fig. 6 Tuning of
clusters—MAE

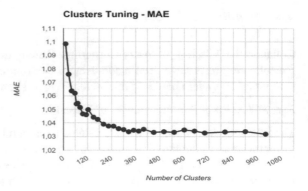

Fig. 7 Tuning of
clusters—RMSE

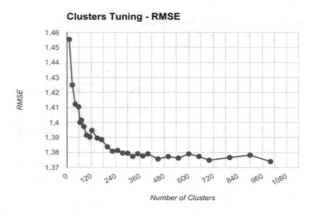

Fig. 8 Comparison of
collaborative filtering with
fuzzy clustering ACO

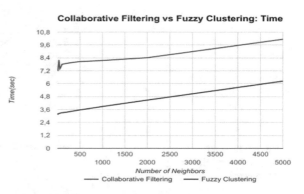

- The quality of the recommendations improves with a high number of neighbors
  K. In particular, results show that it improves for values of K up to 1000, then
  remains constant for values of K up to 5000, where it starts to slightly decrease.
  For completeness, we also considered very high values for K, even though they
  would not be practical (K = 5000 would be not a good choice, especially for a
  system with a dataset of about 10'000 users). For example, in the recommender

system based on collaborative filtering, the quality improvement we get moving from K = 50 (RMSE = 1.282) to K = 5000 (RMSE = 1.238) probably does not justify the increase in the required time to answer queries.

- Focusing on the recommender system based on fuzzy clustering, we can notice that the more the clusters we use, the better the score we get. However, we think the improvement we get moving from, for instance, K = 60 to K = 171.
- The Fuzzy Clustering technique is faster than the Collaborative Filtering one: we can notice a constant improvement of about 4 s, which is independent of the number of neighbors considered. This difference is because the system based on clustering performs kNN using a smaller matrix, while the collaborative filtering one works on the matrix Users X Animes, which is way bigger.

5 Conclusions

To measure the similarity between users we used the cosine similarity between their rating vectors. In both systems, it proved to be better than Pearson similarity. From the graph attached above, we can see that, even if the recommender system based on fuzzy clustering is a bit worse than the collaborative filtering one, it is way faster and therefore it can be considered better than the other one. To achieve the best results, all of the techniques' hyperparameters were tweaked. FCACO outperformed baseline models in terms of efficiency and accuracy due to its capacity to extract features from datasets at varied time steps. This research suggests a viable approach for deep learning approaches by combining multiple architectures for individual benefits, resulting in a large reduction in computational complexity, which would be a useful addition to accurate and reliable crop recommendation prediction. The proposed FCACO recommendation systems model is proven to be effective in recommending a suitable plant. In terms of improving results & evaluate the model technology on an agricultural database, future research should focus on auto encoder-based deep learning processes.

References

1. Hegde G, Hulipalled VR, Simha JB (2020) A study on agriculture commodities price prediction and forecasting. In: 2020 international conference on smart technologies in computing, electrical and electronics (ICSTCEE). IEEE, pp 316–321
2. Mehdizadeh S, Mohammadi B, Pham QB, Khoi DN, Linh NTT (2020) Implementing novel hybrid models to improve indirect measurement of the daily soil temperature: Elman neural network coupled with gravitational search algorithm and ant colony optimization. Measurement 165:108127
3. Moayedi H, Mehrabi M, Bui DT, Pradhan B, Foong LK (2020) Fuzzy-metaheuristic ensembles for spatial assessment of forest fire susceptibility. J Environ Manage 260:109867
4. Lahlouh I, Rerhrhaye F, Elakkary A, Sefiani N (2020) Experimental implementation of a new multi-input multi-output fuzzy-PID controller in a poultry house system. Heliyon 6(8):e04645

5. Omidzade F, Ghodousi H, Shahverdi K (2020) Comparing fuzzy SARSA learning and ant Colony optimization algorithms in water delivery scheduling under water shortage conditions. J Irrig Drain Eng 146(9):04020028
6. Aghelpour P, Bahrami-Pichaghchi H, Kisi O (2020) Comparison of three different bio-inspired algorithms to improve the ability of neuro-fuzzy approach in prediction of agricultural drought, based on three different indexes. Comput Electron Agric 170:105279
7. Shekofteh H, Ramazani F, Shirani H (2017) Optimal feature selection for predicting soil CEC: comparing the hybrid of ant colony organization algorithm and adaptive network-based fuzzy system with multiple linear regression. Geoderma 298:27–34
8. Salam A, Javaid Q, Ahmad M (2021) Bio-inspired cluster-based optimal target identification using multiple unmanned aerial vehicles in smart precision agriculture. Int J Distrib Sens Netw 17(7):15501477211034072
9. Jayaprakash A, KeziSelvaVijila C (2019) Feature selection using ant colony optimization (ACO) and road sign detection and recognition (RSDR) system. Cogn Syst Res 58:123–133
10. Lai C, Shao Q, Chen X, Wang Z, Zhou X, Yang B, Zhang L (2016) Flood risk zoning using a rule mining based on an ant colony algorithm. J Hydrol 542:268–280
11. Sharafati A, Haghbin M, Tiwari NK, Bhagat SK, Al-Ansari N, Chau KW, Yaseen ZM (2021) Performance evaluation of sediment ejector efficiency using hybrid neuro-fuzzy models. Eng Appl Comput Fluid Mech 15(1):627–643
12. Fathi M, Haghi Kashani M, Jameii SM, Mahdipour E (2021) Big data analytics in weather forecasting: a systematic review. Arch Comput Methods Eng 1–29
13. Azad A, Karami H, Farzin S, Mousavi SF, Kisi O (2019) Modeling river water quality parameters using a modified adaptive neuro-fuzzy inference system. Water Sci Eng 12(1):45–54
14. Manavalan R (2020) Automatic identification of diseases in grains crops through computational approaches: a review. Comput Electron Agric 178:105802
15. Maroli A, Narwane VS, Gardas BB (2021) Applications of IoT for achieving sustainability in the agricultural sector: a comprehensive review. J Environ Manage 298:113488
16. Singh SK, Singh RS, Pandey AK, Udmale SS, Chaudhary A (eds) (2020) IoT-based data analytics for the healthcare industry: techniques and applications. Academic Press

Classification and Hazards of Arsenic in Varanasi Region Using Machine Learning

Siddharth Kumar◉, Arghya Chattopadhyay◉, and Jayadeep Pati◉

1 Introduction

Varanasi is one of the oldest, holiest, and sacred city in Hinduism, Jainism, and Buddhism. It is also known as Benares or Kashi which is located on the banks of river Ganga of Uttar Pradesh District. It has two sub Districts Pindra and the other is the main city of Varanasi. According to the census of India 2011 [1], the district has a total population of 3,676,841 persons. Pindra has a population of 627,298 persons whereas the main city of Varanasi has a population of 3,049,543 persons. To revive the glory of Varanasi Fig. 1 by conserving its heritage, culture, and traditions the Government of India under the Capacity Building of Urban Development (CBUD) scheme Varanasi was chosen by the Ministry of Housing and Urban Poverty Allevi-ation (MoHUPA) and the Ministry of Urban Development (MoUD). This will boost tourism, employment, and the quality of life of the residents around the region.

Groundwater aquifers are one of the most valuable and primary sources of water all over the world and is extensively used for agriculture, drinking, and aquaculture in the banks of Varanasi. However, continuous quality monitoring of groundwater samples of Varanasi clearly shows that the As concentration in the water samples from the aquifers exceeds the tolerable level of 10 µg/L advocated by the World Health Organization [3] Consumption of water containing elevated As levels significantly increases the mortality rate as it affects the liver, bladder, cardiovascular system and causes lung cancer [4, 5]. Also, prolonged dermal contact with As affected water can result in acute skin-related conditions. Various studies have shown that continuous

S. Kumar (✉) · J. Pati
Indian Institute of Information Technology, Ranchi 834010, India
e-mail: siddharth.rs@iiitranchi.ac.in

J. Pati
e-mail: jayadeeppati@iiitranchi.ac.in

A. Chattopadhyay
Banaras Hindu University, Varanasi 221005, Uttar Pradesh, India

© The Author(s), under exclusive license to Springer Nature Singapore Pte Ltd. 2022   275
B. Unhelker et al. (eds.), *Applications of Artificial Intelligence and Machine Learning*,
Lecture Notes in Electrical Engineering 925,
https://doi.org/10.1007/978-981-19-4831-2_23

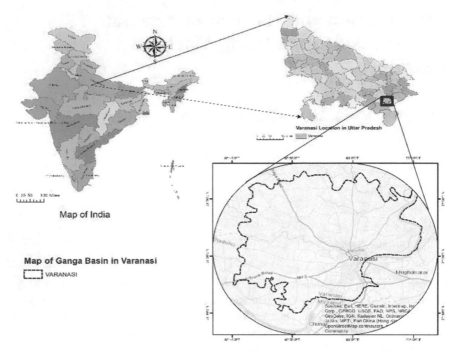

Fig. 1 Location map of the study area [2]

consumption of As in the groundwater can increase infant mortality and affect the nervous system of the children's [6–9].

The major cause of As contamination in aquifers is linked with the deposits of As-rich minerals belonging to the Holocene age which the river Ganga has been depositing in the downstream of the Varanasi region in form of alluvial sediments [10–12]. Due to the reduction of As-bearing iron oxyhydroxides (FeOOH) in presence of Fe and inorganic species which consequently release of As [13, 14, 16] into aquifers.

Most studies have revealed that As contamination in groundwater is related to various factors such as geologic, hydrogeologic, topography, Land use Land cover (LULC), Human factors, and biogeological [11, 15, 17–22]. Testing individual water sources for As poisoning is a very difficult task and involves huge manpower and laboratory testing to determine spatially high-risk regions and populations affected due to As poisoning.

The machine learning approach is accurate to establish complex relations between different attributes of water samples. However, there is no such model which accesses the hazards associated due to As contamination in the region of Varanasi. Therefore, there is a need for a model which can predict the population being affected by the As contamination to provide rapid information for improvement in monitoring and managing public health for the people of the Varanasi region.

Therefore in this study machine learning models have been used to classify water as safe or unsafe for human health. Also to determine the approximate number of people affected with As contamination in the Varanasi region. The work involves the application of different classification algorithms to classify the samples as safe or unsafe namely Simple Logistic, MLP Classifier, and Random Forest. All these models have been analyzed based on multiple evaluation criteria like accuracy, precision, Recall, and Receiver Operating Characteristics (ROC) area.

2 Materials and Methods

2.1 Study Area and Sampling

A total of 62 data points with arsenic contamination were collected along the bank of river Ganga of Varanasi region Fig. 1. Collected samples distilled and filtered before chemical analysis. Different water quality parameters (pH, EC (μScm^{-1}), TDS (ppm), Salinity (ppm), Na$^+$(mEqL^{-1}), K$^+$ (ppm), Ca^{2+} +Mg^{2+} (mEqL^{-1}), SAR, SSP (%), CO$_3{}^{2-}$ (mEqL^{-1}), HCO^{3-} (mEqL^{-1}), RSC (mEqL^{-1}), PO$_4{}^{3-}$ (ppm), Cl$^-$ (mEqL^{-1}), Mn2$^+$ (ppb), Cu^{2+} (ppb), Fe^{2+} (ppm), Zn^{2+} (ppb), SO$_4{}^{2-}$ (ppm), As (III)(ppb)) were determined (Table 1) using standards laid by American Public Health Association [23, 24] Absorption Spectrophotometer (AAS) and Vapour generation atomic absorption spectrometry [25] were used for determining the concentration of iron (Fe), manganese (Mn), zinc (Zn), copper (Cu), and arsenic (As) ions. For the determination of arsenic concentration in water, Vapour Generation Atomic Absorption Spectrometry was used [26].

2.2 Implementation of Machine Learning Algorithms

This section portrays the models used to classify the Arsenic (As) concentration in samples as safe, and unsafe. The methodology used to implement the machine learning models is shown in Fig. 2. For each model, GainRatio Attribute Evaluator [25] along with the Ranker method [26] of the search was chosen for attribute selection. 10 Fold cross-validation test option used to train and test the data sets, [27]. Finally, the models were analyzed on various evaluation criteria like accuracy, precision, recall, and ROC area.

Simple Logistic. Simple logistic regression [28] is used for posterior class with probabilities for all classes in a dataset via a linear function in x and guarantees that the total is 1 or remain between [0, 1]. It uses a vector of weights w_i and bias terms b of all the input features (Eq. 1). The classifier first multiplies each features x_i with its weight w_i and sums it with the bias b to get the expression z.

Table 1 Descriptive statistic and data description

Attribute name	Attribute description	Mode	Min	Median	Max
$Ca^{2+}+Mg^{2+}$ (mEq L^{-1})	Calcium and Magnesium (Milli Equivalent Per Litre)	5.80	1.70	5.15	10.20
CO_3^{2-} (mEq L^{-1})	Carbonate	2.00	1.20	2.00	4.40
Cu^{2+} (ppb)	Copper	22.00	12.00	30.90	74.20
Cl^- (mEq L^{-1})	Chloride	2.40	1.60	2.60	5.00
Fe^{2+} (ppm)	Iron	1.20	0.10	2.10	22.10
EC (μS cm^{-1})	Electrical conductivity (Micro Siemens/cm)	329	209	411	859
K^+ (ppm)	Potassium	2.60	0.70	6.05	44.60
HCO_3^- (mEq L^{-1})	Bicarbonate	3.60	1.80	4.65	9.20
Na^+ (mEq L^{-1})	Sodium	0.14	0.07	0.62	13.40
Mn^{2+} (ppb)	Manganese (Parts Per Billion)	36.00	0.90	26.00	97.00
SAR	Sodium adsorption ratio	0.04	0.04	0.41	9.48
SSP (%)	Soluble Sodium percentage	2.44	0.53	7.53	77.01
PO_4^{3-} (ppm)	Phosphate	3.94	0.45	3.94	14.09
pH	Hydrogen ion concentration	7.35	7.01	7.84	8.71
RSC (mEq L^{-1})	Residual Sodium Carbonate	0.30	0.10	1.15	6.90
Salinity (ppm)	Salt content	248	96	299	647
TDS (ppm)	Total dissolved solid (Parts Per Million)	214	134	266	559
Zn^{2+} (ppb)	Zinc	39.00	10.00	27.50	77.00
SO_4^{2-} (ppm)	Sulphate	48.00	8.00	48.00	98.00
As (III) (ppb)	Arsenic	6.20	3.00	7.52	25.00

$$z = w \cdot x + b \tag{1}$$

This sum of weighted features is applied to the logistic function to calculate the probability between [0, 1] (Eqs. 2 and 3).

$$P(y = 1) = (1 + \exp(-w \cdot x + b))^{-1} \tag{2}$$

$$P(y = 0) = \frac{\exp(-w \cdot x + b)}{1 + \exp(-w \cdot x + b)} \tag{3}$$

MLP Classifier. The multilayer perceptron [31] classifier is a set of layers of perceptron, with the connection of neurons between one layers to the subsequent layer which is termed as feedforward multilayer perceptron (MLP). The inputs are fed forward through the neurons by taking the dot product of the inputs and weights that exist between the input and hidden layer. The multilayer perceptron utilizes activation

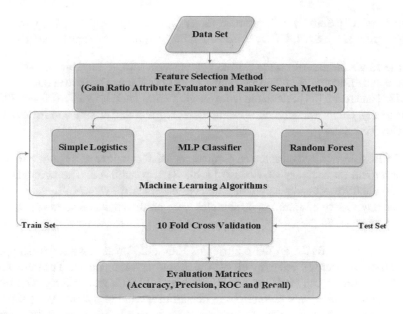

Fig. 2 Methodology for machine learning algorithms

functions at each neuron. The dot product value is passed through this activation function and further passed to the next layer taking the dot product with the weights. This process continues till the output layer is reached which uses these calculations for the classification task.

Random Forest. Random Forest (RF) [32] algorithm is used for classification and regression problems. It constructs several decision trees at the time of training and outputs predictions by combining the result from regression decision trees. Each tree constructed is independent and depends on the input data. Bootstrap aggregation and random feature selection are used for prediction. Firstly, the Random forest algorithm first picks random samples from the training dataset and constructs a decision tree for each sample. To get the prediction voting is performed from every decision tree. Finally, the most voted result is selected as a prediction result.

3   Results and Discussion

3.1   Classification of Arsenic Contamination in Groundwater

The collected samples contains 62 instances which were trained and tested by using MLP classifier, Simple Logistics and Random Forest. Out of 19 parameters the Gain Ratio Evaluator along with ranker search method was used to select 10 relevant parameters $EC(\mu Scm^{-1})$, TDS(ppm), Salinity (ppm), $Na^+(mEqL^{-1})$, K^+(ppm), Ca^{2+} $+Mg^{2+}$ $(mEqL^{-1})$, SAR, $HCO_3^-(mEqL^{-1})$, Fe^{2+} (ppm), Zn^{2+} (ppb), As(III)(ppb)).

Simple Logistic. Based on the selected parameters Simple Logistic classifier was applied and an accuracy of 79.03% (Table 2) was obtained. The percentage of precision [33] and recall [34] calculated from the confusion matrix, was 77.0% and 79.0% respectively. Out of 62 samples, 45 samples were predicted as safe and in actual the samples were safe.

MLP Classifier. Based on selected parameters MLP classifier was applied for classification and an accuracy of 77.41% (Table 3) was achieved. The percentage of precision and Recall calculated from the confusion matrix was 74.9% and 77.40% respectively. Out of 62 samples, 43 samples were predicted as safe, and in actual the samples were safe.

Random Forest. Based on these attributes Random Forest classifier was applied for classification and an accuracy of 79.00% (Table 4) was achieved. The percentage of precision and recall obtained was 76.90% and 79.00% respectively. Out of 62 samples, 44 samples were predicted as safe and in actual the samples were safe.

Experimental results obtained are compared in graph (Figs. 3a–d) for Simple Logistic, MLP Classifier, and Random Forest. Accuracy indicates the ratio of correct predictions to total predictions made. Both Random forest and simple Logistic have an equal accuracy of 79.03% (Fig. 3a) and the MLP classifier has an accuracy of 77.41%. Thus it is perceived from the results that both Random Forest and Simple Logistics performs well as compared to MLP classifier.

Precision indicates the number of true positive outcomes more closely among all positive results. Simple Logistics has the highest precision (Fig. 3b) of 77.00% as compared to Random Forest and MLP Classifier which has a precision of 74.90%,

Table 2 Confusion matrix for simple logistic

		Predicted		
		Safe	Unsafe	Total
Actual	safe	45	2	47
	Unsafe	11	4	15
	Total	56	6	62

Table 3 Confusion matrix for MLP classifier

		Predicted		
		Safe	Unsafe	Total
Actual	safe	43	4	47
	Unsafe	10	5	15
	Total	54	8	62

Table 4 Confusion matrix for random forest

		Predicted		
		Safe	Unsafe	Total
Actual	safe	44	3	47
	Unsafe	10	5	15
	Total	54	8	62

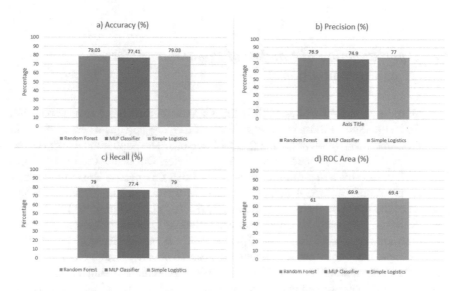

Fig. 3 Accuracy, precision, recall, and ROC area obtained for ML algorithms

and 76.90% respectively. Thus it is perceived from the results that Simple Logistics performs well as compared to others classifiers in terms of precision.

When Recall [35] is considered, both Random Forest and Simple Logistics have a high percentage of Recall value (Fig. 3c) as compared with MLP classifier. Recall signifies the sensitivity of the model. Random Forest and Simple Logistics has a high Recall (Fig. 3c) of 79.00% as compared to the MLP classifier which has a recall of 77.40%. Thus it is perceived from the results that both Random Forest and Simple Logistics perform well as compared to MLP Classifier.

The Receiver Operating Characteristics (ROC) area indicates how well, the classifier separates positive class and negative class samples. The performance of a classifier whose value of ROC near to 1 is considered best and value near to 0 is considered poor. From the result obtained it is observed that the MLP classifier has the highest value (Fig. 3d) of 69.90% as compared to Random Forest and Simple Logistics which has the value of 69.40% and 61.0% respectively.

Finally, from the results, it is concluded that the classifier which has the highest accuracy, precision, recall, and ROC area will be considered best. Both simple logistics and random forest performed equally when accuracy and recall are considered, but when precision and ROC area are considered then it is observed that Simple Logistics performance is better than Random Forest. Thus, Simple Logistics algorithm performance is good in terms of other algorithms and this model can be used to approximate the number of population affected by the As poisoning.

3.2 Approximation of Number of Population Affected by as Contamination

According to the census of India 2011, the Varanasi district has a total population of 3,676,841 persons. Pindra has a population of 627,298 persons whereas the main city of Varanasi has a population of 3,049,543 persons. The population density of the Varanasi district is 2,395 person per square kilometer. Out of the total population, 2,079,790 person lives in a rural area, whereas 1,597,051 person lives in the urban area. As per report of Central Team on Arsenic Mitigation in Rural Drinking Water [36] out of the total population of Varanasi, 2,079,790 people (56.56%) live in rural areas were people mostly dependent on groundwater for their irrigation, domestic needs and cooking. Accordingly, 1,643,034 people are exposed to As contamination in rural areas since they are dependent on groundwater with no supply of piped water. The approx. number of As affected people are calculated using the Eq. 4 [37], which shows a linear functional relationship between accuracy of the predicted As probabilities of the simple logistic model and the population in the area.

$$\text{Population Exposed} = \text{Rural Population} * \text{Predicted Probability} \qquad (4)$$

Thus, the result obtained can be used to guide the policymakers and government to tackle the As contamination in the area. To reduce exposure to As the water sources should be marked and people should be encouraged to switch to other wells or sources of water which is safe. The policymakers and government can devise some plan to supply piped water as well as other As removal techniques of filtering water to the affected areas of the Varanasi Region.

4 Conclusion

It is observed that Ca^{2+}, Mg^{2+} and Na^+ ions are dominated in the region thus water is alkaline. Out of all parameters GainRatio Attribute Evaluator selected EC (μScm^{-1}), TDS (ppm), Salinity (ppm), Na^+ ($mEqL^{-1}$), K^+(ppm), $Ca^{2+} + Mg^{2+}$ ($mEqL^{-1}$), SAR, HCO_3^- ($mEqL^{-1}$), Fe^{2+} (ppm), Zn^{2+} (ppb) which contribute to As prediction in the region. The presence of these parameters in groundwater contributes to the occurrence of As in aquifers and making it unsuitable for human consumption. To approximate the number of person affected by As contamination Simple Logistic classifier was used as its performance was better as compared to MLP Classifier and Random Forest. Out of total population of 3,676,841 persons 1,643,034 people residing in the rural areas are exposed to high As contamination which can damage the liver, bladder, cardiovascular system, skin and causes lung cancer to people living in these areas when exposed for a long time. Arsenic contamination has endangered a huge population of south Asia including India the main cause of occurrence As in still unknown. So, for prediction of As other environmental parameters like LULC, Elevation, Soil properties and other environmental factors can also be considered. Thus, the policymakers and government may devise some plan to tackle the As poisoning in the region of Varanasi.

References

1. Census of India (2011) Uttar Pradesh. India. https://www.censusindia.gov.in/
2. Esri (2021) Location map of the study area [basemap]. Scale 1:10,000,000. Varanasi National Geographic Map, 19 May 2021. http://www.arcgis.com/home/item.html?
3. WHO (2011) Guidelines for drinking-water quality. World Health Organ 216:303–304
4. Chen Y, Graziano JH, Parvez F, Liu M, Slavkovich V, Kalra T, Argos M, Islam T, Ahmed A, Rakibuz-Zaman M et al (2011) Arsenic exposure from drinking water and mortality from cardiovascular disease in Bangladesh: prospective cohort study. Biomed 342(7806):1e11. https://doi.org/10.1136/bmj.d2431
5. Chakraborty M, Mishra AK, Mukherjee A (2021) Influence of hydrogeochemical reaction flow paths on dynamics contrasting groundwater arsenic and manganese distribution across the Ganges River. Chemosphere 132144. https://doi.org/10.1016/j.chemosphere.2021.132144
6. Bhowmick S, Pramanik S, Singh P, Mondal P, Chatterjee D, Nriagu J (2018) Arsenic in groundwater of West Bengal, India: a review of human health risks and assessment of possible intervention options. Sci Total Environ 612:148e169. https://doi.org/10.1016/j.scitotenv.2017.08.216
7. Parvez F, Wasserman GA, Factor-Litvak P, Liu X, Slavkovich V, Siddique AB, Sultana R, Sultana R, Islam T, Levy D et al (2011) Arsenic exposure and motor function among children in Bangladesh. Environ Health Perspect 119(11):1665e1670. https://doi.org/10.1289/ehp.1103548
8. Rahman A, Persson LÅ, Nermell B, El Arifeen S, Ekstrom E-C, Smith AH, Vahter M (2010) Arsenic exposure and risk of spontaneous abortion, stillbirth, and infant mortality. Epidemiology 21(6):797e804. https://doi.org/10.1097/EDE.0b013e3181f56a0d
9. Wasserman GA, Liu X, Parvez F, Ahsan H, Litvak PF, van Geen A, Slavkovich V, Lolacono NJ, Cheng Z, Hussain I, Momotaj H (2004) Water arsenic exposure and children's intellectual function in Araihazar, Bangladesh. Environ Health Perspect 112(13):1329e1333. https://doi.org/10.1289/ehp.6964
10. Mukherjee A, Fryar AE, Thomas WA (2009) Geologic, geomorphic and hydrologic framework and evolution of the Bengal basin, India and Bangladesh. J Asian Earth Sci 34(3):227e244. https://doi.org/10.1016/j.jseaes.2008.05.011
11. Saha D, Sarangam SS, Dwivedi SN, Bhartariya KG (2010) Evaluation of hydrogeochemical processes in arsenic-contaminated alluvial aquifers in parts of Mid-Ganga Basin, Bihar, Eastern India. Environ Earth Sci 61(4):799–811. https://doi.org/10.1007/s12665-009-0392-y

12. Yadav MK, Gupta AK, Ghosal PS, Mukherjee A (2020). Remediation of carcinogenic arsenic by pyroaurite-based green adsorbent: isotherm, kinetic, mechanistic study, and applicability in real-life groundwater. Environ Sci Pollution Res 27(20):24982–24998. https://doi.org/10.1007/s11356-020-08868-0

13. Verma S, Mukherjee A, Mahanta C, Choudhury R, Mitra K (2016) Influence of geology on groundwater-sediment interactions in arsenic enriched tectonomorphic aquifers of the Himalayan Brahmaputra river basin. J Hydrol 540:176e195. https://doi.org/10.1016/j.jhydrol.2016.05.041

14. Drahota P, Falteisek L, Redlich A, Rohovec J, Matousek T, Cepicka I (2013) Microbial effects on the release and attenuation of arsenic in the shallow subsurface of a natural geochemical anomaly. Environ Pollut 180:84e91. https://doi.org/10.1016/j.envpol.2013.05.010

15. Chakraborty M, Sarkar S, Mukherjee A, Shamsudduha M, Ahmed KM, Bhattacharya A, Mitra A (2020) Modeling regional-scale groundwater arsenic hazard in the transboundary Ganges River Delta, India and Bangladesh: infusing physically-based model with machine learning. Sci Total Environ 748:141107. https://doi.org/10.1016/j.scitotenv.2020.141107

16. Mukherjee A, Sarkar S, Chakraborty M, Duttagupta S, Bhattacharya A, Saha D, Gupta S (2021) Occurrence, predictors and hazards of elevated groundwater arsenic across India through field observations and regional-scale AI-based modeling. Sci Total Environ 759:143511.https://doi.org/10.1016/j.scitotenv.2020.143511

17. Shamsudduha M, Uddin A, Saunders JA, Lee M-K (2008) Quaternary stratigraphy, sediment characteristics and geochemistry of arsenic-contaminated alluvial aquifers in the Ganges–Brahmaputra floodplain in central Bangladesh. J Contam Hydrol 99–4:112–136. https://doi.org/10.1016/j.jconhyd.2008.03.010

18. Shamsudduha M, Taylor RG, Chandler R (2015) A generalised regression model of arsenic variations in the shallow groundwater of Bangladesh. Water Resour Res 51:685–703. https://doi.org/10.1002/2013WR014572

19. Burgess WG, Hoque MA, Michael HA, Voss CI, Breit GN, Ahmed KM (2010) Vulnerability of deep groundwater in the Bengal aquifer system to contamination by arsenic. Nat Geosci 3:83–87. https://doi.org/10.1038/ngeo750

20. Biswas A, Nath B, Bhattacharya P, Halder D, Kundu AK, Mandal U, Mukherjee A, Chatterjee D, Jacks G (2012) Testing tubewell platform color as a rapid screening tool for arsenic and manganese in drinking water wells. Environ Sci Technol 46:434–440. https://doi.org/10.1021/es203058a

21. Biswas A, Nath B, Bhattacharya P, Halder D, Kundu AK, Mandal U, Mukherjee A, Chatterjee D, Mörth C-M, Jacks G (2012b) Hydrogeochemical contrast between brown and grey sand aquifers in shallow depth of Bengal Basin: consequences for sustainable drinking water supply. Sci Total Environ 431:402–412. https://doi.org/10.1016/j.scitotenv.2012.05.031

22. Biswas A, Gustafsson JP, Neidhardt H, Halder D, Kundu AK, Chatterjee D, Berner Z (2014) Role of competing ions in the mobilization of arsenic in groundwater of Bengal: insight from surface complexation modeling. Water Res 55:30–39. https://doi.org/10.1016/j.watres.2014.02.002

23. APHA (2005) Standard Methods for the Examination of Water and Wastewater, 21st edn. American Public Health Association, Washington, DC

24. Maite S (2001) Water and wastewater analysis. In: Handbook of methods in environmental studies, vol I. ABD Publishers, Jaipur

25. Chattopadhyay A, Singh AP, Singh SK, Barman A, Patra A, Mondal BP, Banerjee K (2020) Spatial variability of arsenic in Indo-Gangetic basin of Varanasi and its cancer risk assessment. Chemosphere 238:124623. https://doi.org/10.1016/j.chemosphere.2019.124623

26. Van Herreweghe S, Swennen R, Vandecasteele C, Cappuyns V (2003) Solid phase speciation of arsenic by sequential extraction in standard reference materials and industrially contaminated soil samples. Environ Pollut 122:323–342. https://doi.org/10.1016/s0269-7491(02)00332-9

27. Trabelsi M, Meddouri N, Maddouri M (2017) A new feature selection method for nominal classifier based on formal concept analysis. Procedia Comput Sci 112:186–194. https://doi.org/10.1016/j.procs.2017.08.227

28. Hossain MR, Oo AMT, Ali ABM (2013) The effectiveness of feature selection method in solar power prediction. J Renew Energy 2013. https://doi.org/10.1155/2013/952613
29. Fushiki T (2011) Estimation of prediction error by using K-fold cross-validation. Stat Comput 21:137–146. https://doi.org/10.1007/s11222-009-9153-8
30. Sumner M, Frank E, Hall M (2005) Speeding up logistic model tree induction. In: European conference on principles of data mining and knowledge discovery. Springer, Heidelberg, pp 675–683. https://doi.org/10.1007/1156412672
31. Gardner MW, Dorling SR (1998) Artificial neural networks (the multilayer perceptron)—a review of applications in the atmospheric sciences. Atmosp Environ 32(14-15):2627–2636. https://doi.org/10.1016/S1352-2310(97)00447-0
32. Breiman L (2001) Random forests. Mach Learn 45:5–32. https://doi.org/10.1023/A:101093 3404324
33. Stallings WM, Gillmore GM (1971) A note on "accuracy" and "precision." J Educ Meas 8:127–129. https://doi.org/10.1111/j.1745-3984.1971.tb00916.x
34. Tharwat A (2018) Classification assessment methods. Appl Comput Inform
35. Saito T, Rehmsmeier M (2015) The precision-recall plot is more informative than the ROC plot when evaluating binary classifiers on imbalanced datasets. PloS One 10(3). https://doi.org/10. 1371/journal.pone.0118432
36. MDWS Report (2011) Central team on arsenic mitigation in rural drinking water
37. Bindal S, Singh CK (2019) Predicting groundwater arsenic contamination: regions at risk in highest populated state of India. Water Res 159:65–76. https://doi.org/10.1016/j.watres.2019. 04.05

Implementing Reinforcement Learning to Design a Game Bot

Lakshay Narang⊙ and Anshul Tickoo⊙

1 Introduction

1.1 Purpose of Plan

Reinforcement learning is the training of machine learning models to take series of decisions to solve sequence of problems in an optimized manner [6]. This paradigm is often put between the supervised and unsupervised learning. It has a limited feedback mechanism therefore we can have results to test and train model against, but they are scarce. This area of machine learning is highly influenced by behavioral psychology. Let's take an example of dog—a dog can be trained to behave in a certain desired way by the playing a system of rewards and punishment. we give it a treat when it fetches the ball or performs any other desired task. It gets a favorable outcome for a specific behavior. The dog thus learns to associate these rewards and punishment to its's behavioral patterns. This whole process helped in conditioning the behavior of a dog. Similar process can be used to condition the behavior of an artificial or a virtual agent in an environment. For coding purpose this reward—punishment process can be carried out by first creating an agent and then keeping rewards (or positive reinforcement) in the form of +1 and punishment (or negative reinforcement) in the form of −1. For every step that the agent takes towards the goal state (desired task) it receives a +1 and for every step away from the goal state (undesired task) it receives a −1 penalty. But this type of approach can become confusing for the agent in many situations like the example in Fig. 1 where the agent is stuck with all the surrounding states as equally desirable with equal rewards. To curb such situations, we apply q learning algorithm.

L. Narang · A. Tickoo (✉)
Amity School of Engineering and Technology, Amity University, Noida, UP, India
e-mail: anshultickoo@hotmail.com

© The Author(s), under exclusive license to Springer Nature Singapore Pte Ltd. 2022   287
B. Unhelker et al. (eds.), *Applications of Artificial Intelligence and Machine Learning*,
Lecture Notes in Electrical Engineering 925,
https://doi.org/10.1007/978-981-19-4831-2_24

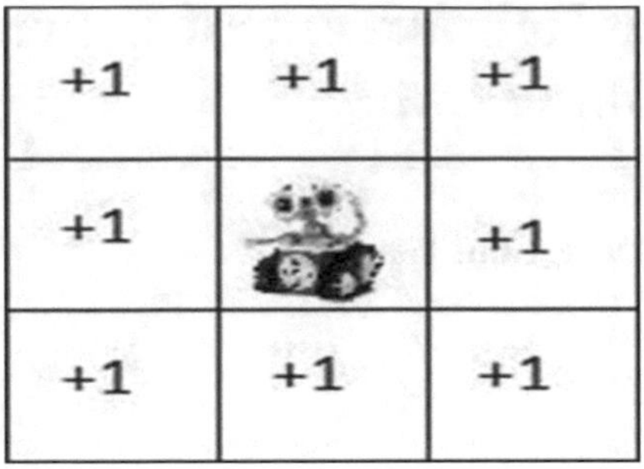

Fig. 1 A bot stuck due to equal reward in all the surrounding states [16]

2   A Dynamic Decision Problem—Bellman Equation

In dynamic programming basically we break a large problem into smaller sub problems. Then we solve the sub problems in an optimal manner recursively and then combine those solutions to give a final solution. It breaks a problem over a long period of time into small problems over small time interval t. These sub parts at a particular time instant can be described as a state. So, it keeps track of states evolving over time. For example, in a travelling salesman problem the current city he is in can be the state. Then action variables like moving forward or backward can control the state of the salesman, going to a new city signifies a new state. These action variables are capable of changing the state of object [1]. The objective of the dynamic programming approach is to fulfill the objective of the function, in the most optimal manner, by defining the rules that describe the actions to be taken to reach the objective. For example, we have a hedge fund the amount of capital in the fund can be a state and the amount of investment is the action variable so the amount of investment that can be made from the fund depends upon the capital. Hence, we can give investment as a function of the capital. Action can be given as a function of the state. This function is known as policy function. The rule that achieves the best possible objective value is defined as the optimal rule. Taking the above example, investing capital has an objective to maximize profits. A function describing the maximum profit or the best possible objective value as a function of a state is called value function. The state at time t is taken as s_t. The initial state at time t = 0 is taken as s_0. The action variable a_t at time t depends upon the current state. Action at can change the current state to next state. So the optimal value to achieve the objective function under the current circumstances is given by the following equation

$$V(s) = \sum_{t=0}^{\infty} \gamma F(s_t, a_t) \qquad (1)$$

where s is the new state that has been achieved by taking action a_t in state s_t. γ is the discount factor with value lying between 0 and 1. Bellman's principal of optimality states that "whatever the initial state and initial decision are, the remaining decisions must constitute an optimal policy with regard to the state resulting from the first decision". Applying bellman's optimality on this equation we get bellman's equation.

$$V(s) = \{F(s, a) + \gamma V(s')\} \tag{2}$$

2.1 Applying Bellman's Equation in Reinforcement Learning

A deterministic environment is where the agent will surely get into the required state when an action is performed, there is no stochasticity so there is no probability of the agent ending up in some other state instead of the same action being taken. In reinforcement learning the agent gets a reward R for every action that it takes and ends up in a new state. How does it decide on which state to end up in? Using bellman equation, it calculates the highest value it can achieve by performing an action and reaching the respective states. Out of all the states that it can end up in, it performs the action to go to the state that maximizes its value function. So bellman equation applied to reinforcement learning is

$$V(s) = \{R(s, a) + \gamma V(s')\} \tag{3}$$

where R is the reward that the agent receives when it is in state s and performs an action a. V(s') is the value of the next future state it ends up in. The above equation is very well valid for a deterministic environment but in reality we hardly have a situation with deterministic environment. To accommodate the randomness or stochasticity of environment we need to apply markov decision process in this equation.

3 Markov Decision Process

A markov decision process is a time stochastic process that follows the markov property. It helps in decision making process in situations where, outcomes are partly controlled by some agent and partly are random in nature due to stochasticity of the environment [8]. The environment is modeled to have agent/agents. The agent can have multiple states S. It can perform actions A on or in the environment to reach another state. The aim of the agent is to reach a goal state, where the required parameter is maximized, through sequence of decision-making processes. A crucial point to understand here is that each state here will be a result of its previous state and that previous state of its own previous state. Storing such vast information will become instantly unfeasible for large environments with multiple agents, multiple states and multiple action to perform. To resolve this issue the markov property comes into play.

3.1 Markov Property

It lays down the principle that—the probability of achieving any particular state in future depends only upon the present state and not the sequence of states that preceded it (past states).

3.2 Defining Markov Decision Process

MDP can be defined as a tuple (A, S, T, R) where:

- S is the set of all possible states agent can get into. At a particular time instant t the agent is in the state s_t.
- A is the set of possible actions that can be taken by the agent. The agent takes an action from the set A to reach a new state s_{t+1}.
- R is the reward function that describes the nature of reward r_t received by the agent on taking action a_t at time to reach state s_{t+1} from state s_t.

$$R : (a_t \in A, s_t \in S) \mapsto r_t \in R \tag{4}$$

- T is the transition rule where $P(s_{t+1}|s_t, a_t)$ is the probability of agent reaching into state s_{t+1} by performing the action at from the state s_t. Hence, we can say that T is the probability distribution of states at time instance $t + 1$ given that a particular action a_t has been performed on a particular state s_t at time instance t.

$$T : (a_t \in A, s_t \in S, s_{t+1} \in S) \mapsto P(s_{t+1}|s_t, a_t) \tag{5}$$

According to the above MDP definition we will again quote Markov property, "The probability of reaching a particular future state $s_t + 1$ depends upon the current state s_t and the action taken, at that instance t, at only and not on any other previous states and actions".

Modifying Bellman Equation to Accommodate Markov Decision Process
The environment around us is full of randomness. There is no guarantee that the action a_t taken by the agent in state s_t will always lead to a fixed particular state $s_t + 1$. It sometimes may lead to some other state due to the changing environment. There is a small probability that the agent ends up in a completely different state. So to account for that and thus accommodate MDP we add another variable-m probability to the bellman equation [2]

$$V(s) = \left(R(s, a) + \gamma \sum P(s, a, s') V(s') \right) \tag{6}$$

where $P(s, a, s')$ is the probability of ending up in state s' when action a is taken in state s.

4 Q Learning Algorithm

Q in Q learning stands for quality. Quality in the context that it gives a value to the quality of decision taken depending upon its usefulness. The target of this algorithm is to learn a policy that guides an agent to take a particular decision under some specific conditions. It is an off-policy algorithm which means that the agent learns by taking actions that are outside the current policy. It upgrades q values and the function learns from completely random actions, there may not be even an exact policy at all. The agent uses greedy policy to learn optimal policy or update policy, and uses a different E-greedy to learn behavior policy. Because of the difference between the upgrade and behavior policy Q learning is an off-policy algorithm [16]. The main task of the algorithm is to learn an optimal policy that maximizes the objective function or in this case the reward. The agent learns by continuously interacting with the environment. The agent takes random actions and ends up in various states, it stores the information in a q table that stores the data in the form (action, state). With every step the q table expands and soon the agent is able to develop a policy on which action to take when it is in a particular state so as to get into a better state that will help in optimizing the objective function. Due to the greedy policy it always chooses the local optima of the available values. Q value can be defined as the value of the quality of action taken to go to another state. Referring to the figure above the agent has an option to choose from 4 options i.e. up, down, left and right. The q values of all the possible actions and their respective states are calculated and then according to the greedy policy to optimize the objective function the agent goes with the action with the maximum q value. This can be represented mathematically by modifying the following equation:

$$V(s) = \left(R(s, a) + \gamma \sum_{s'} P(s, a, s') V(s') \right) \tag{7}$$

where V(s) is the value of the state, R is the reward received for taking action 'a' in the state 's' so it is represented as a function of state and action. P is the probability of ending up in state s' when action a is taken in the state s. So, we take the summation of the product of the probability of the agent ending up in a future state after taking a particular action in the present state. Now for value V(s) we calculate max of all the values possible by taking action 'a' in state 's'. So, each quality value of an action can be represented as:

$$Q(s, a) = +\gamma \sum_{s'} \left(P(s, a, s') \max_{a'} Q(s', a') \right) \tag{8}$$

This gives us the final equation to calculate the q value of an action taken from a state. This equation is used to guide agent to choose action to be taken next. The agent goes with the highest q value function to achieve a local optima, the q algorithm by achieving these local optima then slowly reaches a global optima to maximize the objective function. Given below is Q learning algorithm [16].

Algorithm parameters: step size $\alpha \in (0, I\}$, small $\varepsilon > 0$
Initialize Q(s, a), for all $s \in \delta^+$, $a \in A(s)$, arbitrarily except that Q(terminal, .) =0
Loop for each episode:
Initialize S
Loop for each step of episode:
Choose A from S using policy derived from Q (e.g., ε -greedy)
Take action A, observe R, S'

S←S'
until S is terminal

5 Deep Q Learning

As the name suggests it is the combination of q learning algorithm and the deep learning.

5.1 Need for Deep Q Learning

The q values provide a great framework to train an agent to learn to take decisions. But sometimes situations can arise where there is a large number of states and then further even larger numbers of action possible in every states. If we start making q tables for such large data it may even come upto millions of cells in the table. This will require a lot of resources not just to save the q values to memory but also when we need to search for a specific component, things can get out of hand easily. Hence we use a deep neural network to estimate the q values. The states are given in as input to the network. The network then computes q values for different possible states and then gives out the maximum q value. Deep learning helps increase the potential of the QA learning by continuously updating the weights.

5.2 Process

1. If any past experiences available, they're stored in the memory.
2. The next action to be taken is determined by the final output of the network. The target value is represented by

$$+\gamma \max_{a'} Q(s', a') \tag{9}$$

Table 1 States of Snake game

State	Right	Left	Up	Down
1	0	0.31	0.12	0.87
2	0.98	-0.12	0.01	0.14
3	1	0.10	0.12	0.31
4	0.19	0.14	0.87	-0.12

Since the target function is actually a part of the whole equation thus it keeps on changing and thus is a variable function.

3. Calculate the loss—Loss is calculated by finding the temporal difference and then finding its mean square, which is the difference between the target and the prediction and is given as:

$$TD(a, s) = +\gamma \max_{a'} Q(s', a') - Q(s, a)$$

$$Loss = \frac{1}{2} TD(a, s)^2$$

(10)

4. Backpropagate to adjust the weights and minimize the loss.

Let's take a case study of the snake game to understand this algorithm.

In the game of snake, the goal of the snake is to eat as much particles of food as possible. So, the reward can be the food and punishment is when the snake gets bumped into a wall. Additionally, it can be rewarded at each step for staying alive but such a situation could lead to snake going on about in closed shapes to keep gaining rewards, so we won't consider this. It has 4 possible actions to get out of any state i.e. up, down, left and right. We have made the states by taking into account the direction of the face of the snake, then the direction of the food with respect to snake and finally the direction of any immediate danger to snake (Table 1).

Steps involved in algorithm.

1. After the staring of game the q values are initialized randomly.
2. Then the current state s is recorded.
3. Based on this current state the snake takes action a. The action 'a' could be random or according to the neural network. Generally random actions are taken in the beginning to explore more options in the environment, but later the snake more and more depends upon the neural network to make decision to take action.
4. When an action is performed, the snake collects the reward. After which it gets into a new state. Using the below bellman equation, it adjusts the q value of the state in the q table.

$$New\, Q(S, A) = +\alpha \left[R(s, a) + \gamma max\, Q'(s', a') - Q(s, a) \right]$$

(11)

A buffer is maintained to store a particular number of transitions. It stores the initial state, the action performed, the state reached, the reward received $(R(s, a))$

and whether the game ended or not. Under the process of experience replay this data is sampled to train neural networks.

5.3 Experience Replay

It is an important part of the algorithm, after every step backpropagation is done to minimize loss function. But it wouldn't be practical or feasible to train the network on a single value after every epoch whatever value may it be. Instead of just training the network on the value of the immediate situation of environment, we need to use past experiences also to train the model. Also, a good quality result or value that we achieved in the past may never come up again in the future due to the policies being used. Updating values with only the current experience would result in a very slow convergence of the algorithm. So, it is only beneficial to maintain a buffer to store the values of variables involved in each epoch along with the change observed. These values are stored in the form of a tuple. The buffer has limited memory, hence with new tuples coming in previous ones are orderly removed. For the training purpose, a batch of definite number of tuples is selected at random from the huge data stored in the experience replay memory. This batch is then used to train the model and adjust the weights of the network. The tuples selected from the memory are used to calculate Q values and then use backpropagation to reduce the loss function, which is calculated by squaring the difference of the predicted and the target q value.

$$\text{Loss} = \left(r + \gamma \max_{a'} Q(s', a') - Q(s, a)\right)^2 \tag{12}$$

Algorithm 1: Deep Q-learning with Experience Replay[15]
Initialize replay memory D to capacity N
Initialize action-value function Q with random weights
for episode = 1, M do
Initialise sequence $s_1 = \{x_1\}$ and pre-processed sequenced $\phi_1 = \phi(s_1)$
For t=1, T do
With probability e select a random action a_t
otherwise select $a_t = \phi(s_t)$, a;Θ)
Execute action a_t in emulator and observe reward r_t and image x_{t+1}
Set $s_{t+1} = s_t$,a_t and x_{t+1} and pre-process $\phi_{t+1} = \phi(s_{t+1})$
Store transition (ϕ_t, a_t, r_t, ϕ_{t+1}) in D
Sample random minibatch of transitions (ϕ_j, a_j, r_j, ϕ_{j+1}) from D
Set y_j =
Perform a gradient descent step on ($y_j - Q(\phi_j, a_j ; \Theta))^2$ according to equation 3
end for
end for

---

6 Deep Convolutional Q Learning

It is similar to the deep learning, it has weights neurons, activation function and basically everything that deep learning has, but it additionally deploys other layers called a convolution layer and a pooling layer.

6.1 Brief About Deep Convolutional Learning

In the previous phase we were training a desired model for the game using a set of required values or vectors that we give in as inputs to the network. But in the real situations we are not always given practical or experimental values to work with. Only the view of the current game or task is available to be viewed. For such situations the bot needs to learn from the environment as a human would, i.e., though visual elements. This is where Deep Convolutional learning comes into play. For the purpose of training, we feed in every frame of the game to a convolutional neural network. Using the various layers of the network the raw image is first pre-processed and converted to a form that can be given as input to the neural network. The dimensionality of the image is reduced, and feature extraction is done during the image pre-processing. Important feature like corners objects etc. are detected during this phase. After continuous steps of compression, linearity reduction and feature extraction the image is flattened to enable it's input into the fully connected layer of the neural network (Fig. 2).

Convolution
This layer is used to apply a various number of filters on the image, these filters work to extract features from the image. We use a filter, of size of our choice like 3 * 3, called feature detector. We apply these filters on the image by overlapping them on image pixels and applying convolutional operations to get a feature map. Feature map helps us to extract important features from the images that are important and can help

Fig. 2 Complete process of making a CNN [13]

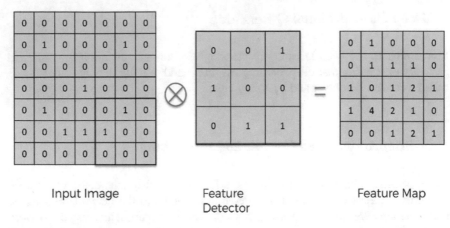

Input Image   Feature   Feature Map
                             Detector

Fig. 3 Feature detection in convolution layer [13]

in image visualisation. We create a number of feature maps using multiple filters. These feature maps form the first layer known as the convolutional layer (Fig. 3).

Max Pooling
The function of this layer is to sequentially reduce the spatial size of the feature map received from the previous layer. Each feature map is independently worked upon by the pooling layer. In max pooling we take a box of size of our choice say 2 * 2 and use this to overlap the image and then out of the four pixels choose the maximum one and write that in the pooled feature map. Pooling is very useful and practical in real world situations because it helps in reducing number of parameters and reducing size of the for computation (Fig. 4).

Flattening
All the pixels of the image are flattened into a column, taken row by row and flattened into a single column in order to feed them as input to the neural network.

Feature Map   Pooled Feature Map

Fig. 4 Pooling of features [13]

Full Connection

The flattened data is used as an input to the neural network which then gives the required output which can be a binary classification or categorical classification or even object detection in the images.

6.2 Steps Involved in Deep Convolutional Q Learning

1. Feed the image of the game screen to the DCQN, all the q values of possible actions available in the state are returned.
2. Actions to be performed is selected using epsilon greedy policy. An action is randomly selected with a probability epsilon.
3. After selection the action is performed to reach a new state and the reward is received. This next state is the pre-processed image of the game screen at next time instance.
4. Transition is then stored in the replay buffer with all it's elements (s, a, s', r).
5. Transitions are chosen at random from the transition buffer and sampled to calculate loss.
6. Perform back propagation to adjust network parameters (weights).
7. After a fixed number of iterations, copy actual network weights to target network weights.
8. Repeat the above steps for N epochs [14].

A3C

A3C stand for Asynchronous advantage actor critic. The A3C algorithm was designed by google deep mind in 2016. This algorithm has been widely accepted over deep networks for Reinforcement learning because of it's ability to give better outputs in shorter durations of time. This algorithm has an advantage because it does not use experience replay memory, which uses more computation and memory for each interaction recorded. Here, totally different process is adopted, multiple agents are executed in a parallel manner in multiple environments asynchronously. These parallel actors are not tied to apply a single exploration policy. In fact, it is beneficial if different exploration policies are being applied by different actors, A3C algorithm depends on this variation in exploration policy by different actors. Experience replay memory is not used in this algorithm because of the different number of actors involved. In the absence of experience replay the usage of different exploration policies help to provide stability to the algorithm giving the required variation for training. Apart from this, the absence of experience replay and presence of multiple actors using multiple exploration policies results in reduction of computation time and opening up of a further option of using on-policy gradient. The A3C stands for Asynchronous, Advantage and Actor Critic.

Asynchronous— As we know till now that A3C does not work like deep Q networks. Instead of a single agent working on a single environment it has multiple agents working on multiple environments in a parallel manner. The experience of each agent in it's environment is different from the other agents in their own respective environments. Each agent has it's own network parameters and exploration policy. There is a global network which gets updated by these various agents. All the agents according to their cycle update the global network individually and do not wait for the other agents to finish. This property of all the agents working in their own timeline and having independent experiences from their environment is characterised as asynchronous.

Actor Critic— This can be considered a way to combine both the q learning and policy gradient. As the name suggests, there is an actor that performs action and then there is a critic that tells the quality of the action. Because of the combination of actions involved in problem, some bad quality actions can be given good q values and create a confusion for the actor. Every step needs to be optimal and have an adequate value. This can be done by having two sperate networks, one network to choose the action that should be taken and then the other network to generate the target value for the action. The actor and the critic are two different networks. Actor calculate the policy or the action, and the critic calculates the value function.

Advantage— To describe simply, the advantage tells about how much better a certain action is than the other actions that can be taken in that particular state. The selected q value is not taken as the max of values calculated. The advantage function not only tells about the received reward but also about how much better the reward turned out to be, than expected. Advantage: $A = Q(s, a) - V(s)$.

In A3C the q values are not directly used so we will use and estimate of advantage, using discounted returns and value function.

Advantage Estimate: $A = \gamma R - V(s)$

Advantage estimate: $A = \sum_{i=0}^{k-1} \gamma r_{t+i} + \gamma V(s_{t+k}; \theta_v) - V(s_t; \theta_v)$

Where γ is the discount function, r is the reward, V is the value function and i, k and t are variables for time where k is bound by a t_{max} i.e. the maximum time (current time).

A3C Algorithm - pseudocode [15]
//Assume global shared parameter vectors Θ and Θ_v, and global shared counter T=0
//Assume thread-specific parameter vectors Θ' and Θ'_v
Initialize thread step counter $t \leftarrow 1$
repeat
Reset gradients: $d\Theta \leftarrow 0$ and $d\Theta_v \leftarrow 0$.
Synchronize thread-specific parameters $\Theta' = \Theta$ and $\Theta'_v = \Theta_v$
$t_{start} = t$
Get state s_t
repeat
Perform a_t according to policy $\pi(a_t|s_t; \Theta')$
Receive reward r_t and new state s_{t+1}
$t \leftarrow t + 1$
$T \leftarrow T+1$
until terminal s_t or $t - t_{start} == t_{max}$
$R =$
for $i \in \{t-1,.....,t_{start}\}$ do
$R \leftarrow r_i + \gamma R$
Accumulate gradients wrt : $d\Theta \leftarrow d\Theta + \nabla_{\Theta'} \log \pi(a_i|s_i;)(R - V(s_i;))$
Accumulate gradients wrt : $d\Theta_v \leftarrow d\Theta_v + (R - V(s_i;))^2 /$
end for
Perform asynchronous update of Θ using $d\Theta$ and of Θ_v using $d\Theta_v$, until $T > T_{max}$

7 Results

In our paper we have used Bellman's equation to solve dynamic programming problems. Then bellman's equation was used to define action-value function and state-value function in both deterministic and non-deterministic environments. The markov decision process was applied on equation to accommodate randomness. On implementing the above-mentioned equations and algorithms using Python programming language, we developed a bot that plays a game just like humans. The screenshots of the developed game are in the figures below (Figs. 5, 6 and 7).

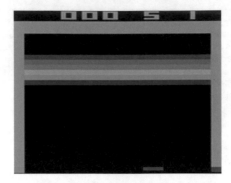

Fig. 5 Atari game breakout1

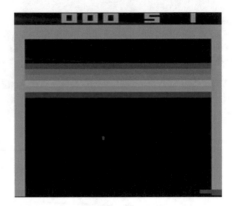

Fig. 6 Atari game breakout2

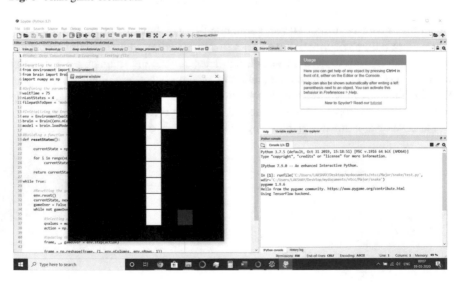

Fig. 7 Snake Game

8 Conclusion and Future Scope

We discussed about the emergence of reinforcement learning the world of AI. How reinforcement learning is being used around the globe for various tasks. Dynamic programming has been useful in breaking down problems into small sequences of problems and then find optimal solutions for all the sub problems. The Q learning algorithm uses action value function in a stochastic environment to calculate rewards other results of performing an action. Combined with DNN and CNN, Q learning algorithm can be a strong sequential decision problem algorithm and overcome some of its limitations. This combination has a lot of applications across various fields. Reinforcement learning has applications in numerous fields. A lot of very successful algorithms were inspired from nature. From the moment a child is born, he starts observing his environment, interacting and starts learning from that, similarly reinforcement learning can make computers learn and see things in the way a human can. It has a lot of potential. It is currently fragile with a lot of limitations, but this also provides opportunities because it's a little less explored. It makes way for us to work on frameworks that can address these inherent limitations. A lot of work is being done in this field but there are always opportunities to expand and innovate. Future technologies depend a lot upon the progress that takes place in this field. In future we may work to build self-aware robots and other artificial technologies that can think and work like a human, reinforcement learning is a crucial component of that master plan.

References

1. Bellman R (1954) The theory of dynamic programming, This paper is the text of an invited address before the annual summer meeting of the American Mathematical Society at Laramie, Wyoming, 2 September
2. Bellman R (1957) Functional Equations in the theory of dynamic programming, VI: a direct convergence proof. Ann Math 65:215–223
3. White DJ (1993) A survey of applications of Markov decision processes. J Oper Res Soc 44(11):1073–1096
4. Barto AG, Sutton RS, Anderson CW (1983) Neuronlike elements that can solve difficult learning control problems. IEEE Trans Syst Man Cybern 13:835–846
5. Boutilier C, Dean T, Hanks S (1999) Decision theoretic planning: structural assumptions and computational leverage. J Artif Intell Res 11:1–94
6. Mahadevan S (1996) Average reward reinforcement learning: foundations, algorithms, and empirical results. Mach Learn 22:159–195
7. Watkins CJCH, Dayan P (1992) Q-Learning. Mach Learn 8(3/4). Special issue on reinforcement learning
8. Witten IH (1977) An adaptive optimal controller for discrete-time Markov environments. Inf Control 34:286–295
9. Ratitch B (2005) On characteristics of Markov decision processes and reinforcement learning in large domains. PhD thesis. The School of Computer Science, McGill University, Montreal
10. Sutton RS (1998) Learning to predict by the methods of temporal differences. Mach Learn 3:9–44

11. van Otterlo M (2009) Markov decision processes: concepts and algorithms
12. Tokic M (2011) Adaptive ε-greedy exploration in reinforcement learning based on value differences
13. Sutton RS, Barto AG (1998) Reinforcement learning: an introduction. MIT Press, Cambridge
14. Matteo H, Joseph M, Hado van H, Tom S, Georg O, Will D, Dan H, Bilal P, Mohammad A, David S (2017) Rainbow: combining improvements in deep reinforcement learning
15. Mnih V et al (2016) Asynchronous methods for deep reinforcement learning. In: Proceedings of the 33rd international conference on machine learning, PMLR, vol 48, pp 1928–1937
16. Deep Q-learning website. https://www.analyticsvidhya.com/blog/2019/04/introduction-deep-q-learning-python/
17. Deep reinforcement learning website. https://julien-vitay.net/deeprl/ActorCritic.html
18. Fayyad J, Jaradat MA, Gruyer D, Najjaran H (2020) Deep learning sensor fusion for autonomous vehicle perception and localization: a review. Sensors 20:4220. https://doi.org/10.3390/s20154220
19. Tingwu W, Xuchan B, Ignasi C, Jerrick H, Yeming W, Eric L, Shunshi Z, Guodong Z, Pieter A, Jimmy B (2019) Benchmarking model-based reinforcement learning. CoRR, abs/1907.02057
20. Hado van H, Matteo H, John A (2019) When to use parametric models in reinforcement learning? In: Wallach HM, Larochelle H, Beygelzimer A, d'Alché-Buc F, Fox EB, Garnett R (eds) Advances in neural information processing systems 32: annual conference on neural information processing systems 2019, NeurIPS 2019, Vancouver, BC, Canada, 8–14 December 2019, pp 14322–14333

Darknet (Tor) Accessing Identification System Using Deep-Wide Cross Network

T. S. Urmila

1 Introduction

The Tor network and platform are low-latency, circuit-based, and privacy-preserving anonymizing platforms. Internet users have been able to get a high level of privacy, and anonymity through this system. One of the main design goals of Tor is to prevent communications from being linked to a single user or communication from being linked to multiple users. Tor anonymizes TCP-based applications such as web browsing, secure shell, or peer-to-peer communication utilizing a distributed overlay network and onion routing. In an Intrusion Detection System (IDS), network traffic is monitored for suspicious activity, and alerts are sent out when it is discovered. An application that detects harmful activities or policy violations on a network or system. In addition to monitoring networks for malicious activity, intrusion detection systems may also trigger false alarms. In this work, an extension of previous works is presented to build a model of Dark net traffic (Tor and Non-Tor) and predict the type of applications that run beneath Dark net traffic by making use of time, and packet-related features. The proposed method lead to 93% accuracy in detecting dark net traffic, based on the selection of the best features. According to the rest of the paper, it is organized as follows. TOR detection system has been discussed previously in Sect. 2. The third section describes the dataset, the machine learning models, the evaluation metrics we used, and the tuned hyper-parameters as well as an overview of SHAP. The results and visualizations are presented in Section and conclusion in Sect. 5.

T. S. Urmila (✉)
Thiagarajar College, Madurai, Tamilnadu, India
e-mail: urmila_cssf@tcarts.in

© The Author(s), under exclusive license to Springer Nature Singapore Pte Ltd. 2022 303
B. Unhelker et al. (eds.), *Applications of Artificial Intelligence and Machine Learning*,
Lecture Notes in Electrical Engineering 925,
https://doi.org/10.1007/978-981-19-4831-2_25

2 Review of Literature

According to Lashkari et al. [1], they used the ISCXTor2016 dataset to identify the type of data within Tor packets using binary classification algorithms and multi-classification algorithms. To classify Tor traffic, AlSabah et al. [2] categorize it into three types: streaming, interactive, and bulk transfer. Specifically, they selected four features to work with: circuit lifespan, data transfer amount, the time between entries, and the number of recent entries. Bayesian networks, functional trees, and logistic tree models were the machine learning models used. Yamansavascilar et al. [3] applied a Random Forest classifier to 111 features in Combination with ISCXVPN2016 and their internal dataset.

Chokravarty et al. [4] present an attack against the Tor network to reveal the identities (IP addresses) of the clients. The research work describes an active traffic analysis attack that deliberately perturbs server traffic characteristics (colluding server) and observes a similar perturbation at the client-side through statistical correlation. Ling et al. [5] present an analysis of Tor traffic using an Intrusion Detection System (IDS). Researchers made use of Suricata and a commercial IDS rule-set (ETPro) for this study. Usman et al. [6], present a comprehensive review of single, hybrid, and ensemble classification algorithms. In the study, the metrics, shortcomings, and datasets were compared among the studies. The future direction of potential research is also discussed. Debmalya Sarkar [7] describes a system that uses deep neural networks (DNN) to detect and classify encrypted Tor traffic. On the UNB-CIC Tor network dataset, the system classified Tor and non-Tor traffic with a 99.89% accuracy rate.

TS Urmila presented [8] an Anomaly and Signature-based collaborative detection scheme by capturing packets in real-time. Filtering and normalization are used to filter out the uninteresting traffic. Correlation-based BAT Feature Selection (CBBFS) is the algorithm that selects relevant features. Through a novel framework, T.S. Urmila proposes [9] recognizing, and labeling the intrusion occurs with Intelligent Intrusion Detection Framework (IIDF). The inspection is suspected to increase performance by including packet headers and payload data.

3 Proposed Scheme

In the proposed model for IDS for monitoring and identifying anomalous traffic in Tor networks, the classification strategy is designed to predict and handle attacks using intrusion detection. A dataset of intrusions is constructed with features including categories and pre-processed to remove missing data, duplicate data, and null values. A feature selection strategy is applied to the cleaned dataset to select the significant features, which reduces the false positive and negative rates. After that, the intrusion prediction is processed with classification techniques.

3.1 Dataset

With CIC (Canadian Institute of Cybersecurity), it was possible to create a dataset that represented real-world traffic. The browser traffic collection was set up with three users, and the communication parts, e.g., chat, mail, FTP, peer-to-peer, etc., with 'n' users. This dataset contact "7" variety type major features categories. Table 1 shows the description of the features.

3.2 Pre-processing

In this part, the work describes the steps involved in pre-processing the dataset. In first step cleaning **Missing values**. Missing values is one of the greatest challenges faced by analysts because making the right decision on how to handle it generates robust data models. In machine learning, missing value gives less accuracy due to the missing values. Next step is **Remove Duplicate Values.** Consequently, duplicate values will outperform the fitted model, so it should be cleared. This dataset has 369 duplicate values out of 67,834. Next, **Remove Unwanted Columns.** This model needs to be cleaned up by removing the unwanted columns. Two columns (Active_STD, IDLE_STD) should be removed because they contained zero values. Next, process is **Encoding Scheme.** Numeric data is required for machine learning in this work. Therefore, string data is converted to integer data here. In Table 2, the encoding value that converts the string into an integer is listed.

End of the Pre-Processing stage is **removing collinear variables.** Those variables that are highly correlated are known as linear variables. Consequently, the model's learnability, interpretability, and generalization performance may be reduced. Establish a threshold for the removal of confounding variables, and then remove one of any pair of variables above it. According to the threshold value, Table 3 shows the collinear removal.

3.3 Feature Selection

To construct efficient IDS with machine learning models, feature selection is crucial. Irrelevant data from the dataset will reduce model accuracy and increase training time. Feature selection strategies are employed to obtain significant features and build the model.

a) Select K-Best
Our dataset is transformed with SelectKBest into a selected subset of features. Following that, machine learning models are trained, tested, and validated using these features. The SelectKBest class just scores the highest-scoring highlights utilizing a

Table 1 Description of the feature

No	Features		Description
1	Identification Features (1) – These features show the identification information in the Data Packet.		
	1.1	Source IP	A field in an IP packet that contains the IP address (Total: 787) of the workstation where it originated.
	12	Source Port	Data is sent over the source port (Total: 17578), which identifies the process that sent it.
	1.3	Destination IP	The dataset contains 1736 different types of Destination IP addresses.
	1.4	Destination Port	In this dataset, there is a total of 2215 many different kinds of Destination ports available for receiving process.
	1.5	Protocol	IP packets include a protocol field that contains an 8-bit number identifying the protocol.
2	**Flow-Based Features (2)**- - A flow is a sequence of unidirectional IP packets with unique source addresses and destination addresses, along with a port number (if TCP or UDP is the transport layer protocol) and protocol signature.		
	2.1	Flow Duration	During a connection, it tracks the total time spent on the server and the client.
	2.2	Flow Bytes/s	In computer networks, it refers to how many bytes are transferred from a server to a client.
	2.3	Flow Packets/s	It is the number of packets transferred from one server to another during a connection.
	2.4	Flow IAT Mean	In a flow, it represents the mean time between two packets.
	2.5	Flow IAT STD	Flow standard deviation is the difference between two packets sent.
	2.6	Flow IAT Max	A maximum time interval between packets sent in a flow is shown here.
	2.7	Flow IAT Min	In the flow, it indicates the Minimum time between packets.
3	**Forward & Backward Features** – A list of forward-looking and backward direction features is presented.		
	3.1	Fwd & Bwd IAT Mean	A measure of the meantime elapsed between sending two packets in both directions.
	3.2	Fwd & Bwd IAT STD	Standard deviation time between two packets in a forward & backward direction can be seen here
	3.3	Fwd & Bwd IAT Max	The Maximum time between two forward & backward packets is indicated by this value
	3.4	Fwd & Bwd IAT Min	It implies a minimum interval between packets sent forward and backward
4	**Active & Idle Status**		
	4.1	Active & Idle Mean	In the meantime, a flow was active before it became active & idle
	4.2	Active & Idle Std	The average time before a flow became active & idle of the standard deviation
	4.3	Active & Idle Max	An active flow's maximum active & idle time
	4.4	Active & Idle Min	Inactivity minimum before a flow becomes active & idle
5	**Class Label** - The class label is the discrete attribute that you want to predict by analyzing other attributes		

Table 2 Encoding scheme

ID	Features	Data types/values	Encoded data type/value	Ranges of the value
1	Source IP	String/10.0.2.15	Integer/1	1–786
2	Destination IP	String/216.58.208.46	Integer/1	1–17,587
3	Label	String/{Tor/Non-Tor}	Integer/{0/1}	0–1

Table 3 Remove collinear variables

ID	Threshold	Removing columns
1	0.90	'Fwd IAT Max', 'Active Max', 'Active Min', 'Idle Max', 'Idle Min'
2	0.80	'Flow IAT Max', 'Flow IAT Min', 'Fwd IAT Mean', 'Bwd IAT Min', 'Active Max', 'Active Min', 'Idle Max', 'Idle Min'
3	0.70	'Flow IAT Max', 'Flow IAT Min', 'Fwd IAT Std', 'Fwd IAT Max', 'Fwd IAT Min', 'Bwd IAT Max', 'Bwd IAT Min', 'Active Max', 'Active Min', 'Idle Mean', 'Idle Max', 'Idle Min'

capacity and after that "removes everything but the highest scoring highlights". It's sort of a cover, but the main thing is the way it uses to score. The f-score is used in this situation by Select K-Best. As a result, class names and processes are determined by 'y'. Here is the recipe that was used: a one-way ANOVA F-test, with KK the number of particular estimates. Using SelectKBest() imprudently could result in the loss of numerous features for some undesirable reasons. With Variance Threshold, the selection of the features is simple. The feature is removed if its variance is below a given threshold. Selecting features using univariate statistical tests in univariate feature selection works by selecting the best features among the available ones. The SelectKBest feature selection selects those features that have the highest scores. Input samples are run through a score function on (X, y) to derive features for training. A training set is matched to data, and then it is transformed. The parameters of this estimator and its contained sub-objects which are estimators are returned if true. In this case, the mask is a numerical index of the features selected. In the transformation operation, zeros in the columns will be removed. In this method, an estimator is employed to determine possible parameters. Once the reduced features have been determined, X represents the selected features.

b) XGB

Machine learning algorithms such as XGBoost are very popular these days. No matter whether it is doing regression or classification. Machine learning algorithms like XGBoost are known to provide better solutions than others. As a result, it has become an important machine-learning algorithm to deal with structured data, since it was launched. Boosting uses the ensemble principle to work sequentially. A combination of weak learners improves prediction accuracy. Based on the outcome of $t - 1$, the model's outcomes are weighed at each instant. Correctly predicted outcomes are weighted lower, while incorrectly predicted outcomes are weighted higher. The weak learner is a bit better than random guessing. Using a decision tree, for instance,

Table 4 Parameters of the XGBoost

Parameter	Description	Values
Learning Rate	It is the step size shrinkage used to prevent overfitting	0.4
Maximum Depth	It determines how deeply each tree is allowed to grow during any boosting round.	20
Estimators	Total Number of trees	50
Loss Function	It's a method of evaluating how well specific algorithm models the given data.	Log loss
Sub Sample	percentage of samples used	7

Table 5 AlexNet model

Type	Input size
Conv 1	$15 \times 67{,}438 \times 1$
Max pool layer, dropout	$15 \times 27 \times 27$
Conv 2	$15 \times 33{,}719 \times 1$
Max pool layer, dropout	$256 \times 15 \times 1$
Conv 3	$15 \times 16{,}859 \times 1$
Max pool layer, dropout	$128 \times 15 \times 1$
Conv 4	$15 \times 8429 \times 1$
Max pool layer, dropout	$64 \times 15 \times 1$
Conv 5	$15 \times 4214 \times 1$
Max pool layer, dropout	$232 \times 15 \times 1$
Fully connected 1	7680
Fully connected 2	7680
Softmax	$1 \times 1 \times 2$

whose predictions are better than 50%. A simple example will help us understand boosting in general. At this point, before constructing the model, it should be familiar with the tuning parameters XGBoost provides and shows in Table 4.

The model uses it to select features from the training dataset, then trains and evaluates it using the selected subset of features on the test dataset. When selecting features by importance, the algorithm can try out multiple thresholds in Table 5.

3.4 Classification

According to some considerations, a dataset is classified as normal or as an attack. Several attack prediction techniques are employed in the work.

a) ANNN (Alex Net Neural Network)

The AlexNet system was primarily designed by Alex Krizhevsky. Geoffrey Hinton and Ilya Sutskever made a CNN, which is a convolutional neural network. The model used an eight-layer CNN called AlexNet. First-ever in computer vision, this network provided evidence that features derived through learning can have greater depth than manually-designed ones. Despite having extremely challenging datasets, ANNN has layers include five convolutional layers and three fully connected layers, as indicated in Table 6. An n-way softmax is used to generate the distribution over the class labels from the output of the last fully connected layer. Convolutional layers two, four, and five are connected only to the kernels of the previous layer which are also on the same GPU. The first convolutional layer filters the 67,438-input dataset with 15 kernels of size $15 \times 67,438 \times 1$. The second convolutional layer takes as input the output of the first convolutional layer and filters it with 33,719 kernels of size $15 \times 33,719 \times 1$. The third, fourth, and fifth convolutional layers are connected without any intervening pooling or normalization layers. The third convolutional layer has 16,859 kernels of size $5 \times 16,859 \times 1$ connected to the (normalized, pooled) outputs of the second convolutional layer. The fourth convolutional layer has 8429 kernels of size $15 \times 8429 \times 1$, and the fifth convolutional layer has 4214 kernels of size $15 \times 4214 \times 1$. The fully connected layers have 7680 neurons each. The ReLU non-linearity is applied to the output of every convolutional and fully-connected layer.

As described in Table 5, each input is routed through network architecture. In this way, the weight parameters learned are more robust and don't over fit too easily. When testing, all of the neurons are used and output is scaled by 0.5 to account for the neurons that are missed during training. Without dropout, AlexNet would overfit substantially without the additional iterations required by dropout.

b) IV3NN

The Inception v3 model [11] is widely used for image recognition and is capable of great accuracy. Several researchers have developed many ideas over the years that culminated in the model. By modifying the previous Inception architectures, the Inception v3 focuses on burning less computational power. TensorFlow provides us with ways to retrain the final Layer of Inception using transfer learning for new

Table 6 Proposed model for IV3NN

Type	Input size
Convolution	$15 \times 67,438 \times 1$
Convolution	$15 \times 128 \times 1$
Pooling	$64 \times 64 \times 1$
$3 \times$ Inception	$15 \times 22,479 \times 1$
$5 \times$ Inception	$15 \times 7493 \times 1$
$2 \times$ Inception	$15 \times 16,859 \times 512$
Convolution	$15 \times 64 \times 1$
Convolution	$15 \times 32 \times 1$
Pooling	$8 \times 8 \times 1$
Linear	$1 \times 1 \times 2048$
Soft max	$1 \times 1 \times 2$

categories. By using an inception-v3 [11] model with the transfer learning method, the last layer is removed and then retrained.

There are $15 \times 67{,}438$ rows of data in this model. First, $15 \times 67{,}438 \times 1$ convolution kernel size is applied to the 67,438 inputs. A second layer filters the output of 128 with $15 \times 128 \times 1$ kernels based on the output of the first convolution layer. This is followed by the application of the pooling layer, which has the dimensions $64 \times 64 \times 1$. In this case, inception was applied with a dimension of $15 \times 22{,}479 \times 1$, 5X inception with a dimension of $15 \times 7493 \times 1$, and secondly, with a dimension of $15 \times 16{,}859 \times 512$. Next, the final two layers are performed using kernels of $15 \times 64 \times 1$ and $15 \times 32 \times 1$. Filters can be considered as a factor of the number of feature mappings that can be extracted, and thus the performance of the model can be improved the more there are. A linear function is applied with a size of $1 \times 1 \times 2048$ and can be chosen from a variety of activation functions on the activation layer. Using the pooling layer, the network's mathematics is reduced while maintaining its main features. The adaptive optimization algorithm SOFTMAX is selected as the optimization algorithm to avoid the inconvenience of manually changing the learning rate.

c) WDCNN

Combining a wide linear model with a deep neural network (for generalization) lets us combine the strengths of both to bring us closer to our goal. Using Deep & Wide Learning is what Google does. An example of a broad component would be a generalized linear model of a wide component

$$y = w^T x + b \tag{1}$$

$$y = Predicted\,Output, b = Bias \tag{2}$$

$$x = \{x_1, x_2, \ldots x_n\} Where,\ X\ is\ Vector\ of\ 'N'\ Features \tag{3}$$

$$w = \{w_1, w_2, w_3 \ldots w_n\} Where,\ w\ is\ the\ Parameter\ of\ the\ Model \tag{4}$$

Raw input features and transformed features are included in the feature set. Cross-product transformation is one of the most important changes, which is defined as

$$\varnothing_k(x) = x_i^{c_{ki}} \tag{5}$$

$$Where, c_{ki} \in \{Intrusion\,or\,No - Intrusion\} \tag{6}$$

$$Where, c_{ki}\,is\,a\,Boolean\,Variable\,i.e., 1 \in Intrusion \tag{7}$$

If the ith feature is part of the kth transformation $\varnothing k$, and No-Intrusion otherwise. An Intrusion-like cross-product transformation consists of binary features if each constituent feature is Intrusion. It is otherwise a No-Intrusion transformation. A generalized linear model that includes nonlinearity is a tool for tracking the interaction between binary features.

In Deep Component the original inputs for categorical features are feature strings (e.g., "Protocol = TCP"). During model training, the embedding vectors are initially initialized randomly and the values are trained to minimize the final loss function. As their hidden layers are fed into a neural network, low-dimensional dense embedding vectors are then fed into the forward pass of neural networks. For each hidden layer, the following computation is performed:

$$a^{(l+1)} = f(W^{(l)} + a^{(l)} + b^{(l)}) \tag{8}$$

where, 'l' is the layer number and 'f' is the activation function, often rectified linear units (ReLUs). ra (l), b (l), and W(l) are the activations, bias, and model weights at the l-th layer. During joint training, the wide and deep parts of each parameter are both considered, along with their weights, simultaneously. Due to the disjoint nature of ensemble training, each model must be larger to achieve an ensemble with any reasonable accuracy. Using mini-batch stochastic optimization, the gradients from the output are backpropagated to the wide and deep portions of a Wide & Deep Model simultaneously. For the wide part of the model, we used Follow the-Regularized-Leader (FTRL) with L1 regularization. The deep part of the model was optimized by AdaGrad. Figure 4 shows the combined model. A logistic regression model predicts that:

$$P(Y = 1|X) = \sigma(W_{wide}^T[X, \varnothing(X)] + W_{deep}^T a^{(l_f)} + b) \tag{9}$$

$$X = Features, y = PredictedOutput, \sigma = sigmoid function, b = bias \tag{10}$$

$$(x) = CrossProductionTrasformationof X, a^{(l_f)} = FinalActivations \tag{11}$$

$$W_{wide} = WegihtofWideModel \& W_{deep} = WegihtofDeepModel \tag{12}$$

A huge dataset is used to train the Wide and Deep models. It is necessary to retrain the model whenever new training data is released. It is costly and takes a long time to retrain every time because a new version of the model has to be ordered after data arrives. Our solution to this challenge is to use a warm-start system that uses the linear model weights and embedding from the previous model to initialize the new model (Fig. 1).

WDCNN is based on the same general architecture as CNN. It consists of some filter stages and one classification stage. Convolutional kernels at the start of the filter stage have a wide range ($15 \times 67,438 \times 1$) and the following ones have a narrow range. When compared with small kernels, the first convolutional layer's wide kernels can better suppress high-frequency noise. With a layer-by-layer convolutional kernel,

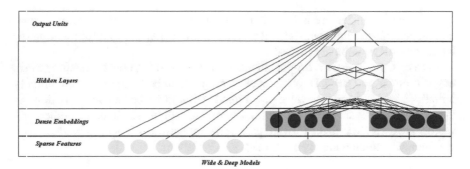

Fig. 1 Wide & deep models

the networks become deeper, which improves the representations of the input signals and the network's performance. To speed up the training process, batch normalization $(15 \times 512 \times 1)$ is implemented right after the convolutional layers and the fully connected layer.

4 Performance Evaluation

The section includes the performance evaluation for the feature selection and classification techniques to evaluate the performance while prediction.

a) Constant Features
The constant features are the3 important features are shown in Table 7.

b) Selected Features
Table 8 shows the selected features based on two feature selection algorithms. Such as SKB, XGB.

Table 7 Constant features

No	Constant features
1	Source IP, Source Port, Destination IP, Destination Port, Protocol

Table 8 Selected Features

No	Algorithm	Selected features
1	SKB	Flow Bytes/s, Flow Packets/s, Flow IAT Mean, Flow IAT Std, Flow IAT Max, Fwd IAT Mean, Bwd IAT Mean, Bwd IAT Std, Bwd IAT Min, Idle Mean
2	XGB	Flow Duration, Flow Packets/s, FlowIAT Mean, Flow IAT Std, Flow IAT Min, Fwd IAT Mean, Fwd IAT Std, Fwd IAT Min, Bwd IAT Std, Active Mean

$$n_{11} = True Positive, n_{00} = True Negative \qquad (13)$$

$$n_{01} = False Positive \quad n_{10} = False Negative \qquad (14)$$

Table 9 and 10 shows the Truth Table & classification metrics.

c) Proportion Correctly Classified (PCC): These metrics show how many tests set is correctly classified. To estimate the PCC of a test, it should calculate the proportion of true positive and true negative in all evaluated cases. Mathematically this can be stated as the following equation.

$$PCC = \frac{n_{11} + n_{00}}{n_{00} + n_{10} + n_{01} + n_{11}} \qquad (15)$$

d) Proportion Incorrectly Classified (PIC): These metrics are computed based on the proportion of instances misclassified over the whole set of instances. Mathematically this can be stated as the following equation. Figure 2 shows the PCC & PIC values of the Proposed Classifiers.

$$PIC = \frac{n_{10} + n_{01}}{n_{00} + n_{10} + n_{01} + n_{11}} \qquad (16)$$

Table 9 TP, FP, TN & FN values

FSA	Alg	TP(n_{11})	TN(n_{00})	FP(n_{01})	FN(n_{10})
SKB	ANNN	6002	48,789	2042	11,001
	IV3NN	5989	52,878	2055	6912
	WDCNN	7002	54,789	1042	5001
XGB	ANNN	5832	52,354	2212	7436
	IV3NN	6876	55,432	1168	4358
	WDCNN	7233	56,345	811	3445

Table 10 Classification metrics

FSA	Alg	PCC	PIC	FDR	FOR
SKB	ANNN	0.8077218	0.1922782	0.04017234	0.2538538
	IV3NN	0.8678097	0.1321903	0.03740921	0.25546992
	WDCNN	0.9109149	0.0890851	0.01866347	0.12953754
XGB	ANNN	0.8577704	0.1422296	0.04053806	0.27498757
	IV3NN	0.9185364	0.0814636	0.02063604	0.14520139
	WDCNN	0.9372586	0.0627414	0.01418924	0.10082049

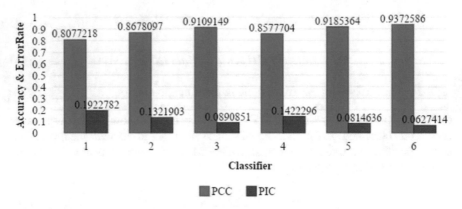

Fig. 2 Accuracy and error rate

e) False Discovery Rate: FDR is the expected ratio of the number of false-positive classifications (false discoveries) to the total number of positive classifications (rejections of the null). The total number of rejections of the null includes both the number of false positives (FP) and true positives (TP). Figure 3 shows the false discovery rate.

$$FDR = \frac{n_{01}}{(n_{01} + n_{00})} \tag{17}$$

f) False Omission Rate: The false omission rate is the proportion of the individuals with a negative test result for which the true condition is positive. Figure 4 shows the false omission rate.

$$FOR = \frac{n_{01}}{(n_{01} + n_{11})} \tag{18}$$

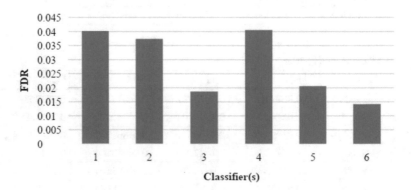

Fig. 3 False discovery rate

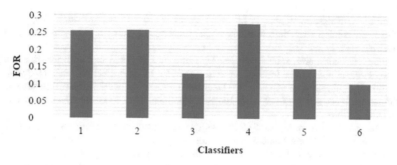

Fig. 4 False omission rate

5 Conclusion

Researchers are most attracted to IDS to reduce their trouble analyzing intrusion datasets. This work examines how machine learning techniques can increase the accuracy of Tor networks with intrusions. To prepare the dataset for training, the unwanted columns are removed to make it clean for the significant features. By utilizing classification techniques, one can reduce the false positives and false negatives in an IDS. Utilizing XGB's feature selection strategy, the classifier can be trained more quickly and more accurately by selecting significant features. Based on the proposed classifier WDCNN, better results are achieved for accuracy, error rate, false discovery rate, and false omission rate.

References

1. Lashkari AH, Gil DG, Mamun M, Ghorbani A (2017) Characterization of tor traffic using time-based features. In: 3rd international conference on information systems security and privacy (ICISSP 2017), pp 253–262
2. Al-Sabah M, Bauer K, Goldberg I (2012) Enhancing tor's performance using real-time traffic classification. In: Proceedings of the 2012 ACM conference on computer and communications security, pp 73–84
3. Yamansavascilar B, Guvensan MA, Yavuz AG, Karsligil ME (2017) Application identification via network traffic classification. In: International conference on computing, networking and communications (ICNC), pp 843–848
4. Chakravarty S, Barbera MV, Portokalidis G, Polychronakis M, Keromytis AD (2014) On the effectiveness of traffic analysis against anonymity networks using flow records. In: International conference on passive and active network measurement, pp 247–257
5. Ling Z, Luo J, Wu K, Yu W, Fu X (2014) Toward Discovery of malicious traffic over tor. In: IEEE conference on computer communications, pp 1402–1410
6. Musa US, Chakraborty S, Abdullahi MM, Maini T (2021) A review on intrusion detection system using machine learning techniques. In: International conference on computing, communication, and intelligent systems (ICCCIS)
7. Sarkar D, Vinod P, Yerima Y (2020) Detection of Tor traffic using deep learning. In: 17th ACS/IEEE international conference on computer systems and applications (AICCSA)

8. Urmila TS, Balasubramanian R (2017) A novel framework for intrusion detection using distributed collaboration detection scheme in packet header data. Int J Comput Netw Commun (IJCNC) 9(4)

9. Urmila TS, Balasubramanian R (2018) Dynamic multi-layered intrusion identification and recognition using artificial intelligence framework. Int J Comput Sci Inf Secur (IJCSIS) 17(2)

10. Urmila TS, Balasubramanian R (2015) Decision tree-based network packet classification algorithms. In: Proceedings of the UGC sponsored national conference on advanced networking and applications

11. Szegedy C, Vanhoucke V (2015) Rethinking the inception architecture for computer vision. In: 2016 IEEE conference on computer vision and pattern recognition (CVPR)

12. Yadav MK, Sharma KP (2021) Intrusion detection system using machine learning algorithms: a comparative study. In: 2nd international conference on secure cyber computing and communications (ICSCCC)

13. Shah A, Clachar S, Minimair M, Cook D (2020) Building multiclass classification baselines for anomaly-based network intrusion detection systems. In: IEEE 7th international conference on data science and advanced analytics (DSAA)

14. Ghorbani AA, Lu W, Tavallaee M (2010) Network intrusion detection and prevention. Adv Inf Secur. ISSN 1568-2633

15. Yan M, Chen Y (2021) Intrusion detection based on improved density peak clustering for imbalanced data on sensor-cloud systems. J Syst Archit 118

16. Alsudani MQ, Abbdal Reflish SH (2021) A new hybrid teaching-learning based Optimization -Extreme learning Machine model-based Intrusion-Detection system. Proc Mater Today

Energy Efficient Dual Probability-Based Function of Wireless Sensor Network for Internet of Things

Nikhil Ranjan, Parmalik Kumar, and Ashish Pathak

1 Introduction

The growth of the internet is a surprise in every field of communications. The internet of things is a bidirectional communication application in terms of local or device to device communication. In the internet of things, wireless sensors play a vital role in data quality and reliable service. The maximization of the life of the sensor network depends on the utilization of the energy factor [1]. The IoT devices play the role of edge network in the scenario of the wireless sensor network. The integration of sensors network and IoT devices support the property of being dynamic and easy to access [2, 3]. The processing of edge networks provides support of cloud-assisted services through the internet of things in the way of the network's data storage and computational capability. However, with the multiple integrations of the gateway with sensor network and cloud network, the edge network suffers from the problem of bandwidth and energy during the mode of transmission and data receiving [4–6]. The dynamic nature and support system of the sensor network utilized most of the battery-operated energy, the maximum consumption of energy compromised the life of IoT communicating devices. Despite serval energy model and lightweight protocol of energy management in wireless sensor network energy is a significant issue on the internet of things-based communication models—the overall efficiency of IoT based communication based on the life of sensor network [7]. The various research scholar proposed an energy-based model for the proper utilization of energy factors in integrating gateways. The reported survey suggests that the cluster-based

N. Ranjan (✉)
Sarvepalli Radhakrishnan University, Bhopal, M.P., India
e-mail: nikhilranjan101@gmail.com

P. Kumar
Lakshmi Narain College of Technology, Bhopal, M.P., India

A. Pathak
Freemaster Education Centre and Software Development Pvt. Ltd., Bhopal, M.P., India

© The Author(s), under exclusive license to Springer Nature Singapore Pte Ltd. 2022   317
B. Unhelker et al. (eds.), *Applications of Artificial Intelligence and Machine Learning*,
Lecture Notes in Electrical Engineering 925,
https://doi.org/10.1007/978-981-19-4831-2_26

routing protocol improves the sensor network's life and enhances the duty cycle of the internet of things. The machine learning-based algorithms apply the case of switching and harvesting the energy factors for the IoT devices [8]. The process of energy harvesting also increases the processing of multiple gateway integration with wireless sensor networks. In current research trends, most authors focus on a lightweight and low-cost routing protocol for the transmission and receiving of data in the scenario of multiple integrations of wireless sensor networks and the internet of things. Furthermore, cost and path optimization using swarm intelligence and heuristic-based function also improves energy efficiency, and the quality of services in IoT enables communication [8, 9]. Other ways to harvest the energy level in the internet of things in the mode of switching are active and passive. The processing of active and passive save the mode of energy reduces the loss of data packet and increases the life of the network and multiple gateways. The probabilistic based function contributes more to the management of energy in a wireless sensor network. The incremental enhancement of probabilistic function derives a dual probability-based function for energy management as an integral function of IoT devices [10]. The concept of dual probability function measures the level of energy in different levels of network and coordinates with sink node for proper communication. The levelling of energy is categorized into three sections: low level, middle level, and high level. The low level of energy processing cannot involve in the process of communication. The middle and high-level energy scheme integrates with gateways of cloud and IoT [11]. The proposed model enhances the service and reliability of IoT enabled communication systems. The methodology of dual probability function cooperates with compressed sensing methods for energy minimization in cloud-assisted IoT networks [12]. The rest of paper organized as in Sect. 2 related work in Sect. 3 proposed model and algorithm in Sect. 4 describe the experimental analysis and finally conclude in Sect. 5.

2 Related Work

The advancement of internet technology derived the concept of the intelligent Internet is called the Internet of things. The Internet of things provides services in all eras of society. The things deal with electronic communication objects connected through the Internet. The acceptability of IoTs is increasing day to day due to easy installation and low-cost maintenance. IoTs change the scenario of remote area data accessing for remote area data such as temperature, pressure, weather and fire event in the forest used sensors networks. The sensors collect the information and transmit it to the base station with IoT devices. Now a day the IoTs application integrates with cloud-based services. Cloud-based services are deployed over intelligent devices [12]. The cloud services support the static infrastructure; IoTs integrates these services with dynamic infrastructure. The dynamicity of the cloud enhances the reachability of IoTs services with the last person in the universe. The IoTs connects real-world objects and embeds the intelligence in the system to smartly process the object specific information and take good autonomous decisions [13]. Thus, IoTs can give birth to enormous valuable applications and services that we never imagined before. With technological advancement, the devices processing power and storage capabilities significantly increased while their sizes reduced. These intelligent devices are usually equipped with different types of sensors and actuators. Also, these devices can connect and communicate over the Internet that can enable a new range of opportunities [14]. Moreover, the physical objects are increasingly equipped with RFID tags or other electronic bar codes that can be scanned by smart devices, e.g., smartphones or small embedded RFID scanners. IoT is the Internet's extending and expanding to the physical world, and its related properties include focus, content, collection, computing, communications and connectivity scenarios. These properties show the seamless connection that between people and objects or between the objects and objects [15, 16]. In [1] authors describe the methods of energy efficiency for wireless sensor networks. The proposed algorithm uses particle swarm optimization to select cluster heads during communication with other cluster heads. The proposed algorithm reduces the utilization of energy and boosts the life of smart things. In [2] authors proposed the methods of energy optimization based on machine learning algorithms. The proposed algorithm enhances the performance of energy in the scenario of the internet of things. The methods applied on the mode of even and odd basis in-network and routing balanced the process of energy optimization. Finally, in [3] authors proposed the methods for energy optimization based on an intelligent fuzzy-based network. The proposed algorithms describe the process of packet-based operation, maintain the switching process, and reduce the impact of energy in IoT communication. [4] authors proposed the method of the dynamic stochastic optimization process for balancing the energy factors in IoTs communication. The design model focusses on the computational efficiency of energy and others parameters for the process of routing. The proposed algorithm is very efficient with certain limitations. The authors [5] proposed a neural network-based optimization model to control data storage and communication traffic, and energy in a wireless sensor network. The applied neural network model selects the optimal sensor node for the communication of the internet of things. The experimental results of the proposed methods suggest that the proposed algorithm is very efficient for wireless sensor networks. In [6] authors proposed the methods based on multi-objective constraints of energy optimization. The process of optimization resolves the issue of optimal node selection in an extensive wireless sensor network. The proposed algorithm uses a memetic algorithm for the sub-group of wireless networks. In [7] authors design the sensor network based on the human body. The human body is a good transmission section of the wireless network. The authors also applied the probability function methods to estimate the energy level of the sensors node. In [8], the authors investigate energy level performance in the internet of things using clustering algorithms and swarm intelligence algorithms such as ant colony optimization. The investigation process suggests that a clustering process is a handy approach for optimising energy in the internet of things. Furthermore, the ant colony optimization algorithm helps to estimate the optimal node selection in a sensor network. In [9] authors proposed the chicken-based optimization algorithm for the clustering of the wireless sensor network. The proposed algorithm is very efficient for the selection of optimal nodes in wireless sensor network environments.

3 Proposed Methodology

The proposed methodology describes in three parts first one dual probability function, second part describe the node distribution in IoT communication and finally in third part describe the integration with cloud network. The dual probability function (DPF) is derived function of probability and its estimate the level of energy in dense network in forward as reverse. The system model describes the approach of integration factor to cloud network. The major contribution of this algorithm is estimation of energy in three levels and reduces the cost function of dense IoT devices.

3.1 Dual Probability Function (DPF)

The derived probabilistic function applies on the source node to measure the level of energy for the categorization of sensors nodes to integrate with gateway. The DP function estimates the level of energy in all condition as directed by communication devices [2, 5].

$P - DP(n)$: Level of energy.

sink probablity $(level - Dp(n))$ of a sensor's node n with respect to another sensor's node

$$nlevel - prop_k(p, o) = \max\{n - probablity(o), n(p, o)\} \tag{1}$$

where $n(p, o)$ is the similar probability between p and o.

sub level probablity (slp) of a sensor's node n

$$slp_k(p) = \left(\frac{1}{k} \sum_{o \in N_{(p,k)}} level - prob_k(p, o) \right)^{-1}, \tag{2}$$

where $N_{(p,k)}$ is the set of n node of similar probability of energy of n.

sink node energy of a sensor's node n

$$DP\_OT_k(p) = \frac{1}{k} \sum_{0 \in N_{(p,k)}} \frac{slp_k(o)}{slp_k(p)} \tag{3}$$

The value of K is sub group of network and N is total number of nodes.

3.2 System Model

Given $n_t \in \mathfrak{R}^{GP}$ collected at level energy $e \in E$, the goal of DP and DP_E is to assign an DP_OT value to n_t, for the value of energy $E < P$ of the n nodes that have been measured up to level energy E. All n sensor level and their corresponding DP_OT values in sink connection of subgroups. Hence measure the energy value of extended energy DP_OT values of new nodes can be calculated. the goal of DP and DP_E is to detect same level for the whole network's communication and not just for the n last sensor nodes where the available sink connection of subgroups is limited to P. DP_E is an extension to DP.

DP—Dual Probability
DP_OT—Dual Probability Outer Sensor Node
DP_E—Extension to Dual Probability

3.3   Algorithm 1: DP_E Estimation

Input: a sensors noden_t at level energy e
Output: DP_OT value $DP\_OT_{(n_t)}$
Estimate $N_{(n_t,k)}$ and $p - probab(n_t)$
for all $n \in N_{(n_t,k)}$ do
Estimate $level - prob(p_t, o)$ using Equation (1)
end for
$$S_{node} \leftarrow Pn_{(n_t,k)}\{the\ set\ of\ sink\ node\ n_t\}$$
for all $n \in S_{node}$ and $q \in P_{(o,k)}$ do
Node $k - prob(o)$ and $level - prob(q, o)$
if $o\ N_{(q,k)}$ then
$$S_{node} \leftarrow S_{node} \cup \{q\}$$
end if
end for
for all $o \in S_{node}$ do
Node $slp(o)$ and $DP\_OT(\{Pn_{o,k}\})$
end for
Estimate $slp_{(n_t)}$ and DP_OTn
return DP_OT

3.4   Algorithm 2: DP (Dual Probability)

Input: set of sensor nodes $N = \{n_1, \ldots \ldots, n_n\}$
sink connection of subgroups size limit of L
Output: set of $DP\_OT = \{DP\_OT(n_1), \ldots \ldots., DP\_OT(n_n)\}$ nodes
$i \leftarrow 0$; {energy level}
for all $n_t \in n$ do
$DP\_OT(n_t) \leftarrow$ connection (n_t)
if Number of sensor nodes in sink connection of subgroups
$$(sc^i, N^i) \leftarrow GP - Probal\ (n^i)$$
for all $sc_j^i \in n^i$ do
Estimate $k - probab\ (sc_j^i), slp(v_j^i), DP\_OT(v_j^i)$
end for
return DP_OT

The design layout of network models describes in figure q and Fig. 2. The processing of node mention as PD as selection node and BS is base node or sink node.

Integration of IoTs with dual probability-based function with application with internet gateway. The measure probability value DP and DP_OT creates different level of energy group for the integration of application.

If $DP\_OT = 0$), the energy level of all sensor's node is same level and direct connect each node with sink nodes.

If $DP\_OT > 0$) the value of energy level of DP is lower and creates more similar probability-based node connection.

If $DP\_OT < 0$), the value of energy level is average and some node are not alive so gateway only maintain a connection.

3.5 Algorithm 3: Integration of Sink Nodes

Input: Sensor nodes set $s = \{n_i, i = 1, \dots \dots, n\}$
Output:S - the selected sensor nodes subset
Begin
Define the initial value of request $R = \emptyset$
Calculate $GP(DP\_OT; \; ni)$ for each sensor nodes, $i = 1, \dots \dots \dots ., n$
$n_f = n$; Select the sensor nodes ni
Then, set $S \leftarrow S\{ni\}; S \leftarrow S \cup \{ni\}; n_s = n_s - 1$
while$s \neq \emptyset$ do
Measure DP_OT in (2) to find n_i where $i \in \{1,2, \dots \dots, ns\}$;
$n_s = n_s - 1$;
if $(DP\_OT > 0)$then
$S \leftarrow S \cup \{n_i\}$
end
end
Set level of energy of each group selected sensor nodes.
return S

Figure 3 represents the real scenario of the communication of wireless sensors network with IoTs and cloud-based services. I have deployed network send the information to sink node (Base station) and base station integrate with the waterway through IoTs elements. The mention DPF represents the value of energy level for the selection of sink node to communicate to things.

Integration of IoTs with dual probability-based function with application with internet gateway. The measure probability value DP and DP_OT creates different level of energy group for the integration of application.

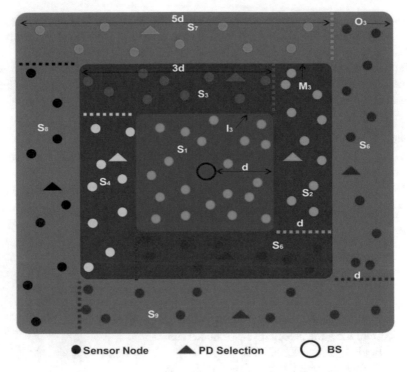

Fig. 1 Scenario of dense sensor network for the selection of PD node and process of BS (base station)

1. If $DP\_OT = 0$), the energy level of all sensor's node is same level and direct connect each node with sink nodes
2. If $DP\_OT > 0$) the value of energy level of DP is lower and creates more similar probability-based node connection.
3. If $DP\_OT < 0$), the value of energy level is average and some nodes are not alive so gateway only maintain a connection.

4 Experimental Results

To evaluate the performance of the proposed model for cloud integration with IoT simulated in MATLAB environments. The distribution of nodes mentioned in Fig. 1 and Fig. 2—the number of sensor node selections in a random fashion and fixed patterns. The selection of nodes cannot impact on design model—the process of simulation conducted on the scenario of 50,75,100 and 5000 nodes. The traffic of data is CBR 512. The dual probability function (PDF) controls and manage the level of energy value. Initially, all sensors node assigns the range value of energy is 12–14 J. The data are collected from the sink node is every 120 s. The transmission interval of data is 10 s. The total simulation time is 600 s. The process includes aggregated connectivity data, representing the link quality between any two sensor nodes and between sensor nodes and the base station. In our simulations of the DPF protocol, we selected a time interval of 120, such that at least 60% of data are transmitted successfully to the base station. In our simulations, the integration factor deals with algorithm three and sets the condition value DP_OT = 0, DP_OT > 0 and DP_OT < 0. in the case of DP_OT < 0, the data transmission process is terminated. Analyzed the performance of the proposed model and existing model estimates these parameters, energy utilization parameter, data utility and data error correction during the transmission [2, 7].

RESULT ANALYSIS

See Tables 1, 2 and 3.

The Fig. 4 describes the analysis of results in terms of proposed algorithm as well as existing algorithm in variation of sensor node as 5, 10, 15, 20, 25, 30, 35, 40, 45. In the case of node variation the results of proposed algorithm is increases instead of existing algorithms.

Figure 5 describe the process of energy utilization in mode of variable nodes such as 5, 10, 15, 20, 25, 30, 35, 40, 45 sequentially decreased percentage 2, 7, 15, 25 and 33%. The proposed algorithm gains maximum utilization of energy factor instead of existing algorithm.

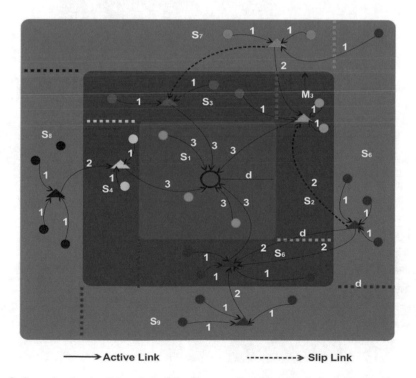

—————▶ Active Link   ------------▶ Slip Link

Fig. 2 Scenario of active link and slip link of inner node and outer node for the base station

Fig. 3 Proposed system model of IoT enable cloud data storage

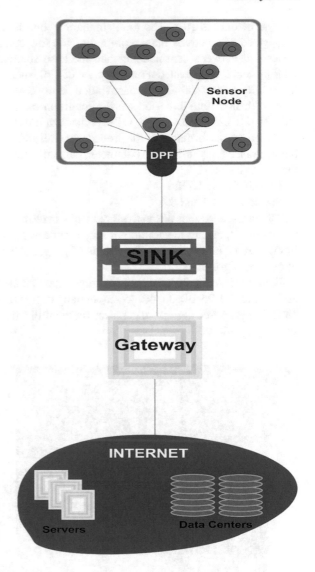

Table 1 Comparative result of data utility using reference [1], reference [5] and proposed techniques

Number of nodes	5	10	15	20	25	30	35	40	45
Reference [1]	−110	−50	0	−50	−150	−200	−250	−275	−280
Reference [5]	−20	25	35	30	−10	−50	−100	−200	−270
Proposed	25	60	60	80	70	40	10	−100	−250

Table 2 Comparative result of energy consumed in transmission using reference [1], reference [5] and proposed techniques

Number of nodes	5	10	15	20	25	30	35	40	45
Reference [1]	300	500	800	1200	1500	1900	2800	3200	3500
Reference [5]	400	600	1000	1500	2000	2400	3200	3500	3800
Proposed	200	400	600	1000	1200	1600	2500	2800	3000

Table 3 Comparative result of data reconstruction error using reference [1], reference [5] and proposed techniques

Number of nodes	5	10	15	20	25	30	35	40	45
Reference [1]	12	11	10	9	8	8	7	5	4
Reference [5]	15	14	13	12	10	9	8	7	5
Proposed	11	10	9	8	7	6	6	4	2

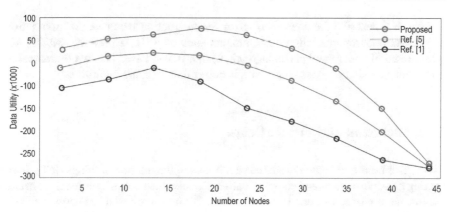

Fig. 4 Comparative result analysis between data utility with number of nodes (0–45)

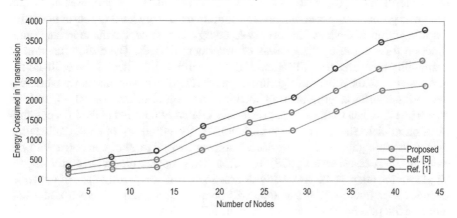

Fig. 5 Comparative analysis between energy consumed and number of nodes (0–45)

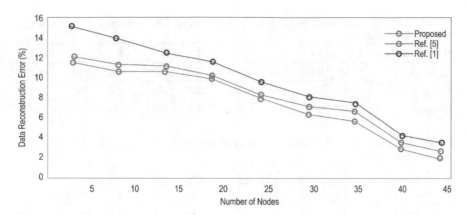

Fig. 6 Comparative result analysis of data reconstruction and number of nodes (0–45)

Figure 6 describe the analysis of error reconstruction parameter with variation of sensors node in simulation environments such as 5, 10, 15, 20, 25, 30, 35, 40, 45 sequentially decreased percentage 0.2, 0, 0.5, 0.8 and 1%. The rate of deceasing error indicates that proposed algorithm is better than existing algorithm.

5   Conclusion and Future Scope

The sensor node and networks are the backbones of the internet of things (IoT). Integrating multiple devices and communication models consumes a high energy rate and expires the life of sensor nodes and networks. We have proposed an energy-efficient integration model for the internet of things. The proposed algorithm reduces the energy level during the transmission and receiving of data. The proposed algorithm designs a probability-based function to estimate the energy level of sensor nodes for the integration of devices. The estimated energy factors decide the inner and outer function for connecting the gateway of intelligent devices. The design function of probability measures the utilization of data quality of IoTs device. To evaluate the performance of the proposed algorithm, use MATLAB software and design different network scenarios in terms of small, medium and large scale, such as 50, 75, 100 and 500. The UDP data traffic is used to estimate a delivery function of the IoTs communication model. The proposed model increases the efficiency of energy utilization during the integration of IoT devices. The proposed algorithm compares with the existing compressive sensing (CP) algorithm. The proposed algorithm performs only single source node. If the node of the source increases, the performance of networks decreases. In future, the single node will proceed in a group of sensor nodes and enhance the possibility of energy factors in IoT devices.

References

1. Senthil GA, Raaza A, Kumar N (2021) Internet of things energy efficient cluster-based routing using hybrid particle swarm optimization for wireless sensor network
2. Chen JIZ, Lai K-L (2020) Machine learning based energy management at Internet of Things network nodes. J Trends Comput Sci Smart Technol (3):127–133
3. Lavanya R, Kavitha MG, Radha D (2021) Intelligent fuzzy enabled wireless sensor network energy optimization scheme using IoT
4. Sundhari, Meenaakshi RP, Jaikumar K (2020) IoT assisted hierarchical computation strategic making (HCSM) and dynamic stochastic optimization technique (DSOT) for energy optimization in wireless sensor networks for smart city monitoring. Comput Commun 150:226–234
5. Govindaraj S, Deepa SN (2020) Network energy optimization of IOTs in wireless sensor networks using capsule neural network learning model. Wireless Pers Commun 115(3):2415–2436
6. Ahmad M, Shah B, Ullah A, Moreira F, Alfandi O, Ali G, Hameed A (2021) Optimal clustering in wireless sensor networks for the Internet of things based on memetic algorithm: memeWSN. Wirel Commun Mobile Comput 2021
7. Thabit AA, Mahmoud MS, Alkhayyat A, Abbasi QH (2019) Energy harvesting Internet of Things health-based paradigm: towards outage probability reduction through inter–wireless body area network cooperation. Int J Distrib Sensor Netw 15(10):1550147719879870
8. Kumar S, Solanki VK, Choudhary SK, Selamat A, Crespo RG (2020) Comparative study on ant colony optimization (ACO) and K-means clustering approaches for jobs scheduling and energy optimization model in internet of things (IoT). Int J Interact Multimedia Artif Intell 6(1)
9. Osamy W, El-Sawy AA, Salim A (2020) CSOCA: chicken swarm optimization based clustering algorithm for wireless sensor networks. IEEE Access 8:60676–60688
10. Bijarbooneh FH, Du W, Ngai EC-H, Fu X, Liu J (2016) Cloud-assisted data fusion and sensor selection for internet of things. IEEE, pp 257–268
11. Kaur K, Dhand T, Kumar N, ZeaDally S (2017) Container-as-a-service at the edge: trade-off between energy efficiency and service availability at fog nano data centers. IEEE, pp 48–56
12. Nguyen TD, Khan JY, Ngo DT (2016) Energy harvested roadside IEEE 802.15.4 wireless sensor networks for IoT applications. Ad Hoc Netw, 1–14
13. Wu S, Niu J, Chou W, Guizani M (2016) Delay-aware energy optimization for flooding in duty-cycled wireless sensor networks. IEEE Trans Wirel Commun 8449–8462
14. Shen H, Bai G (2016) Routing in wireless multimedia sensor networks: a survey and challenges ahead. J Netw Comput Appl 30–49
15. Chakchouk N (2015) A survey on opportunistic routing in wireless communication networks. IEEE Commun Surv Tutor 2214–2251
16. Jiang J, Han G, Guo H, Shu L, Rodrigues JJPC (2015) Geographic multipath routing based on geospatial division in duty-cycled underwater wireless sensor networks. J Netw Comput Appl 1–15

OFDMA Based UAVs Communication for Ensuring QoS

Muhammet Ali Karabulut, A. F. M. Shahen Shah, Md Baharul Islam, and Muhammad Ehsan Rana

1 Introduction

An attractive alternative to wireless services has been recognized in recent years by aerial communications platforms. Unmanned aerial vehicles (UAVs), offer human protection concerning the use of human vehicle missions. A wide range of military and civil tasks, including monitoring environment and meteorology, forest fire management, agriculture monitoring, support for surveillance, search and rescue missions, radar locality, border monitoring and aerial photography, etc. are supported by UAVs [1–5]. The UAV ad hoc networks also termed as the FANETs, as shown in Fig. 1, enable UAV-to-infrastructure (UAV2I) communication and UAV-to-UAV (UAV2UAV) communication. UAVs are being touted as one of the most promising technologies for next-generation wireless networks. In a similar spirit, artificial intelligence (AI) is fast evolving and proving to be highly successful, thanks to the vast quantity of data accessible. Therefore, a large portion of the research community has begun to integrate intelligence into the heart of UAV networks by using AI algorithms to solve a variety of drone-related issues. The impact of the height of UAV

M. A. Karabulut
Department of Electrical and Electronics Engineering, Kafkas University, Kars, Turkey
e-mail: mali.karabulut@kafkas.edu.tr

A. F. M. Shahen Shah
Department of Electronics and Communication Engineering, Yildiz Technical University, Istanbul, Turkey
e-mail: shah@yildiz.edu.tr

M. B. Islam
School of Data Science and Engineering, American University of Malta, Bormla, Malta

M. E. Rana (✉)
School of Computing, Asia Pacific University of Technology and Innovation, Kuala Lumpur, Malaysia
e-mail: muhd_ehsanrana@apu.edu.my

on coverage, link reliability, performance, and service availability are key issues for designing the FANETs. Recent studies [6–8] have shown interest in this issue. To our knowledge, there has never been a study of the relationship between UAV height and coverage that incorporates more accurate wireless link statistics. In this paper, the height of the UAV is considered in the analysis.

The EDCAF of IEEE 802.11 is used to support QoS [9–11]. There are four access categories (ACs), including AC0 (Voice), AC1 (Video), AC2 (Best effort) and AC3 (Background), according to IEEE 802.11 EDCAF [11, 12]. The priority among ACs is defined by various EDCA parameters, for example, the maximum and minimum contention window size (ς_{max} and ς_{min}), the arbitration interframe space number (AIFSN), etc.

The CSMA/CA approach is worthy if traffic is low. Nevertheless, data rates decrease and delay increases because of greater collision when there is high traffic. In contrast, OFDMA can provide a greater data throughput and reduced latency in high traffic [13]. OFDM in the PHY layer brings frequency as well as time multiplexing together, that provides high spectrum efficiency and solves the problem of hidden nodes. OFDMA appears to be an option for high data rate communication [14–17]. To exploit the benefit of OFDMA, we develop an OFDM analysis framework for UAV communications systems. This paper contributions are summarized below.

The paper offers an analytical study to investigate IEEE 802.11 EDCAF performance for OFDMA based UAVs communication. Parameters that could influence FANET performance are taken into account. Markov's chain model-based theoretical study derives a relationship between EDCA parameters and performance metrics. A new channel contention method that combines EDCAF and OFDMA is developed. The first step is to do EDCA-based prioritized channel contention. Then OFDMA is used for channel access and transmission. The proposed OFDM-based protocol has been analyzed with simulation results. It has been observed that the proposed OFDMA-based protocol has a high data rate when there is high traffic. The protocol we recommend provides consistent communication by reducing packet drop rate (PDR) and greater throughput. In addition, our proposed protocol reduces latency and meets the strict 100 ms latency requirements.

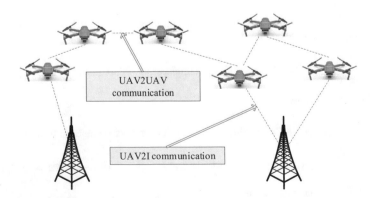

Fig. 1 Basic FANETs structure

The rest of this paper has the following structure: Sect. 2.1 describes the EDCAF of IEEE 80,211 and the analytical analysis of the Markov Model Chain. The throughput and the PDR analysis are carried out in Sect. 2.3. Section 3 presents delay analysis. Simulation results are presented in Sect. 4. Section 5 provides a conclusion with future research directions.

2 System Model

The height of UAV that maximizes area, where it can provide a reliable wireless connection, is formulated. For this purpose, first of all, the area of coverage is demarcated according to the concept of probability of outage. As shown in Fig. 2, a land-air communication network where the UAV is positioned at an elevation of h meters, provides a wireless connection to the nodes on the ground is considered. r is the space between terrestrial node and the UAV's projection onto the ground.

For a given G node, the probability of outage of terrestrial aerial communication link is expressed as:

$$
\begin{aligned}
P_{out}(h, r) &= P\left(\frac{AP_T}{N_0 R^\alpha} \leq \gamma_{th}\right) \\
&= 1 - Q\left(\sqrt{2(h/r)}, \sqrt{2\gamma_{th}(1 + (h/r)R^\alpha/(AP_T/N_0))}\right)
\end{aligned}
\tag{1}
$$

where $P(.)$ denotes probability, $Q(.,.)$ is Marcum function, γ_{th} is a threshold value of SNR. A is a constant describing the impact of parameters. P_T is the transmitter's radiated power, N_0 is the power of noise. α is the path loss exponent.

It is assumed that total number of subcarriers available is 64, where only 52 are utilized to map. To carry the pilot signal, four subcarriers among 52 subcarriers are designated. To forecast channel and check alterations in the signal, which is transmitted, pilot symbols shall be used. Table 1 indicates the IEEE 802.11 OFDMA parameters [18].

Fig. 2 Terrestrial aerial communication with UAV

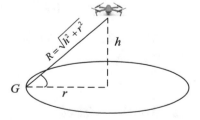

Table 1 OFDMA parameters in IEEE 802.11 standards

Parameter	Value
Total number of subcarriers	64
Pilot subcarriers	4
Data subcarriers	48
Null subcarriers	12
Frequency spacing of subcarriers	312.5 kHz
Bandwidth	20 MHz

Table 2 EDCA parameters of each ACs

AC	ς_{min}	ς_{max}	AIFSN
0	ς_{min}	ς_{max}	2
1	ς_{min}	ς_{max}	3
2	$(\varsigma_{min} + 1)/2 - 1$	ς_{min}	6
3	$(\varsigma_{min} + 1)/4 - 1$	$(\varsigma_{min} + 1)/2 - 1$	9

2.1 IEEE 802.11 EDCAF

DCF does not distinguish between different traffic classes in IEEE 802.11. Alternatively, QoS on MAC layer supports IEEE 802.11 EDCAF [19–21]. Every AC queue serves as an autonomous DCF STA with EDCAF employing their EDCA parameters to compete for transmission opportunities. The priorities of users are divided between 4 various first-in-first-out queues referred to as categories of access (ACk) (k = 0, 1, 2, 3). Table 2 gives EDCA parameters for each AC. Figure 3 shows the priority mechanism for each STA. 4 transmission queues and 4 independent EDCAF are available for the various ACs. Each AC queue employs distinct AIFS, ς_{max} and ς_{min} because of its different EDCA parameters for each AC. EDCAF uses arbitration inter-frame space (AIFS) as a different IFS for implementation of priority for transmission. AIFS represents idle intermission after a busy intermission that can be obtained directly from value of AIFSN of AC shown as

$$T_{AIFS}[AC] = AIFSN[AC] \times T_{slot} + T_{SIFS} \tag{2}$$

where T_{SIFS} and T_{slot} are short interframe space (SIFS) and duration of an empty slot time, respectively.

2.2 Markov Model

A FANET which consists of N is deliberated and UAV are randomly distributed on the street. Let s(t) be backoff stage for a certain UAV at time slot and b(t) be backoff

Fig. 3 EDCA mechanism

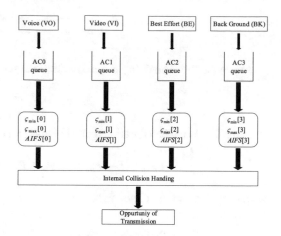

counter respectively which are stochastic processes. b_c is backoff counter value where $b_c \in (0, \varsigma[AC] - 1)$. Initially, at each packet transfer, the value of b_c is chosen evenly between 0 and $\varsigma_0[AC] - 1$. When channel is free, value of b_c is reduced by 1. b_c stops at the current backoff value while channel is busy and resumes when channel is still free. If b_c is zero, it transfers the packet. Although a failing transmission is present, the ς is multiplied by 2 and the backoff stage is increased by 1. The number of unsuccessful transmissions for packet is dependent on ς. ς is doubled after each failed transmission, equal to an end value known as maximum contention window ($\varsigma_{max} \leq m \, r_{min_{max}}$) where m_r is the maximum retransmission limit.

$b_{i,b_c} = \lim\limits_{t \to \infty} P\{s(t) = i, b(t) = b_c\}, (i \in (0, m_r), b_c \in (0, \varsigma_i[AC] - 1))$ is Markov chain's stationary distribution [22]. Let P_t be the chance of packet transmission in a time slot that is randomly selected. P_t can be given as

$$P_t = \frac{2}{(1 + \varsigma[AC] + m_r \varsigma[AC]/2)/N_{sc}} \tag{3}$$

where N_{sc} is the number of subcarriers. If 1 of the remaining $N - 1$ UAVs in transmission range (R_t) transfers a packet, a collision will occur. P_c can be given as

$$P_c = 1 - (1 - P_t)^{N-1} \tag{4}$$

If any node of N UAVs delivers a packet in a time slot, channel will be busy. Hence, P_b can be expressed as

$$P_b = 1 - P_t^N \tag{5}$$

Let P_s be the chance of successful delivery that can be shown as

$$P_s = \frac{N P_t (1 - P_t)^{N-1}}{P_b} \tag{6}$$

Let P_c be the probability of packet entrance with an average arrival rate λ_r

$$P_c = 1 - e^{-\lambda_r T_{ms}}. \tag{7}$$

Here, T_{ms} is a mean duration that a UAV spends for individual Markov chain state

$$T_{ms} = (1 - P_b)T_s + P_b P_s T_s + P_b(1 - P_s)T_c \tag{8}$$

where, T_s and T_c are defined in below.

$$T_s = T_{AIFS}[AC] + 3T_{SIFS} + T_{RTS} + T_{ACK} + \frac{L}{R_d} + T_{del} \tag{9}$$

$$T_c = T_{AIFS}[AC] + T_{SIFS} + T_{RTS} + T_{del} \tag{10}$$

where T_{RTS}, T_{CTS} and T_{ACK} are the spell for the request to send (RTS), clear to send (CTS) and acknowledgement (ACK), respectively. T_{del}, R_d and L represents propagation delay time duration, data transmission rate and size of data, respectively.

2.3 Throughput and PDR Analysis

S is the normalized performance that represents average slot data and average slot data transmission time that can be given as

$$S = \frac{P_s P_b R_d T_d}{T_{ms}} \tag{11}$$

where, T_d is the average data transmission duration.

If after the maximum transmission limit a packet is failing to transmit it then the packet is dropped. Therefore, PDR can be expressed as

$$PDR = (1 - P_s)^{m_r}. \tag{12}$$

3 Delay Analysis

Frame delay is the time it takes between the generation of a frame to its successful delivery. Delay $E[Del]$ can be given as:

$$E[Del] = E[C](E[B] + T_c + T_w) + E[B] + T_s, \tag{13}$$

where mean collision's number of a frame has until it is successfully sent is T_c, $E[B]$ is the backoff latency of a UAV node before it may access the channel. After a collision, T_w is the time it takes for a UAV to perceive the channel again which can be written as

$$T_w = T_{SIFS}. \tag{14}$$

$E[C]$ can be given as from P_s

$$E[C] = \frac{1}{P_s} - 1. \tag{15}$$

When the channel is busy, $E[B]$ is determined by the backoff counter value and the counter's stopped time. If at c backoff counter and to touch 0, b_c slot times are required without taking into consideration the counter paused. The mean duration can be expressed as

$$
\begin{aligned}
E[\phi] &= \sum_{c=0}^{c-1} cb_c = \sum_{c=0}^{c-1} c \frac{\varsigma - c}{\varsigma} b_0 \\
&= \frac{(\varsigma - 1)(\varsigma + 2)}{(\varsigma + 1)N_{sc}}.
\end{aligned}
\tag{16}
$$

$E[\Delta]$ is the average time a UAV node counter freeze, $E[N_\Delta]$ is the average time that a UAV node listened transmission from other UAVs before the backoff counter is zero, and before transmission, $E[\Xi]$ is the average number of consecutive idle periods. The connection can be described as follows:

$$E[N_\Delta] = \frac{E[\phi]}{max(E[\Xi], 1)} - 1, \tag{17}$$

$$E[\Delta] = E[N_\Delta](P_s T_s + (1 - P_s)T_c), \tag{18}$$

$$E[\Xi] = \frac{1}{P_b} - 1. \tag{19}$$

By using (16) to (19), $E[B]$ can be given as

$$E[B] = E[\phi] + E[N_\Delta](P_s T_s + (1 - P_s)T_c), \tag{20}$$

Here, T_s and T_c can be obtained by using (15) to (19). Thus, $E[Del]$ can be calculated by substituting (14), (15), and (20) into (13).

4 Simulation Results

This section assesses performance of proposed OFDMA based UAVs communication to support QoS. UAVs are allocated randomly. Simulations are performed in MATLAB. Parameter values employed in the simulation are presented in Table 3.

It is perceived from Fig. 4 that the probability of outage decreases with altitude and increases with r. With increasing r, the path loss increases because a longer link length is required. However, increasing the height also results in a larger elevation angle, which increases the effect of multipath scattering.

Table 3 Parameter values assigned in simulation

Parameter	Value
$T_{slot}, T_{SIFS}, T_{AIFS}, T_{del}$ (μs)	30, 10, 64, 3
$L_{RTS}, L_{CTS}, L_{ACK}$ (bytes)	26, 20, 14
L, L_h (bytes)	1024, 50
R_t (m), R_d (Mbps)	100–500, 11
N, m_r	0–25, 5
$\varsigma min min min_{min}$	3, 3, 7,15
$\varsigma max max max_{max}$	7, 15, 1023, 1023
$AIFS[0], AIFS[1], AIFS[2], AIFS[3]$	2, 3, 6, 9

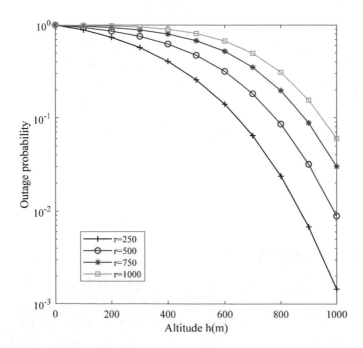

Fig. 4 Outage probability against altitude

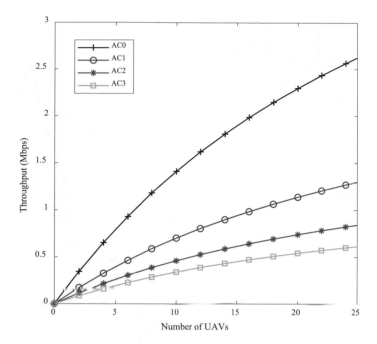

Fig. 5 Throughput versus number of UAVs

The throughput of different ACs is displayed in Fig. 5 in contrast with the number of UAVs. AC0 has the highest throughput because AC0 is expected to wait less time because of its EDCA parameters. AC1 has lower throughput than AC0 and AC2 has lower throughput than AC1. The lowest is AC3 because it has higher backoff due to the value of EDCA parameters. However, this mechanism ensures QoS.

Figure 6 shows the PDR for different ACs against the number of UAVs. The reliability of data transfer hinges on PDR. When PDR decreases, transmission reliability increases. PDR is growing with the number of UAVs because more packets cause more collisions. AC0 has the highest PDR, then AC1 and AC2, and AC3 has the lowest PDR. Due to the value of EDCA parameters, backoff of AC0 is low which increases PDR as well. In AC0, since there is less backoff, packets transmitted more frequently, thus increasing collision probability. Nevertheless, AC3 has the highest contention time that reduces the collision probability. The PDR is also falling when collision probability is decreasing.

For each AC, Fig. 7 illustrates the average delay vs the UAVs number. The average latency increases as the number of UAVs grows because more packets compete for transmission, making the channel crowded and requiring a longer backoff, resulting in a backoff delay. Moreover, collision probability will also increase with the increase of UAV's number that will increase the delay.

It is also apparent that the delay for AC0 is lowest, while the delay is highest for AC3. Delay from AC0 to AC3 is increased due to the EDCAF values of IEEE 802.11

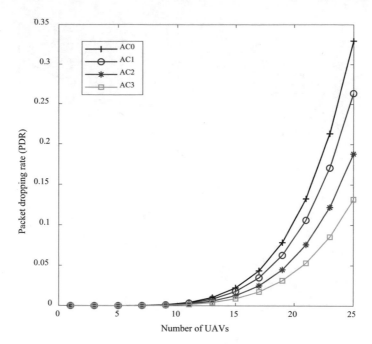

Fig. 6 The PDR vs. UAVs number

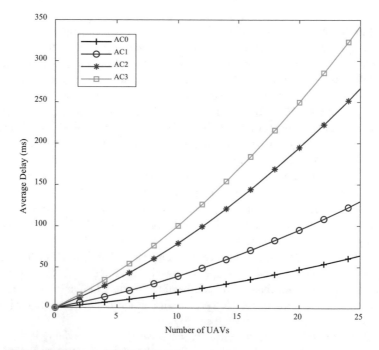

Fig. 7 The average delay vs. UAVs number

EDCAF. As AC0 is low for $\varsigma[AC]$ and AIFSN[AC], the contend period for AC0 is less. This means that the delay for AC0 is lower. In addition, the value of $\varsigma[AC]$ and AIFSN[AC] increases from AC0 to AC3 and increases the delay between AC0 and AC3. However, because of using OFDMA delay is lower for all ACs. Even in the high traffic AC0 has the delay lower than 100 ms.

5 Conclusion

This study develops an analytical model for evaluating IEEE 802.11 EDCAF performance for OFDMA-based UAV communication. A new channel contention method that combines EDCAF and OFDMA is proposed. The first step is to do EDCA-based prioritized channel contention. Then OFDMA is used for channel access and transmission. A 2D Markov chain model is used to assess the model's performance, which considers all major aspects that might impact performance. The UAV's altitude is considered. Due to OFDMA, the simulation results demonstrate that the system enhances data throughput while lowering packet loss rate and latency. Moreover, the proposed system also ensures QoS due to EDCAF. We are planning to study the physical layer in the future study. Researchers have begun to incorporate artificial intelligence into UAV networks by using AI algorithms to solve a variety of drone-related challenges which includes future research works.

References

1. Mozaffari M, Saad W, Bennis M, Nam Y, Debbah M (2019) A Tutorial on UAVs for wireless networks: applications, challenges, and open problems. IEEE Commun Surv Tutor
2. Mozaffari M, Saad W, Bennis M, Debbah M (2016) Unmanned aerial vehicle with underlaid device-to-device communications: performance and tradeoffs. IEEE Trans Wirel Commun 15(6):3949–3963
3. Shah AFMS, Ilhan H, Tureli U (2019) Designing and analysis of IEEE 802.11 MAC for UAVs ad hoc networks. In: Proceedings of IEEE 10th IEEE annual ubiquitous computing, electronics & mobile communication conference (UEMCON), New York, USA, October 2019, pp 0934–0939
4. Zeng Y, Zhang R, Lim TJ (2016) Wireless communications with unmanned aerial vehicles: opportunities and challenges. IEEE Commun Mag 54(5):36–42
5. Bekmezci I, Sahingoz OK, Temel Ş (2013) Flying ad-hoc networks (FANETs): a survey. Ad Hoc Netw 11(3):1254–1270
6. Košmerl J, Vilhar A (2014) Base stations placement optimization in wireless networks for emergency communications. In: IEEE international conference on communications workshops (ICC), pp 200–205
7. Al-Hourani A, Kandeepan S, Lardner S (2014) Optimal LAP altitude for maximum coverage. IEEE Wirel Commun Lett 3(6):569–572
8. Mozaffari M, Saad W, Bennis M, Debbah M (2015) Drone small cells in the clouds: design, deployment and performance analysis. In: IEEE global communications conference (GLOBECOM), pp 1–6

9. IEEE Standard for Information technology--Telecommunications and information exchange between systems Local and metropolitan area networks--Specific requirements - Part 11: Wireless LAN Medium Access Control (MAC) and Physical Layer (PHY) Specifications. In: IEEE Std 802.11–2016 (Revision of IEEE Std 802.11–2012), pp 1–3534, 14 December 2016
10. Shah AFMS, Ilhan H, Tureli U (2019) qCB-MAC: QoS aware cluster-based MAC protocol for VANETs. In: Arai K, Kapoor S, Bhatia R (eds) Intelligent computing, vol 857. Advances in intelligent systems and computing. Springer, Cham, pp 685–695
11. Shah AFMS, Karabulut MA, Ilhan H (2018) Performance modeling and analysis of the IEEE 802.11 EDCAF for VANETs. In: Intelligent systems and applications. Springer, Cham, pp 34–46
12. Gosteau J, Kamoun M, Simoens S, Pellati P (2004) Analytical developments on QoS enhancements provided by IEEE 802.11 EDCA. In: Proceedings of IEEE international conference on communications, vol 7, pp 4197–4201
13. Karabulut MA, Shah AFMS, Ilhan H (2022) A novel MIMO-OFDM based MAC protocol for VANETs. IEEE Trans Intell Transp Syst. https://doi.org/10.1109/TITS.2022.3180697
14. Ferdous H, Murshed M (2011) Ad hoc operations of enhanced IEEE 11 with multiuser dynamic OFDMA under saturation load. In: Proceedings of the IEEE WCNC, March 2011, pp 309–314
15. Kwon H, Seo H, Kim S, Lee BG (2009) Generalized CSMA/CA for OFDMA systems: protocol design, throughput analysis, implementation issues. IEEE Trans Wireless Commun 8(8):4176–4187
16. Xiong H, Bodanese E (2009) A scheme to support concurrent transmissions in OFDMA based ad hoc networks. In: Proceedings of the IEEE VTC Fall, September 2009, pp 1–5
17. Rashtchi R, Gohary R, Yanikomeroglu H (2012) Joint routing, scheduling and power allocation in OFDMA wireless ad hoc networks. In: Proceedings of the IEEE ICC, June 2012, pp 5483–5487
18. Karabulut MA, Shah AFMS, Ilhan H (2020) OEC-MAC: a novel OFDMA based efficient cooperative MAC protocol for VANETS. IEEE Access 8:94665–94677
19. IEEE Standard for Information technology-- Local and metropolitan area networks-- Specific requirements-- Part 11: Wireless LAN Medium Access Control (MAC) and Physical Layer (PHY) Specifications Amendment 6: Wireless Access in Vehicular Environments, IEEE 802.11p-2010, 15 July 2010
20. Shankar R, Muthaiya AT, Mathew Janvier L, Dananjayan P (2011) Quality of service enhancement for converging traffic in EDCA based IEEE 802.11. In: Proceedings of IEEE international conf. on process automation, control and computing, Coimbatore, pp 1–6
21. Mukherjee S, Peng XH, Gao Q (2009) QoS performances of IEEE 802.11 EDCA and DCF: a testbed approach. In: Proceedings of IEEE 5th international conference on wireless communications, networking and mobile computing, Beijing, pp 1–5
22. Karabulut MA, Shah A, Lhan HI (2020) Performance modeling and analysis of the IEEE 802.11 MAC protocol for VANETs. J Fac Eng Archit Gazi Univ 35(3):1575–1587

Personalization and Prediction System Based on Learner Assessment Attributes Using CNN in E-learning Environment

J. I. Christy Eunaicy, V. Sundaravadivelu, and S. Suguna

1 Introduction

E-Learning that is customized to meet the needs of individual students is called personalized e-Learning. To facilitate these levels of learning, content has to become more accurate, clearer, and concise, so it can be consumed by the student more easily. Data is also contributing to vast improvements in e-Learning, as it aids organizations in improving a deeper understanding of which content and learning approaches are most effective down to the individual level. Personalization can be done online. The achievement of expressing target study plans and certain skill gaps of the learners would lead to the personalization of their requirements with more experience of personalized learning. If a learner had the basic skills related to a particular topic, a system would be able to recognize that they might be able to skip a few modules to take a more comprehensive and less linear path of learning [1].

The proposed work obtains learner log information to monitor and assess the learner's learning performance. The log information is pre-processed to get the user identification, session identification, and learning path for the construction of custom features for the evaluation. The pre-processed log information is employed to derive the custom features based on the learning methods of the learners. The custom features are involved in the evaluation of every individual learner for their performance assessment and labeling. Based on the learner's assessment feature vector the classifiers are trained to label the learners as good and Average learners.

J. I. Christy Eunaicy (✉)
Thiagarajar College, Madurai, India
e-mail: christyeunaisy_caitsf@tcarts.in

V. Sundaravadivelu
Aringar Anna Government Arts College, Villupuram, India

S. Suguna
Sri Meenakshi Government Arts College for Women, Madurai, India

B. Unhelker et al. (eds.), *Applications of Artificial Intelligence and Machine Learning*,
Lecture Notes in Electrical Engineering 925,
https://doi.org/10.1007/978-981-19-4831-2_28
343

The work contains 5 main sections which include the introduction of the work as Sect. 1. The background and the related study of the work are discussed in Sect. 2. Section 3 contains the proposed scheme in detail with the sub-sections of the data collection layer, pre-processing layer, feature construction layer, and prediction layer. Results and discussion is described with illustration in Sect. 4. Finally, the work is concluded with the achievement of the objective is noted in Sect. 5.

2   Review of Literature

Dianshuang et al. [2] proposed this work to depict the perplexing learning exercises and student profiles extensively; this work developed fuzzy tree-organized learning action models, as well as student profiles models. PLORS is a framework proposed by Imran et al. [3] that enables students to determine which learning objects within a course are most helpful for them. An affiliation rule mining procedure is used to identify the relationship between Los and the proposal instrument. Abelardo et al. [4] suggest that combining oneself report information with information generated from an observation of the commitment of learners with web-based learning occasions provides a more thoughtful understanding and explanation of why a few students attain more impressive levels of scholarly accomplishment. In a survey of current web-based e-learning scenarios, Suguna and SundaraVadivelu [4] assess their level of effectiveness. Identify and describe the service-oriented architecture of e-learning systems. According to Khribi [6], association rules are used to determine relationships between variables (features) in e-learning systems so many algorithms may generate recommendations based on recommending systems. It is most commonly used to create custom algorithms and their improvements.

The Agarwal intelligent agents [7] are also worthy of mention here. Usually, e-learning systems achieve all their functions when implemented. In an environment like this, software agents can carry out autonomous actions to achieve a goal. The eLearning environment that [8] provides personalized web content recommendations and motivates the learners to participate actively in their education process. Learners' navigation patterns can be identified through web mining to identify content that they frequently visit. Personalized web content can be created using this pattern. The paper [9] proposed a technique called educational mining to predict student's performance. In [10] the proposed system provides learners with higher-quality learning objects. Learners with similar profiles determine personalization based on the learning objects they visit. It is the purpose of this article [11] to identify overlapping points between KM and e-learning phases so that meaningful and effective methods of ontology and metadata standards may be used to efficiently structure and transfer personalized course knowledge.

In paper [12] focused on the learner's learning style and strategy that fits every individual learner for the recommendation. The main of the work is to reduce the number of failures in exams for the preparation of the IDCL certificate and to improve the performance of the learner during exams. The study [13] analyzing the performance of the student by obtaining academic, behavioral, and non-behavioral features of the students to improve their achievement in academic. In the article [14] a survey of the systematic review of the e-learning and organizational learning for the current findings is described. The paper [15] proposed a personalization of the e-learning based on the Felder Silverman Learning Style Model (FSLSM) to recommend the learners with their learning style. The work [16] aims on suggesting the content to the learner by understanding the user behavior and requirement of the contents.

3 Proposed Scheme

In the proposed approach, the learner's details are combined based on the learner's profile and dynamically determined based on monitoring of the learner's characteristics. Phases of the proposed method include Data Collection, Data Pre-Processing, User Profiling, Feature Construction, and Classification of the Users. Figure 1 shows the complete flow of the current work.

3.1 Initial and Collection Phase

a) Initial Phase
The dataset for the work is obtained from the e-learning system working in a local server and it is tested with 100 Students (learners) for the initialization of the dynamic dataset. The Website has the content of the courses like C, C++, Java and the learners

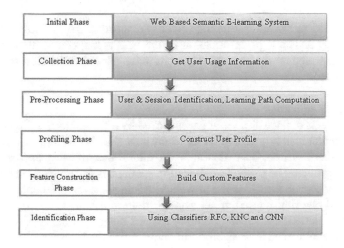

Fig. 1 The flow of the proposed scheme

are allowed to learn their study plan. The learner's log (User IP, Username, Session Time, etc.) information is extracted after they complete learning.

b) Collection Phase

Making a suitable target data set and gathering adequate usage data is the key to a successful mining process. E-Learning environment usage data are reclaimed from log files and client data, similar to other web-based applications. For the proposed work, the access logs of the web server are used. The learning portal's access log contains a record of all activities carried out on the webserver. The IP address, URL, referrer URL, the response code, the size of the files, the date, and the time stamp are all included.

3.2 Pre-processing Phase

In the pre-processing phase, the raw dataset and processed to eliminate duplicates, null values, missing values and converted into the common format.

i. User Identification

In this work, the IP address is employed to identify the user. The log file information of the learner is obtained with the user login detail. The user ID in the login table is compared with the user ID of the log information and obtains the log file fields of the particular learner. For getting the username, the system used the *user login* detail table, which can be maintained by the server. Attributes in this table are *Date, User ID, IP, and Time.*

$$If\ log.userid == user.userid\ \textbf{then}\ get\ (log\_file\_fields)$$

The user log information from the server is mapped to the user login details are shown in Fig. 2. The login information of the user is mapped to the user log file information. Based on the mapped information the user identification from the log file is extracted.

ii. Session Identification

A user session is defined as a set of pages stayed by a similar user within the period of one particular visit to a website. A user may have single or several sessions during a period. Once a user has been identified, the clickstream of every user is portioned into logical groups. The method of allocating sessions is called Sessionization. This section takes session identification in weblog mining as a research object, proposes an improved algorithm based on an average time threshold value. By computing the average intervals dynamically among request records in the session, adjusting the time threshold value individually, and compared to the traditional algorithm that describes a uniform threshold value for all user's web pages, the algorithm in this work can recognize the long session more precisely.

User Login   Log File

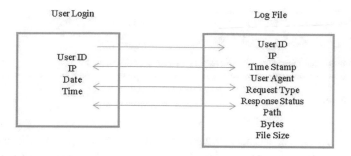

Fig. 2 User identification

iii. Learning Path

This work evaluates the learning paths with the learning domain. The learner's learning path is constructed to evaluate the learner's visiting path. The path is constructed with the URI is read for learning sessions. The learning path is build using the learning domain is shown in the following format:

$$Course->Subject->Topic->Subtopic$$

The sample learning path for the learning domain based on the above format is shown in the following Table 1.

Table 1 User profile based on the learning domain

Learner ID	Date	Learning path	Starting session	End session	Absent time
LID003	02/07/2021	Computer Science C Programming Introduction to C Overview of C	12:09:34 pm	02:03:54 pm	0
LID006	02/07/2021	Computer Science C Programming Introduction to C History of C	10:23:12 am	10:43:21 am	0
LID006	02/07/2021	Computer Science C-Programming Introduction to C Facts about C	10:53:33 am	11:16:21 am	0

Computer Science→C Programming→Introduction to C→History of C

Computer Science→JAVA→Introduction→Java Overview

Computer Science→C Programming→Basic Syntax of C→Tokens

Fig. 3 Learning path

Figure 3 depicts the learning path based on the domain chosen by the learner and the subject, topic, and subtopic of the learning course. After obtaining the user log information, the profile construction is carried out for the prediction of the user's learning patterns.

3.3 Profiling Phase

To personalize user queries and store answers from data, user-specific DATA is important. Users are given unique IDs based on log information and login information obtained by interacting among the pages. By identifying the resource's connection with the learner's page usage, a learner's learning path can be evaluated. Regarding the information that the active user session is extracted from the Weblog file, which only shows the last visited pages in the active user session. If these pages are too similar or contain terms that do not match the recommendation procedure, then the recommendation procedure will need to be changed. As illustrated in Table 1, this domain is based on user profiles.

3.4 Feature Construction Phase

This Feature Construction Phase includes the assessment part of the learner's activities. The feature construction phase includes the various features identified from the user profile constructed in the previous phase. The following features are the custom features constructed from the User Profile.

a) **Learner Learning Capacity (*LLC*)**

This Learner Learning Capacity (LLC) is the extent to which learners are capable of studying the chapter or topic during the time allotted for their study.

b) **Presence of Learner (*PoL*)**

To compute PoL, learners are asked to demonstrate their presence while they learn their study plan. The presence of the learner is evaluated to identify whether the learner is in presence through the no. of alarms missed while learning their study plan. The presence of the learner is evaluated using:

$$PoL = \frac{Missed\ Alarm\ while\ Studing\ Subject}{Total\ Alarm\ in\ the\ Subject} \tag{1}$$

c) **Subject Complexity (*SC*)**

A subject complexity is based on a subject or topics that the learner studies repeatedly. Subject Complexity is computed based on the following equation:

$$SC = \sum_{n}^{i=1} \frac{Repeated\ Topics\ Learned}{No.\ of\ Topics}, n \in No.\ of\ Subjects \tag{2}$$

d) **Completion of Levels (*CoL*)**

This metric is computed based on how the learner completed their study plan levels. Learning levels were computed during assessments for achievement reports. To calculate level completion, using the following equation:

$$CoL = \frac{\sum Completed\ Levels}{\sum Assigned\ Levels} \tag{3}$$

e) **Learner Absence Rate (*LAR*)**

The learner absence rate is calculated by looking at how many days the learner is not logged in the E-learning Portal. Using this equation as a basis for evaluation:

$$LAR = \frac{\sum Day\ of\ Not\ Logged\ in}{\sum Logged\ Day} \tag{4}$$

$$where,\ Total\ Logged\ day = Current\ Date - First\ Login\ Date$$

f) **Subject Effectiveness Rate**

In this section, the effectiveness of the subject is evaluated based on its quality. Evaluation of subject quality is based on the total number of subjects learned which one is not found on the repeated subject list. The SER equation is as follows:

$$SER = \frac{No.\ of\ Learned\ Subject\ !\ in\ Repeated\ Reading\ Subject}{No.\ of\ Subject} * 100 \tag{5}$$

g) **Login Failure Rate**

To secure the E-learning system, the login authentication constraints are determined based on the failure rate of the learner's login. This equation evaluates login failure rate:

Table 2 Sample feature vector of '5' learners for 20 days

Learner's	LLC	PL	SC	CoL	LAR	SER	LFR	APR (%)	CCR (%)
L1	45	0.4	5	1	3	6	3	40	34
L2	30	0.2	1	3	2	2	1	10	12
L3	42	0.6	2	2	3	5	5	30	23
L4	31	0.1	1	4	1	2	2	12	13
L5	52	0.9	4	1	6	4	7	35	24

$$LFR = \frac{\sum Login\ Attempt\ Failure}{\sum Total\ Login\ of\ Learners} * 100 \qquad (6)$$

h) **Access Privacy Rate Based on Login Mode**

The feature is evaluated the access rate of privacy based on the login mode of the learners by knowing whether the learner accessing his system or another system. Based on this criterion the access privacy rate is calculated by the following equation:

$$APR = \frac{\sum Own\ System\ Login\ of\ Learner}{\sum Total\ Login\ of\ Learners} * 100 \qquad (7)$$

i) **Chapter Complexity Ratio**

Chapter Complexity is based on the Subject Complexity that calculates how many subjects have complexity through studying, the repeated time of study complexity rate is taken as the maximum value. According to the following equation, chapter (topic) complexity is evaluated for the course content:

$$Chapter\ Complexity\ Ratio = \frac{MRT}{No\ of\ SubChapter} . * 100 \qquad (8)$$

$$Where\ MRT \in Maximum\ time\ of\ Repeated\ Reading\ Subject$$

Table 2 shows the information of the Sample feature vector of 10 Learners from the features was describe above in order.

3.5 Identification Phase

The constructed features from the user profiling are employed to the learner classification (Good/Poor) for the automatic identification based on the user's learning pattern. Thus, the work employs this layer to predict the user's learning pattern from their learning Custom Constructed Features. The CCF is converted into the training sets. In the learning process, a training dataset is used to fit the parameters of the model. To build up a machine learning model, a training dataset is constructed from the features, whereas the test set is used to validate the ML model. Test data set is excluded from the training set. The feature set is assigned to training and testing for 80% and learning for 20%.

3.5.1 KNC

Tuned (K-Nearest Neighbour) KNN classifiers are most commonly used for classification by choosing a K-value and distance metric. A variable 'K' affects the estimation of the conditional class probability because local regions are used by the distance of the K-th nearest neighbor to the query. Different values of 'K' will yield different conditional class probabilities. Since data sparsity and ambiguous, noisy, or mislabeled points occur when 'K' is very small, the local estimate is very poor. Increasing 'K' and adding a large area around the query will smooth out the estimate even further. As the number of outliers from other classes are increasing, the estimate can easily be considered to be over smoothed and the classification performance will degrade.

Table 3 shows the parameters of the proposed model. By updating K-Neighbor from 5 to 10, this work improves performance. Weight is the next parameter that is used for improving the prediction. It is set the value to 'Distance' while weight points by the inverse of their distance. Closer neighbors to a query point will have more influence than further away from neighbors. For N-point generalized problems, the KD-Tree is suitable. Leaf size '50' is passed to KD-Tree. The power parameter for the Minkowski metric. In the case of $p = 1$, this would be the equivalent of using manhattan_distance and using euclidean_distance in the case of $p = 2$.

3.5.2 RFC

A Random Forest analyzes the test results based on the decision trees which class received the most votes. Due to their ease of implementation and efficiency of computation, random forests are frequently used. It is possible to make a random forest more complex (by adding more trees) while still avoiding the risk of overfitting. A random forest uses the collective wisdom of the crowd to leverage the information provided by multiple decision trees, which makes it often more accurate than a single decision tree. The reason is that each model has certain strengths and weaknesses that it can use to predict certain outcomes. The fact that there is only one correct prediction and numerous incorrect predictions leads to the mutual reinforcement of the models that yield the correct predictions, while these models tend to cancel. The bagging method is used to create thousands of decision trees with minimal correlations in

Table 3 Tuned parameter for KNC

Parameter	Default value	Tuned value
N-Neighbors (NN)	5	10
Weight (W)	Uniform	Distance
Algorithm (Alg)	Auto	KD-Tree
Leaf Size (Leaf_Size)	30	50
Power (P)	2	2

Table 4 Tuned parameter for RFC

Parameter	Default value	Tuned value
Estimators	100	200
Criterion	Gini	Entropy
Max depth	None	5
Max features	Auto	SQRT
Max leaf nodes	None	10

random forests. To train each tree, a random subset of the training data is selected. Additionally, the model restricts randomly which variables may be used at every split. Hence, the trees grown are dissimilar, but they still retain certain predictive power.

Table 4 shows the parameters of the proposed model. There is a parameter called an Estimator in the forest algorithm, which is set to 200. Entropy is set as the criteria value to assess split quality. A tree's maximum depth is set to five. Maximum features for the best split are set into the 'SQRT'. Trees are grown with the maximum number of leaf nodes set to '10'.

3.5.3 CNN

Convolutional Neural Networks (ConvNet/CNN) are Deep Learning algorithms capable of processing large datasets. Different aspects of the dataset should be assigned a weight (learnable weights and biases) and they should be able to be distinguished from each other. ConvNets have architecture analogous to that of Neuronal connections in the Human Brain and are inspired by the organization of the Visual Cortex. Receptive Fields are restricted regions of the visual field in which neurons respond to stimuli. The visual field is covered by a collection of such fields. CNN's have multiple layers, such as input, convolutional, MaxPool, flatten, and density layers.

Table 5 shows the parameters of the proposed model. In this work, the input Shape of the neural network would be 10X 500, have three Convolutional '1' Dimensions of 32, 64, and 128 Filters. And MaxPool Layer with 6, 4, 1 Shapes. Flatten and the dense layer will be 128 and 64. The Epochs is 200, Kernel is Uniform and Optimizer will be set into Adam and the output layer is 2.

4   Result and Discussion

This section includes the results and discussion for the prediction phase of the learner. The evaluation of the performance metrics are as follows:

Table 5 Tuned parameter for RFC

Parameter	Filters	Size of the parameter
Input shape	–	10X 500
Conv1D	32	(8,2,32)
MaxPool	–	(6,2,32)
Conv1D	64	(5,2,64
MaxPool	–	(4,2,64)
Conv1D	128	(3,2,128)
MaxPool	–	(1,2,1)
Flatten	–	(128,2)
Dense	–	(64,2)
Softmax	–	(1,2)
Epochs	–	200
Kernel		Uniform
Optimizer		Adam
Output Layer		2

4.1 Environmental Setup

The Proposed Work Developed Using Python 3.7.3 in Anaconda 4.10.3 with Pandas and SciKit-learn Libraries. Development Hardware Environment includes Core i7 Processor, 16 GB Ram with 1 TB Hard disk. Server Configuration is AMD 7351P, 16C/32 T–2.9 GHz Turbo, DDR4 RAM 128 GB, and 2 × 8 TB HDD Storage.

4.2 Evaluation Result

a) Accuracy
Accuracy is the percentage of correctly predicted outcomes compared to all predicted outcomes.

$$Accuracy(ACC) = \frac{TP + TN}{TP + TN + FP + FN} \tag{9}$$

The accuracy evaluation for the classifiers is depicted in Fig. 4. The figure illustrates that the proposed CNN provides a high rate of accuracy while predicting.

b) Classification Error
The classified results are affected by errors of omission. These are sites that are classified as references whose class has been left out of the classified map (or omitted).

$$Classification\ Error(CE) = \frac{FP + FN}{TP + TN + FP + FN} \tag{10}$$

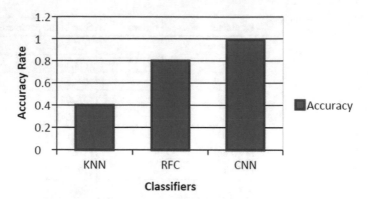

Fig. 4 Accuracy evaluation of KNN, RFC, and CNN

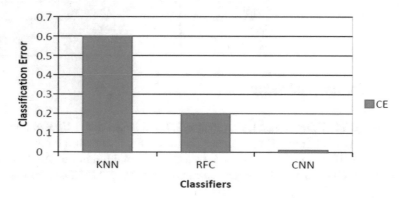

Fig. 5 Classification error of KNN, RFC, and CNN

The error rate evaluation for the classifiers while prediction is shown in Fig. 5. The proposed CNN possesses less error rate than other methods.

c) Sensitivity
True positive rate (also referred to as sensitivity) is the percentage of people with the disease with a positive test result.

$$Sensitivity = \frac{TP}{FN + TP} \qquad (11)$$

d) Specificity
Specificity is determined by the percentage of people without the disease who will have a negative test result (also called True Negative Rate).

$$Specificity = \frac{TN}{TN + FP} \qquad (12)$$

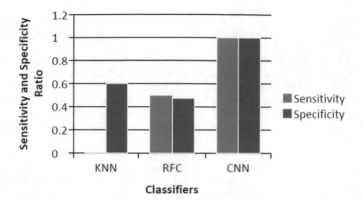

Fig. 6 Sensitivity and specificity ratio evaluation

For the proposed CNN in Fig. 6 shows the sensitivity and specificity ratios are high compared to other algorithms.

5 Conclusion

The proposed architecture is elaborated with its layers containing the process of the Learner's phase. This framework includes the entire process flow of the learner's phase. In this framework, the layers containing personalization are included for the automatic recommendations. In the personalization system, user profiling with the weblog information for the prediction of learner's achievement during an assessment is described. After that, in the prediction layer, the user profile is employed to find the learner's assessment with aspects such as the presence of the learner, learner-learning capacity, and subject complexity. Based on the feature construction the user's pattern is evaluated for the classification. The features set is split into training and testing set for the user classification. The performance results show that the proposed technique CNN provides high results for accuracy, sensitivity, and specificity than others and consumes less error rate for prediction.

References

1. Web Reference. https://elearningindustry.com/benefits-of-artifcial-intelligence-in-personali zed-learning
2. Wu D, Lu J, Zhang G (2015) A fuzzy tree matching-based personalized eLearning recommender system. IEEE Trans Fuzzy Syst 23(6):2412–2426
3. Imran H, Belghis-Zadeh M, Chang T, Graf KS (2016) PLORS: a personalized learning object recommender system. Vietnam J Comput Sci 3(1):3–13

4. Pardo A, Han F, Ellis RA (2017) Combining university student self-regulated learning indicators and engagement with online learning events to predict academic performance. IEEE Trans Learn Technol 10(1):82–92
5. Suguna S, Sundaravadivelu V (2018) An extensive survey on personalized and secured e-learning systems using data mining process. Int J Res Anal Rev (IJRAR) 6(1):664–667
6. Khribi MK, Jemni M, Nasraoui O (2008) Automatic recommendations for e-learning person-alization based on web usage mining techniques and information retrieval. In: Eighth IEEE international conference on advanced learning technologies, pp 241–245
7. Agarwal R, Deo A, Das S (2004) Intelligent agents in e-learning. SIGSOFT Softw Eng Notes 29(2):1–9
8. Herath D, Jayaratne L (2017) A personalized web content recommendation system for E-learners in E-learning environment. In: National information technology conference (NITC)
9. Thai-Nghe N, Drumond L, Krohn-Grimberghe A, SchmidtThieme L (2010) Recommender system for predicting student performance. Proc Comput Sci 1(2):2811–2819
10. Imran H, Belghis-Zadeh M, Chang T-W, Kinshuk, Graf S (2016) PLORS: a personalized learning object recommender system. Vietnam J Comput Sci 3(1):3–13
11. Cakulaa S, Sedleniecea M (2013) Development of a personalized e-learning model using methods of ontology. In: ICTE in regional development, pp 113–120
12. Ali NA, Eassa F, Hamed E (2019) Personalized learning style for adaptive E-learning system. Int J Adv Trends Comput Sci Eng 8:223–230
13. Jenila Livingston LM, Merlin Livingston LM, Agnel Livingston LGX, Annie Portia A (2019) Personalized tutoring system for E-learning. In: 2019 international conference on recent advances in energy-efficient computing and communication (ICRAECC), pp 1–4
14. Giannakos MN, Mikalef P, Pappas IO (2021) Systematic literature review of E-learning capabilities to enhance organizational learning. Inf Syst Front
15. Sihombing J, Laksitowening K, Darwiyanto E (2020) Personalized E-learning content based on Felder-Silverman learning style model, pp 1–6
16. Joshi N, Gupta R (2020) A personalized web-based E-learning recommendation system to enhance and user learning experience. Int J Recent Technol Eng (IJRTE) 9(1):1186–1195. ISSN 2277-3878

Prognosis of Clinical Depression with Resting State Functionality Connectivity using Machine Learning

S. Saranya⬡ and N. Kavitha⬡

1 Introduction

The term "Health" refers to a condition of total mental, physical, and social well-being that is the most distinguishing feature of humans [1]. Healthcare informatics is extremely important for the advancement of numerous healthcare industries. The main goal of Healthcare Informatics is to improve the healthcare process by combining improved efficiency, quality, and other important factors. Healthcare Informatics aims to improve the effectiveness of patient care delivery on a global scale.

Machine Learning is a popular discipline in the computer business, with Healthcare Informatics being one of the most challenging areas [2]. Machine learning allows people to learn without having to be explicitly programmed. Pattern recognition and computational learning theory in Artificial Intelligence have determined its evolution (AI). The basic goal of machine learning is to utilize it to make predictions and create algorithms that can learn over time.

One of the significant psychiatric disorder named, Major Depressive Disorder (MDD) is characterized by the existence of depressed mood or a loss of interest or fondness in daily activities for a duration of two weeks or more [3]. Many studies have depicted that the presence of low mood, worthlessness, anxiety, cognitive impairments, presence of altered physical activity levels increases with severity of the disease. In some core region of brain, there are structural and functional changes due to these symptoms of MDD which are detected by magnetic resonance imaging (MRI).

S. Saranya (✉)
Nehru Arts and Science College, Coimbatore, India
e-mail: ssaranyamphil@gmail.com

Department of Computer Science, Dr.N.G.P. Arts and Science College, Coimbatore, India

N. Kavitha
Nehru Arts and Science College, Coimbatore, India

MDD is the most common psychiatric disorder which has affected more than 322 million people according to the World Health Organization's statistics. To investigate the structural and functional alterations in brain of MDD patients, MRI are used which are reliable and contributes to clinical findings and treatments. The structural and functional modification of brain caused by MDD is shown in Fig. 1 [4].

The Structural MRI captures the anatomical alteration of brain, symbolized as volume differences caused by MDD. It also measures the changes in gray matter of several brain regions which also aid in predicting MDD. In functional MRI, the neuronal activity and the abnormal brain activity is assessed. The local blood flow of the functional brain area is increased significantly by the neuronal activity enhancement which is caused by MDD. Due to the increased blood flow the ratio of deoxygenated hemoglobin/oxyhemoglobin is decreased, resulting in changes of Blood Oxygen Level Dependent (BOLD) signal as neurophysiologic indicators.

Even in the resting state (absence of externally prompted task) there is a fundamental activity of brain, which results in number of resting-state conditions in the brain. The default mode network is one the most easily pictured and most studied network during rest state [5]. Resting state functional magnetic resonance imaging (rsfMRI) is a tool for examining the different types of brain disorders by evaluating the regional interaction of brain in the resting state [4]. The fMRI measures the changes in blood flow known as BOLD signal in the default mode network. In presence of any psychiatric disease, the BOLD signal fluctuates and the brain's functional organization is also altered.

Functional connectivity in neuro imaging is defined as the associations between spatially distant neuro physiological events. Recently many researchers have found that functional connectivity of brain can be more extensively analyzed with dynamic

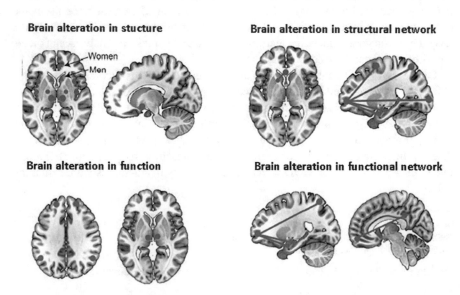

Fig. 1 Structural and functional brain map details for major depression disorder

temporal information [7]. Unlike structural connectivity, the functional connectivity changes over a short time in order of seconds which can be captured by rsfMRI. In this study, an automated system is built by using the high-order functional connectivity network information of resting state of the brain which holds rich dynamic time data is used to discriminate MDD individuals and healthy controls by using supervised machine learning methods.

2 Related Work

Kun Qian et al. [6] developed an automatic detection method for predicting major depressive disorder by wearable devices united with artificial intelligence. They recorded the impulsive activities of a person by a wearable device and the higher representations were extracted from the recorded information by bag-of-behavior-words method. Lastly support vector machine algorithms were used to predict whether a patient is affected with MDD and were able to achieve a better accuracy.

To identify MDD at starting phase an Electroencephalogram (EEG) was used to record the electrical activities of brain and the data was decomposed by Wavelet Transformation (WT) in [7]. The important features were selected by using Random Forest (RF) and Ant Colony Optimization (ACO) algorithm. To classify patients with MDD, the data after feature selection was given to k-nearest neighbors (KNN) and SVM classifiers. Accuracy was calculated with all the features and also after feature selection. There was significant improvement in accuracy in both SVM and KNN algorithms after feature selection.

Hao Guo [8] developed a model based on the fact that patients with MDD possess an abnormal resting state functional brain network metrics. The data of resting state functional brain networks were used. Feature selection was done by considering the threshold of statistical significance. The importance of various features was estimated by sensitivity analysis method. They obtained a highest average accuracy of the data with 28 features by neural network algorithm and Support vector machine algorithm with radial basis kernel.

Support Vector Machine (SVM) was applied to MRI measures of brain to categorize adults with MDD in [9] which has been a long term goal to predict mental disorders using MRI. With an equally balanced data of MDD patients and healthy individuals, the algorithm classified 74% of records accurately. The significant fractional anisotropy values (FA) for classifying MDD patients and healthy individuals was derived when Diffusion tensor imaging (DTI) data was fed to Support Vector Machine. This study also revealed that prediction information is not localized in any part of brain; rather it is spread across brain network.

Wajid Mumtaz et al. [10, 11] used a machine learning framework to construct a model to predict MDD at an early stage. The framework was fed EEG-derived synchronization likelihood (SL) parameters that were thought to distinguish MDD patients from healthy controls. The support vector machine (SVM), logistic regression (LR), and Nave Bayesian (NB) machine learning methods were employed to construct a prediction model, with SVM providing the best accuracy.

Single resting-state functional magnetic resonance imaging scans combined with an unsupervised machine learning technique were utilized to accurately predict MDD in the absence of clinical information by Ling-Li Zeng et al. In this investigation, magnetic resonance imaging scans of 54 MDD patients and healthy controls with similar demographics were employed. When the voxels were clustered into subgenual and pregenual regions, the largest margin was found. This clustering provided sufficient information for the classification of MDD patients and healthy controls, with 92.5% group-level clustering consistency.

Runa Bhaumik [12] used resting state functional magnetic resonance imaging (rs-fMRI) to develop a model that distinguishes between those who have MDD and those who do not. In their investigation, they applied the Support Vector Machine on a remitted population. For feature selection, the multivariate Least Absolute Shrinkage and Selection Operator (LASSO) and Elastic Net feature selection methods were used. The best classification accuracy was achieved using SVM classification with Elastic Net feature selection. The use of rs-fMRI connectivity as a recognized neurobiological signal for early diagnosis of MDD patients is also claimed in this work.

The anomalies in brain anatomy of patients with MDD were used to differentiate patients with MDD in [13]. They developed a novel hybrid model combining relevance vector machines and support vector machines for classification, complete with feature selection and characterization. T1-weighted 'structural' scan data from 62 patients with MDD and matched controls were used. The missing data was utilized to determine the classifiers' generalization ability and predictive accuracy. When compared to feature characterization, feature selection performed a critical impact in boosting the overall accuracy of this model.

Based on multiple brain network properties Matthew D. Sacchet et al. [14] proposed a model to segregate individuals with MDD and healthy controls by Support Vector Machines. Diffusion-weighted imaging was used to trail the neural pathways of human brain. The properties of fiber networks were obtained from graph theory. Some graph metrics gave vital information for the classification. The abnormal connectivity at specific nodes of the network was recognized by local graph analysis. The graph metric had a vital role in the classification of MDD and healthy individuals.

3 Proposed Methodology

Major Depressive Disorder is a heterogeneous disorder with a definite depressive mood for at least duration of two weeks with a high life time prevalence of 17% [15]. Brain's functional organization and how it differs from normal functionality is learned through rs-fMRI. The fMRI connectivity measures the correlations in BOLD Signal difference within the spatially distributed brain network.

Machine learning comprises many methods to develop prediction models for depression prediction from observed data to make predictions on new data. In this study Enhanced K Nearest Neighbor Algorithm is used for effective prediction of depression [16].

The proposed approach has two stages. The first stage is the preprocessing stage and the second stage is the prediction stage. Principal Component Analysis, a dimensionality reduction algorithm is used in the first phase and in the second stage depression is predicted using classification algorithm and to improve the performance of the algorithm two approaches of KNN algorithm, KD-Tree and Ball tree are used.

3.1 Dimensionality Reduction Using Principal Component Analysis

Principal Component Analysis (PCA) is used to reduce the dimension of complex data. It reveals the underlying structure of data by analytical solutions there by making a classifier system more effective. In this study, Regional volumes of brain which are affected by depression are considered as features in rsfMRI. PCA is used to reduce the number of dimensions, to a great extent which helps in increasing the calculation speed [17]. Principal Component Analysis reduces the dimensions of the features and focuses on maximizing the variance of the BOLD signal from other components.

Consider a dataset of dimension "ρ", with "m" number of principal axes where m lies $1 \le m \le \rho$. The orthonormal axes in the projected space where the variance are maximum is given by $T_1, T_2, \ldots T_m$ The covariance of m leading eigen vectors is given by the covariance matrix

$$S = \frac{1}{N} \sum (x_l - \mu)^T (x_i - \mu) \tag{1}$$

where $x_i \epsilon X$ is the sample mean and N is the number of samples. Then

$$ST_i = \lambda_i T_1$$

where λ_i is the ith eigen vector.

The Principal Component of a given vector is given by x_{i_-}

$$y = [y_1, y_2, \ldots, y_m] = [T_1^T x, T_2^T, \ldots \ldots T_m^T] = T_x^T \tag{2}$$

The subspace of the first m principal, where $m < \rho$ is the major assumption of PCA, contains the greatest information of the observation vectors. All original data can be represented by an m-dimensional main component vector [18].

3.2   K-Nearest Neighbor Algorithm for Depression Prediction

K-Nearest Neighbor (KNN) is one of the classification algorithm and its performance competes with the complex algorithm [19, 20]. It is used extensively in various fields such as text categorization, pattern recognition, event recognition and object recognition applications. In KNN the training dataset determines the parameters so it is a non-parametric algorithm [21] and the label of the closest specified training samples are used in predicting the label of test data.

Consider X as the training sample set which is divided into L classes [21]

$$X = [X_1, X_2, \ldots, X_i, \ldots X_L]^T \qquad (3)$$

where x_i is a matrix of class α_i there are N_i samples in each class α_i, which has n features. N, the total number of samples is given by

$$N = N_1 + N_2 + \ldots + N_L \qquad (4)$$

A test sample $y, y = (y_1, y_2, \ldots, y_n)^T$, is classified based on the concept of KNN as

$$y \in \alpha_i : 1 \leq I \leq L., \qquad (5)$$

$$\exists [d^* = \min(d_1, d, \ldots, d_i, \ldots d_L), d_j = \text{SEP}(x_i, y), x_i \in y] \qquad (6)$$

where d_i is a measure obtained from the separation function SEP (x_i, y), of n-dimension. Jaccard similarity, Manhattan distance, Euclidean distance, cosine similarity and many other separation functions are there to find the similarity or distance. Because distance or similarity measures are so critical in deciding the final classification outcome, picking the right distance metric for the data is crucial.

In the proposed work difference distance metrics are evaluated for the better performance. Jaccard is one of the similarity metrics which is most suitable for image similarity. In this work, rsfMRI images are used to classify the MDD individuals and healthy controls. To find the deformity in the medical images, Jaccard coefficient is extensively used by the researchers [22]. Jaccard's coefficient of similarity (JCS) is given by the formula

$$\text{JCS} = S_c/(S_x + S_y) \qquad (7)$$

where (S_x and S_y) are the number of sizes of the populations of X and Y respectively, and S_c is the size of population present in both samples.

In the traditional KNN model, to make a new prediction, the algorithm calculates the distance between the new data and the other data, and the data points which are most similar are taken up for prediction based on the voting process. The depression data set is an image dataset, which consist of large number of data points this repeated distance calculation takes much time for prediction. To speed up the process certain tuning can be done in the algorithm level. Ball tree and KD tree are the two approaches were the KNN algorithm can be tuned for the better performance.

KD-Tree and Ball tree are used, when there are multidimensional data points and spatial division of data points are done on these data points. When used for depression prediction, the spatial information of BOLD signal is predicted with faster response time and better accuracy.

4 Experimental Evaluation

The proposed methodology is developed using Python programming language in Anaconda environment, which is an interactive IDE for reproducing code, results and explanations is used in implementing the proposed method for the prediction of MDD. The sklearn module is imported to introduce the KNN Classifier in this work.

A dataset from Openfmri is used in this experimental evaluation. 19 Individual (11women, nine men, with an average age of 34.15) with MDD are registered. Depressive individuals were not under any medication and had depression at the time of scanning. 20 Never Depressed (ND) individuals (11 female, nine men, and average age of 28.5) with no past history of depression or any other psychiatric disorders were used for the evaluation of Depression. There were no current alcohol influence in the individuals during the time of scanning. All of the participants were right-handed, had no metal implanted in body, and had no medication or controls affecting blood flow, brain function or neurological condition.

4.1 Performance Evaluation Metrics

The performance of binary classification problem is generally assessed by Classification Accuracy, Recall, Precision, Confusion Matrix and F1-Score. The details of True Positives (TP), True Negatives (TN), False Positives (FP) and False Negatives (FN) are obtained from the confusion matrix, where

TP—Total Depressed patients correctly classified.
TN—Total Healthy Controls correctly classified.
FP—Total Healthy Control incorrectly classified as Depressed.
FN—Total Depressed patients incorrectly classified as Healthy Controls

Accuracy is the ratio of correctly predicted MDD and HC's to the total number of cases. It is calculated by

$$\text{Accuracy} = (TP + TN) / (TP + TN + FP + FN) \qquad (8)$$

Precision is defined as the classifier's expertise, not to label an instance positive that is actually negative. It is also known as the ratio of number of correctly classified depressed patients to the total of individuals predicted as depressed. It is denoted by

$$\text{Precision} = TP / (TP + FP) \tag{9}$$

Recall is defined as the classifier's proficiency in finding all positive instances. It is the ratio of number of correctly classified depressed patients to the total number of depressed patients in the study. It is expressed as

$$\text{Recall} = TP / (TP + FN) \tag{10}$$

F1 score is a measure of a model's accuracy on a dataset. It calculates the harmonic average of the precision and recall metrics, which is denoted by

$$F1 - \text{Score} = 2 * (\text{Precision} * \text{Recall}) / (\text{Precision} + \text{Recall}) \tag{11}$$

The prediction results for the various performance parameters are shown in tables below for the various distance metrics and with number of neighbors as eight. Tables 1 and 2 shows the train data and test data results respectively; KNN with the Jaccard distance metrics shows the highest accuracy. Tables 3 and 4 shows the results of Ball tree and K-D tree with performance tuning. Ball tree and K-D tree are used as tuning algorithm. For prediction depression using fmri images ball tree shows better performance.

Table 1 Training data

Distance metric	Accuracy (%)	Precision	Recall	F1 score
Euclidean	75.1	0.68	0.85	0.85
Manhattan	74.2	0.67	0.77	0.71
Cosine	74.2	0.73	0.62	0.67
Jaccard	87.7	0.85	0.92	0.96

Table 2 Test data

Distance metric	Accuracy (%)	Precision	Recall	F1 score
Euclidean	50	1	0.33	0.5
Manhattan	62.5	0.8	0.67	0.73
Cosine	62.5	1	0.5	0.67
Jaccard	75	0.75	1	0.86

Table 3 BALL TREE (Test data)—number of neighbors—8

Distance metric	Accuracy (%)	Precision	Recall	F1 score
Euclidean	64.3	1	0.5	0.67
Manhattan	70.1	0.75	0.5	0.6

Table 4 KD-TREE (Test data) number of neighbors—8

Distance metric	Accuracy (%)	Precision	Recall	F1 score
Euclidean	62.5	1	0.5	0.67
Manhattan	65.2	0.75	0.5	0.6

5 Conclusion

This research illuminates the use of machine learning models to predict the Major depressive disorder using rs-fMRI images. In this study we propose a novel approach to distinguish individuals with a history of MDD from healthy controls with logically good accuracy by using KNN algorithm with various distance metrics with performance optimization. The performance of the KNN classifier for different neighbors and different similarity measures is calculated by various metrics like accuracy, precision, recall, and f-measure. This method can be supportive for clinicians to predict the MDD at early stage and patients can be treated as early as possible. In future work we planned to use more dataset to predict the depression with various emotional factors to improve the accuracy.

References

1. Jadad AR, O'Grady L (2008) How should health be defined? BMJ Br Med J (Online) 337
2. Jordan MI, Mitchell TM (2015) Machine learning: trends, perspectives, and prospects. Science 349:255–260
3. Randy B, KrienenFenna Y, Thomas BT (2013) Opportunities and limitations of intrinsic functional connectivity MRI. Nat Neurosci 16:832–837. https://doi.org/10.1038/nn.3423
4. Sharaev MG, Zavyalova VV, Ushakov VL, Kartashov SI, Velichkovsky BM (2016) Effective connectivity within the default mode network: dynamic causal modeling of resting-state fMRI data. Front Hum Neurosci 10:14. https://doi.org/10.3389/fnhum.2016.00014. Accessed 1 Feb 2016
5. Qian K, Kuromiya H, Ren Z, Schmitt M, Zhang Z, Nakamura T (2019) Automatic detection of major depressive disorder via a bag-of-behaviour-words approach. In: Proceedings of the third international symposium on image computing and digital medicine, August 2019. https://doi.org/10.1145/3364836.3364851
6. Bandopadhyay S, Nag S, Saha S, Ghosh A (2020) Identification of major depressive disorder: using significant features of EEG signals obtained by random forest and ant colony optimization methods. In: Proceedings of the 2020 4th international conference on intelligent systems, metaheuristics & swarm intelligence, March 2020, pp 65–70. https://doi.org/10.1145/3396474.3396480

7. Guo H, Cao X, Liu Z, Li H, Chen J, Zhang K (2013) Machine learning classifier using abnormal brain network topological metrics in major depressive disorder. Neuro Rep 24(1):51. https://doi.org/10.1097/WNR.0b013e32835ca23a

8. Schnyer DM, Clasen PC, Gonzalez C, Beevers CG (2017) Evaluating the diagnostic utility of applying a machine learning algorithm to diffusion tensor MRI measures in individuals with major depressive disorder, March 2017. https://doi.org/10.1016/j.pscychresns.2017.03.003

9. Mumtaz W, Ali SSA, Yasin MAM, Malik AS (2017) A machine learning framework involving EEG based functional connectivity to diagnose major depressive disorder (MDD). Med Biol Eng Comput 56(2):233–246. https://doi.org/10.1007/s11517-017-1685-z

10. Zeng L-L, Shen H, Liu L, Hu D (2014) Unsupervised classification of major depression using functional connectivity MRI. Hum Brain Map 35:1630–1641. https://doi.org/10.1002/hbm.22278

11. Bhaumik R, Jenkins LM, Gowins JR, Jacobs RH, Barba A, Bhaumika DK, Langenecker SA (2017) Multivariate pattern analysis strategies in detection of remitted major depressive disorder using resting state functional connectivity. Neuro Image Clin 16:390–398. https://doi.org/10.1016/j.nicl.2016.02.018

12. Mwangi B, Ebmeier KP, Matthews K, Douglas Steele J (2012) Multi-centre diagnostic classification of individual structural neuroimaging scans from patients with major depressive disorder. Brain J Neurol 135(Pt 5):1508–1521. https://doi.org/10.1093/brain/aws084

13. Sacchet MD, Prasad G, Foland-Ross LC, Thompson PM, Gotlib IH (2015) Support vector machine classification of major depressive disorder using diffusion-weighted neuro imaging and graph theory. Front Psychiat 6:21. https://doi.org/10.3389/fpsyt.2015.00021

14. Thara DK, Prema Sudha BG, Xiong F (2019) Auto-detection of epileptic seizure events using deep neural network with different feature scaling techniques. Elsevier Pattern Recogn Lett 128:544–550. https://doi.org/10.1016/j.patrec.2019.10.029

15. Kataria A, Singh MD (2013) A review of data classification using K-nearest neighbor algorithm. Int J Emerg Technol Adv Eng 3(6):354–360

16. Vishnuvarthanan G, Pallikonda Rajasekaran M, Anitha Vishnuvarthanan N, Arun Prasath T, Kannan M (2017) Tumor detection in T1, T2, FLAIR and MPR brain images using a combination of optimization and fuzzy clustering improved by seed-based region growing algorithm. Int J Imag Syst Technol

17. Kluyver T, Ragan-Kelley B, Pérez F, Granger BE, Bussonnier M, Frederic J, Kelley K, Hamrick JB, Grout J, Corlay S et al (2016) Jupyter notebooks-a publishing format for reproducible computational workflows. In: ELPUB, pp 87–90

18. Rotzek M, Koitka S, Friedrich C (2017) Linguistic metadata augmented classifiers at the CLEF 2017 task for early detection of depression. 2017 presented at: conference labs of the evaluation forum, Dublin, Ireland, 11–14 September 2017

19. Lepping RJ, Atchley RA, Chrysikou E, Martin LE, Clair AA, Ingram RE, Kyle Simmons W, Savage CR (2016) Neural processing of emotional musical and nonmusical stimuli in depression. Plos One. https://doi.org/10.1371/journal.pone.0163631

20. Tadesse MM, Lin H, Xu B, Yang L (2018) Personality predictions based on user behavior on the facebook social media platform. IEEE Access 6:61959–61969

21. Almeida H, Briand A, Meurs M-J (2017) Detecting early risk of depression from social media user-generated content. In: Proceedings of the CLEF, pp 1–10

22. Friedrich MJ (2017) Depression is the leading cause of disability around the world. JAMA 317(15):1517

Medical Diagnosis Using Image-Based Deep Learning and Supervised Hashing Approach

Aman Dureja and Payal Pahwa

1 Introduction

In the past years of development in medical and clinical diagnosis, a large amount of data has occupied the storage bins and the data centers by the availability of computers and various media devices, which is of use to doctors and the researchers and they are greatly benefiting from this advancement [1]. The underlying problem that has emerged despite of great advantage of advancement in storage technology is the image retrieval. The efficiency and accuracy of image retrieval is the major concerning issue because of large data repositories a large amount of processing must take place which may result in less accurate or less efficient retrieval [2, 3]. To ease the task of image retrieval and increase the efficiency and accuracy of retrieved images of the concerned body organ, many algorithms have been proposed [4, 5]. There are various methods that are used earlier for retrieval of medical images that were based on decoded images with text.

Nowadays, Content based Medical Image Retrieval (CBMIR) have increased emphasis in the retrieval of medical images in the field of bioinformatics, educational institutions etc. for retrieval of images and classification tasks.

CBMIR main aim to retrieving out the best similar images by analyzing the images contents from the databases. In the earlier stages, the histogram-based features approach was utilized for retrieval of medical images features. The histogram displays the no. of pixels in an image at each power value. Using this power value of histogram, the matches is done with the predetermined histogram stored in the databases. However, the problem with the histogram-based features was that the performance

A. Dureja (✉)
Department of Computer Science and Engineering, USICT, GGSIPU, New Delhi, India
e-mail: amandureja@gmail.com

Department of Information Technology, BPIT, Rohini, New Delhi, India

P. Pahwa
Department of Computer Science and Engineering, BPIT, Rohini, New Delhi, India

© The Author(s), under exclusive license to Springer Nature Singapore Pte Ltd. 2022   367
B. Unhelker et al. (eds.), *Applications of Artificial Intelligence and Machine Learning*,
Lecture Notes in Electrical Engineering 925,
https://doi.org/10.1007/978-981-19-4831-2_30

is frequently limited for retrieval of huge databases. To solve this issue, texture-based features [6] method was introduced for retrieval of medical images. The Local Binary Pattern (LBP) was used in texture-based features retrieval that have exhibited extremely assuring effects in medical diagnosis.

As of late, there is a development in the field of profound learning, the Convolutional Neural Network (CNN) more pulled in this field of removing features dependent on the convolutional strategy by presenting the convolutional layers, pooling layers and completely associated layers in the Deep CNN organizations. Different models of CNN have been proposed by numerous scientists [7, 8] that give the better and are exceptionally successful to recover the fundamental features for the characterization and Image recovery measure. The normal CNN models utilized these days are LeNet-5, AlexNet, ZFNet, VGG, GoogleNet (v1), Res-Net, Inception-v4, Res-next.

The primary benefit of utilizing the CNN is that there is no compelling reason to separate features utilizing handmade things that have restrictions.

The element extraction of robotized methods is useful for recovering clinical Images [9]. These removed features are extremely high-dimensional in nature. To store these features into the data sets lessens the speed and effectiveness of access and expands the computational expense of recovery of Images. The initiation work additionally assists with expanding the non-linearity of the organizations [10]. It additionally assumes a significant part in expanding the proficiency of the CNN organizations. For recovery of clinical Images productively and quick from the put away information bases, the methods of creating parallel codes can be utilized by utilizing hash work [11], that are useful for producing smaller paired hash codes for expanding computational force and execution of the organizations.

Through this paper, a novel structure of image retrieval is proposed. In the initial step of features extraction, a famous feature learning model is utilized. Image features and semantic data with dimensional vectors is extricated utilizing the model of CNN which is joined with the hash code.

The primary point of the proposed system is to expand the judgment capacity of produced hash codes (twofold) which addresses that the comparative paired hash codes should have comparative images and the other way around.

Because of the proficient preparing of the CNN [2] Fig. 1, the profound learning models can be applied to the clinical Images for eliminating the hole that is there uniting the graphic information recorded by the clinical imaging machines and the graphic information comprehended by individual's reasoning force and seen by vision.

In this paper, the powers of CNN are utilized to build up a system that can efficiently bridge the semantic gap and also lay a strong basis for the task of content or more precisely feature based or texture based medical image retrieval [12, 13]. In the underlying stage, the CNN [14, 15] model will be prepared to characterize the clinical Images and in the later stage, the learned capacities of the CNN will be utilized for highlight based clinical Image retrieval [2].

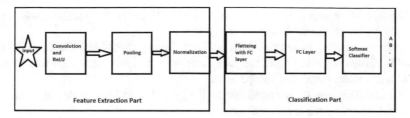

Fig. 1 Block diagram of architecture of a CNN [7]

1.1 Content Based Medical Image Retrieval (CBMIR)

In this paper, the more specifically concentrate on the content based medical image acquisition which is based on Content Based Medical Image retrieval which is the extension of existing technology that is the Content Based Image Retrieval (CBIR), which was enhanced for its use in medical imaging. The only problem that faces in using CBMIR is that of the semantic gap which is found between two information areas. The first information area is of the pictorial information clicked by the imaging machine and second information area is of the Image formation taking place in human brain as a mere result of human imaginative power. In simpler words, a human can imagine something different than what is captured by the imaging machine while diagnosing the patient, which is the major concern in the CBMIR system. To improve the working of CBMIR system, a lot of features descriptors have been developed that can represent image from a higher-level perspective. Scale Invariant Features Transform (SIFT), features descriptors [17, 18] were developed including many more. Also, a Bag of Words model was developed using SIFT or SURF (Speed up Robust Features). Despite these modifications in features extraction layers, the semantic gap was not able to bridge up. Another method of improving the CBMIR system is integrating it with deep learning constructs which can directly learn the features from images without the use of descriptors or any fabricated features. CBMIR was also equipped with support vector machines to filter and eliminate the irrelevant images so that the size of the database gets reduced, and some amount of efficiency, accuracy and precision gets increased [2].

1.2 Deep Learning with DCNN

Convolutional Neural Networks (CNN) [19, 20] based systems with deep learning techniques have been already used in the medical research and application areas such as for classification of medical images related to lung diseases to classify and retrieve lung patterns, lung textures and airway detection. DCNN models have been used for Computer Aided Diagnosis (CAD) [21–24].

The major working procedure discussed for CBMIR in this paper is given below:

- The transfer learning model specially with CNN and hash function (hash layer) is applied to retrieval of medical images.
- A new framework is adopted with adding 5 Convolutional layer, 3 pooling layer and 2 fully connected layer and one hash layer and a contrastive loss function. which is mainly into two components.
- The first component is the order task utilizing the pre-trained model on the clinical dataset to create the mid-level features.
- In second parts, added the hash layer for age of n-digit hash codes which will help of make quick and productive image retrieval task.
- The nitty gritty bit by bit two calculations are introduced for preparing the organization, the first is features extraction and image retrieval algorithm and other is the execution of hash function and fast image retrieval algorithm [16].

The remaining part of this paper is organized as: Sect. 2 discussed the proposed framework briefly. In Sect. 3 step by step algorithms are discussed and important results are presented in Sect. 4 and 5, and conclusion is given in Sect. 6 in this paper.

2 Methodology

In this proposed work of recovering clinical Images for the analysis, a system that depends on the grouping of Images and highlight extraction from that arrangement is utilized. To achieve this, the basic design which comprise of information, convolution and pooling layers, completely associated layers and the basic yield layer is utilized and two new pursuit methodologies are applied on the removed features for making the cycle quick and proficient.

2.1 Proposed Framework

In this paper, for retrieval of medical images the new network is introduced with 5 convolutional layers, 3 pooling layers and 2 fully connected layers are used. The main work of convolutional layers is to extract high level features and pooling layer is used here to find the local maxima in the neighborhood for decreasing the size of the features. The fully connected layer is utilized for making classification task with SoftMax layer so that based on probabilities, the classes could be find. At the end, the output is fed into the contrastive loss function. The main purpose of introducing the loss function is to minimize the distance between the features of the similar image pairs and maximize it for dissimilar pairs. With the proposed network, the transfer learning approach is also used for checking the efficiency of medical images retrieved from the new proposed model and with transfer learning approach.

3 Algorithms

3.1 Algorithm 1: Training the Network

The most common algorithm that is used in training of neural networks is the back-propagation algorithms which can be integrated with some or the other optimization technique. Stochastic Gradient descent (SGD) can be used as an optimization technique with the backpropagation. The major challenge of working of SGD is that it will take the calculated gradients as input and will keep on minimizing the objective function. So there is a need to calculate all the gradients related to all the parameters first and then feed them inputs to SGD. There are associated parameters only with the convolution layer and no associated parameters with the pooling layers.

The algorithm used here is the 3 steps algorithm consisting of forward flow, backward flow and the calculation of gradients of parameters of convolution layer.

Step 1: Forward flow
In this part, the message is transferred to compute all z's where z is the function of input x. Equation 1 gives the notation:

$$z^{k+1} = f(z^k)z^{k+1} = f(z^k) \tag{1}$$

where k is the layer index and $z = f(x_i)$.

Step 2: Backward flow
In this section a retrogressive message is moved to register every one of the subsidiaries for example Δ's the place where Δ is the subsidiary of cost work w.r.t y's, so we address it in numerical documentation as:

$$\Delta_i^k = \frac{dX}{dz_i^k}$$
$$\sum_j \frac{dX}{dz_i^k} \cdot \frac{dz_i^{k+1}}{dz_i^k} = \sum_j \Delta_i^{k+1}\left(\frac{dz_i^{k+1}}{dz_i^k}\right) \tag{2}$$

where i is the unit index of layers, l is the input sample, X signifies the loss-function. This equation is recursive so SGD will recursively minimize the loss function.

Step 3: Derivative of parameters of Convolution Layer
The mathematical equation for calculating the derivatives of parameters of any layer is given below:

$$\frac{dE}{d\Theta^k} = \sum_i \frac{dE}{dz_i^{k+1}} \cdot \frac{di^{k+1}}{d\Theta^k}$$
$$\frac{dE}{d\Theta k} = \sum_i \Delta i^{k+1}\left(\frac{dzi^{k+1}}{d\Theta^k}\right) \tag{3}$$

3.2   Feature Extraction and Image Retrieval

The output of transfer learning model is fed to the two fully connected layers, the FC layer 8 has 1000 nodes, depending on the size of domain dataset. The output of this layer will be the deep features. These features are put to the hash layer that is helpful to generate hash code for making fast comparison.

A hash is basically a function that receives the data, applies some mathematical constraints and rules and outputs a value. The value that the hash function will output will always be of fixed size. The hash layer in this system model consists of 2 sublayers. The first sublayer will govern the task of hash calculation techniques and the last layer will binarize the continuous values to produce binary hash. The task of functional layer is carrying out the mathematical computations. For these mathematical computations, sigmoid function is used. If say, L is hash layer then, the sigmoid function will be:

$$L = \text{sigmoid}(WT.f + b) = V\,q. \qquad (4)$$

where, f is the features vector such that $f \in V$ m. and V m is the random vector whose number is q′ or can say that q is the length of the hashing code that will be generated.

b is the bias and $b \in V$ q, $W \in V$ q × m.

The below algorithm shows the implementation of hash function as hash layer and retrieval of fast images in proposed framework.

3.3   Algorithm 2: Implementing Hash Function and Fast Image Retrieval

The algorithm proceeds as a two-phases and are a 7-steps process to increase the speed, efficiency, and accuracy of the image retrieval system. The image retrieval process is improved by adopting rough to smooth image retrieval strategies.

Phase 1

First retrieve the candidates that have similar binary activations from the hash layer or in simple words the set of candidates where the hamming distance between $\text{sign}(D_q)$ and $\text{sign}(D_i)$ is smaller than threshold. D_q the sigmoidal output of query image and D_i is the sigmoidal output of images in the dataset.

Step 1: D_i is computed by the pretrained model by using the activation function discussed above.

Step 2: D_q is computed by applying the sigmoidal function on the features of query image.

Step 3: Compute $D_i = \text{sign}(D_i)$ for every image in the image dataset as

$$D_i = \text{sign}(D_i) = \begin{cases} 1, if\ Dq \geq 0.5 \\ 0, otherwise \end{cases} \quad (5)$$

where Di is the binary hash code of image in the image dataset and all the Di will be stored in the vector say I_h. such that $I_h = \{D_1, D_2, \ldots Dn\}$.

Step 4: Compute Di = sign (Dq) as

$$Dq = \begin{cases} 1, if\ Dq \geq 0.5 \\ 0, otherwise \end{cases} \quad (6)$$

where Dq is the binary hash code of query image.

Step 5: If assume that $I = \{I1, I2, \ldots In\}$ is the image dataset with n images and $I_h = \{D_1, D_2, \ldots Dn\}$ are the corresponding binary hash codes of images in image dataset, and Iq is the query image with Dq is the binary hash code of query image, then, if (t (Dq, Di) < = Ω) return(a ∈ I).

where 't' is the hamming distance, 'Ω' is the threshold value and 'a' is an element of set I.

Step 6: Store each a in a vector p. Now p contains the rough level images.

Through these 6 steps, the rough image retrieval process is completed. For finding the k-top images from the rough images databases, apply the phase 2 of algorithm 2, which is show below.

Loss Function

Assume G mirrors the grayscale space; the reason for the whole space; System is intended to investigate interpretation from G to k-bit parallel code: f: G → {−1, 1} k, which permits indistinguishable pictures to be named as Related twofold codes to that. Provided the pair of photos I_1, I_2 and G, the related network contribution is b_1, b_2 and, in general, b_1. The loss function for this interest point is defined as follows:

$$L(b_1, b_2, y) = \frac{1}{2}(1 - y)D(b_1, b_2) + \frac{1}{2}ymax(m - D(b_1, b_2)) \quad (7)$$

We characterize y = 0 if two pictures are comparable; something else, y = 1, where D(,) is the distinction between two yield vectors and m > 0 is an edge limit boundary. At the point when D m is more noteworthy than one, the main segment in (1) punishes indistinguishable pictures projecting to different yield sections, while the subsequent thing punishes unique pictures projecting to comparable yield vectors. The accompanying misfortune work is acquired for a preparation set comprising N sets of pictures:

$$L = \sum_{i=1}^{N} L(b_{i,1}, b_{i,2}, y) \quad (8)$$
$$where, i \in (1, \ldots N), j \in (1, 2)$$

A regularization model is stretched out to the misfortune work, and the Euclidean distance is used to restrict the scope of boundaries and guarantee the organization yields look like the parallel hash codes, which will be favorable to the subsequent binarization. The proposed loss function looks like this:

$$L(b_1, b_2, y) = \frac{1}{2}(1 - y)\|b_1 - b_2\|_2^2 + \frac{1}{2}ymax\left(m - \|b_1 - b_2\|_2^2, 0\right)$$
$$+\alpha(\||b_1| - 1\|_1 + \||b_2| - 1\|_1)$$
(9)

where L with addendum r is the unwinding misfortune work, 1 signifies an every one of the one vector, ‖.‖ demonstrates the vector's L1 standard and is the regularization term's weight boundary. The regularization term may restrict the vectors' components to one.

We lean toward the L1 standard over the higher-request standard since it has a lower figuring cost and rates up the preparation cycle. The above condition permits us to communicate the complete misfortune work as follows:

$$L_r = \sum_{i=1}^{N} \frac{1}{2}(1 - y_i)\|b_{i,1} - b_{i,2}\|_2^2 + \frac{1}{2}y_imax\left(m - \|b_{i,1} - b_{i,2}\|_2^2, 0\right)$$
$$+\alpha(\||b_{i,1}| - 1\|_1 + \||b_{i,2}| - 1\|_1)$$
(10)

Paired hash coding is just accomplished after network preparing by parallel processing the vector b, which would be sign (b).

Phase 2

In this stage, the estimation of the Euclidian distance is performed between the element vector of inquiry picture and highlight vector of each chose competitor 'a' from the vector p delivered by the harsh hunt. The more modest the Euclidean distance, the more will be the similitude between the inquiry picture and the ith competitor of the vector p.

In this interaction, managing the entire picture dataset isn't performed, yet the smooth hunt is in effect just completed on the vector p, the vector which conveys the unpleasant pictures recovered in stage 1. On the off chance that estimation of Euclidean distances between the question picture highlight vectors and highlight vectors of each picture of the area dataset is applied, then, at that point it would require some investment, energy, computational force. Along these lines, to diminish the assignment generally first and foremost by utilizing the hash like capacities for harsh picture recovery is performed and afterward refine the hunt to recover the k-top pictures by applying smooth pursuit on unpleasant pictures dataset/information bases.

If Y is the Euclidean distance, then,

Step 7: For every i = 1, 2, …. N calculate $Y_i = \|U_q - U_i\|$,

Where U_q and U_i denote features vectors of query image and the rough image candidate in the vector k respectively.

The Yi will be calculated for each element in the vector k to retrieve the smoothest and most similar images. And top k ranked images are retrieved through this process.

4 Result and Discussion

The proposed clinical diagnosis model is carried out and executed utilizing the foundation of the deep libraries of Keras and Tensorflow with the framework setup Ryzen 7 processor, 16 Gigs RAM and Radeon RX 5700 GPU processor. The standard dataset of CT Images (Coronavirus-19) from Kaggle online platform was utilized.

4.1 Description of Dataset

In this paper the standard medical dataset is utilized from standard repository Kaggle, named as Coronavirus-19 CT Sweeps which contains chest CT Images of both Coronavirus-19 patients and non-Coronavirus-19 contaminated individuals [34]. The dataset contains 5000 ordinary Images and 7000 Coronavirus-19 contaminated images and are utilized for training the model with two classifications that are utilized for discovery purposes. The area showing the infection of Covid was perceived and determined by the clinical specialists and gathered by differentiation of two classes. The two classes are shown in Figs. 2 and 3.

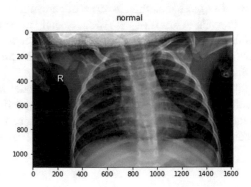

Fig. 2 Normal lungs

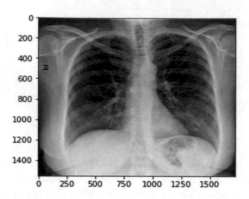

Fig. 3 Covid-19 infected lungs

5 Experimental Results

In this experiment, 3500 images normal and 4900 images of Covid-19 corrupted, for training data are randomly picked and for the test purpose out of the total 12,000 Images are used [34].

During training phase, two classes namely Covid-19 and normal images are used as an image pair. During training phases, two comparable classes of images were used to evaluate the performance. The results are obtained by applying the proposed model are represented with point operations.

Table 1 shows the image pair datasets for training and testing images for analysis. Using the input test data, the proposed method can predict the Chest's diseases.

The accuracy level of suggested method in terms of percentage is shown in Table 2. It shows the obtained results in terms of Precision, Recall and Specificity.

The obtained results are compared with another classifier technique which is based on deep neural network algorithm [33] and shown in Table 2. Our proposed method gives good results than existing methods.

In above table (Table 2), the absolute assessment of the model with various execution measure estimations like PPV, NPV, Precision, Recall, Specificity is shown. The findings show that the Precision attained up to 94.5% with recall 98.2%, specificity 91.2%, Positive Predictive Value (PPV) 93.6% and Negative Predictive Value (NPV) 95.0%.

Table 1 Training & testing dataset images for analysis

Phase of networks	Proposed classifier		
	Normal	Covid-19	Total images
Training	3500	4900	8400
Testing	1500	2100	3600

Table 2 (a) Performance results proposed method and (b) Performance results of [33].

CT Images	Image Types	Precision	Recall	Specificity
	Normal			
	Covid – 19	94.5%	98.2%	91.2%
(b)				
	Normal			
	Covid – 19	93.6%	93.9%	91.4%

6 Conclusion

The method proposed in this paper significantly means to help advanced pathology to help clinical experts in diagnosing the basic cases. The proposed clinical Image retrieval technique dependent on profound convolutional neural organizations incorporated with two degrees of highest-level image retrieval systems (harsh search and smooth search) contributes for quick, productive, and more exact strategy for clinical diagnostics by retrieving most comparative images dependent on learning of the model. The proposed model with highlight dimensionality decrease showed the better order on account of Ordinary Chest CT and Coronavirus-19 Chest CT datasets when contrasted and other characterization calculation. The methodology utilized here gives better outcomes and shows that the manual marking time is diminished. Different exploration has been done on the recognition of Chest Coronavirus-19, yet the procedure proposed here gives the better outcomes as far as accuracy, specificity, and recall. Accuracy is demonstrated to improve by 94.5%, specificity by 91.2% and recall by 98.2%. The significant advantage is focused on the clinical professionals and specialists to help them in the advanced pathology and finding by utilizing this proposed model and calculation. This will accordingly be advantageous to the specialists in the recognition of Coronavirus-19.

References

1. Khened M, Kollerathu VA, Krishnamurthi G (2019) Fully convolutional multi-scale residual DenseNets for cardiac segmentation and automated cardiac diagnosis using ensemble of classifiers. Med Image Anal 51:21–45. ISSN 1361-8415
2. Bakkouri I, Afdel K (2019) Multi-scale CNN based on region proposals for efficient breast abnormality recognition. Multimed Tools Appl 78:12939
3. Qayyum A, Anwar SM, Awais M, Majid M (2017) Medical image retrieval using deep convolutional neural network. Neurocomputing
4. Mizotin M, Benois-Pineau J, Allard M, Catheline G (2012) Feature-based brain MRI retrieval for Alzheimer disease diagnosis. In: 2012 19th IEEE international conference on image processing, pp 1241–1244
5. Rahman MM, Antani SK, Thoma GR (2011) A learning-based similarity fusion and filtering approach for biomedical image retrieval using SVM classification and relevance feedback. IEEE Trans Inf Technol Biomed 15(4):640–646
6. Zhang F et al (2016) Dictionary pruning with visual word significance for medical image retrieval. Neurocomputing 177:75–88
7. Jia Y et al (2014) Caffe: convolutional architecture for fast features embedding. In: Proceedings of the ACM international conference on multimedia, pp 675–678
8. Khatami A, Babaie M, Khosravi A, Tizhoosh HR, Salaken SM, Nahavandi S (2017) A deep-structural medical image classification for a Radon-based image retrieval. In: Proceedings of the IEEE 30th Canadian conference on electrical and computer engineering (CCECE), Windsor, ON, USA, April/May 2017, pp 1–4
9. Lu J, Liong VE, Zhou J (2017) Deep hashing for scalable image search. IEEE Trans Image Process 26(5):2352–2367
10. Dureja A, Pahwa P (2019) Analysis of non-linear activation functions for classification tasks using convolutional neural networks. Recent Patents Comput Sci 12:156. https://doi.org/10.2174/2213275911666181025143029
11. Liong VE, Lu J, Wang G, Moulin P, Zhou J (2015) Deep hashing for compact binary codes learning. In: 2015 IEEE conference on computer vision and pattern recognition (CVPR), pp 2475–2483
12. Wu C, Li Y, Zhao Z et al (2019) Image classification method rationally utilizing spatial information of the image. Multimedia Tools Appl 78:19181
13. Dureja A, Pahwa P (2018) Image retrieval techniques: a survey. Int J Eng Technol (IJET, UAE) 7(12):215–219
14. Xia R, Pan Y, Lai H, Liu C, Yan S (2014) Supervised hashing for image retrieval via image representation learning. In: Proceedings of the national conference on artificial intelligence, vol 3, pp 2156–2163
15. Baskar D, Jayanthi VS, Jayanthi AN (2019) An efficient classification approach for detection of Alzheimer's disease from biomedical imaging modalities. Multimed Tools Appl 78:12883
16. Hu Y, Modat M, Gibson E, Li W, Ghavami N, Bonmati E, Wang G, Bandula S, Moore CM, Emberton M, Ourselin S, Alison Noble J, Barratt DC, Vercauteren T (2018) Weakly-supervised convolutional neural networks for multimodal image registration. Med Image Anal 49:1–13. ISSN 1361-8415

17. Lowe G (2004) SIFT-The scale invariant features transform. Int J 2:91–110
18. Bay H, Tuytelaars T, Van Gool L (2006) SURF: speeded up robust features. In: Leonardis A, Bischof H, Pinz A (eds) Computer vision – ECCV 2006, ECCV 2006. Lecture notes in computer science, vol 3951
19. Krizhevsky A, Sutskever I, Hinton GE (2012) ImageNet classification with deep convolutional neural networks. In: Advances in neural information processing systems
20. Simonyan K, Zisserman A (2015) Very deep convolutional networks for large-scale image recognition. In: ICLR
21. Yan Z et al (2016) Multi-instance deep learning: discover discriminative local anatomies for bodypart recognition. IEEE Trans Med Imaging 35(5):1332–1343
22. Anthimopoulos M, Christodoulidis S, Ebner L, Christe A, Mougiakakou S (2016) Lung pattern classification for interstitial lung diseases using a deep convolutional neural network. IEEE Trans Med Imaging 35(5):1207–1216
23. van Tulder G, de Bruijne M (2016) Combining generative and discriminative representation learning for lung CT analysis with convolutional restricted Boltzmann machines. IEEE Trans Med Imaging 35(5):1262–1272
24. Moeskops P et al (2016) Automatic segmentation of MR brain images with a convolutional neural network. IEEE Trans Med Imaging 35(5):1252–1261
25. Seetharaman K, Sathiamoorthy S (2016) A unified learning framework for content based medical image retrieval using a statistical model. J King Saud Univ-Comput Inf Sci 28(1):110–124
26. Glorot X, Bordes A, Bengio Y (2011) Deep sparse rectifier neural networks. In: AIStats, p 275
27. Glorot X, Bengio Y (2010) Understanding the difficulty of training deep feedforward neural networks. In: AIStats, pp 249–256
28. LeCun Y, Bottou L, Bengio Y, Haffner P (1998) Gradient-based learning applied to document recognition. Proc IEEE 86(11):2278–2324
29. Gionis A, Indyk P, Motwani R et al (1999) Similarity search in high dimensions via hashing. In: VLDB, vol 99, pp 518–529
30. Weiss Y, Torralba A, Fergus R (2009) Spectral hashing. In: Advances in neural information processing systems, pp 1753–1760
31. Gong Y, Lazebnik S, Gordo A, Perronnin F (2013) Iterative quantization: a procrustean approach to learning binary codes for large-scale image retrieval. IEEE Trans Pattern Anal Mach Intell 35(12):2916–2929
32. Lin K, Yang H-F, Hsiao J-H, Chen C-S (2015) Deep learning of binary hash codes for fast image retrieval. In: Proceedings of the IEEE conference on computer vision and pattern recognition workshops, pp 27–35
33. Kermany D, Goldbaum M, Cai et al (2018) Identifying medical diagnoses and treatable diseases by image-based deep learning. Cell 172:1122–1131
34. COVID-19 CT scans (2021). https://www.kaggle.com/andrewmvd/covid19-ct-scans

Ontological Representation and Analysis for Smart Education

Bikram Pratim Bhuyan and Shelly Garg

1 Introduction

In current scenario, technology has proven to be very helpful in providing ease into day to day life. Every person has witnessed the power of technology. It can be applied in various fields such as smart education, smart technologies and smart systems. Many of the new approaches have been developed so far. Those services can fulfill the requirement of on and off campus students with a modern era setup. Nowadays in pandemic we have seen that every student is attending his classes from home. With technology enhancement, hardware and software market has clearly seen a boom in the market of smart universities. This growth will be more in the coming years. When we talk about smart technology, a new concept comes into the picture that is smart education. When we see smart education market a tremendous growth has been observed. In 2019, around 23.2 billion dollar investments were seen which is expected to increase up to USD 56.5 billion in 2024, a growth of 19.5% is expected [12]. The major factor of contribution are the smart technologies connecting various devices together. Many e-Learning solutions are adopted which uses the smart technologies such as artificial intelligence, concepts of machine learning etc. in the field of smart learning. Some of the key major players in the field are Blackboard, IBM, Adobe, Oracle, Saba Software, Microsoft which are developed in US. Many other companies are contributing in this field like Samsung from south Korea, SAP from Germany, BenQ from Taiwan, Hawaii from china etc. Many companies have come forward with the technology solution which can be provided in the field of smart education. In general, on these platforms two type of learning's are provided which is Synchronous and Asynchronous learning. Therefore, this area has been a source of attraction to many researchers. To get clear understandings we need to

B. P. Bhuyan (✉)
Department of Informatics, School of Computer Science, University of Petroleum and Energy Studies, Dehradun, India
e-mail: bikram23bhuyan@gmail.com

S. Garg
Department of Virtualization, School of Computer Science, University of Petroleum and Energy Studies, Dehradun, India

© The Author(s), under exclusive license to Springer Nature Singapore Pte Ltd. 2022   381
B. Unhelker et al. (eds.), *Applications of Artificial Intelligence and Machine Learning*,
Lecture Notes in Electrical Engineering 925,
https://doi.org/10.1007/978-981-19-4831-2_31

know what are the hardware, software services related strategies and activities are required to be built which can provide more and more information to the next era of smart classrooms [13].

2   Literature Survey

Several years ago, a new concept was introduced which is smart classrooms. Where a need was seen that in future everything will be accomplished via technology applications [14]. The basic concept of smart classroom is to reorder the school infrastructure with a vision of focused learning's and business model development which considers the overall development of individual students with learning needs. Many companies have developed a transformation strategy which is providing a transition from procedural ways to the digital way of educating young students with a vision of meaningful engagement [15]. Specially in the Covid-19 scenerio, use of smart classroom is the need of the hour [16]. Papers which discuss the need of artificial intelligence [17] can be used as a focal point for these issues [18].

2.1   First Generations

Smart classroom concept has been divided into set of generations. The vision of early smart classroom concept was to do delivery of education in a synchronous manner. The major learning was delivered to the locals to the students who can come to attend face to face teaching method and online teaching methodology where various students are being located at remote locations. Shie et al. [2] shows a method where more than one natural modalities are being used by teaching to interact with remote students to achieve a same effect as that in face to face classroom. This method allows students located at different locations can participate in the synchronous mode of education. Xie et al. [3] described an additional method on top of the traditional approach. Where teachers are no longer required to stick to the traditional ways of using laptops or mouse or keyboards for delivery of education. This method introduces the tele-education method which is based on intelligent concept in the nature. At the same time, they provide an experience of just like real time classroom. Uskov et al. [4] developed an interesting method where 2-way education with the usage of high definition quality of audio and videos are introduced. This method introduces the real time two-way communication between remote and local students. Remote students may be located at different locations in the world. This method highly provide an ease to the local as well as remote students for imparting of education.

2.2 Second Generations

After the popularity obtained by first generation of smart classroom concept, new concepts are being introduced in the second generation of smart classroom concept. In this era of generation, mobile technology was also actively introduced for imparting the smart education. Various mobile devices used by learners, users and students with a communication between traditional and new smart classroom concepts environment are discussed. In this generation Vladimir et al. [1] developed smart class room concept based on ontology for understanding and analysis purpose. This method helps in identification of hardware, services, software, pedagogy, learning and teaching methods related activities. Driscoll et al. [5] introduces the smart class concept based on the context aware technology (CASC). In this responses are generated based on the lecturer timetable and already established policies. Low cost identification location based system is developed. It uses a central scheduling policy to determine the activity. Yau et al. [6] RCSM a smart classroom solution comprises of re configurable context sensitive middle ware. In this method, PDA are used with a knowledge of light, location, noise and mobility. This concept is applied on each student with a vision of collaborative learning. Huang et al. [7] focuses on optimization of the delivery content presentations. It also aim at to provide ease for probe of various learning resources available. It also focuses on teaching and learning concept go hand in hand with management of class layout. Many universities offer the distant education so, Pishva et al. [8] define an intelligent classroom approach where students can experience a real time classroom experience despite of enrolled in distant education course. In this approach amalgamation of smart technology is introduced such as computer vision, voice-recognition, artificial intelligence concepts where agents are used to provide real time tele-education experience. Glogoric et al. [9] proposes the use of internet of things in the concepts of smart classroom, where behaviours are observed and analyzed in the ordinary classrooms.

3 Basic Definitions Used

Following the terminologies used in [1], we denote the end users of the system i.e. the individual students by $\{I_1, I_2, \ldots, I_n\}$. Each of the levels of smartness of the smart system is denoted by $\{s_1, s_2, \ldots, s_n\}$; goals and objectives of the system is represented by $\{g_1, g_2, \ldots, g_n\}$ and $\{o_1, o_2, \ldots, o_n\}$ respectively. Hardware and software used is symbolized by $\{h_1, h_2, \ldots, h_n\}$ and $\{so_1, so_2, \ldots, so_n\}$. Finally activities and pedagogy utilized is signified by $\{a_1, a_2, \ldots, a_n\}$ and $\{p_1, p_2, \ldots, p_n\}$. We form a set of $\{S, G, O, H, SO, A, P\}$ to accumulate the properties of the system (see notations used) (Table 1).

Table 1 Notation and symbols used in this paper

Symbol	Description
I	Set of students
S	Set of smartness levels
G	Set of goals
O	Set of objectives
H	Set of hardware
SO	Set of software
A	Set of activities
P	Set of pedagogy
$(S, G, O, H, SO, A, P, \varpi)$	Meet semi-lattice
ξ	Mapping function
I^θ	Familiar systems for a set of students $N \subseteq I$
$\{S, G, O, H, SO, A, P\}^\alpha$	Familiar students for a set of systems $M \in (\{S, G, O, H, SO, A, P\}, \varpi)$
(N, M)	Tuple for the set of students with the system properties

Definition 1. *A mapping function is defined from the set of students to the set of properties of the system as*

$$\xi : I \rightarrow \{S, G, O, H, SO, A, P\} \tag{1}$$

The mapping function acts as a binary operator from the set comprising of students to the system properties [10, 11]. The Table 2 represents a graphical representation of the concepts discussed so far.

Table 2 Students with system properties

Student	S_1	S_n	G_1	G_n	O_1	O_n	H_1	H_n	SO_1	SO_n	A_1	A_n	P_1	P_n	...
I_1	y		y	y		y			y	y		y		y	
I_2	y	y		y	y		y	y	y				y	y	y
I_3		y		y	y	y	y			y	y			y	
I_4	y		y				y	y			y	y	y	y	
I_5	y	y	y	y		y					y	y		y	y
I_6		y		y	y		y		y		y	y			
I_7			y	y			y	y			y		y	y	
I_8	y	y		y	y				y	y		y		y	
...		y	y	y	y	y			y	y		y		y	
I_n	y	y		y	y	y	y			y	y	y		y	y

We now define a familiarity operator as I^α to represent the common set of system properties and $\{S, G, O, H, SO, A, P\}^\alpha$ to represent common set of students reflecting the properties. Before formally defining the same, we first design Algorithm 1 to cluster the students with common properties.

Algorithm 1 lets us to amalgamate the tuples of the students set and the systems set to fully closed set relationship.

Algorithm 1. Clustering students with the system

Group students with the system properties to form a tuple $(I, \{S, G, O, H, SO, A, P\})$
Input: Set of Students: I and System properties: $(I, \{S, G, O, H, SO, A, P\})$
Output: Set of tuples (CT)
1: **procedure** CREATE–TUPLE
2:     Initialize N_0 and M_0 to ϕ.
3:     **for** each student $N_i \in I$ **do**
4:         Compute N_i^α and $N_i^{\alpha\alpha}$
5:         **if** $N_i^\alpha == N_i^{\alpha\alpha}$
6:             $N \leftarrow N_i^{\alpha\alpha}$
7:             $M \leftarrow N_i^{\alpha}$
8:             Form tuple (N, M)
9:             Remove duplicates if any.
10:    **end for**
11: **end procedure**

Definition 2. *The familiarity for the set of students is be represented by*

$$I^\theta = \xi_{x \in X} \varpi(x) \quad for \quad N \sqsubseteq I \tag{2}$$

Definition 3. *The familiarity operator for the system properties can be represented by*

$$\{S, G, O, H, SO, A, P\}^\theta = \{x \in X | y \sqsubseteq \varpi(x)\} for \quad y \in (\{S, G, O, H, SO, A, P\}, \xi) \tag{3}$$

Theorem 1. *The tuples formed under the familiarity operator is closed under the same operator.*

We are now at a position to formally design the smart system ontological knowledge representation in the next section.

4   Ontological Knowledge Representation

We start with the creation of a lattice as shown in Algorithm 2. The lattice formed of the closed tuples result in a meet semi lattice.

Now, we are in a position to define a smart education system formally.

Algorithm 2. Lattice creation with tuples

Form a lattice from the tuples generated from Algorithm 1
Input: Set of tuples(N,M)
Output: Meet semi-Lattice $(S, G, O, H, SO, A, P, \varpi)$

1: **procedure** CREATE–LATTICE
2:     Sort the tuples with respect to $\mid N \mid$ in a non-increasing order list.
3:     Insert ϕ as the bottom item.
4:     **for** each tuple (N, M) $\in$ the ordered list **do**
5:         Insert N_i in the lattice following the partially ordered relation.
6:     **end for**
7:     Insert $\{I_1, I_2, \ldots, I_n\}$ in the lattice.
8:     **for** each Student item $\in$ N **do**
9:         Add the corresponding M in the tuple.
10:    **end for**
11:    Finalize the Meet semi-Lattice $(S, G, O, H, SO, A, P, \varpi)$
12: **end procedure**

Definition 4. *A smart system can be defined as a triple:*

$(I, (S, G, O, H, SO, A, P, \varpi), \varrho)$ *where;*

I is the set of students; $(S, G, O, H, SO, A, P, \varpi)$ is a meet-semilattice comprising of the systems defined for the students and $\xi : I \rightarrow \{S, G, O, H, SO, A, P\}$ represents the mapping.

Let us understand the concept of smart system with an example. Table 3 is used as a toy example comprising of ten students bearing one individual property of each system properties.

Algorithm 1 is used to create the following tuple as shown in Table 4. It is seen that 31 unique tuples are generated.

Now, we implement Algorithm 2, to create the meet semi-lattice as shown in Fig. 1.

Table 3 Students with system properties

Student	S_a	G_b	O_c	H_d	SO_e	A_f	P_g
I_1	y		y	y		y	
I_2	y	y		y	y		y
I_3		y		y	y	y	y
I_4	y		y				y
I_5	y	y	y	y			y
I_6		y		y	y		y
I_7			y	y			y
I_8	y	y		y	y		
I_9		y	y	y	y	y	
I_{10}	y	y		y	y	y	y

Table 4 Tuples

$Tuple_{id}$	N (students)	M (system properties)
T_1	$\{\phi\}$	$\{S_a, G_b, O_c, H_d, SO_e, A_f, P_g\}$
T_2	$\{I_1\}$	$\{S_a, O_c, H_d, A_f\}$
T_3	$\{I_9\}$	$\{G_b, O_c, H_d, SO_e, A_f\}$
T_4	$\{I_{10}\}$	$\{S_a, G_b, H_d, SO_e, A_f, P_g\}$
T_5	$\{I_3, I_{10}\}$	$\{G_b, H_d, SO_e, A_f, P_g\}$
T_6	$\{I_1, I_{10}\}$	$\{S_a, H_d, A_f\}$
T_7	$\{I_3, I_9, I_{10}\}$	$\{G_b, H_d, SO_e, A_f\}$
T_8	$\{I_1, I_9\}$	$\{O_c, H_d, A_f\}$
T_9	$\{I_5\}$	$\{S_a, G_b, O_c, H_d, P_g\}$
T_{10}	$\{I_1, I_5\}$	$\{S_a, O_d, H_f\}$
T_{11}	$\{I_4, I_5\}$	$\{S_a, O_c, P_g\}$
T_{12}	$\{I_5, I_7\}$	$\{O_c, H_d, P_g\}$
T_{13}	$\{I_1, I_5, I_7, I_9\}$	$\{O_c, H_d\}$
T_{14}	$\{I_5, I_9\}$	$\{G_b, O_c, H_d\}$
T_{15}	$\{I_4, I_5, I_7\}$	$\{O_c, P_g\}$
T_{16}	$\{I_1, I_4, I_5\}$	$\{S_a, O_c\}$
T_{17}	$\{I_2, I_{10}\}$	$\{S_a, G_b, H_d, SO_e, P_g\}$
T_{18}	$\{I_2, I_3, I_6, I_{10}\}$	$\{G_b, H_d, SO_e, P_g\}$
T_{19}	$\{I_2, I_8, I_{10}\}$	$\{S_a, G_b, H_d, SO_e\}$
T_{20}	$\{I_2, I_5, I_{10}\}$	$\{S_a, G_b, H_d, P_g\}$
T_{21}	$\{I_2, I_5, I_8, I_{10}\}$	$\{S_a, H_d\}$
T_{22}	$\{I_2, I_4, I_5, I_{10}\}$	$\{S_a, P_g\}$
T_{23}	$\{I_2, I_3, I_5, I_6, I_7, I_{10}\}$	$\{P_g, H_d\}$
T_{24}	$\{I_2, I_3, I_5, I_6, I_{10}\}$	$\{G_b, P_g, H_d\}$
T_{25}	$\{I_1, I_3, I_9, I_{10}\}$	$\{H_d, A_f\}$
T_{26}	$\{I_1, I_4, I_5, I_7, I_9\}$	$\{O_c\}$
T_{27}	$\{I_2, I_3, I_6, I_8, I_9, I_{10}\}$	$\{G_b, H_d, SO_e\}$
T_{28}	$\{I_1, I_2, I_4, I_5, I_8, I_{10}\}$	$\{S_a\}$
T_{29}	$\{I_2, I_3, I_4, I_5, I_6, I_7, I_{10}\}$	$\{P_g\}$
T_{30}	$\{I_2, I_3, I_5, I_6, I_8, I_9, I_{10}\}$	$\{G_b, H_d\}$
T_{31}	$\{I_1, I_2, I_3, I_4, I_5, I_6, I_7, I_8, I_9, I_{10}\}$	$\{\phi\}$

5 Discussion

The asymptotic time complexity of Algorithm 1 is $O(NM)$ and that of Algorithm 2 is $O(N \mid H \mid M)$; where $\mid H \mid$ is the meet semi lattice size. Implementing Algorithm 1 in the demo example shown in Table 3, we notice the creation of 31 tuples shown in Table 4. Now, we can put forward a threshold on the minimum number of student set in the tuple so as to extract the tuples with at least a minimum count of students. Suppose we put a threshold of 5, we get eight tuples namely $T_{23}, T_{24}, T_{26}, T_{27}, T_{28}, T_{29}, T_{30}, T_{31}$. On analysing individual tuples say T_{27}, we find that the set of students $\{I_2, I_3, I_6, I_8, I_9, I_{10},\}$ share the system properties goals, hardware and software. Hence we can put-up a holistic approach to place these stu-

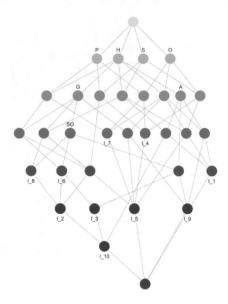

Fig. 1 Lattice

dents under a common platform. Also from the lattice, we infer some knowledge on the system properties like $A \to H$ and $SO \to G \to H$ which means that if a student bear the property of Software then automatically that student will have the property Goal, also of goal property is observed, then Hardware property is seen. Hence inferential learning [] is also observed herein.

6   Conclusion

In this paper, smart schooling is discussed in detail, and the essential terminology that provide the basis for algorithms to build tuples and lattices are defined. Binary inputs generate student and system groups, which in turn develop the inferential semi-lattice, which is used to study and refine knowledge representation. A toy example's complexity and implementation are also analysed. The concept of generalised learning may be expanded in future research. The use of inferential statistics as well as an ontological purview through the tools of machine learning and artificial intelligence is a work to be explored in future.

References

1. Uskov VL, Bakken JP, Pandey A (2015) The ontology of next generation smart classrooms. In: Smart education and smart e-learning. Springer, Cham, pp 3–14
2. Shi Y, Xie W, Xu G, Shi R, Chen E, Mao Y, Liu F (2003) The smart classroom: merging technologies for seamless tele-education. IEEE Pervasive Comput 2(02):47–55
3. Xie W, Shi Y, Xu G, Xie D (2001) Smart classroom-an intelligent environment for tele-education. In: Pacific-rim conference on multimedia. Springer, Heidelberg, pp 662–668
4. Uskov V, Uskov A (2005) Streaming media-based education: outcomes and findings of a four-year research and teaching project. Adv Technol Learn 2(2):45–57
5. O'Driscoll C, Mithileash M, Mtenzi F, Wu B (2008) Deploying a context aware smart classroom
6. Stephen SY, Karim SKF, Ahmed SI, Wang Y, Wang B (2003) Smart classroom: enhancing collaboartive learning using pervasive computing technology. In: Proceedings of American Society of Engineering Education 2003 annual conference
7. Huang R, Hu Y, Yang J, Xiao G (2012) The functions of smart classroom in smart learning age. Open Educ Res 18(2):22–27
8. Pishva D, Nishantha GGD (2008) Smart classrooms for distance education and their adoption to multiple classroom architecture. J Netw 3(5):54–64
9. Gligorić N, Uzelac A, Krco S (2012) Smart classroom: real-time feedback on lecture quality. In: 2012 IEEE international conference on pervasive computing and communications workshops. IEEE, pp 391—394
10. Bhuyan BP, Karmakar A, Hazarika SM (2018) Bounding stability in formal concept analysis. In: Advanced computational and communication paradigms. Springer, Singapore, pp 545–552
11. Bhuyan BP (2017) Relative similarity and stability in FCA pattern structures using game theory. In: 2017 2nd international conference on communication systems, computing and IT applications (CSCITA), 7 Apr 2017. IEEE, pp 207–212
12. Kwet M, Prinsloo P (2020) The 'smart' classroom: a new frontier in the age of the smart university. Teach High Educ 25(4):510–526
13. Saini MK, Goel N (2019) How smart are smart classrooms? A review of smart classroom technologies. ACM Comput Surv (CSUR) 52(6):1–28
14. Yang J, Pan H, Zhou W, Huang R (2018) Evaluation of smart classroom from the perspective of infusing technology into pedagogy. Smart Learn Environ 5(1):1–11
15. MacLeod J, Yang HH, Zhu S, Li Y (2018) Understanding students' preferences toward the smart classroom learning environment: development and validation of an instrument. Comput Educ 122:80–91
16. Cox AM (2021) Exploring the impact of Artificial Intelligence and robots on higher education through literature-based design fictions. Int J Educ Technol High Educ 18(1):1–19
17. Westman S, Kauttonen J, Klemetti A, Korhonen N, Manninen M, Mononen A, Paananen H (2021) Artificial intelligence for career guidance-current requirements and prospects for the future. IAFOR J Educ 9(4)
18. Liu H, Tan W, Li H, Gong J, Liu X (2021) Application of artificial intelligence technology in the teaching of mechanical education courses in universities. J Phys Conf Ser 1992(4):042065

Empirical Analysis of Diabetes Prediction Using Machine Learning Techniques

Nikita Poria and Arunima Jaiswal

1 Introduction

Before understanding diabetes and how it develops, we should know how things work without diabetes. The body requires energy for its proper functioning, and the major source of energy for the body is carbohydrates. Even diabetic people need carbohydrates. When we eat food containing carbohydrates, it gets broken down into glucose by the body. The glucose is then used by the body for energy. The body makes the glucose travel through the bloodstream to the cells. Insulin is a hormone that helps the glucose in moving from the blood into the cells [1]. Insulin is produced by the beta cells in the pancreas. Sometimes because of some reason, the pancreas is unable to create enough insulin, or the cells are unable to respond to the insulin correctly, which causes the glucose to stay in the bloodstream and increase the blood sugar level. This increase in glucose (sugar) level for an extended duration of time is categorized by diabetes. A few diabetes symptoms include excessive urination, feeling hungry all the time, feeling excessive thirst, and sometimes feeling fatigued [2]. Diabetes is considered a serious health issue that needs early treatment. If not treated on time, it can lead to many other serious health issues or can possibly result in death. It may also lead to long-lasting disabilities or complications like heart attack, stroke, renal failure, eye issues, and so on. There are mainly two sorts of Diabetes Mellitus [3]. The inability of the pancreas to produce enough insulin due to the loss of beta cells causes type 1 diabetes. Beta cells are lost as a result of the autoimmune response. So far, the reason behind this autoimmune reaction is unclear. Diabetes Type 2 happens because of body resistance to insulin. Insulin resistance is a situation in which the

N. Poria (✉) · A. Jaiswal
Department of Computer Science and Engineering, Indira Gandhi Delhi Technical University for Women, Delhi, India
e-mail: nikita090btcse19@igdtuw.ac.in

A. Jaiswal
e-mail: arunimajaiswal@igdtuw.ac.in

© The Author(s), under exclusive license to Springer Nature Singapore Pte Ltd. 2022   391
B. Unhelker et al. (eds.), *Applications of Artificial Intelligence and Machine Learning*,
Lecture Notes in Electrical Engineering 925,
https://doi.org/10.1007/978-981-19-4831-2_32

cells are not able to react to the insulin correctly. As the illness worsens, it leads to a deficiency in insulin levels. One of the most often causes seen is excessive body weight combined with a lack of workout. There is also diabetes during pregnancy [4] which happens when the blood sugar levels of women reach a high level. It's not necessary to have a history of diabetes to have it. Those with gestational diabetes have a greater danger of type 2 diabetes in later years of life. Diabetes is one of the most expensive chronic illnesses [5]. It creates a substantial economic burden on society. There is no known cure for diabetes. But one can control the blood sugar levels by following a diabetes-healthy lifestyle. There is a pressing requirement for the early diagnosis and prediction of symptoms so that the treatment can be done on time and more lives can be saved. For the early diagnosis, we can make use of machine learning algorithms. Machine learning has quickly entered the field of healthcare [6]. It is used to discover patterns from the given dataset and provides excellent disease prediction capabilities [7]. In this research paper, we did a comparative analysis on the following machine learning techniques K-Nearest Neighbors, Extra Tree, Random Forest, Naive Bayes, Bagged Decision Tree, Adaptive Boosting, Stochastic Gradient Boosting, Support Vector Machine, and MLP Classifier. The results were validated on the PIMA Indian Diabetes Dataset using accuracy and F1 score as the performance criteria. It was found that KNN gave the best accuracy as compared to other machine learning techniques. We believe that our research will help clinicians predict diabetes more accurately.

In the subsequent sections of the paper, Related work is discussed in Sect. 2. In Sect. 3, all the applied techniques have been discussed in brief. Section 4 describes the methodology used in the paper. Section 5 includes results and analysis and, Sect. 6 talks about the conclusion and future scope of this work.

2 Related Work

Diabetes mellitus is a chronic condition that affects people all over the world [8]. As diabetes is a very dangerous disease the scientists for the many past decades are researching its early diagnosis so that more lives can be saved. For the research paper, we reviewed few research papers and, we will be discussing them below. In many of the research studies on the prediction of diabetes, the authors have used the PIMA Indian Diabetes Dataset [9–11]. In paper [12], the authors experimented with three different algorithms on the PIMA Indians of Arizona diabetes dataset to build the prediction model. They used Logistic Regression, Support Vector Machine, and Random Forest. According to their calculations, they found out that Random Forest is the ideal algorithm from the other two for the prediction of diabetes. They also showed the importance of the glucose level feature in the dataset. If people want to keep diabetes away, they should try to keep their glucose level down and, people with past diabetic history should have a proper diet. In paper [13], the authors tried to determine the accuracies of few data mining techniques used for prediction. They used the Pima Indian Diabetes Dataset and then developed five predictive models on it. They found from their research that the C5.0 decision tree and the logistic regression gave equally good accuracies, following which Naive Bayes was the second good and then ANN and at the end SVM that performed the worst based on the accuracy. In paper [14], the authors discuss the importance of early diagnosis of diabetes. They try to predict the risk of diabetes using different machine learning techniques. They implemented logistic regression, Decision Tree, Linear SVM, Random Forest, Gradient boosting on the standard PIMA Indian Diabetes Dataset. They used 80% of the dataset for the training and the rest 20% of the dataset for the testing. At the end of their research, they found out that the gradient boosting classifier had an accuracy of 79%, which was better than the rest of the classifiers. In paper [15], the authors tried to improve the accuracy using the bootstrapping re-sampling technique. They, later, applied three algorithms, naive Bayes, KNN, and the decision tree, to find out which algorithm performed the best based on the accuracy perimeter. They did their research on a diabetes dataset that had 768 records and was obtained from UCI. After their research, they found out that the decision tree algorithm gave the best accuracy in both the bootstrapping and without bootstrapping models. After bootstrapping, the best accuracy they got was 94.4% from the Decision Tree algorithm. In paper [16], the author did a comparison between the various machine learning techniques, logistic regression, decision tree, and ANN. They did a questionnaire on family history of diabetes, demographic characteristics, anthropometric measurements, and lifestyle of people to gather the dataset. They did their evaluation based on three parameters, accuracy, sensitivity, and specificity. They found out with their research that season three gave the best accuracy of 77.87%. In paper [17], the authors employed four machine learning algorithms on a dataset that they collected from diagnostic of Medical Centre Chittagong, Bangladesh. The dataset had data of 200 patients having 16 attributes in total. The machine learning algorithm applied to the dataset were KNN, SVM, Naive Bayes, and Decision Tree. According to their research, the decision tree gave the best accuracy of 73.5% which, was significantly better than the accuracies given by the other algorithms.

3 Application of Techniques

In this section, we have briefed about the various machine learning techniques applied in this paper. The following Table 1 contains the details of the machine learning techniques used.

Table 1 Techniques implemented on the dataset

Techniques	Details
K-Nearest Neighbor Classifier	• It is a machine learning algorithm that is based on distance • It is used to tackle classification and regression problems • The final classification output is determined based on the distance between the test samples and the training sample [18]
Naive Bayes	• Classification problems that are based on the Bayes theorem are addressed using the Naive Bayes algorithm • Mostly, it is used in text classification that requires a large training dataset [19] • Because the technique is quick and efficient, it is utilised to make real-time predictions • When it comes to solving multi-class problems, it is effective
SVM	• Both the regression and the classification problems can be solved using it but it is most often employed for classification problems • It is based on the statistical learning theory [20] • Linear SVM and Non-linear SVM are the two types of SVM. The linear SVM is used on data that can be divided into two classes using a straight line, whereas the non-linear SVM is used on data that cannot be classified using a straight line
Random Forest Classification Method	• Decision Tree Forest is another name for the Random Forest • It's a well-known decision tree-based ensemble model • Random Forest models are found to be more accurate than the decision tree models • We generate a vast number of decision trees in a random forest, and each decision tree is fed with each observation. The most common outcome for each observation is the final output [21] • It can be used for both classification and regression • It provides great accuracy through cross-validation • It's capable of working with datasets with higher dimensions
Extremely Randomized Trees Classifier	• It is a type of ensemble learning technique that generates a classification result by combining the results of numerous de-correlated decision trees in a "forest."

(continued)

Table 1 (continued)

Techniques	Details
Adaptive Boosting	• It is an ensemble approach for creating a strong classifier from a group of weaker ones [22] • It is used to boost the performance of weak learners • It is best used to boost the performance of the decision tree algorithms on binary classification problems
Stochastic gradient boosting	• It is one of the variations of boosting • A subsample of the training data is randomly selected (without replacement) from the whole training dataset at each cycle. The randomly selected subsample is then utilized to fit the base learner instead of the entire sample • It's used for regression, classification, and other tasks, which produces a prediction model in the form of an ensemble of weak prediction models
Bagged Decision Tree	• Bagging is a bootstrap ensemble method that works well with algorithms whose variance is high. One such example is the decision tree
Multi-layer Perceptron or MLP	• It is the most often used type of neural network [23] • It is primarily made up of numerous layers of the perceptron • It is appropriate for classification prediction problems in which inputs are classified or labeled • It is also suitable for regression prediction problems where a real-valued quantity is predicted given a set of inputs

4 Dataset Details

4.1 Understanding the Dataset

The dataset used in the paper is the PIMA Indian Diabetes Dataset which was originally·provided by "The National Institute of Diabetes and Digestive and Kidney Diseases [24]". In this dataset, we have been provided with a total of 768 training instances where each training instance has nine features (See Table 2). All patients in the dataset are females of at least 21 years old. The features include diabetes pedigree function, number of pregnancies, the concentration of plasma glucose, the thickness of skin, serum insulin, body mass index, blood pressure, age, and outcome. The outcome attribute of the dataset tells whether a person has diabetes or not. It stores a binary value where 0 depicts that the person is non-diabetic and 1 depicts that the person is diabetic. In this paper, we are doing two-way text classification into categories: positive and negative. Out of 768 instances, there are 500 negative instances and 268 positive instances. There are 111 instances with no history of pregnancy and 657 instances with the number of pregnancies lying between 1–17. The minimum

Table 2 PIMA Indian diabetes dataset has been used

Pregnancies	Age	Diabetes pedigree function	Skin thickness	Insulin	BMI	Blood pressure	Glucose	Outcome
6	50	0.627	35	0	33.6	72	148	1
1	31	0.351	29	0	26.6	66	85	0
8	32	0.672	0	0	23.3	64	183	1
1	21	0.167	23	94	28.1	66	89	0
0	33	2.288	35	168	43.1	40	137	1
5	30	0.201	0	0	25.6	74	116	0

age of the patient in the dataset is 21 and the maximum age in the dataset is 81. We have used the accuracy and F1 score parameters for the evaluation of the classifiers used.

4.2 Cleaning and Preprocessing of Dataset

Firstly, we checked for the null values in our dataset. Then, we found out that there were no null values. In the next step, we looked for potential missing values which are not null but 0. All the missing values were replaced with the mean of the respective columns. In the second step, we split the given data set into a training dataset and a test dataset. The training and test dataset constitutes 70% and 30% respectively.

4.3 Training Model Using Different Classifiers

Python was used for implementing the machine learning techniques. The various machine learning classifiers, namely K-Nearest Neighbors, Support Vector Machine, Extra Tree, Random Forest, Naive Bayes, Bagged Decision Tree, Adaptive Boosting, Stochastic Gradient Boosting, MLP Classifier were implemented on the Pima Indian Diabetes Dataset. All the machine learning techniques were compared with each other based on their accuracy and F1 score performance.

4.4 Evaluation of Classifiers Performance Metrics

We have evaluated the models based on their accuracy, sensitivity, precision, specificity, and F1 Score measures.

Fig. 1 Proposed methodology

$$Accuracy = (TN + TP)/(TN + FN + TP + FP)$$

$$Precision = TP/(FP + TP)$$

$$Specificity = TN/(FP + TN)$$

$$Sensitivity = TP/(FN + TP)$$

$$F1Score = (2 * Recall * Precision)/(Precision + Recall)$$

where TN stands for true negatives, TP stands for true positives, FP stands for false positives and FN denotes false negatives. The model having the highest sensitivity, specificity, and accuracy will be the best predictive model [25]. Figure 1 shows the steps followed.

5 Results

In this section, we have briefed about the results obtained upon application of aforesaid machine learning techniques on the Pima dataset for diabetes prediction. In Table 3 we have shown the results obtained by the various machine learning techniques with the default hyperparameters. It was observed that the Extra trees classifier gave the best accuracy of 80% with default hyperparameters followed by Bagged Decision Tree, Random Forest, Naive Bayes, MLP Classifier, Adaptive Boost, Stochastic Gradient Boosting, KNN, and Sigmoid SVM. It was also observed that the Extra trees classifier gave the best F1 score of 68% with default hyperparameters followed by Bagged Decision Tree, Random Forest, Stochastic Gradient Boosting, Naive Bayes, Adaptive Boost, MLP Classifier, KNN, and Sigmoid SVM (Fig. 2).

In Table 4 we have shown the results obtained by the various machine learning techniques after Hyper Parameter Tuning using GridSearchCV. It was observed that the KNN gave the best accuracy of 81% after Hyper Parameter Tuning followed by Extra trees, Sigmoid SVM, Bagged Decision Tree, Random Forest, Naive Bayes, Adaptive Boost, MLP Classifier, Stochastic Gradient Boosting. It was also observed that the KNN gave the best F1 score of 69% after Hyper Parameter Tuning followed

Table 3 Performance with default hyperparameters

Classification algorithms	Accuracy (%)	F1-Score (%)
K-Nearest Neighbors	75	59
Sigmoid SVM	68	50
Bagged Decision Trees	79	66
Random Forest classifier	79	64
Extra Trees	80	68
AdaBoost	77	62
Stochastic Gradient Boosting	77	64
MLP Classifier	78	60
Naive Bayes	78	63

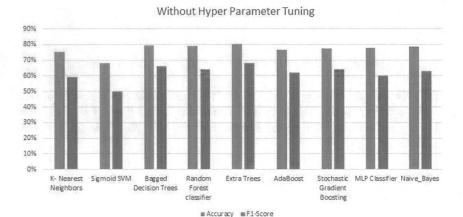

Fig. 2 Performance of techniques used with default hyperparameters

by Extra trees, Random Forest, Stochastic Gradient Boosting, Naive Bayes, Adaptive Boost, Sigmoid SVM, Bagged Decision Tree, and MLP Classifier (Figs. 3, 4 and 5).

6   Conclusion and Future Scope

Diabetes is a deadliest disease. It is required to do the early diagnosis of it so that the treatment can be done on time and more lives can be saved. In this paper, we have done the empirical analysis of diabetes using various machine learning techniques. The techniques used were K-Nearest Neighbors, Support Vector Machine, Extra Tree, Random Forest, Naive Bayes, Bagged Decision Tree, Adaptive Boosting, Stochastic Gradient Boosting, and MLP Classifier. It was observed that the KNN gave the best accuracy of 81% after Hyper Parameter Tuning using GridSearchCV. In the future,

Table 4 Performance after hyper parameter tuning using GridSearchCV

Classification algorithms	Accuracy (%)	F1-Score (%)
K- Nearest Neighbors	81	69
Sigmoid SVM	80	62
Bagged Decision Trees	79	60
Random Forest classifier	79	65
Extra Trees	80	65
AdaBoost	76	62
Stochastic Gradient Boosting	68	65
MLP Classifier	71	57
Naive Bayes	78	62

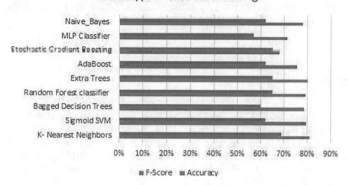

Fig. 3 Performance of techniques used after Hyper Parameter Tuning using GridSearchCV

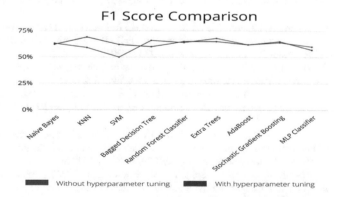

Fig. 4 Comparison of techniques based on F1-Score parameter before and after hyperparameter tuning

Fig. 5 Comparison of
techniques based on
Accuracy parameter before
and after hyperparameter
tuning

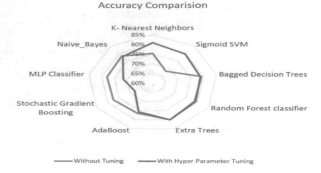

the performance and efficiency of the models can be tested on other datasets or self-collected real-time datasets as well by using other soft computing, machine learning, deep learning techniques, or hybrid models.

References

1. Indoria P, Rathore Y (2018) A survey: detection and prediction of diabetes using machine learning techniques. Int J Eng Res Technol (IJERT) 07(03)
2. Tigga N, Garg S (2020) Prediction of type 2 diabetes using machine learning classification methods. Procedia Comput Sci 167:706–716. https://doi.org/10.1016/j.procs.2020.03.336
3. Alehegn M, Joshi R, Mulay P (2018) Analysis and prediction of diabetes mellitus using machine learning algorithm. Int J Pure Appl Math 118:871–878
4. Kampmann U, Madsen LR, Skajaa GO, Iversen DS, Moeller N, Ovesen P (2015) Gestational diabetes: a clinical update. World J Diabetes 6(8):1065–1072. https://doi.org/10.4239/wjd.v6.i8.1065
5. Moucheraud C, Lenz C, Latkovic M, Wirtz VJ (2019) The costs of diabetes treatment in low- and middle-income countries: a systematic review. BMJ Glob Health 4(1):e001258. https://doi.org/10.1136/bmjgh-2018-001258. Accessed 27 Feb 2019
6. Malik S, Harous S, El-Sayed H (2021) Comparative analysis of machine learning algorithms for early prediction of diabetes mellitus in women. In: Chikhi S, Amine A, Chaoui A, Saidouni D, Kholladi M (eds) Modelling and implementation of complex systems, MISC 2020. Lecture notes in networks and systems, vol 156. Springer, Cham. https://doi.org/10.1007/978-3-030-58861-8_7
7. Shailaja K, Seetharamulu B, Jabbar MA (2018) Machine learning in healthcare: a review. In: 2018 second international conference on electronics, communication and aerospace technology (ICECA), pp 910–914. https://doi.org/10.1109/ICECA.2018.8474918
8. Ahmad HF, Mukhtar H, Alaqail H, Seliaman M, Alhumam A (2021) Investigating health-related features and their impact on the prediction of diabetes using machine learning. Appl Sci 11:1173. https://doi.org/10.3390/app11031173
9. Sisodia D, Sisodia DS (2018) Prediction of diabetes using classification algorithms. Procedia Comput Sci 132:1578–1585. https://doi.org/10.1016/j.procs.2018.05.122. ISSN 1877-0509
10. Tripathi G, Kumar R (2020) Early prediction of diabetes mellitus using machine learning. In: 2020 8th international conference on reliability, Infocom technologies and optimization (trends and future directions) (ICRITO), pp 1009–1014. https://doi.org/10.1109/ICRITO48877.2020.9197832.

11. Pradeep Kandhasamy J, Balamurali S (2015) Performance analysis of classifier models to predict diabetes mellitus. Procedia Comput Sci 47:45–51. https://doi.org/10.1016/j.procs.2015. 03.182. ISSN 1877-0509

12. Dutta D, Paul D, Ghosh P (2018) Analysing feature importances for diabetes prediction using machine learning. In: 2018 IEEE 9th annual information technology, electronics and mobile communication conference (IEMCON), pp 924–928. https://doi.org/10.1109/IEMCON.2018. 8614871

13. Varma K, Panda B (2019). Issue 6 www.jetir.org. ISSN 2349-5162

14. Singh A (2020) Performance analysis of diabetes prediction by using different machine learning algorithms. Int J Sci Res (IJSR) 9(7): 1472–1476. https://www.ijsr.net/search_index_results_p aperid.php?id=SR20722143245

15. Saru S, Subashree S (2019) Analysis and prediction of diabetes using machine learning. Int J Emerg Technol Innov Eng 5(4). SSRN: https://ssrn.com/abstract=3368308

16. Meng XH, Huang YX, Rao DP, Zhang Q, Liu Q (2013) Comparison of three data mining models for predicting diabetes or prediabetes by risk factors. Kaohsiung J Med Sci 29(2):93–99. https:// doi.org/10.1016/j.kjms.2012.08.016. Epub 2012 Oct 16. PMID 233478

17. Faruque MF, Asaduzzaman, Sarker IH (2019) Performance analysis of machine learning techniques to predict diabetes mellitus. In: 2019 international conference on electrical, computer and communication engineering (ECCE), pp 1–4. https://doi.org/10.1109/ECACE.2019.867 9365.

18. Guo G, Wang H, Bell D, Bi Y (2004) KNN model-based approach in classification

19. Kaviani P, Dhotre S (2017) Short survey on naive bayes algorithm. Int J Adv Res Comput Sci Manage 04

20. Zhang Y (2012) Support vector machine classification algorithm and its application. In: Liu C, Wang L, Yang A (eds) Information computing and applications, ICICA 2012. Communications in computer and information science, vol 308. Springer, Heidelberg. https://doi.org/10.1007/ 978-3-642-34041-3_27

21. Ali J, Khan R, Ahmad N, Maqsood I (2012). Random forests and decision trees. Int J Comput Sci Issues (IJCSI) 9

22. Tu C, Liu H, Xu B (2017) AdaBoost typical Algorithm and its application research. MATEC Web Conf 139:00222. https://doi.org/10.1051/matecconf/201713900222

23. Marius P, Balas V, Perescu-Popescu L, Mastorakis N (2009). Multilayer perceptron and neural networks. WSEAS Trans Circ Syst 8

24. https://www.kaggle.com/uciml/pima-indians-diabetes-database

25. Hossin M, Sulaiman MN (2015) A review on evaluation metrics for data classification evaluations. Int J Data Mining Knowl Manage Process 5:01–11. https://doi.org/10.5121/ijdkp.2015. 5201

An Energy Efficient Smart Street Lamp with Fog-Enabled Machine Learning Based IoT Computing Environments

J. Angela Jennifa Sujana, R. Vennita Raj, and V. K. Raja Priya

1 Introduction

Fog and Cloud computing paradigms provide the resources, computation, storage and communication which are routed over the Internet backbone. Fog computing or fog networking is the distributed decentralized infrastructure that uses edge devices to process an instant connection locally [1]. Both of these technologies are emerging over different industries. But for latency-sensitive applications [2] or real-time response applications cloud computing lags behind fog computation where resources are deployed closer to the user. Fog computing provides innovative solutions to network latency and response time. Cloud computing is coalesced with these emerging technologies through service and infrastructure [3].

A Smart Street Lighting control system is to automate the function of streetlights based on the intelligent network function. The intelligence in street light, states the efficient power consumption with greater illuminance heeding by climate conditions and vehicle movements. The system requires real time results where Fog computing comes in, making the resources close to the user. The real-time monitoring and amending illuminance help in increasing the safety measures, anti-glare and response on risky situations. The higher in prediction accuracy leads the higher in Quality-of-Service. Thus, the objective of delivering superhuman accuracy increases the prediction time, which is crucial for real-time applications. As the process at Edge nodes reduces the response time, enabling edge intelligence with self-learned edges obtains real-time high accuracy results [4]. In smart street light applications, the edges (Street Light) are assigned some autonomous function to reduce the response time at critical situations.

J. Angela Jennifa Sujana
Department of AI&DS, MSEC, Sivakasi, India

R. Vennita Raj · V. K. Raja Priya (✉)
Department of IT, MSEC, Sivakasi, India
e-mail: rajapriya@mepcoeng.ac.in

In this work, proposed a Fog based smart street light system for the automatic diagnosis of motion of pedestrians and vehicles using machine learning and IoT. The contrivance of the learning of traffic prediction enhances the energy conservation in smart street lights. The data generated from different IoT devices are efficiently managed by the Fog server. The proposed objectives of the project are:

- To design Smart Street Lamp (SSL) by developing the Energy Conservation Unit, this incorporates energy conservation in the lamp and aids in 'Save electricity'.
- To develop a framework in Fog Server to integrate Edge-Fog-Cloud for real-time data analysis.
- To enhance the energy conservation model of the street lamps by using AI techniques to make intelligent decisions on a real time basis.

2 Related Work

Fog computing can provide low-latency network connections within IoT devices and decision environments by making edge devices closer to IoT devices than with cloud servers. Charith Perera et al. [5] reviewed various approaches to tackle the Fog Computing domain for building sustainable smart cities. Maryam et al. [6] discussed a-state-of-the-art review on fog computing approaches. The Fog computing environment is classified as service-based, resource-based and application based. Each of these classes was evaluated with various metrics such as scalability, security, energy, cost, response time, latency and throughput. The important aspect of a smart lighting system is remote management with less human intervention. The dynamic and flexible web interface is designed as a central web server in [7]. The smart street light system consists of several street light groups with each streetlight group coordinator (SGC) and streetlight group members (SGMs). The system is constructed as a tree topology with streetlight group coordinator (SGC) as a network coordinator. The smartness in the street lighting system is built-up by using sensors. The footstep power generation mechanism is introduced in [9] using the piezoelectric sensor. The presence or the level of light is detected using Light Dependent Resistor (LDR) in [10, 14]. The control of light illumination further reduces the power consumption of street lights. LDR usually senses day and night, which turns the street light to ON state when dark and OFF state during the day.

In addition to smartness, the intelligence of the system is enhanced with machine learning. The fusion of machine learning algorithms with IoT devices helps with the smart usage of massive amounts of data in [11]. Generally, the intelligence over the system is attained with the ability of learning and decision-making. The implemented algorithm is adapted through knowledge from the training process and efficiency is analysed through testing or validation. Besides intelligence, machine learning algorithms can also predict the defect over the system in [12]. The requirement of manpower in the street lamp fault diagnosis system is overwhelmed with extreme learning machine (ELM). The fault of the system is diagnosed by the value of the output voltage at different frequencies.

The predictive based model improved Bayesian neural network (IBNN) is introduced to notify the controller in [13]. The model attains greater accuracy with less consumption time. The process of data in forms of prediction or classification is difficult when there is a massive amount of data. Thus the 'Energy on-demand' cloud based smart street light management system is introduced in [8]. The smart street lighting system is managed based on a cloud computing model. The state of the lights is changed with the cloud-based controller. However, the cloud-based network lags in the performance with high latency. To overcome the latency issues and reliable network system the proposed system introduces a fog-based computing model to achieve greater performances in prediction accuracy, power consumption, network bandwidth and latency.

3 Smart Street Lamp

The Smart Street Lamp (SSL) comprises: 1) Energy Conservation Unit (ECU) to design the energy conserving model for conserving the energy; 2) communication network between server and massive street lights; 3) Master Light Controller (MLC) to predict the lightness by creating uncorrelated forest of trees.

3.1 Energy Conserving Unit

The versatile illuminance optimizes the light output and conserves energy. The intellect street lamp is equipped with sensors such as motion detection, range finder and photoresistor. The fall off solar radiation detected by photoresistor, will set off the glow of light at required rate. The range finder measures the distance of oncoming traffic/vehicles by ultrasonic waves. The nearby traffic/vehicles intimate the requirement of efficient illumination to avert accidents. Further the movement is predicted with the motion detection sensor. The interactive results from sensors assist in street light asset tracking using GPS. The energy conservation mainly comes with the dimness state of street lights when there is a lack of street activity.

3.2 Communication Network

Smart Street Lamp (SSL) communicates with servers to send/receive information through the network. There are various types of network technologies Bluetooth, Wi Fi, ZigBee, GSM, 4G etc. The data communication network may encounter limitations: rate of transfer of data within the network, reliability in efficient management of the system, security vulnerabilities, wide coverage of communication over the network, latency over the change of state of street lamp control from the server and low power consumption.

The system adopts the Wi-Fi technology due to its benefit over convenience with multiple user connection, expandability, high security, low latency and low power consumption. In addition, Smart Street Lamp (SSL) is a public service which ensures a safe environment for pedestrians and travellers. The act of street lights as WiFi points enhances the safety and protection over the travellers in case of emergency.

3.3 Master Light Controller

The Master Light Controller tracks the states of the street lamps periodically from the Energy Conservation Unit (ECU). MLCs implement the flexible management platform by constructing machine learning models in the fog server. The learning undergoes the training of a random subset of individual models from the dataset and makes a decision by aggregating the votes from all the subset decision trees. The machine learning construction undergoes various steps:

i.   Use bagging and randomly select n subsets
ii. Train n decision trees
iii. Collecting the votes from each decision tree
iv. Aggregates the votes from all decision tree
v.   Make a final prediction with the label with maximum votes. The prediction is more accurate with a group of decision trees than the individual decision tree.

4 System Architecture

The efficient Fog-IoT enabled computing model for street light automatically diagnoses the traffic flow and implements an energy conserving model using machine learning. The generic framework for end-to-end integration of Edge-Fog-Cloud provides fast and accurate delivery of results (see Fig. 1).

4.1 Street Lamp Sensor Network

The Street Lamp contains various sensors such as Passive Infrared Sensor (PIR), Light Dependent Sensor (LDR), Ultrasonic Sensor and environment sensors. The environmental sensors include Temperature sensor, Humidity sensor and Pressure sensor. The sensor network senses the data and sends the data to the fog computing environment where the Worker nodes process the data as a job request.

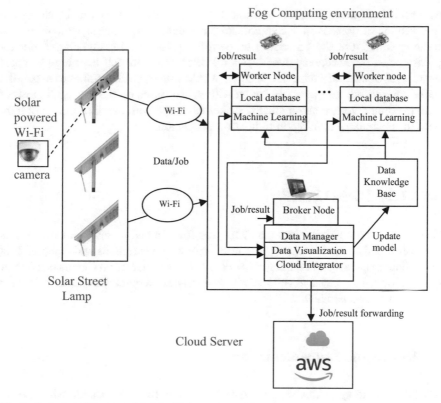

Fig. 1 System architecture of fog enabled smart street lamp

4.2 Fog Computing Framework

The Fog Computing framework improves the computing paradigm of user demands closer to the end user. The architecture comprises of the various nodes:

i. **Broker node:** This module receives data or job requests from the sensor network. The digital information is protected from unauthorized users by the Security Management module where it maintains the unique identity of the connected IoT devices. The Data Management module processes the raw data with data mining analytics such as data preprocessing. The data derived from different data streams, offers several insights with trends and patterns by Data Visualization. The interplay between fog and cloud server handled by Cloud Integrator module.

ii. **Worker node:** Worker nodes comprise a special-purpose computing system like Raspberry Pi as an embedded controller. The sensor readings are locally published to the database. The input data is processed and filtered with various mining techniques. The Machine Learning model analyzes and makes predictions with the input data which is shared with the network.

Machine Learning Module: Machine Learning learns the data and reduces the intervention of humans in decision making by identifying patterns from data. The training introduces the procedure of performing tasks without being explicitly programmed. The accuracy over decision is increased with self-learning from past experience or historical data. The training of data with more decision trees in parallel makes the system easy over prediction with votes. Based on the aggregation result with maximum vote, it automatically makes decisions with the illuminance of light with the continuous training on traffic flow prediction.

4.3   Cloud Data Center

The Cloud server comes into action when the fog computing environment services are overloaded and in increasing response latency. The enormous and complex data collections may encounter the fog network to harness resources on Cloud Data Center (CDC). Thus, the extensive volume of data is processed quickly in the cloud server with location independent.

5   Design and Implementation

The fog computing model takes the data from street lights associated with sensors and sends back the results of the level of brightness. This is implemented with the machine learning module with data mining pre-processing techniques.

Below representation describes the pseudo code for the implementation of the Energy Conservation Unit (ECU). The simplified experimental setup with LED resembles the 'Smart Street Lamp' (see Fig. 2).

Fig. 2 Experimental setup

PSEUDOCODE
ECU (Energy Conserving Unit) construction:
    //Define the level of brightness
    LEVEL1 0
    LEVEL2 102
    LEVEL3 153
    LEVEL4 204
    LEVEL5 255
    Repeat forever:
    Read the intensity of light value
    Read the duration of signal from objects
    Calculate the distance of the object
    if intensity of light >= 300
        then turn on LED with LEVEL2 brightness
    else if intensity of light >= 300 AND motion detection == 1
        then turn on LED with LEVEL3 brightness
    else if intensity of light >= 300 AND distance <= 70cm
        then turn on LED with LEVEL4 brightness
    else if intensity of light >= 300 AND motion detection == 1AND distance <=70cm
        then turn on LED with LEVEL5 brightness
    else turn off LED

Build Decision Tree (T, F, C)
Input: Training dataset $T = (X_i, Y_i)_{i=1}^n$
C= count of trees, F = number of features, k = number of classes in Y, S = size of subspace and P_i = probability of getting the final output as i^{th} class.
Output: A set of decision trees
Begin:
for $1 \le j \le C$ **do**
    Obtain subset T_p from T using Bagging
    Randomly select S, subset of features F
    for $1 \le f \le S$ **do**
        Compute Gini index, G.I.$(attr_f) = 1 - \sum_1^k P_i^2$
    end for
    choose the splitting attribute as the one with minimum value of G.I.
    continue in split until maximum tree construction
end for
Final Y = MajorityVote $(Y_i)_1^C$

6 Performance Evaluation

The proposed system is implemented and deployed in a fog computing environment. The model has been used for real time analytics on the various levels of brightness on street lights using machine learning techniques. We have analyzed the accuracy of the model with 0.9666666666666667. The following Table 1 gives the first five rows of sample dataset of the Smart Street Lamp. The evaluation result of the model classification performance over various metrics such as precision, recall, f1-score and support predict high accuracy (see Fig. 3). Also developed the learning curve of Random Forest Classifier (see Fig. 4).

Precision metric calculates the ratio of actual number of incorrect predictions classified correctly to the total number of nodes with incorrect predictions.

Recall metric calculates the ratio number of incorrect prediction nodes classified correctly to the total number of nodes classified correctly as incorrect predictions or

Table 1 Sample of smart street lamp dataset

SI. no	LDR	Ultrasonic	PIR	Level of brightness
0	128	54	1	1
1	310	798	0	2
2	313	52	0	4
3	304	70	1	3
4	288	61	0	1
5	306	36	1	5

```
Anaconda Prompt (anaconda3) - python ML_SSL.py                              —  □  ×
     LDR   Ultrasonic   PIR   level of brightness
0    128         54      1                 1
1    310        798      0                 2
2    313         52      0                 4
3    304         70      1                 3
4    288         61      0                 1
Confusion Matrix:
[[15  0  0  0  0]
 [ 0  4  0  0  0]
 [ 0  0  4  0  0]
 [ 0  1  0  4  0]
 [ 0  0  0  0  2]]
Classification Report:
              precision    recall  f1-score   support

           1       1.00      1.00      1.00        15
           2       0.80      1.00      0.89         4
           3       1.00      1.00      1.00         4
           4       1.00      0.80      0.89         5
           5       1.00      1.00      1.00         2

    accuracy                           0.97        30
   macro avg       0.96      0.96      0.96        30
weighted avg       0.97      0.97      0.97        30

Accuracy: 0.9666666666666667
```

Fig. 3 Evaluation of model over classification metrics (precision, recall, f1-score, support)

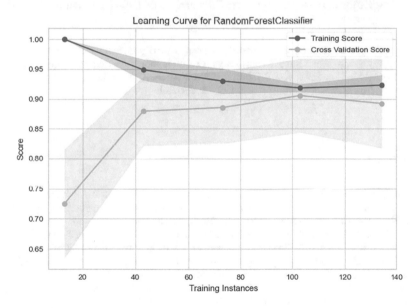

Fig. 4 Learning curve of random forest classification model

incorrectly as true prediction. F-Score is calculated as the harmonic mean of Precision and Recall metrics.

$$Precision = \frac{t_p}{t_p + f_p} \qquad (1)$$

$$Recall = \frac{t_p}{t_p + f_n} \tag{2}$$

$$F - Score = 2 \times \frac{Precision \times Recall}{Precision + Recall} \tag{3}$$

7 Conclusion

The proposed Smart Street Lamp model based on fog computing automatically made decisions on the level of brightness without human intervention. The deployment of the machine learning models on fog server reduces latency, power consumption and network bandwidth. The proposed system also reduces the human resources of maintenance over the system with periodic inspection. In the future, the proposed system can be further extended with the implementation of various applications such as agriculture, in smart cities—traffic management, environmental monitoring, parking and so on.

Acknowledgements The authors acknowledge the support and grant for this work under the project titled "Energy Aware Smart Street Lamp Design for Autonomous Maintenance using AI and Fog Computing" (vide No.: SP/YO/2019/1304(G), dated 20.05.2020) from Department of Science & Technology, SEED Division, Ministry of Science & Technology.

References

1. Sanchez-Corcuera R, Nunez-Macros A, Sesma-Solance J, Bilbao-Jayo A, Mulero R, Zulaika U, Azkune G, Almeida A (2019) Smart cities survey: technologies, application domains and challenges for the cities of the future. Int J Distrib Sensor Netw 15(6):1–36
2. Tuli S, Basumatary N, Gill SS, Kahani M, Arya RC, Wander GS, Buyya R (2020) Healthfog: an ensemble deep learning based smart healthcare system for automatic diagnosis of heart diseases in integrated IoT and fog computing environments. Future Gener Comput Syst 104:187–200
3. Vinueza Naranjo PG, Pooranian Z, Shojafar M, Conti M, Buyya R (2019) FOCAN: a Fog-supported smart city network architecture for management of applications in the Internet of Everything environments. J Parallel Distrib Comput 132:274–283
4. Khan LU, Yaqoob I, Tran NH, Ahsan Kazmi SM, Dang TN, Hong CS (2020) Edge computing enabled smart cities: a comprehensive survey. IEEE Internet Things J 7(10):10200–10232
5. Perara C, Qin Y, Estrella JC, Reiff-marganiec S, Vasilakos AV (2017) Fog computing for sustainable smart cities: a survey. ACM Comput Surv 50:1–43
6. Songhorabadi M, Rahimi M, Farid AMM, Kashani MH (2020) Fog computing approaches in smart cities: a state-of-the-art review. networking and internet architecture (cs.NI); distributed, parallel, and cluster computing (cs.DC)
7. Daely PT, Reda HT, Satrya GB, Kim JW, Shin SY (2017) Design of smart LED streetlight system for smart city with web-based management system. IEEE Sens J 17(18):6100–6110
8. Umar M, Gill SH, Shaikh RA, Rizwan M (2020) Cloud-based energy efficient smart street lighting system. Indian J Sci Technol 13(23):2311–2318

9. Ahmad H, Naseer K, Asif M, Alam MF(2019) Smart street light system powered by foot-steps. In: Hwang SO (Editor of ETRI Journal) International conference on green and human information technology (ICGHIT), vol 1. IEEE, Kuala Lumpur, pp 122–124
10. Manitha PV, Anandaraman, Sudharsan S, Manikumaran K, Aswathaman K (2017) Design and development of enhanced road safety mechanism using smart roads and energy optimized solar street lights. In: International conference on energy, communication, data analytics and soft computing. IEEE, pp 1650–1654
11. Sharma K, Nandal R (2019) A literature study on machine learning fusion with IOT. In: Cletus Babu S (ed) Third international conference on trends in electronics and informatics. IEEE, pp 1440–1445
12. Lee Y, Zhang H, Rosa J (2019) Street lamp fault diagnosis system based on extreme learning machine. IOP Conf Ser Mater Sci Eng 490:1–10
13. Suresh M, Anharasi M, PraveenKumar SV, Mohamed Hasvak A (2021) An intelligent smart street light system with predictive model. In: International conference on system, computation, automation and networking (ICSCAN)
14. Velaga NR, Kumar A (2012) Techno-economic evaluation of the feasibility of a smart street light system: a case study of rural India. Proc Soc Behav Sci 1220–1224

A Comparative Study of Machine Learning and Deep Learning Techniques on X-ray Images for Pneumonia

Amisha Jangra and Arunima Jaiswal

1 Introduction

Pneumonia is a life-threatening inflammatory condition of lung(s) which primarily affects the tiny air sacs called alveoli. The alveoli may get filled with purulent material leading to infection in one or both the lungs. Statistics for Pneumonia presented by a research paper for The Lancet [1] states that India accounts for 20% of the deaths for childhood pneumonia worldwide. The number of infections in adults are alarming as well. In addition to physical examination, diagnosis is often based upon Chest X-ray, blood sample and sputum culture. It is evident that India is highly burdened with number of infections as well as deaths. One of the major causes of this can be poor healthcare facilities in rural India leading to delay in diagnosis.

With the success of Deep learning [2] and Machine Learning techniques in classification and analysis of medical images, these techniques are emerging of prominent use in medical diagnosis. Using these techniques, we can detect a particular disease by examining a specific pattern in the healthcare records e.g., X-ray images of the patients and check for anomalies to predict results.

Though there are 2 sub-categories of Pneumonia as bacterial and viral but in our research, we have focused on the binary classification of data as Pneumonia or Normal. In this work, we will compare some existing Machine learning and Deep learning algorithms implemented using Python to compare their accuracy for classification of images as normal or pneumonia. For the research we have implemented 4 Machine learning techniques namely Random Forest (RF), Support Vector Machine (SVM), K-Nearest-Neighbors (KNN), Voting classifier algorithm

A. Jangra (✉) · A. Jaiswal
Department of Computer Science and Engineering, Indira Gandhi Delhi Technical University for Women, Delhi, India
e-mail: amisha114btcse19@igdtuw.ac.in

A. Jaiswal
e-mail: arunimajaiswal@igdtuw.ac.in

© The Author(s), under exclusive license to Springer Nature Singapore Pte Ltd. 2022   415
B. Unhelker et al. (eds.), *Applications of Artificial Intelligence and Machine Learning*,
Lecture Notes in Electrical Engineering 925,
https://doi.org/10.1007/978-981-19-4831-2_34

and 3 Deep learning techniques namely Multi-Layer Perceptron (MLP), Convolution Neural Network (CNN) and MobileNet technique.

In the next section i.e., Related work we talk about some of the related research on the steps of pneumonia detection, statistics and cure. In the section Dataset description, we describe the dataset and preliminary work i.e., Exploratory data analysis and data preprocessing for different techniques used. In the section application of techniques, we brief about the implementation of the techniques and obtained results for each. In the section Results and Discussion, we compare the accuracies and confusion matrices of the models implemented. Finally, we conclude the results and discuss future scope in the Conclusion and future scope section.

2 Related Work

Pneumonia has been one of the 10 major causes of death worldwide, as stated by WHO [3]. As discussed in a recent research paper [4] childhood pneumonia [5] is a major cause of death for children below age 5, and the risk of infections is not limited to children as evident from another study [6]. Pneumonia is a fatal problem in India, especially in rural areas where detection of disease is delayed due to lack of proper resources. In a research paper [5] it is stated that with improvement in socioeconomic factors and various government initiatives the mortality has substantially decreased over the past years in India but morbidity is still a cause of concern.

As a part of systematic literature review in order to identify published data and research on the similar lines of our research we found some resourceful studies conducted in past to deal with this fatal disease. In recent times there has been an inclination in computer-aided diagnosis [2] of diseases to facilitate the early diagnosis of various fatal diseases like cancer [7], liver disease [8], neurological disorders [9], pneumonia [10] etc., various studies have also been conducted [11–14] and [15] using deep learning to detect pneumonia from chest x-ray images. In a paper [11], two computer-aided detection problems namely thoraco-abdominal lymph node detection and interstitial lung disease (ILD) classification are used to evaluate CNN performance. In a paper [12], authors have developed CheXNeXt, a CNN to detect 14 distinct pathologies which includes pneumonia as well. In another paper [13], authors have implemented three-dimensional CNNs to access Chest CT and were able to achieve sensitivity as high as 91% at 2 false positives per scan on their chosen dataset. In paper [15], authors have used CNN models Xception and Vgg16 for pneumonia detection and compared both the models on basis of accuracies. One of the recent researches [16] also records exceptional accuracies by using various deep learning-based approaches for diagnosis of COVID-19, in this paper authors evaluated accuracy for various fine-tuned state-of-art deep learning models pretrained on ImageNet, and DenseNet121 comes out to be the most precise model. In paper [17], authors have compared various generic machine learning algorithms for Image classification. In paper [18], authors have compared traditional machine learning techniques with emerging deep learning techniques for image classification. It is also

Fig. 1 Distribution of different categories of data in train dataset

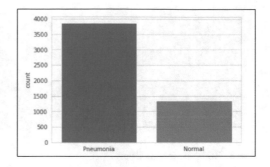

evident that early diagnosis [19] can help reduce the risk of mortality and morbidity [20] in the pneumonia patients.

3 Dataset Description

The dataset used for the research is chest X-ray Images (Pneumonia) by Paul Mooney [21], which is also publicly available on the Kaggle platform. The dataset comprises of 3 folders viz. train, test, val which further contain subfolders for each category i.e., Pneumonia or Normal. There are a total of 5,856 X-ray jpeg images (anterior–posterior), out of which there are 234 images of Normal category and 390 images of Pneumonia category in the test directory, 1341 images of Normal category and 3875 images of Pneumonia category in the train directory, 8 images of Normal category and 8 images of Pneumonia category in the val directory. The data was taken as a part of regular clinical checkup comprising of x-ray images of pediatric patients between the age 1 to 5 years old from Guangzhou Women and Children's Medical center. The dataset is well compiled by removing low quality and unreadable scans with expertise of 2 physicians. Followed by grading error checks by a third expert.

As a part of data visualization and preprocessing it can be seen in Fig. 1 the dataset seems imbalanced as number of x-ray images for Pneumonia are much greater as compared to Normal category. Therefore, there is a need to balance the dataset which we perform in the data preprocessing step as described in the next section.

4 Exploratory Data Analysis (EDA) and Data Preprocessing

In this section we discuss the data preprocessing done for each technique implemented.

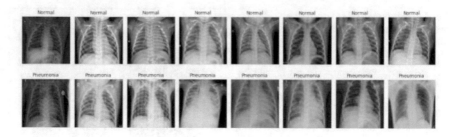

Fig. 2 Raw data in train dataset

4.1 Convolution Neural Network

This section includes brief discussion about the dataset and further about the data preprocessing for implementing convolution neural network technique.

EDA
In Fig. 2 few raw X-ray images from the dataset are shown corresponding to each category. As shown in Fig. 2 the X-rays of infected patients are blurry (ground-glass opacity) as compared to normal patients.

Data Preprocessing
In order to reduce the effect of illumination while training we implement gray scale normalization on the dataset by dividing the train data, test data and validation data by 255 as CNN converges better for such data followed by resizing the data for model implementation.

Next step would be data augmentation to deal with the imbalance in dataset and the problem of overfitting. For the CNN model, we randomly rotated some images, randomly zoomed, randomly shifted few images horizontally and vertically, flipped some of the images as well. After applying data augmentation, we have expanded our dataset which would help us to create a more robust model.

4.2 Multi-layer Perceptron

In this section we discuss the data preprocessing for implementing multi-layer perceptron technique on our dataset.

Data Preprocessing
As implied in EDA there was an imbalance in data, therefore for the purpose of training we transform 400 images from each category (to balance data) from train folder to feature vector. Then we perform data augmentation in order to build a robust model using ImageDataGenerator method. Further we assign labels. And

finally preprocessing is completed by transforming the shape of dataset as required for training.

4.3 MobileNet

In this section we discuss the data preprocessing involved for implementing MobileNet technique on our dataset.

Data Preprocessing
In data preprocessing step we initially combine all images for pneumonia class and normal class followed by dividing the images into 80:10:10 ratio as train, test, validation respectively for both classes and shuffling them. In the next step we normalize the images and assign labels. Further, we implement data augmentation. And our dataset is ready for training.

4.4 Machine Learning Techniques

In this section we discuss the data preprocessing for implementing the selected machine learning techniques.

Data Preprocessing
As learnt from EDA that there exists an imbalance in dataset i.e., for one class (pneumonia) the images are nearly equal to 3 times for other class (normal). To implement Machine Learning algorithms, we prepare image set and label set for 400 images of each category from training set, then resize and convert the images from image set to a feature vector as required for model implementation. The features mainly obtained are variance, standard deviation, mean, skewness, entropy and kurtosis of the images followed by applying Sobel filter and canny edge detection to the feature vector. The feature vector is further split into 20% test data and 80% train data using train_test_split method.

5 Application of Techniques

In this section we discuss the implementation aspect of the machine learning and deep learning techniques used.

5.1 Deep Learning Models

Convolution Neural Networks
CNN [22] is one of the most popular deep neural networks. CNNs are widely used to analyze visual imagery specifically designed to process pixel data by assigning weights and biases to various aspects of image.

After performing data augmentation on the data, we implement the model by adding convo layers followed by batch normalization to standardize activations of prior layers and optimization which is followed by pooling layers and dropout for generalization purpose. In the next step the output of all layers is flattened to feed to the fully connected layer. Finally, we compile our model using rmsprop optimizer.

After compilation we train our model using the fit method. Lastly, we find the accuracy and loss of the model on the test data. Accuracy comes out to be 87.77% and loss is 28.72%. Confusion matrix (see Fig. 5) for the implemented model is shown in the result section.

MobileNet
MobileNet is a streamlined architecture that uses depth wise separable convolutions to construct lightweight deep CNNs. After data preprocessing, we implement the pretrained CNN architecture i.e., MobileNet by setting weights none and input_shape. In next step we apply average pooling on the spatial dimensions followed by adding a logistic layer. Then we compile our model using adam optimizer. After compilation we train our model using the fit method. Finally, we obtain test accuracy as 94.70%. Confusion matrix (see Fig. 11) for the implemented model is shown in the result section.

Multi-layer Perceptron
MLP is a classic feed forward artificial neural network, the core component of Deep learning. MLP uses at least 3 layers viz. input layer, hidden layer and output layer.

Followed by data preprocessing, model is initialized as sequential followed by adding three Dense hidden layers with activation relu, followed by a dense layer with sigmoid activation as we are performing binary classification.

The test accuracy for MLP model comes out to be 72.91% which is very different from the train accuracy signaling our model to be overfit. Confusion matrix (see Fig. 10) for the implemented model is shown in the result section. As a result, the model did not generalize with the test data. As predicted by f1-score the model performed well with class 1 i.e., Pneumonia but results were inaccurate for class 0 i.e., normal.

5.2 Machine Learning Models

Random Forest
A Random Forest merges a collection of independent decision trees to get a more accurate and stable prediction. RandomForestClassifier is used to implement Random

Forest model. Firstly, we set the 2 key hyperparameters i.e., 'n_estimators' and 'max_depth' followed by fitting the model to the training set. Then using predict and confusion_matrix method we construct the confusion matrix as shown in Fig. 6. The accuracy for model implemented on test data comes out to be 80.6%. And the accuracy came out to be 82.95% after performing 10-fold cross validation.

Support Vector Machines

A SVM is a classifier that finds an optimal hyperplane that maximizes the margin between the 2 classes. SVC is used to implement Support Vector Machine algorithm. Firstly, we initialize the model followed by fitting the model to the training set. Then using predict and confusion_matrix method we construct the confusion matrix as shown in Fig. 7. The accuracy for model implemented on test data comes out to be 81.25%. And the accuracy came out to be 79.875% after performing 10-fold cross validation.

Voting Classifier

A Machine Learning model that trains on ensemble of various models and predicts a class based on highest majority of voting.

VotingClassifier is used to implement the model. Firstly, we set the key hyperparameter i.e., 'estimators' followed by fitting the model to the training set. Then using predict and confusion_matrix method we construct the confusion matrix as shown in Fig. 8. The accuracy for model implemented on test data comes out to be 80.62%. And the accuracy came out to be 79.25% after performing 10-fold cross validation.

K-Nearest Neighbors

KNN is one the most basic classification algorithms. KNN belongs to supervised learning domain.

KNeighborsClassifier is used to implement K-Nearest Neighbors model. Firstly, we set one of the key hyperparameter i.e., 'n_neighbors' followed by fitting the model to the training set using fit method. Then using predict and confusion_matrix method we construct the confusion matrix as shown in Fig. 9. The accuracy for model implemented on test data comes out to be 56.25% which is very low as compared to other ML models implemented. And the accuracy came out to be 50.125% after performing 10-fold cross validation.

6 Results and Discussion

After observing performance of various Machine Learning and Deep Learning models following results were obtained:

As noted from Table 1 that Support Vector Machine gave best accuracy for the considered dataset, but after 10-fold cross validation Random Forest technique turns out to be most accurate (Fig. 3).

Table 1 Comparing ML techniques

S No.	Name of the model	Accuracy (%)	Accuracy (%) (With 10-fold cross validation)
1	Random forest	80.6	82.95
2	Support vector machine	81.25	79.875
3	Voting classifier	80.62	79.25
4	K-Nearest neighbor	56.25	50.125

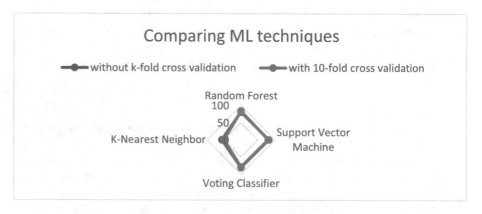

Fig. 3 Comparing machine learning techniques on the basis of accuracies with and without k-fold cross validation

Table 2 Comparing DL techniques

S No.	Name of the model	Accuracy (%)
1	Convolution neural network	87.77
2	MobileNet	94.70
3	Multi-Layer perceptron	72.91

From Table 2 it is evident that Deep learning technique MobileNet is more accurate for the considered dataset as compared to others. Therefore, on further comparison it can be noted that Deep learning techniques gave best statistical results (see Fig. 4) out of all implemented models as the nature of task was image recognition. Another thing to keep in mind is that the data preprocessing done was different for different models implemented as per the model's requirement.

The confusion matrices (see Figs. 5, 6, 7, 8, 9, 10 and 11) also depict that in terms of performance deep learning techniques show better results as compared to machine learning techniques supporting our results.

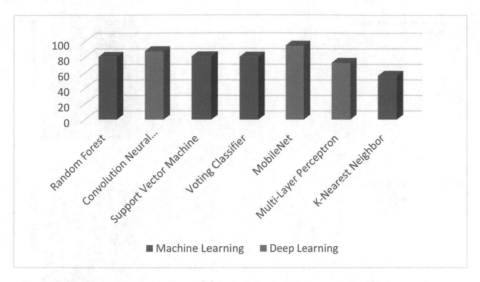

Fig. 4 Comparing machine learning and deep learning techniques on the basis of accuracies

Fig. 5 Confusion matrix
(CNN)

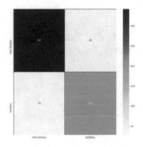

Fig. 6 Confusion matrix
(Random Forest)

Fig. 7 Confusion matrix
(SVM)

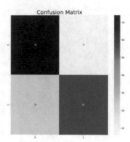

Fig. 8 Confusion matrix
(Voting Classifier)

Fig. 9 Confusion matrix
(KNN)

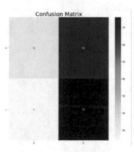

Fig. 10 Confusion matrix
(MLP)

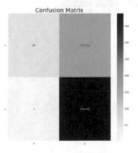

Fig. 11 Confusion matrix
(MobileNet)

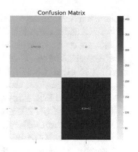

7   Conclusion and Future Scope

It can be concluded that Deep Learning techniques gave better accuracy and confusion matrix for the considered dataset.

In future more techniques can be incorporated to widen the research and in other medical diagnosis. Also, in future the best techniques can be used in the form of mobile and web apps [23] to make the diagnosis accessible for general public.

Also, the paper aims to help facilitate early diagnosis of the disease especially in rural areas to prevent fatalities but presence of expert radiologists and physicians is important for proper diagnosis of the disease.

References

1. Pandey A, Galvani AP (2020) The burden of childhood pneumonia in India and prospects for control. Lancet Child Adolesc Health 4(9): 643–645. ISSN: 2352-4642
2. Litjens G et al (2017) A survey on deep learning in medical image analysis. Med Image Anal 42:60–88. ISSN 1361-8415, https://doi.org/10.1016/j.media.2017.07.005, https://www.scienc edirect.com/science/article/pii/S1361841517301135
3. WHO (2020) The top 10 causes of death, World Health Organization
4. Marangu D, Zar HJ (2019) Childhood pneumonia in low-and-middle-income countries: An update. Paediatr Respir Rev 32:3–9. ISSN 1526-0542, https://doi.org/10.1016/j.prrv.2019. 06.001, https://www.sciencedirect.com/science/article/pii/S1526054219300594
5. Wahl B et al (2020) National, regional, and state-level pneumonia and severe pneumonia morbidity in children in India: modelled estimates for 2000 and 2015. Lancet Child Adolesc Health 4(9):678–687. ISSN: 2352-4642
6. Ghoshal AG (2016) Burden of pneumonia in the community. JAPI
7. Jeyaraj PR, Samuel Nadar ER (2019) Computer-assisted medical image classification for early diagnosis of oral cancer employing deep learning algorithm. J Cancer Res Clin Oncol 145:829–837. https://doi.org/10.1007/s00432-018-02834-7
8. Reddy DS, Bharath R, Rajalakshmi P (2018) A novel computer-aided diagnosis framework using deep learning for classification of fatty liver disease in ultrasound imaging. In: 2018 IEEE 20th International Conference on e-Health Networking, Applications and Services (Healthcom), pp 1–5. https://doi.org/10.1109/HealthCom.2018.8531118
9. Gautam R, Sharma M (2020) Prevalence and diagnosis of neurological disorders using different deep learning techniques: a meta-analysis. J Med Syst 44:49. https://doi.org/10.1007/s10916-019-1519-7

10. Hossain S, Rahman R, Ahmed MS, Islam MS (2020) Pneumonia detection by analyzing Xray images using MobileNET, ResNET architecture and long short-term memory. In: 2020 30th International Conference on Computer Theory and Applications (ICCTA), pp 60–64. https://doi.org/10.1109/ICCTA52020.2020.9477664

11. Shin H et al (2016) Deep convolutional neural networks for computer-aided detection: CNN architectures, dataset characteristics and transfer learning. IEEE Trans Med Imaging 35(5):1285–1298. https://doi.org/10.1109/TMI.2016.2528162

12. Rajpurkar P et al (2018) Deep learning for chest radiograph diagnosis: a retrospective comparison of the CheXNeXt algorithm to practicing radiologists. PLoS Med. https://doi.org/10.1371/journal.pmed.1002686

13. Pezeshk A, Hamidian S, Petrick N, Sahiner B (2019) 3-D convolutional neural networks for automatic detection of pulmonary nodules in chest CT. IEEE J Biomed Health Inform 23(5):2080–2090. https://doi.org/10.1109/JBHI.2018.2879449

14. El Asnaoui K, Chawki Y, Idri A (2021) Automated methods for detection and classification pneumonia based on X-Ray images using deep learning. In: Maleh Y, Baddi Y, Alazab M, Tawalbeh L, Romdhani I (eds) Artificial Intelligence and Blockchain for Future Cybersecurity Applications. Studies in Big Data, vol 90, pp 257–284. Springer, Cham. https://doi.org/10.1007/978-3-030-74575-2_14

15. Ayan E, Ünver HM (2019) Diagnosis of pneumonia from chest X-Ray images using deep learning. In: 2019 Scientific Meeting on Electrical–Electronics & Biomedical Engineering and Computer Science (EBBT), pp 1–5. https://doi.org/10.1109/EBBT.2019.8741582

16. KC K et al (2021) Evaluation of deep learning-based approaches for COVID-19 classification based on chest X-ray images. SIViP 15:959–966. https://doi.org/10.1007/s11760-020-01820-2

17. Marée R, Geurts P, Visimberga G, Piater J, Wehenkel L (2004) A comparison of generic machine learning algorithms for image classification. In: Coenen F, Preece A, Macintosh A (eds) Research and Development in Intelligent Systems, pp 169–182 XX. SGAI 2003. Springer, London. https://doi.org/10.1007/978-0-85729-412-8_13

18. Lai Y (2019) J Phys Conf Ser 1314:012148

19. Vincent JL, De Souza Barros D, Cianferoni, S (2010) Diagnosis, management and prevention of ventilator-associated pneumonia. Drugs 70:1927–1944. https://doi.org/10.2165/11538080-000000000-00000

20. Kallander K, Burgess DH, Qazi SA (2016) Early identification and treatment of pneumonia: a call to action. Lancet Glob Health 4(1):e12–e13. https://doi.org/10.1016/S2214-109X(15)00272-7

21. https://www.kaggle.com/paultimothymooney/chest-xray-pneumonia. Accessed 02 Aug 2021

22. Albawi S, Mohammed TA, Al-Zawi S (2017) Understanding of a convolutional neural network. In: 2017 International Conference on Engineering and Technology (ICET), pp 1–6.https://doi.org/10.1109/ICEngTechnol.2017.8308186

23. Sait U, Shivakumar S, KV GL, Kumar T, Ravishankar VD, Bhalla K (2019) A mobile application for early diagnosis of pneumonia in the rural context. In: 2019 IEEE Global Humanitarian Technology Conference (GHTC), pp 1–5. https://doi.org/10.1109/GHTC46095.2019.9033048

Analysis of Covid-19 Fake News on Indian Dataset Using Logistic Regression and Decision Tree Classifiers

Rajiv Ranjan, Akanksha Srivastava, and Utkarsh Uday Singh

1 Introduction

The amount of what we read on the web and probably "dependable" news locales are reliable? It is generally simple for anybody to post what they want and although that can be acceptable. As the biggest proportion of our lives has passed online through social media platforms [1–3], a gigantic number of people everyday chase out and appreciate the news from social media rather than conventional news affiliations. Regardless of the benefits given by online media, the level of nature of stories through social media is lower. Since it is sensible to supply news on the internet and distant quickly with no issue to diffuse through online media. Gigantic of fake news, i.e., those reports with an intentioned wrong suggestion are passed on online for an assortment of purposes like money related and political extension. The open spread of fake news can conflictingly influence individuals and society. Social media's capacity to allow users to talk about and share contemplations and talk over issues like larger part run the government, preparing and prosperity. Such stages are additionally utilized with a negative methodology by specific elements ordinarily for a money-related issues [4] and in different cases for making one-sided suppositions, controlling outlooks, and propagating parody or insanity. It is normally referred to as fake news. Today, we have faith in what we see on the sites or social media and don't seek after to check if the given data is valid or not [5]. It is hard to separate the fake and true news physically. Our perspective is shaped by the data we absorb. It is becoming increasingly clear that users have reacted absurdly to news that later turns out to be false. The graph is given below Fig. 1 is obtained from Google Trends [6]

R. Ranjan (✉) · U. U. Singh
Plaksha University, Mohali, India
e-mail: rajiv.ranjan3105@gmail.com

U. U. Singh
e-mail: shivamutkarsh14@gmail.com

A. Srivastava
Delhi Technical Campus, Greater Noida, India
e-mail: srivastavaakanksha717@gmail.com

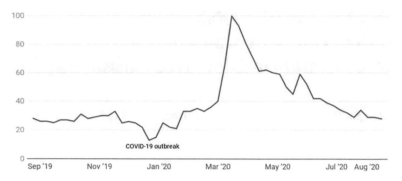

Fig. 1 Covid 19 fake news

where the keyword "Covid 19 fake news" was considered as the search term. A late case is the spread of the new Corona Virus, where false reports spread on the internet about the start, about nature and direction of the contamination [7]. Here we have proposed a solution where the current fake news three-sided course of action had a distinct perspective where both supervised and unsupervised learning algorithms are used for the representation of the text [8]. Here, we proposed a response for the fake news analysis on self made dataset using the Logistic Regression, Decision Tree Classifier, Random Forest Classifier and Gradient Boosting Classifier [9, 10].

For the remainder of the article, Sect. 2 focuses on the Literature Review. Materials and Methods are illustrated in Sects. 3 and 4 describes the Dataset which is collected by us. Evaluation Metrics are presented in Sect. 5. Results are presented in Sect. 6 and conclusions and directions for future work are presented in Sect. 7.

2   Literature Review

Several researchers have aimed to solve this challenge in various ways to test which method works and get desirable results. A few studies have been discussed are: Wenlin Han demonstrated the detection of fake news on social media using machine learning performance scoring [11]. In this proposed framework, they alluded to some common views of machine learning, for example-Deception modeling, clustering, Naive Bayes scoring for accurate detection of TF-IDF(Term Frequency-Inverse Document Frequency) and PCFG(Probabilistic Context Free-Grammar) with convolution & recurrent neural network models that are examined to account for execution with conventional machine learning techniques. Rohit Kumar Kaliyar demonstrated Fake News Detection Using A Deep Neural Network [12] where he proposed the framework in which he used natural language management techniques, deep learning, and machine learning to run this model and then test results with a more precise results. Ranojoy Barua has launched an application that uses machine learning technology to detect fake news articles [13]. The framework is based on machine learning methods, such as Long Present Moments (LPM) and Gated Recurrent Unit (GRU) to describe data in the spam or raw data. Karishnu Poddar proposed the idea of correlating different Machine Learning models to accurately identify fake news [14]. It's core is to solve the problem of the classification of false information. Rahul Mandical proposed the idea of using machine learning with Naive Bayes and Passive-Aggressive classifiers to detect fake news [15]. These classifiers are used to load models that rely on the TF-IDF vectorizer, where TF-IDF is the Term Frequency Inverse Document Frequency and it's value increases. The word appeared repeatedly in the report, but the semantic importance of the word was lost.

3 Materials and Methods

In this section proposed system, algorithm, data set, and results are discussed.

3.1 Proposed Framework

In this section, detailed information about the analysis of fake news framework is explained. In the proposed method, as shown in Fig. 2 represents the developed current documents that demonstrate supervised learning in different ways to classify messages from different fields as true or false. Several reputable websites can publish real news and various websites (such as PolitiFact) are used to check the reality. For experiments two datasets are created which contain news related to Covid-19.

Data Extraction- When data is processed from a database, the process is called Data Extraction. This is important because obtaining good-quality data is considered an important step.

Preprocessing the data- Data preprocessing is a vital step to build a good model. For simplicity, the columns "title", "subject" are removed and retained the text and date column for further processing.

Feature Extraction- The way toward recognizing significant highlights or properties of the information is called Feature extraction. It may be utilized to diminish the quantity of traits that portray the information. Here, TF-IDF is being utilized to include extraction. It is a critical method utilized for data recovery to address how significant a particular word or expression is to a given document.

Test-Train Split- The test-train split strategy is utilized to assess the exhibition of machine learning algorithms whenever they are utilized to make predictions about information that won't be utilized to prepare the model. We have trained the model on train-test split ratio of 50:50, 60:40, 70:30 and 80:20. However, best result achieved on 70:30 train-test split with an accuracy of 95%.

Classification using Decision Tree and Logistic Regression- Logistic regression is a classification algorithm, utilized when the worth of the objective variable is unmitigated. Logistic regression is normally utilized when the data being referred to

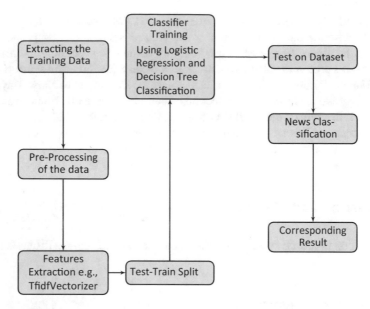

Fig. 2 Flowchart of training algorithm and classification of news articles

has binary output, so when it has a place within any event some class or is either a 0 or 1. The decision tree creates a regression model in a tree structure. The data set is divided into smaller subsets and the related decision trees are constantly evolving. The final product is a tree with decision nodes and leaf nodes.

3.2 Collecting Data

A brief description of the datasets used in this research work is provided below. The datasets we utilized in this work are created by us. The dataset utilized in this paper is one of the novelty of our proposed approach. There are two sections for the information procurement measure, "fake news" and "true news". The dataset used for this work are in csv format named true.csv and fake.csv (https://github.com/Aks 121/Fake-News-Analysis-on-Indian-Dataset.git).

3.3 Algorithms

The step-wise procedure for detecting fake news is described in the algorithm. The algorithm represents the Fake News Detection generated by different models applied on the test set. In this, we calculated the output value of fake and true news. If the output of the test set is 1 then the sample belongs to class 1 (true) otherwise to class 0 (fake).

Algorithm Algorithm for Fake News Detection
___
     Input Loading the data i.e fake.csv and true.csv
     Output Classification of test set
Step 1: Inserting a column "class" as target
Step 2: Merging True and Fake Dataframes
Step 3: Removing columns which are not required
Step 4: Random shuffling the dataframe
Step 5: Creating a function wordopt to process the text
Step 6: Defining dependent and independent variable
Step 7: Splitting Training and Testing
Step 8: Convert text to vectors
Step 9: Building the model

if $Output = 1$ **then**
    test set Class ϵ 1($true$)
else
       test set Class ϵ 0($fake$)
end if

Logistic Regression- As textual content dependent on a huge rundown of capacities with a binary output (true/false) are grouped together. Logistic Regression is used because it offers the instinctual circumstance to illustrate issues into twofold or various classes. Numerically, the logistic regression hypothesis function are regularly characterized as:

$$g(z) = \frac{1}{1 + e^{-z}} \qquad (1)$$

Logistic regression makes use of a sigmoid feature to trade the honour a probability regard. The motive here is to limit the cost ability to reap an ideal probability.

Decision Tree Classifier- One of the most common algorithms used in classification is the Decision Tree Classifier algorithm. Decision Tree tackles the issue of machine learning by changing the data into a tree representation. Each internal node of the tree representation indicates a characteristic and each leaf node signifies a class name. Decision tree algorithms are regularly utilized for taking care of both regression and classification problems. A decision tree could overfit when there are an oversized wide variety of sparse functions and in this manner carry out ineffectively on the testing records.

Random Forest Classifier- Random Forest (RF) is machine learning technique that is used to solve regression and classification problems. RF comprises an enormous number of Decision Trees working particularly to foresee a result of a class where a definite expectation is predicted on a classification that got majority votes. The error rate is less in the random forest when contrasted with different models because of low correlation among trees [16–18].

Gradient Boosting Classifier- The Gradient Boosting approach Ensemble Learning [19, 20] joins various base classifiers to shape a solid predicting model. One of the advantages of ensemble learning is that it doesn't need base classifiers to have high accuracy to get the general high accuracy of the predicting model [21, 22]. By ensemble learning, an unpredictable issue can be disintegrated into numerous sub issues that are simpler to unwind and comprehend. Boosting, e.g., gradient boosting, and bagging are two methodologies normally utilized in ensemble learning. In this work, we proposed to utilize a gradient boosting approach to rumor detection. Boosting is an ensemble modeling technique that is used to build a robust classifier from the number of weak classifiers.

4 Datasets

In this work, two dataset(fake and true) are created which contains news related to Covid-19, for example- 5G is the cause of the coronavirus pandemic, Hydroxychloroquine-The virus cure, Drinking alcohol will prevent coronavirus etc.

4.1 Dataset Information

The datasets used in this work are created by us. The dataset utilized in this paper is one of the novelty of our proposed approach. There are two datasets one for fake news and one for genuine news. Both the datasets are combined together using Panda's built-in function. Final dataset is balanced because both categories have the approximate same number of examples. The data incorporates both fake and genuine news stories identified with Covid-19. The genuine news stories distributed contain a true depiction of events while the fake news sites claims that aren't lined up with realities. Dataset for fake news can be gathered from more than one sources like news agencies, webpages, different social media websites, Twitter, Facebook, Instagram, and others. The similarity of cases from the clinical space for a large number of those articles can be checked with various sites, for example- times of india, indiatimes.com, and thehindu.com.

4.2 Dataset Analysis

This section presents a definite investigation of the dataset. Online news are gathered from sources like social media websites, the home page of stories office sites, or fact-checking sites. Online news is often accumulated from different sources, like press association home pages, search engines, and social media sites. Here dataset creation was done by us. We added news and labeled them manually. We added the labels into CSV files and stored them. In this paper, the dataset (fake.csv and true.csv file) is used (https://github.com/Aks121/Fake-News-Analysis-onIndian-Dataset.git). This dataset has records from various articles found on the internet and their attributes are text, title, subject and date. Only two features (text, date) are used to detect fake news in this work. Label zero is assigned to represent unreliable news (or fake), while one is assigned to real news.

5 Evaluation Metrics

To assess the performance of algorithms for fake news detection, different evaluation metrics are utilized. In this segment, we audit the first broadly utilized measurements for fake news recognition. Most existing methodologies consider the fake news issue as an order issue that predicts if a report is fake or not is True Positive (TP), True Negative (TN), False Negative (FN), False Positive (FP).

Confusion Matrix
The confusion matrix is a table that is regularly used to depict the classification model on a set of test data, where the true value of is known. The confusion matrix is a summary of the prediction results in the classification problem.

		Predicted Class	
		Class = Yes	Class = No
Actual Class	Class = Yes	TP	FN
	Class = No	FP	TN

By planning it as a classification problem, we will characterize following metrics,

5.1 Recall

Recall represents the entire number of positive classifications out of true class. For this situation, it shows the measure of articles predicted as true out of all the true articles [23].

$$Recall = \frac{TP}{TP + FN} \tag{2}$$

5.2 Precision

On the other hand, the precision score tends to the extent of TP to all or any events predicted as true. For this circumstance, precision represents the proportion of the articles that are separate as true out of all the positively forecasted (true) articles [24].

$$Precison = \frac{TP}{TP + FP} \tag{3}$$

5.3 F1-score

F1-score shows the compromise between recall and precision. It calculates the mean of the two. In this way, it takes both the FP and thusly the FN perceptions under consideration [25]. F1-score can be determined utilizing the given equation:

$$F1 - score = \frac{2 * Precision * Recall}{Precision + Recall} \tag{4}$$

5.4 Accuracy

Accuracy is regularly the most used metric representing the proportion of efficiently predicted observations, either true or fake. To calculate the accuracy of model performance, the given equation can be used:

$$Accuracy = \frac{TP + TN}{TP + TN + FP + FN} \tag{5}$$

6 Results

Using the above Algorithm, implementation has been done with Vector feature-Tf-Idf vector at Word level.

Confusion Matrices
After applying extracted feature (Tf-Idf) on four different classifiers (Logistic Regression, Random Forest Classifier, Decision Tree Classifier and Gradient Boosting Classifier), the confusion matrix with actual and predicted sets are acknowledged shown below in Tables 1, 2, 3 and 4:

Table 1 Confusion matrix for logistic regression using Tf-Idf features-

Total = 60	Logistic regression	
	Fake (Predicate)	True (Predicate)
Fake (Actual)	26	7
True (Actual)	3	24

Table 2 Confusion matrix for random forest classifier using Tf-Idf features-

Total = 60	Random forest	
	Fake (Predicate)	True (Predicate)
Fake (Actual)	23	10
True (Actual)	4	23

Table 3 Confusion matrix for decision tree classification using Tf-Idf features-

Total = 60	Decision tree	
	Fake (Predicate)	True (Predicate)
Fake(Actual)	21	12
True(Actual)	5	22

Table 4 Confusion matrix for gradient boosting classifier using Tf-Idf features-

Total = 60	Gradient boosting classifier	
	Fake (Predicate)	True (Predicate)
Fake (Actual)	24	9
True (Actual)	5	22

The Table 5, given underneath sums up the precision accomplished by every algorithm on the datasets.

Table 5 Comparison result

Classifier	Precision	Recall	F1-score	Accuracy
Logistic Regression (LR)	90%	91%	95%	95%
Random Forest Classifier (RF)	84%	81%	83%	83%
Decision Tree Classification (DT)	88%	87%	87%	87%
Gradient Boosting Classifier (GB)	89%	89%	88%	88%

Fig. 3 Accuracy results of
all the algorithms

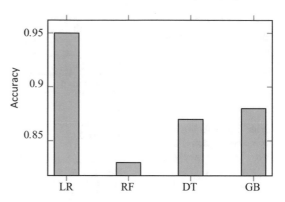

The classifiers are broke down subject to Precision, Recall, and F-Measure and
Accuracy. Logistic Regression has the most elevated F1 score, 95%, and subsequently
the best classification quality as demonstrated by Table. The precision attained by
Gradient Boosting Classifier is 89%, whereas the precision for Decision Tree Clas-
sification is 88%. The absolute difference between Decision Tree Classification and
Gradient Boosting Classifier is 1% which is not significant. The Random Forest
Classifier performed average than all other algorithms as shown in Fig. 3.

7 Conclusion

With the growing obviousness of social media, an ever increasing number of people
eat up news from social media as opposed to customary news-casting. In any case,
web-based media has additionally been used to expand fake information which has
solid adverse consequences on singular clients and more extensive society. Considering the changing scene of the trendy business world, the difficulty of phony news has
become very warrants genuine endeavors from security analysts. Presently arranging
news physically needs top to bottom information on the space and the ability to spot
peculiarity inside the content. We have thought of a response to the question of
ordering fake news. The information we utilized here is gathered online and contains
news stories. It is a simple however successful way to deal with licensed clients
to distinguish fake news. In this paper, We likewise further analyzed the self made
datasets, evaluation metrics, and promising future direction in counterfeit information revelation research moreover, stretch out the area to various applications. The
main point of the research is to spot designs to differentiate fake articles from true
news.

As future work, we plan to make a machine learning model to detect fake news
on regional languages. The generalised machine learning model to detect fake news
can be developed for many regional languages in future work.

References

1. Meel P, Vishwakarma DK (2020) Fake news, rumor, information pollution in social media and web: a contemporary survey of state-of-the-arts, challenges and opportunities. Expert Syst Appl 153:112986
2. Savyan PV, Bhanu SMS (2017) Behavior profiling of reactions in Facebook posts for anomaly detection. In: 2017 ninth international conference on advanced computing (ICoAC). IEEE, pp 220–226
3. Varshney D, Vishwakarma DK (2020) A review on rumour prediction and veracity assessment in online social network. Expert Syst Appl 168:114208
4. Lazer DMJ, Baum MA, Benkler Y (2018) The science of fake news. Science 359(6380):1094–1096
5. Yang Y (2018) TI-CNN: convolutional neural networks for fake news detection
6. Google Trends (2020) Google Trends. https://www.haasia.com/internet-searches-on-fake-news-misinformation-and-disinformationpeak-during-covid-19/, Accessed 21 Aug 2020
7. Hua J, Shaw R (2020) Coronaviruss (covid-19) "infodemic" and emerging issues through a data lens: "the case of China". Int J Environ Res Public Health 17(7):2309
8. Ruchansky N, Seo S, Liu Y (2017) Csi: a hybrid deep model for fake news detection. In: Proceedings of the 2017 ACM on conference on information and knowledge management, Singapore, pp 797–806
9. Kaliyar RK, Goswami A, Narang P, Sinha S (2020) FNDNet a deep convolutional neural network for fake news detection. Cogn Syst Res 61:32–44
10. Dewan P, Kumaraguru P (2017) Facebook Inspector (FbI): towards automatic real-time detection of malicious content on facebook. Soc Netw Anal Min 7(1):15
11. Han W, Mehta V (2019) Fake news detection in social networks using machine learning and deep learning: performance evaluation. IEEE
12. Kaliyar RK (2018) Fake news detection using a deep neural network. IEEE
13. Barua R, Maity R, Minj D, Barua T, Layek AK (2018) F-NAD: an application for fake news article detection using machine learning techniques. IEEE
14. Poddar K, Bessie G, Amali D, Umadevi KS (2019) Comparison of various machine learning models for accurate detection of fake news. IEEE
15. Mandical RR, Mamatha N, Shivakumar N, Monica R, Krishna AN (2020) Identification of fake news using machine learning. IEEE
16. Gregorutti B, Michel B, Saint-Pierre P (2017) Correlation and variable importance in random forests. Stat Comput 27(3):659–678
17. Rampersad G, Althiyabi T (2020) Fake news: acceptance by demographics and culture on social media. J Inf Technol Politics 2020(17):1–11
18. Meel P, Mishra M, Vishwakarma D, Dinesh K (2019) A contemporary survey of machine learning techniques for fake news identification. In: Proceedings of the international conference on advances in electronics, electrical & computational intelligence (ICAEEC)
19. Zhou Z, Feng J (2017) Deep forest: towards an alternative to deepneural networks. In: Proceeding of the 26th international joint conference on artificial intelligence (IJCAI 2017), pp 3553–3559
20. Meel P, Vishwakarma DK (2021) A temporal ensembling based semi-supervised ConvNet for the detection of fake news articles. Expert Syst Appl 177:11500
21. Yuan X, Xie L, Abouelenien M (2018) A regularized ensemble frame-work of deep learning for cancer detection from multi-class. Imbalanced Train Data Pattern Recognit. 77:160–172
22. Meel P, Vishwakarma DK (2021) HAN, image captioning, and forensics ensemble multimodal fake news detection. Inf Sci 567:23–41

23. Shu K, Sliva A, Wang S, Tang J, Liu H (2017) Fake news detection on social media. ACM SIGKDD Explor Newsl 19(1):22–36
24. Vosoughi S, Roy D, Aral S (2018) The spread of true and false news online. Science 359(6380):1146–1151
25. Allcott H, Gentzkow M (2017) Social media and fake news in the 2016 election. J Econ Perspect 31(2):211–236

A Multilingual iChatbot for Voice Based Conversation

Dhanishtha Patil and Amit Barve

1 Introduction

The main goal of a chatbot is to make conversation between humans and machines. The machine is thus trained such that it is very well contextually aware of the input query asked by the user. As it becomes contextually aware the chatbot model glances throughout the dataset to find the relevant response in a human-like way. A simple chatbot is easier to build as compared to complex bots and developers should analyse, understand and consider the stability, scalability and flexibility issues that come in a way while building such systems along with a high level of intention in human language [1]. Unlike other customer services the chatbot system is available 24/7 and 365 days, and is helpful and useful as it provides an enlightening answer, upholding a framework of conversation and being inarticulate. The motive of chatbots is to reinforce and implant real-time websites to maintain their relations with customers. Machine Learning-Based chatbots are very intelligent and can handle different kinds of queries efficiently than human intelligence [2]. The chatter-bots or chatbot can be utilized securely as an open-source and is usually implemented in popular web services. As a whole, inquiry chat-bots are built using deep-learning algorithms that examine a user's queries and understand them [3]. The study of chatbot using CiteSpace and Bibliometrics shows that the old research data available on chatbots and conversational agents is fragmented so it is possible to explore various aspects in this field in future era [4], taking this in consideration several algorithms researched may be used to develop the self-learning chatbot, as well as the problems that have been experienced since the first chatbot was created [5]. Chatbot incorporates an important feature which is human machine automation and ability to get the context

D. Patil (✉)
Ramrao Adik Institute of Technology, Nerul, Navi Mumbai, India
e-mail: dhanishthapatil1804@gmail.com

A. Barve
Parul Institute of Engineering and Technology, Parul University, Vadodara, India

© The Author(s), under exclusive license to Springer Nature Singapore Pte Ltd. 2022   439
B. Unhelker et al. (eds.), *Applications of Artificial Intelligence and Machine Learning*,
Lecture Notes in Electrical Engineering 925,
https://doi.org/10.1007/978-981-19-4831-2_36

of the entire user command based on tests that are being conducted for producing the output [6]. A chatbot is a system that is user-friendly and connects to the user in a human-like way. AI-based conversational agents are very popular in various settings and offer a number of time- and cost-saving opportunities [7]. Chatbots are of multiple categories and are again sub- categorized and even integrated into a few different patterns [8]. There is wide classification of the existing approaches and the variety of chatbots categorized by its techniques [9]. Chatbot technology implementation can be done with different programming languages and databases [10]. Also, a variety of platforms and techniques can be used for conversational chatbots to get developed [11], in which one of the examples is the echo platform [12], taking this into consideration we tried to implement all the aspects of deep learning which made the proposed system more reliable, a framework which is designed using theoretical designs can result in satisfying the required capabilities of that particular industry for which it is developed [13]. The proposed system is thus made in such a way that the dataset can be updated according to the requirements of the intended industry for which it has flourished. This chatbot can be integrated with the website of that particular industry for which it is developed [14]. Users will just have to ask their queries either by typing them or using the speech recognition button. Users can type their queries in Marathi, English, Hindi, Gujarati, Punjabi, Telugu, and Japanese as it is multilingual. Chatbot will analyze the entered input and give an appropriate response to that input using the provided dataset. The proposed system also replies in the same language in which the user has asked the query. This way it will be easier and more viable for the users to get answers to their queries as they don't personally need to visit the office for that [7]. For a variety of application fields, chatbots are used [4], In addition to marketing, entertainment, customer service with basic activities, such as general aid, social chatbot, health and education are its expanse of utilization. In the current scenario when it is inconvenient for the students to visit the college/institute most of the students rely on the website of the college/institute. The paper comprises of literature survey, description for novel multilingual university dataset which is in both international and national languages, proposed methods of preprocessing done for the readiness of model, the model and its working, incorporation of voice recognition system, incorporation of Graphical User Interface along with analysis of accuracy for features and all languages.

2 Related Work

Ever after the first chatbots like Eliza [15] to the latest virtual chatbots assistants like Siri [16], have modernized the way with great attainment [17]. Eliza, which was developed at MIT by Joseph Weizenbaum in 1966, is the very first widely known chatbot. Eliza is the chatbot working on the pattern matching algorithms. Basically, it act as a system that represents an actual human being and was developed to show the superficiality between human and computer communication [15]. The system is made intelligent by providing a legit dataset to it. The chatbot is trained well using this dataset created by human experts, this increases the overall accuracy of the system. Many researchers felt that the software would have a significant effect on the lives of many individuals, especially those with psychological disorders [5]. An addition to ELIZA, a chatbot created in 1972 with a personality called PARRY which passed the Turing test [5]. A.L.I.C.E [18], which was the first personality program focused on AIML, won the Loebner and bestowed at the annual Turing Test competitions in 2000, 2001 and 2004 as "the most human computer" [19]. At present, A.L.I.C.E. Bot has more than 40,000 types of information, when there were just about 200 in the original ELIZA. Many chatbots were created for the Loebner Prize beginning in 1991.The competition was its oldest Turing Test contest to find the most human-like chatbot considered by the judges. Over recent years, Mitsuku [20] and Rose [21] have been the two chatbot champions The next move was to build automated personal assistants such as Apple Siri, IBM Watson, Amazon Alexa etc. Many chatbots have been established for tech companies, whereas a diverse range of less known chatbots are applicable for its analysis and their purposes [22]. Chatbots are then characterized as text-based, turn-based, task-fulfilling programs incorporated in existing frameworks with clarifying characteristics. Capabilities of the chatbot judged and deduced were helping dialog context resolution performance, handling mistakes in conversation, engaging in small talk, gracefully concluding and terminating conversation [23]. The modern phenomenon is gaining popularity for online chatting particularly where the consumer can communicate with customer service from anywhere at any moment which is a great part for the latest business [24]. The comprehensive and historical review of the global organization's interest in chatbots is extensive, and it stimulates the use of chatbots. Since the general design of modern chatbots and platforms for their production is improving, their usefulness in a number of sectors is expanding [25].

3 Dataset Description

3.1 Multi-linguistic Dataset

Our owned designed and constructed multilingual dataset, which includes languages like Marathi, Hindi, English, Gujarati, Punjabi, Telugu, and Japanese also more can be added further is used, which is one of the key features of ichatbot. The multi-linguistic dataset is in the JavaScript Object Notation (JSON) format [26], with fields for questions and responses, and values for different patterns that the user may ask. This multi-linguistic data is structured in such a manner that if a user asks a question in his chosen language, the response of chatbot will likewise be in the user's preferred language. The questions and responses in each labeled segment are in seven different languages. The segment format of the dataset in Fig. 1 is of English and Hindi language which includes the Thank you tag respectively. It is composed of three sections which are Tags, Questions and Responses. The tag is the query's frame

```
{ "tag": "thanks",
  "patterns": ["Thanks", "Thank you", "That's helpful", "Awesome, thanks", "Thanks for helping me","okay"],
  "responses": ["Happy to help!", "Any time!", "My pleasure"]
},
{"tag": "शुकरीया",
  "patterns": ["शुकरीया", "यह सहायक है", "बहुत बढ़िया", "मदद करने के लिए शुकरीया"],
  "responses": ["मदद करने के लिए  हमेशा खुश!", "हमेशा आपकी सेवा में!", "आपका स्वागत है"]
}
```

Fig. 1 Instance of our dataset for Thank you tag of English and Hindi Language

of reference. It specifies the subject of the query. Questions are several patterns in which a chatbot user can ask a question of the respective tags.

Each question's segment has responses set from which the chatbot can return any of the relevant responses. They're employed together to train the model consisting of a variety of language's patterns and respective responses.

3.2 Preprocessing with NLP and NLTK

While implementing a chatbot different functions and techniques of Natural Language Processing (NLP) and Natural Language Toolkit (NLTK) are used. Certainly, NLP techniques and tools have been known for its enhancement of data and from the original plain text, NLP completely automates the method of required optimization [27]. NLTK has been known as an amazing tool for functioning in computational linguistics with python along with Python it also provides practical knowledge for natural language processing [28]. After loading dataset the first step done as a preprocessing is Tokenization; it is a process in which breaking down a vast amount of data into smaller tokens. The output of the word tokenizer is converted to a data frame for improved text interpretation in subsequent applications; punctuation is frequently eliminated from our corpus because it is of little use, by analyzing our data with a for-loop within a function. Lemmatization is applied on the data in which transformed words to lemma form in order to lessen all of the canonical words. Transform all tokenized words into lowercase by iterating over multiple words using for-loop within a function to apply the "lower" function to each word. Here all words with their respective tags are saved in the form of a pickle file. Frequency of words in each text in this stage is analyzed and the main objective is to convert each word into a vector so that it may be readily utilized as an input to a deep learning model. NLTK helps with data processing in the form of a pipeline whereby the data then is collected in the form of a bag of words which is used for feature extraction and even used as an input for training the model.

4 Proposed Method

In this method there are three significant modules, the input data is preprocessed following the intents and entity generation and the last is response generation. The proposed ichatbot is regarding the intelligent chatbot technology based on natural language communication, as the world is becoming technically developed. The system built is of the conversation category in chatbots. Self-learning bots are one of the sub- categorized systems which we have proposed Deep learning is the ambiguous perspective to make this kind of bots. For the proposed system we used a "bag of words" algorithm which helped us to increase the accuracy. A neural model is fabricated using TensorFlow for training and the bag-of-words algorithm is implemented which examines by collecting data and designs the vocabulary, manufacturing document vector, and then storing words. Even the algorithm extracts features that are further used to train the model which is suitable for a simplified stack of layers with about one input tensor and one output tensor on every layer.

As shown in the below Fig. 2, the architecture consists of different modules like intent classification, entity recognition, response generator and graphic user interface. Intents are the end-objectives of the user, these intentions are passed by the user to chatbot which is used to classify all the intents in the input. The modifier that the client provides to define their problem is referred to as the entity. Entity recognition is used to recognize the input and find the solution to the input given. Responses are generated from the database. The Response Generator explores the keywords, and the specific keywords are searched through the database. And after detecting the keyword relative response is generated by the Response Selector from the response database and is displayed on the user interface.

4.1 Deep Neural Network Model

When the user gives a query, this query will be passed through the query preprocessing and modulation process and converted into a bag of words and will be given as input

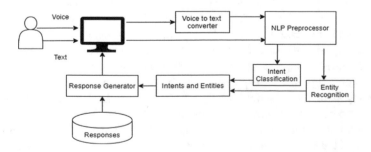

Fig. 2 Overall architecture diagram of proposed iChatBot

to the model. For accurate analysis of the input user query and generation of the most relevant response, we implement the feed forward Model. The feed forward neural network model is a framework for processing complicated data inputs that features a one-way information flow and no feedback loops. The multi-layer perceptron, a neural network consisting mainly of several perceptrons that can train to compute non-linearly separable functions, is used to approximate a function. The proposed model is constructed using in-built functions and methods of deep learning in which Keras offers high-level neural networks and dynamic library of deep learning on control of TensorFlow [29]. Model will return all the tags that have probabilities greater than the error threshold defined, then the response model will select the response which will be more accurate and have greater ranking given by the response System.

Cross Entropy loss function is used to find the distance between the output probability values from the softmax and the ground truth labels. The objective here is to reduce this distance. During training, weights and biases are tuned to minimize this cross entropy loss. Adam Optimizer an extension of the stochastic gradient descent is used to minimize the loss function for noisy problems, to manage sparse gradients.

$$Categorical\ Cross\ Entropy\ Loss = -\sum_{i=1}^{n} y_c, \log(p_c, l) \qquad (1)$$

where, N is the total classes, y is the classification label for particular class correctly classified or not, and p is the probability for classification c for class label l.

Training accuracy is the level of precision you acquire when you apply the model to the training data and is calculated by given formula.

$$Accuracy = \frac{Total\ Correct\ Predictions}{Total\ Predictions} \qquad (2)$$

4.2   Speech-to-Text Recognition

Speech recognition is also known as automatic speech recognition and computer speech recognition. The speech-to-text recognition is added with the help of google speech recognition library [30]. With microphone function from library speech_recognition is used to input the voice note from the user thereafter converting the speech to text using recognize_google function the query is passed to the model for generating appropriate response. If the function didn't receive any query or in case of any error, an exception message of "Sorry, I did not get that" is displayed. The user can not only enquire via text but can also enquire via audio, just has to tap on the microphone placed at the right corner of the GUI and say/enquire something. Feature is added so as to ease the efforts of users, as people mostly prefer to ask their queries via audio.

4.3 Graphical User Interface

The purpose of the user interface is more focused and expected since it is an interactive part and plays an important role in chatbots [9]. User interface helps to take users' input inquiries and process it through preprocessing and response prediction in the backend and acts as a link between backend and user. GUI is also integrated with the proposed ichatbot model which can be embedded onto the website. The GUI is implemented by using the Tkinter module which is to build the structure of the desktop application and then will capture the user message. The model then predicts the tag of the user's message, and will randomly select the response from intents through list of responses in our multilingual dataset and returns it to GUI. Components on GUI are designed by functions button text, scrollbar, etc using specific parameters. And the default message of "Hey there now I can listen to you, for speaking just tap on microphone" is displayed using the ChatLog library insert function.

5 Implementation Details

The feed forward neural network with best accuracy amongst tested, is of two hidden layers. The feed forward neural network architecture proposed consists of four dense layers of which the first input layer consists of 64 neurons, 2 hidden layers with 258 and 356 neurons respectively and an output layer. ReLU activation function is used in input as well as hidden layers and a softmax function in the output. Dropout of 0.3 and

0.4 has been added in order to reduce overfitting while training the model. Categorical cross entropy is used as loss function and Adam optimization. It is trained with the multilingual dataset which is in the form of a bag of words on 350 epochs, and a batch size of 95 as given in Table 1 accuracy reached 96.2% after 350 epochs and loss of 0.1331.

Table 1 Performance measures of iChatbot at different epochs

Epochs	Accuracy	Loss
50	86.89	0.5001
100	91.36	0.2985
150	92.65	0.1995
200	93.52	0.1805
250	94.92	0.1689
300	95.98	0.1543
350	96.28	0.1331

6 Result Analysis

The proposed iChatbot have accuracy 96%. The graphical user interface embedded on the website will be the source through which the user can have a live chat as shown in Fig. 3. The user can just click onto the logo embedded on either side of the website and a dialog box pops up where the user can ask or type the inquiry. iChatbot offers two options for the user to place inquiry either text or speech form. If any user asks their query via text then the iChatbot responds to it from the database that the system has provided to it. In another way user ask queries through audio by clicking on the microphone image provided. iChatbot converts the audio query into text and replies to the user in text format. Thus, the iChatbot can help various organizations and institutions to provide quality service and satisfy the user with less human efforts.

Two feed forward neural network architecture are compared, the first with one hidden layer and the second with two hidden layers, comparison of both is shown in Fig. 4. Inferred from the analysis, the feed forward neural network with two hidden layers is having great accuracy compared to the one having one hidden layer.

Table 2 shows a question asked through text from greeting tag in seven different languages and their response whereas Table 3 shows questions asked through voice for Thank you tag and their respective replies

In Fig. 5 describes testing done for evaluating accuracy of tag and response predicted for both text and voice base queries which resulted in efficiency of chatbot. A dataset is designed with both text and voice base queries for analyzing the accuracy of tag prediction, it consists of around 160 rows including 110 text questions and 50 voice inquiries. The predicted tag and response were compared with the actual response and respective tag. It was observed that 6 out of 110 inquiries predicted

Fig. 3 Graphical user interface of iChatBot

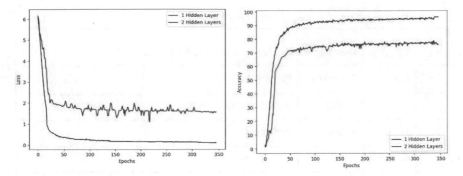

Fig. 4 Performance measures of iChatbot

Table 2 Results for responses generated for greeting tag through typing

Sr no.	Language	Question	Response generated by the Chatbot
1	English	Hello	Hi nice to see you I can speak in the following languages: Marathi, English, Hindi, Punjabi, Gujarati, Telugu, Japanese
2	Hindi	नमस्ते	आपक○ दे॑ खकर बहु॒त अच्छ○ा लगा मे॑○ं नम◌लखत भा○ष○ाओं मे॑○ ब○ल सकत○ा हं मराठ○ी अं○गे॑ज़ी १5○ंदी पं○जाबी गु॒जर○ाती ते॑लु॒गु॒ ज○ापानी
3	Marathi	नमस्कार	तुम्हाला पाहन मला आनंद झाला मी खालील भाषांमध्ये बोलु शकतो मराठी इ○ं ग्रजी १5○ंदी पं○जाबी गु॒जर○ाती ते॑लगु॒ जपानी
4	Gujarati	નમસ્તે	તમને જ○ઇને હ○ાય સરસ, હु○ં ન○ીયને ○ી ભાષાઓમ○ां૦ મરાઠ○ી અ○ं○ेજી ઇ॒ंग્ल○ी પુ○જાબી ગु॒જરાતી ગु॒જરાતી તેલ,ग॒म०૦ બ૦લ૦ી શ॒ક॒○ં છ○ં
5	Punjabi	ਹਾਇ ਉਥੇ	ਹਾਇ ਤੁ॒ ਹਾਨ॒ਤੁੰ ਬਹੁ॒ ਚੁੰਗਾ ਲਾਗਆ ਮੈਂ ਹੇਠ ਲਿਖੀਆਂ ਤ○ਸ○ਵ○ਂ ਮਰ○ਠੀ ਇ॒ੰ○ਗਲਸ ਹੁੱਦੀ ਪੁੰਜ○ਬੀ ਗु॒जਰ○ਤੀ ਤੇਲਗ ਜਪਾਨੀ ਬਵਚ ਬੋਲ ਸਕਦਾ ਹ○
6	Telugu	హలో	హాయ్ మరిమ్మ ఎరి చ○డటం ఆనందంగా ఉందరి నేనం ఈ8 కోరంది భూతపలలో మాట్ల డగలను క్రొంತी ఇ○గ్లీష్ హ○ందరి పంజాబీ గుజరాతీ తెలుగు జపనీస్
7	Japanese	こんにちは	こんにちは、私は次の言語で話すことができます。マラティ英語ヒンディー語パンジャブ語グジュラティテルグ語日本語

the erroneous tag, while 3 out of 50 voice queries had failed predictions. Hence, total testing accuracy achieved through analyzing tag prediction is 94.9% while for individual classes of text and voice predictions it is 94.5 and 94% respectively.

The pie chart in Fig. 5 was principally conducted for evaluation for individual language question's responses. Dataset of 50 questions for each language along with respective response was designed effectively, further calculating total correct responses received from a fixed set in different seven languages the above graph was

Table 3 Results for responses generated for greeting tag through Voice

Sr no.	Language	Question	Response generated by the Chatbot
1	English	Thank you	Happy to help!
2	Hindi	शुकरीया	य5 मेरा सौभाग्य 5े
3	Marathi	धन्यवाद	आपले स्वागत आ5े
4	Gujarati	આભાર	તમારૂં સ્વાગત છે
5	Punjabi	ਧੁੰਨਵਾਦ	ਮਦਦ ਕਰਕੇ ਖੁਸੀ
6	Telugu	ధనయ వాదాలు	ఇది నాకు సంతోషమే
7	Japanese	ありがとう	どういたしまして

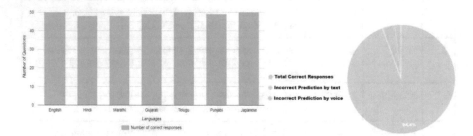

Fig. 5 Analysis of responses for each language and Feature testing based on results

plotted. It is observed that three of the total languages had 100% accurate results and the remaining between 96–98% precise results and average accuracy for overall testing is 98.28%.

We had a small evaluation analysis in which ten students actively participated with the bot and provided information shown in Fig. 6. These were students who had gone through the admissions procedure and had to wait long in the college or were not able to reach college to have their questions answered. Depending on their needs, each user asked a different amount of questions and shows which responses are appropriate and which are unsatisfying. Analyzing experiment into account, a total of 159 questions were asked, with 150 of them being satisfactory, resulting in an accuracy of 94.33%.

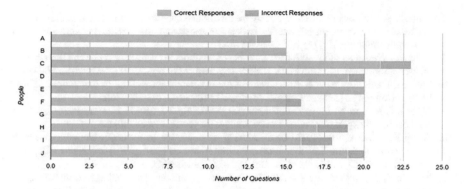

Fig. 6 Analysis of responses by views of people

7 Conclusion

In this paper, we proposed a self-learning voiced based intelligent chatbot with the idea of building a multilingual conversation that could be used to identify the user input and respond accordingly. We utilize various natural language processing techniques, 'Bag of words' algorithm and feed forward neural network. We created a database that would contain answers to each and every query in seven languages. iChatbots reaches a large number of people at the same time and more effective than humans. The proposed iChatbot design is simple, user-friendly and the responses given by the chatbot are understood easily by an individual. In upcoming years as technology grows rapidly chatbot can provide a response through speech in any language asked. Other than live chat responses we can develop it to support short message service also on user's mobile.

References

1. Rahman AM, Mamun AA, Islam A (2017) Programming challenges of chatbot: current and future prospective. In: IEEE region 10 humanitarian technology conference, Dhaka
2. Khanna A, Pandey B, Vashishta K, Kalia K, Pradeepkumar B, Das T (2015) A study of today's AI through chatbots and rediscovery of machine intelligence. Int J u-and e-Serv Sci Technol 8(7):277–284
3. Nuruzzaman M, Hussain OK (2018) A survey on chatbot implementation in customer service industry through deep neural networks. In: IEEE 15th international conference on e-business engineering
4. Io HN, Lee CB (2017) Chatbots and conversational agents: a bibliometric analysis. In: IEEE international conference on industrial engineering and engineering management (IEEM), Singapore
5. Shum H, He X, Li D (2018) From Eliza to XiaoIce: challenges and opportunities with social chatbots. Front Inf Technol Electronic Eng 19: 10–26. https://doi.org/10.1631/FITEE.1700826

6. Angelov S, Lazarova M (2019) E-commerce distributed chatbot system. In: Proceedings of the 9th balkan conference on informatics association for computing machinery, New York, NY, USA. https://doi.org/10.1145/3351556.3351587
7. Adam M, Wessel M, Benlian A (2020) AI-based chatbots in customer service and their effects on user compliance. Electron Mark 31:427–445. https://doi.org/10.1007/s12525-020-00414-7
8. Baez M, Daniel F, Casati F, Benatallah B (2021) Chatbot integration in few patterns. IEEE Internet Comput 25(3):52–59. https://doi.org/10.1109/MIC.2020.3024605
9. Thies IM, Menon N, Magapu S, Subramony M, O'Neill J (2017) How do you want your chatbot? in: an exploratory wizard-of-oz study with young, Urban Indians
10. Setiaji B, Wibowo FW (2016) Chatbot using a knowledge in database: human-to-machine conversation modeling. In: 7th international conference on intelligent systems, modelling and simulation (ISMS), Bangkok. https://doi.org/10.1109/ISMS.2016.53
11. Hussain S, Sianaki OA, Ababneh N (2019) A survey on conversational agents/chatbots classification and design techniques. In: Barolli L, Takizawa M, Xhafa F, Enokido T (eds) Web, artificial intelligence and network applications. advances in intelligent systems and computing, vol 927. Springer, Cham. https://doi.org/10.1007/978-3-030-15035-8_93
12. Argal A, Gupta S, Modi A, Pandey P, Shim S, Choo C (2018) Intelligent travel chatbot for predictive recommendation in echo platform. In: IEEE 8th annual computing and communication workshop and conference (CCWC), Las Vegas. https://doi.org/10.1109/CCWC.2018.830 1732
13. Følstad A, Brandtzaeg PB (2020) Users' experiences with chatbots: findings from a questionnaire study. Qual User Exp 5:3. https://doi.org/10.1007/s41233-020-00033-2
14. Satu MS, Parvez MH, Al-Mamun S (2015) Review of integrated applications with AIML based chatbot. In: International conference on computer and information engineering (ICCIE), Rajshahi. https://doi.org/10.1109/CCIE.2015.7399324
15. Weizenbaum J (1966) ELIZA—a computer program for the study of natural language communication between man and machine. Commun ACM. https://doi.org/10.1145/365153. 365168
16. Simone N (2020) To believe in Siri: A critical analysis of AI voice assistants 32:1–17
17. Coheur L (2020) from eliza to siri and beyond. In: Lesot MJ, et al (eds) Information processing and management of uncertainty in knowledge-based systems. IPMU. Communications in computer and information science, vol 1237. Springer, Cham. https://doi.org/10.1007/978-3-030-50146-4_3
18. Bayan A, Eric A (2015) ALICE chatbot: trials and outputs. Computación y Sistemas
19. Wallace RS (2009) The anatomy of A.L.I.C.E.. In: Epstein R, Roberts G, Beber G (eds) Parsing the turing test. Springer, Dordrecht. https://doi.org/10.1007/978-1-4020-6710-5_13
20. Koperniak M, Dąbrowska M, Mańczak-Wohlfeld E (2014) A linguistic analysis of the conversations with two chatbots: elbot and mitsuku
21. Park M, Aiken M, Salvador L (2018) How do humans interact with chatbots? an analysis of transcripts. Int J Manag Inf Technol 14, 3338–3350
22. Adamopoulou E, Moussiades L (2020) An overview of chatbot technology. In: Maglogiannis I, Iliadis L, Pimenidis E (eds) Artificial intelligence applications and innovations. AIAI. IFIP advances in information and communication technology, vol 584. Springer, Cham (2020). https://doi.org/10.1007/978-3-030-49186-4_31
23. Jain M, Kumar P, Kota R, Patel SN (2018) Evaluating and informing the design of chatbots. In: Proceedings of the 2018 designing interactive systems conference (DIS 2018). Association for Computing Machinery, New York, pp 895–906. https://doi.org/10.1145/3196709.3196735
24. D'silva GM, Thakare S, More S, Kuriakose J (2017) Real world smart chatbot for customer care using a software as a service (SaaS) architecture. In: 2017 international conference on I-SMAC (IoT in social, mobile, analytics and cloud) (I-SMAC), Palladam
25. Wei C, Yu Z, Fong S (2018) How to build a chatbot: chatbot framework and its capabilities. In: Proceedings of the 2018 10th international conference on machine learning and computing
26. Baazizi MA, Colazzo D, Ghelli G et al (2019) Parametric schema inference for massive JSON datasets. VLDB J 28:497–521. https://doi.org/10.1007/s00778-018-0532-7

27. Nazir F, Butt WH, Anwar MW, Khattak MAK (2017) The applications of natural language processing (nlp) for software requirement engineering - a systematic literature review. In: Kim K, Joukov N (eds) Information science and applications
28. Lobur M, Romanyuk A, Romanyshyn M (2011) Using NLTK for educational and scientific purposes. In: 11th international conference the experience of designing and application of CAD systems in microelectronics (CADSM), Polyana-Svalyava
29. Manaswi NK (2018) Understanding and working with keras deep learning with applications using python. Apress, Berkeley. https://doi.org/10.1007/978-1-4842-3516-4_2
30. Shakhovska N, Basystiuk O, Shakhovska K (2019) Development of the speech-to-text chatbot interface based on Google API. In: CEUR workshop proceedings, pp 212–221

Analysis and Forecasting of COVID-19 Pandemic on Indian Health Care System During Summers 2021

Vidhi Vig◉ and Anmol Kaur◉

1 Introduction

COVID-19 has wreaked havoc globally as health and government officials struggle to cope with this catastrophe. The ailment first spread in Wuhan, a city in China. Nearly two months after the world's first case, COVID-19 was declared a pandemic [1]. India encountered its first COVID-19 positive case on 30 January, 2020 in Kerala. To stop the further propagation of the virus, the Government of India decided to implement a nationwide lockdown on 23 March, 2020. This nationwide lockdown helped the Government to scale up the healthcare facilities. As on 28 July, 2021 India has reported 31,526,628 confirmed cases, 30,694,134 recovered cases, 422,695 fatalities and 409,799 active cases because of COVID-19 [2].

Figure 1 depicts the cumulative number of active cases in India from May, 2021 to July, 2021. India observed second COVID-19 wave in the months of April and May, which justifies the sudden peak in the active cases as shown in Fig. 1. The second wave was way more powerful than the first one due to consistent mutation of the virus [3]. As more people started testing positive for COVID-19, India experienced a shortage in the supply of oxygen which is crucial to the patients [4]. The healthcare system witnessed a significant rise in the amount of oxygen beds in the months following the second wave during which India went through a crippling shortage of oxygen and ICU beds. As of July 23, 2021, there are over 4.25 lakh oxygen-supported isolated beds, 57,518 ventilators and 1,573 PSA oxygen generation plants are to be set up in public health facilities across India.

The pandemic has stretched the global healthcare system to its limits. Figure 2 depicts the rise and fall in the number of Intensive Care Units (ICU), oxygen supported beds and ventilators within the public healthcare system [5]. The data

V. Vig (✉) · A. Kaur
Department of Computer Science, Shri Guru Tegh Bahadur Khalsa College, University of Delhi, New Delhi, India
e-mail: vidhi.ipu@gmail.com

B. Unhelker et al. (eds.), *Applications of Artificial Intelligence and Machine Learning*,
Lecture Notes in Electrical Engineering 925,
https://doi.org/10.1007/978-981-19-4831-2_37

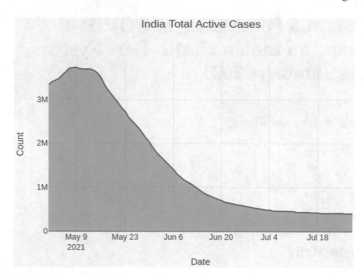

Fig. 1 Total number of COVID-19 active cases in India from May, 2021 to July, 2021

provided the quantity of ICU, oxygen supported beds and ventilators available in the states recorded on 21 April, 2020 and 28 January, 2021. Figure 2 gives a view of the struggling healthcare system during the pandemic. As a developing country, 80% of the Indian population lives in rural areas where access to quality healthcare is scarce. Even though private healthcare facilities also provide treatment for COVID-19, the cost of private healthcare is a barrier for many Indians [6]. They can only rely on the healthcare facilities provided by the government. Therefore, it is important to identify the trend of the pandemic and use the resources efficiently and sustainably. This study aims to forecast overall active cases in India and in turn help the government to provide health facilities without straining the health system, especially during extreme summers.

It has been known that there is a strong inter connection between the weather and the respiratory viral infections. Still, a very limited amount of climatic variables has been researched with respect to the COVID-19 virus. Earlier it was found that the virus can sustain for good 14 days in the host body before becoming too weak to change the host. It was further proposed that temperature and humidity can heavily affect the lifecycle of the virus. However, its exact pattern and outspread is still unknown.

The aim of the current study is to empirically investigate the effect of higher temperatures and to learn the proliferation of the virus using ARIMA model. Papers on the kindred research have been explored and studied however, none of these explored the same for hot Indian summers.

The structure of the paper is as follows. Section 2 provides a literature review of the studies that have analyzed environmental influence on the spread of COVID-19 and used time series models to predict the propagation of the virus. Section 3 discusses the materials and methods used in the paper. Section 4 discusses the outcome of the

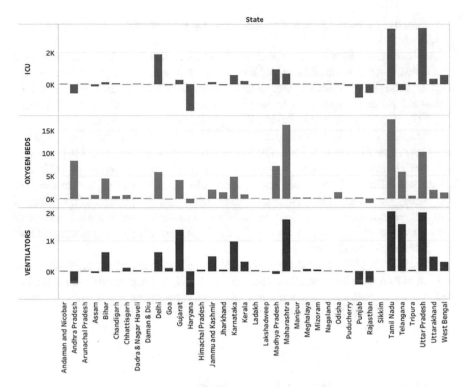

Fig. 2 The rise and fall in the number of ICU, oxygen supported beds and ventilators (April 2020–January 2021)

experiment and its possible meaning. Lastly, Sect. 5 provides the conclusion drawn from the analysis.

2 Literature Review

In [7], the authors predicted the necessity of healthcare facilities in Telangana, India. The study implemented SIR, FB-Prophet and ARIMA models on the data extracted from the Government of Telangana bulletins. They found that even though SIR model is more intuitive and explainable, it uses the method of trial and error. The FB-Prophet model outperforms the SIR model as its prediction process is simple and precise.

The study presented in [8] found a big flare-up of COVID-19 in India utilizing the data scraped from [2] and information available on COVID-19 publicized by ICMR. They also estimated the number of patients that recovered, tested positive and died due to COVID-19 using ARIMA (1, 2, 0) model. Their findings uncover that the overall confirmed cases were expanding at the rate of 3.48%, recuperated cases at

4.09%, active cases at 2.92% and casualty cases at the rate of 3.51% day by day over the country.

Another study [9] used the time series ARIMA model to forecast the confirmed and active cases in India and some of its affected states until June 2020. The authors then used the predicted active cases to estimate the healthcare requirements. Findings reveal that India will require to prepare for 106,006 isolation beds, 12,471 ICU beds and 6236 ventilators to accommodate the patients at the end of June.

The influence of weather conditions on COVID-19 seems to be related and has been investigated by researchers. The systematic review presented in [10] explored the interrelationship between temperature and humidity and observed that COVID-19 do get influenced by humid and warmer climates. However, the certainty of this theory was found weak. Another systematic review [11] analyzed the effect of ambient environmental conditions on COVID-19 mortality and affirmed that lower temperature contribute to an increment in cases. Again this fact was not found true for all the reviewed studies.

The study in [12] found that the transmission of flu viruses are relatively slow in high humidity and hot weather. The studies in [13, 14] suggest that the spread of COVID-19 may slow down in regions with high temperatures. But on the other hand some studies [15–17] came to the conclusion that there's no affiliation of COVID-19 with temperature.

3 Research Dataset and Methodology

The study has drawn the data from the publicly available API https://www.covid19in dia.org. Since, the work is focused specifically on assessing the impact of the virus in extreme summers of India, the data was collected and analyzed ranged from May, 2021, to July, 2021. The data thus extracted was then investigated for relevance and duplicity.

ARIMA is a comprehensive form of ARMA (Autoregressive Moving Average) which is further a comprehensive from of AR (auto regressive) and MA (moving average) models. The time series model ARIMA is depicted using (p, d, q) where element 'p' belongs to the auto-regressive component of the model and represents the number of terms included. The element 'd' stands for the degree of differencing required to stabilize the series if it isn't stationary. The element 'q' indicates the order of the number of terms included in moving average (MA) model. Further, the time series ARIMA model [18] was implemented on the total active cases recorded for the same.

The entire process of data cleaning and implementation of ARIMA model was done using R programming language in R studio [19]. R studio is a free and open source platform that provided extensive packages for data analysis and forecasting.

4   Results and Discussion

Initially, the stationarity of the data was determined by Augmented Dickey Fuller (ADF) test [20]. The null hypothesis for the test determines if the data is non-stationary. On implementation it was observed that the p-value came out to be 0.99 indicating that the data was not stationary. It is also known that the model with the least AICc value is deemed to be the best model. Consequently, Autocorrelation function (ACF) and partial autocorrelation function (PACF) plots were then used to determine the optimal model to fit the data using AICc.

Akaike information criterion (AIC) investigates the quality of every model with rest to each other, therefore, it is considered a strong variable in model selection. With the AICc value of 1970.95 and RMSE of 1965, the study found that ARIMA (0, 2, 0) was the best model for forecasting the active cases in India (for the current dataset).

Figure 3 below, shows the forecasted total active cases of COVID-19 in India. The black line in the graph indicates the active cases recorded from May 2021 to July 2021 in India and the blue line indicates the forecasted active cases for the upcoming 15 days.

The trend in August however, seems to be following the trend of June and July months. This further supports the claim made by the study where months with higher temperatures tends to follow the same growth patterns. It may also be observed that the criterion temperature alone cannot be assumed to be the reason for such a trend. As the disease progresses and more and more data is available for research. This knowledge along with other sociopolitical reforms play a major role in setting trends. However, based on the current forecasts, India will have around 469,619 active cases by mid-August.

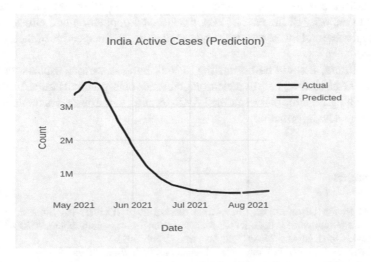

Fig. 3   Forecast of total active cases in India from 29 July, 2021 to 12 August, 2021

The COVID-19 pandemic revealed the underlying problems in the health system of India. Before the pandemic, India had 1 doctor per 1404 people and an average of 0.55 government hospital beds per 1000 population (as of 2019) [21]. These estimates are profoundly low than what is recommended by WHO. India needs to address the shortcomings and under-investment in its health system to combat the pandemic and studies predicting the trends using efficient modelling can definitely come handy.

5 Conclusions

The present study carried out an experimental investigation in India over the summer to estimate rise in COVID-19 active cases. It is a common sight for respiratory ailments to peak in the winters and afterwards subside in the summers. Even so, throughout the months of June and July, India saw a decrease in the number of cases. However, despite the extreme heat and humidity, the pandemic has continued to spread rapidly globally. These disparities push the need to better understand the pandemic.

Statistical modeling helps to analyze the trend and gives a broader view of the pandemic. Such studies will further help the healthcare workers and government to determine the daily caseloads, quantity of ICU, ventilators and isolation beds required, thereby preparing for the worst. The current study investigates and explore the healthcare system in India and the pattern of the COVID-19 pandemic during summers. However, ARIMA model has its limitations, the model's accuracy often depends on the size of the data. Therefore, the forecasted values may contrast from reality owing to the inconsistent nature of the pandemic.

It has further been observed that these factors alone cannot justify the liability to vary the disease transmission. Moreover, the claim laid by these studies are weak and fail to majority of the cases. Thus, the affected population and cities/countries must emphasize and invest on health and social policies, irrespective of their climatic condition.

In the future, it could be fascinating to look into the various aspects that could influence the pandemic, such as meteorological and socio-political issues. In addition, analyzing the pandemic using seasonal ARIMA and other time series methods could reveal some new information.

References

1. World Health Organization. Coronavirus disease 2019 (COVID-19) Situation Report – 82. https://www.who.int/docs/default-source/coronaviruse/situation-reports/20200411-sitrep-82-covid-19.pdf?sfvrsn=74a5d15_2, Accessed 29 July 2021
2. Coronavirus Outbreak in India – covid19india.org. https://www.covid19india.org/, Accessed 29 July 2021

3. Ranjan R, Sharma A, Verma MK (2021) Characterization of the second wave of COVID-19 in India. medRxiv. https://doi.org/10.1101/2021.04.17.21255665

4. Bonnet L, Carle A, Muret J (2021) In the light of COVID-19 oxygen crisis, why should we optimise our oxygen use? Anaesth Crit Care Pain Med 40:100932. https://doi.org/10.1016/j.accpm.2021.100932

5. Ministry of Health and Family Welfare (2021) Rajya Sabha Unstarred Question No.102. https://pqars.nic.in/annex/253/A102.pdf, Accessed 29 July 2021

6. Pandey K (2021) Is India's public health infrastructure ready to tackle the second COVID-19 wave? Here's what data says. https://www.downtoearth.org.in/news/health/is-india-s-pub lic-health-infrastructure-ready-to-tackle-the-second-covid-19-wave-here-s-what-data-says-76320, Accessed 29 July 2021

7. Darapaneni N, Mahita GM, Paduri AR, Talupuri SK, Konanki V, Galande S, Tondapu CH (2021) Predicting hospital beds utilization for COVID-19 in Telangana, India. In: 2021 IEEE international IOT, electronics and mechatronics conference (IEMTRONICS). IEEE, pp 1–6

8. Athe R, Dwivedi R, Modem N (2020) Is India ready to address its biggest public health challenge? forecasts from the publicly available data on COVID-19. Int J Public Health 10:226–230. https://doi.org/10.5530/ijmedph.2020.4.47

9. Tyagi R, Bramhankar M, Pandey M, Kishore M (2020) COVID 19: Real-time forecasts of confirmed cases, active cases, and health infrastructure requirements for India and its states using the ARIMA model. medRxiv. https://doi.org/10.1101/2020.05.17.20104588

10. Mecenas P, Bastos R, Vallinoto A, Normando D (2020) Effects of temperature and humidity on the spread of COVID-19: a systematic review. PLoS ONE 15:e0238339. https://doi.org/10.1371/journal.pone.0238339

11. Romero Starke K, Mauer R, Karskens E, Pretzsch A, Reissig D, Nienhaus A, Seidler A, Seidler A (2021) The effect of ambient environmental conditions on COVID-19 mortality: a systematic review. Int J Environ Res Public Health 18:6665. https://doi.org/10.3390/ijerph18126665

12. Lowen A, Mubareka S, Steel J, Palese P (2007) Influenza virus transmission is dependent on relative humidity and temperature. PLoS Pathog 3:e151. https://doi.org/10.1371/journal.ppat.0030151

13. Wang J, Tang K, Feng K, Lv W (2020) High temperature and high humidity reduce the transmission of COVID-19. SSRN Electron J. https://doi.org/10.2139/ssrn.3551767

14. Demongeot J, Flet-Berliac Y, Seligmann H (2020) Temperature decreases spread parameters of the new Covid-19 case dynamics. Biology 9:94. https://doi.org/10.3390/biology9050094

15. O'Reilly K, Auzenbergs M, Jafari Y, Liu Y, Flasche S, Lowe R (2020) Effective transmission across the globe: the role of climate in COVID-19 mitigation strategies. Lancet Planet Health 4:e172. https://doi.org/10.1016/S2542-5196(20)30106-6

16. Yao Y, Pan J, Liu Z, Meng X, Wang W, Kan H, Wang W (2020) No association of COVID-19 transmission with temperature or UV radiation in Chinese cities. Eur Respir J 55:2000517. https://doi.org/10.1183/13993003.00517-2020

17. Sahafizadeh E, Sartoli S (2020) Rising summer temperatures do not reduce the reproduction number of COVID-19. J Travel Med 28:189. https://doi.org/10.1093/jtm/taaa189

18. Box G, Jenkins G, Reinsel G, Ljung G (2015) Time series analysis: forecasting and control. John Wiley & Sons, Hoboken

19. R Core Team: R (2013) A language and environment for statistical computing

20. Dickey D, Fuller W (1981) Likelihood ratio statistics for autoregressive time series with a unit root. Econometrica 49:1057. https://doi.org/10.2307/1912517

21. Singh P, Ravi S, Chakraborty S (2021) Is India's health infrastructure equipped to handle an epidemic? In: Brookings. https://www.brookings.edu/blog/up-front/2020/03/24/is-indias-hea lth-infrastructure-equipped-to-handle-an-epidemic/, Accessed 29 July 2021

A Novel Approach to Image Forgery Detection Techniques in Real World Applications

Dhanishtha Patil, Kajal Patil, and Vaibhav Narawade

1 Introduction

In the past decades, the era of digitalization has made it easier to use a variety of data sources, including images, to conduct scientific experiments, derive valuable insights for business and technology, identify repetitive patterns of data to develop the machine learning models, a popular field in today's world [1]. Among all the varied sources of data available, digital image seems to be one of an emerging category of producing a high impact which will accelerate the research of the particular domain [2].

Significant advancements in computer engineering have culminated in very high resolution capture equipment and substantial breakthroughs in image processing, making picture tampering and forgeries more accessible and straightforward. Some people distort the contents of a picture by modifying parts of it, resulting in image forgery, which is becoming more hard to trace using modern image processing tools [3, 4]. With the aid of advanced computer graphics algorithms and picture editing softwares like Photoscape and Acorn, photographs may now be manipulated without leaving any imprints [5]. To modify picture characteristics such as pixels, size, and resolution, a variety of applications is available [6].

Given the accessibility of powerful media editing, analysis, and production tools, as well as the increasing processing capability of contemporary computers, image altering and generation is now straightforward even for unskilled users. This trend is only likely to continue, resulting in more automated and precise processes becoming available to the general public [1, 7]. Image forgery can be defined as "erroneously and maliciously altering a picture." The concept of image forgery dates back to 1840.

D. Patil (✉) · K. Patil · V. Narawade
Ramrao Adik Institute of Technology, Nerul, Navi Mumbai, India
e-mail: dhanishthapatil1804@gmail.com

V. Narawade
e-mail: vaibhav.narawade@rait.ac.in

461

B. Unhelker et al. (eds.), *Applications of Artificial Intelligence and Machine Learning*,
Lecture Notes in Electrical Engineering 925,
https://doi.org/10.1007/978-981-19-4831-2_38

The earliest altered image was created by Hippolyte Bayard, a French photographer, in which Bayard admitted to attempting himself [8]. The certainty that the picture has not been altered or accessed is known as image integrity. When the integrity of an associated sensitive picture is breached in particularly applications that are critical, such as radioactive and military, a critical procedure might be jeopardised, as a consequence [9].

To counter the issue, digital forensics emerged as a crucial and trending field with a motive of ensuring the detection of digital tampering and manipulation. Because of their capacity to recreate the evidence left by cyber assaults, forensic tools have become a critical tool for Information Assurance [10]. As a result, pictures no longer have the status of being the only definitive record of events. In the lack of an authentication or watermark, digital forensics describes numerous statistical approaches for identifying evidence of digital manipulation [11].

Active and blind or passive image validation methods are available. Digital watermarking and digital signatures, for example, to detect active counterfeiting, include a valid authentication feature in the picture content before delivering it over an unprotected public channel [5]. The active forging approach uses a digital icon and requires pre-processing to create a watermark and embedded or signatures before compiling the picture. Although watermarking is a live intruding disclosure approach in which a certification shape is implanted into the image, most imaging devices lack a watermarking or check module [12]. This data might be added throughout the picture collection process, or afterwards with the help of an appropriate tool. Passive image forgery detection strategies, on the other hand, to detect fabrication, no previous information of the input image is required. Alternatively, these approaches recognize forgeries by looking for changes in the image's fundamental characteristics that may have been introduced during the modification process [13]. Copy-move forgery is copying a piece or region from one position in a photo and transferring it to another site inside the same image in order to conceal or imitate an item or a collection of items and create a fictitious perspective [10]. To hide an existing item in a picture, make a replica of the entity, copy-move forgery is used [2].

Based on the study of pre-existing literature, it was discovered that there are numerous domains in which image manipulation malpractices are carried out, each with a varied immoral, unethical motive. Some of them are briefly explained as follows:

Medical Diagnosis forgery:
Image falsification is a major problem in the healthcare industry. The diagnosis will be incorrect, and the patient will be in life-threatening danger if a visual diagnostic report is manipulated and the attacker employs any kind of forgeries to expand the cancerous region. If a healthcare framework includes an image forgery detection system, it can identify the counterfeit before the diagnostic process begins if medical data is stolen or changed, for example, the patient may experience social humiliation or be let down, while others may gain an unfair advantage. As a result, in a smart healthcare architecture, there should be a system that can verify whether medical data is corrupted during transmission by hackers or intruders [11].

Deepfake:
DeepFakes are artificial intelligence systems that utilise deep learning to replace the likeness of one person in video and other digital media with that of another. Deepfake technology has raised worries that it may be used to produce fake news and deceptive films. The most challenging challenge in the realm of picture forgery detection is detecting fake faces. Phony face pictures may be used to construct fake identities on social networking sites, allowing for the illicit theft of personal information. The false picture generator, for example, may be used to create photos of celebrities with improper material, which might be dangerous.

Social media image manipulation:
In today's digital world, social media platforms play a critical role in news dissemination. They have, however, been disseminating false pictures. On social media platforms like Twitter, forged photos cause deception and negative user sentiments. As a result, identifying fraudulent photos on social media sites has become a pressing requirement [17]. Not every material shared on social media is exactly what it claims to be. People upload and disseminate incorrect information for a variety of reasons. These might include a) personal goals (to acquire recognition or celebrity) or b) intents to sway public opinion or promote specific points of view (marketing, propaganda, etc.) [13]. Image forgery can now be done with a variety of technological tools, computer software, and web applications. As a result, the usage of modified photographs in news portals and, more specifically, social media has proliferated [14].

Criminal investigation evidence forgery:
The reliability of photos is critical in a variety of fields, including forensics, criminal investigations, surveillance systems, and intelligence agencies. Image manipulation is quite prevalent [18]. Manipulating the images is very much possible using the software accessible to the criminals. Because digital photographs contain vital information and are used in various sectors such as authenticating digital photographs is becoming increasingly important in legal issues of investigation and serve as a crucial source of proofs [19]. Image tampering, which is a critical component of a criminal investigation, can lead to legal authorities making defective or improper judgment [15].

Art forgery:
Even for veteran art historians, authenticating paintings may be challenging. It might be difficult to determine the authenticity of an artwork. Forgery detection has traditionally relied on the discriminating talents of "connoisseurs," who can determine credibility based on a prominent art piece, lifestyle, and circumstances [20]. There has been a lot of development in utilizing digital feature extraction techniques to characterize an artist's style. Studies have been able to categorize test paintings as originals or forgeries based on these findings [16].

2   Literature Survey

Forgery detection is a crucial issue in visual forensics, with a large body of literature on the subject and numerous deep learning techniques have dominated in recent years. General-purpose extremely deep architectures produce similar good outcomes in favourable circumstances when given a sufficiently big training set. The goal is to detect patterns that identify transitional zones that are out of the ordinary in comparison to the backdrop in order to locate probable forgeries. This concept is also explored in [21], which uses a hybrid CNN- LSTM architecture to train a binary mask for forgery localisation. However, to train the net, these approaches require comprehensive ground truth maps, which may not be available or exact. Researchers demonstrated and developed numerous methods for verifying picture integrity and detecting various types of image fraud.

Shen et al. [22] proposed a technique for detecting forgeries based on textural characteristics extracted from Grayscale Co-Occurrence Matrices (GCOM) and DCT. The CASIA-V1 dataset had a detection rate of 98%, whereas the CASIA-V2 dataset had a detection rate of 97%. In the quaternion DCT (QDCT) domain, Li et al. [23] proposed a forgery detection method based on Markov. For the CASIA-V1 dataset, the detection rate was 95.217%, while for the CASIA-V2 dataset, it was 92.38%. The CASIA-V1 dataset had a detection accuracy of 97%, whereas the CASIA-V2 dataset had a detection accuracy of 97.5%. The CASIAV1 dataset had a detection accuracy of 99.16%, whereas the CASIA-V2 dataset had a detection accuracy of 97.52% [24]. The CASIA-V1 dataset had a detection accuracy of 98.3%, whereas the CASIA-V2 dataset had a detection accuracy of 99.5% [25]. The Curvelet transform and the LBP were employed by Al Hammidi et al. [26] to identify forgeries, and the findings were 93.4% accurate. For the CoMoFoD dataset, the findings obtained a detection accuracy of 92.22% [27]. He et al. [28] developed a technique for detecting forgeries in both the DCT and DWT domains based on extracting Markov characteristics. For the CASIA-V1 dataset, the findings obtained a detection rate of 93.33%. Kakar et al. [29] proposed a forgery detection technique based on DCT and DWT domain Markov characteristics. For the CASIA-V2 dataset, the findings obtained a 95.5% detection rate. Using a pyramid model and Zernike moments, Ouyang et al. [30] proposed a Copy Move Forgery Detection (CMFD) method. The findings revealed that for scaling ranges of 50 to 200%, the given technique had the best influence on the arbitrary rotation angle. A new block-based CMFD matching method was proposed by Lai et al. [31]. The findings revealed that the proposed technique is more resistant to JPEG picture compression assaults and the addition of Gaussian or salt pepper noises than the traditional algorithm. Saleh et al. [32] proposed a multiscale weber local descriptor-based forgery detection technique. Vaishnavi and Subashini utilised the Random Sampling Consensus method and local symmetry characteristics (RANSAC). For MICC-F600 datasets, the findings reached 90.2% detection accuracy [33]. Rao and Ni used CNN to achieve a detection rate of 98.04% for the CASIA-V1 dataset and 97.83% for the CASIA-V2 dataset. The findings obtained

95% identification accuracy for the MICC-F600 dataset when Agarwal and Verma used the pretrained Visual Geometry Group-net (VGGNet) deep learning model [34].

3 Proposed Methodology

Recent research in computer vision, however, has revealed that CNNs are taking advantage of the rapid rise in the amount of annotated data and significant advances, topics have been made. Convolutional Neural Networks shown in Fig. 1 are scaled up for increased accuracy if more resources are available [35]. The CNN accumulates hierarchical visual characteristics from training dataset, that are subsequently utilised to differentiate between tainted and genuine pictures. Extensive testing has shown that the proposed algorithm outperforms existing technologies [36]. In this paper we have investigated model scaling in depth and attempted to find a model that carefully balances network depth, width, and resolution to improve performance for Image Forgery.

3.1 Dataset

Deep learning-based approaches beat other technologies and solutions for the MICC group dataset, the most well-known and practical datasets for testing copy- move forgery detection methods [10] developed by Amerini et al. MICC data include pictures that have a visible manipulation impact. In MICC F2000 picture's tampered regions are randomly picked regions from the that data itself and level of forgery for particular image in MICC F2000 is around 1.12% of the whole image of 2048 × 1536 pixels which 700 are tampered and 1300 arc original images [36, 37]. For model training and validation, MICC F2000 is utilized, also there are 260 forged picture sets in the CMFD [38]. The CoMoFod and MICC F220 datasets are used for testing. The input images are converted to RGB and scaled to a predetermined size

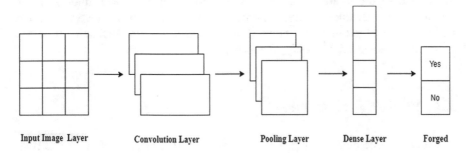

Input Image Layer   Convolution Layer   Pooling Layer   Dense Layer   Forged

Fig. 1 Basic CNN architecture

Scaled Image	Tampered Image 1	Tampered Image 2	Tampered Image 3

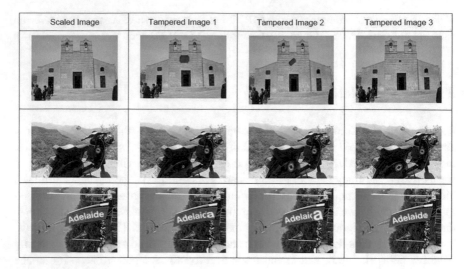

Fig. 2 Images from dataset MICC F2000

without cropping any sections during the data pre-processing step. Figure 2 shows some images from the training dataset that is MICC F2000.

3.2 VGG-19 Architecture

The VGG-19 model is a convolutional model with a wide range of image classification success. Its design consists of three small convolutional filters (3×3), the convolution stride is fixed at 1 pixel, and the padding is similarly fixed at 1. On the supplied dataset, a batch size of 16 is used to train the network. Validation is done with the same batch size. The activation function of the ReLU is implemented. A dropout of 0.5 is used. The RMSprop optimizer is used with a learning rate of 1e–4. Only vertical variations are allowed by the RMSprop optimizer. When training the model, the Adam optimiser was also utilised to get to the global minima. When training, if you get stuck in local minima, the Adam optimiser will help you get out and attain global minima. For training, we utilised 15 epochs. Validation accuracy of 97.50% is achieved with only 15 epochs.

3.3 VGG-16 Architecture

Instead of a wide range of hyper-parameters, VGG16 concentrated on 3×3 filter convolution layers with a stride 1 and always retained the same padding and maxpool layer of 2×2 filter stride 2. Throughout the design, the convolution and max pool layers are positioned in the same way. It has two completely linked layers in the end [39]. The network is trained with a batch size of 30 on the provided dataset. ReLU's activation function has been implemented. A dropout of 0.5 is used. The RMSprop optimizer is used with a learning rate of 1e–3 and a decay factor of 0.75. The Adam optimizer to attain global minima. We used 15 epochs for training. The accuracy of validation is 97.75%.

3.4 EfficientNet

EfficientNet is a convolutional neural network design that performs better on ImageNet than existing CNNs in terms of effectiveness and precision, while lowering parameter size and FLOPS by an order of magnitude and scaling approach that uses a compound coefficient to evenly scale all depth, breadth, and resolution parameters [40]. The drop connect rate is set for better regularisation and rebuilt top layers after loading EfficientNetB0, 0.2 dropout rate is chosen. Also the GlobalAverage-Pooling2D layer is added, since by imposing correspondences between feature maps and categories, convolution structure is created. In addition, the batch normalisation layer is used in the initial training and dropout to set the input unit to 0 at a frequency of at each step throughout the training time, which helps avoid overfitting, and the dense layer with softmax activation is used in the final training and dropout. We used Adam Optimizer to train the model with a learning rate of 1e-3 and a decay of 0.75. The batch size of 16 and the number of epoch 15 are the same. Validation accuracy of 96% is achieved after 15 epochs. In the second training phase top 20 layers are not froze except the batch normalization layers and further the model is trained on 15 epochs and adam optimizer with learning rate of 1e-4 which in final obtained accuracy of 98.2%. The training was done on Google Colab with NVIDIA Tesla P100 GPU and 12 GB RAM.

4 Results

4.1 Evaluation Metrics

A confusion matrix is a table that describes the results of categorisation problem prediction. Count values are used to sum and break down the number of successful and failed projections by class (Table 1).

True positive (TP): The model's estimated value corresponds to the origanality of the image. It can be deduced it is identified that the image as forged.

False negative (FN): The expected output is a false negative, in which the image in input is wrongly labelled negative, despite the fact that the image is unforged.

True negative (TN): The anticipated outcome is a genuine negative when the model

Table 1 General confusion matrix

	Predicted forged(1)	Predicted unforged(0)
Forged(1)	TP	FN
Unforged(0)	FP	TN

Table 2 Confusion matrix for EfficientNetB0

	Predicted forged(1)	Predicted unforged(0)
Forged(1)	249	7
Unforged(0)	0	144

Table 3 Confusion matrix for VGG 16

	Predicted forged(1)	Predicted unforged(0)
Forged(1)	248	8
Unforged(0)	1	143

Table 4 Confusion matrix for VGG 19

	Predicted forged(1)	Predicted unforged(0)
Forged(1)	247	9
Unforged(0)	1	143

predicted value matches to the reality that the picture is unforged. It may be inferred that the machine classified properly.

False positive (FP): Image is incorrectly categorized as an unforged image, despite the fact that it is a forged image (Tables 2, 3 and 4).

From above observations, TNR, TPR is high and FNR, FPR is low. So our model is not in overfit or underfit.

In comparison research, the performance of data classification algorithms is evaluated using metrics such as classification accuracy, sensitivity or recall, precision, and Matthew Correlation Coefficient after the confusion matrix is constructed (MCC) (Tables 5, 6 and Figs. 3, 4, 5).

Table 5 Rate comparison

	TPR%	FPR%	FNR%	TNR%
EfficientNetB0	100.00	4.64	0.00	95.36
VGG 16	99.59	5.30	0.40	94.70
VGG 19	99.59	5.92	0.40	94.07

Table 6 Performance comparison

	Accuracy	Recall	Precision	F1 Score	MCC
EfficientNetB0	0.9825	1.00	0.9727	0.9861	0.9631
VGG 16	0.9775	0.9960	0.9688	0.9822	0.9524
VGG 19	0.9750	0.9960	0.9648	0.9802	0.9473

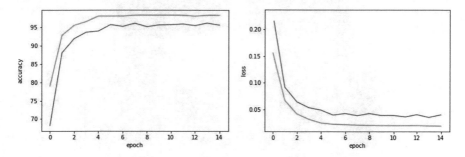

Fig. 3 Performance curves for first phase for EfficientNetB0

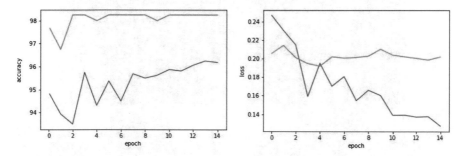

Fig. 4 Performance curves for final phase for EfficientNetB0

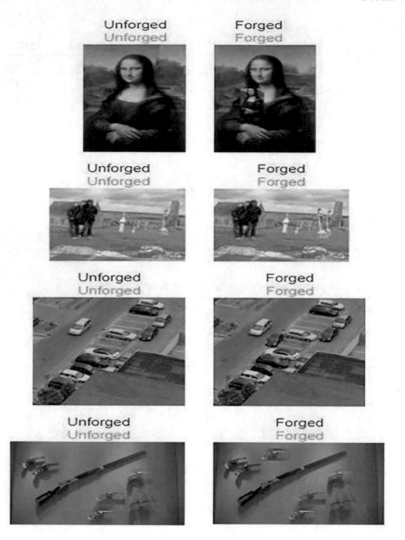

Fig. 5 Final results for most of real time based applications of Image Forgery in fields of Art, Social Media, Criminal Investigation

5   Conclusion

The focus of this research is to show how picture data counterfeiting methodologies may be employed to perform image tampering and falsification. One of the most rapidly developing fields of research is forgery detection utilising passive forgery detection approaches. Various fields being affected by image forgery have also been scrutinized. Analysis of the performance studies on forged images from various realistic domains where image forgery is do8ne to depict the usefulness of this method.

The main objective was was to identify the fabrication using copy move technique a variety of CNN architectures and compare them in terms of how accurate their results are. Amongst the Convolutional Neural Network architectures, EfficientNetB0 achieved the highest validation accuracy of over 98%. The study can further be expanded to develop various methodologies to detect audio forgery and an efficient system can be established.

References

1. Sencar HT, Memon N (eds) (2013) digital image forensics. Springer-Verlag New York. https://doi.org/10.1007/978-1-4614-0757-7
2. Walia S, Kumar K (2018) Digital image forgery detection: a systematic scrutiny. Aust J Forensic Sci 51:1–39. https://doi.org/10.1080/00450618.2018.1424241
3. Kasban H, Nassar S (2020) An efficient approach for forgery detection in digital images using Hilbert Huang transform. Appl Soft Comput 97:106728. https://doi.org/10.1016/j.asoc.2020.106728
4. The 2015 IEEE RIVF International Conference on Computing. Communication Technologies Research, Innovation, and Vision for Future (RIVF)
5. Bharti CN, Tandel P (2016) A survey of image forgery detection techniques. In: International conference on wireless communications, signal processing and networking (WiSPNET). IEEE. https://doi.org/10.1109/wispnet.2016.7566257
6. Meena KB, Tyagi V (2019) Image forgery detection: survey and future directions. In: Shukla RK, Agrawal J, Sharma S, Tomer GS (eds) Data, engineering and applications. Springer, Singapore. https://doi.org/10.1007/978-981-13-6351-1_14
7. Garfinkel SL (2010) Digital forensics research: the next 10 years. Digit Invest 7:S64–S73. https://doi.org/10.1016/j.diin.2010.05.009
8. Taylor JRB, Baradarani A, Maev RG (2015) Art Forgery Detection via craquelure pattern matching+. In: Garain U, Shafait F (eds) Computational forensics. IWCF 2012, IWCF 2014. Lecture notes in computer science, vol 8915. Springer, Cham. https://doi.org/10.1007/978-3-31920125-2_15
9. Meena KB, Tyagi V (2020) A copy-move image forgery detection technique based on tetrolet transform. J Inf Secur Appl 52:102481. https://doi.org/10.1016/j.jisa.2020.102481
10. Elaskily MA, Elnemr HA, Sedik A et al (2020) A novel deep learning framework for copy-move forgery detection in images. Multimedia Tools Appl 79:19167–19192. https://doi.org/10.1007/s11042-020-08751-7
11. Ghoneim A, Muhammad G, Amin SU, Gupta B (2018) Medical image forgery detection for smart healthcare. IEEE Commun Mag 56(4):33–37. https://doi.org/10.1109/MCOM.2018.1700817
12. Hsu C-C, Zhuang Y-X, Lee C-Y (2020) Deep fake image detection based on pairwise learning. Appl Sci 10:370. https://doi.org/10.3390/app10010370
13. Zampoglou M, Papadopoulos S, Kompatsiaris Y, Bouwmeester R, Spangenberg J (2016) Web and social media image forensics for news professionals. SMN@ICWSM
14. Rahman MM, Tajrin J, Hasnat A, Uzzaman N, Atiqur Rahaman GM (2019) Novel social media image forgery detection database. In: 22nd international conference on computer and information technology (ICCIT), pp 1–6. 1109/ICCIT48885.2019.9038557
15. Sadeghi S, Dadkhah S, Jalab HA et al (2018) State of the art in passive digital image forgery detection: copy-move image forgery. Pattern Anal Appl 21:291–306. https://doi.org/10.1007/s100440170678-8

16. Buchana P, Cazan I, Diaz-Granados M, Juefei-Xu F, Savvides M (2016) Simultaneous forgery identification and localization in paintings using advanced correlation filters. In: IEEE international conference on image processing (ICIP), pp 146–150. https://doi.org/10.1109/ICIP.2016. 7532336

17. Thakur A, Neeru J (2018) Machine learning based saliency algorithm for image forgery classification and localization. In: First international conference on secure cyber computing and communication (ICSCCC), pp 451–456. https://doi.org/10.1109/ICSCCC.2018.8703287

18. Gardella M, Musé P, Morel J-M, Colom M (2021) Forgery detection in digital images by multi-scale noise estimation. J Imaging 7:119. https://doi.org/10.3390/jimaging7070119

19. Duan S, Shujian Y, Principe JC (2022) Modularizing deep learning via pairwise learning with kernels. IEEE Trans Neural Netw Learn Syst 33(4):1441–1451. https://doi.org/10.1109/TNNLS.2020.3042346

20. Polatkan G, Jafarpour S, Brasoveanu A, Hughes S, Daubechies I (2020) Detection of forgery in paintings using supervised learning. In: 16th IEEE international conference on image processing (ICIP), 2009. Cosine Transform, KSII Transaction Internet Information System, vol 14, no7, pp 2981–2996. https://doi.org/10.3837/tiis.2020.07.014

21. Bappy JH, Simons C, Lakshmanan BS, Manjunath AK, Chowdhury R (2019) Hybrid LSTM and encoder–decoder architecture for detection of image forgeries. IEEE Trans Image Process 28(7):3286–3300. https://doi.org/10.1109/TIP.2019.2895466

22. Shen X, Shen H, Chen L (2016) Splicing, image forgery detection using textural features based on the grey level co-occurrence matrices. IET Image Process 11:44–53. https://doi.org/10.1049/iet-ipr.2016.0238

23. Li C, Ma Q, Xiao L, Zhang A (2017) Image splicing detection based on Markov in QDCT domain. Neurocomputing 228:29–36. https://doi.org/10.1016/j.neucom.2016.04.068

24. Wang J, Liu R, Wang H, Wu B, Shi YQ (2020) Quaternion Markov: splicing detection for color images based on quaternion discrete. Cosine Transf KSII Trans Internet Inf Syst 14(7):2981–2996. https://doi.org/10.3837/tiis.2020.07.014

25. Jaiswal AK, Srivastava R (2020) A technique for image splicing detection using hybrid feature set. Multimedia Tools Appl 79(17–18):11837–11860. https://doi.org/10.1007/s11042-019-08480-6

26. Al-Hammadi M, Ghulam M, Muhammad H, George B (2013) Curvelet transform and local texture based image forgery detection. 8034:503–512. https://doi.org/10.1007/978-3-642-41939-3_49

27. N.K. Rathore, N.K. Jain, P.K. Shukla, U.S. Rawat, R. Dubey.: Image forgery detection using singular value decomposition with some attacks", Natl.Acad. Sci. Lett. (2020) http://dx.doi.org/https://doi.org/10.1007/s40009-020-00998-w.

28. He Z, Lu W, Sun W, Huang J (2012) Digital image splicing detection based on Markov features in DCT and DWT domain. Pattern Recogn 45(12):4292–4299. ISSN 0031–3203, https://doi.org/10.1016/j.patcog.2012.05.014

29. Kakar P, Sudha N, Ser W (2011) Exposing digital image forgeries in motion blur. IEEE Trans Multimedia 13(3):443–452. https://doi.org/10.1109/TMM.2011.2121056

30. Ouyang J, Liu Y, Liao M (2019) Robust copy-move forgery detection method using pyramid model and Zernike moments. Multimed Tools Appl 78:10207–10225. https://doi.org/10.1007/s11042-018-6605-1

31. Lai Y, Huang T, Lin J et al (2018) An improved block-based matching algorithm of copy-move forgery detection. Multimedia Tools Appl 77:15093. https://doi.org/10.1007/s11042-017-5094-y

32. Saleh SQ, Hussain M, Muhammad G, Bebis G (2013) Evaluation of image forgery detection using multi-scale weber local descriptors. In: Bebis G, et al (eds) Advances in visual computing. ISVC lecture notes in computer science, vol 8034. Springer, Berlin, Heidelberg. https://doi.org/10.1007/978-3-642-41939-3_40

33. Vaishnavi D, Subashini TS (2019) Application of local invariant symmetry features to detect and localize image copy move forgeries. J Inf Secur Appl 44:23–31. https://doi.org/10.1016/j.jisa.2018.11.001

34. Jawadul B, Cody S, Lakshmanan BS, Manjunath Amit K, Chowdhury R (2019) Hybrid LSTM and encoder–decoder architecture for detection of image forgeries. IEEE Trans Image Process 28(7):3286–3300. https://doi.org/10.1109/TIP.2019.2895466
35. Gu J, Wang Z, Kuen J, Ma L, Shahroudy A, Shuai B, Liu T, Wang X, Wang G, Cai J, Chen T (2016) Recent advances in convolutional neural networks. Pattern Recogn 77:354–377
36. Barad ZJ, Goswami MM (2020) Image forgery detection using deep learning: a survey. In: 6th international conference on advanced computing and communication systems (ICACCS). https://doi.org/10.1109/ICACCS48705.2020.9074408
37. Soni B, Das PK, Thounaojam DM (2018) CMFD: a detailed review of block based and key feature based techniques in image copy-move forgery detection. IET Image Process 12(2):167–178. https://doi.org/10.1049/iet-ipr.2017.0441
38. Tralic D, Zupancic I, Grgic S, Grgic M (2013) CoMoFoD — New database for copy-move forgery detection. In: Proceedings ELMAR, pp 49–54
39. Srikanth T (2019) Transfer learning using VGG-16 with deep convolutional neural network for classifying images. Int J Sci Res Publ (IJSRP) 9(10):143–150. https://doi.org/10.29322/IJSRP.9.10.2019.p9420
40. Tan M, Le QV (2018) Efficientnet: rethinking model scaling for convolutional neural networks. arXiv:1905.11946

Modified Bat Algorithm for Balancing Load of Optimal Virtual Machines in Cloud Computing Environment

Gaurav Raj, Shabnam Sharma, and Aditya Prakash

1 Introduction

Optimization is the process of obtaining the best possible solution for any particular problem while satisfying underlying constraints. While solving any problem and obtaining its solution, the optimization of the solution is considered as one of the major concerns. This concern becomes more important when the solution to any combinatorial problem is to be obtained. To obtain the optimal result for combinatorial optimization problems, the finite sets of results are selected among all feasible results, which satisfy certain constraints. Generally, the problems are related to finding the optimal path in vehicular routing, for the job and task scheduling, like preparing the time-table for trains in railways, solving knapsack problems, or any other engineering problem. Various approaches have been adopted so far to provide the solution to these problems, which include Iterative Improvement, Simulated Annealing, Evolutionary Computation, Variable Neighborhood Search, Iterative Local Search, Meta-Heuristic Techniques and Heuristic Techniques. Currently, there are a variety of optimization techniques available to solve a different kind of problems. Bat Algorithm is a meta-heuristic technique, used for optimization and has gained popularity in past years. It has proven to be beneficial and result oriented in different fields. Bat Algorithm was formulated in 2010 and worked based on echolocation. In BA, artificial bats emit a pulse in different directions and wait for the echo to receive. Once the echo of all emitted pulses is received, artificial bats compare the computed fitness value. The solution, which has minimum/maximum fitness value,

G. Raj (✉)
Sharda University, Noida, India
e-mail: gaurav.raj@sharda.ac.in

S. Sharma
CMR University, Bangalore, India

A. Prakash
iNurture Education Solutions Pvt. Ltd., Bangalore, India

is considered as the best solution. Cloud Computing gained popularity due to its pay per use facility and enabled users to access different services. Cloud consists of virtual machines, which enables cloud users to use cloud services. Whenever a user requests for the execution of any task/job, it is assigned to a virtual machine. The decision regarding which virtual machine will process the request, depends upon the load balancer. The load balancing algorithm's task is to select the most appropriate virtual machine for the execution of the task. Different optimization techniques have been used so far for efficient load balancing. Bat Algorithm has proved its applicability in this area as well. In this research work, the nature-inspired bat algorithm is used to maintain the balance distribution among available virtual machines in 'cloud'. To this end, an improved/modified version of the bat algorithm is deployed. This modified version of bat algorithm works by considering distance as a fitness value.

This work has been divided into various sections. The second section highlights existing studies conducted by numerous researchers in the field of Cloud Computing and Bat Algorithm. The third section describes the Modified Bat Algorithm for load balancing in Cloud Computing. The fourth section focuses on conducting the comparative analysis of the proposed algorithm, followed by the conclusion and future work in the last section of this research work.

2 Related Work

Nuaimi et al. [1] carried out a review and analysis of existing load balancing techniques and the associated challenges. The author described the pros and cons of numerous load-balancing algorithms. Moreover, the author also categorized these algorithms based on various factors, like replication, speed, heterogeneity, SPOF, network overhead, spatially distributed, implementation complexity, and fault tolerance. Chang et al. [2] proposed a resource scheduling algorithm, that possesses dynamic load balance qualities. Khan et al. [3] mentioned different cloud computing research areas, which need researchers' immediate attention to enhance the productivity of cloud. Afzal et al. [4] classified load unbalancing as multi variant and multi constraint problem of cloud computing. Hsieh and Chiang Hsieh [5] proposed a three-phase dynamic data replication algorithm, which was validated on Hadoop, based cloud environment. Jyoti et al. [6] reviewed existing load balancing and service brokering techniques of cloud computing environment.

Panda and Jana [7] proposed a probability-based load balancing algorithm. Dey and Gunasekhar [8] presented comprehensive study of virtual machine migration techniques, along with their challenges and advantages. Pan et al. [9] developed an online load balancing algorithm to improve overall user experience and system performance.

Mathur et al. [10] proposed layered architecture to balance the load and resource allocation in the cloud computing environment. Chawla and Ghumman [11] proposed a novel package-based load balancing algorithm in the cloud computing environment. Ghomi et al. [12] categorized load-balancing techniques broadly into six categories. These categories were Agent-based, Hadoop Map Reduce, workflow specific, General, application-oriented, Natural Phenomena-based, and network-aware. Raj et al. [13] proposed an improvised bat algorithm, and it worked on the concept of min-min, max–min, and alpha–beta pruning algorithm, for task scheduling in cloud computing. Sharma et al. [14] reviewed the existing work done by numerous researchers who had developed different types of nature-inspired algorithms. Ghose et al. [15] elaborated pursuit strategies adopted by bats while capturing preys. Chen et al. [16] proposed a novel variant of bat algorithm, i.e. guidable bat algorithm which integrated doppler effect. Mirjalili et al. [17] developed a binary bat algorithm, as a standard bat algorithm only optimizes continuous problems.

Li and Zhou [18] proposed a variant of bat algorithm, i.e., complex valued bat algorithm. Yang [19] developed a meta-heuristic approach for optimizing solutions, which was inspired by bats and named as bat algorithm. Kurdi [20] proposed a locust inspired scheduling algorithm for the cloud computing environment.

Khoda et al. [21] proposed an intelligent computational offloading system for 5G mobile devices. Sayantani et al. [22] used a bio-inspired cognitive model to schedule the tasks for IoT applications in a heterogeneous cloud computing environment. Asma et al. [23] proposed a mobility aware optimal resource allocation architecture, namely, Mobi-Het for big data task execution in a mobile cloud computing environment. Shirjini et al. [24] analyzed the stability and convergence capability of the bat algorithm. Lin et al. [25] proposed two algorithms, namely, the time balancing algorithm and main resource load balancing algorithm. Gopinath and Vasudevan [26] analyzed existing load balancing techniques of the cloud computing environment. Jodayree et al. [27] proposed a rule-based workload dynamic balancing algorithm based on predictions of incoming jobs and named Cicada. Kumar et al. [28] proposed a dynamic load balancing algorithm for cloud computing environment. Kumar et al. [29] proposed a cloud architecture which worked on the principle of elasticity.

This work aims to apply the biological characteristics of bats to enhance the functionality of Bat Algorithm. This research paper, therefore, explores the flight behavior of bats. Bats adopt either the Constant Bearing strategy or the Constant Absolute Target Detection strategy. In this work, the main focus has been to balance the load among virtual machines of the Cloud Computing environment using CATD strategy. To this end, a new variant of the Bat Algorithm, which is inspired by CATD strategy of bats, is proposed and applied to solve load balancing. A new technique of determining fitness value has been deployed here, i.e., fitness value is computed considering 'range between target and bat' as a parameter. A detailed explanation of the proposed algorithm is given in the subsequent sections.

3 Modified Bat Algorithm for Load Balancing in Cloud Computing

This section describes the proposed algorithm for selecting the optimal virtual machine for the execution of tasks in the Cloud Computing environment while balancing the load. In order to find the best solution, the distance between artificial bats and solutions is calculated in 'magical way'. The advancements introduced in standard bat algorithm, revolve around either parameter initialization or parameter update or hybridization with other techniques or mapping continuous space problems to binary space problems. In these advancements to BA, very few authors have adopted different strategies to compute distance. In this research paper, a different way of computing distance has been suggested, which is equivalent to the strategy adopted by real bats while targeting their target (prey). Further, the modified bat algorithm has been applied for balancing the load in 'cloud' by using CATD inspired Bat Algorithm. The steps for balancing the load in Cloud Computing using Modified Bat Algorithm are described in pseudo-code.

Data: Input number of bats, N and Number of Virtual Machines, V.
Set min_freq, max_freq, velocity, pulse emission rate, loudness and position for entire bat population.
Result: Selection of best suited virtual machine
Begin
For i = 1 to V
Compute the fitness value of every available virtual machine V.
1. *Deploy 'N' number of bats, where each bat is responsible for computing the fit ness value of V which is present in search space.*
2. *Every bat will emit pulse and received echo for the computation of distance.*
3. *Detect the presence of any obstacle.*
   If present, delay, β, and attenuation, α, will affect the solution, as per following equation.
   *Echoi= (Pulse_Emittedi * rand1) +rand2 otherwise*
   Echoi= Pulse_Emittedi+rand2
4. *Compute the similarity among sound produced Pulse_Emittedi and Echoi using mathematical function, cross correlation and compute delay samples.*
   [Correlation] xcorr (Pulse_Emittedi, Echoi)
   DelaySample= Lags(find(Correlation==maximum(Correlation)))
5. *Distance can be computed, using DelaySample and TimeSample.*
6. *Select solution as a best which is hving minimum fitness value. end for*
   Select first virtual machine as the best local solution, having minimum fitness value. for i=1 to V
   if VMi == visited VM(i, :)
   Increment variable and check for other VM's assigned load. end if
   To balance the load, the task is to be assigned to under loaded virtual machine, even in the presence of virtual machine which is having lesser fitness value than selected virtual machine
   if count(i,:)>threshold
   Compute the fitness value, again and select the optimal virtual machine. end if
End

4   Comparative Analysis of Proposed Algorithm

In order to evaluate the performance of modified bat algorithm while selecting best suited virtual machine (optimal virtual machine), results are contrasted with the results of standard bat algorithm, when applied for balancing the load in cloud computing environment.

A modified bat algorithm is applied to balance the load among the available virtual machines. During this evaluation, it is noticed that the optimal virtual machine may get overloaded, while other virtual machines are under loaded. This variant of modified bat algorithm is named as Overloaded Optimal Virtual Machine (OOVM). In order to maintain the balance between overloaded and underloaded virtual machines, an advancement is introduced in modified bat algorithm, where the selected optimal

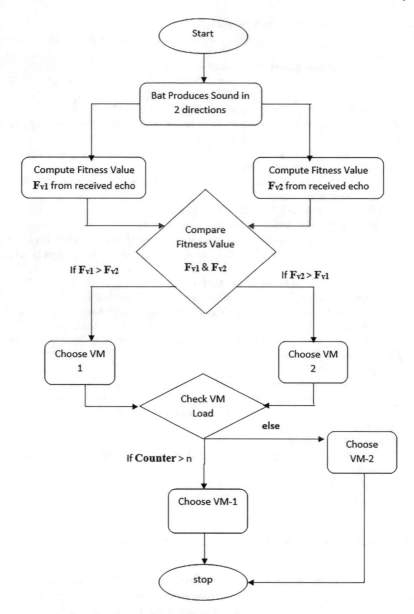

Fig. 1 Flowchart of modified bat algorithm

virtual machine is not assigned any task for evaluation, if its existing task count exceeds the threshold value. This variant of modified bat algorithm is named as Balanced Virtual Machine (BVM). The results of modified bat algorithm for an over-loaded optimal virtual machine, on the basis of execution time and cost are shown in Tables 1 and 2 respectively. The results of modified bat algorithm for a balanced

Table 1 Performance evaluation of MBA-OOVM on the basis of execution time

VM	Parameters	Standard bat algorithm			Modified bat algorithm (OOVM)		
		Bat population					
		10	15	20	10	15	20
10	Best	4.030	4.503	4.823	3.976	4.250	4.633
	Median	4.070	4.379	4.832	4.058	4.337	4.728
	Worst	4.223	4.982	4.901	4.123	4.407	4.804
	Mean	4.079	4.632	4.862	4.057	4.337	4.727
	SD	0.040	0.043	0.047	0.039	0.037	0.038
15	Best	4.151	4.343	4.386	4.187	4.552	4.684
	Median	4.241	4.432	4.993	4.273	4.645	4.780
	Worst	4.349	4.423	4.999	4.342	4.720	4.857
	Mean	4.213	4.446	4.343	4.272	4.644	4.779
	SD	0.041	0.040	0.048	0.041	0.046	0.046
20	Best	4.233	4.777	4.983	4.234	4.889	5.142
	Median	4.124	4.982	4.783	4.321	4.989	5.248
	Worst	4.381	5.012	5.012	4.391	5.070	5.333
	Mean	4.230	4.992	4.784	4.320	4.988	5.247
	SD	0.042	0.049	0.049	0.042	0.046	0.047

virtual machine on the basis of execution time and cost are shown in Tables 3 and 4 respectively.

Considering 20 VM's with the deployment of 10 bats, there will be a trade-off between SD and cost. SD will increase and the cost will decrease. If the number of bats is increased from 10 to 15 for 20 VM's, then the SD will decrease and cost will increase. For 20 bats, SD and cost both increases. So, for 20 VM's, 15 bats are suitable to obtain optimal results. While evaluating the results of modified bat algorithm-overloaded optimal virtual machine variant, it was observed that it had lesser cost and less variation in the values of SD in comparison to the results obtained using standard bat algorithm for solving the same problem, if 10 bats are used for 10 virtual machines. In the case of 15 virtual machines, one should prefer the deployment of 15 bats, as it aims at lesser SD values at a lesser cost. For 20 VM's, cost increases as the number of bats increases. So, an optimal solution can be obtained by deploying 15 bats. As depicted in Table 2, the cost of selecting optimal virtual machine is lesser than the standard bat algorithm.

While evaluating the performance on the basis of execution time, it is observed that the variation among the optimal results obtained reduced and the difference between the execution time of standard bat algorithm and modified bat algorithm-OOVM version became almost negligible up to three decimal points. It is evident from the results that the Modified Bat Algorithm-OOVM produced more optimal results in comparison to the standard bat algorithm, while considering cost as the

Table 2 Performance evaluation of MBA-OOVM on the basis of cost

VM	Parameters	Standard bat algorithm			Modified bat algorithm (OOVM)		
		Bat Population					
		10	15	20	10	15	20
10	Best	155.263	165.977	180.914	152.263	162.977	177.914
	Median	156.746	167.111	182.150	153.746	164.111	179.150
	Worst	158.147	168.097	183.225	155.147	165.097	180.225
	Mean	156.757	167.093	182.131	153.757	164.093	179.131
	SD	0.100	0.082	0.070	0.092	0.089	0.077
15	Best	163.492	177.761	182.905	160.502	174.771	179.915
	Median	165.053	178.975	184.154	162.063	175.985	181.164
	Worst	166.528	180.031	185.241	163.538	177.041	182.251
	Mean	165.065	178.956	184.135	162.075	175.966	181.145
	SD	1.048	0.068	0.070	0.099	0.067	0.070
20	Best	165.356	190.915	200.815	161.516	187.075	196.975
	Median	166.934	192.220	202.187	163.094	188.380	198.347
	Worst	168.426	193.354	203.380	164.586	189.514	199.540
	Mean	166.946	192.199	202.166	163.106	188.359	198.326
	SD	1.060	0.074	0.077	1.060	0.071	0.067

Table 3 Performance evaluation of MBA-BVM on the basis of execution time

VM	Parameters	Standard bat algorithm			Modified bat algorithm (BVM)		
		Bat population					
		10	15	20	10	15	20
10	Best	4.030	4.503	4.823	4.045	4.351	4.723
	Median	4.070	4.379	4.832	4.078	4.391	4.794
	Worst	4.223	4.982	4.901	4.732	4. 412	4.801
	Mean	4.079	4.632	4.862	4.067	4.311	4.872
	SD	0.040	0.043	0.047	0.038	0.038	0.037
15	Best	4.151	4.343	4.386	4.173	4.442	4.785
	Median	4.241	4.432	4.993	4.363	4.691	4.673
	Worst	4.349	4.423	4.999	4.413	4.812	4.866
	Mean	4.213	4.446	4.343	4.176	4.773	4.671
	SD	0.041	0.040	0.048	0.043	0.043	0.043
20	Best	4.233	4.777	4.983	4.332	4.777	4.992
	Median	4.124	4.982	4.783	4.213	4.671	4.675
	Worst	4.381	5.012	5.012	4.399	5.016	5.031
	Mean	4.230	4.992	4.784	4.312	4.773	4.673
	SD	0.042	0.049	0.049	0.048	0.046	0.047

Table 4 Performance evaluation of MBA-BVM on the basis of cost

VM	Parameters	Standard bat algorithm			Modified bat algorithm (BVM)		
		Bat population					
		10	15	20	10	15	20
10	Best	155.263	165.977	180.914	149.293	159.157	174.944
	Median	156.746	167.111	182.150	150.776	160.291	176.180
	Worst	158.147	168.097	183.225	152.177	161.277	177.255
	Mean	156.757	167.093	182.131	150.787	160.273	176.161
	SD	0.100	0.082	0.070	0.954	0.815	0.647
15	Best	163.492	177.761	182.905	156.672	170.941	177.495
	Median	165.053	178.975	184.154	158.233	172.155	178.744
	Worst	166.528	180.031	185.241	159.708	173.211	179.831
	Mean	165.065	178.956	184.135	158.245	172.136	178.725
	SD	1.048	0.068	0.070	1.011	0.585	0.564

factor. The results which are computed on the basis of best, worst, median, mean and standard deviation values by varying the number of bats present in bat population and number of virtual machines, are represented in graphs below. The results are shown on the basis of execution time in Figs. 2, 3 and 4. Figure 2 represents the varying values of performance evaluation parameters for 10 virtual machines and varying bat population from 10, 15 and 20. Similarly, Figs. 3 and 4 represent the results for 15 and 20 virtual machines, for varying bat population, respectively. Further, with the inclusion of threshold value to limit the over utilization of the optimal virtual machine, BVM version of Modified Bat Algorithm has been introduced. The results obtained using a modified bat algorithm for balanced virtual machine are better in comparison to the results obtained using standard bat algorithm, as depicted in Table 4.

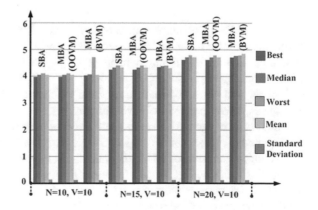

Fig. 2 Comparison result graph on the basis of execution time of SBA, MBA-OOVM and MBA-BVM for V = 10 and N varying between [10, 15, 20]

Fig. 3 Comparison result
graph on the basis of
execution time of SBA,
MBA-OOVM and MBA-
BVM for V = 15 and N
varying between [10, 15, 20]

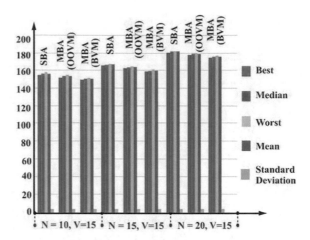

Fig. 4 Comparison result
graph on the basis of
execution time of SBA,
MBA-OOVM and MBA-
BVM for V = 20 and N
varying between [10, 15, 20]

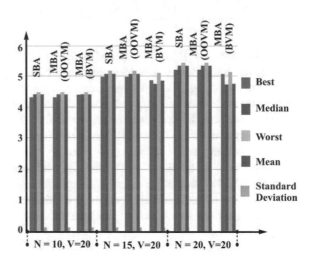

For 10 VM's, if the number of bats deployed is increased, the time to obtain the optimal result also increases, but the standard deviation among the feasible solution decreases. For 15 VM's, deployment of 15 bats will be suitable while applying standard bat algorithm and modified bat algorithm-BVM. For 20 VM's, deployment of 15 bats will obtain optimal results and balance the load among all the virtual machines present in the cloud environment.

The results evaluated on the basis of the cost incurred during the entire process, are depicted in Figs. 5, 6 and 7. Figure 5 represents the varying values of performance evaluation parameters for 10 virtual machines and varying bat population from 10, 15 and 20. Similarly, Figs. 6 and 7 represent the results for 15 and 20 virtual machines for varying bat population, respectively. Fig. 5 shows the graphical representation of the comparative results of the cost required to select the optimal virtual machine for the execution of jobs/tasks by appointing 10 bats for 10 virtual machines. Results prove that deployment of more number of bats will not improve the results and selection of an optimal virtual machine can be done by deploying only 10 bats. Figure 6 depicts that for 15 virtual machines, deployment of 15 bats will serve the purpose. If we deploy more number of bats, it will not improve the performance, but will lead to an increase in cost. Figure 7 depicts that for 20 virtual machines, 15 bats are sufficient to select the optimal virtual machine. In earlier related literature works, main focus was to ensure the optimal utilization of virtual machines and balancing the load among those machines, in Cloud Computing environment. Consequently, number of jobs submitted to 'best/optimal' virtual machines, may overburden them. In order to avoid such situation and to ensure that the jobs are assigned to all virtual machines in a balanced way, two new variants of Bat Algorithm are designed. It is clearly evident from the results that the best mean value obtained using Modified Bat Algorithm-BVM had outperformed Modified Bat Algorithm-OOVM and Standard Bat Algorithm. Except in the case of 15 bats and 15 virtual machines, the results of Standard Bat Algorithm and Modified Bat Algorithm-BVM are quite similar.

While evaluating the performance on the basis of execution time, it is observed that the variation among the optimal results obtained reduced and the difference between the execution time of standard bat algorithm and modified bat algorithm-OOVM version became almost negligible up to three decimal points. It is evident from the results that the Modified Bat Algorithm-OOVM produced more optimal results in comparison to the standard bat algorithm, while considering cost as the factor. For 10 VM's, if the number of bats deployed is increased, the time to obtain the optimal result also increases, but the SD among the feasible solution decreases. For 15 VM's, deployment of 15 bats will be suitable while applying standard bat algorithm and modified bat algorithm-BVM. For 20 VM's, deployment of 15 bats will obtain optimal results and balance the load among all the virtual machines present in the cloud environment. Results prove that deployment of a greater number of bats will not improve the results and selection of an optimal virtual machine can be done by deploying only 10 bats. In earlier related literature works, main focus was

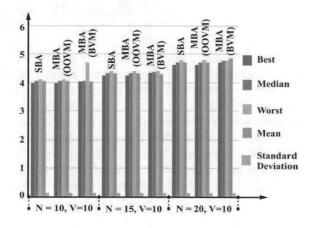

Fig. 5 Comparison result graph on the basis of cost of SBA, MBA-OOVM and MBA-BVM for V = 10 and N varying between [10, 15, 20]

Fig. 6 Comparison result graph on the basis of cost of SBA, MBA-OOVM and MBA-BVM for V = 15 and N varying between [10, 15, 20]

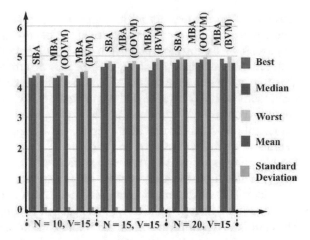

Fig. 7 Comparison result graph on the basis of cost of SBA, MBA-OOVM and MBA-BVM for V = 20 and N varying between [10, 15, 20]

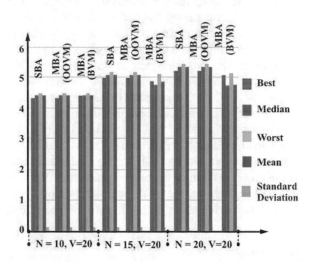

to ensure the optimal utilization of virtual machines and balancing the load among those machines, in Cloud Computing environment. Consequently, number of jobs submitted to 'best/optimal' virtual machines, may overburden them. In order to avoid such situation and to ensure that the jobs are assigned to all virtual machines in a balanced way, two new variants of Bat Algorithm are designed. It is clearly evident from the results that the best mean value obtained using Modified Bat Algorithm-BVM had outperformed Modified Bat Algorithm-OOVM and Standard Bat Algorithm. Except in the case of 15 bats and 15 virtual machines, the results of Standard Bat Algorithm and Modified Bat Algorithm-BVM are quite similar.

5 Conclusion and Future Work

A novel variant of bat algorithm, which is inspired by bat's flight behavior, has been designed for solving combinatorial problems. The different flight behavior adopted by Microchiroptera bats has been studied and modelled mathematically. The motive of this research is to provide an efficient technique for balancing the load of virtual machines available on Cloud and ensuring that no optimal virtual machine should be overloaded. In order to ensure the same, standard bat algorithm is modified by incorporating the strategy that real bats use while estimating the distance between itself and its prey and while capturing the target. It improved the performance of the algorithm and the modified bat algorithm has been applied for solving the load balancing problem of cloud computing. The results computed have proven the applicability of a modified bat algorithm to balance the load on 'cloud' and generated more optimal results. Further, to improve the performance of the algorithm and increase its applicability to other fields, the standard bat algorithm can be hybridized with other newly developed meta-heuristic techniques. Real Bats have the capability to jam the pulse emitted by other bats or receive the echo and target the prey of other real bats. This astonishing feature also motivates to develop another bat algorithm variant. Moreover, real bats adopt different pursuit strategies depending on the movement of prey's, depending on the prey to capture and many additional factors. One can research on these areas to propose a new bat algorithm with improved performance. There can also be potential research to develop other variants of standard bat algorithm or to improve the performance of the proposed algorithm, other biological features of the bat can also be explored.

References

1. Nuaimi KA, Mohamed N, Nuaimi MA, Al-Jaroodi J (2012) A survey of load balancing in cloud computing: challenges and algorithms. In: Second symposium on network cloud computing and application, pp 137–142
2. Chang H, Tang X (2011) A load-balance based resource-scheduling algorithm under cloud computing environment. In: New horizons in web-based learning - ICWL 2010 workshops, vol 6537, pp 85–90
3. Khan RZ, Ahmad MO (2016) Load balancing challenges in cloud computing: a survey. In: Proceedings of the international conference on signal, networks, computing, and systems, vol 396, pp 25–32
4. Afzal S, Kavitha G (2019) Load balancing in cloud computing–a hierarchical taxonomical classification. J Cloud Comput 8(1):22
5. Hsieh HC, Chiang ML (2019) The incremental load balance cloud algorithm by using dynamic data deployment. J Grid Comput 17(3):553–575
6. Jyoti Amrita, Shrimali Manish, Tiwari Shailesh, Singh Harivans Pratap (2020) Cloud computing using load balancing and service broker policy for IT service: a taxonomy and survey. J Ambient Intell Hum Comput 11(11):4785–4814
7. Panda SK, Jana PK (2019) Load balanced task scheduling for cloud computing: A probabilistic approach. Knowl Inf Syst 61(3):1607–1631

8. Dey NS, Gunasekhar T (2019) A comprehensive survey of load balancing strategies using hadoop queue scheduling and virtual machine migration. IEEE Access 7:92259–92284
9. Pan J, Ren P, Tang L (2015) Research on heuristic based load balancing algorithms in cloud computing. Intell Data Anal Appl 370:417–426
10. Mathur H, Tazi SN, Bayal RK (2016) Cloud load balancing and resource allocation. In: Proceedings of the second international conference on computer and communication technology, vol 380, pp 745–753
11. Chawla A, Ghumman NS (2018) Package-based approach for load balancing in cloud computing. Big Data Anal 654:71–77
12. Ghomi E, Rahmani A, Qader N (2017) Load-balancing algorithms in cloud computing: a survey. J Netw Comput Appl 88:50–71
13. Raj B, Ranjan P, Rizvi N, Pranav P, Paul S (2018) Improvised bat algorithm for load balancing-based task scheduling. In: Progress in Intelligent Computer Technology: Theory, Practice, and Applications, vol 518, pp 521–530
14. Sharma S, Luhach AK, Jyoti K (2018) Research & analysis of advancements in BAT algorithm. In: IEEE 3rd intrnational conference on computer for sustainable global development, INDIACom, pp 2391–2396
15. Ghose K, Horiuchi TK, Krishnaprasad PS, Moss CF (2006) Echolocating bats use a nearly time-optimal strategy to intercept prey. PLoS Biol 4(5)
16. Chen YT, Shieh CS, Horng MF, Liao BY, Pan JS, Tsai MT (2014) A guidable bat algorithm based on doppler effect to improve solving efficiency for optimization problems. Comp Collective Intell Tech Appl 8733:373–383
17. Mirjalili S, Mirjalili SM, Yang X-S (2014) Binary bat algorithm. Neural Comp. and Appl. 25(3):663–681
18. Li L, Zhou Y (2014) A novel complex-valued bat algorithm. Neural Comput Appl 25(6):1369–1381
19. Yang XS (2010) A new metaheuristic bat-inspired algorithm. Nat Insp Cooper Strat Optim 284:65–74
20. Kurdi HA, Alismail SM, Hassan MM (2018) LACE: a locust-inspired scheduling algorithm to reduce energy consumption in cloud datacenters. IEEE Access 6:35435–35448
21. Khoda ME, Razzaque MA, Almogren A, Hassan MM, Alamri A, Alelaiwi A (2016) Efficient computation offloading decision in mobile cloud computing over 5G network. Mob Netw Appl 21(5):777–792
22. Sayantani B, Karuppiah M, Selvakumar K, Li KC, Islam SKH, Hassan MM, Bhuiyan MZA (2018) An intelligent/cognitive model of task scheduling for IoT applications in cloud computing environment. Fut Gener Comput Syst 88:254–261
23. Asma E, Razzaque MA, Hassan MM, Alamri A, Fortino G (2018) A mobility-aware optimal resource allocation architecture for big data task execution on mobile cloud in smart cities. IEEE Commun Mag 56(2):110–117
24. Shirjini MF, Nikanjam A, Shoorehdeli MA (2020) Stability analysis of the particle dynamics in bat algorithm: standard and modified versions. Eng Comput 37:1–12
25. Lin W, Peng G, Bian X, Xu S, Chang V, Li Y (2019) Scheduling algorithms for heterogeneous cloud environment: main resource load balancing algorithm and time balancing algorithm. J Grid Comput 17(4):699–726
26. Gopinath PG, Vasudevan SK (2015) An in-depth analysis and study of Load balancing techniques in the cloud computing environment. Procedia Comput Sci 50:427–432
27. Jodayree M, Abaza M, Tan Q (2019) A predictive workload balancing algorithm in cloud services. Procedia Comput Sci 159:902–912
28. Kumar M, Sharma SC (2017) Dynamic load balancing algorithm for balancing the workload among virtual machine in cloud computing. Procedia Comput Sci 115:322–329
29. Kumar M, Dubey K, Sharma SC (2018) Elastic and flexible deadline constraint load balancing algorithm for cloud computing. Procedia Comput Sci 125:717–724

Forecasting Floods Using Classification Based Machine Learning Models

Vikas Mittal, T. V. Vijay Kumar, and Aayush Goel

1 Introduction

India being the seventh largest country in the world consists of regions having drastically varying geographic and climatic conditions [12]. Therefore, various natural hazards like earthquakes, tsunamis, cyclones and floods pose recurrent threats to the second largest populated country in the world [22]. Amongst these natural hazards, floods and tsunamis have caused maximum damage [36]. Frequent floods in the second largest flood affected nation in the world have also resulted in malnutrition and stunted growth among children [24, 35, 36]. From 1953 to 2016, the Central Water Commission, reported the deaths of more than one lakh people, and an economic loss of more than 347,000 crore rupees, due to floods in the country [6]. Every year, floods in India severely threaten its people, infrastructure and its economy. Therefore, to overcome the challenges posed by floods, efficient flood mitigation strategies are required. Such strategies primarily focus on structural and non-structural measures [23]. The structural measures focus on the construction of hazard resistant buildings, dikes, dams etc. on the river. Whereas, non-structural measures design early warning systems that focus on forecasting future floods and use communication systems to disseminate the flood related warnings to the first responders. Early forecasting of floods provides more time to first responders and helps them in better planning and preparation. Machine learning (*ML*), a branch of artificial intelligence, can be used to design flood forecasting models using historical data. In this paper, such ML based flood forecasting models have been designed. Historical data with longer lead times, monthly average of precipitation and temperature for twelve flood affected districts of Northern Bihar for the period (1991–2002) obtained from the Climatic Research

V. Mittal (✉) · T. V. Vijay Kumar
School of Computer and Systems Sciences, Jawaharlal Nehru University, New Delhi, India
e-mail: vikas.mittal.10@gmail.com

A. Goel
Bharati Vidyapeeth's College of Engineering, New Delhi, India

© The Author(s), under exclusive license to Springer Nature Singapore Pte Ltd. 2022   489
B. Unhelker et al. (eds.), *Applications of Artificial Intelligence and Machine Learning*,
Lecture Notes in Electrical Engineering 925,
https://doi.org/10.1007/978-981-19-4831-2_40

Unit, University of East Anglia [1, 18] has been used. *ML* techniques viz. Artificial Neural Network (*ANN*), k-Nearest Neighbor (*KNN*), Logistic Regression (*LR*), Naive Bayes (*NB*), Random Forest (*RF*) and Support Vector Machine (*SVM*) have been applied on this data for forecasting floods.

This paper is organized as follows: An overview of the related work is presented in Sect. 2. *ML* based flood forecasting model using six *ML* techniques is explained in Sect. 3. In Sect. 4, experimental results and performance of the *ML* techniques is analyzed. Section 5 is the conclusion.

2 Related Work

Flood forecasting (*FF*) models can be broadly classified into three major categories: physical models, conceptual models and data-driven models [7, 13, 31]. Physical models are based on the principle of physics and hydrology. These models use water equations characterizing physical properties of the catchment area. Water equations use various physical parameters like coefficient of channel roughness, river geometry etc. to predict water levels of the river [25]. Conceptual models for forecasting floods use large amounts of hydrological and meteorological data to calibrate the model parameters. Interpretation of model parameters used by the physical and conceptual models require domain expertise and is a complex process [25]. Data driven flood forecasting models apply *ML* techniques for forecasting floods using historical data related to floods. These models are less complex and do not require domain expertise for interpretation of the physical parameters. Therefore, *ML* based data driven models are widely used for forecasting floods [9, 17, 19, 21, 25, 28, 33, 34].

In [9], an *ANN* model using back propagation was proposed for predicting hourly runoff for Govindpur basin on Brahmani river in Odisha, India. Nodes in this *ANN* model were arranged in three layers and their weights were optimized using the back propagation technique. In [33], a feed forward neural network (*FFANN*) model consisting of three layers was proposed. This *FFANN* model forecast floods with a lead time of upto 3 h in the Bhasta river region of Maharashtra, India. This model was trained using the error backpropagation, conjugate gradient and cascade correlation algorithm on the storm hydrograph data. In [25], the *FFANN* model, using levenberg–marquardt (*LM*) back propagation, for forecasting floods with a lead time of upto 5 h in Kushabhadra branch of the Mahanadi delta in Odisha, India was proposed. In [21], a neurofuzzy model called the adaptive neurofuzzy inference system (*ANFIS*) was proposed for forecasting floods in the Kolar basin, Madhya Pradesh, India, by combining the features of the fuzzy inference system (*FIS*) and the neural networks. Further, for longer lead time, the neurofuzzy model performed comparatively better than the *ANN* and the *FIS* models [20]. In [17], Takagi Sugeno (*T-S*) fuzzy inference system (*FIS*) [32] was modified as a Threshold Subtractive Clustering based Takagi Sugeno (*TSC-T-S*) to forecast rare (high to very high river flow) and frequent hydrological events in the upper Narmada basin with lead times of upto 6 h. In [34], the Wavelet-Bootstrap-*ANN* (*WBANN*) flood forecasting model using the wavelet transform and neural networks was proposed. In this model, a wavelet transform was used to decompose the five year hourly monsoon period data into sub-components. Further, these sub-components were used for training and forecasting future floods with a lead time of upto 10 h in the Mahanadi river basin, Maharashtra, India. In [28], a Wavelet-Genetic-*ANN* (*WGANN*) based flood forecasting model with a lead time of upto 24 h for the Kosi and Gandak rivers of Bihar, India, was proposed. In this model, time series data was decomposed into sub-components using the wavelet transform and initial parameters of the *ANN* were optimized using the genetic algorithm. In [19], a *SVM* based low precision flood forecasting model using meteorological parameters was proposed. This model forecast floods in urban areas with a lead time of upto 48 h.

Flood forecasting models discussed above have insufficient lead times and were of very low precision. In order to improve this lead time and precision, this paper focuses on the designing of *ML* based flood forecasting models using the monthly means of precipitation and temperature data.

3 ML Based Flood Forecasting Model

Floods are an annual feature in Bihar. Every year, alongwith heavy rainfall during the monsoon period, heavy discharge of sediments from mountains of Nepal leads to rise in the water levels of rivers like Adhwara, Bagmati, Burhi Gandak, Gandak, Kamla Balan, Kosi and Mahananda, resulting in floods in the plains of Northern Bihar. River basins of these rivers are shown in Fig. 1. More than 73% area comprising these river basins are prone to floods [30, 37]. These rivers cause floods in East Champaran, Mujaffarpur, Samastipur, Khagaria, Bhagalpur, Madhubani, Patna, Katihar, West Champaran, Sitamarhi, Darbhanga and Begusarai districts of Northern Bihar. Flood affected districts of Northern Bihar from 1991 to 2002 are mapped in Fig. 2. Darker shades in Fig. 2, depict the higher flood occurrences while lighter shades depict the lower number of flood occurrences.

In order to design *ML* based flood forecasting models for flood affected districts of Northern Bihar using historical meteorological data, the monthly mean data of precipitation and temperature for the period (1991–2002) was obtained from the Climatic Research Unit, University of East Anglia [1, 18] and labeled using the flood information available on the state's flood management portal (FMISC, http:// www.fmis.bih.nic.in/). This dataset has a total 1728 instances, out of which 201 instances represent flood occurrences. Further, labeled datasets were normalized to a common scale using the min–max scaler as varying ranges of the features may lead to poor classification performance [29]. Min–max scaler [14] is defined by Eq. 1.

$$X' = \frac{X - X_{min}}{X_{max} - X_{min}} \tag{1}$$

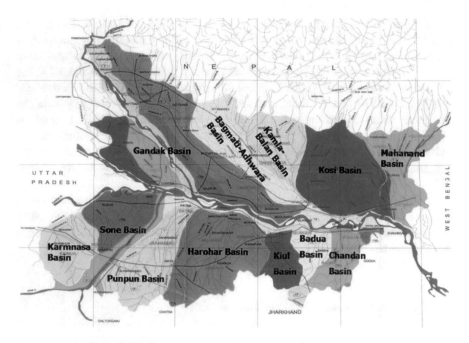

Fig. 1 River basins in Bihar (Source: http://wrd.bih.nic.in)

where X_{max} and X_{min} denotes the maximum and minimum values respectively of the feature (X). The value of X' varies between 0 and 1. Classification techniques viz. Artificial Neural Network (ANN) [10], k-Nearest Neighbor $(k\text{-}NN)$ [11, 15], Logistic Regression (LR) [5, 26], Naive Bayes (NB) [27], Random Forest (RF) [3, 4] and Support Vector Machine (SVM) [8, 38] have been used to design flood forecasting models that use the normalized flood forecasting dataset. These models are briefly discussed next.

The classification based ML techniques, used in this paper, for designing flood forecasting models are briefly discussed below:

Logistic Regression (LR): Logistic Regression is a classification technique which classifies data points into different classes by using a logistic function [5, 26]. The Logistic function used by logistic regression is a sigmoid function and the logistic function $L_\theta(X)$ is defined as: $L_\theta(X) = $ sigmoid (Z), where Z is a linear function of input features X and Z is defined as: $Z = \beta_0 + \beta_1 X$, where β_0 is the intercept and β_1 is the weight.

Support Vector Machine (SVM): Support Vector Machine (SVM) is a machine learning classifier which classifies data points using a hyperplane [8, 38]. The SVM classifier aims to maximize the distance of the data points from the hyper plane in N-dimensional space that distinctly classify the data points.

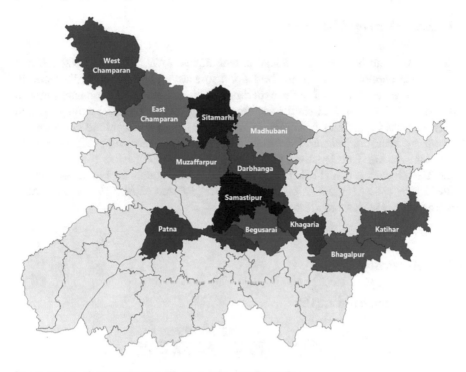

Fig. 2 Flood affected districts in Northern Bihar (1991–2002)

k-Nearest Neighbors (*KNN*): *KNN* is a classification technique which uses the k-nearest neighbors of the data point to be classified for deciding the class of the data point. Plurality of vote from the k-nearest neighbors is used to make the decision [11, 15]. Distance measures like Euclidean distance, Manhattan Distance etc. are used to find the nearest neighbors.

Naive Bayes (*NB*): Naive Bayes *(NB)* is a probabilistic model which uses Bayes Theorem for classification of data points into discrete classes [27]. It is based on the naive assumption that attributes are independent of each other.

Random Forest (*RF*): Random Forest is an ensemble classifier. It uses multiple decision trees to classify a data point where each decision tree classifies and votes for the output class for the given data point [3, 4]. Multiple decision trees are built on random data re-sampled using the bootstrapping technique.

Artificial Neural Network (*ANN*): *ANN* is a classification technique which comprises a network of computational nodes called neurons [10]. Each node in the network is connected with the other node in the network through synapses. Each connection/synapse in the network has a weight, which is optimized using the backpropagation algorithm.

Results of these *ML* models are compared on various performance metrics like accuracy, precision, recall, F-measure and *AUC-ROC*.

4 Experimental Results

The above mentioned *ML* classification techniques: *ANN, KNN, LR, NB, RF* and *SVM* were applied on the normalized flood forecasting (*FF*) dataset discussed in Sect. 3. The details related to the experimentation and the simulations are given in the Table 1. Experimental results are obtained for each *ML* model using stratified *fivefold* cross validation [2]. Minimum, maximum, mean and standard deviation for Accuracy, Precision, Recall, F-measure and *AUC-ROC* were computed, across the five folds, using True Positives (*TP*), False Positives (*FP*), False Negatives (*FN*) and True Negatives (*TN*) [16]. The comparison of ML models for Flood Forecasting (*FF*): *ANN_FF, KNN_FF, LR_FF, NB_FF, RF_FF* and *SVM_FF*, based on these performance metrics, are discussed below:

4.1 Accuracy

Accuracy is computed as [16]:

$$Accuracy = \frac{TP + TN}{TP + FP + FN + TN} \tag{2}$$

Accuracy of the above mentioned *ML* models is given in Table 2. It can be noted from Table 2 that mean accuracy of the *ANN_FF, KNN_FF, LR_FF, RF_FF* and *SVM_FF* models are almost similar having ranges between 0.893 and 0.899. The *SVM* model has the highest mean accuracy while the *NB_FF* model has the least mean accuracy. Except the *NB_FF* model, the maximum accuracy of all models exceeds 0.9. The minimum accuracy of the *ANN_FF*, the *LR_FF* and the *RF_FF* models are almost similar. Amongst all the *ML* models, the *SVM_FF* model has the least standard deviation while the *KNN_FF* model has the highest standard deviation.

Table 1 Experimental setup

Operating system	Windows 10
Processor	Intel i7@2.80 GHz
RAM	16 GB
Tool	Python 3.7.7
Features	Monthly average of temperature and precipitation
Number of folds	5
Learning rate in ANN	0.05
No. of decision trees in RF	20
Value of k in k-NN	5

Table 2 Accuracy of all ML based FF models

Model	Accuracy			
	Min	Max	Mean	SD
ANN_FF	0.882	0.905	0.896	0.009
KNN_FF	0.875	0.91	0.897	0.013
LR_FF	0.887	0.91	0.898	0.008
NB_FF	0.844	0.864	0.852	0.007
RF_FF	0.884	0.901	0.893	0.006
SVM_FF	0.89	0.908	0.899	0.005

Table 3 Precision of all *ML* based *FF* models

Model	Precision			
	Min	Max	Mean	SD
ANN_FF	0.5	0.652	0.603	0.054
KNN_FF	0.457	0.667	0.58	0.077
LR_FF	0.529	0.846	0.651	0.11
NB_FF	0.4	0.443	0.425	0.015
RF_FF	0.5	0.636	0.565	0.043
SVM_FF	0.6	0.833	0.698	0.089

4.2 Precision

Precision is computed as [16]:

$$Precision = \frac{TP}{TP + FP} \tag{3}$$

Precision of the above mentioned *ML* models is given in Table 3. It can be noted from Table 3 that the mean precision value of the *SVM_FF* model is highest and the mean precision value of the *NB_FF* model is lowest amongst all *ML* models. Standard deviations of the *NB_FF* model is also the lowest amongst all *ML* models. The maximum precision value and the standard deviation of the *LR_FF* model is the highest amongst all *ML* models. Furthermore, the mean precision value of the *SVM_FF* model is comparatively better than the maximum precision value of all the *ML* models except the *LR_FF* model.

4.3 Recall

Recall is computed as [16]:

Table 4 Recall of all *ML* based *FF* models

Model	Recall			
	Min	Max	Mean	SD
ANN_FF	0.225	0.425	0.328	0.068
KNN_FF	0.375	0.55	0.463	0.065
LR_FF	0.22	0.4	0.279	0.065
NB_FF	0.634	0.875	0.772	0.098
RF_FF	0.293	0.475	0.369	0.06
SVM_FF	0.2	0.375	0.249	0.066

$$Recall = \frac{TP}{TP + FN} \tag{4}$$

Recall of the above mentioned *ML* models is given in Table 4. It can be noted from Table 4 that the mean recall value of the *NB_FF* model is the highest and the mean recall value of the *SVM_FF* model is the lowest amongst all the *ML* models. Further, the standard deviation of the *NB_FF* models is comparatively high amongst all the *ML* models. If recall is the key performance metric then the *NB_FF* model can be used for forecasting floods.

4.4 F-measure

F-measure is computed as [16]:

$$F - Measure = \frac{2 \times Precsion \times Recall}{Precision + Recall} \tag{5}$$

As observed above, the mean accuracy of all the models are comparable. The mean precision value of the *SVM_FF* model is the highest and its mean recall value is the lowest amongst all the *ML* models. Whereas, the mean precision value of the *NB_FF* model is the lowest and its mean recall value is the highest amongst all the *ML* models. Therefore, the performance of the *ML* models has been evaluated using the F-measure and the *AUC-ROC*. The F-measure of all models is given in Table 5. It can be noted from Table 5, that the mean F-measure value of *NB_FF* model is the highest whereas the mean F-measure value of the *SVM_FF* model is the least amongst all *ML* models. Further, the standard deviation of the *NB_FF* model is the least amongst all the *ML* models and the standard deviation of the *SVM_FF* model is comparatively high. Therefore, it can be inferred that performance of the *NB_FF* model is comparatively better than any other *ML* model when the F-measure is used as the key performance metric.

4.5 AUC-ROC

AUC-ROC is the area under the ROC curve plotted between the True Positive Rate (TPR) and the False Positive Rate (FPR). TPR and FPR are defined as [16]:

$$TPR = \frac{TP}{TP + FN} \tag{6}$$

$$FPR = \frac{FP}{FP + TN} \tag{7}$$

To assess the overall performance of the ML models across various thresholds, the AUC-ROC has been computed. The AUC-ROC of above mentioned models is given in Table 6. It can be noted from Table 6 that the mean AUC-ROC value of the NB_FF model is the highest and the mean AUC-ROC value of the KNN_FF model is the lowest amongst all the ML models. Further, the NB_FF models has a lower standard deviation than all other ML models except the ANN_FF model.

It can be noted from the above performance metrics that the SVM_FF model performs comparatively better than any other ML model in terms of accuracy and precision. Whereas, the NB_FF model outperforms all ML models in terms of recall, F-measure and AUC-ROC values.

Table 5 F-measure of all ML based FF models

Model	F-Measure			
	Min	Max	Mean	SD
ANN_FF	0.327	0.507	0.421	0.064
KNN_FF	0.427	0.563	0.51	0.053
LR_FF	0.316	0.478	0.385	0.061
NB_FF	0.491	0.576	0.546	0.032
RF_FF	0.387	0.514	0.442	0.041
SVM_FF	0.314	0.462	0.359	0.058

Table 6 AUC-ROC of all ML based FF models

Model	AUC-ROC			
	Min	Max	Mean	SD
ANN_FF	0.91	0.932	0.919	0.008
KNN_FF	0.818	0.916	0.864	0.036
LR_FF	0.898	0.931	0.915	0.011
NB_FF	0.909	0.934	0.924	0.009
RF_FF	0.89	0.926	0.913	0.013
SVM_FF	0.857	0.913	0.893	0.019

5   Conclusion

This paper focused on designing *ML* based flood forecasting models with the aim to achieve large lead times for designing proactive and preventive flood mitigation strategies. classification based *ML* models like *NB, LR, SVM, KNN, RF and ANN* were applied on the meteorological data (precipitation and temperature) of the twelve most flood affected districts of Northern Bihar (India) during the period 1991–2002. These models were trained and tested using the stratified fivefold cross validation. Performance of these models were then compared on five performance parameters namely accuracy, precision, recall, *F*-measure and *AUC-ROC*. Amongst these classification models, the *SVM_FF* model performed comparatively better in terms of accuracy and precision whereas, the *NB_FF* model performed comparatively better in terms of recall, *F*-Measure and *AUC-ROC* value. Based on the key performance parameters, *SVM_FF* or *NB_FF* can be used for forecasting future floods in the plains of Northern Bihar.

References

1. Allen RG, Pereira LS, Raes D, Smith M (1998) Crop evapotranspiration: Guidelines for computing crop water requirements, FAO Irrigation and drainage paper 56. Italy, Rome
2. Berrar D (2018) Cross-validation. Encycl Bioinf Comput Biol ABC Bioinf 1–3:542–545
3. Biau G (2012) Analysis of random forests model. J Mach Learn Res 13:1063–1095
4. Breiman L (2001) Random forests. Mach Learn 45:5–32
5. Cabrera AF (1994) Logistic regression analysis in higher education: an applied perspective. In: Smart JC (ed) Higher Education Handbook of Theory and Research, vol 10, pp 225–256
6. Central Water Commission (2018) Flood Damage Statistics (Statewise and for the Country as a whole) for the Period 1953 to 2016; Central Water Commission (CWC), Flood Forecast Monitoring Directorate, Government of India (3), 37. http://www.indiaenvironmentportal.org.in/content/456110/flood-damage-statistics-statewise-and-for-the-country-as-a-whole-for-the-period-1953-to-2016/
7. Devia GK, Ganasri BP, Dwarakish GS (2015) A review on hydrological models. Aquatic Procedia 4(Icwrcoe):1001–1007
8. Evgeniou T, Pontil M (2001) Support vector machines: theory and applications. In: Paliouras G, Karkaletsis V, Spyropoulos CD (eds) Machine learning and its applications. ACAI 1999. Lecture Notes in Computer Science, vol. 2049, pp 249–257
9. Ghose DK (2018) Measuring discharge using back-propagation neural network: a case study on brahmani river basin. Intell Eng Inf, 591–598
10. Gurney K (1997) An introduction to neural networks. CRC Press, Boca Raton
11. Guo G, Wang H, Bell D, Bi Y, Greer K (2003) KNN model-based approach in classification. In: Meersman R, Tari Z, Schmidt DC (eds) On the move to meaningful internet systems 2003: CoopIS, DOA, and ODBASE. OTM 2003. Lecture Notes in Computer Science, vol 2888, pp 986–996
12. Jain SK, Agarwal PK, Singh VP (2007) Physical environment of India. In: Jain SK, Agarwal PK, Singh VP (eds) Hydrology and water resources of India, vol 57

13. Jain SK, Mani P, Jain SK, Prakash P, Singh VP, Tullos D, Dimri AP (2018) A Brief review of flood forecasting techniques and their applications. Int J River Basin Manag 16(3):329–344
14. Jayalakshmi T, Santhakumaran A (2011) Statistical normalization and back propagationfor classification. Int J Comput Theory Eng 3(1):89–93
15. Laaksonen J, Oja E (1996) Classification with learning k-nearest neighbors. In: IEEE International conference on neural networks - conference proceedings, vol 3, pp 1480–1483
16. Liu Y, Zhou Y, Wen S, Tang C (2014) A strategy on selecting performance metrics for classifier evaluation. Int J Mob Comput Multimedia Commun 6(4):20–35
17. Lohani AK, Goel NK, Bhatia KKS (2014) Improving real time flood forecasting using fuzzy inference system. J Hydrol 509:25–41
18. Mitchell TD, Jones PD (2005) An improved method of constructing a database of monthly climate observations and associated high-resolution grids. Int J Climatol 25(6):693–712
19. Nayak MA, Ghosh S (2013) Prediction of extreme rainfall event using weather pattern recognition and support vector machine classifier. Theoret Appl Climatol 114(3–4):583–603
20. Nayak PC, Sudheer KP, Ramasastri KS (2005) Fuzzy computing based rainfall-runoff model for real time flood forecasting. Hydrol Process 19(4):955–968
21. Nayak PC, Sudheer KP, Rangan DM, Ramasastri KS (2005) Short-term flood forecasting with a neurofuzzy model. Water Resour Res 41(4):1–16
22. NDMA (2017) National Disaster Management Authority (NDMA). In Ndma. Retrieved from http://ndma.gov.in
23. NIDM (2002) Disaster Management- Terminology 8. 1–4. https://nidm.gov.in/PDF/Disaster_terminology.pdf
24. NIDM (2018) Safety Tips for Floods, Cyclones & Tsunamis. Retrieved from https://nidm.gov.in/safety_flood.asp
25. Panda RK, Pramanik N, Bala B (2010) Simulation of river stage using artificial neural network and MIKE 11 hydrodynamic model. Comput Geosci 36(6):735–745
26. Peng CJ, Lee KUKL, Ingersoll GM (2002) An introduction to logistic regression analysis and reporting. J Educ Res 96(1):3–14
27. Rish, I (2001) An empirical study of the naive Bayes classifier, vol 3, no 22, pp 4863–4869
28. Sahay RR, Srivastava A (2014) Predicting monsoon floods in rivers embedding wavelet transform, genetic algorithm and neural network. Water Res Manag 28(2):301–317
29. Singh D, Singh B (2020) Investigating the impact of data normalization on classification performance. Appl Soft Comput 97(B):105524
30. Singh SK (2013) Flood management information system. In: Government of Bihar. http://fmis.bih.nic.in/
31. Sitterson J, Knightes C, Parmar R, Wolfe K, Muche M, Avant B (2017) An overview of rainfall-runoff model types an overview of rainfall-runoff model types
32. Takagi T, Sugeno M (1985) Fuzzy identification of systems and its applications to fault diagnosis systems. IEEE Trans Syst Man Cybern 15(1):116–132
33. Thirumalaiah K, Deo MC (1998) Real-time flood forecasting using neural networks. Comput-Aided Civil Infrastruct Eng 13(2):101–111
34. Tiwari MK, Chatterjee C (2010) Development of an accurate and reliable hourly flood forecasting model using wavelet–bootstrap–ANN (WBANN) hybrid approach. J Hydrol 394(3–4):458–470
35. UNDRR (2020) Human cost of disasters. In: Human Cost of Disasters
36. Wallemacq P, Below R, McLean D (2018) UNISDR, CRED.: economic losses, poverty & disasters (1998–2017). CRED. https://www.cred.be/unisdr-and-cred-report-economic-losses-poverty-disasters-1998-2017
37. Water Resource Department, Govt. of Bihar (2021) Flood Control and Drainage: Introduction. http://wrd.bih.nic.in
38. Zhang Y (2012) Support vector machine classification algorithm and its application. In: Communications in computer and information science, CCIS(PART 2), vol 308, pp 179–186 (2012)

Multilayer Perceptron Optimization Approaches for Detecting Spam on Social Media Based on Recursive Feature Elimination

Puneet Garg and Shailendra Narayan Singh

1 Introduction

Social media platforms are gradually becoming the primary information source and public events and people all over the world active in social networks. About 62% of US adults currently receive information and news via online social networks, according to a survey. Social network proliferation is developed on shared activities and public user comments. Nonetheless, for spreading fake news, misleading ads, perpetuating political agendas, biasing product ideals and even triggering social chaos, the social media platforms have become a common medium [1].

Social media platforms are capable of influencing as well as the virtual and the physical worlds. Therefore, it is a significant necessity to secure these networks. Both restrictions and technical controls are usually required to prevent spammers and increase user confidence in those platforms [2].

While extensive study of spam has been carried out in the perspective of the emails, a security report depicts that the nature of the attacks and the quantity of spam it carries are still a problem, particularly with the widespread and emergence of online social networks [3]. Since spam affects the performance, effectiveness and reliability of an Online Social Network, which exploit account holders to countless security threats. Therefore, reducing or removing the number of spam profiles is crucial [4].

Since the inception of the internet, various types of measures and techniques have been built to avoid the spam. The emails had enormous amount of spammers' and spoofing assaults. The enormous techniques have been applied to identify spam over the email and social media events. Different types of four spams are characterized

P. Garg (✉) · S. N. Singh
Amity University, Noida 201301, India
e-mail: puneetgarg15797@gmail.com

S. N. Singh
e-mail: snsingh36@amity.edu

by inconsistent, high prosperity in online social networks. Spammers, who intend to publicize their post victim links, propagate their attacks events more frequently over dissimilar online social networks [5].

A user faces multiple search results issues, which share repetitive and unnecessary details due to the continued information dissemination. At times, this can be quite disappointing, as a user has to navigate through all the information in the process of having an abstract view of the matter. Since URLs, abbreviations, modern language definitions and informal language are used predominantly, spam detection on online social networks is a challenging task [6].

In this work, identifying spammers on Twitter [7] is primary focus. The attributes for the spam identification, being considered are content-based and user behavior-based. Correlation based preprocessing along with the simple Genetic algorithm for searching among the attribute set for preprocessing is also applied on the collected dataset from [8] and then applied Recursive Feature Elimination method for feature selection. The three Multilayer Perceptron optimization approaches namely Stochastic Gradient Descent (SGD), Limited-memory Broyden–Fletcher–Goldfarb–Shanno (L-BFGS), and Adam, are evaluated to identify a suitable Multilayer Perceptron approach for distinguishing spammers and non-spammers via Recursive Feature Elimination. This paper is arranged as follows: Sect. 2 contains the literature Review, Sect. 3 contains the methodology, Sect. 4 contains the experimental results and evaluation, and the final section summarizes and concludes the results along with the future scope.

2 Literature Review

Paul Heymann studied that the profiles of Twitter that send spam posts and how Twitter spammers manipulate links [9] for fake advertising, phishing, and spreading malware. Their study suggested that, instead of generating new accounts, spammers spread the spam by misusing hacked account.

Faraz Ahmed proposed that expanding the interpretation of content-based filtering [10] by considering the content of web pages linked to e-mail messages. Separating the conventional content-based email spam considers nature of email messages and applies machine-learning techniques to differentiate between spam and hams. Specifically, the use of substance-based spam separation released an endless arms race among spammers and channel engineers, allowing the ability of spammers to persistently alter the content of spam messages in ways that could circumvent the current channels. They present a paradigm for removing linked URLs in spam messages and explain the relation between those pages and the messages. They later use a machine learning technique to isolate grouping rules from the webpage that are related to the position of spam. They stated that the use of data from linked sites, as depicted by Spam Assassin, would comfortably complement current spam grouping systems.

Marco Ribeiro reviewed the spam filtering techniques already in use [11]. Spam has been a difficult problem recently over digital communications and the internet.

Against this issue numerous approaches are introduced. They presented the conventional learning-based approaches for assessment, classification and contrast. Several innovative anti-spam approaches were tested and contrasted. Saadat Nazirova examined and used features [12] derived from user profiles, social network followers and followees, tweet content, and the temporal nature of user activity to classify polluters of content.

Kyumin Lee argued that the social media platforms provide a way [13] for the users to monitor their friends' contacts. Developing social network popularity helps entire users to collect massive quantities of personal information and data about their friends. Unfortunately, the abundance of data as well as the ease of access to user data and information may draw malicious group attention. That is why spammers infiltrate these networks because the effort to repair and treat them was enormous. There have been ongoing strategies to identify malicious emails and spammers in respect of this matter, which spam account users look for new ways of targeting these platforms.

Nasim Eshraqi addressed a scalable method for predicting fake account groups [14], with the appropriate actor reported. The key approach is a supervised pipeline that classifies the entire collection of accounts as either malicious or legitimate. A generic pattern encoding algorithms are proposed in this work, which allow users to collapse use generated text into a small space on which to evaluate statistical characteristics. Cao Xiao proposed a novel approach for differentiating non-spam vs. spam posts [15] in social media and provides a great deal of insight into twitter spam user activity that is feasible on twitter for a limited period of time, motivated by the need to search out and classify spam content in social media networks.

Isa Inuwa-Dutse predicted the correlation between classification trends and prediction features [16] for spam messages from unwanted websites. After assessing the functional importance for spam detection the program applies successful classification algorithms in a good way. The outcomes of their work shall support people who are engaged on the social networks for efficient needs such as marketing, company and communication setting up. This work also looked at current approaches for identifying spam accounts on the twitter. Spammer identification tools may be material or user-based mechanisms and mechanisms for classifying spammers.

Meet Rajdev addressed that the users on social media prefer to believe in sharing information quickly [17] connected with certain incidents during emergencies or natural disasters and send retweets in the intention of reaching many other users. There are deceptive users who post disinformation and trend of massive dissemination, such as false and spam messages. This work performs a case analysis of Hurricane Sandy and Moore Tornado of 2013. This work considers the actions of malicious users, explores the properties of false, legitimate and spam messages, suggests hierarchical and flat classification methods and identifies all real and fake spam messages, thus distinguishing between them.

Amira Soliman proposed a novel graph-based approach for identifying the spam [18]. This system is unsupervised, thereby omitting the need for labelled training data and training cost. This approach can effectively identify spam in large scale online social networks by using graph clustering approach to analyze user behaviors. This approach further updates identified communities on an ongoing basis to comply with diverse online social networks where events and interactions are continually evolving. This system is capable of identifying spam more accurately, and has a false positive rate that is less than half of the rate achieved by state-of-the-art methods.

Surendra Sedhai suggested a semi-supervised framework for spam detection on twitter [19]. Several spam detection approaches on twitter are aimed at blocking a recognized user posting spam messages. The proposed structure consists of two major modules operating in real time mode, namely model update module in batch mode and spam detection module. The information required by the detection module is revised in batch mode based on the number of tweets depicted in the previous time frame.

Sreekanth Madisetty developed a neural network approach for identification of spam on twitter [20] and proposed approach found that deep learning methods operate more efficiently than feature-based approaches for the HSspam-14 dataset. Zulfikar Alom developed spam detection frameworks involving data gathering, analysis, assessment and rating of attributes [21]. It is noted that the top attributes found by the selection method (i.e., data gaining) yield much increased performance.

Until now, traditional classifiers and neural networks have been used to identify spam over the social networks. In this work, the three Multilayer Perceptron optimization approaches namely Stochastic Gradient Descent (SGD), Limited-memory Broyden–Fletcher–Goldfarb–Shanno (L-BFGS), and Adam are used along with recursive feature elimination, and evaluated to find the best approach amongst the three approaches.

3 Methodology

The proposed spam detection model consists of four phases named Data collection and description, Feature Selection, Multilayer Perceptron, and Evaluation. It is represented using Fig. 1.

3.1 Data Collection and Description

The dataset from [8] is obtained, consisting of 37 content features and 21 user-behavior features from the Twitter. The size of the labeled dataset is 1535 users.

3.2 Feature Selection

Recursive Feature Elimination (RFE) is a selection algorithm for wrapper-type features. It operates by removing attributes recursively and creating a model on

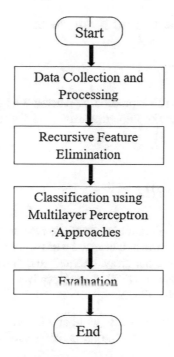

Fig. 1 Proposed spam detection model

those remaining attributes. The model accuracy is used to determine which features make significant contribution to the target attribute prediction.

3.3 Multilayer Perceptron

The three Multilayer Perceptron optimization approaches namely Stochastic Gradient Descent (SGD), Limited-memory Broyden–Fletcher–Goldfarb–Shanno (L-BFGS), and Adam, were applied by splitting the dataset as 75% training data and 25% test data, after applying the Recursive Feature Elimination.

3.4 Evaluation

Performance factors namely Confusion matrix, Accuracy, TP rate, Precision, FP rate, and F-Score are used for the assessment of the three optimization approaches used.

4 Experimental Results and Evaluation

4.1 Tool Used

The collected dataset from [8], is preprocessed using WEKA. WEKA is an open and free Data Mining software. For implementation, the sklearn libraries and packages in Python are used.

4.2 Performance Metrics

Confusion Matrix: This is the simplest metric to evaluate the efficiency of the problems of classification where the output may be of two or more classes. A confusion matrix is just a two-dimensional array. "Actual" and "Predicted" as well as "True Positives (TP)", "False Positives (FP)", "True Negatives (TN)" and "False Negatives (FN)". It is represented using Fig. 2.

TP Rate: True Positive Rate is the number of users correctly categorized as spammers (yes) and non-spammers (no). It is represented using the formula:

$$TPR = \frac{Tp}{Tp + Fn} \tag{1}$$

FP Rate: False Positive Rate is the number of users wrongly categorized as spammers (yes) and non-spammers (no). It is represented using the formula:

$$FPR = \frac{Fp}{Fp + Tn} \tag{2}$$

Precision: Precision (also known as positive predictive value) is the ratio of relevant users with the retrieved users. It is represented using the formula:

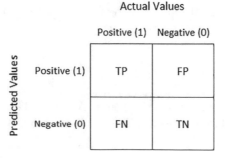

Fig. 2 Confusion matrix

$$P = \frac{Tp}{Tp + Fp} \tag{3}$$

Recall: Recall (also known as sensitivity) is the ratio of the total amount of relevant users that are actually retrieved.

$$R = \frac{Tp}{Tp + Fn} \tag{4}$$

F-Measure: The results of both recall and precision are used to calculate F-Measure as follows:

$$F - Measure = 2 * \frac{PR}{P + R} \tag{5}$$

4.3 Results

Tables 1, 3, and 5 depict values of performance metrics like Accuracy, TP rate, Precision, FP rate, and F-Score. The value of these metrics are computed using pycm in python after applying the Stochastic Gradient Descent (SGD), Limited-memory Broyden–Fletcher–Goldfarb–Shanno (L-BFGS), and Adam Multilayer Perceptron optimization approaches.

Tables 2, 4, and 6 contain confusion matrices of Stochastic Gradient Descent (SGD), Limited-memory Broyden–Fletcher Goldfarb–Shanno (L-BFGS), and Adam Multilayer Perceptron optimization approaches. The classifiers information used for the performance analysis is contained in confusion matrix. It contains both actual information as well as predicted information by the classifier.

Table 1 SGD-MLP values with different performance metrics

No. of iterations for convergence: 850					
TP rate	FP rate	Precision	F-score	Accuracy	Class
0.97070	0.06306	0.97426	0.95060	0.96094	Yes
0.93694	0.02930	0.92857	0.95421	0.96094	No
0.95382	0.04618	0.95142	0.95240	0.96094	**Weighted avg.**

Table 2 Confusion matrix of SGD-MLP

a	b	Classified as
265	7	a = yes
8	104	b = no

Table 3 Adam-MLP values with different performance metrics

No. of iterations for convergence: 700					
TP rate	FP rate	Precision	F-score	Accuracy	Class
0.99634	0.03604	0.98551	0.98970	0.98698	Yes
0.96396	0.00366	0.99074	0.97839	0.98698	No
0.98015	0.01985	0.98812	0.98404	0.98698	**Weighted avg.**

Table 4 Confusion matrix of Adam-MLP

a	b	Classified as
272	4	a = yes
1	107	b = no

Table 5 L-BFGS-MLP values with different performance metrics

No. of iterations for convergence: 100					
TP rate	FP rate	Precision	F-score	Accuracy	Class
0.98535	0.01802	0.99262	0.97737	0.98438	Yes
0.98198	0.01465	0.96460	0.98479	0.98438	No
0.98366	0.01634	0.97861	0.98108	0.98438	**Weighted Avg.**

Table 6 Confusion matrix of L-BFGS-MLP

a	b	Classified as
269	2	a = yes
4	109	b = no

SGD and Adam optimization approaches are good at predicting the spam users but they misclassified few non-spam instances as spam users. SGD and Adam optimization approaches, both are gradient descent based, but the Adam approach yields accuracy of around 98% as that of SGD's 96%. Both SGD and Adam optimization approaches required relatively high number of iterations for convergence, as compared to L-BFGS optimization approach. L-BFGS optimization approach is found to be the best among the three approaches, as its TP rate for both spam as well as non-spam users is approximately same and yields accuracy of around 98%.

5　Conclusion and Future Scope

Social media platforms are the largest information and content sharing network in the world. Spam messages, posts, comments, and links are also circulating at a huge transmission rate along with the merits of the significant amount of information on these platforms. In this work, three different Multilayer Perceptron optimization approaches namely Stochastic Gradient Descent (SGD), Limited-memory Broyden–Fletcher–Goldfarb–Shanno (L-BFGS), and Adam are used along with Recursive feature elimination algorithm. The Weka tool is used for preprocessing on the dataset. The above-mentioned classifiers are implemented & feature selection method in Python using the sklearn libraries. Although, L-BFGS optimization approach required less no of iterations for convergence as in contrast with SGD and Adam optimization approaches. The experimental results depict that the Adam optimization approach yields better accuracy in contrast with SGD and L-BFGS approach but L-BFGS yields the same TP rate of 98% for Spammers as well as Non-Spammers.

For future works, the approach being proposed can also be evaluated for a bigger dataset, sparse datasets, and other social networking platforms like Facebook, Instagram, LinkedIn, and so on. Further, these optimization approaches can also be applied and evaluated on a Big Data application, which could not be done owing to the hardware constraints. NLP techniques and Context knowledge can also be integrated in proposed approach so as to take individual message, post or tweet into account or increase the accuracy.

References

1. Allcott H, Gentzkow M (2017) Social media and fake news in the 2016 election. J Econ Perspect 31(2):211–236
2. Goolsby R, Shanley L, Lovell A (2013) On cybersecurity, crowdsourcing, and social cyberattack. Office of Naval Research Arlington, VA
3. Haley K, Wood P (2015) Internet security threat report. Symantec, Mountain View
4. Herzallah W, Faris H, Adwan O (2018) Feature engineering for detecting spammers on twitter: modelling and analysis. J Inf Sci 44(2):230–47. http://www.springer.com/lncs, Accessed 21 Nov 2016
5. Rathore S, Sharma PK, Loia V, Jeong YS, Park JH (2017) Social network security: issues, challenges, threats, and solutions. Inf Sci 1(421):43–69
6. Stringhini G, Kruegel C, Vigna G (2010) Detecting spammers on social networks. In: Proceedings of the 26th annual computer security applications conference, pp 1–9
7. Twitter. https://twitter.com/, Accessed 26 Feb 2021
8. TransientObject (Priya Narayana Subramanian) GitHub. https://github.com/TransientObject/SpamDetectionTwitter/tree/master/data_analysis/arff_files/, Accessed 25 Apr 2020
9. Heymann P, Koutrika G, Garcia-Molina H (2007) Fighting spam on social web sites: a survey of approaches and future challenges. IEEE Internet Comput 11(6):36–45
10. Ahmed F, Abulaish M (2012) An mcl-based approach for spam profile detection in online social networks. In: 2012 IEEE 11th international conference on trust, security and privacy in computing and communications, 25 June 2012. IEEE, pp 602–608
11. Ribeiro MT, Guerra PH, Vilela L, Veloso A, Guedes D, Meira Jr W, Chaves MH, Steding-Jessen K, Hoepers C (2011) Spam detection using web page content: a new battleground. In: Proceedings of the 8th annual collaboration, electronic messaging, anti-abuse and spam conference, 1 September 2011, pp 83–91
12. Saadat N (2011) Survey on spam filtering techniques. Commun Netw 29:2011

13. Lee K, Caverlee J, Webb S (2010) Uncovering social spammers: social honeypots+ machine learning. In: Proceedings of the 33rd international ACM SIGIR conference on Research and development in information retrieval, 19 July 2010, pp 435–442
14. Eshraqi N, Jalali M, Moattar MH (2015) Spam detection in social networks: a review. In: 2015 international congress on technology, communication and knowledge (ICTCK), 11 November 2015. IEEE, pp 148–152
15. Xiao C, Freeman DM, Hwa T (2015) Detecting clusters of fake accounts in online social networks. In: Proceedings of the 8th ACM workshop on artificial intelligence and security, 16 October 2015, pp 91–101
16. Inuwa-Dutse I, Liptrott M, Korkontzelos I (2018) Detection of spam-posting accounts on Twitter. Neurocomputing 13(315):496–511
17. Rajdev M, Lee K (2015) Fake and spam messages: detecting misinformation during natural disasters on social media. In: 2015 IEEE/WIC/ACM international conference on web intelligence and intelligent agent technology (WI-IAT), 6 December 2015, vol 1. IEEE, pp 17–20
18. Soliman A, Girdzijauskas S (2016) Adaptive graph-based algorithms for spam detection in social networks
19. Sedhai S, Sun A (2017) Semi-supervised spam detection in Twitter stream. IEEE Trans Comput Social Syst 5(1):169–175
20. Madisetty S, Desarkar MS (2018) A neural network-based ensemble approach for spam detection in Twitter. IEEE Trans Comput Social Syst 5(4):973–984
21. Alom Z, Carminati B, Ferrari E (2018) Detecting spam accounts on twitter. In: 2018 IEEE/ACM international conference on advances in social networks analysis and mining (ASONAM), 28 August 2018. IEEE, pp 1191–1198

Convolution Neural Network Based Classification of Plant Leaf Disease Images

K. Jaspin, Shirley Selvan, Princy Salomy Packianathan, and Preetha Kumar

1 Introduction

The primary occupation in our country is agriculture and India ranks second in the agricultural output worldwide. Here, farmers cultivate an excellent diversity of crops. But there has been a fall in agricultural production due to various plant diseases. The first step in plant disease identification is to look for signs of the infection which varies with the infecting organism. In this paper, we propose a solution to determine leaf diseases, a system that identifies leaf diseases of the tomato plant and pepper bell plant through Deep learning methodology. Several diseases affect the tomato plant in which four diseases Mosaic Virus [15], Target Spot, Bacterial Spot, and Early Blight can effectively be identified by our proposed work. Karnataka, Kerala, and Tamil Nadu are the leading producers of Pepper Bells in India. In our proposed work, we have identified the Bacterial Spot disease of Pepper Bell that results in water-soaked lesions that dry out and turn brown forming on the underside of the leaves. It leads to the disfiguration of the leaves and fruit. In severe cases, the plants may die. Deep Learning has become a recent trend, excelling in the fields of image processing [16] and data analysis.

2 Related Work

Sachin et al., (2019) [1] proposed an efficient soybean disease identification method [1], which made use of the transfer learning approach. The networks AlexNet,

K. Jaspin (✉) · P. S. Packianathan · P. Kumar
St. Joseph's Institute of Technology, Tamil Nadu, Chennai 600 119, India
e-mail: jaspink@stjosephstechnology.ac.in

S. Selvan
St. Joseph's College of Engineering, Tamil Nadu, Chennai 600 119, India

© The Author(s), under exclusive license to Springer Nature Singapore Pte Ltd. 2022   511
B. Unhelker et al. (eds.), *Applications of Artificial Intelligence and Machine Learning*,
Lecture Notes in Electrical Engineering 925,
https://doi.org/10.1007/978-981-19-4831-2_42

GoogleNet, VGG16, and DenseNet101were trained using 1200 village plants' image datasets of both healthy and diseased soybean leaves. They used five-fold cross-validation for analyzing the performance of the networks. The networks proved accuracies of 95, 96.4, 96.4, 92.1, 93.6% respectively. T. Rumpf et al. (2010) [2] proposed a method of early detection and classification of sugar beet diseases [17] with SVM based on the Hyperspectral Reflectance. The hyperspectral data were recorded from leaves inoculated with certain bacterial pathogens for 21 days after inoculation and healthy leaves [24]. A total of nine different spectral vegetation indices related to some physiological parameters were used as features for the automatic classification. The accuracies varied depends upon the type and various disease stages of disease [26], ranging between 65 and 90%.

R. Zhou et al. (2014) [3] had come together and presented an image algorithm for detection of Cercospora Leaf Spot in sugar beet on real field conditions [18]. The first framework was based on robust template matching by orientation code matching (OCM) [21], which involved a single leaf from the beet plant exposed to successive tracking under the conditions of severe illumination changes and non-rigid plant movements. The second framework made use of the pattern recognition method of SVM to improve further classification of diseases. Lin Yuan et al. (2017) [4] proposed habitat monitoring for evaluating the potential occurrence and distribution of wheat plant leaf diseases & pests in Hebei province using Worldview 2 and Landsat 8 satellite data. They employed an approach called FLDA which was most effective in crop disease monitoring and pest detection. This approach could give an accuracy of about 82% compared to 71% while considering only the vegetative indices. K. P. Waidyarathne et al. (2014) [5] made an attempt to classify the Weligama Coconut Leaf Wilt Disease (WCLWD) using visual symptoms through computational modelling. The result of their work revealed a correspondence of 73.45% with expert decisions on disease severity classification.

M. Neumann et al. (2014) [6] proposed a system to detect five different diseases, in beet leaves with the use of Erosion band features using images taken in cell phone camera. They had evaluated 1st and 2nd order features to classify the leaf spots texture. S. S. Patil et al. (2014) [7] proposed a method using SVM classifier to Identify and classify the cotton leaf diseases. K-means clustering [29] was employed to segment the image & colour, texture-based features, and shape were extracted. The SVM classifier was used for identifying the diseases. S. Phadikar et al. (2015) [8] classified diseases of rice plant using feature selection and rule generation techniques. They proposed a segmentation algorithm based on Fermi-energy which isolated the infected region from the background. Important features were extracted through rough set theory (RST) [19] to minimize the loss of information and to reduce the complexity of the classifier. M. Ranjan et al. (2015) [9] extracted HSV features after segmentation of the diseased region in leaves of cotton plants. The ANN was trained by selecting the feature values that would effectively differentiate the healthy and diseased plant leaf images with overall accuracy of 80% in comparison to its other counterparts.

Usama Mokhtar et al. (2015) [10] presented a method to detect Tomato Leaves Diseases using SVM. They had made use of the Gray Level Co-occurrence Matrix

(GLCM) for the purpose of detection and identification of state of tomato leaf, infected or healthy. For testing and training, 800 different images were used. This proposed work could achieve a classification accuracy of about 99.83%, using the linear kernel function. Prashant R. Rothe et al. (2019) [11] had given an intelligent system to identify and classify the Cotton leaf diseases [27] such as Bacterial leaf blight, Alternaria, and Myrothecium [28]. Colour, texture, and shape-based features were cultured and fed to the Back Propagation neural network (BPNN) for assimilation. They had collected the datasets from CICR Nagpur and from the actual fields of Wardha and Buldhana districts. Their methodology could achieve an accuracy of 95.48%, for categorization. Yang Lu et al. (2017) [12] proposed a deep CNN technique for a novel rice disease classification method [25]. Jie Tian et al. (2012) [13] designed an Improved KPCA/GA-SVM Classification Model [30] to detect the apple leaf disease. Their system employed the Genetic algorithm (GA) [10] for supporting SVM classifier in parameter determination. To identify the best features among them, they used a feature selection method based on Kernel principal component analysis (KPCA) [23]. It was observed that the proposed KPCA/GA-SVM model could achieve recognition ratios of 94.05, 97.96, and 98.14%, for apple rust and apple Alternaria leaf spot and apple mosaic virus, respectively.

Bock et al. (2011) [14] were the designers of a new methodology for the "Detection and Measurement of Plant Disease Symptoms Using visible wavelength photography and image analysis". This was done using thresholding technique.

3 Materials and Methodology

3.1 Materials

Dataset. The required data for our proposed system is collected from Kaggle dataset. As shown in Table 1, we have collected 1750 sample leaf images that consist of both diseased and healthy leaf images of tomato and pepper bell plant. Among 1750 total sample leaf images, there are 1050 diseased leaf images and 420 healthy leaf images. We have labelled pepper bell bacterial spot disease as class 1, mosaic virus disease as class 2, target spot disease as class 3, tomato bacterial spot as class 4, early blight disease as class 5 [20], healthy pepper bell as class 6 and healthy tomato as class 7. These input images are augmented to the configuration of 150 × 150 pixels. Both diseased and healthy sample leaf images of tomato and pepper bell plant can be seen in Table 1.

Table 1 Training and testing dataset of VGG16

S.No	Disease class	Training samples	Testing samples	Total
1	Pepper bell bacterial spot	210	40	250
2	Mosaic virus	210	40	250
3	Target spot	210	40	250
4	Tomato bacterial spot	210	40	250
5	Early blight	210	40	250
6	Healthy pepper bell	210	40	250
7	Healthy tomato leaf	210	40	250
	Total images	**1470**	**280**	**1750**

3.2 Methodology

The proposed retrained VGG16 CNN is used to classify test images from a validation dataset. We in our work have made use of the pre-trained deep learning models namely VGG16, ResNet50, and AlexNet. By using transfer learning method, the pre-trained VGG16 model shown in Fig. 1 is retrained. The proposed retrained model presented in Fig. 2 is being trained on a new dataset.

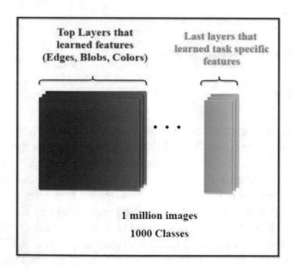

Fig. 1 Pretrained existing VGG16 model

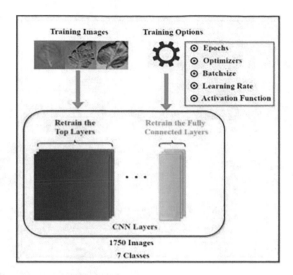

Fig. 2 Retrained proposed VGG16 model

4 Proposed Work

4.1 Proposed VGG16 Architecture

In our proposed system, we have made use of VGG16, the pre-trained deep learning model. The inner configuration of the retrained model of VGG16 is described in Fig. 3 which has 13 convolution layers, 3 dense layers (fully connected layers) and 5 max-pooling layers which sum up to 21 layers, but only 16 weight layers [22]. The last layer of the CNN model is the SoftMax layer.

System Design. The proposed method of leaf disease detection system uses VGG16 model to identify and classify five different leaf diseases of tomato and pepper bell plants, the block diagram of which is presented in Fig. 4. After splitting the dataset, we have loaded the pre-trained VGG16, AlexNet, and ResNet50 models using Keras. After retraining the models, the network is trained on a new dataset that can help the model to accurately predict presence or absence of a disease. We have set the maximum number of epochs to 350, 500, and 600 for the three models respectively. Then, the model is being trained on a new dataset. We have implemented VGG16, AlexNet, and ResNet50 models using various optimizers such as RMSProp, Nadam, Adam, SGD, Gradient Descent, Adagrad, Adadelta, Adamax, and Momentum.

After training the network on our dataset, classification of the different diseased class categories has been done. The name of the identified disease is displayed as output to the user. Our classification of diseased leaves achieves an accuracy of approximately about 97.4%, which is an acceptably higher rate.

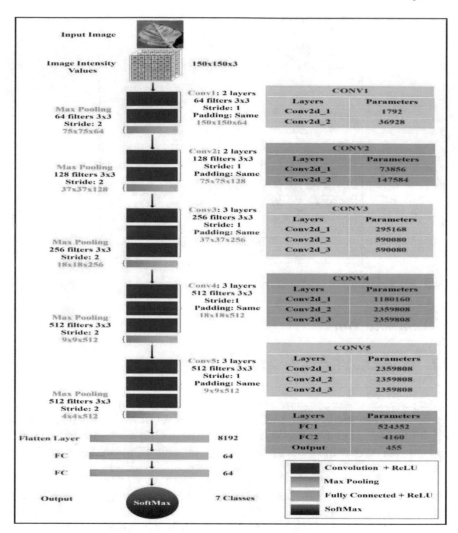

Fig. 3 Architecture of proposed retrained VGG16 CNN model

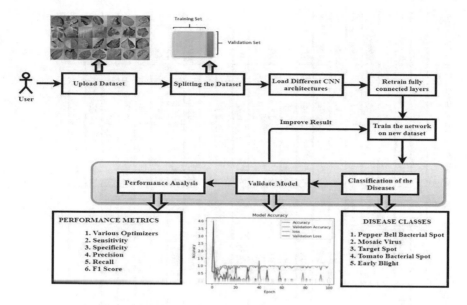

Fig. 4 Block diagram of our proposed leaf detection system

5 Experimental Results

We have evaluated our retrained VGG16 model based on the testing accuracy, confusion matrix, sensitivity, specificity, recall, f1 score, receiver characteristic (ROC) curve and precision.

5.1 Accuracy, Loss and Execution Time for Different CNN's Using Various Optimizers

In our proposed work, we make a comparison between VGG16, AlexNet, and ResNet50 using various optimizers namely RMSProp, Nadam, Adam, SGD, Adagrad, Adadelta, Adamax, Momentum and Gradient Descent (GD). Accuracy, Loss, and Execution Time of each optimization technique in classification of each disease are being tabulated in Table 2. From Table 2, we have observed that out of nine optimizers, RMSProp has given the best-optimized result and out of three CNN's, VGG16 has given higher accuracies for each disease. Therefore, we have implemented our system using VGG16 model with RMSProp optimizer. Our proposed system achieves an overall accuracy of about 97.4%, which is an acceptable higher rate.

Table 2 Performance analysis, training accuracy & loss and validation accuracy & loss of different optimization techniques used in seven different class categories

S. No	DL model	Performance metrics							
		Optimizer	Training accuracy	Validation accuracy	Training loss	Validation loss	Precision	F1 score	Recall
Hyper Parameters: LR:0.001, Batch size:4, Epochs:350, Activation Function: Softmax									
1	VGG16	RMSProp	0.925	0.974	0.075	0.026	0.97	0.96	0.96
		Adam	0.906	0.936	0.094	0.064	0.94	0.94	0.94
		Nadam	0.916	0.94	0.084	0.06	0.94	0.94	0.94
		SGD	0.907	0.925	0.093	0.075	0.93	0.92	0.93
		Adagrad	0.884	0.92	0.116	0.08	0.92	0.92	0.92
		Adadelta	0.89	0.91	0.11	0.09	0.91	0.91	0.91
		Adamax	0.883	0.903	0.117	0.097	0.9	0.9	0.9
		Momentum	0.908	0.876	0.092	0.124	0.88	0.87	0.87
		GD	0.869	0.906	0.131	0.094	0.91	0.91	0.91
Hyper Parameters: LR:0.001, Batch size:4, Epochs:600, Activation Function: Softmax									
2	RES NET50	RMSProp	0.902	0.935	0.098	0.065	0.94	0.93	0.93
		Adam	0.91	0.917	0.09	0.083	0.92	0.91	0.91
		Nadam	0.95	0.913	0.05	0.087	0.91	0.91	0.91
		SGD	0.93	0.902	0.07	0.098	0.9	0.9	0.9
		Adagrad	0.89	0.886	0.11	0.114	0.89	0.88	0.88
		Adadelta	0.902	0.88	0.098	0.12	0.88	0.88	0.88
		Adamax	0.84	0.862	0.16	0.138	0.86	0.86	0.86
		Momentum	0.89	0.857	0.11	0.143	0.86	0.85	0.85
		GD	0.82	0.85	0.18	0.15	0.86	0.86	0.85

(continued)

Table 2 (continued)

S. No	DL model	Performance metrics								
		Optimizer	Training accuracy	Validation accuracy	Training loss	Validation loss	Precision	F1 score	Recall
Hyper Parameters: LR:0.001,Batch size:4, Epochs:500, Activation Function: Softmax									
3	ALEX NET	RMSProp	0.936	0.902	0.064	0.1	0.9	0.9	0.9
		Adam	0.884	0.896	0.116	0.104	0.9	0.89	0.89
		Nadam	0.912	0.88	0.088	0.12	0.88	0.88	0.88
		SGD	0.86	0.859	0.14	0.141	0.86	0.85	0.85
		Adagrad	0.836	0.884	0.164	0.116	0.88	0.89	0.88
		Adadelta	0.843	0.887	0.157	0.113	0.89	0.89	0.88
		Adamax	0.91	0.856	0.09	0.144	0.86	0.85	0.86
		Momentum	0.903	0.86	0.097	0.14	0.86	0.86	0.85
		GD	0.82	0.847	0.18	0.153	0.85	0.84	0.85

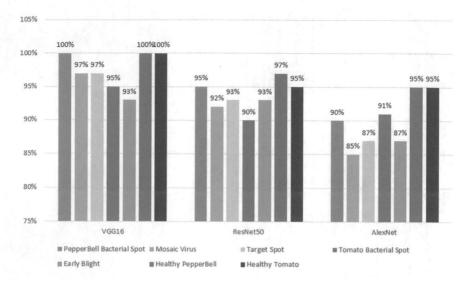

Fig. 5 Classification result of leaf disease detection using VGG16, ResNet50 and AlexNet

5.2 Performance Analysis of Alex Net, ResNet50 and VGG16 CNN Models in Classification

In this work, 1470 samples are trained and 280 samples are tested based on each class. A total of 40 samples are tested in each disease class category. Accuracy of the overall classification of each disease classes using different CNN models can be seen in Fig. 5. The classification accuracy (average accuracy of each disease class category) for the VGG16 CNN model is 97.4%.

5.3 Model Validation

VGG16 produces better classification results when compared with Alexnet, ResNet 50. We have built the model but would like to validate it by including datasets. The training and validation loss, training and validation accuracy curves can be seen in Figs. 6, 7, 8, 9 and 10 respectively. The graph is plotted for Training accuracy, Training loss, and Validation loss on the Y-axis and 350 epochs on the X-axis. The accuracies achieved for Pepper bell bacterial spot, Mosaic Virus, Target Spot, Tomato Bacterial Spot, Early Blight detection are about 100%, 97%, 97%, 95% and 95% respectively.

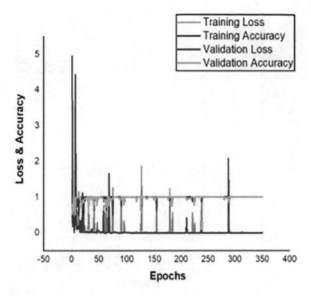

Fig. 6 Model accuracy for pepper bell bacterial spot detection

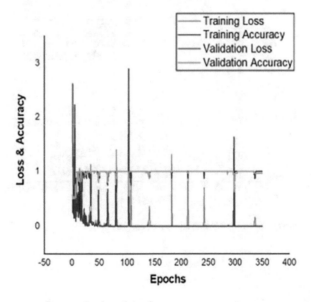

Fig. 7 Model accuracy for mosaic virus detection

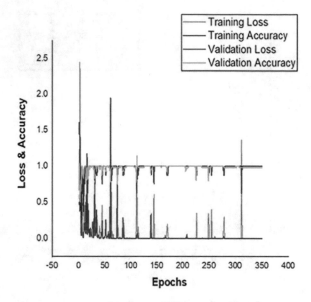

Fig. 8 Model accuracy for target spot detection

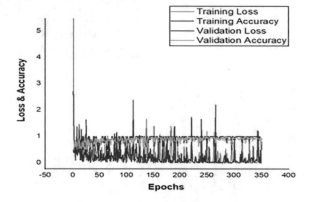

Fig. 9 Model accuracy for tomato bacterial spot detection

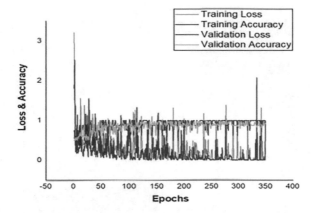

Fig. 10 Model accuracy for early blight detection

5.4 Performance Measure Indices

The models are tested and trained using 280 and 1470 images respectively. A total of 210 and 40 images are trained and tested for each disease class category. In ROC the area under the curve for Pepper bell bacterial spot detection as seen in Fig. 11.is 1.00 which shows that the model predictions are 100% correct, for Mosaic Virus detection is 0.97 as in Fig. 12, for Target spot detection is 0.97 as in Fig. 13 which shows that the model predictions are 97% correct, for Tomato bacterial spot detection is 0.95 which shows that the model predictions are 95% correct in Fig. 14 and for Early Blight detection is 0.92 as seen in Fig. 15.

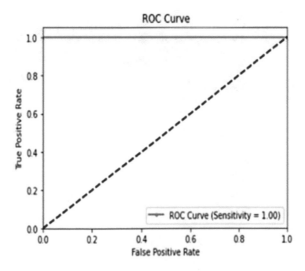

Fig. 11 ROC curve for pepper bell bacterial spot detection

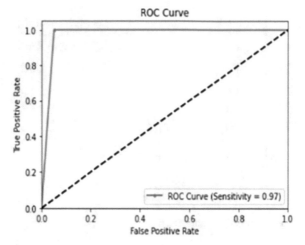

Fig. 12 ROC curve for mosaic virus detection

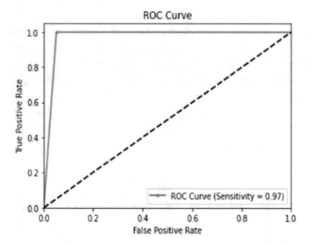

Fig. 13 ROC curve for target spot detection

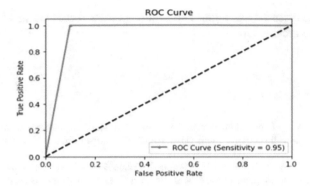

Fig. 14 ROC curve for tomato bacterial spot detection

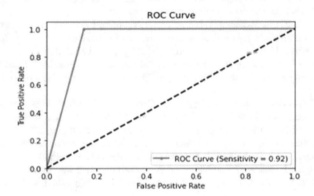

Fig. 15 ROC curve for early blight detection

6 Conclusions

Agriculture has a part in the entire life of humans and the economy. The economy largely depends on agricultural productivity. Since, plants disease is quite natural, plant leaf disease detection plays a key role in the agriculture field. If proper detection is not taken in those area, then it brings out serious consequences on plants due to which respective product quantity, quality, or productivity will be affected. Some automatic techniques are used to detect plant diseases as it decreases intense to monitor in big farms and to detect symptoms of plant diseases earlier. Hence, we have proposed a better system towards the culmination of the above-stated problem. Finally, we have implemented our work using the VGG16 CNN model to increase classification accuracy. Thus, we have proposed a system using CNN for the classification of diseased and non-diseased from the given dataset (during training and testing).Our system achieves an accuracy of more than 97% with other variations of diseases. In Future we plan to implement semantic segmentation using deep learning to segment the diseased leaf region and to attempt to increase classification accuracy by varying the architecture.

References

1. Jadhav S (2019) Convolutional neural networks for leaf image-based plant disease classification. IAES Int J Artif Intell (IJ-AI) 8(4):328. https://doi.org/10.11591/ijai.v8.i4.pp3 28-341
2. Rumpf T, Mahlein A-K, Steiner U, Oerke E-C, Dehne H-W, Plümer L (2010) Early detection and classification of plant diseases with Support Vector Machines based on hyperspectral reflectance. Comput Electron Agric 74:91–99. https://doi.org/10.1016/j.compag.2010.06.009
3. Zhou R, Kaneko S, Tanaka F, Kayamori M, Shimizu M (2014) 'Dis- ease detection of Cercospora Leaf Spot in sugar beet by robust template matching.' Comput Electron Agric 108:58–70. https://doi.org/10.1016/j.compag.2014.07.004
4. Yuan L, Bao Z, Zhang H, Zhang Y, Liang X (2017) Habitat monitoring to evaluate crop disease and pest distributions based on multi-source satellite remote sensing imagery. Optik Int J Light Electron Optics 145:66–73. https://doi.org/10.1016/j.ijleo.2017.06.071
5. Waidyarathne P, Samarasinghe S (2014) Artificial neural networks to identify naturally existing disease severity status. Neural Comput Appl 25:1031–1041. https://doi.org/10.1007/s00521-014-1572-6
6. Neumann M, Hallau L, Klatt B, Kersting K, Bauckhage C (2014) Erosion band features for cell phone image based plant disease classification. In: 2014 22nd international conference on pattern recognition, Stockholm, pp 3315–3320. https://doi.org/10.1109/ICPR.2014.571
7. Patil SS, Suhas KC (2014) Identification and classification of cotton leaf spot diseases using SVM CLASSIFIER. Int J Eng Res Technol (IJERT) 03(04):1511–1544
8. Phadikar S, Sil J, Das A (2013) Rice diseases classification using feature selection and rule generation techniques. Comput Electron Agric 90:76–85. https://doi.org/10.1016/j.compag.2012.11.001
9. Ranjan M, Weginwar MR, NehaJoshi P, Ingole AB (2015) Detection and classification of leaf disease using artificial neural network'. Int. J. Tech. Res. Appl. 3(3):331–333
10. Mokhtar U, El-Bendary N, Hassenian AE (2015) SVM-based detection of tomato leaves diseases. Adv Intell Syst Comput Intell Syst 323:641–652
11. Rothe, P. R. and R. V. Kshirsagar. "Cotton leaf disease identification using pattern recognition techniques," 2015 International Conference on Pervasive Computing (ICPC) (2019): 1–6.
12. Yang L, Yi S, Zeng N, Liu Y, Zhang Y (2017) Identification of rice diseases using deep convolutional neural networks. Neurocomputing 267:378–384
13. Tian J, Hu Q, Ma X, Han M (2012) An improved KPCA/GA-SVM classification model for plant leaf disease recognition. J Comput Inf Syst 8(18):7737–7745
14. Bock C (2011) Detection and measurement of plant disease symptoms using visible-wavelength photography and image analysis. CAB Rev Perspect Agric Veter Sci Nutr Nat Res 6. https://doi.org/10.1079/PAVSNNR20116027
15. Mim TT, Sheikh MH, Shampa RA, Reza MS, Islam MS (2019) Leaves diseases detection of tomato using image processing. In: 2019 8th international conference system modeling and advancement in research trends (SMART)
16. Iliadis L, et al (eds) (2020) Proceedings of the 21st EANN (engineering applications of neural networks) 2020 conference: proceedings of the EANN 2020, vol 2 Springer Nature, Heidelberg
17. Sowmiya M, Thilagavathi C (2020) Leaf disease detection of soybean plant using machine learning algorithms. Int J Innov Technol Explor Eng (IJITEE) 9(3). ISSN: 2278-3075
18. https://plantvillage.psu.edu/topics/pepper-bell/infos
19. Phadikar S, Sil J, Das AK (2013) Rice diseases classification using feature selection and rule generation techniques. Comput Electron Agric 90:76–85

20. Jadhav SB, Udupi VR, Patil SB (2020) Identification of plant diseases using convolutional neural networks. Int J Inf Technol 13:2461–2470
21. Zhou R, Kaneko S, Tanaka F, Kayamori M, Shimizu M (2015) Image-based field monitoring of Cercospora leaf spot in sugar beet by robust template matching and pattern recognition. Comput Electron Agric 116:65–79
22. Paulson A, Ravishankar S (2020) AI based indigenous medicinal plant identification. In: 2020 advanced computing and communication technologies for high performance applications (ACCTHPA). IEEE
23. Thampi SM, et al (eds) (2015) Advances in signal processing and intelligent recognition systems: 4th international symposium SIRS 2018, Bangalore, India, 19–22 September 2018, Revised Selected Papers, vol. 968. Springer Science and Business Media LLC, Heidelberg
24. Manavalan R (2020) Automatic identification of diseases in grains crops through computational approaches: a review. Comput Electron Agric 178:105802
25. Anami BS, Malvade NN, Palaiah S (2020) Deep learning approach for recognition and classification of yield affecting paddy crop stresses using field images. Artif Intell Agric 4:12–20
26. Rupnik R, Kukar M, Vračar P, Košir D, Pevec D, Bosnić Z (2019) AgroDSS: a decision support system for agriculture and farming. Comput Electron Agric 161:260–271
27. Rothe PR, Kshirsagar RV (2015) Cotton leaf disease identification using pattern recognition techniques. In: 2015 international conference on pervasive computing (ICPC)
28. Gupta S, et al (2019) A hybrid machine learning and dynamic nonlinear framework for determination of optimum portfolio structure. In: Innovations in computer science and engineering. Springer, Singapore, pp 437–448
29. Dhingra G, Kumar V, Joshi HD (2017) Study of digital image processing techniques for leaf disease detection and classification. Multimedia Tools Appl 77:19951–20000
30. Durga NK, Anuradha G (2019) Plant disease identification using SVM and ANN algorithms. Int J Recent Technol Eng (IJRTE) 7(5S4). ISSN: 2277–3878

Predicting Deflagration and Detonation in Detonation Tube

Samira Namazi, Ljiljana Brankovic, Behdad Moghtaderi, and Jafar Zanganeh

1 Introduction

Accidental fires and explosions in underground coal mines pose a continuous threat to miners' lives and welfare, expensive equipment, and timely coal delivery. Over years, there were several reports of accidental explosions which took human lives and caused a massive property damage. For example, in 1942, the deadliest coal mine explosion in human history (1549 victims) occurred in Benxi Hu Colliery, China [1]. More recently, in 2018, there was a fatal accident in Czech Republic that killed thirteen and injured ten people [2]. In the same year, there was a fire in North Goonyella mine in Australia [3].

To control the accidental fires and explosions, the mine safety operators can use a number of mitigation measures. Many of these countermeasures are proven technologies for mitigation of fires and explosions in process industries and can be potentially adapted without much difficulty, but they require mine safety operators to understand characteristics of methane explosion trough the pipes and mining tunnels. For example, in order to design an accurate capture duct with nearly zero error in VAM Abatement System, they need to know the maximum overpressure expected from an explosion. Similarly, in order to design a sufficiently strong capture duct and deploy

S. Namazi (✉) · L. Brankovic
School of Electrical Engineering and Computing, FENBE, The University of Newcastle,
Newcastle, Australia
e-mail: Samira.Namazi@uon.edu.au

L. Brankovic
e-mail: Ljiljana.Brankovic@newcastle.edu.au

B. Moghtaderi · J. Zanganeh
Frontier Energy Technologies Centre, FENBE, The University of Newcastle, Newcastle, Australia
e-mail: Behdad.Moghtaderi@newcastle.edu.au

J. Zanganeh
e-mail: Jafar.Zanganeh@newcastle.edu.au

© The Author(s), under exclusive license to Springer Nature Singapore Pte Ltd. 2022   529
B. Unhelker et al. (eds.), *Applications of Artificial Intelligence and Machine Learning*,
Lecture Notes in Electrical Engineering 925,
https://doi.org/10.1007/978-981-19-4831-2_43

an appropriate fast response countermeasure, they need to know the maximum flame front velocity.

In this paper, we apply data mining approach to predict the maximum pressure and maximum velocity during the explosion in a detonation tube. We use five different classification algorithms as follows: Decision Tree (DT), Random Forest (RF), Support Vector Machine with Sequential Minimal Optimization (SMO), Naïve Bayes (NB) and AdaBoostM1.

The paper is organized as follows. In Sect. 2, we present a literature review. In Sect. 3, we describe the experiment and the dataset. In Sect. 4, we compare the results obtained by five different methods. We provide the discussion of results and conclusion in the last section.

2 Literature Review

There are quite a few papers where authors studied explosion in coal mines. However, in many of them no data mining technique was used [4–8].

The most relevant research for this paper was IJCRS'15 Data Challenge, a data mining competition associated with 2015 International Joint Conference. This challenge focused on predicting methane outbreaks in Polish Long Wall coal mine [4]. Among 50 proposed methods, the five winning approaches that were able to achieve the highest accuracy were, respectively: 1) Random Forest (generic approach) [9], 2) Selective Naïve Bayes with automatic variable construction [6], 3) Fast Greedy Backward-Forward Search [7], 4) Recurrent Neural Network with LSTM [8], and 5) Support Vector Machines with grid search [4].

More recently, new studies have been conducted, which apply data mining to predicting explosions in coal mines. In particular, these studies used Fuzzy Analytic Hierarchy Process (FAHP) and Bayesian Network [10], Adaptive Weighted Least Squares Support Vector Machine (AWLS-SVM) [11], Support Vector Regression (SVR) [12], Artificial Neural Network (ANN) [13] and, Uniform Manifold Approximation and Projection (UMAP) and Long Short-Term Memory (UMAP-LSTM) [14].

There are also several studies focusing on predicting concentration, dispersion and emission of methane in coal mines, using Artificial Neural Network (ANN) [15–17], Deep Belief Network (DBN) [18], and Multilayer perceptron (MLP) Network [19].

Our work described in this paper focuses on predicting deflagration and detonation in detonation tubes rather than coal mines. Most pervious experimental studies of the properties of methane-mixture explosions were conducted in small-scale tubes, using high energy ignition sources. Additionally, these studies did not report on the value of pressure during deflagration. Table 1 shows a historical summary of previous experiments on methane air mixture explosion in cylindrical vessels. More research is needed to identify the variables that have the most impact on the deflagration to detonation transition in longer tubes using low ignition energy.

Table 1 Summary of experiments on methane air mixture explosion cylindrical tubes

Authors	Experiments
Sir Frederick Augustus Abel (1869) [20]	First to observe pressure development of gasses in tubes. Scales: 5 m long and 50 mm diameter
M. Berthelot and P. Vieille (1881) [21]	First to systematically document deflagration and detonation in tubes of hydrocarbon gases
E. Mallard and Henry Le Chatelier (1881, 1883) [22], E. Jouguet (1913) [23]	First theories of flame velocity in gases
W. Mason and R.V. Wheeler (1917, 1920) [24, 25]	First to study properties of methane explosions in tubes; detected a gradual increase of flame velocity. Scales: 5 m long and 50 mm diameter
S. M. Kogarko (1958) [26]	Observed deflagration to detonation transition in the range of 6.3–13.5% of methane-air mixture. Scales: 11.2 m long and 305 mm diameter
Wolanski et al. (1981) [27]	Observed transition from deflagration to detonation in the range of 8–14.5% of methane in the methane-air mixture. Scales: 6.35 cm square tube and 9.42 MJ/m2 ignition
Knystautas et al. (1982) [28]	Discover the deflagration to detonation transition. Scales: longer than 0.24 m
Phylaktou et al. (1990) [29]	Observed the flame speed and pressure rate increase in a vertical closed tube. Scales: 1.64 m long, 21.6 L/D ratio, 16 J ignition energy
Kindracki et al. (2007) [30]	Observed the maximum pressure and flame speed in the middle of tube and the effect of ignitor location on the result. Scales: 1.325 m long and 128.5 mm diameter
Wei et al. (2009) [31]	Observed that the experimental maximum pressure was less than theoretical estimation, while the total pressures were very close. Scales: 30 m long and 0.5 m diameter
Li et al. (2012) [32]	Discovered a significant impact of closed pipe on the deflagration to detonation transmission. Scales: 12 m long, 80 mm diameter and 10 kJ ignition energy

3 Experiment Description

Data used in our machine learning study were obtained in an industrial scale research project which was carried out at the University of Newcastle, Australia, 2014–2018 (VAM Abatement Safety Project). Figure 1 shows the general view of the detonation tube used to examine the characteristics of the methane-air mixture fire and explosion. The total length of the detonation tube is 30 m, the diameter is 0.5 m and there is a silencer at the open end of the tube to reduce the explosion noise. The detonation tube is constructed in a modular fashion and consists of 11 spools. In each of these spools,

Fig. 1 Experiment setup at UoN [33]

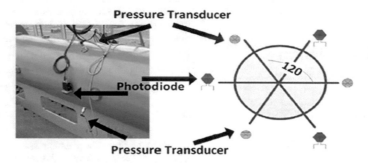

Fig. 2 Pressure transducer and photodiode locations [33]

explosion pressure and flame front velocity are measured by 3 pressure transducers and 3 photodiodes located at the middle of the spool on the same plane (see Fig. 2).

The methane explosions encompass two phenomena: deflagration and detonation. Flame deflagration is characterised by low pressure and subsonic velocity. Under some conditions, the flame deflagration may transition into detonation, which comes with supersonic velocity and high pressure. The transition from deflagration to detonation is known as DDT. Prediction of the occurrence of DDT is not easy. A typical pressure and flame velocity variation profile for methane fire and explosion in a pipe closed at one end is shown in Fig. 3.

The aim of this study is to predict the "dangerous" value of pressure and "supersonic" value of flame front velocity in the detonation tube. Velocity is supersonic if it is greater than or equal to the speed of sound, which is approximately 343.2 m/s. Since velocity and pressure in the experimental tube are directly related to each other, we determine the bound between "Dangerous" and "Not-dangerous" pressure from the data itself. We fit a curve to describe maximum pressure as quadratic function of maximum velocity (Eq. 1, Fig. 4). Using this function, we find that the maximum pressure corresponding to maximum velocity of 343.2 m/s is around 240 kPa.

$$MP = 0.001(MV) + 0.034MV + 105.3 \tag{1}$$

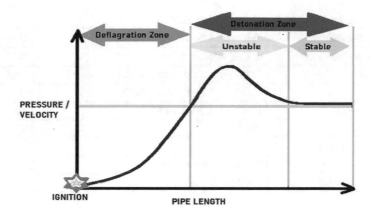

Fig. 3 Flame propagation and explosion pressure rise development zones

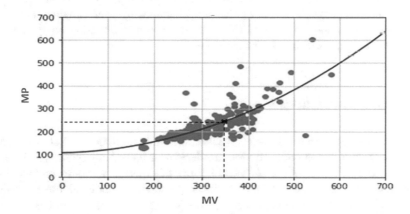

Fig. 4 The scatter plot and curve fitted between maximum pressure and maximum velocity

Our dataset contains a row for each valid experiment (244 rows in total) and seven columns corresponding to seven variables. Five of the columns correspond to the conditions of each experiment as independent variables and the final two correspond to maximum pressure and maximum velocity attained in the experiment as dependent variables. To determine the maximum pressure and maximum velocity for each experiment, we first compute the average of three readings in each spool and each point in time, and then we compute the maximum over all spools and all time points. Table 2 shows the variables and their domains.

This dataset is divided into two datasets, dataset A with 215 experiments for 50 mJ ignition energy and dataset B with 29 experiments with a varying ignition energy. Datasets A and B differ significantly in a way in which the experiments were conducted, regarding how the methane was distributed between the two injections site.

Table 2 Independent variable and dependent variable

Independent variable	**Values** (number of experiments)	Unit
Methane concentration	**0, 1.25, 2.5, 5, 7, 7.5, 8, 9.5, 11** (1), **15** (6)	%
Coal dust	**0** (109), **0.01** (15), **0.02** (3), **10** (51), **30**(66)	g/m^3
Length	**3** (111), **6** (91), **12** (31), **25** (11)	m
Ignition energy	**50 mJ, 1 kJ, 5 kJ, 10 kJ**	mJ/kJ
Temperature	**[10, 41]**	°C
Dependent variable		
Maximum pressure	Dangerous, Not dangerous	
Maximum velocity	Subsonic, Supersonic	

Dataset A comprises seven variables: methane concentration (M), coal dust concentration (C), reactive section length (L), temperature (T), maximum pressure (MP), and maximum velocity (MV). Dataset B includes all seven variables contained in dataset A, plus ignition energy (IE). In the above, reactive section length refers to the area in the pipe which contains methane.

3.1 Study Limitations

Careful examination of the datasets reveals some study limitations. First, some variables, such as methane concentration and length of the tube, exhibit significantly unbalanced distribution of values. Vast majority of experiments were conducted for short tube length and low methane concentration. On the other hand, there is only a handful of experiments on longer tube lengths (such as 25 m) and/or higher methane concentration (such as 8 and 11%). For more information, we refer the reader to Table 2, where the number of experiments for some of the values of independent variables are given in parenthesis and Figs. 5 and 6, which show the distribution of number of experiments over different methane concentrations and tube lengths.

Secondly, some of the independent variables exhibit non-trivial correlation which is undesirable [24]. For example, the methane concentration and temperature have correlation coefficient of 0.41, length and temperature 0.23, and length and coal concentration 0.20. All values are summarized in Table 3. The relationship between methane concentration and temperature is further illustrated in Fig. 7.

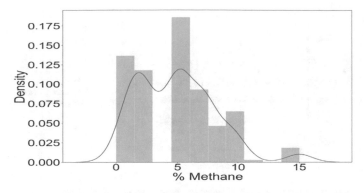

Fig. 5 The distribution of number of experiments for different methane concentration

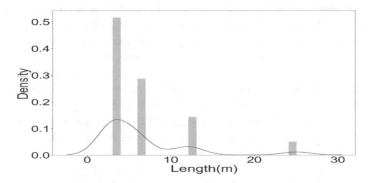

Fig. 6 The distribution of number of experiments for different lengths of the tube

Table 3 Correlation between independent and dependent variables

	M	C	L	T	IE	MP	MV
M	1	-0.08	0.05	0.41	0.25	0.49	0.51
C	-0.08	1	-0.20	-0.04	-0.07	-0.05	0.01
L	0.05	-0.20	1	0.23	-0.01	0.13	-0.00
T	0.41	-0.04	0.23	1	0.06	0.23	0.16
IE	0.25	-0.07	-0.01	0.06	1	0.02	0.05
MP	0.49	-0.05	0.13	0.23	0.02	1	0.76
MV	0.51	0.01	-0.00	0.16	0.05	0.76	1

3.2 Preprocessing

Real-world data is typically incomplete, inconsistent, and erroneous. Our dataset is no exception. We identified some missing values in our dataset, for example, some of the experiments did not have the temperature recorded. Similarly, among 1,000,000

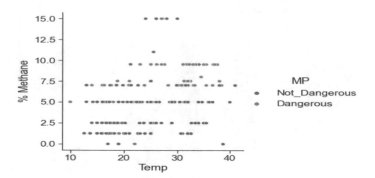

Fig. 7 The scatter plot of methane concentration versus temperature

pressure readings, several missing values were found. To handle the missing temperature values, we filled it in manually with values obtained official Australian government meteorological website (www.bom.gov.au), based on the place, date, and time of each experiment. This approach was time consuming, but we were able to use it as we had relatively small number of missing values. For missing pressure values, we used a mean of the values recorded immediately before and after the missing value.

As expected from a real-world data, our raw dataset was noisy and massive. In order to visualize the data, we used moving average technique to reduce the noise and smooth out the dataset. Figure 8 shows the moving average of the mean pressure over time, for each of the spools.

Next, we transformed the data by representing each experiment with seven values as described in Sect. 3. Finally, we converted dependent variables from numerical into binary to facilitate classification.

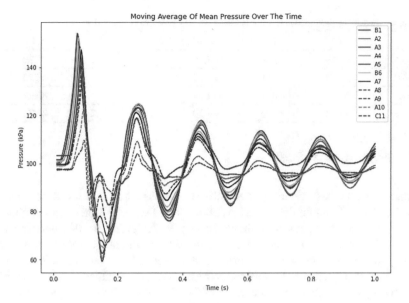

Fig. 8 Moving average of mean pressure over the eleven spools (Experiment 002: M = 1.25%, L = 6 m, C = 0, IE = 50 mJ)

3.3 Tools Used in Research

In this research we used both Python programming language with large standard libraries and Weka, a free data analysis software designed by University of Waikato in New Zealand that supports all standard machine learning task such as preprocessing, classification, regression, clustering, visualization, and feature selection [34].

4 Classification Results

We applied five classification techniques including Decision Tree (DT), Random Forest (RF), Support Vector Machine with Sequential Minimal Optimization (SMO), Naïve Bayes (NB) and AdaBoostM1 with Tree Stumps as underlying classifier. We used tenfold cross validation and compared the methods based on the accuracy (percentage of correctly classified instances).

Our results are summarized in Table 4. Each column in the table corresponds to a classification method. Each row corresponds to either dataset A or dataset B, and either maximum pressure or maximum velocity. For example, row MP-A shows classification accuracies for predicting maximum pressure in dataset A.

The results indicate that the best classification model in most cases is AdaBoostM1, followed by Decision Tree.

Table 4 Results of classification models based on 10 - fold cross validation

	DT	RF	AdaBoostM1	SMO	NB
MV-A	85.58	85.58	85.58	73.02	81.39
MV-B	86.20	82.75	89.65	79.31	86.20
MP-A	80.93	81.86	84.65	73.95	80.93
MP-B	89.65	82.75	86.20	79.31	82.75

4.1 Maximum Velocity

In dataset A, the best accuracy of 85.58% was achieved by three methods, namely Decision Tree, Random Forest and AdaBoostM1. Decision Tree for predicting the maximum velocity in set A is shown in Fig. 9. According to this decision tree, supersonic velocity of the flame front occurs only in the range 5–11% of methane concentration. For methane concentration in the range 7.5–11%, the velocity of the flame front is predicted to be supersonic regardless of the values of other variables. For methane concentration in the range of 5–7.5%, flame front achieves supersonic velocity when there is a sufficient length of the tube (greater than 6 m), and/or sufficient coal concentration (greater than 10 g/m^3). This finding is consistent with the theoretical understanding of DDT process, except that explosive methane concentration range is deemed to be 5–15%. This discrepancy is not unexpected, considering that our dataset is unbalanced regarding methane concentration, and contains only 7 experiments with methane concentration over 11%.

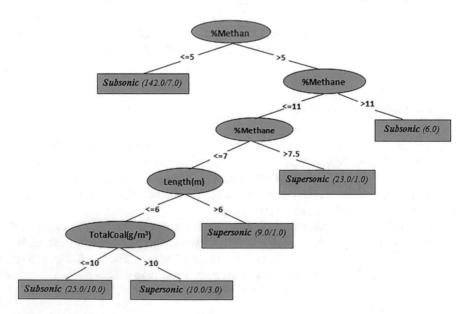

Fig. 9 The decision tree for predicting the maximum velocity, dataset A

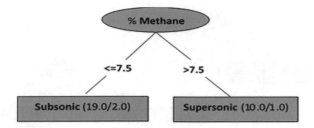

Fig. 10 The decision tree for predicting the maximum velocity, dataset B

Dataset B (the set with a varied ignition energy), contains much smaller number of experiments than dataset A. For dataset B, AdaBoostM1 achieved the highest accuracy (89.65%) among all the models. While this accuracy is higher than the best one achieved for dataset A, it may be due to overfitting. Decision Tree model uses a single attribute (methane concentration) for predicting the flame front velocity (see Fig. 10). It is worth noting that all the experiments in dataset B have the same tube length (6 m). Unlike in dataset A, here very few experiments with methane concentration less than or equal to 7.5% achieve supersonic velocity, even for high concentration of coal dust (30 g/m^3). Closer inspection of the dataset revealed that for methane concentration of 7.5%, high coal dust concentration (30 g/m^3) appears to lead to supersonic flame front velocity but due to a small number of experiments was not discovered by the decision tree builder.

4.2 Maximum Pressure

In dataset A, the best solution is delivered by AdaBoostM1 with accuracy of 84.65%. The model indicates that dangerous pressure occurs for methane concentration in the range 7–9.5% of methane concentration, as well as the range 5–7%, providing that coal dust concentration is higher than 0.02 g/m^3 and temperature is higher than 30.3 °C. As shown in Fig. 11, it is worth mentioning that the original decision tree model for predicting maximum pressure appeared to be over-fitted which was solved by bounding the minimum number of experiments per tree leaf to 7.

In dataset B, decision tree was identical to the decision tree for predicting supersonic velocity of the flame front and achieved the highest accuracy (89.65%) among all models (see Fig. 12).

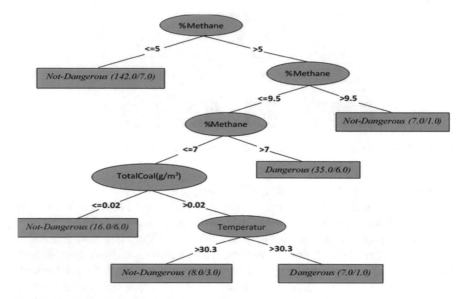

Fig. 11 The decision tree for predicting maximum pressure, dataset A

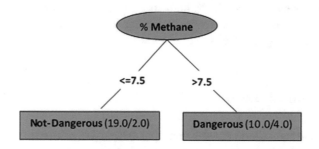

Fig. 12 The decision tree for predicting maximum pressure, dataset B

5 Discussion of the Results

In this paper, we used five different techniques to classify the maximum pressure and maximum flame velocity in a closed end detonation tube. We observed that the tree-based methods have the highest accuracy. The AdaBoostM1 performed the best in all but one case, while Decision Tree achieved the highest accuracy in the remaining case.

We evaluated the five methods for two datasets: dataset A with a 50 mJ ignition energy, and dataset B with varying ignition energy. As expected, for dataset A, classification of maximum flame velocity is more accurate than classification of maximum pressure. Moreover, the decision tree model for classifying maximum velocity is more consistent with theoretical understanding of DDT phenomenon than classification of maximum pressure. This is most likely due to the reflection of pressure waves in the detonation tube. A study by Qingzhao Li et al., showed that in the closed pipe, the reflected pressure wave has a considerable impact on the properties of gas explosion flame [35].

In dataset B, due to very small number of experiments (29), decision trees for maximum pressure and flame velocity are identical and less informative than for dataset A.

Overall, this study identifies a detonation methane concentration range of 5 to 11%. It is worth noting that previous studies as well as theoretical understanding of this field identified DDT methane concentration range to be 5% to 15%. We note that our data sets contain only a very small number of experiments with methane concentration of 11% or greater (Table 2), which could explain this discrepancy.

6 Conclusion and Future Work

In this paper, we applied five different classification techniques to predict dangerous pressure and supersonic flame front velocity in the experimental tube based on methane concentration, coal dust concentration, length of tube, ignition energy and temperature. The tree-based methods such as AdaBoostM1 show more accurate predictions. The dangerous pressure and supersonic velocity occur for methane concentration in the range 5 to 11%.

The most significant result of our study is identifying tube length and coal dust concentration as deciding factors for predicting DDT for methane concentrations in the range (5, 7%). Additionally, we identified coal dust concentration and temperature as most influential parameters for raising maximum pressure when methane concentration is in the rang (5, 7%).

In the future work, we intend to perform feature extraction to reduce the number of variables in order to enable nonlinear regression and other methods to predict the value of maximum pressure and value of maximum flame front velocity.

References

1. Dhillon BS (2010) Mine safety: a modern approach. Springer science & business media, Heidelberg
2. Santora M (2018) 13 Dead in Czech Coal Mine Explosion. https://www.latimes.com/world/la-fg-czech-coal-mine-20181221story.html
3. Ninness J (2018) Coal mine fire building at North Goonyella. https://www.amsj.com.au/coal-minefire-building-at-north-goonyella/
4. Wang Y, Qi Y, Gan X, Pei B, Wen X, Ji W (2020) Influences of coal dust components on the explosibility of hybrid mixtures of methane and coal dust. J Loss Prev Process Ind 67:104222
5. Zhu Y, et al (2020) Investigation of methane-air explosions and its destruction at longwall coalface in underground coalmines. Energy Sources Part A Rec Utilizat Environ Effects, 1–18

6. Zhu Y, Wang D, Shao Z, Xu C, Li M, Zhang Y (2021) Characteristics of methane-air explosions in large-scale tunnels with different structures. Tunn Undergr Space Technol 109:103767
7. Gamezo VN, Oran ES, Kunka L, Kaplan CR (201) Numerical analysis of gas explosions in coal mines. Alpha Foundation for the Improvement of MineTexas A and M
8. Qi Y, Gan X, Li Z, Li L, Wang Y, Ji W (2021) Variation and prediction methods of the explosion characteristic parameters of coal dust/gas mixtures. Energies 14(2):264
9. Zagorecki A (2015) Prediction of methane outbreaks in coal mines from multivariate time series using random forest. In: Rough Sets, fuzzy sets, data mining, and granular computing. Springer, Heidelberg, pp 494–500.
10. Li M, Wang H, Wang D, Shao Z, He S (2020) Risk assessment of gas explosion in coal mines based on fuzzy AHP and bayesian network. Process Saf Environ Prot 135:207–218
11. Sun Z, Li D (2020) Coal mine gas safety evaluation based on adaptive weighted least squares support vector machine and improved dempster-shafer evidence theory. Disc Dyn Nat Soc 2020:1–12
12. Liu L, Liu J, Zhou Q, Qu M (2021) An SVR-based machine learning model depicting the propagation of gas explosion disaster hazards. Arab J Sci Eng 46:1–12
13. Soomro AH, Jilani MT (2020) Application of IoT and artificial neural networks (ANN) for monitoring of underground coal mines. In: 2020 international conference on information science and communication technology (ICISCT). IEEE, pp 1–8
14. Kumari K et al (2021) UMAP and LSTM based fire status and explosibility prediction for sealed-off area in underground coal mine. Process Saf Environ Prot 146:837–852
15. Mathatho S, Owolawi PA, Tu C (2020) An artificial neural network and principle component analysis based model for methane level prediction in underground coal mines. In: Proceedings of the 2nd international conference on intelligent and innovative computing applications, pp 1–7
16. Tutak M, Brodny J (2019) Predicting methane concentration in longwall regions using artificial neural networks. Int J Environ Res Public Health 16(8):1406
17. Mishra DP, Panigrahi DC, Kumar P, Kumar A, Sinha PK (2021) Assessment of relative impacts of various geo-mining factors on methane dispersion for safety in gassy underground coal mines: an artificial neural networks approach. Neural Comput Appl 33(1):181–190
18. Wu X, Zhao Z, Wang L (2019) Deep belief network based coal mine methane sensor data classification. J Phys Conf Ser 1302(3):032013
19. Tutak M, Brodny J (2019) Forecasting methane emissions from hard coal mines including the methane drainage process. Energies 12(20):3840
20. Abel FA (1869) XIV. Contributions to the history of explosive agents. Philos Trans R Soc Lond 159:489–516
21. Berthelot M (1882) Sur la vitesse de propagation des phénomènes explosifs dans les gaz. CR Acad Sci 95:151–157
22. Mallard E, Le Chatelier H (1881) Sur les vitesses de propagation de l'inflammation dans les mélanges gazeux explosifs. Comptes Rendus Hebdomadaires des Séances de l'Académie des Sciences 93:145–148
23. Jouguet E (1913) Sur la propagation des dtflagrations et sur les limites d'inflammabilitt. CR Acad Sci 136:1058–1061
24. Mason W, Wheeler RV (1917) XCII.—the "uniform movement" during the propagation of flame. J Chem Soc Trans 111:1044–1057
25. Mason W, Wheeler RV (1920) V.—the propagation of flame in mixtures of methane and air. Part I. horizontal propagation. J Chem Soc Trans 117:36–47
26. Kogarko S (1958) Detonation of methane-air mixtures and the detonation limits of hydrocarbon-air mixtures in a large-diameter pipe. Soviet Phys-Tech Phys 3(9):1904–1916
27. Wolański P, Kauffman C, Sichel M, Nicholls J (1981) Detonation of methane-air mixtures. In: Symposium (international) on combustion, vol 18, no 1. Elsevier. Amsterdam, pp 1651–1660
28. Knystautas R, Lee J, Guirao C (1982) The critical tube diameter for detonation failure in hydrocarbon-air mixtures. Combust Flame 48:63–83

29. Phylaktou H, Andrews G, Herath P (1990) Fast flame speeds and rates of pressure rise in the initial period of gas explosions in large L/D cylindrical enclosures. J Loss Prev Process Ind 3(4):355–364
30. Kindracki J, Kobiera A, Rarata G, Wolanski P (2007) Influence of ignition position and obstacles on explosion development in methane–air mixture in closed vessels. J Loss Prev Process Ind 20(4–6):551–561
31. Wu HW, Gillies A, Oberholzer J, Davis R (2009) Australian sealing practice and use of risk assessment criteria-ACARP Project C17015. In: Proceedings of the Queensland mining industry health and safety conference, Townsville, Queensland, Australia, 23–26 August 2009
32. Li Q, Lin B, Jian C (2012) Investigation on the interactions of gas explosion flame and reflected pressure waves in closed pipes. Combust Sci Technol 184(12):2154–2162
33. Ajrash MJ, Zanganeh J, Moghtaderi B (2017) Deflagration of premixed methane–air in a large scale detonation tube. Process Saf Environ Prot 109:374–386
34. Holmes G, Donkin A, Witten IH (1994) Weka: a machine learning workbench. In: Proceedings of ANZIIS'94-Australian New Zealand intelligent information systems conference. IEEE, , pp 357–361
35. Pyatnitskii LN (2019) Flame propagation and acoustics. Comb Explos Shock Waves 55(6):633–643

Movie Recommendation Based on Fully Connected Neural Network with Matrix Factorization

Vineet Shrivastava and Suresh Kumar

1 Introduction

The fast development of the web has resulted in a new generation of Information sorting. The World Wide Web (WWW) provides a novel way of communication that exceeds the traditional way of communication [1]. It has a great impact on daily life. The World Wide Web has revolutionized the way info was gathered, discussed, provided, used, and stored.. Still, large quantity of information over the internet remains unorganized, so naive users face challenge while searching for the required information. This requires the help of experts to search for their preference over large and complex website [2].

To support users for dealing with a large quantity of information, several companies have built RS to assist in searching their preferences [3]. Research in the domain of RS was going on almost half-century, but interest remains higher due to plenty of practical applications and issues rich domain. Amazon Prime, Netflix are examples of big companies that uses online RS for suggesting movies to the users [4].

RS has a balanced economy of e-commerce websites like Netflix and Amazon which have been created as a prominent part of websites [5]. Some website's profit is illustrated in Table 1.

RS provides a suggestion list, where users can accept or reject according to their interest and user can also provide feedback about the suggestion list. User's feedback and their actions are preserved in the recommender database for developing a new recommendation list for the next interaction of users with the system. High-quality personalized recommendations increase more dimension to user experience. Web personalized RS was utilized to give varied kinds of customized information to the

V. Shrivastava (✉) · S. Kumar
Department of Computer Science and Engineering, Manav Rachna International Institute of
Research and Studies, Faridabad, India
e-mail: vineet_shrivastava@manavrachna.net

© The Author(s), under exclusive license to Springer Nature Singapore Pte Ltd. 2022   545
B. Unhelker et al. (eds.), *Applications of Artificial Intelligence and Machine Learning*,
Lecture Notes in Electrical Engineering 925,
https://doi.org/10.1007/978-981-19-4831-2_44

Table 1 Benefits of companies with RS

Choicestream	28% of peoples buy music based on their likes
Amazon	35% of sales from recommendation
Google news	Recommendations produce 38% more click-troughs
Netflix	2/3 of watched movies are recommended

users [6]. For the case of movie recommendation, the issue of choosing good films gets increases as time passes.

A movie is a combination of visual arts and entertainment. Generally, posters of the movie give an idea of the movie to the audience. Posters act as a key element to accelerate the hyper of the audience before and after the release of the movie. Most of the people decide to book the movies by seeing the posters [7]. The mood of the movie can be easily predicted by looking at the typography of the poster. The decision process is direct and does not need to review reading. Hence, in addition to standard movies, a recommendation algorithm is preferred to process posters of movies to predict similar movies which are provided as a recommended list to the users [8]. With content-based filtering, RS does not require information about other users and provides a recommendation list based on the interest of a specific user. Collaborative filtering is used to predict user interest without knowing domain knowledge of previous preference. In this paper, a hybrid technique is developed for movie recommendation. User interest is captured, and a recommended list of similar movies is provided. To find similarities among low dimensional data, Matrix factorization is used.

2   Review of Existing Works

The body of work paying attention to the movie recommendation system is diverse and vast, spanning a couple of interconnected disciplines. The evaluation of the reviews measures the different important research work related to the movie recommendation process that happened in the past.

Visualizing posters of movies can provide better learning and may result in a better recommendation. In their paper, Zhao L. et al. [9] developed a new framework for movie RS facilitates the inclusion of visual features to support recommendation tasks. The authors focused on high and low-level visual features from still frames and movie posters for further enhancement of RS. Generally, visual information taken from the still frame is a better measurement of similarities among movies. Additionally, the utilization of a linear combination of the visual feature was capable of learning bias most accurately.

Initially, for developing a hybrid recommendation system, Singhal A. et al. [10], used various deep learning (DL) techniques to convert current advances in the recommendation system. Here review was done based on a hybrid system, content-dependent system, and collaborative system. This study discussed the impact of DL integrated recommendation system with various application domains. Finally, this study concludes with the impact and importance of DL over the recommended system.

For the recommendations of movies on mobile applications, Ibrahim M. et al. [11], developed a Movie recommendation tool for the user provided with valuable services. Reviews were examined by a deep semantic analyzer depending on RNN (Recurrent Neural Network) along with UMA (User Movie Attention) to create emotion. The RS examines multivariate and creates the most prominent movie recommendation list related to users' tastes on a mobile application in an effective manner.

Optimization of RS can provide refined results for recommendation and can achieve by combining various conventional techniques with parameter optimization algorithms. In their work, Aljunid M. F. et al. [12], developed a movie RS depending on ALS (Alternating Least Square) with the use of Apache Spark. The work concentrates on parameter selection of the ALS algorithm combined with Collaborative filtering affects the performance of developing a movie recommender engine. The design was evaluated with the use of various metrics like RMSE (Root Mean Square Error) for prediction and execution time for training design.

A movie genre is category based on similarities in the narrative. Reddy S. et al. [13], created a movie RS based on the type of genres, users curious to look at. In the paper authors use content dependent filtering with genre correlation. In case the user provides a higher rating to a film of a certain genre, then a movie of the equivalent genre will be highly recommended to that specific user.

Social media plays a vital part in extracting the taste of users. In their work, Virk H. K. et al. [14] developed a movie recommendation system considering its significance in social media because of its feature of suggesting a list of movies to users depending on their taste. It extracts valuable information like attractiveness and popularity from the movie database for the recommendation process. Additionally, collaborative, content-dependent, and hybrid filtering used to build a system that provides more concern for recommended movies.

Social media personalization is also significant in the context of business. In Personalization, the recommendations are focused on the client. Adeniyi D. A. et al. [15], have used overspread information in their work that quickly determines one's favorite movie from numerous movies. The authors developed a Personalized RS with the help of the K Neighbor Algorithm that plays a major role, particularly when users have no final decision. Additionally, Subramaniyaswamy V. et al. [16], have developed a recommendation engine based on the idea of collaborative filtering. User ratings were utilized to connect other users having similar tastes and recommendations were provided without common entities. Euclidean distance scores were computed to predict the neighbours. User with minimum Euclidean distance score was predicted. Results proved that the proposed system was more reliable, precise, and develop numerous personalized movie recommendations when compared to existing models.

In terms of combining various methods in personalization, Rajarajeswari S. et al. [17] have developed movie RS utilizing content-dependent and collaborative filtering for form hybrid model. Later experiments were conducted on benchmark datasets to predict computation time and accuracy of the hybrid model. The results show a better prediction than the previously defined work.

Neural Networks provides more learning ability to a system, Utilizing the same concept, Yi B. et al. [18], developed a DL dependent collaborative filtering framework, called DMF (Deep Matrix Factorization), which could add some information type effectively and handily.

A special kind of neural network known as autoencoder is utilized for the purpose of recommendation where the feedback is not involved. An autoencoder is a specific kind of artificial neural network used to discover effective information coding in an unsupervised way. In their work, Zhao J. et al. [19], have developed a hybrid initialization technique depending on the autoencoder neural network (NN) and attribute mapping to resolve the issues of feedback.

3 Proposed Technique

3.1 Dataset Description

MovieLens dataset are basic datasets used to train recommendation models. It is received from the GroupLens website. The main data file includes a tab-separated list along with user-id, item-id, timestamp, and rating as 4 attributes.

Datasets are characterized as:

100,000 examination (scale from 1–5) of 1682 films by 943 users.

Every user has examined a minimum of 20 films, which results in a sparsity of 6.3%.

Which means that 6.3% of user-item rating has values. Though we filled the missing rating with 0, it should not be assumed as 0. They are considered as empty entries. Dataset is split into training and testing dataset by omitting 10 ratings given by the user from the training set and positioning them over test set. This kind of movies are found from Internet Movie Database. Several demographic information about users like age, professional, sex exist. But this type of information is not utilized. "rating.csv" consists of time, rating, movie-id, user-id information, and "link.csv" consist of TMDB-id, IMDB-id. Integrating these 2 attributes will result in retrieval of IMDB-id information which gets a poster of all movies from Movie database website using its API.

3.2 Feature Analysis

Movies dataset holds info regarding the various fields related to movies. The fields are movie id, ratings, tags, genre, movie title, timestamps, and movie title. These fields or features are utilized for the evaluation of RS. Pandas library and mat plot library are utilized for removing the undesired fields from the dataset and plotting the graph making use of the characteristics for everyone to offer much better visualization of the analysis. Certain conditions were applied over the dataset to make an overall analysis. This analysis helps the researcher to get a view about the useful fields for the implementation of a RS.

3.3 Content-Based Filtering

This kind of filtering concentrates on the properties of items. This filtering intends to predict significant characteristics of items that are simply identifiable which serves to create a profile. Similarity measurement among the items means measuring the similarity between the profiles.

Weighting Calculation Based on Score
Here weight calculation is done with the use of IMDB. Where IMDB provides a rating scale that permits the users to rate the film on a scale of 1–5 or 1–10. IMDB denotes that rating submitted by users which are filtered and weighted in several ways to create weighted mean for every film. IMDB uses the following formula to compute its weighted rating:

$$W = \frac{S.u + D.n}{u + n} \tag{1}$$

where w = weighted rating.
S = average for movies as from 1–15 (mean) = (rating).
u = number of votes for movies.
n = minimized votes needed to be listed in top 250.
D = mean vote across the whole report.

IMDb has 100 features assembled in a way through a similar process, 10,000 votes are needed to qualify for the recommended list.

Finding Best Items by Genres
Here, based on the weighted mean best movie is predicted related to several genres which can be provided as a recommendation list to that user.

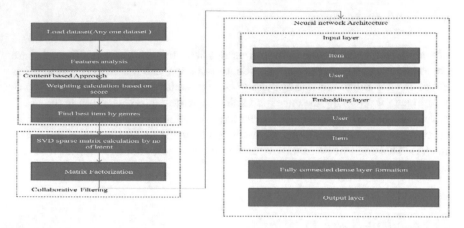

Fig. 1 Overall flow of movie recommendation system

3.4 Collaborative Filtering (CF)

It is a technique used to find the interest of users depending on information about other user's preferences. For example, if X has a similar opinion as Y for an issue, then it assumed that mostly X have a similar opinion as Y on various issues. There is a various collaborative filtering system. Most popular CF are user-dependent CF and item-dependent CF.

SVD Sparse Matrix Calculation
Collaborative filtering finds unknown outcomes by developing a user-item matrix of preference for items by users. Similarity among user's profiles is measured by matching the user-item matrix with the interest of users. The neighborhood is formed by users having similar. Suppose if the user is not able to relate with items, then a recommendation list is provided to that user by a positive rating given to the item by neighborhood users. Collaborative filtering in RS can be utilized for prediction. Prediction means rating value s_{ji} of item i for user j. There are 2 kinds of collaborative filtering they are (Fig. 2):

- Memory-dependent collaborative filtering
- Model-dependent collaborative filtering

Here we utilize model-based collaborative filtering. This technique depends on the previous rating to know about the model which utilizes data mining or machine learning techniques. Numerous approaches were used to categorize items and users depending on the model. Among those approaches Artificial Neural Network best suits for categorization.

Matrix Factorization (MF)

Here, items and users are indicated as latent vectors in shared latent t-dimensional space R^t, where user j is denoted as latent vector $v_j \in R^t$ and item i is indicated as latent vector $v_i \in R^t$. Predicting whether user j would like the item I is provided by inner product among their latent representations, $\hat{s}_{ji} = v_j^K u_i$. To use MF for CF, a matrix of rating needs to know the latent indication of items and users. General methods is to reduce regularized square error loss related to user factors $V = (v_j)_{j=1}^J$ and item factor $V = (v_i)_{i=1}^I$.

$$min_{V,U} \sum_{j,i} \left(s_{ji} - v_j^K u_i^2 \right) + \lambda_v ||v_j||^2 + \lambda_u ||v_i||^2 \qquad (2)$$

where λ_v and λ_u are regularization parameters. $s_{ji} > 0$ if user j rated item i, and $s_{ji} = 0$ otherwise. MF is generalized as a probabilistic model by positioning zero mean spherical Gaussian prior on latent factors of users and things, that is briefed as succeeding generative process,

1. For every user j, draw user latent vector $v_j \sim M(0, \lambda_v^{-1} F_t)$;
2. For every item I, draw item latent vector $v_i \sim M(0, \lambda_u^{-1} F_t)$;
3. For every user-item pair (j,i), draw rating $s_{ji} \sim M\left(v_j^K u_i, D_{ji}^{-1}\right)$;

Where D_{ji} serves as confidence parameter for s_{ji}. If D_{ji} is large then s_{ji} is more trusted. Usually, $D_{ji} = b$ if $s_{ji} > 0$ and $D_{ji} = a$ if $s_{ji} - 0$, b and a were tuning parameters fulfilling ba $> = 0$. In this way, MF deals with unnoticed ratings. MF can simply expand to include biases for various contexts, items, and users to receive a more powerful latent factor design.

3.5 Deep Neural Network

In order to explore textual info regularizing generation of latent look for owners and things in MF must have brand new DNN (Deep Neural Network) to learn distribution representation of files of a bottom up manner which is similar to word embedding done in sentiment analysis [20]. Primarily, DNN study illustration for every text-piece in a document of items/users with embedding layer is done. Afterward semantics of all the text pieces as well as their links had been adaptively encoded in document representation with a completely connected dense layer. The framework of DNN is illustrated in Fig. 1. Which contains the embedding layer and fully connected dense layer.

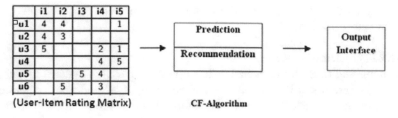

	i1	i2	i3	i4	i5
u1	4	4			1
u2	4	3			
u3	5			2	1
u4				4	5
u5			5	4	
u6		5		3	

(User-Item Rating Matrix)

CF-Algorithm

Fig. 2 Techniques of collaborative filtering

Input Layer

This layer consists of a user and an item matrix. The user input matrix with a latent feature produces a user vector. Similarly, the Item input matrix with latent feature produces the item vector. Both user and item vector are given as input to the next embedding layer.

Embedding Layers

To mine semantics data from of the dataset, every content portion is denoted as a sequence of word embedding. Here, content portion might be reviews, paragraphs, or sentences. R is represented as content-portion with m words and every word is mapped to global vector, then it will be (Table 2):

$$R = \left[\vec{f_1} \parallel \vec{f_2} \parallel \vec{f_3} \parallel \cdots \parallel \vec{f_m} \right] \tag{3}$$

Where vector $\vec{f}_j$ denotes the vector of the j-th word. Vector of word embedding was focused on controlling the order of words in r. Subsequently, it could overcome shortages of bag-of-words methods. For document, if there were K text-pieces aligned in temporal order, we receive sequence as $R = (R_1, R_2, R_2, \ldots R_K)$, where R_K is embedding –dependent subsequence of k-th text-piece.

Fully Connected Dense Layer

The dense level is nothing though an overall level of neurons in NN. Every neuron in the level obtains feedback from neurons of the prior layer in by doing this they connected heavily. This means is every neuron is attached to various other neurons in the prior level. The layer has weight matrix W, bias vector is denoted as 'a', and the previous layer activation is denoted as b.

Output Layer

In this layer, high-level features mined by fully connected dense layer are projected on t-dimensional space with the use of non-linear projection:

$$\theta = \tanh(P * g + a) \tag{4}$$

Table 2 Proposed deep neural network architecture

Layer (type)	Output Shape	Param #	Connected to
1. Item (Input Layer)	(None, 1)	0	
2. User (Input Layer)	(None, 1)	0	
3. Movie-Embedding (Embedding)	(None, 1, 13)	182,351	Item[0][0]
4. User-Embedding (Embedding)	(None, 1, 10)	71,210	User[0][0]
5. FlattenMovies (Flatten)	(None, 13)	0	Movie-Embedding[0][0]
6. FlattenUsers (Flatten)	(None, 10)	0	User-Embedding[0][0]
7. dropout_1 (Dropout)	(None, 13)	0	Flatten Movies [0][0]
8. dropout_2 (Dropout)	(None, 10)	0	Flatten Users [0][0]
9. Concat (Merge)	(None, 23)	0	dropout_1[0][0], dropout_1[0][0]
10.FullyConnected-1 (Dense)	(None, 50)	1200	Concat[0][0]
11.FullyConnected-2 (Dense)	(None, 20)	1020	FullyConnected-1[0][0]
12.FullyConnected-3 (Dense)	(None, 10)	210	FullyConnected-2[0][0]
13.Activation (Dense)	(None, 1)	11	FullyConnected-3[0][0]
(a) Total params: 256,002			
(b) Trainable params: 256,002			
(c) non-trainable params: 0			

where P is t × e projection matrix, a is a bias vector for P, g is the output of a fully connected layer. Furthermore, we use dropout to stay away from projection out of over-fitting. The concept is to drop neurons randomly during training. If dropout_rate is q, the probability of retaining neuron is 1-q. For updation of every parameter, the only portion of model parameter P and a pay to the projection of θ would be improved. In this way, it could avoid difficult co-adaptions of neurons on the training data.

It is observed that, from the benefits of the fast development of DL technology, it's simple to utilize a few high-level API. Similarly, like [21], back-propagation is utilized to explain the abstract model for proposed DNN and also to train the model parameter by specifying loss function, output, and input of the model.

4 Performance Analysis

The performance of the proposed work is examined on the MovieLens dataset. Collaborative and content dependent filtering techniques are used together to provide recommended movie list to specific users depending on their interest. For this purpose, it makes use of the concept of SVD and matrix factorization to examine the performance of proposed movie RS in terms of Mean Square Error (MSE) and Means Absolute Error (MAE) (Tables 3 and 4).

Table 3 Matrix factorization

Latent features	RMSE values
1	1.9435224404259104
2	1.2860963726502623
5	1.3295593001659105
20	0.6362012238538575
40	2.7297190856220594
60	2.111120943394768
100	3.8120449511589753
200	3.904324878539917

Table 4 Sample recommendation list for user 1

Id	Ratings	movieId	Title	Genres
0	5.543971	4738	Happy Accidents (2000)	Romance|Sci-Fi
1	4.742045	6699	Once Upon a Time in the Midlands (2002)	Drama
2	4.695138	5622	Charly (2002)	Comedy|Drama|Romance
3	4.324161	1159	Love in Bloom (1935)	Romance
4	4.316933	1142	Get Over It (1996)	Drama
5	4.27286	287	Nina Takes a Lover (1994)	Comedy|Romance
6	4.269008	1201	Good, the Bad and the Ugly, The (Buono, il bru…	Action|Adventure|Western
7	4.244108	2605	Entrapment (1999)	Crime|Thriller
8	4.076005	2796	Funny Farm (1988)	Comedy
9	3.892875	312	Stuart Saves His Family (1995)	Comedy

5 Conclusion

In various movie RS, ranking or suggestion is performed based on ratings or like. Moreover, system mine information from 1 or 2 websites only. Votes, rating, or semantics for the ranking movie is not reliable since they are not able to provide good recommendation services and there exist a large gap between reviews and statistical information of movie websites. So, it is not reliable to use information from websites since one website uses a qualitative score illustrating the high popularity of movies whereas another website shows low popularity for the same movie. To overcome this drawback, in this paper deep neural network is used with a hybrid combination of content and collaborative dependent filtering to build a movie recommendation system to provide individual recommendation lists for every user based on their preference. Experiments were carried out on MovieLens dataset to evaluate the performance of proposed movie RS in terms of MAE and MSE. The proposed hybrid technique attains a MAE value of 0.6532, matrix factorization with latent features attains an MAE value of 0.6659. Similarly, the proposed hybrid technique achieves MSE of 0.8458, and matrix factorization with latent features achieves MSE of 0.8785. Result proves that the proposed movie RS outperforms other RS. In future the work can be further extended for multiple datasets that can give more precise recommendations and various other metric parameters can be measured for providing more validity to the results.

Conflict of Interest Vineet Shrivastava and Suresh Kumar declared that this manuscript has not been submitted to, nor is under review at, another journal or other publishing venue and the authors have no affiliation with any organization with a direct or indirect financial interest in the subject matter discussed in the manuscript.

References

1. Wei J, He J, Chen K, Zhou Y, Tang Z (2017) Collaborative filtering and deep learning based recommendation system for cold start items. Expert Syst Appl 69:29–39
2. Hwangbo H, Kim YS, Cha KJ (2018) Recommendation system development for fashion retail e-commerce. Electron Commer Res Appl 28:94–101
3. Su J (2016) Content based recommendation system. Google Patents
4. Jiang L, Cheng Y, Yang L, Li J, Yan H, Wang X (2019) A trust-based collaborative filtering algorithm for E-commerce recommendation system. J Ambient Intell Humaniz Comput 10:3023–3034
5. Hsiao JH, Liu N, Li J (2019) E-commerce recommendation system and method. Google Patents
6. Singh T, Nayyar A, Solanki A (2020) Multilingual opinion mining movie recommendation system using RNN. In: Proceedings of first international conference on computing, communications, and cyber-security (IC4S 2019), pp 589–605
7. Cui Z, Xu X, Xue F, Cai X, Cao Y, Zhang W et al (2020) Personalized recommendation system based on collaborative filtering for IoT scenarios. IEEE Trans Services Comput 13:685–695

8. Wang M, Liu X, Jing L (2020) Deep learning based recommedation system: a review of recent works. In: 2020 IEEE 5th information technology and mechatronics engineering conference (ITOEC), pp 1245–1252
9. Zhao L, Lu Z, Pan SJ, Yang Q (2016) Matrix factorization+ for movie recommendation. In: IJCAI, pp 3945–3951
10. Singhal A, Sinha P, Pant R (2017) Use of deep learning in modern recommendation system: a summary of recent works. arXiv preprint arXiv:1712.07525
11. Ibrahim M, Bajwa IS, Ul-Amin R, Kasi B (2019) A neural network-inspired approach for improved and true movie recommendations. Comput Intell Neurosci 2019:1–19
12. Aljunid MF, Manjaiah D (2019) Movie recommender system based on collaborative filtering using apache spark. In: Data management, analytics and innovation. Springer, Heidelberg, pp 283–295
13. Reddy S, Nalluri S, Kunisetti S, Ashok S, Venkatesh B (2019) Content-based movie recommendation system using genre correlation. In: Smart intelligent computing and applications. Springer, Heidelberg, pp 391–397
14. Virk HK, Singh EM, Singh A (2015) Analysis and design of hybrid online movie recommender system. Int J Innov Eng Technol (IJIET) 5:1–5
15. Adeniyi DA, Wei Z, Yongquan Y (2016) Automated web usage data mining and recommendation system using K-Nearest Neighbor (KNN) classification method. Appl Comput Inf 12:90–108
16. Subramaniyaswamy V, Logesh R, Chandrashekhar M, Challa A, Vijayakumar V (2017) A personalised movie recommendation system based on collaborative filtering. Int J High Perform Comput Netw 10:54–63
17. Rajarajeswari S, Naik S, Srikant S, Prakash MS, Uday P (2019) Movie recommendation system. In: Emerging research in computing, information, communication and applications. Springer, Heidelberg, pp 329–340
18. Yi B, Shen X, Liu H, Zhang Z, Zhang W, Liu S et al (2019) Deep matrix factorization with implicit feedback embedding for recommendation system. IEEE Trans Ind Inf 15:4591–4601
19. Zhao J, Geng X, Zhou J, Sun Q, Xiao Y, Zhang Z et al (2019) Attribute mapping and autoencoder neural network based matrix factorization initialization for recommendation systems. Knowl-Based Syst 166:132–139
20. Tang D, Qin B, Liu T (2015) Deep learning for sentiment analysis: successful approaches and future challenges. Wiley Interdisc Rev Data Min Knowl Disc 5:292–303
21. Keras K (2019) Deep learning library for theano and tensorflow

PropFND: Propagation Based Fake News Detection

Pawan Kumar Verma⊙ **and Prateek Agrawal**⊙

1 Introduction

In the recent years large number of people are using social media websites not only for maintaining social connections and entertainment but they also use these sites for dissemination of news all over the world. Tech giants like Twitter and Facebook reinvented the technique that how news can be routed across the world within a second [1]. Social media websites are also favorite place for people where they express and share their views related to politics, religion and entertainment etc. Apart from all these advantages some people are using this platform for dissemination of fake news. The term fake news includes several terms like news manipulation, misinformation, rumor and disinformation [2]. This term is not new but it became famous after social media websites being used by large number of users. These fake news affects people politically, socially, financially and by so many other ways. For example, in May 2016 "Bill and Hillary Clinton were reported to be using a Pizza restaurant as a front for a pedophile sex ring" was tweeted on twitter and it was fake news but as an affect a 28-year-old guy entered a pizza joint with an assault rifle to investigate the claim. With God's grace there was no injury but he was arrested [3]. Also, people are using digital platform to spread fake news to create chaos regarding health. After the whole world is suffering from COVID-19, few people are taking its advantage by sharing many rumors all over the world. The International Fact-Checking Network found more than 3,500 false claims related to COVID-19 in less than two months. As the propagation of news on social media in many-to-many fashion, the detection of fake news becomes complicated task.

P. K. Verma
Department of Computer Engineering & Applications, GLA University, Mathura Uttar Pradesh,, India

P. K. Verma · P. Agrawal (✉)
Department of Computer Science & Engineering, Lovely Professional University, Phagwara, Punjab, India
e-mail: dr.agrawal.prateek@gmail.com

The potential and risk of fake news is increasing everyday [4, 5]. Therefore, large number of researchers are working in the field of fake news detection. Also, large number of researchers are using user and text-based features for spotting fake news. However, some of them are also using propagation-based features and performance of their model outperforms as compared to baseline approaches. Vosoughi et al. [6] compared the propagation pattern of real and fake news and concluded that fake news propagates much longer, quicker and deeper as compared to real news.

This paper clearly shows that combination of propagation and user profile-based features improves the accuracy of fake news detection as compared to independent propagation-based features. It includes the explanation of experimental setup and model accuracy i.e., 93.81%. This paper is organized in six sections. Section 2 explains some literature work in the field of fake news detection. Section 3 explains the basic working of classifiers which are used in this paper. Sections 4 and 5 explains the working of proposed model and implementation respectively. At the end of the paper Sect. 6 explains the conclusion and future direction for further research.

2 Literature Survey

Currently researchers are using machine learning and deep learning approaches for predictions in various fields. Therefore, we also extracted different features and fed into the classifiers for classification of news as real or fake. These features are classified into three categories; user profile based, text based and propagation-based features [7, 8].

2.1 Text Based Detection

Sadia et al. [14] analyzed all possible linguistic features which can be used for classification of normal document from misleading documents with the accuracy (F-measure) of 96.6%. Victoria et al. [15] used combined features of absurdity, punctuation and grammar on twelve contemporary news topics of four different areas for the detection of satirical news. They proposed an algorithm based on SVM and achieved the precision of 90%. Veronica et al. [16] built fake news detection model which was based on the linguistic features and achieved the accuracy of 76%. They also presented the comparative analysis of manual and automatic fake news detection result.

2.2 User Profile-Based Detection

Zhou et al. [9] analyzed few users profile features for spammers detection and for this analysis they used Sina Weibo and Tencent Weibo micro-blogging sites. Mateen et al. [10] used both user profile and content-based features to improve accuracy. Ersahin et al. [11] used user profile-based features with Entropy Minimization Discretization (EMD) and obtained 90.9% accuracy using Naive Bayes classifier. Wanda et al. [12] used modified CNN model for the classification of fake profile. They used user profile-based features like name, follower count, location, language and achieved the precision of 94.00%. Akyon et al. [13] analyzed the performance of various models on two user account (fake and automated) datasets from Instagram social media website and obtained the accuracy of 96%.

2.3 Propagation Based Detection

Vosoughi et al. [6] analyzed the propagation pattern over 1500 users and concluded that real news spreads approximately six times slower than fake news. Kwon et al. [17] used temporal diffusion for classification of rumors and non-rumors with precision and recall ranging from 87 to 92%. Castillo et al. [18] used message, user, topic and propagation-based features for the analysis of credibility and concluded that features based on the propagation plays vital role in the classifier's performance. Federico et al. [19] extracted user profile, user activity, network and content-based features from dataset and concluded that network-based features are good for fake news detection in terms of high accuracy and early detection. Meyers et al. [20] applied propagation-based features on Random Forest Classifier and achieved the accuracy of 87%. They also used graph based i.e., Geometric Deep Learning approach on same features and achieved 73.3% accuracy.

In literature survey we observed that as compared to user profile and text-based analysis there is less research has been done in propagation based fake news detection. We also noticed following insights in above detection techniques:

1. Text based approach uses both syntactic and semantic approach for classification of news. Syntactic approach majorly focuses on linguistic pattern of news like number of special characters, number of nouns, adjectives and semantic approach considers global/local context of word in the piece of text. Some time it is difficult to detect fake news because the person who generates fake news can easily handle these patterns during writing of fake news.
2. Features based on user profile and propagation does not need any additional information other than profile information. Therefore, there is no effort needed for feature extraction.

After seeing these observations, we proposed a PropFND model which uses the combination of propagation and user profile-based features for the classification of news.

3 Classifier Background

3.1 Naive Bayes (NB)

This ML model is used for large volume of data. And it is fast and simple classification algorithm. It is based on Bayes Theorem. This algorithm can deal with binary as well as multi-class classification. It has three types; GaussianNB, MultinomialNB and BernoulliNB. These three types of NB are used for different use cases. NB algorithm assumes that each feature is independent and participates equal contribution for obtaining the outcome.

$$P(X|Y) = \frac{P(Y|X) * P(X)}{P(Y)}$$

where
P(X|Y) Probability of X occurring given evidence Y has already occurred
P(Y|X) Probability of Y occurring given evidence X has already occurred
P(X) Probability of X occurring
P(Y) Probability of Y occurring

3.2 Support Vector Machine (SVM)

It is an example of supervised learning method which uses hyper-plan for classification of data. SVM is used for linear as well as non-linear classification with the help of kernel. For the improvement of model performance parameter tuning is performed. In case of SVM we have four tuning parameters; kernel, regularization, gamma and margin. In case of high dimensional space this algorithm is very useful, but the process of parameter tuning is computationally intensive.

3.3 Random Forest (RF)

It is considered under supervised classification algorithm. As the name suggests, it is a collection of several decision trees in random manner. Accuracy of this algorithm is directly related to the number of trees connected in this forest. The advantages of this algorithm is that; it can be used for both classification and regression problem and in case of sufficient number of trees in forest over-fitting problem won't occur.

3.4 *Artificial Neural Network (ANN)*

ANN is a collection of simple and highly connected processing elements and form a computing system. This computing system responses dynamically according to the external input. ANN tries to imitate the neural network of human body. In case of pattern recognition and data classification, ANN is very useful. ANN uses large number of neurons which are arranged in multiple layers. In this multi-layer architecture, there are mainly three layers; input layer, output layer and hidden layers. Each layer is fully connected with other layer. At the beginning of ANN, input is multiplied by some weight and generates output accordingly; if the actual output is different from predicted output, error value is calculated and modifies the weight. Updation of weight is performed till the minimum error. For this task two techniques are used; forward propagation and back propagation.

4 Proposed Model

The proposed model consists three layers: (i) data pre-processing, (ii) feature engineering and (iii) model building and tuning; as shown in Fig. 1. This section demonstrates the working of model using algorithm and explanation.

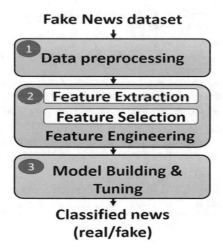

Fig. 1 PropFND working flow

4.1   PropFND Model Algorithm

This section explains the working of PropFND model using algorithm 1. The steps involved in the algorithm are as follows:

1. Initially we selected one dataset among publicly available various fake news datasets as per our requirement as shown in line 1. After that we performed some basic pre-processing operation like missing value imputation, feature scaling, outlier removal etc. in line number 2.
2. After data pre-processing, we performed feature engineering which involves feature extraction as shown in line number 4 & 5 and feature selection as shown in line number 8.
3. Using extensive experiments, we evaluated various models for classification and finally we selected best model for final classification which gives the improved accuracy as in line 9.

Algorithm 1: Algorithm for PropFND model

Data: FakeNewsDataset
Result: News classification using PropFND Model
// Stage 1: Data preprocessing
1 dataset ← dataCollection(*FakeNewsDataset*) // Fake news dataset selection
2 dataset ← preprocess(*dataset*) // Dataset preparation
// Stage 2: Feature Engineering
3 **Feature Extraction**
4   F_1 ← FeatureExtraction(*dataset*) // Extraction of available features in dataset
5   F_2 ← ManualFeatureExtraction(*dataset*) // Generation of additional features from already available features
6   FinalFeatures ← Union(F_1, F_2) // Combining two feature sets
7 **Feature Selection**
8   features ← FeatureSelection(*FinalFeatures*) // Selection of important features using feature selection technique
// Stage 3: Model Building & Tuning
9 Labelled_News ← BestModel(*features*) // Features passed to the multiple classifiers for classification of news as real or fake

---

4.2   PropFND Model Design

Layer 1: Data Pre-processing Layer: The structure of dataset always depends on problem statement. Therefore, in this phase we used the fake news dataset from public repository and performed some basic pre-processing tasks for the conversion of data into required format. This pre-processing task also handles the problem of incomplete, noisy and inconsistent data.

Layer 2: Feature Engineering Layer: After obtaining the dataset in required format we extracted and selected essential features for further execution. Feature extraction and selection phases consist following operations:

1. *Feature extraction* phase extracts already available features from the dataset and also generates some manual features from available features.
2. *Feature selection* phase selects the best feature which is responsible for the better accuracy. Therefore, in our model we used Mutually Informed Correlation Coefficient (MICC) which is the combination of Pearson Correlation Coefficient and Mutual Information [21]. Pearson Correlation Coefficient measures the feature-to-feature dependency and on the other hand Mutual Information measures the feature to output feature dependency. The objective behind the use of MICC is to take the advantage of both type of features which are exceptionally correlated with other features and output feature.

Layer 3: Model Building and Tuning Layer: In this phase we fed selected features into several classifiers for classification. We also performed some model tuning operations for improved result. Section 5 explains all the best possible values that improves the final prediction.

5 Implementation and Result

In this section we explain all the information related to dataset, performance metrics and implementation of PropFND model.

5.1 Experimental Setup

We implemented PropFND model using Python programming. We used Jupyter notebook for code development and debugging. For the execution of this proposed model, we used Intel Core i7 9th generation with 16 GB RAM, 256 HDD and 256 SSD. And we used Windows 10 operating system.

5.2 Dataset

In this research we used FakeNewsNet dataset for the training and testing of proposed model [22]. Dataset construction consists following steps:

1. Labelled news articles are collected from two fact checking organizations [23, 24].
2. Keywords are extracted from headings of labelled news articles and based on those keywords tweets and retweets are extracted.

Finally, this dataset contains news articles in the form of text/image and publishing date of article and it also contains information related to tweets/retweets and user information.

5.3 Performance Metrics

As we are focusing on binary classification (real and fake), we used confusion matrix for the evaluation of our proposed model. Terms used in confusion matrix are shown in Table 1 are associated with the confusion matrix:

Accuracy is the ratio of total number of correct predictions done by proposed model to the total number of predictions.

$$Accuracy = \frac{T_P + T_N}{T_P + T_N + F_P + F_N}$$

Precision is also known as positive predictive value. It is the ratio of true positive value to the summation of true positive and false positive value. It explains that how many positive predictions were actually positive.

$$Precision = \frac{T_P}{T_P + F_P}$$

Recall is also called sensitivity. It explains that among all positive results how many results predicted as positive.

Table 1 Terms associated in confusion matrix

Term	Description
True positive (T_P)	Number of news that classified as real are actually real
True negative (T_N)	Number of news that classified are fake are actually fake
False positive (F_P)	Number of news which are predicted as real but actually they are fake
False negative (F_N)	Number of news which are predicted as fake but actually they are real

$$Recall = \frac{T_P}{T_P + F_N}$$

F1-score is harmonic mean of precision and recall.

$$F1 - score = \frac{2}{Recall^{-1} + Precision^{-1}}$$

5.4 PropFND Model

Data pre-processing layer fills missing values based on the feature property and remove some outliers from our dataset. In this layer we also performed data normalization for improved result. For these tasks we used inbuilt libraries of Python.

Feature extraction and selection layer reads pre-processed data from previous layer and performed feature extraction and selection. In feature extraction process we extracted features related to news propagation and user profile. The objective behind the feature extraction is to reduce the dimension of dataset for fast processing and improved result. Thereafter we ignored irrelevant features which were responsible for decreasing the accuracy of model. Therefore, we used MICC feature selection scheme for the improvement of model accuracy.

$$MICC(i) = \alpha * MI(i) - (1 - \alpha) * Sum(PCC(1, 1 : dim))$$

where $MICC(i)$ is the mutually informed correlation coefficient value and $MI(i)$ is mutual information of ith feature, α represents the weight, PCC is the Pearson Correlation Coefficient and dim is the number of features present in the dataset.

Model building and tuning layer reads selected features and feeds into several classifiers and perform extensive experiments with different parameter values. We performed hyper parameter tuning for the selection of best parameter value. After that we performed tenfold cross-validation for the improved result. Finally observed that SVM with some hyperparameter tuning gives the best result among several classifiers. Table 2 shows the final parameter values which gives the improved result.

Table 2 Hyperparameter tuning of SVM model

Parameter name	Value
Kernel	Linear
Regularization	100
Kernel coefficient	0.0001
Degree	3
Dataset split	80% train, 20% test

Table 3 Accuracy comparison on various classifier

Classifier	Accuracy (%)	Precision (%)	Recall (%)	F1-score (%)
NB	82.52	82.11	83.45	83.02
SVM	93.81	92.94	92.51	93.03
RF	91.84	90.54	91.65	91.84
NN	92.13	91.15	91.84	90.74

Table 4 Accuracy analysis with different training–testing split

Classifier	70–30%	80–20%	90–10%
NB	82.31%	82.52%	81.89%
SVM	93.11%	93.81%	93.49%
RF	91.84%	92.03%	91.42%
NN	91.31%	92.13%	91.87%

5.5   Result

We performed extensive experiments on various classifiers and analyzed the performance of each model. Initially we split the dataset in the ratio of 80%-20% for training and testing purpose respectively. After the execution we observed that Naive Bayes gives the minimum accuracy of 82.52% and SVM gives the maximum accuracy of 93.81% among various models as shown in Table 3. For the generalized accuracy analysis, we tested the performance of each classifier with different training/testing dataset size. Table 4 shows SVM accuracy varies between 93 to 94%.

5.6   Comparative Study

For the performance comparison we compared our model performance with existing model proposed by Meyers et al. [20]. Table 5 shows that after adding the user profile-based features and MICC feature selection technique the accuracy of news classification improves by 6.81% from existing model executed on same dataset.

Table 5 Comparison between existing and proposed model

Parameters	Meyers et al. [20] model	PropFND model
Features	Propagation based	Propagation and profile based
Dataset	FakeNewsNet	FakeNewsNet
Feature selection	Manual analysis	MICC
Classifier	NN	SVM
Accuracy	87%	93.81%

6 Conclusion and Future Work

In this paper we proposed a PropFND model for the fake news detection using propagation and user profile-based features. We used FakeNewsNet dataset and split into 80% training and 20% testing purpose. Thereafter we passed essential features to the modified NN for classification of news as real or fake. After the implementation of PropFND model we obtained the accuracy of 93.81%.

Inspite of good accuracy this model has one limitation. During the selection of feature, we used one feature that counts the number of unique users involved in the spreading of news in starting few hours, this feature clearly says that the model can label the news after few hours. Therefore, the model cannot give the labelled result in early stage.

References

1. Center PR (2018) News use across social media platforms 2018. https://www.journalism.org/2018/09/10/news-use-across-social-media-platforms-2018/. Accessed 07 June 2021
2. Lazer DM et al (2018) The science of fake news. Science 359(6380):1094–1096
3. What to Know About Pizzagate, the Fake News Story With Real Consequences (2016). https://time.com/4590255/pizzagate-fake-news-what-to-know/. Accessed 07 June 2021
4. Guardian T (2018) Bolsonaro business backers accused of illegal whatsapp fake news campaign. https://www.theguardian.com/world/2018/oct/18/brazil-jairbolsonaro-whatsapp-fake-news-campaign. Accessed 07 Aug 2021
5. Marwick A, Lewis R (2017) Media manipulation and disinformation online. Data & Society Research Institute, New York
6. Vosoughi S, Roy D, Aral S (2018) The spread of true and false news online. Science 359(6380):1146–1151
7. Shu K, Sliva A, Wang S, Tang J, Liu H (2017) Fake news detection on social media: a data mining perspective. ACM SIGKDD Explor Newsl 19(1):22–36
8. Zhou X, Zafarani R (2018) Fake news: a survey of research, detection methods, and opportunities. arXiv:1812.00315
9. Zhou Y, Chen K, Song L, Yang X, He J (2012) Feature analysis of spammers in social networks with active honeypots: a case study of chinese microblogging networks. In: International conference on advances in social networks analysis and mining. IEEE/ACM, Istanbul, pp 728–729

10. Mateen M, Iqbal MA, Aleem M, Islam MA (2017) A hybrid approach for spam detection for twitter. In: 14th international bhurban conference on applied sciences and technology (IBCAST). IEEE, Islamabad, pp 466–471
11. Ersahin B, Aktas O, Kilinc D, Akyol C (2017) Twitter fake account detection. In: International conference on computer science and engineering (UBMK). IEEE, Antalya, pp 388–392
12. Wanda P, Jie HJ (2020) Deepprofile: finding fake profile in online social network using dynamic CNN. J Inf Secur Appl 52:1–13
13. Akyon FC, EsatKalfaoglu M (2019) Instagram fake and automated account detection. In: Innovations in intelligent systems and applications conference (ASYU). IEEE, Izmir, pp 1–7
14. Afroz S, Brennan M, Greenstadt R (2012) Detecting hoaxes, frauds, and deception in writing style online. In: Proceedings of IEEE symposium security and privacy (SP), pp 461–475
15. Rubin V, Conroy N, Chen Y, Cornwell S (2016) Fake news or truth? using satirical cues to detect potentially misleading news. In: Proceedings of computational approaches to deception detection, pp 7–17
16. Pérez-Rosas V, Kleinberg B, Lefevre A, Mihalcea R (2017) Automatic detection of fake news. ArXiv arXiv:1708.07104
17. Kwon S, Cha M, Jung K, Chen W, Wang Y (2013) Prominent features of rumor propagation in online social media. In: 2013 IEEE 13th international conference on data mining. IEEE, pp 1103–1108
18. Castillo C, Mendoza M, Poblete B (2011) Information credibility on twitter. In: Proceedings of the 20th international conference on world wide web. ACM, pp 675–684
19. Federico M, Fabrizio F, Davide E, Damon M (2019) Fake news detection on social media using geometric deep learning. arXiv preprint arXiv:1902.06673
20. Meyers M, Weiss G, Spanakis G (2020) Fake news detection on twitter using propagation structures. In: Disinformation in open online media. Springer, Heidelberg, pp. 138–158
21. Guha R, Ghosh KK, Bhowmik S, Sarkar R (2020) Mutually informed correlation coefficient (MICC) - a new filter based feature selection method. In: 2020 IEEE calcutta conference (CALCON), pp 54–58
22. Shu K, Mahudeswaran D, Wang S, Lee D, Liu H (2018) Fakenewsnet: a data repository with news content, social context and dynamic information for studying fake news on social media
23. Politifact. The Poynter Institute. https://www.politifact.com/. Accessed 25 June 2021
24. Snopes. https://www.snopes.com/. Accessed 25 June 2021

Depression Detection Using Spatial Images of Multichannel EEG Data

Akriti Goswami, Shreya Poddar, Ayush Mehrotra, and Gunjan Ansari

1 Introduction

Depression is a mental illness that affects a person's thought process, moods, and daily activities. It affects an individual's mind as well as body. Depression affects how an individual perceives things and acts on stimulus which is given to the nervous system. Everyone feels down or low in their lives sometimes but when this emotion lasts for a longer time, it affects general cognitive function of the brain. Early recognition and treatment of depression can improve the negative impacts of the disorder. Depression detection would be a helpful step in eradicating depression in its early stages and prevention of any life or material loss due to depression. The previous studies have shown that the brain functions of a depressed and normal individual acts differently. The electroencephalogram (EEG) has been a prevalent approach for examining brain activities in depression. Electroencephalography is a non-invasive procedure, in which signals are obtained from electrodes placed on the scalp. The two Brain Computer Interface (BCI) methods that use EEG datasets are linear and nonlinear. Linear methods include Linear Discriminant Analysis and Non-linear methods includes Support vector Machines and Neural Networks.

In the past, deep learning architectures have made very little progress in neuroscience and biomedical domains due to unavailability of dataset. But, in today's era, there has been a significant improvement in learning and diagnosis of neural disorders including Alzheimer, seizure, epilepsy, Parkinson, depression, emotional states and other abnormality diseases due to the rapid enhancement of available neurological data. Advanced neurocomputing, machine learning and deep learning techniques

A. Goswami · S. Poddar · A. Mehrotra (✉) · G. Ansari (✉)
Department of Information Technology, JSS Academy of Technical Education, Noida 201301, India
e-mail: ayush.mehrotra900@gmail.com

G. Ansari
e-mail: gunjanansari@jssaten.ac.in

© The Author(s), under exclusive license to Springer Nature Singapore Pte Ltd. 2022   569
B. Unhelker et al. (eds.), *Applications of Artificial Intelligence and Machine Learning*,
Lecture Notes in Electrical Engineering 925,
https://doi.org/10.1007/978-981-19-4831-2_46

have been used for EEG based diagnosis of various neurological disorders [8]. The first study conducted in utilizing the CNN for analysis of EEG signals was conducted by Acharya et al. [1, 2] for classification of epilepsy into normal, seizure and preictal classes. In the classification of normal and epileptic subjects, highest accuracy of 99.7% has been achieved for a three-class classification problem using nonlinear features. Various studies have been done for effectively scoring sleep stages where CNN has been used to identify hidden features in EEG. For automated detection of Parkinson's disease, authors developed a thirteen-layer CNN architecture [14]. They achieved a promising performance of 88.25% accuracy, 84.71% sensitivity, and 91.77% specificity.

Betul Ay et al. [3] proposed a deep hybrid model developed using (CNN) and Long-Short Term Memory (LSTM) architectures for depression detection using EEG signals. The LSTM network [13] is effective in learning long-term dependencies present in the CNN architecture. The feature maps were obtained by applying raw EEG signals from the CNN model which was then fed as input to LSTM to perform sequence learning on these signals. The EEG signals were acquired from 15 normal and 15 depressed subjects. They collected EEG signals from both the left and right hemispheres of the brain and the accuracy achieved was 97.6% and 99.12% respectively. The drawback of their research was that they used few subjects and the model was computationally intensive. Inspired from their study, Sandheep P et al. [15] proposed a CNN model to detect depression. They collected data from 30 normal and 30 depressed patients and achieved accuracy of 99.31% from the right-brain hemisphere and 96.31% from the left hemisphere of the brain after tenfold cross-validation. Hanshu et al. [16] conducted a study to detect depression using a pervasive prefrontal-lobe three-electrode EEG system at Fp1, Fp2, and Fpz electrode sites. They used four classification methods (Support Vector Machine, K-Nearest Neighbor, Classification Trees, and Artificial Neural Network) and the performance was evaluated using tenfold cross validation. The data was taken from 213 (92 depressed patients and 121 normal controls) subjects. The accuracies achieved in their work were 72.56% using SVM classifier, 76.83% using KNN, 68.29% using Decision Trees and 72.56% using ANN.

Xiaowei Li et al. [11] employed a deep learning approach for detection of mild depression based on functional connectivity for EEG data. They obtained functional connectivity matrices from five EEG bands and applied CNN to it. Their study concluded that the recognition performance of coherence was superior to the other functional connectivity matrices (correlation, phase lag index and phase locking value) by giving classification accuracy of 80.74%. Various studies have proved that the temporal region of the brain is affected mostly in depressed patients. Shalini Mahato et al. [12] detected depression and its severity using six (FT7, FT8, T7, T8, TP7, TP8) channel EEG data extracted from the temporal region of the brain. Their model could achieve the highest classification accuracy of 96.02% using SVM and 79.19% using ReliefF as feature selection.

There are various EEG features in the frequency domain, time domain, and time frequency domain. EEG time-domain features can be used to determine time-series properties that differ between distinct emotional states [7]. In the proposed work, time series EEG data is used and the approach employed in this work is inspired by the work of Pouya Bashivan et al. [4] on learning representations from EEG data with Deep Recurrent-CNN for modelling cognitive events. In the proposed approach, instead of representing the low-level EEG features as a vector, the data is transformed into topology-preserving multi-spectral images. This robust data representation in the form of images is used to train the deep recurrent-CNN to provide an effective solution for depression detection. In the experimental setup of this work, after validating the model, a web-based application has been created that would take EEG data of a patient as input and predicts whether the patient is suffering from depression or not. The developed application provides a cost-effective solution to medical health care providers for detecting depression at an early stage.

2 Materials and Methods

2.1 About Dataset

In this study, the depression and normal signals are the same as used by Cavanagh et al. [6] to prove that anxiety and mood dimensions are associated with unique aspects of EEG responses to punishment and reward respectively. Participants were chosen from introductory psychology classes based on their Beck Depression Inventory (BDI) scores in a mass survey for creating this dataset. All the participants had to provide a written informed consent approved by the University of Arizona before participating in this survey. The data is in the form of csv files with raw waveform signals from 66 Ag/AgCl probes places around the scalp. Figure 1 shows 2D visualizations of the channels and their locations. The sampling rate is 500 Hz and the band-pass filter is 0.5–100 Hz.

2.2 EEG Classification System

The proposed work consists of the following modules for developing a fast, efficient and robust EEG classification system as shown in Fig. 2.

Preprocessing. The resting EEG data is entirely raw and has few readings of the signals of the human scalp which are not required for the model. So, preprocessing is the primary step of EEG analysis as the data which is needed for classification should be clean and noise free [5] for better prediction. Preprocessing steps followed in this work includes re-referencing, band pass filtering, down-sampling, artifact and bad channels removal and independent component analysis.

The primary goal of re-referencing is to demonstrate the voltage at the EEG scalp channels in terms of a new, different reference. It can be made up of any recorded

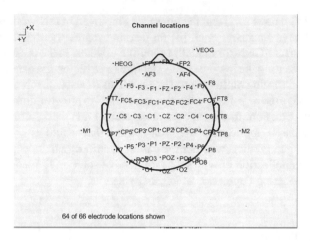

Fig. 1 EEG channel locations

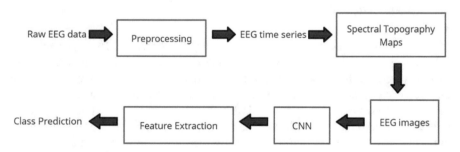

Fig. 2 EEG classification system

channel or a composite of multiple channels. The signal of the new reference is removed from each EEG channel during re-referencing. In this approach, average mastoids M1 and M2 are used since the main focus is on central areas of the scalp and also mastoids record less cortical signals from the brain. References become symmetric due to averaging across the left and the right ears. Subsequently, the frequency of the time series EEG data is changed by performing down-sampling. As signals with unnecessary high sampling rates slows the operations of computing and also takes up huge memory, resampling is included while preprocessing.

Further, artifacts have been identified and removed from the data [10]. Artifacts in EEG data can be due to various reasons like improper measuring instrument setup or human error. Artifacts imitate cognitive activities which can result in wrong prediction of depression. These artifacts are unwanted signals of different amplitudes and waveforms. The resting EEG data consists of 67 channels initially out of which 'CB1', 'CB2', 'HEOG', 'VEOG', 'EKG','M1','M2', 'Status' channels were explicitly removed. Lastly, independent component analysis is employed to reduce the statistical dependence between components of the signal that means to separate the mixed signals from each other. It also removes the extra noises from the data which survived from the previous stages of the preprocessing.

Constructing Images from EEG Time Series. Multichannel EEG time-series data can be represented in a robust way to maintain the topology of the EEG data spatially, spectrally and temporally [4]. This approach basically preserves the structure of the network among different electrodes by forming succeeding 2D images that are multi-spectral which typical EEG analysis approaches generally neglect. The method is capable of locating features with the least amount of value change owing to dispro-portion and differences induced by dimension changes. The technique appears fair because there will be no disruption in network structures as EEG data is represented in low-level vector form. The depiction of data will be of high dimension which can be used for further classifications.

Another goal for this technique is to use the well-known deep learning classes, which self-train themselves based on the inputs they get. This eases out the upcoming learning of this study. So, for spectral analysis of EEG data and to perform Fast Fourier Transform (FFT), data was split into intersecting 1-s 'frames'. Later, to have better periodic extension in FFT, for each trial, signals were windowed. In the approach used in this work, the Hanning window is applied on these one second frames. This time series data is later converted to frequency domain data using FFT by transforming the domain that is the x axis of a signal from time to frequency. FFT improves the performance of CNN. The data is converted to the frequency domain because the most prominent feature resides in this domain and also frequency analysis is considered to be one of the most powerful analysis for EEG. So, frequency resolution was calculated as shown in Eq. (1) where Fs is sample frequency, n is number of data points used in the FFT

$$\text{Frequency resolution} = \frac{Fs}{n} \tag{1}$$

Next, FFT amplitudes is binded into three frequency bands that are affected mostly in depressed individuals, i.e. delta (0–4 Hz), theta (4–8 Hz) and alpha (8–12 Hz). This resulted in three scalar values for each probe per frame. After frequency binding, 2D Azimuthal Projection was performed where the values collected from the previous steps are considered as three values of RGB color channels and later transformed into two-dimensional measurements. The values are considered as RGB color channels to preserve the spectral dimensions. To project 3D locations of electrodes onto a 2D surface, Azimuthal Equidistant Projection is used. This particular transformation also preserves the relative distance between the neighboring electrodes. The values within the electrodes are also calculated by performing interpolation. Topographical activity maps for each band were formed which are then merged to three channel images. These images are further used for training the CNN model discussed in the later section.

Convnet Architecture. In the proposed structure, a 4-layer CNN has been applied. Convolutional layer helps to learn EEG features, a single max pooling layer is used

Table 1 CNN architecture (Total params:64,267)

Layer (type)	Output shape	Param #
conv2d (Conv2D)	(None, 28, 28, 32)	896
activation (Activation)	(None, 28, 28, 32)	0
conv2d_1 (Conv2D)	(None, 26, 26, 32)	9248
activation_1 (Activation)	(None, 26, 26, 32)	0
max_pooling2d (MaxPooling2D)	(None, 13, 13, 32)	0
flatten (flatten)	(None, 5408)	0
dense (Dense)	(None, 10)	54,090
activation_2 (Activation)	(None, 10)	0
dense_1 (Dense)	(None, 3)	33
activation_3 (Activation)	(None, 3)	0

for dimensionality reduction or down-sampling, and a fully-connected layer at last for classification. Table 1 shows the parameter settings of the sequential CNN architecture used in this work. The input size of the image is 28×28. All convolutional layers use small receptive fields of size 3×3 and stride of 1 pixel with ReLU activation functions. The ReLU function f(x) is as follows:

$$f(x) = \max(x, 0) \tag{2}$$

The padding in the convolutional layer is kept the same to preserve the spatial resolution after convolution [4]. Two convolution layers are stacked together followed by the max pooling layer across time frames. Max pooling is performed over a 2×2 window and stride of 2 pixels. Finally, a fully-connected network is appended followed by a softmax layer.

3 Experimental Results

3.1 Experimental Setup

In this approach, the preprocessing is performed using MATLAB, a proprietary multi-platform programming language and numeric computing environment developed by MathWorks. Matlab EEGlab plugin [9] tools have been used for functions like re-referencing, changing sampling rates, filtering data, automatic channel rejections, remove baseline and run independent component analysis. The experimental study utilized MNE package in Python for reading, visualization and running independent component analysis. Numpy library in Python was used for implementing FFT and hanning functions. The keras library of tensorflow was used to implement various functions of the CNN model. Later, an interactive interface was designed using Flask framework to build a user-friendly application for depression detection.

3.2 Results

The data used in this model consists of 29 subjects. This EEG data is raw and is labelled into three classes out of which 7 were normal, 10 currently depressed and 12 past depressed. In our work, 75% data was used for training and the rest 25% was used for validating the model. After the raw data was preprocessed and all artifacts were eliminated, the data was split using a random splitting technique to prevent the process from being biased towards any one attribute of the data. The images generated for high level representation of EEG data were of dimensions 28 × 28 with frame duration as 1.0 s and overlapping value as 0.5 s. These generated images were fed as input to the CNN model where RMSprop optimizer was used, it balances the step size automatically by recognizing how large or small the gradient is in order to handle the issue of exploding and vanishing gradient. Table 2 shows the parameters setting used in our model for depression detection. It contains information about learning rates and decay, batch sizes set, optimizers and label settings.

Figure 3 shows the graphical representation of the CNN model's performance. The irregularity in the validation loss and accuracy arises due to the insufficient amount of data available for the validation of the model.

The model's performance during training and validation sets is shown in Table 3. Here the total number of epochs were 100, out of which the performance of randomly selected ten epochs are tabulated below. It can be observed that the value of training loss decreases from 1.0722 to 0.4164. Increase in training accuracy from 0.4168 to 0.8370 was also noted. For validation, there was decrease in validation loss from 1.0562 to 0.8135 and increase in validation accuracy from 0.4115 to 0.6721.

To increase the model's interaction with the user, a web application has been developed. There are two components of this application: one for static data and the other for dynamic data. The input option in the static area allow users to upload EEG

Table 2 Parameter settings

Parameters settings	Values
Batch size	32
Decay	1e-6
Epochs	100
Learning rate	0.0001
Loss function	Categorical cross entropy
Metrics	Accuracy
Optimizer	RMSprop
Labels	0 Controlled, 1 Current MDD, 2 Past MD

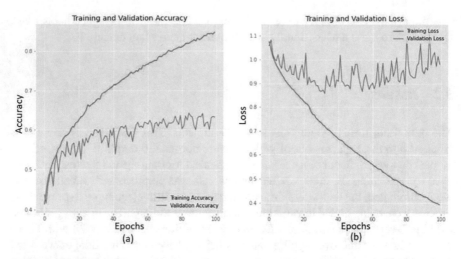

Fig. 3 Performance graph for CNN model **a** Accuracy graph **b** Loss graph

Table 3 Training performance of CNN model using EEG signals for various epochs

Epochs (nth)	Time (Second)	Training loss	Training accuracy	Validation loss	Validation accuracy
1	37	1.0722	0.4168	1.0562	0.4115
25	24	0.7264	0.6838	0.8464	0.6289
39	25	0.6496	0.7208	0.8135	0.6505
46	26	0.6159	0.7424	0.8207	0.6508
55	26	0.5762	0.7583	0.8164	0.6555
62	22	0.5474	0.7773	0.8315	0.6572
75	22	0.4984	0.7988	0.8354	0.6663
84	25	0.4687	0.8131	0.8406	0.6721
97	27	0.4273	0.8324	0.8625	0.6645
100	28	0.4164	0.8370	0.9259	0.6455
Average	26.2	0.59	0.73	0.86	0.63

files from their local systems. This model is channel independent, which means data of any appropriate number of electrodes can be uploaded. The EEG system should contain a maximum number of 67 electrodes. The data format is also taken into consideration. Users can upload EEG data of formats like.set and.fdt or.vhdr,.vmrk and.dat or.edf.

In the dynamic section, the number of seconds during which real time EEG data is to be generated should be entered as an input, which could later be replaced with a real-time input fed from EEG device. The lab streaming layer was used to test

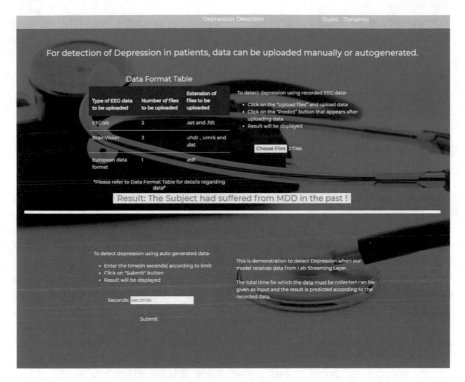

Fig. 4 GUI of proposed model for depression detection

the capability of the system for recording and analyzing real-time EEG data. The predicted result is displayed on the screen as shown in Fig. 4.

3.3 Discussion

In this paper, a robust approach for detecting depression from multi-channel EEG time-series data has been proposed. The association between EEG signals and depression, as well as the most significant frequency bands, were investigated. Different brain regions contributing to depression were studied and accordingly electrodes positioned on different locations of the scalp were considered for the analysis. To generate images for CNN input, top three frequency bands with values in the ranges associated with depression detection were evaluated. The validation accuracy achieved by this model is around 63.0%. Pouya et al. [4] proposed an efficient method to preserve EEG data spatially, spectrally and temporally but their work was limited on learning representations from image sequences. However, in this work, we tried to extend their approach to provide an effective solution using high level representation of EEG data. The application proposed in this work can prove to be highly beneficial to medical practitioners as it provides early detection of depression in real-time with limited resources.

4 Conclusion and Future Work

This paper proposed the utility of deep neural network and CNN for depression detection. The proposed model self learns the features and recognizes prominent ones for algorithm's training, thus saving computational time consumed in feature selection and extraction. Based on the findings obtained from the experimental analysis on the small set of available EEG data, it can be inferred that the proposed model can be used for computer-assisted depression diagnosis with reasonable accuracy. CNN utilized in the proposed work performs parameter sharing and its special convolution and pooling algorithms make it globally appealing. The model also employed an image representation of EEG data to save information about the electrode networks. Instead of representing data in low-level formats, high-level formats were used as input for the CNN model. The raw time series data was analyzed and preprocessed, with all undesired artifacts removed to improve the model performance.

The main drawback of this study is that only few subjects (7 normal, 10 currently depressed and 12 past depressed) have been used. In future, this work can be extended collecting data of more subjects and focus can be given to specific regions of the brain to improve validation and testing accuracy. Additionally, CNN model can be combined with other classifiers such as LSTM, SVM to increase the prediction accuracy. In order to determine the brain regions responsible for mental diseases, functional connectivity analysis can be considered. To better understand the prominent frequency band, this analysis can be performed independently on each frequency band data. The matrices obtained by the functional connectivity study, which yield different outcomes for different mental states, can be contemplated in future works. The model can also be used for detecting depression in early stages by using reports generated by electroencephalogram test. This work is focused to be an asset to institutions bound on fighting depression and problems related to mental health.

Acknowledgements All data and code for this experiment are available on the PRED+CT website, http://www.predictsite.com, accession no. d003. The concept of this work is supported by INMAS(Institute of Nuclear Medicine and Allied Sciences) Lab, Defense Research and Development Organisation, Ministry of Defense, India.

References

1. Acharya UR, Oh SL, Hagiwara Y, Tan JH, Adeli H (2018) Deep CNN for the automated detection and diagnosis of seizure using EEG signals. Comput Biol Med 100:270–278. https://doi.org/10.1016/j.compbio-med.2017.09.017

2. Acharya UR, Vinitha Sree S, Swapna G, Martis RJ, Suri JS (2013) Automated EEG analysis of epilepsy: a review. Knowl-Based Syst 45:147–165. https://doi.org/10.1016/j.knosys.2013.02.014

3. Ay B, Yildirim O, Talo M et al (2019) Automated depression detection using deep representation and sequence learning with EEG signals. J Med Syst 43:205. https://doi.org/10.1007/s10916-019-1345-y

4. Bashivan P, Rish I, Yeasin M, Codella N (2016) Learning representations from EEG with deep recurrent-convolutional neural networks. CoRR, abs/1511.06448

5. Bigdely-Shamlo N, Mullen T, Kothe C, Su KM, Robbins KA (2015) The PREP pipeline: standardized preprocessing for large-scale EEG analysis. Front Neuroinform 9:16. https://doi.org/10.3389/fninf.2015.00016

6. Cavanagh JF, Bismark AW, Frank MJ, Allen J (2019) Multiple dissociations between comorbid depression and anxiety on reward and punishment processing: evidence from computationally informed EEG. Comput Psychiat 3:1–17. https://doi.org/10.1162/cpsy_a_00024

7. Chao H, Zhi H, Dong L, Liu Y (2018) Recognition of emotions using multichannel EEG data and DBN-GC-based ensemble deep learning framework. Comput Intell Neurosci 2018:1–11. https://doi.org/10.1155/2018/9750904

8. Craik A et al (2019) Deep learning for electroencephalogram (EEG) classification tasks: a review. J Neural Eng 16(031001):1–38. https://doi.org/10.1088/1741-2552/ab0ab5

9. Delorme A, Makeig S (2004) EEGLAB: an open source toolbox for analysis of single-trial EEG dynamics including independent component analysis. J Neurosci Methods 134(1):9–21. https://doi.org/10.1016/j.jneumeth.2003.10.009

10. Jiang X, Bian GB, Tian Z (2019) Removal of artifacts from EEG signals: a review. Sensors (Basel, Switzerland) 19(5):987–1005. https://doi.org/10.3390/s19050987

11. Li X, La R, Wang Y, Hu B, Zhang X (2020) A deep learning approach for mild depression recognition based on functional connectivity using electroencephalography. Front Neurosci 14(192):1–20. https://doi.org/10.3389/fnins.2020.00192

12. Mahato S, Goyal N, Ram D et al (2020) Detection of depression and scaling of severity using six channel EEG data. J Med Syst 44(118):2–13. https://doi.org/10.1007/s10916-020-01573-y

13. Ng J, Hausknecht M, Vijayanarasimhan S, Vinyals O, Monga R, Toderici G (2015) Be-yond short snippets: deep networks for video classification. In: IEEE conference on computer vision and pattern recognition (CVPR). IEEE, Boston, pp 4694–4702. https://doi.org/10.1109/CVPR.2015.7299101

14. Oh SL, Hagiwara Y, Raghavendra U et al (2020) A deep learning approach for Parkinson's disease diagnosis from EEG signals. Neural Comput Appl 32:10927–10933. https://doi.org/10.1007/s00521-018-3689-5

15. Sandheep P, Vineeth S, Poulose M, Subha D (2019) Performance analysis of deep learning CNN in classification of depression EEG signals. In: TENCON 2019 - 2019 IEEE region 10 conference. IEEE, Kochi, pp 1339–1344. https://doi.org/10.1109/TENCON.2019.8929254

16. Cai H, Han J, Chen Y, Sha X, Wang Z, Hu B, Yang J, Feng L, Ding Z, Chen Y, Gutknecht J (2018) A pervasive approach to eeg-based depression detection. Complexity 2018:1–13. https://doi.org/10.1155/2018/523802

Feature Selection for HRV to Optimized Meticulous Presaging of Heart Disease Using LSTM Algorithm

Ritu Aggarwal ⑩ and Suneet Kumar ⑩

1 Introduction

Heart disease (HD) contrarily influences various individuals throughout the world. [1]. HD cases are high in the United States (US) [2]. The HD primary diagnoses methods are breath brevity, body weakness, and so forth [3]. The coronary illness conclusion strategies were not solid at an early age due to numerous causes [4]. Discovery and therapy of HD are troublesome where present-day determination innovation and clinical specialists. To identify the disease the past medical history of a patient is required. HRV is related to heartbeats. It is measured by fluctuation in heartbeats at given time intervals. HRV focus on the neurocardiac functions that are generated by the nonlinear autonomic nervous system process.

1.1 HRV Metrics

HRV utilizes time-domain, Frequency domain measures, and non-linear measurements. Time–space records of HRV evaluate the measure of changeability in estimations of the interbeat span (IBI) which is the time-frame between progressive pulses. These qualities might be communicated in unique units or as the regular calculation (ln) of unique units to accomplish a more ordinary conveyance. Recurrence area estimations gauge the conveyance of outright or relative force into four recurrence groups. The team of the European culture of cardiology isolated pulse motions into super low-recurrence (ULF), extremely low-recurrence (VLF),[5] Low-recurrence (LF), and High-recurrence (HF) groups. Force is the signal energy found inside a recurrence band. The ULF (≤ 0.003 Hz) files variances in IBI with a period from 5 min

R. Aggarwal (✉) · S. Kumar
Maharishi Markandeshwar Engineering College, Mullana, Ambala, Haryana, India
e-mail: errituaggarwal@gmail.com

to 24 h and is estimated utilizing 24 h chronicles. The VLF band (0.0033–0.04 Hz) is included rhythms with periods somewhere in the range of 25 and 300 s. The LF band (0.04–0.15 Hz) is contained rhythms with periods somewhere in the range of 7 and 25 s and is influenced by breathing from $\sim$ 3 to 9 bpm inside a 5 min example, there are 12–45 complete times of motions. The HF or respiratory band (0.15–0.40 Hz) is impacted by breathing from 9 to 24 bpm [6]. The proportion of LF to HF force may assess the proportion between thoughtful sensory system and parasympathetic sensory system movement under controlled conditions. Complete force is the amount of energy in the VLF, ULF, LF& HF groups for 24 h. The VLF, LF, and HF groups for the momentary account. HRV time–space boundaries are SDNN(standard deviation of NN span), SDANN(standard deviation of the normal NN stretches for every 5 min portion of a 24 h HRV recording), pNN50 (level of progressive RR spans that contrast by more than 50 ms), RMSSD (Root mean square of progressive RR span contrasts). Non-linear measurements permit us to measure the unconventionality of a period series. HRT portrays momentary vacillations in a sinus cycle length that follow unconstrained ventricular premature complexes (VPCs). In ordinary subjects, sinus rate at first and consequently decelerates contrasted and the pre-VPC rate, before getting back to benchmark. Two phases of HRT, the early sinus rate speed increase and late deceleration, are evaluated by 2 parameters named Turbulence onset (TO) and turbulence slope (TS). It is determined as [3].

$$TO(onset) = \frac{(RR_1 + RR_2) - (RR_2 + RR_{-1})}{(RR_2 + RR_1)} \times 100[\%] \tag{1}$$

TO are a turbulence onset and RR1 and RR2 in the intervals to precede the VPC coupling interval. The turbulence is measured for R-R intervals within 15 sinus rhythms for VPC [11]. In this proposed work the HRV dataset is implemented For SFFS using the LSTM to select the relevant features set. Long short term memory (LSTM) is a repetitive neural organization (RNN) that is reasonable for preparing and foreseeing significant functions with moderately long spans and delay in time series [7]. Be that as it may, insights to clinical industry, the time between various hospitalizations of patients is unique, and the earlier traditional LSTM can't adequately gain proficiency with the significant symptoms of patient's medical condition, which restricts the down to earth use of LSTM in clinical issues [9]. In this paper, worked on the LSTM by smoothing the unpredictable time between various clinical stages of the patient to get the temporal vector. TFV deals with the irregular time interval between the multiple data because it is used as a forgetting threshold input. It will improve the performance of the prediction model. In Sect. 1 introduction is discussed related to HRV and cardiac disease, Sect. 2 literature review discussed related existing work in feature selection using HRV, in Sect. 3 feature selection techniques discussed the proposed methodology used, in Sect. 4 machine learning classifiers, Sect. 5 Material, and Methods, Sect. 6 Results, and discussion, Sect. 7 conclusion.

2   Literature Review

Aravind Natarajan, et al. [1] in this proposed work the researcher used the HRV parameters for showing the stress levels using the different machine methods. With the help of patient saliva easily predict their stress level. Features used to show a variation of recurrence plot that are DET, REC, Lmax, Lmean, etc. it is the time domain features such as NN50, pNN50, NN50, RMSSD, STD HR, mean HR, and DFA, etc. with the help of these features of HRV in stress levels the classification accuracy obtained. Güzel Aydın et al. [2] in this study proposed a disease metabolic syndrome that describes the chances of increasing CD and their risk is more. İt is associated with autonomic nervous system dysfunction. to find out the metabolic syndrome used the HRV. HRV is used for those patients that have hypertension. With the help of HRV parameters measured the RR intervals. For ambulatory elec-trocardiography, RR intervals were recorded. Different parameters related to HRV such as LF, SDNN, RMSSd, PNN50, TP, etc. Rajendra Acharya et al. [3] in this study the researchers proposed the features selection model using the HRV dataset and extract the relevant feature set. In males and females achieve accuracy 69.3 and 80.9%. It measured the ECG signal with high frequency. P. Karthikeyan et al. [4] in this researcher designed a model to detect stress in patients. It implemented the clas-sification for feature extraction on the ECG signals. For the HRV dataset proposed the 60 subjects that differentiate the male and female ratio. DWT is used to detect and denoising the signals and derives a detection algorithm for preprocessing the signals. For sampling, the HRV signals the LSP is used that easily rectified the signal. With the help of machine learning classifier easily detects and obtained the classification accuracy. C. -W. Sung et al. [5] the author study the biosignal and recording system for HRV analysis. This analysis indicates the ANS disorders. This work used specific parameters for the detection of diseases such as SDNN, HF, LF/HF, and Samp. EN. Dalmeida et al. [6] In this study comparative analysis is described for detection of HRV features to diagnose the stress in patients. For achieved accuracy implemented this on different machine learning algorithms. İn this model feature selection and extraction were done by wearable devices and achieved a score in terms of sensi-tivity 80%.Sumit et al. [7] in this proposed work implements the algorithms DNN for improvement in results by the methods used Talos the dataset used for their work is guts malady datasets and achieved an accuracy of 90.76%. Bindhika et al. the author proposed a method for HD using different machine learning techniques. It used a random forest approach to calculate the results of heart rate for different samples. N. Kumar et al. [12] designed an efficient automated disease diagnosis model using various ML models. The focus is majorly made on three crucial diseases like heart disease, coronavirus, and diabetes. In this technique, an android app is used where the data is entered as input and using pre-trained ML models, the analysis is then performed in a real-time database and finally, the result of disease detection is again shown in the android app. The computation for prediction is done using Logistic Regression. General experimental results divulge that this technique performs much better than various competitive ML models in terms of accuracy as well as F-measure by 1.4765% and 1.2782, respectively, for the COVID-19 dataset. For the diabetes dataset, this technique performs well than other competitive ML models in terms of accuracy and F-measure by 1.8274% and 1.7264, respectively.

3   Feature Selection Techniques

This study proposed features that were selected according to the relevancy of classifying features and tasks. Some of the features selection that is used in this proposed are following:-

3.1   Sequential Forward Selection

SFS is based on the greedy approach in which the best relevant best new feature is selected from the set of features iteratively. It starts with zero features and to maximize its value a cross-validated score is applied to estimate and trained it by a single feature. The best features are selected from the next triplets of features, after that the best triplet is.

3.2   Sequential Floating Forward/Backward Selection (SFFS and SFBS)

In LRS the value is fixed for L and R respectively, these floating methods verified the data values. During the search, the dimensionality of the subset for searching could be floating values up and down. According to feature selection, two methods are used:

   Sequential floating Forward Selection and Sequential floating backward Selection.

3.3   SFFS (Sequential Floating Forward Selection)

It is similar to the above approach SFS, except that it selects the optimal features sequentially for each step. Then repeat until the worst features are selected.it removes the worst features from the optimal set [4].

3.4 Sequential Floating Backward Selection

In this approach SFBS, it starts with the full feature set. It moves backward in each step until the objective function increase [8].

3.5 Sequential Backward Selection

SBS is a traditional feature selection algorithm, which concluded the element space into subspace include with minimum latency in classifier execution and decreases the model execution time. At times, SBS can work on the prescient ability of the model if a model dealing with overfitting issue SBS successively wipe out highlights from the full element space until the new component subspace have enough of highlights [4]. To figure out what highlight ought to dispose of from include space at each progression needed to characterize a component of basis J to limit. The basis is processed by a measure that is just be the distinction in the execution of the classifier previously, then after the fact the end of a particular element. The component that is wiped out at each progression can be characterized as the element that boosts the rule [1]. The Pseudo-code of SBS calculation can be outlined into 4 stages: 1: algorithm starting with k = d, the d is dimensional of highlight full space Xd 0.2: Eliminate include x-, that expands the rule x- = arg max J(Xk − x) Where xₑX k 3: Eliminate highlight x-from include space: Xk−1: = X k − x-; k: = k − 1. 4: Finish if k came to the required features, if not repeat stage 2.

3.6 LSTM

LSTM model is an updated version of RNN. In this, the previous state is get by the current output. These models overcome the negatives in traditional RNN. The algorithm depicts the disadvantages of traditional RNN, in manners, for example, utilizing angle plummet when managing the issue of long terms conditions [4]. LSTM has been effective in applications, for example, handwriting acknowledgment [5], regular language preparing machine interpretation, and speech recognition. Regular RNN ascertains its ht intermittent secret state, and the yt yield relies upon the previous secret state ht − 1 just as current xt contribution as below equations where A, U, V address the weight measurements between the present secret state ht and the former secret state ht − 1, current state, and yield, individually, while g(.) and f(.) address the element-wise AF. RNNs can use data about the past circumstance, yet the current state depends upon the past data, yet additionally on the approaching setting data. To defeat this issue, bidirectional RNNs were created [14]. Be that as it may, during preparing, bidirectional RNNs experience the will effects of the evaporating and detonating issue while handling the drawn-out conditions. The disappearing and

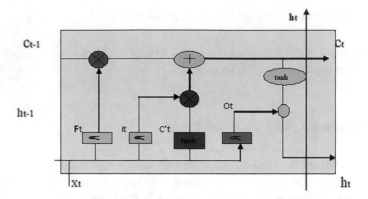

Fig. 1 LSTM network

detonating issue implies that BRNN is not reasonable for use in circumstances with longer conditions. This is a major question in intermittent organizations [8]. To beat this issue an unrivaled RNN structure was proposed, by and large, known as long short term memory (LSTM). As shown by Fig. 1 LSTM Network.

$$ht = (Aht - 1 + Uxt) \tag{2}$$

$$yt = (Ght) \tag{3}$$

4 Machine Learning Classifiers

The various ML classifiers are used to implement this proposed study. These are the following:

4.1 KNN

In a given region the closest neighboring trained points performed classification. The number of neighboring labeled points with examples present at a given location is based on the classification of new test data [3]. To find the different values of k and p-value the best KNN classification model.

4.2 SVM

It locates Hyperplane, where the numbers of features are present and represented by N.it, which is in N-dimensional space. It classifies the data according to the corresponding class. It has hyperparameters by which the performance of algorithms is affected [7].

4.3 Random Forest (RF)

It is an ensemble-based learning calculation comprising of randomly produced DT classifiers, the consequences of which are collected to acquire obtained better predictive performance [10].

4.4 Gradient Boosting (GB)

It is a machine learning technique that is used for classification and regression. it produces the ensembling of weak production models such as decision tree is considered as weak learner after that applying algorithm is called GB tress. Gradient Boosting (GB) is a group-based calculation made out of numerous decision trees l [6]. The optimal value for this boundary was observed to be 0.14 as other learning rate upsides of 1, 0.5, 0.25, 0.1, 0.05, and 0.01 were likewise tried in the grid search. A Naïve Bayes probabilistic algorithm was used as the standard model for execution examination between the other more unpredictable calculations. The setup for this model was kept as basic as conceivable by using the boundaries in their default esteems as introduced by the Gaussian NB python model. Besides, to decide if there were factual contrasts between the researched models and the gauge model, a One-Way ANOVA measurable test with Tukey's post hoc examination was performed on the mean AUROC scores. The invalid and substitute theory formed was: Hypothesis 1 (H1). Invalid Hypothesis: The mean AUROC score for the looked at 2 models is equivalent. Theory 2 (H2). Elective Hypothesis: The mean AUROC score [9].

4.5 NB

Naïve Bayes based on the probabilistic classifiers. İt is implemented by applying the Bayes theorem. It is highly scalable. It requires the number of parameters or features in a learning problem [13].

5 Materials and Methods

In this proposed work 419 instances are taken from the HRV dataset, this dataset has a real dataset. Some of the labels were done from the physionet library. This dataset selected 30 features using the feature selection technique SFFS [4]. in the given dataset the person who is suffering from heart disease or not could be identified by the resultant value that 3 for normal means no heart disease and 1,2 have abnormal means heart disease is there [5].

5.1 Proposed Methodology

This section proposed a model LSTM and Gradient boosting classifier to detect the heart disease by relevant feature selection technique (SFFS) using HRV dataset which has 419 instances. As shown proposed figure by which the input values will send to the module and the applying the LSTM for 178 samples and GB to improve accuracy and learn deep learning and relative information among the features as shown by Fig. 2 proposed Methodology and described the research workflow.

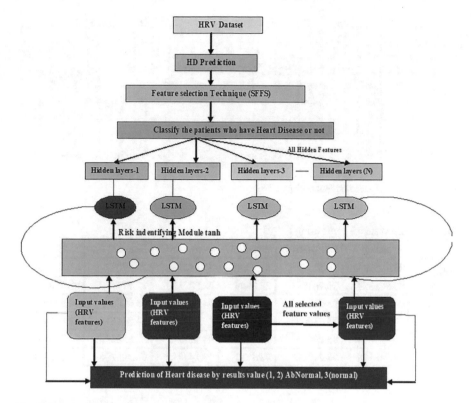

Fig. 2 Proposed methodology

6 Results and Discussion

The HRV dataset has 419 instances and 34 attributes then select the best 30 features. Which are as follows. The CV score obtained for SFFS is 0.9269662921348315. These are features that have been selected for SFFS.

('td_mean', 'td_median', 'td_SDNN', 'td_SDANN', 'td_NNx', 'td_pNNx', 'td_RMSSD', 'td_SDNNi', 'td_meanHR', 'td_sdHR', 'td_HRVTi', 'td_TINN', 'fd_aVLF', 'fd_aLF', 'fd_aTotal', 'fd_nLF', 'fd_nHF', 'fd_LFHF', 'fd_Poincare_SD1', 'fd_Poincare_SD2', 'fd_Nonlinear_sampen', 'fd_Nonlinear_alpha"fd_Nonlinear_alpha1', 'fd_Nonlinear_alpha2', 'tf_aVLF', 'tf_aLF', 'tf_nLF', 'tf_nHF', 'tf_LFHF', 'tf_rLFHF') as shown in Fig. 3 Number of Selected Feature by SFFS.

The performance and number of selected features by the given HRV dataset with their results were obtained.

The results obtained by given Table 1 Performance Results using LSTM. LSTM give best results in terms of all the performance metrics.

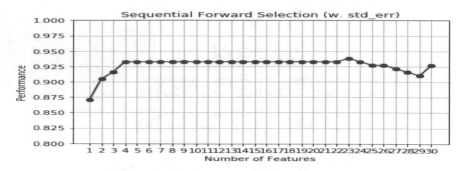

Fig. 3 Number of selected feature by SFFS

Table 1 Performance Results using LSTM

	Accuracy	Precision	Recall	F-measure	Kappa	AUC
RF	85.185185	88.571429	85.1852	84.852735	70.37037	85.185185
KNN	88.888889	89.761571	88.8889	88.827586	77.777778	88.888889
SVM	94.444444	95.00000	94.4444	94.427245	88.888889	94.444444
NB	57.407407	68.75000	57.4074	49.818182	14.814815	57.407407
GB	94.444444	95.00000	94.4444	94.427245	88.888889	94.444444
LSTM	98.148148	98.214286	98.1481	98.147513	96.296296	98.148148

As by given Fig. 4, the Accuracy obtained over LSTM for feature selection(SFFS) is: **Accuracy** 98.148148, **Precision** 98.214286, **Recall** 98.1481, **F-measure** 98.147513, **Kappa** 96.296296, **and AUC** 98.148148.

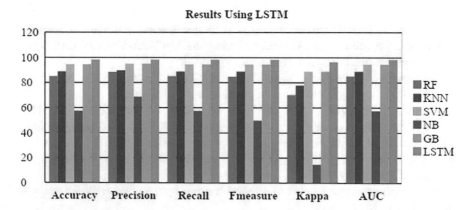

Fig. 4 Various performance metrics results for ML classifier over Deep learning

Fig. 5 LSTM AUC curve

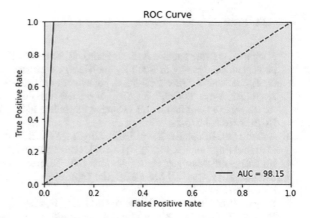

This ROC curve represents by the confusion matrix. The percentage obtained for LSTM for AUC is 98.148% as per the confusion matrix obtained and shown by above Fig. 5 LSTM AUC curve.

[[26 1]
[0 27]]

7 Conclusion

This study developed a Deep learning approach for the study of high dimensional information. LSTM DL approach was used the classify the features using SFFS. With high-dimensional information examination utilizing DL performed preferred as far as precision over past ML and deep learning structures. The HRV dataset taken from Max hospital mohali. LSTM is helpful for high-dimensional data analysis and can deal with exceptionally huge datasets, with low computational intricacy and high exactness. LSTM achieved Accuracy 98.148148, Precision 98.214286, Recall 98.1481, F-measure 98.147513, Kappa 96.296296, and AUC 98.148148 with 30 features. The primary approach is to improve the existing work and build a novel model which has an outcome for more in terms of their accuracy rate. Deep learning always gives the best results and classifies more data features. In the future generalize this model for other feature selection algorithms and datasets. in the future implement some other deep learning techniques to improve accuracy and better results and also computed and make it multimodal system by using different datasets.

References

1. Natarajan A, Pantelopoulos A, Emir-Farinas H, Natarajan P (2020) Heart rate variability with photoplethysmography in 8 million individuals: a cross-sectional study. Lancet Dig Health 2(12), e650–e657. ISSN 2589–7500. https://doi.org/10.1016/S2589-7500(20)30246-6
2. Aydin SG, Kaya T, Guler H (2016) Heart rate variability (HRV) based feature extraction for congestive heart failure. Int J Comput Electr Eng 8(4):272–279. https://doi.org/10.17706/IJCEE.2016.8.4.272-279
3. Acharya UR, Joseph KP, Kannathal N, Lim CM, Suri JS (2006) Heart rate variability: a review. Med Biol Eng Comput 44(12):1031–1051. https://doi.org/10.1007/s11517-006-0119-0
4. Karthikeyan P, Murugappan M, Yaacob S (2013) Detection of human stress using short-term ECG and HRV signals. J Mech Med Biol 13(2). https://doi.org/10.1142/S0219519413500383, PubMed: 1350038
5. Sung C-W, Shieh J, Chang W, Lee Y, Lyu J, Ong H, Chen W, Huang C, Chen W, Jaw F (2020) Machine learning analysis of heart rate variability for the detection of seizures in comatose cardiac arrest survivors. IEEE Access 8:160515–160525. https://doi.org/10.1109/ACCESS.2020.3020742
6. Dalmeida KM, Masala GL (2021) HRV features as viable physiological markers for stress detection using wearable devices. Sensors 21(8), 2873. https://doi.org/10.3390/s21082873
7. Younis H, Anwar MW, Khan MUG, Sikandar A, Bajwa UI (2021) A new sequential forward feature selection (SFFS) algorithm for mining best topological and biological features to predict protein complexes from protein–protein interaction networks (PPINs). Interdisc Sci Comput Life Sci 13(3):371–388. https://doi.org/10.1007/s12539-021-00433-8
8. Lee J, Park D, Lee C (2017) Feature selection algorithm for intrusions detection system using sequential forward search and random forestclassifier. KSII Trans Internet Inf Syst 11(10):5132–5148. https://doi.org/10.3837/tiis.2017.10.024
9. Ali L, Rahman A, Khan A, Zhou M, Javeed A, Khan JA (2019) An automated diagnostic system for heart disease prediction based on x2 Statistical model and optimally configured deep neural network
10. Solanki Y, Sharma S (2019) Analysis and prediction of heart health using deep learning approach. In: 2019 international journal on computer science and engineering 7(8):309–315. E-ISSN: 2347–2693
11. Karayilan T, Kilic O (2017) Prediction of heart disease using neural network. In: International conference on computer science and engineering, pp 719–723
12. Kumar N, Das NN, Gupta D, Gupta K, Bindra J (2021) Efficient automated disease diagnosis using machine learning models. J Healthcare Eng 2021:13. Article ID 9983652
13. Aggarwal R, Kumar S (2022) An automated perception and prediction of heart disease based on machine learning. In: AIP conference proceedings, vol 2424, no 1. AIP Publishing LLC, p 020001
14. Aggarwal R, Kumar S (2022) An enhanced fusion approach for meticulous presaging of HD detection using deep learning. In: IEEE international conference on distributed computing and electrical circuits and electronics (ICDCECE), pp 1–4. https://doi.org/10.1109/ICDCECE53908.2022.9793141

Determining the Most Effective Machine Learning Techniques for Detecting Phishing Websites

S. M. Mahamudul Hasan, Nirjas Mohammad Jakilim, Md. Forhad Rabbi, and Rumel M. S. Rahman Pir

1 Introduction

Phishing is one of the most serious cybersecurity threats since it includes creating fake websites that seem to be legitimate [1]. In this assault, the user inputs important information such as credit card numbers, passwords, and so on to a bogus website that seems to be real. The sectors most severely affected by this attack include online payment services, e-commerce, and social media [2, 3]. Phishing attacks make use of the aesthetic resemblance between fake and legitimate websites [4]. Phishing has taken on several forms throughout history, including legal, educational, and awareness efforts [5]. Phishing attacks use several methods to access sensitive information, including link manipulation, filter evasion, website forgery, covert redirection, and social engineering [6, 7]. It is estimated that internet-based theft, fraud, and exploitation would account for an astonishing $4.2 billion in financial losses in 2020, according to the FBI's Internet Crime Complaint Center 2020 report [8]. During this attack, the attacker mainly produces a website that is identical to the genuine web page in appearance. The phishing web page's URL is subsequently sent to thousands of Internet users through email and other contact forms [9]. Typically, the false email content creates panic, urgency, or promises money in exchange for the recipient taking immediate action. The fake email will prompt customers to change their PIN to prevent their debit/credit card suspension. When a user changes their sensitive credentials inadvertently, cyber thieves get the user's information [10]. Phishing attacks are not just used to get information, they have also become the primary technique for

S. M. Mahamudul Hasan (✉) · N. M. Jakilim · Md. Forhad Rabbi
Shahjalal University of Science and Technology, Sylhet, Bangladesh
e-mail: smmahamudul32@student.sust.edu

N. M. Jakilim
e-mail: nirjas01@student.sust.edu

Md. Forhad Rabbi
e-mail: frabbi-cse@sust.edu

R. M. S. Rahman Pir
Leading University, Sylhet, Bangladesh
e-mail: rumelpir@lus.ac.bd

© The Author(s), under exclusive license to Springer Nature Singapore Pte Ltd. 2022
B. Unhelker et al. (eds.), *Applications of Artificial Intelligence and Machine Learning*,
Lecture Notes in Electrical Engineering 925,
https://doi.org/10.1007/978-981-19-4831-2_48

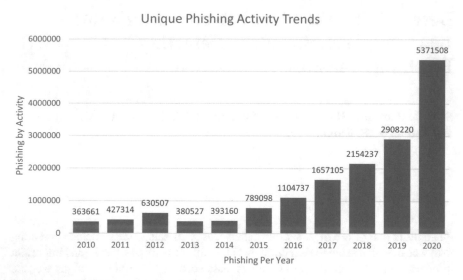

Fig. 1 Unique phishing activity trends

distributing other kinds of harmful software, such as ransomware. 91% of current cyber-attacks begin with phishing emails [11].

Phishing assaults account for more than half of all cyber fraud affecting Internet users. More than 245,771 distinct phishing websites were identified in January 2021, according to the APWG study. Monthly attack growth rose by 1477% during a ten-year period from 2010 to 2020. (363661 Phishing attacks in 2010 and an average of 5371508 attacks in 2020). From 2010 through 2020, Fig. 1 depicts the rise of phishing assaults [12].

Numerous variations of phishing attacks have occurred throughout history [13]. The attacker may be motivated by identity theft, financial gain, or celebrity. Scientists and researchers face significant challenges when it comes to identifying and blocking phishing attacks [14]. Even the most seasoned and informed users may face an assault. Phishing attacks usually include the transmission of a fake email purporting to be from a reputable company or organization, requesting sensitive information such as a bank login or password. Phishing communications are delivered through email, SMS, instant messaging, social media, and voice over internet protocol (VoIP) [15]. However, the most frequent form of attack is through email. 65% of phishing efforts include the use of malicious URLs in emails [16]. Due to the fact that phishing websites are only up for a limited time, the phisher may flee immediately after committing the crime. That's why automated systems are needed to prevent them as soon as possible. To determine if a website is phishing or not, a variety of methods have been developed. Attackers may use several methods at different phases of the attack cycle. Among other things, network security, user education, user authentication, server-side filters, client-side tools, and classifiers are some of the methods that are available. While each kind of Phishing attack is unique, the bulk of them have certain characteristics and patterns [17]. Due to the essential role of machine learning techniques for identifying patterns in data, it has become possible to identify a large number of common Phishing characteristics, as well as to recognize Phishing websites [18]. Specifically, the purpose of this paper is to examine and evaluate a number of machine learning methods for identifying fake websites. Logistic Regression, Random Forest, Support Vector Machine, KNN, Naive Bayes, and the XGBoost classifier were among the machine learning methods we investigated.

The remainder of this research paper is divided into the following sections: The second section covers similar studies on phishing websites. Section 3 explored our suggested method of work. In Sect. 4, we provide a short description of the dataset. In Sect. 5, we examined different machine learning techniques to detect phishing and summarized the experiments and their outcomes. The conclusion and future study are described in Sect. 6.

2 Related Work

There are various methods of phishing detection and they are list-based and machine-learning-based etc. The most often used detection technique is list-based. Whitelists consist of legitimate websites, while blacklists consist of phishing websites. Phishing detection systems that are list-based rely on these lists to identify phishing attempts. C Whittaker, B Ryner, M Nazif [19] compiled and published a whitelist of all URLs accessed through the Login user interface. When a user visits a website, the system informs the user if their data is incompatible with the site. This method may be used by a user for the first time when they visit an authorized website. SL Pfleeger, G Bloom [20] developed an automatic whitelist of user-approved websites. They used two stages of feature extraction, which are domain-IP address matching and source code connections. They got a true positive rate of 86.02% for all observations, whereas false negatives accounted for 1.48%. Zhang et al. [21] identified phishing by developing CANTINA that uses TF-IDF techniques. It is a content-based approach. The keywords are put into Google. When a website appears in search results, it earns the confidence of users. CANTINA Plus is equipped with fifteen HTML-based features. Despite the algorithm's 92% accuracy rating, it can give a substantial number of false positives. Islam et al. [22] created a categorization system for communication by reviewing the website titles and messages. The technique is designed to minimize false positives. The research gathered information on URL-specific characteristics such as length, subdomain names, slashes, and dots. Rule mining was utilized to develop detection rules as a priority. In testing, 93% of phishing URLs were correctly identified.

To categorize phishing attempts more correctly and effectively, an adaptive self-structuring neural network was employed by Rami et al. [23]. It includes seventeen features, some of which are dependent on third-party services. As a result, real-time execution is sluggish. While it has the potential to enhance accuracy, it is not currently used. It manages noisy data by using a small dataset of 1400 elements. Others combine artificial intelligence and image processing. For image/visual-based applications, the internet domain (web history) is needed to recognize phishing attempts. Basit et al. [24] proposed an approach that circumvents these constraints. They categorized features according to whether they included hyperlinks, third-party material, or masked URLs. By using third-party services, the system's accuracy is increased to 99.55%, while detection time and latency are decreased. NLP receives little attention in the scholarly literature (NLP). In a recent study, Peng et al. [25] used natural language processing to detect phishing emails. This program scans the plain text content of emails for harmful intent. It gathers queries and responses via the use of natural language processing (NLP). Phishing attempts are detected using a custom-built blacklist of word pairs. The algorithm was trained on 5009 phishing emails and 5000 real emails prior to becoming public. Their experimental research demonstrated a 95% accuracy rate.

3 Proposed Approach

The method we use to identify phishing websites is one that is based on machine learning. Our model incorporates a variety of machine learning techniques, including logistic regression, KNN, decision trees, Random Forest, support vector machines, and gradient boosting. We collected the dataset from kaggle [26] and highlighted the dataset's vector in our model (Fig. 2).

Then, we utilized this dataset for training six machine learning classification models to identify the characteristics of a phishing website. To implement the machine models, we have used the sci-kit learn library [27]. We also utilized their tweaking parameters to hyparametertune the models to get the best results for a particular model. Finally, we compare the results and find out the best-performing models and evaluate the reasons for their good performances. Our categorization algorithm detects Phishing websites with an accuracy of about 97%. And our model is capable of detecting approximately 97.48% of the genuine phishing sites.

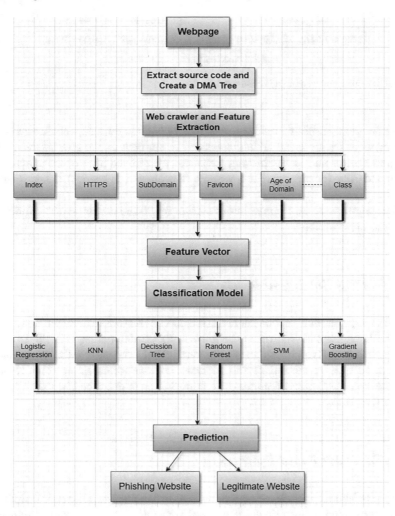

Fig. 2 Diagram of our approach

4 Dataset

One of the most important challenges we faced during our research was the scarcity of phishing databases. There have been a large number of academic papers on the subject of phishing detection. However, none of them have made the datasets used in their research available to the general public. The absence of a common feature set that captures the features of a phishing website also makes it more difficult to collect useful data, which makes it more difficult to build a useable dataset in the first place. Numerous academics carefully examined and benchmarked the dataset used in our analysis. We collected the dataset from kaggle. [26] which contains about

Fig. 3 URL components of a legitimate website

11,054 sample websites with 32 features. Nearly 6080 legitimate websites and 4974 phishing websites are included in the dataset. In our dataset, we give a score of 1 to genuine websites and −1 to phishing websites. We utilized 30% of the samples for testing and 70% for training. Each website is evaluated to see if it is legitimate or fake.

We categorize the 32 features of our dataset into three categories. The following categories are described:

4.1   Address Bar Based Features

The following elements are considered in the address bar-based features. Long URL, Short URL, Symbol@, Redirecting/, Prefix Suffix-, Subdomains, HTTPS, Domain-RegLen, Favicon, NonStdPort, HTTPSDomainURL, Request URL, and Anchor URL are all used in the index. The address bar-based features are often referred to as the link URL-based features. The address bar is described in the following manner (Fig. 3):

4.2   Abnormal Based Features

There are 9 features that define abnormal-based characteristics. LinksInScriptTags, Server Form Handler, Info Email, Abnormal URL, Website Forwarding, StatusBar-Cust, Disable Right Click, Using popup Window, and IframeRedirection are some of them.

4.3   Domain Based Features

There are 8 features that describe domain-based features. These are the following: Domain Age, DNSRecording, Website Traffic, Links Pointing to Page, PageRank, Google Index, Status Report, and class.

If the URL-based feature has a value of −1, it is a phishing website. If the value is 0, the website is suspect. If the URL value includes one, it indicates that the website is genuine.

5 Result and Analysis

We utilized 10-fold cross-validation to evaluate the model's overall performance in our trials. 10 sub-samples were drawn from the original data set using a random number generator. Three samples are tested (30%), while the other samples are used to train model-based categorization algorithms. Because phishing detection is categorical in nature, we must employ a binary classification model to identify phishing assaults. "−1" denotes a phishing sample, while "1" denotes a genuine sample. We identified phishing websites using a variety of machine learning models, including logistic regression, random forest, KNN, SVM, gradient boosting, and decision trees.

We assessed these models' accuracy, precision, recall, F1 score, and confusion matrix, and then utilized a variety of feature selection and hyperparameter tweaking techniques to get the best possible results. The precision, recall, and F1 scores and accuracy of different models, as well as their overall performance, are summarized in Table 1. The accuracy of different models, as well as their overall performance, is compared in Fig. 5. The Random Forest has been proven to be extremely accurate, reasonably resistant to noise and outliers, simple to build and comprehend, and capable of implicit feature selection in our studies. In Fig. 4, we show the learning

Table 1 Evaluation of all the models

Algorithms	Precission	Recall	F1 score	Accuracy %
Logistic Regression	0.92	0.93	0.93	92.76
KNN	0.52	0.55	0.54	60.45
Decision Tree	0.95	0.95	0.95	94.75
Random Forest	0.97	0.96	0.97	97.17
SVM	0.50	0.28	0.36	56.04
Gradient Boosting	0.94	0.95	0.95	94.75

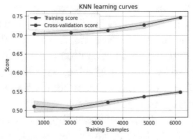
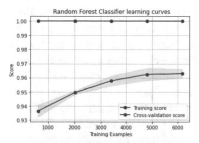

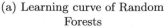

(a) Learning curve of Random Forests   (b) Learning curve of KNN

Fig. 4 Learning curves of RF and KNN

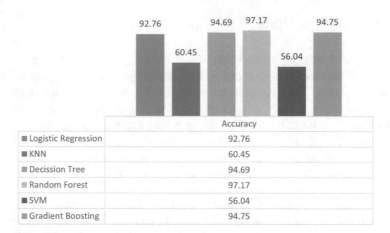

	Accuracy
■ Logistic Regression	92.76
■ KNN	60.45
■ Decission Tree	94.69
■ Random Forest	97.17
■ SVM	56.04
■ Gradient Boosting	94.75

Fig. 5 Accuracies of the models

curves of random forest. The Random Forest offers a number of benefits over the decision tree, the most important being its resistance to noise.

By increasing the number of trees in each woodland inside the forest, the Random Woodland decreases variance. The primary drawback of Random Forests was the large number of hyperparameters that required tuning to attain optimum performance. Additionally, it adds a random aspect to both the training and testing data, which may not be appropriate for all data sets and circumstances. The study could establish the optimal classification accuracy of the KNN by using k = 10. There is no one-size-fits-all k value for KNN classification. In Fig. 4, we show the learning curves of the KNN classifier. Due to the huge number of neighbors, it is computationally costly to develop a solution. Additionally, we discovered that a few neighbors provide the most flexible fit, with low bias but a high variation, while a large number of neighbors produces a smoother decision boundary with a lower variance but greater bias.

Logistic regression is predicted to be 92.76% accurate. Additionally, our system accurately detects about 93.50% of true positives in the confusion matrix. We can evaluate the model's correctness throughout both the training and testing stages by examining the training and cross-validation scores. The actual accuracy of the KNN model is just 55.28%, and the model cannot identify 44.71% of phishing websites. This accuracy is poor, and the total performance of the KNN model is 60.45%. When the accuracy of decision tree tests is considered, the cross-validation score performs well. The decision tree classifier has a true positive score of 95.32%, but a false positive score of just 4.67%. With an overall accuracy of 94.69%, the model is very accurate. The Random forest has the greatest accuracy among all the models. At level 1, the training score of the model is negligible. Cross-validation surpasses single-validation. This model has a true positive rate of 97.48% and a false negative rate of 2.51%. The system works optimally 97.17% of the time. The support vector machine model performs the least well of all the models. The accuracy is optimal 56.0% of the time. It is unable to identify any phishing websites effectively using this. The training score is the lowest, but the cross-validation score is close to the training score overall. As a result, the model is unable to function properly. In our model, the gradient boosting method works well. It detects true positives at a rate of 95.53% and false positives at a rate of just 4.46%. The model's total accuracy is 94.75%.

The primary benefit of Gradient Boost over other techniques such as decision trees and support vector machines is its speed. Additionally, it has a regularization parameter that significantly lowers variance. To further enhance the generalizability of this approach, the learning rate, and subsamples from features such as random forests are combined with the regularization parameter. Compared to Logistic Regression and Random Forests, Gradient Boost is more difficult to comprehend, visualize, and change. Numerous hyperparameters may be adjusted to improve overall performance. Gradient Boost is an enticing technique to use when both speed and accuracy are required. Despite this, more resources are needed to train the model, since model tweaking takes additional effort and skill on the part of the user to get statistically significant results.

The decision tree outperforms the KNN, Logistic regression, and SVM in our model. Due to the huge volume of data and the diversity of characteristics included therein, the decision tree works well in this scenario. The Decision Tree has two nodes. The Decision Node is the first of these nodes, followed by the Leaf Node. In contrast to decision nodes, which are used to make choices and include many branches, leaf nodes reflect the result of those choices and contain no further branches that branch out to other locations. The judgments or tests are done in light of the dataset's characteristics.

For SVM, we just apply the linear kernel model. Our prior experience indicates that the linear kernel does not perform well on this dataset. Consequently, SVM is ineffective at detecting phishing websites. It is unable to properly detect any phishing websites. Despite the size of our dataset, the SVM technique is not designed for big data sets. When the data set has a high level of noise and the target classes overlap, SVM performs poorly. It is unusual for the SVM to perform poorly when a single data point has more features than there are training data samples. Therefore, SVMs fail in our model.

6 Conclusion

We developed and tested six phishing website classifiers on a dataset of 6080 legitimate websites and 4974 phishing websites in this study. Classifiers such as Logistic Regression, Decision Trees, Support Vector Machines, Random Forests, KNNs, and Gradient Boosting are examined. Our classifiers, Random Forest and Gradient Boost, perform well in terms of computation time and accuracy, as shown in Tables 1. Experimental findings indicate that logistic regression works best for the identification of phishing websites. The suggested method has reasonably high accuracy in identifying phishing websites as it was obtained for random forest classification. It has more than 97.41% true positive rate and 2.58% false positive rate. Moreover, our method's accuracy, precision, and f1 score are 97.17, 97.80, and 97.59%, respectively. We have also examined the area under the classification model and the learning curve for all the models to discover a better measure of accuracy. Our experiment calculated training and cross-validation scores independently for all classification models used to categorize correct web pages.

Our model could not make use of support vector machines because SVM uses the linear kernel model. That's why when the input is noisy and the target classes overlap, SVM performs poorly. When a single data point has more features than the amount of training data samples, it fails. As a consequence, SVMs don't perform optimally in our model. Utilizing a different SVM kernel may be beneficial. Also, using a polynomial, Sigmoid, or RBF kernel may increase accuracy. Logistic regression is predicted to be 92.76% accurate. The KNN accuracy is poor, and the total performance of the KNN model is 60.45%. When the accuracy of decision tree tests is considered, the cross-validation score performs well. The decision tree classifier has a true positive score of 95.32%, but a false positive score of just 4.67%. With an overall accuracy of 94.69%, the model is very accurate.

References

1. Shaikh AN, Shabut AM, Alamgir Hossain M (2016) A literature review on phishing crime, prevention review and investigation of gaps. In: 2016 10th international conference on software, knowledge, information management & applications (SKIMA). IEEE
2. Scheau C, Arsene A, Dinca G (2016) Phishing and e-commerce: an information security management problem. J Def Resources Manage 7(1):12
3. Sarjiyus O, Oye ND, Baha BY (2019) Improved online security framework for e-banking services in Nigeria: a real world perspective. J Sci Res Rep 1–14
4. Mohammad RM, Thabtah F, McCluskey L (2015) Tutorial and critical analysis of phishing websites methods. Comput Sci Rev 17:1–24
5. Adebowale MA et al (2019) Intelligent web-phishing detection and protection scheme using integrated features of Images, frames and text. Expert Syst Appl 115:300–313
6. Ali A (2016) Social engineering: phishing latest and future techniques. Accessed 10 Mar 2015
7. Goel D, Jain AK (2018) Mobile phishing attacks and defence mechanisms: state of art and open research challenges. Comput Secur 73:519–544
8. FBI releases the internet crime complaint center 2020 internet crime report, including COVID-19 scam statistics. https://www.fbi.gov/news/pressrel/press-releases/fbi-releases-the-interne-crime-complaint-center-2020-internet-crime-report-including-covid-19-scam-statistics
9. Jain AK, Gupta BB (2016) A novel approach to protect against phishing attacks at client side using auto-updated white-list. EURASIP J Inf Secur 2016(1):1–11
10. Dhamija R, Doug Tygar J, Hearst M (2006) Why phishing works. In: Proceedings of the SIGCHI conference on Human Factors in computing systems
11. 91% of all cyber attacks begin with a phishing email to an unexpected victim. https://www2.deloitte.com/my/en/pages/risk/articles/91-percent-of-all-cyber-attacks-begin-with-a-phishing-email-to-an-unexpected-victim.html
12. Phishing activity trends reports. https://apwg.org/trendsreports/
13. Charoen D (2011) Phishing: a field experiment. Int J Comput Sci Secur (IJCSS) 5(2):277
14. Jakobsson M, Myers S (eds) Phishing and countermeasures. Understanding the increasing problem of electronic identity theft. Wiley, Hoboken
15. Ramzan Z (2010) Phishing attacks and countermeasures. In: Handbook of information and communication security, pp 433–448

16. Must-know phishing statistics. https://www.tessian.com/blog/phishing-statistics-2020/
17. Jain AK, Gupta BB (2021) A survey of phishing attack techniques, defence mechanisms and open research challenges. Enterp Inf Syst 1–39
18. Passos IC, Mwangi B, Kapczinski F (2016) Big data analytics and machine learning: 2015 and beyond. Lancet Psychiat 3(1):13–15
19. Whittaker C, Ryner B, Nazif M (2010) Large-scale automatic classification of phishing pages
20. Pfleeger SL, Bloom G (2005) Canning spam: proposed solutions to unwanted email. IEEE Secur Priv 3(2):40–47
21. Zhang Y, Hong JI, Cranor LF (2007) Cantina: a content-based approach to detecting phishing web sites. In: Proceedings of the 16th international conference on World Wide Web
22. Islam R, Abawajy J (2013) A multi-tier phishing detection and filtering approach. J Netw Comput Appl 36(1):324–335
23. Mohammad RM, Thabtah F, McCluskey L (2014) Predicting phishing websites based on self-structuring neural network. Neural Comput Appl 25(2):443–458
24. Basit A et al (2020) A comprehensive survey of AI-enabled phishing attacks detection techniques. Telecommun Syst 1–16
25. Peng T, Harris I, Sawa Y (2018) Detecting phishing attacks using natural language processing and machine learning. In: 2018 IEEE 12th international conference on semantic computing (ICSC). IEEE
26. Phishing website detector. https://www.kaggle.com/eswarchandt/phishing-website-detector
27. Pedregosa F, Varoquaux G, Gramfort A, Michel V, Thirion B, Grisel O, Blondel M, Prettenhofer P, Weiss R, Dubourg V, Vanderplas J, Duchesnay E (2011) Scikit-learn: machine learning in Python. J Mach Learn Res 12:2825–2830

Performance Analysis of Computational Task Offloading Using Deep Reinforcement Learning

S. Almelu, S. Veenadhari, and Kamini Maheshwar

1 Introduction

There are many fields in which deep learning and machine learning techniques are applied to provide the fruitful, secured and accurate results. These techniques are used in the streams like natural language processing, fog computing, mobile edge computing, etc. Internet of things are becoming an important part of technologies to be used, like for smart city, telecom etc. Numerous applications are developing for the mobile users which are increasing dynamically; similarly multiple uses of internet of things are applied in different areas to modify the environment of internet [1]. These devices consume energy as they perform various tasks and thus their battery life also reduces with the increase in workload of the devices, so this is one of the problems with such devices. It becomes necessary to reduce the workload and increase the battery life of IoT devices and for this the researchers have developed many techniques or proposed several models to increase the energy efficiency of such devices, and deep learning is also one of the techniques which is applied in the process of task offloading in IoT devices [2].

Task offloading can be defined as the process of reducing the workload of the task offered in any particular device and enhancing the energy efficiency of IoT devices. Task offloading is applied in fog computing, edge computing, fog networks, mobile edge computing, vehicular cloud computing systems, cloud edge collaborative systems, vehicular edge computing networks and systems, etc. Mobile devices as known that not able to make fine grained offloading decisions in real time environment, or these devices are wireless and connected through cloud servers so it is become difficult for them to make the decision while mobile clouds are used in the wireless communication process. Thus, to make the decision for appropriate

S. Almelu (✉) · S. Veenadhari · K. Maheshwar
Department of CSE, RNTU, Bhopal, M.P., India
e-mail: salmelu.uit@gmail.com

offloading task in mobile devices, deep learning techniques are used which are somewhere proven beneficial [3]. Similarly, if we discuss about use of IoT devices at large scale like for developing smart homes, smart cities, several industries, transportation, etc. then it is important that physical size of these devices should be smaller and computational cost must be low. with these qualities it is also the need of the devices that their battery life and energy efficiency is to be maintained. Offloading of data is required because it reduces the burden of computing in devices and balances the energy which is to be consumed while performing multiple tasks. Task offloading using edge computing, fog computing, deep learning techniques and many more other techniques significantly increased the computational flexibility and low latency in IoT devices. In urban areas or in smart cities for vehicles also intelligent transportation systems are developed to understand the complex real time traffic problems [4]. Thus, for reliable functionality of the devices and improving the computational process by reducing the energy consumption and decision making in task offloading deep learning techniques are applied in this paper to provide energy efficient and enhanced battery life of the IoT devices used in the environment of internet.

2   Related Work

Mohamed K. Hussein [1] In this paper author introduces two different schedulers Ant Colony Optimization and Particle Swarn Optimization, and presented separate task offloading algorithms to make balance the workload over the fog nodes as fog computing performs better when compared to cloud computing, and the performed experiments resulted that Ant Colony Optimization improves the efficiency in time and also make balance in the task over fog nodes. Mohammad Aazam [2], three tier IoT fog-cloud model is proposed which executes through firstly in IoT nodes, secondly in fog and lastly in cloud. This paper introduces performance and energy consumption by performing the experiments on real data sets based on three different scenarios: fog only, cloud only and fog-cloud only and concluded that fog-cloud experiments provide effective results in task offloading with less energy consumption. Xu Chen [3] In this article author introduced another worldview of resource efficient edge computing for the arising smart IoT applications. He devise this technique to such an extent that an insightful IoT user can well help its computationally escalated task by appropriate undertaking offloading across the local device, close by nearby device, furthermore, the edge cloud in nearness. Nan Cheng [4] here researcher proposed SAGIN (space air ground integrated network) an architecture based on edge or cloud computing for offloading the computational problems in UAV edge servers. Here he used RL- based computational task offloading approach for comprehensive computation tasks. Ke Zhang [5] he started by creating the best possible scenario offloading scheme with MEC server collection as a community and the determination of transmission mode in a deep Q-learning strategy. Then, in the context of task communication failure, he focuses on efficient offloading and proposes an adaptive redundant offloading algorithm to ensure offloading efficiency while improving system utility. Kaoru Ota [6] He developed an elastic model for varying deep learning models in edge computing for the Internet of Things. He also devised an effective online algorithm for optimizing the edge computing model's service capability. Finally, he puts the IoT deep learning model to the test in a real-world edge computing environment. He also compares the edge computing approach to more conventional approaches. He chose ten different options, Deep learning networks are represented by CNN models, and gather data of a medium size and computational complexity overhead resulting from deep functional learning the apps The outcome of the performance review demonstrate that the solutions will improve the situation the number of tasks that have been implemented in edge servers guaranteed Quality of Service (QoS) specifications. Minghui Min [7]. In this paper, author has proposed a reinforcement learning (RL) based offloading scheme for an IoT system with EH that uses the current battery level, the previous radio transmission rate to each edge device, and the expected amount of harvested energy to pick the edge device and offloading rate. Shuai Yu [8] suggested a new deep imitation in this paper, Edge-cloud computing offloading based on deep learning (DIL) MEC networks have a structure. One of the key goals the framework's aim is to reduce offloading by expense in time-varying network environments behavioral cloning at its best. Xiaolong Xu [9] To make the process go faster, this paper proposes a heuristic offloading approach for deep learning tasks with transmission delays. Here author builds a mechanism for offloading within the CU-DU architecture and on the basis, examine the offloading period. The process of offloading deep learning tasks is then described in detail. Chandan Pradhan [10] presented a computation-offloading problem for IoT applications in an uplink xL-MIMO C-RAN with a latency restriction. The nonconvex optimization problem that minimises the total transmit power of the IoTDs thus meeting the latency requirement is discovered. Guanjin Qu [11] proposed an edge offloading framework for addressing task offloading decision-making in heterogeneous IoT-edge-cloud computing settings. The DMRO framework combines a task offloading decision model depending on a distributed deep reinforcement-learning algorithm and a training initial parameter model relying on deep meta learning, both of which intended to deal with the issue of neural network mobility. The work concluded that the DMRO architecture outperforms full offloading approaches and traditional reinforcement learning-based methods in terms of task offloading decisions. Furthermore, because of the usage of Meta parameters, the model has increased adaptability and speedy environmental ability to learn. S. Aljanabi, A. Chalechale [12] proposed a model to identify the ideal selection on when and where to offload a task; to a specific fog node or to the cloud server. As a Markov decision process, the issue is investigated and assessed. Two decision-makers have been examined in the suggested MDP, where Operators can determine which fog node they want to offload their duties to, and fog nodes can choose to offload some jobs to other fog nodes or to cloud servers to balance the activities across fog nodes. To acquire optimal policy, a Q-learning-based method is developed to handle large-size state space and action space. When compared to earlier research, measured values show that the suggested technique provides superior load balancing and minimizes delay time. Mingfeng Huang [13] built a cloud-MEC collaborating computing platform and develop a CTOSO scheme on top of it to fulfill the low latency, high dimensionality, and highly reliable needs of developing applications. Firstly, researchers offer a unique service orchestration methodology for orchestrating data as high-quality operations at the edge layer using software defined network technology, therefore significantly lowering network transmission load. Simultaneously, the CTOSO scheme describes a target optimization function regarding the communication energy usage, data processing energy usage, task wait-time models, and pre-estimates the task offloading cost based on the optimization function, before making a distinguishable offloading decision relying on the estimation result and task attributes. The test results reveal that the CTOSO system described in this study has a significant impact on the latency and energy performance of large data-based implementations. M. Aazam et al. [14] reviewed various offloading approaches that have recently been described in the literature, with an emphasis on fog or edge computing in the cloud-IoT ecosystem. They also discussed some of the factors that have been recommended for deciding whether to offload to the cloud or the fog. They looked at some of the techniques that are being used to facilitate offloading, such as wireless communication. They also highlighted various offloading circumstances documented in the research, as well as newly invented offloading methodologies for such cases. Lastly, they discussed some of the major research difficulties involved with work offloading in fog computing.

3   Overview of Computational Offloading

Another technique for dealing with the limited resources in mobile and smart gadgets is computation offloading. The benefits of code offloading include improved power administration, reduced storage requirements, and improved application productivity. In the Smartphone evaluating realm, computation offloading has been a hot topic, with the majority of recommendations offloading tasks to the cloud. Because offloading to the cloud isn't always possible or practical [15]. The computational offloading concept is applied among smart devices and server to reduce the power consumption and workload while executing tasks. When smart devices compute sensitive and intensive data then it requires more computational resources. If they perform these tasks locally then it may suffer from more power consumption. Therefore, these tasks are offloaded to edge servers via wireless transmission. Mobile Edge Computing (MEC) programming framework is an effective to offload duties to edge machines, making it easier to build flexible and scalable edge-based mobile applications. When there is a computer-powered-restricted atmosphere, computational offloading is a cloud computing method that is utilized to run programme and offer contents. Complex jobs necessitate more processing power. The QoS of the delivered contents/programs is reduced if the destination gadget has inadequate computational power. Offloading is a feasible remedy in this situation. As a result, rather than doing complicated operations directly, the target device outsources them to a cloud server.

3.1   Cost of Computation Offloading

The energy usage for consumer n's task m contains two elements, namely the part for data transfer and the part for data reception, when users offload their duties to the edge server. The energy utilization for data transmission is denoted by e_{NM}^T, and the energy usage for data reception is denoted by e_{NM}^R. The system utilities value is given by when mobile customer n offloads its job m to the edge server [16].

$$d_{NM}^{CLO} = l_{NM}^{IN} + \frac{\psi_1}{F^{CLO}} + \frac{\psi_2}{c^{up,MAX}} + \frac{\psi_3}{c^{DO,MAX}} \tag{1}$$

$C^{up,MAX}$ and $C^{DO,MAX}$ are the overall uplink and whole downlink bandwidth variables, etc. The CPU rate of the edge server is variable F^{CLO}. The overall price for consumer n to offload assignment m to the edge server is denoted as:

$$e_{NM}^{CLO} = e_{NM}^T + e_{NM}^T + \lambda d_{NM}^{CLO} \tag{2}$$

Here λ is the discount element. It's worth noting that the cost e_{NM}^{CLO} includes the energy costs of transferring and obtaining the job, as well as the server's utilities cost for completing it. The delayed in computational offloading is then modelled. The system employed c_N^{UP} to represent the assigned bandwidth to consumer n for transferring its offloaded assignment to the edge server, and c_N^{DO} to represent the disbursed bandwidth to users n for obtaining the task's outcome data from the edge server. It's worth noting that the implementation and recommended method work in the case where distinct mobile customers have varying quantities of jobs to perform [17].

$$t_{NM}^{CLO} = \frac{l_{NM}^{IN} n_{NM}^C}{F^{CLO}} \tag{3}$$

where n_{NM}^C is the quantity of operating phases for every bit of input data and F^{CLO} is the edge processing rate. In short, the overall latency of customer n when it implements MEC may be calculated using the offloading selections $[x_{NM}]$.

$$t_N^{CLO} = \sum_{M=1}^{m} \left(t_{NM}^{UP} + t_{NM}^{do} + t_{NM}^{CLO} \right), X_{NM}, \forall N \tag{4}$$

The edge-server can only begin processing user n's task m once the edge has acquired the task in its entirety, and that the edge-server can only begin sending back the final output after the finished work M has been completed.

3.2   Cost of Local Computing

Furthermore, the situation in which user N decides to perform its energy domestically [18]. The symbol E_N^{LOC} is used to represent the energy usage per data bit. As a result, the energy usage of user n for doing task M domestically is represented by:

$$e_{NM}^{LOC} = l_{NM}^{IN} E_N^{LOC} \tag{5}$$

Likewise, it is referred to as T^{LOE} as user n's regional processing period per data bit. The total processing period required for customer n to complete job m domestically is:

$$t_{NM}^{LOE} = [l_{NM}^{IN} + l_{NM}^{OUT}] T^{LOE} \tag{6}$$

The overall delay for customer n to accomplish its duties regionally is provided by given user n's offloading choice [x_{NM}].

$$t_N^{LOE} = \sum_{M=1}^{m} t_{NM}^{LOE} [1 - x_{NM}] \tag{7}$$

4   Problem Formulation

In IoT networks, the emerging of edge computing has brought many research challenges, such as dynamic computation offloading scheme design, resource (e.g., computing resource, spectrum resource) allocation, and transmit power control. Moreover, they cannot be solved independently, such as designing an optimal computation offloading scheme has to consider the limited resource of the gateway and the transmit power of the users. Computation task offloading scheme has been investigated to access to multi resources efficiently in edge computing networks, especially in the MEC system. A joint optimization of task offloading scheduling and transmit power allocation scheme has been proposed in the MEC system with single mobile user [18].

By jointly optimizing every customer n's offloading judgment, the optimization challenge is defined with the goal of minimizing the total delay in completing all customers' activities and the accompanying energy usage. For consumer n's task transmitting and reception, use [x_{NM}] and the bandwidth allocations c_N^{UP} and c_N^{DO}. Client n's overall latency in completing all of its duties can be given by, based on our prior modelling.

$$t_N^{OVE} = MAX[t_N^{LOE}, t_N^{CLO}] \tag{8}$$

As a result, the recommended bandwidth allocation is as follows:

$$\text{Bandwidth allocation} = MIN \sum_{N-1}^{n} [\sum_{M-1}^{n} (e_{NM}^{LOC}(1-x_{NM}) + e_{NM}^{CLO} x_{NM})(T)_N t_N^{OVE}]$$
$$\sum_{N-1}^{n} c_N^{UP} \leq c^{UP,MAX}$$
$$\sum_{N-1}^{n} c_N^{DO} \leq c^{DO,MAX}$$
$$c_N^{UP}, c_N^{DO} \leq 0, \forall N \tag{9}$$
$$x_{NM} \in [0,1], \forall N, M$$
$$[x_{NM}], [c_N^{UP}], [c_N^{DO}]$$

The weight among energy usage and overall processing latency is represented by the parameter n. The overall uplink bandwidth allotment for all customers cannot exceeds the optimum band-width, $c^{UP,MAX}$ according to constraint the overall downlink band-width allotment for all customers cannot exceeds the optimum bandwidth $c^{DO,MAX}$ due to a constraint.

5 Computational Offloading Scheme Using Deep Reinforcement Learning

The gateway groups IoT users into separate clusters based on their unique features and user priorities after centralized user clustering. Edge computing is assigned to the maximum priority cluster, whereas local computing is assigned to the cluster with the lowest priority. This section proposes the DRL-solved optimum distributed computation offloading technique for the remaining clusters. The fundamental formulas of reinforcement learning are demonstrated first, followed by the creation of a computation offloading scheme using an MDP to simulate the computation offloading process. To find the appropriate computation offloading strategy for the MDP issue with large-state space, a deep reinforcement learning based on Q-learning is proposed for computation offloading, as depicted in Fig. 1.

Reinforcement learning is the process through interaction with the environment which enables agents to constantly learn optimal strategies. During this situation, each Internet of things users is regarded as an agent, whereas the environment encompasses everything else in the IoT network. The time slots structure of the computation offloading system is assumed.

Each agent IoT user u_i recognizes the state s_k from State space S at times k, the action a_k is taken in action space A, i.e., by choosing various transmission power, which mode is taken in a computer mode, based on the policy α. As a result, a new state S_{k+1} changes, and a R_k reward is acquired when the environment changes. The system cost SC_k, is represented by sum of energy utilization and latency. A reinforcement learning problem usually represented as following terms:

Action: Action, a_k, is the discrete transmit power of each IoT user, while others in edge computing actually are transmitting power.

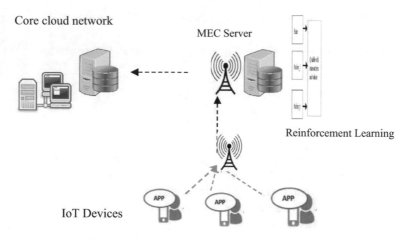

Fig. 1 Proposed task offloading flow chart

State: Environmental exploration is termed as state. The channel gain, task queue, and storage space can be considered to decide the state.

Reward: The reward signal's purpose is to motivate the learning algorithm to achieve the optimization problem's goal. The negative reward is used in this article to reduce system costs while try to make the best selections about power allocation and computation offloading. To be clear, reward shaping is used to make the reward more informative and to speed up the training process.

The proposed work is performed in following steps:

Step 1: Input IoT user list along with their tasks.
Step 2: Generate cluster of IoT users.
Step 3: Determine Priority levels of multiple IoT users in each cluster.
Step 4: Train deep reinforcement network using Q-learning algorithm.
Step 5: Task Offloading according to above steps.
Step 6: Evaluate performance parameters.

6 Result Analysis

The proposed technique is built in MATLAB R-2020a to evaluate its effectiveness. The simulations were run on a PC with an Intel i5, 3.7 GHz processor and 8 GB of RAM.

Fig. 2 Energy Consumption
with respect to number of
iterations of mobile users

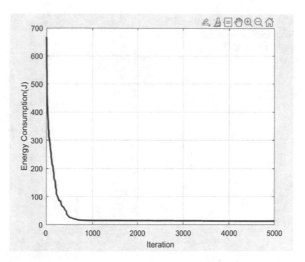

6.1 Energy Consumption Versus the Number of Iteration

The maximum power pmax is taken to be 5w in this situation of mobile users.
Figure 2 depicts the deep reinforcement learning's performance throughout a range
of iterations. It has been discovered that as the number of iterations of reinforcement
learning increases, the energy consumption of the suggested methodology reduces.

6.2 Energy Consumption Versus the Maximum Power

We assess the energy consumption under varied maximum power pmax, i.e., pmax
= 4w, 5w, and 6w, as shown in Fig. 3, for different numbers of mobile users. The
energy consumption is shown to be proportional to the maximum power pmax, i.e.,
the greater the pmax, the higher the energy consumption. This is because, with a
greater maximum power pmax, mobile users' average transmitting power is higher,
resulting in increased energy usage.

Fig. 3 Energy Consumption
with respect to maximum
power of mobile users

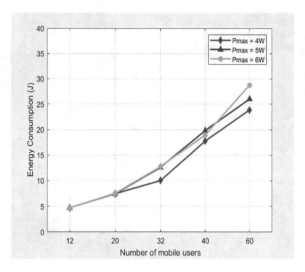

6.3   Welfare Versus Number of Mobile Users

As shown in Fig. 4, the maximum power pmax was set to be 5 w. The graph represents
that as the number of users increases in network, the welfare increases while adopting
DRL.

Fig. 4 Welfare with respect
to number of mobile users

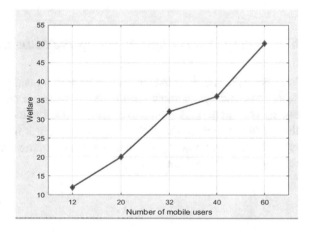

Fig. 5 Welfare with respect to different request workload of mobile users

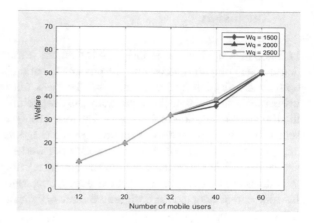

6.4 Welfare Versus Different Request Workload

As shown in Fig. 5, under different request workload Wq, i.e., Wq = 1500, 2000, 2500, welfare was evaluated for mobile user scenario. The graph shows that as the Wq increases the welfare increases.

6.5 Comparative State-Of-Art

To achieve efficacy over existing work, an optimal task offloading policies is validated in this section. The clustering of the IoT users into various groups has been proposed by making use of clustering optimizing algorithm which is framed as initial step for the task offloading scheme designing. The reduction in the energy is achieved by making use of distribution computation task offloading algorithm. Diverse contributions from various authors towards the work have been described in the Table 1. In this the result was compared with existing works with some features and results.

Table 1 Comparative analysis of computational offloading applications

Ref	Method	Multi-user	Multi-task	Average response time
[1]	Ant colony optimization	No	No	~85 ms
[1]	Particle swarm optimization	No	No	~90 ms
[13]	Deep Reinforcement learning with markov decision process	No	Yes	–
[19]	Boltzmann machines learning	No	Yes	~20 s
Proposed	Reinforcement learning with clustering and q-learning	Yes	Yes	~80 ms

7 Conclusion

In this paper, application of deep reinforcement learning is analyzed in the ultra-dense edge computing network. The power allocation problem among IoT users is optimized using reinforcement learning. The task offloading problem is solved by using mixed-integer non-linear program, the deep reinforcement learning algorithm, to offload the request for mobile users. The result analysis was presented for dynamic scenario system and observed optimal power usage during offloading. The result analysis was also compared with some existing works and with their features. This shows the efficiency of the deep reinforcement learning with Q-learning. In order to address the bottlenecks of practical issues, deep reinforcement learning with Q-learning can be adopted for designing framework for future offloading planning for IoT or Edge applications.

References

1. Hussein MK, Mousa MH (2020) Efficient task offloading for IoT-Based applications in fog computing using ant colony optimization. IEEE Access. 8:37191–37201. https://doi.org/10.1109/ACCESS.2020.2975741
2. Aazam M, Islam SU, Lone ST, Abbas A (2020) Cloud of Things (CoT): cloud-fog-iot task offloading for sustainable internet of things. IEEE Trans Sustain Comput. 7:87–98. https://doi.org/10.1109/TSUSC.2020.3028615
3. Chen X, Shi Q, Yang L, Xu J (2018) ThriftyEdge: resource-efficient edge computing for intelligent IoT applications. IEEE Networks. 32:61–65. https://doi.org/10.1109/MNET.2018.1700145
4. Cheng X, Lyu F, Quan W (2019) Space/aerial-assisted computing offloading for IoT applications: a learning-based approach. IEEE J Sel Areas Commun 37:1117–1129. https://doi.org/10.1109/JSAC.2019.2906789
5. Zhang K, Zhu Y, Leng S (2019) Deep learning empowered task offloading for mobile edge computing in urban informatics. IEEE Internet Things J 6:7635–7647. https://doi.org/10.1109/JIOT.2019.2903191
6. Li H, Ota K, Dong M (2018) Learning IoT in edge: deep learning for the internet of things with edge computing. IEEE Netw 32(1):96–101. https://doi.org/10.1109/MNET.2018.1700202
7. Min M, Xiao L, Chen Y (2019) Learning-based computation offloading for IoT devices with energy harvesting. IEEE Trans Veh Technol 68:1930–1941. https://doi.org/10.1109/TVT.2018.2890685
8. Yu S, Chen X, Yang L (2020) Intelligent edge: leveraging deep imitation learning for mobile edge computation offloading. IEEE Wirel Commun 27:92–99. https://doi.org/10.1109/MWC.001.1900232
9. Xu X, Li D, Dai Z (2019) A heuristic offloading method for deep learning edge services in 5G networks. IEEE Access. 7:67734–67744. https://doi.org/10.1109/ACCESS.2019.2918585
10. Pradhan C, Li A, She C et al (2020) Computation offloading for IoT in C-RAN: optimization and deep learning. IEEE Trans Commun 68:4565–4579. https://doi.org/10.1109/TCOMM.2020.2983142
11. Qu G, Wu H, Li R, Jiao P (2021) DMRO: a deep meta reinforcement learning-based task offloading framework for edge-cloud computing. IEEE Trans Netw Serv Manag 18, 3448–3459. https://doi.org/10.1109/tnsm.2021.3087258.

12. Aljanabi S, Chalechale A (2021) Improving IoT services using a hybrid fog-cloud offloading. IEEE Access 9:13775–13788. https://doi.org/10.1109/ACCESS.2021.3052458
13. Huang M, Liu W, Wang T, Liu A, Zhang S (2020) A cloud-MEC collaborative task offloading scheme with service orchestration. IEEE Internet Things J 7(7):5792–5805. https://doi.org/10.1109/JIOT.2019.2952767
14. Aazam M, Zeadally S, Harras KA (2018) Offloading in fog computing for IoT: review, enabling technologies, and research opportunities. Fut Gener Comput Syst 87:278–289
15. Wu H, Zhang Z, Guan C, Wolter K, Xu M (2020) Collaborate edge and cloud computing with distributed deep learning for smart city internet of things. IEEE Internet Things J 7(9):8099–8110. https://doi.org/10.1109/JIOT.2020.2996784
16. Han Y, Zhao Z, Mo J, Shu C, Min G (2019) Efficient task offloading with dependency guarantees in ultra-dense edge networks. In: IEEE global communications conference (GLOBECOM), pp 1–6. https://doi.org/10.1109/GLOBECOM38437.2019.9013142
17. Chen M, Liang B, Dong M (2016) Joint offloading decision and resource allocation for multi-user multi-task mobile cloud. In: IEEE international conference on communications (ICC), pp 1–6. https://doi.org/10.1109/ICC.2016.7510999.
18. Huang L, Feng X, Zhang C, Qian L, Wu Y (2019) Deep reinforcement learning-based joint task offloading and bandwidth allocation for multi-user mobile edge computing. Digital Commun Netw 5(1):10–17. https://doi.org/10.1016/j.dcan.2018.10.003
19. Alelaiwi A (2019) An efficient method of computation offloading in an edge cloud platform. J Para Distrib Comput 127:58–64

GyanSagAR 1.0: An AR Tool for K-12 Educational Assistance

Shweta Taneja, Nidhi Sharma, Arshita Bhatt, and Khushboo Gupta

1 Introduction

The conventional teaching and learning methods lead to rote learning of students. That needs to be made more interesting and exciting. Moreover, the Covid-19 pandemic has also transformed our lives. In a very short span of time, various educational institutions were required to shift to an online platform such as Zoom, Microsoft teams, Google Classroom, etc. to continue the curriculum.

But as time progressed, there has been rising disinterest in students to learning by watching online videos or lectures. In fact, studies have indicated students suffering from loss of focus, disengagement, and Zoom fatigue from attending multiple online lectures. This has led to a direct impact on student's learning which is prominently visible by falling grades and no desire to learn. Students with special needs are finding it difficult to learn or understand subjects. This has added to the woes of increasing anxiety, distress, and other psychological problem in students.

1.1 Motivation of Work

Following are some points that motivated us to undertake this work:

- Receding curiosity to learn and explore: With remote education and learning, there's been no physical experiment that gave rise to disengagement from peers and teachers, loss of focus, social isolation, and lack of general interest in students.

S. Taneja (✉) · A. Bhatt · K. Gupta
Department of Computer Science, Bhagwan Parshuram Institute of Technology, GGSIPU, Delhi, India
e-mail: shwetataneja@bpitindia.com

N. Sharma
Department of Applied Science, VIPS Technical Campus, Shalimar Bagh, Delhi, India

© The Author(s), under exclusive license to Springer Nature Singapore Pte Ltd. 2022
B. Unhelker et al. (eds.), *Applications of Artificial Intelligence and Machine Learning*,
Lecture Notes in Electrical Engineering 925,
https://doi.org/10.1007/978-981-19-4831-2_50

- Superficial understanding of the subject: with no experiments to perform, STEM subjects are a bit difficult to grasp, this led to only surface knowledge and cramming of formulas by students, instead of seeing and experimenting with the theory.
- Learning limited to the curriculum: with a shorter semester span, schools and colleges focus on completing the defined curriculum in a limited time. This has led to depletion in the exploration of new topics and ideas among students, reduced intelligent discussion, etc.

1.2 Augmented Reality

While the use of AR/VR technologies is not new in the education domain, it wasn't openly available to everyone due to the lack of proper gadgets and tools. But with many open-source libraries such as AR Foundation, Vuforia, Unity Engine and AR supporting Smartphone devices has led to accessibility to a wider audience. The history of Augmented Reality dates back to the 1960s on the concept of a video immersion project named Sensorama. Later in the 1960s, Ivan Sutherland designed 3D modeling and visual simulation software, the Sketchpad. He is usually termed as the inventor of Augmented Reality. Fast-forwarding to 2012 when google launched google glasses which is available to all gave a boost to the technology and made more people aware of it. Since then, we can see an immense amount of development in AR like Image Detection, Plane Detection, and occlusion, depth-sensing, etc. [1].

Therefore, looking at manifold advantages of AR in education we propose our solution, GyanSagAR 1.0, an AR education tool for K-12 educational assistance.

Previously in [2], we developed the application providing content on area and volume from mathematics and organic chemistry with a minimal user interface. In this paper, we have extended our work supporting common STEM subjects such as physics, chemistry, mathematics, and biology which will help students to visualize various concepts. The user interface is designed keeping the accessibility and ease of usage in mind. Through this instrument, we aim to arouse the interest of the student in learning subjects in an enjoyable manner along with conventional teaching methods like reading and making notes watching videos, etc. so that students can actively participate in their learning journey instead of passively grasping and cramming subject.

2   Literature Review

An effort to contain COVID-19 infection prompted many countries to bring full closure to public places such as offices, educational institutions, parks, gyms cinemas, shopping centers, etc. This unprecedented lockdown has impacted billions of people worldwide, including students. While MOOCs and videos are successful examples of online and remote education but they were not structured to serve all requirements of the syllabus and curriculum of various institutions around the world. They caused school and college authorities to hastily adopt and adapt as per online tools for teaching. A study conducted by Bozkurt et al., which highlights the overall action and reaction of online education in pandemic time. The study conducted for 31 nations with a population share of 62.7% of the whole world population, highlighted the difference between planned remote education and pandemic imposed remote education practices. The study also highlighted social injustice, inequity, and the digital divide have been exacerbated during the pandemic due to lectures and learning material be served to students via online medium [3]. A similar analysis conducted by Onyema et al. suggested that even though online education and remote learning was introduced in good faith to continue education, it became a glaring example of inefficacy of the online education system due to poor infrastructures including, network, power, inaccessibility, and unavailability issues and poor digital skills [4]. However, students pursuing their undergraduate degree, felt that the physical classrooms-based learning and attending MOOCs, were better than learning through online education, they felt that there has been a gradual improvisation by professors for their online teaching skills since the beginning of the pandemic and online education is quite useful right now. But they also felt that online education is stressful and affecting their health and social life, as per the study conducted by Chakraborty et al. for about 350 + students [5].

Specially abled students and those who fall under the autism spectrum suffer from the added difficulty of online education, as per the study conducted by Buchnat et al. [6]. Though it was observed that the online content accessibility is still not favorable for these students but in classroom-based learning, a teacher had access to various tools and techniques to connect with the child and adapt lectures according to an individual's need but this option is not available via online mode. The curriculum design for every student was unique to their personality but catering to this through online mode has become an over-burden for a teacher. As per analysis conducted by Copeland et al. the number of cases where students have been increasingly feeling stressed out, losing focus on studies, suffering from attention problems and fatigue, and another psychological syndrome have increased [7].

While the above literature provides a view of the impact on education due to the pandemic, we explored the alternative to curb some of these impacts and along that direction, we found AR/VR as a powerful tool. As per a systemic review for applications of AR in education and game-based learning done by Alper et al. it was revealed that applying game-based activities in the learning process increased the academic performance of students in terms of exam scores and behaviors and unlike other information technologies, AR interfaces offer an infinite interaction between real and virtual worlds, a concrete interface metaphor, and a means of transition between real and virtual worlds. Through the review, it was discovered that AR technology has great potential to provide effective learning support both in-school and out-of-school activities and usage of this technology can't not only be applied to STEM subjects but also to learning languages [8]. In fact, a major finding reported by Papakostas et al. in this domain was of the improvement of learners' spatial ability using AR in educational settings, and the noted challenge is the need for more learning content. [9]. One such study by del Cerro Velázquez et al. explores usages of an AR tool for learning Mathematical functions and finds it to be an effective tool in teaching the subject and also showed that the students had a positive perspective on the use of the tool which managed to capture their attention and increase their motivation from the beginning [10]. In an another work, an app based on AR was developed for learning chemistry in schools [17].

3   Implementation

In GyanSagAR 1.0, we propose a model of a mobile implementation to teach concepts of STEM subjects such as Biology, Chemistry, Maths, and Physics. This model is created using Unity3D and ARFoundation.

3.1   Unity

Unity is a cross-platform game engine developed by Unity Technologies to create mesmerizing experiences which provide an entire suite of solutions for various industry-related problems and can also serve as an aide to existing solutions to make them more efficient and creative. Unity came into existence in 2006, since then it has won several awards, including the 2010 Wall Street Journal Technology Innovation Award and the 2009 Gamasutra Top 5 Gaming Companies. In addition, the company became a responsible party to democratize the game development process [11]. Unity was developed with the purpose of providing a toolset for game development and animation, but recently automotive companies have started to use it to simulate Product Lifecycle Management (PLM) prior to physical implementation, and architecture firms engage in building information modeling (BIM) to integrate physical infrastructures, such as public utilities, into the design of an app [12].

3.2   AR Foundation

ARFoundation is a collection of various software development kits for Augmented Reality. It is created by Unity to bring different features of various SDKs under one hood and make them more compatible with Unity. Some of the packages it includes are ARCore, ARKit, Vuforia, Lumin SDK, Windows 10 SDK.
   Some of its common features are:

- Plane Detection: It detects the size and type of plane i.e., horizontal or vertical. After detection, we can use that detected plane to put our gameobjects and interact.

- Image Detection: It scans the live input feed and if the image on the screen matches with the database of images stored, it can spawn anything on that image or just show details of the image captured according to the what developer has coded.
- Anchors: It is a term used for relating the real world's position with the virtual world's position. To keep any object in its place even when we move or close our camera, we use anchors which calculate its real-world position and stays there for the given session.
- Raycast: A raycast is essentially a ray that is sent from a position in 3D or 2D space and moves in a specific direction [13].

3.3 Sections of GyanSagAR 1.0

The GyanSagAR 1.0 tool has 4 sections, each section dedicated to a subject. Navigating to these sections individually opens up a window to choose the concept the user wants to learn and visualize in 3D.

1. **Physics section**: In this subject, we have included Kepler's Law for planetary motion (Fig. 1) wherein we depict a 3D model following the Law's along with a text description to show refer to and read the laws.

2. **Mathematics section**: This section contains two topics, namely Area and Mensuration and Trigonometry (Fig. 2). In Area and Mensuration concept, the user can

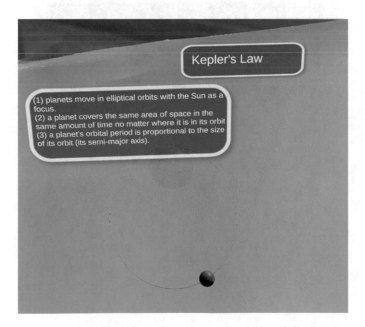

Fig. 1 Kepler's laws of planetary motion

visualize the 3D objects such as cubes, spheres, etc. with their dimensions set per their choice and learn various formulas related to the object. In Trigonometry, the user can visualize sine waveform and along with it the entire set of trigonometric functions are displayed for ease of understanding.

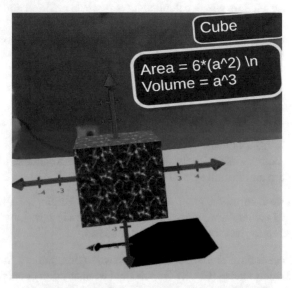

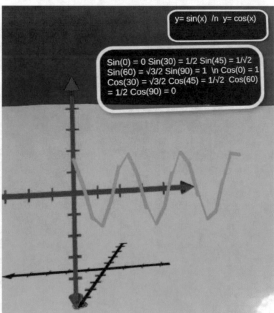

Fig. 2 Concepts of mathematics

3. **Chemistry section**: This section is different from the ones implemented above as the above objects were spawned on plain surfaces where this section requires a card with a certain texture to spawn the atom on it. There are a number of in-organic formulas given for the user to choose from where they may select one. For one kind of atoms cards of different textures are used and atoms are spawned on their unique card. Bringing the two cards closer enough will activate a molecule formed using the respective atoms and taking the card farther from each other destroys the molecule (Fig. 3). This is symbolic of the bond formed in atoms when electrons are shared among atoms to give stability and an overall neutral charge to the formed molecule.

4. **Biology section**: This section contains the concept of a human digestive organ system (Fig. 4), which covers the gastrointestinal tract starting from mouth to large intestine and major organs such as stomach, liver, pancreas, etc. that facilitate the digestion of food we ingest. Selecting each organ, by clicking on it displays a card with information snippets of that organ.

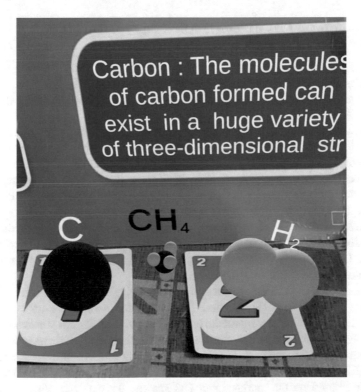

Fig. 3 Inorganic chemistry formula

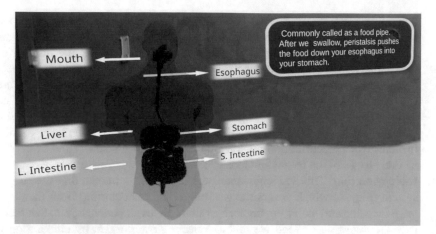

Fig. 4 Human digestive system

3.4 Workflow of GyanSagAR 1.0

To start using the tool, the user is asked to scan the surface first by following the action present on-screen, and as soon as the surface is scanned, the plane detection modules provided by ARFoundation instantiate a virtual plane on top of it to show the detected surface.

Then system waits for the user's response which can be activated by tapping on the icon for the detected surface on the screen which further verifies whether the tap detected is on the detected plane or not. If the tap on the icon is detected, Plane Manager calculates the corresponding position of the tap in the real world and spawns a virtual reality object as per the topic selected by the user, otherwise, the system waits for the correct response from the user (Fig. 5).

In the Mathematics section, the user has the freedom to select the width, height, and depth of the 3D geometrical shape. To cater to this functionality, Object Creator creates a Game Object at runtime using the dimensions specified by the user and instantiates it as per the user's input.

In the case of the Chemistry section, we are using image detection and image processing modules of ARFoundation. Here user needs to scan the image, which is processed using the processing model and sent to the database for verification, if the same image is found we send the data to the next module else we erase the current detected image and move back to the image scanning process. If the scanned image is verified, image processing algorithms of ARFoundation calculate the position and orientation based on the marker input and spawns the expected virtual object on top of it, in the current example it will be spawning atoms or molecules of organic chemistry (Fig. 6).

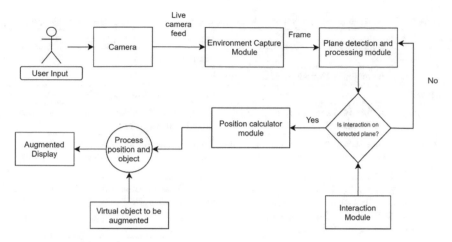

Fig. 5 GyanSagAR 1.0 Workflow

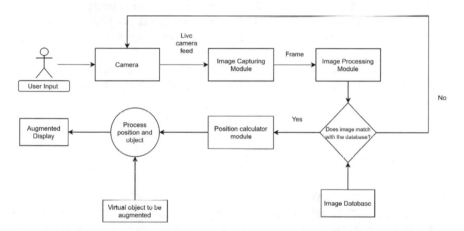

Fig. 6 Chemistry section flowchart diagram depicting detection of playing card or any pre-trained image surface to spawn Game Object of the molecule on top of it

4 Comparative Analysis

There are commonly available commercial and non-commercial implementations that demonstrate the varied usability of AR in education. Commercial tools such as Osmo [13], which develops educational games for young kids, Arloon plants [14] an app by which students can explore interactive plants to learn about structure and parts and QuiverVision [15] is an easy-to-use, augmented reality app triggered by images that can be scanned to activate the AR experience.

Non-commercial tools, such as AR Lab is used to analyze the impact of AR technology in mainstream learning of Chemistry subject in which the authors presented

Table 1 Comparison between GyanSagAR 1.0 vs. similar tools in AR in the Education domain

GyanSagAR 1.0	Other AR education assistive apps
Wider audience	Limited audience
Suitable for the majority population of the country	Not suitable
Open source	Closed source and expensive
Complete Learning Package as it targets actual concepts taught by schools	Support single concepts like botany topics, number theory, etc.
Developed to assist in educational learning of the student and arouse interest in mainstream topics, Topic first learning	Teaches simple concepts such as adding numbers, how a plant grows i.e., Fun first learning
Besides supporting basic functionalities as in other apps, this tool supports added features such as height and width adjustment, options to visualize various compounds, etc	Such functionalities are not found in mainstream apps

the result that the mean result was achieved by implementation using student group is similar to that of students using the lab equipment, hence providing the evidence that the tool has high usability in the schools with no funding for setting up the lab [16].

On the other hand, GyanSagAR 1.0 overcomes many shortcomings for the above-mentioned implementations and key differences are mentioned in the table below (Table 1).

5   Conclusion and Future Scope

Our proposed GyanSagAR 1.0 tool has been developed using the AR concept in education. It has many advantages. Targeting tough topics in STEM subjects that often require visualization skills, such as the motion of planets in the solar system, molecular structure of compounds, 3D shapes visualization, etc. this tool helps students to clear doubts at grain level and fuels academic exploration. Keeping in mind the accessibility needs of students, the user interface is designed with easy-to-use and interactive features. This tool promotes unrestricted learning for all strata of learners, which comes without the confines of the curriculum.

AR is a promising field in various subject areas and this tool can be expanded to many other education assistive solutions. Currently, this tool supports visualization of game objects on certain pre-recognized solid surfaces such as playing cards, whereas this functionality can be expanded to handwriting recognition wherein the user may provide the handwritten name of elements or shapes on top of which game objects can be scoped. Furthermore, this tool is developed keeping in mind the perspective of a learner providing wholesome resources for them, in the future, it can be extended for educators as well, providing functionalities such as creating assignments, generating reports, and personalized student dashboards. Since the tool is based on game-based learning, each subject or environment can be expanded to a multi-player environment where different learners can interact with the same environment.

References

1. Laghari AA, Jumani AK, Kumar K, Chhajro MA (2021) Systematic analysis of virtual reality & augmented reality. Int J Inf Eng Electron Bus 13(1):36–43
2. Devfolio webpage. https://devfolio.co/submissions/gyansagar-b695, and https://devfolio.co/submissions/gyansagar-53a1, Accessed 28 Sept 2021
3. Bozkurt A, Jung I, Xiao J, Vladimirschi V, Schuwer R, Egorov G, Paskevicius M (2020) A global outlook to the interruption of education due to COVID-19 pandemic: Navigating in a time of uncertainty and crisis. Asian J Dist Educ 15(1):1–126
4. Onyema EM, Eucheria NC, Obafemi FA, Sen S, Atonye FG, Sharma A, Alsayed AO (2020) Impact of Coronavirus pandemic on education. J Educ Pract 11(13):108–121
5. Chakraborty P, Mittal P, Gupta MS, Yadav S, Arora A (2021) Opinion of students on online education during the COVID-19 pandemic. Human Behav Emerg Technol 3:357–365
6. Buchnat M, Wojciechowska A (2020) Online education of students with mild intellectual disability and autism spectrum disorder during the COVID-19 pandemic. Interdyscyplinarne Konteksty Pedagogiki Specjalnej 29:149–171
7. Copeland WE, McGinnis E, Bai Y, Adams Z, Nardone H, Devadanam V, Hudziak JJ (2021) Impact of COVID-19 pandemic on college student mental health and wellness. J Am Acad Child Adolesc Psychiatry 60(1):134–141
8. Alper A, Öztaş EŞ, Atun H, Çınar D, Moyenga M (2021) A systematic literature review towards the research of game-based learning with augmented reality. Int J Technol Educ Sci 5(2):224–244
9. Papakostas C, Troussas C, Krouska A, Sgouropoulou C (2021) Exploration of augmented reality in spatial abilities training: a systematic literature review for the last decade. Inf Educ 20(1):107–130
10. del Cerro Velázquez F, Morales Méndez G (2021) Application in augmented reality for learning mathematical functions: a study for the development of spatial intelligence in secondary education students. Mathematics 9(4):369
11. Canossa A (2013) Interview with nicholas francis and thomas hagen from unity technologies. In: Game analytics. Springer, London, pp 137–142
12. Unity Technologies homepage. https://unity.com/, Accessed 28 Sept 2021
13. AR Foundation homepage. https://developers.google.com/ar/develop/unity-arf#:~:text=AR%20Foundation%20is%20a%20cross,for%20the%20AR%20Foundation%20framework, Accessed 28 Sept 2021
14. Osmo application. https://www.playosmo.com/en/, Accessed 28 Sept 2021
15. Arloon plants application. http://www.arloon.com/apps/plants/, Accessed 28 Sept 2021
16. QuiverVision application. https://quivervision.com/, Accessed 28 Sept 2021
17. da Silva BR, Zuchi JH, Vicente LK, Rauta LRP, Nunes MB, Pancracio VAS, Junior WB (2019) Ar lab: Augmented reality app for chemistry education. In:Nuevas Ideas en Informática Educativa. Proceedings of the international congress of educational informatics, Arequipa, Peru, vol 15, pp 71–77

Performance Analysis of Energy Efficient Optimization Algorithms for Cluster Based Routing Protocol for Heterogeneous WSN

Kamini Maheshwar, S. Veenadhari, and S. Almelu

1 Introduction

Wireless sensor network consists of wireless sensor nodes which stores information and used to exchange the information for the communication process. These networks are playing very significant role as they are used in numerous fields like monitoring or particular area which includes health care monitoring, industrial monitoring, pollution monitoring, detecting threats, traffic management and control and many more. Wireless sensor networks are heterogeneous and homogenous, heterogeneous networks wireless sensor networks are comprised of sensing nodes with enhanced configuration for complicated tasks like clustering, routing etc. [1–3]. These wireless sensor networks store energy in their nodes which is consumed in performing the communication between the nodes [4]. Consumption of energy takes place in the sensor nodes by, sensing the data, processing it and finally exchanging it. Thus, it is necessary that wireless sensor networks should be energy efficient so that lifetime of the networks can be enhanced by using multiple techniques. One of the most significant problems in prolonging the life of a WSN is developing an energy-efficient routing protocol [5, 6]. Network scale can have an impact on lifetime; the network's stability becomes heavily essential as the scale grows. Grouping is one of the most effective ways to improve energy efficiency and the network's lifetime. K-means clustering separates information into K-clusters with more resemblance inside groupings but less similarities across clusters, and is one of the most widely used data clustering methods. CH is a node that collects information from sensor nodes in the cluster and delivers it to the base station. Clustering has shown to be one of the most effective strategies for boosting scalability and developing an energy-efficient WSN routing algorithm [7–11]. Several sensors can even con-serve energy

K. Maheshwar (✉) · S. Veenadhari · S. Almelu
Department of CSE, RNTU, Bhopal, M.P, India
e-mail: kaminimaheshwar@gmail.com

© The Author(s), under exclusive license to Springer Nature Singapore Pte Ltd. 2022   631
B. Unhelker et al. (eds.), *Applications of Artificial Intelligence and Machine Learning*,
Lecture Notes in Electrical Engineering 925,
https://doi.org/10.1007/978-981-19-4831-2_51

by adjusting each node's sample rate. These techniques use quantitative optimization techniques or heuristic models to alter sample rates for topological modification, covering protection, or localization. Furthermore, as previously noted, model-based WSN management has several downsides. The effectiveness of sensor nodes implemented in the real world, on the other hand, is heavily dependent on the environment in which they operate. The quality of the obtained sensor data is directly affected by environmental conditions (e.g., location, etc.). Machine Learning (ML) methods are well-known for their self-experiencing characterized by the fact that they do not need reprogramming. Machine Learning is an effective technique that provides a computational method that is scalable, dependable, and cost-effective [12–15]. There are three forms of machine learning: Supervised learning, unsupervised learning, and reinforcement learning. It was noticed that techniques based on machine learning are beneficial in understanding the core issues of wireless sensor networks. The ML algorithm is effective in increasing the existence of the network, which, in turn, enhances the WSNs and how to make them appropriate for forecast the amount of energy that can be harvested in a given time frame. The WSNs' performance can also be increased by machine learning technique. WSNs need dynamic routing due to the variable nature of the sensor network. ML techniques can help with this and improve the system's efficiency.

2 Related Work

YuchaoChang [4] proposed a model termed as MLPGA algorithm that is based on determining optimal number of chromosomes in evolutionary algorithm. The model is designed in two-tier architecture. The clustering is done using k-mean clustering. Shashi Bhushan [5] the proposed method introduces a new concept called hybridization of population initialization. To cluster WSN, a hybrid method combining GA (as stated in Algorithm 3) and K-means is suggested in this paper. To ensure high-quality CHs, GA's initial population is seeded with K-means. The fitness module is based on factors such as intra-cluster distance, inter-cluster distance, and cluster number. Deyu Lin [9] focused on two main factors for energy efficiency: One is reducing energy consumption and second is managing energy consumption. The clustering algorithm is based on game theory. Dual cluster heads are formed in this algorithm for energy efficiency. JunfengXie [11] attempted to give readers a basic grasp of how machine learning algorithms work and when they can be applied to SDN challenges. Serious research difficulties and regarding the study paths in ML-based SDN include elevated quality training datasets, decentralized multi-controller platforms, increasing network security, cross-layer network optimization, and progressively implemented SDN. Sahoo et al. [12] suggested a particle swarm optimization (PSO) approach paired through energy efficient clustering and sinking (PSO-ECSM) to deal with the CH selection difficulty and the sink mobility issue. The efficiency of the PSO-ECSM is determined using comprehensive simulations. As per simulation results was performed on stability period, partially node dead, lifetime of the network, and performance, respectively. Nigam et al. [13] presented an upgraded algorithm dubbed ESO-LEACH to overcome the difficulties with LEACH, such as the non-uniformity of the number of cluster heads and the disdain for the nodes' residual energy. In this paper, meta-heuristic particle swarm augmentation is used to cluster the sensor nodes at first. In this paper, meta-heuristic particle swarm augmentation is used to cluster the sensor nodes at first. In the presented ESO-LEACH, the idea of upgraded nodes and an updated set of rules for CH election are employed to reduce the algorithm's randomness. Khatoon et al. [14] leverages the multiagent randomized parallel search approach of particle swarm optimization to construct a clustering algorithm that addresses both mobility and energy efficiency issues. Using particle swarm optimization, a multi-objective fitness function is used to build clusters. Nayyar et al. [15] represented an interesting energy-efficient ACO-based multipath routing technique for WSNs in this investigation, i.e., IEEMARP. Maheshwari et al. [16] used a combination of BOA and ACO to reduce total energy consumption and networking longevity. Wang et al. [17] proposed an algorithm termed as GECR to compute the total energy consumed by all sensor nodes throughout, where the technique integrates a clustering scheme and a routing approach in the same chromosomes. To find the ideal number of clusters and cluster heads, Bhushan et al. [18] suggested protocol uses range among clusters, length inside clusters, and a number of cluster heads. To lengthen the network's survival time, a fuzzy expanded grey wolf optimization algorithm-oriented threshold sensitive energy efficient clustering protocol is considered by Mittal et al. [19]. Daneshvar et al. [20] described a new clustering algorithm that uses the grey wolf optimizer to choose CHs. Zivkovic et al. [21] augmented GWO swarm intelligence metaheuristics to solve the clustering issue in WSNs in the research that will be conducted. Oluwasegun Julius Aroba [22] proposed a machine learning algorithm-based model termed as DEEC GAUSS for optimizing positioning and energy efficiency in wireless sensor network nodes. B. R. Al-Kaseem et al. [23] proposed a system that produced clusters based on heuristic data via sensor nodes, and this study proposes an efficient route optimization strategy focused on it. Abidoye and Kabaso [24] suggested as a fog-based technique for WSNs termed as EEHFC. EEHFC demonstrated a hierarchical routing framework for data transfer via fog nodes from typical sensor nodes to data centers. Vially Kazadi Mutombo [25] presented an EER-RL, an EER-based energy-efficient IoT routing protocol relying on reinforcement learning. Seyyedali Hosseinalipour [26] presented fog learning, a new approach for distributing machine learning model training across massive networks of heterogeneous devices.

3 Methodology

Clustering has proven to be one of the most powerful methods for increasing network scalability and designing a WSN routing protocol that is energy efficient. Moreover, model-based WSN management has some drawbacks, as mentioned herein. The routing algorithm is designed to follow various QoS criteria to provide better efficiency and to increase the lifetime of WSN challenges and problems considered by the network. This section, therefore, discussed about some optimization algorithms that contributes in energy efficiency of WSN and to provide application-specific assurance for Quality of Service (QoS).

3.1 Network Model

The following is the model-based presumption employed in this study:

N sensor nodes are placed at random throughout the sensing region, which is A = N * N in size. Both the sensor nodes and the base stations are situated random.

Every node in the network has a unique ID identifier and the similar beginning energies. The nodes have a definite amount of energy, but the Base Station has an endless amount.

The connection is symmetric. Depending on the acquired signal strength, the node can estimate the span between the transmitter and itself.

Every node only requires one primetime to connect with its parent node, and in that timeframe, every node can only accept or transmit one data packet and accompanying control packet.

The transmit power of the node can be adjusted based on the interaction distance.

3.2 Energy Consumption Model

The data exchange consumes the amount of energy consumed by sensor nodes. In this research, we solely consider the energy usage cost of data transmitting and merging information. The energy usage of transmitting and receiving is formalized in the calculations below in Eq. (1) [16]:

$$E_{tx}(m, s) = \begin{cases} m E_{selects} + m\varepsilon_{fs}s^2, & s < s_0 \\ m E_{selects} + m\varepsilon_{amp}s^4 & s \geq s_0 \end{cases} \tag{1}$$

$$E_{rx}(m) = m E_{selects} \tag{2}$$

where, m = data length, s = data transmission distance or span, Eselects = energy usage during transmitting and receiving of unit length data, ε_{fs} and ε_{amp} = amplifier energy usage of free space model and multiple path attenuation model.

The free space model is employed when the length s between the transmission and receiving nodes is smaller than the energy usage model cutoff, and the transmission range is attenuated as s^2. Instead, the multi-path attenuation framework is employed, using s^4 as the transmitted power. The energy needed for nodes to merge m-length information is calculated as follows:

$$E_u(m) = m E_{da} \tag{3}$$

where, E_{da} = The amount of energy it takes to merge a unit length of information.

3.3   Bio-inspired Optimization Algorithms

Cluster optimization is one of the very useful technique in wireless sensor network that was introduced to optimize the cluster size and selection of optimal candidate of cluster head. The bio-inspired algorithms are used to evaluate the optimal parameters for cluster head selection that is based on intelligence and capabilities. The fundamental challenge in a wireless sensor network is sensor node deployed, covering capability, and movement plan; yet, the connectivity challenge in a wireless sensor node is dependent on a deployment sensor node. Even as scale of the problem grows larger, the optimizing strategies increase dramatically. As a consequence, an optimization algorithm which uses minimal memory and processing while producing excellent results is preferable, particularly for sensor network implementations. Traditional analytic procedures are substantially inefficient; thus, bioinspired optimization techniques offer a good alternative. General architecture of bio-inspired algorithm is presented in Fig. 1. Some of bio-inspired algorithms are discussed below:

Particle swarm optimization-based clustering algorithm: This algorithm automatically compute the optimal number of cluster without much intervention. It employs a swarm of agents (particles) that move around in the search space in search of the optimum answer. In search space, each particle changes its "flying" based on its own flying experience as well as that of other particles and hence the name "Particle swarm optimization based clustering algorithm".

Ant colony optimization-based clustering algorithm: A way to solve optimisation problems based on the way that ants indirectly communicate directions to each other. In this algorithm some random solutions are generated first represent the word ant and the ant path represent the solution.

Genetic optimization-based clustering algorithm: Genetic based algorithm technique based on the principle of genetic and natural selection. It is frequently used to find the optimal and near optimal solutions to difficult problem which otherwise would tale a lifetime to solve.

Gray wolf optimization-based clustering algorithm: This algorithm predicts the solution based on the mathematical model in which a particle or object has better knowledge about the solution. In this algorithm, best solutions are saved and used to update other particle's position for search optimal results.

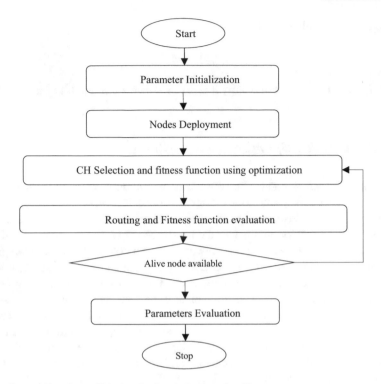

Fig. 1 General flowchart of bio-inspired optimization algorithms

3.4 Parameters for Designing Bio-Inspired Optimization Algorithm

Residual Energy of the CH: Inside a system, the cluster head conducts a variety of functions such as data collection from regular sensors and transfer of data to the base station. Because the cluster head takes a huge amount of energy therefore the nodes with the highest residual energy is selected to be a cluster head. The remaining cncrgy (F_1) is expressed mathematically below in Eq. (4).

$$F_1 = \sum_{i=1}^{M} \frac{1}{E_{ch(i)}} \tag{4}$$

where $E_{ch(i)}$ is the ith remaining energy of the cluster head.

The Range Among Sensor Nodes: It specifies the spacing in between standard sensor nodes as well as its own cluster head. The spacing of the transmitting link has the greatest impact on the node's energy loss. Once the selected node has a shorter transmission range to the base stations, the node's energy usage was reduced. The Eq. (5) expresses the range between the regular sensors and cluster head (F_2).

$$F_2 = \sum_{j=1}^{M} \left(\sum_{i=1}^{I_j} \frac{range(R_i, CH_j)}{I_j} \right) \tag{5}$$

Here, the range between the sensors (i) and cluster head CH_j is represented as $range(R_i, CH_j)$, and the quantity of sensor nodes that belongs to cluster head is denoted as I_j.

The Range Among the Cluster Head and Base Station: It specifies the range between the cluster head and the base station. The node's energy usage is proportional to the range travelled along the transmission link. For example, if the base station is located far off the cluster head, it will consume more energy for information transfer. As a result, the rapid decrease of cluster head might occur as a result of rising energy usage. As a result, throughout transmission, the node closest to the base station is favored. The objective function (F_3) of range in between cluster head and the base station is expressed by the following Eq. (6).

$$F_3 = \sum_{i=1}^{M} range(CH_j, BS) \tag{6}$$

The Degree of the Node: It specifies the quantity of sensor nodes assigned to each cluster head. Cluster heads with fewer sensor-nodes are chosen as cluster heads with more cluster members incur losses in a shorter period of time. Equation (7) expresses the node degree (F_4).

$$F_4 = \sum_{i=1}^{M} I_i \tag{7}$$

The Centrality of Nodes: Node centrality (F_5) expresses how far a node is from its neighbors and is stated in Eq. (8).

$$F_5 = \sum_{i=1}^{M} \frac{\sqrt{\sum_{j \in N} Range^2(i, j) / N(i)}}{Dimension\ of\ Network} \tag{8}$$

Here, N(i) is the quantity of nodes in neighborhood of CH_i.

Table 1 Simulation scenario for WBAN

Simulation scenario	Values
Area	100 * 100 m
WBAN sensor nodes	100
Initial energy of network	0.5 J
Packet size	2000, 4000

4 Discussion

4.1 Simulation Scenario

The simulation scenario is deployed with sensor nodes in area of 100 * 100 m as mentioned in Table 1. For energy evaluation, energy consumption model/radio model is considered to analyze some bio-inspired algorithms. The main objective of this paper is to evaluate and reduce the overall energy consumption of each node in the network. So, the cluster-based routing is developed with bio-inspired optimization based cluster head selection and routing between the nodes. The inputs given to bio-inspired optimization for better CH selection are residual energy, node degree, distance to the neighbors, distance to the BS, etc. For performance evaluation of the proposed model, the paper presents the simulated the scenario on the MATLAB platform under different conditions and with different parameters. These parameters are discussed below.

Throughput: Another important parameter is the throughput that is calculated on the successful delivery of data packets to the sink node at a particular time. The routing protocols of WBAN are dedicated to maximizing the throughput.

Network Longevity: In WBAN routing algorithms, the most important parameter is network longevity as sensor nodes are battery-operated. This is evaluated by counting the alive and dead nodes after every round or after a particular period.

4.2 Result Analysis

Heterogeneous WSN areas are simulated for multiple sensor nodes in a particular area. The configuration of the simulation is presented in Table 1. Random location of sink node is selected with unlimited energy whereas sensor nodes are deployed with limited energy with different energy levels. The proposed scheme is implemented for variable rounds of iterations for different packet sizes (2000 and 4000). In this paper, we have presented the comparative analysis of optimization algorithms. So, Tables 2 and 3 represents the result analysis under 2000 packet size and 4000 packet size respectively of some well-known bio-inspired optimization algorithms such as PSO [12], ACO [16], GA [12] and GWO [21].

Table 2 Comparative performance analysis with 2000 packet size

	Network Longevity (in rounds)	Throughput (packets)
LEACH [16]	1500	40,000
PSO-ECSM [12]	19,071	637,880
ICRPSO [12]	17,360	568,457
PSOBS [12]	15,222	486,712
DEEC [16]	1700	40,000
ACO [16]	8500	350,000
DCH-GA [12]	10,466	308,695

Table 3 Comparative performance analysis with 4000 packet size

Algorithms	Network Longevity (in rounds)	Throughput (packets)
LEACH	1570	13,000
LEACH-PSO [21]	3,880	20,000
LEACH-EEGWO [21]	3900	21,000

From Fig. 2 it can be concluded that the PSO optimized cluster routing protocols outperforms better as compared to others. In PSO [12], the network lifetime was observed to be 19,071 rounds, GA [12] achieves 10,466 rounds of lifetime and ACO [16] achieves 8500 rounds of network lifetime. Similarly, from Fig. 3 it was observed that PSO [12], GA [12] and ACO [16] achieved 637880 packets, 308695 packets and 350,000 packets of throughput respectively. So, from this analysis it can be said that there is future research scope with PSO optimization algorithm because by using PSO optimization maximum output can be achieved. Another analysis is presented on 4000 packets and its result is shown in Table 3. From Figs. 4 and 5 shows another comparison between PSO and GWO optimization algorithms. The result analysis shows just similar results of PSO and GWO. Therefore, from this analysis it can be concluded that PSO and GWO can be opted in future as an optimization algorithm for routing in cluster based WSN.

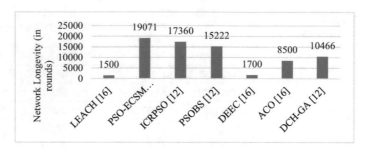

Fig. 2 Network longevity analysis with 2000 packet size

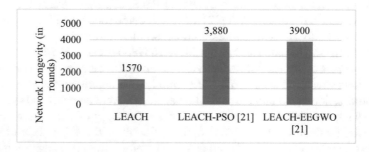

Fig. 3 Throughput analysis with 2000 packet size

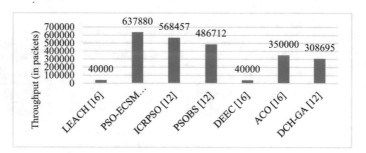

Fig. 4 Network longevity analysis with 4000 packet size

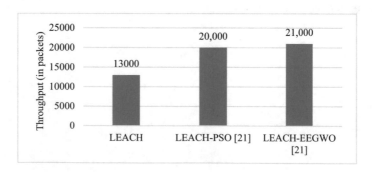

Fig. 5 Throughput analysis with 4000 packet size

5 Conclusion

One of the most frequently used topologies for WSN is clustering for energy efficiency, load balancing and achieving quality of service (QoS). This paper summarizes the objectives for future research works in field of cluster-based routing for WSNs. This paper is dedicated to analyze the effectiveness of bio-inspired optimization algorithms for clustering of sensor nodes in WSN applications. The comparative analysis was performed in this paper among PSO, GA, ACO and GWO. Out of which PSO and GWO outperforms better as compared to others. So, for future research work, PSO can be used for designing more efficient routing protocol for WSN application such as Internet of things (IoT).

Recently, bio-inspired computing had attracted the interest of researchers due to its capabilities of intelligence and adaptive nature. Besides that, those technologies must be enhanced as well. For node mobility management, these novel models were based on bio-inspired computation must be implemented. There's also a function for node mobility, which aims to boost network coverage while simultaneously increasing network lifetime and improving data detail timeliness and dependability. Upcoming scientific investigations can look on how to reduce power usage for coverage options with gaps, as well as how to control sensor node tactical strategy to repair network coverage and extend lifetime of the network. Other factors, including an energy usage model for network nodes and their movement approach, must also be considered when constructing movement strategies.

References

1. Gautam N, Vig R (2014) Energy efficient approach through clustering and data filtering in WSN. In: Proceedings 2014 international conference advanced computing communication informatics, ICACCI, pp 2142–2148. https://doi.org/10.1109/ICACCI.2014.6968467
2. Al-Aboody NA, Al-Raweshidy HS (2016) Grey wolf optimization-based energy-efficient routing protocol for heterogeneous wireless sensor networks. In: International symposium computing bus intelligence ISCBI, pp 101–107. https://doi.org/10.1109/ISCBI.2016.7743266
3. Yan J, Zhou M, Ding Z (2016) Recent advances in energy-efficient routing protocols for wireless sensor networks: a review. IEEE Access 4:5673–5686. https://doi.org/10.1109/ACCESS.2016.2598719
4. Chang Y, Yuan X, Li B (2019) Machine-learning-based parallel genetic algorithms for multi-objective optimization in ultra-reliable low-latency WSNs. IEEE Access 7:4913–4926. https://doi.org/10.1109/ACCESS.2018.2885934
5. Bhushan S, Pal R, Antoshchuk SG (2018) Energy efficient clustering protocol for heterogeneous wireless sensor network: a hybrid approach using GA and K-means. In: IEEE 2nd international conference data stream mining process DSMP, pp 381–385. https://doi.org/10.1109/DSMP.2018.8478538
6. Lata S, Mehfuz S (2019) Machine learning based energy efficient wireless sensor network. In: International conference power electron control automation ICPECA. https://doi.org/10.1109/ICPECA47973.2019.8975526

7. Kang J, Kim J, Kim M, Sohn M (2020) Machine learning-based energy-saving framework for environmental states-adaptive wireless sensor network. IEEE Access 8:69359–69367. https://doi.org/10.1109/ACCESS.2020.2986507

8. Bharot N, Suraparaju V, Gupta S (2019) DDoS attack detection and clustering of attacked and non-attacked VMs using SOM in cloud network. In: Singh M, Gupta P, Tyagi V, Flusser J, Ören T, Kashyap R (eds) Advances in Computing and Data Sciences. ICACDS 2019. CCIS, vol 1046. Springer, Singapore. https://doi.org/10.1007/978-981-13-9942-8_35

9. Lin D, Wang Q (2019) An energy-efficient clustering algorithm combined game theory and dual-cluster-head mechanism for WSNs. IEEE Access 7:49894–49905. https://doi.org/10.1109/ACCESS.2019.2911190

10. Kumar N, Sandeep, Bhutani P, Mishra P (2012) U-LEACH: a novel routing protocol for heterogeneous wireless sensor networks. In: International conference communication information computing technology ICCICT. https://doi.org/10.1109/ICCICT.2012.6398214

11. Xie J, Richard YuF, Huang T (2019) A survey of machine learning techniques applied to software defined networking (SDN): research issues and challenges. IEEE Commun Surv Tutorials 21:393–430. https://doi.org/10.1109/COMST.2018.2866942

12. Sahoo BM, Amgoth T, Pandey HM (2020) Particle swarm optimization-based energy efficient clustering and sink mobility in heterogeneous wireless sensor network. Ad Hoc Networks 106:102237. https://doi.org/10.1016/J.ADHOC.2020.102237

13. Nigam GK, Dabas C (2018) ESO-LEACH: PSO based energy efficient clustering in LEACH. J King Saud Univ Comput Inf Sci. https://doi.org/10.1016/J.JKSUCI.2018.08.002

14. Khatoon N, Amritanjali (2017) Mobility aware energy efficient clustering for MANET: a bio-inspired approach with particle swarm optimization. Wirel Commun Mob Comput 2017:1–12. https://doi.org/10.1155/2017/1903190

15. Nayyar A, Singh R (2019) IEEMARP- a novel energy efficient multipath routing protocol based on ant colony optimization (ACO) for dynamic sensor networks. Multimed Tools Appl 79:35221–35252. https://doi.org/10.1007/S11042-019-7627-Z

16. Maheshwari P, Sharma AK, Verma K (2021) Energy efficient cluster-based routing protocol for WSN using butterfly optimization algorithm and ant colony optimization. Ad Hoc Netw 110:102317. https://doi.org/10.1016/J.ADHOC.2020.102317

17. Wang T, Zhang G, Yang X, Vajdi A (2018) Genetic algorithm for energy-efficient clustering and routing in wireless sensor networks. J Syst Softw 146:196–214. https://doi.org/10.1016/J.JSS.2018.09.067

18. Bhushan S, Pal R, Antoshchuk SG (2018) Energy efficient clustering protocol for heterogeneous wireless sensor network: a hybrid approach using GA and K-means. In: International conference data stream mining process DSMP, pp 381–385. https://doi.org/10.1109/DSMP.2018.8478538

19. Mittal N (2020) An energy efficient stable clustering approach using fuzzy type-2 bat flower pollinator for wireless sensor networks. Wirel Pers Commun 1122(112):1137–1163. https://doi.org/10.1007/S11277-020-07094-8

20. Daneshvar SMMH, Mohajer PAA, Mazinani SM (2019) Energy-efficient routing in WSN: a centralized cluster-based approach via grey wolf optimizer. IEEE Access 7:170019–170031. https://doi.org/10.1109/ACCESS.2019.2955993

21. Zivkovic M, Bacanin N, Zivkovic T et al (2020) Enhanced grey wolf algorithm for energy efficient wireless sensor networks. In: Zooming innovation consumer technology conference ZINC, pp 87–92. https://doi.org/10.1109/ZINC50678.2020.9161788

22. Aroba OJ, Naicker N, Adeliyi T (2021) An innovative hyperheuristic, gaussian clustering scheme for energy-efficient optimization in wireless sensor networks. J Sens 1–12. https://doi.org/10.1155/2021/6666742

23. AL-Kaseem BR, Taha ZK, Abdulmajeed SW, Al-Raweshidy HS (2021) Optimized energy efficient path planning strategy in WSN with multiple mobile sinks. IEEE Access 9:828833. https://doi.org/10.1109/access.2021.3087086

24. Abidoye AP, Kabaso B (2021) Energy-efficient hierarchical routing in wireless sensor networks based on fog computing. EURASIP J Wirel Commun Netw. https://doi.org/10.1186/s13638-020-01835-w

25. Mutombo VK, Lee S, Lee J, Hong J (2021) EER-RL: energy-efficient routing based on reinforcement learning. Mob Inf Syst. https://doi.org/10.1155/2021/5589145
26. Hosseinalipour S, Brinton CG, Aggarwal V, Dai H, Chiang M (2020) From federated to fog learning: distributed machine learning over heterogeneous wireless networks. IEEE Commun Mag 58(12):41–47. https://doi.org/10.1109/MCOM.001.2000410

Applied Multivariate Regression Model for Improvement of Performance in Labor Demand Forecast

Hai Pham Van and Nguyen Dang Khoa

1 Introduction

Forecasting the labor demand in the future figures out the lack of redundancy of labor among sectors, provinces, locals, regions, degrees, for matching policies of human resources with its labor effectiveness. Labor forecast is an important task, emphasized through various works. In related works, Yas A. Alsultanny used data mining with Naïve Bayes Classifiers and created the Decision rules technique, which they recommended in predicting labor forecast [1]. Ross Gruetzemacheret et. al studied labor displacement in the advances of AI, with 90 and 99% of human tasks in the range between 10 and 15 years, respectively [2]. A. Luz et al. also studied effect from local labour demand on immigrant employment [3]. Studies of labor demand forecast impact using multivariate regression method for prediction of historical data [4, 5]. Some studies combined with the time series prediction method [6–8] in order to forecast demand for labor economic sectors. Yalcinkaya, A. et al. studied Maximum likelihood estimators of the model parameters in multiple linear regression obtained genetic algorithm, which they proved to outperform traditional algorithms in most cases, and suggested using GA to obtain maximum likelihood estimators in specific cases [4]. Pan, Y. et al. used multiple linear regression and life-cycle cost analysis for cost-effective evaluation of pavement maintenance, showing the ability to apply multiple regression to establish decision – making systems [5].

Multivariate regression is an extension of multiple regression, which can al-low more than one response for each input, making it better than multiple regression for predictions. Recent technological advancements and developments have led to

H. P. Van (✉)
School of Information Technology and Communication, Hanoi University of Science and Technology, Hai Ba Trung, Hanoi, Vietnam
e-mail: haipv@soict.hust.edu.vn

N. D. Khoa
Hanoi University of Science and Technology, Hai Ba Trung, Hanoi, Vietnam

© The Author(s), under exclusive license to Springer Nature Singapore Pte Ltd. 2022　　645
B. Unhelker et al. (eds.), *Applications of Artificial Intelligence and Machine Learning*,
Lecture Notes in Electrical Engineering 925,
https://doi.org/10.1007/978-981-19-4831-2_52

a dramatic increase in the amount of high-dimensional data for proper and efficient multivariate regression methods. Xiaoxi H. et al. extended the scope of multivariate regression with sparse reduced rank regression and subspace assisted regression with row sparsity. The study has been enhanced model with improved interpretability of regression models [6]. Consonni, V. et al. (2021) described the regression toolbox for multivariate regression using MATLAB, which majorly contributed to the improvement of multivariate regression application in general [7]. L. Lucy and Z. Julie (2020) studied minimax D-optimal designs for multivariate regression models with multifactors, hence using multivariate regression to introduce a design robust against small departures from an assumed error covariance matrix [8]. W. Yihe and Z.S. Dave used nonparametric empirical Bayes approach to large-scale multivariate regression, suggesting methods to improve multivariate regression with big data input [9]. Some approach was used by P.V. Hai to enhance uncertainty model [10]. In multiple variables in forecast under uncertainty. L.H. Son et al. proposed a new method for Hospital Cost Analysis using genetic algorithm and artificial neural network, suggesting the application of genetic algorithm in building decision – making systems that allow more variable and more accuracy [12]. P.V. Hai and N.T. Dong presented the Hybrid Louvain-Clustering model using a knowledge graph to cluster contents based on user behaviors in a social network [13]. The result is a model representing all multi-dimensional user relationships of contents based on users' behaviors. D.X. Truong and P.V. Hai presented the Bayesian graph deep learning framework for the case of classified mixed node random block models to classify the topic of social posts as nodes by creating a homogeneous graph with links between them, showing improved performance of the Bayesian formulation in topic classification in social during the training process [14]. N.T. Dong and P.V. Hai also proposed a new graph deep learn-ing model associated with knowledge graph with to prediction model the latent fea-ture of user and item, supplying the principle of organizing interactions as a graph, combines information from social network and all kind of relations in the heterogene-ous knowledge graph [15].

This paper proposes a model using multivariate multiple regressions, dealing with time series for forecasting labor demand. The approach is to propose multivariate regression approach by dealing time series data. To do an experiment in a real case study show that the multivariate regression model enhances significant performance for forecasting labor demand.

2 The Proposed Model

2.1 Applied Regression Function to the Proposed Model for Labor Forecast

In labor forecast, the relationships of labor demands are influenced by many factors [variables] such as GDP, export, import investment, science and technology, wages …

Hence, dependent variable Y depends on various explanatory variables. The random overall regression function of the proposed model with k variables can be expressed as follows:

$$Y_t = \beta_1 + \beta_2 X_{2t} + \beta_3 X_{3t} + \ldots + \beta_k X_{kt} + u_t \tag{1}$$

where β_1 is the cutting coefficient, $t = 1,2,3, \ldots, n$

- $\beta_2, \beta_3, \ldots, \beta_k$ are particular regression coefficient,
- u_t is a parameter estimation
- t is t^{th} observation
- n is the whole scale of overall

It supposes that n observations, each observation consists of k values $(Y_i, X_{2i}, \ldots, X_{ki})$ with $i = 1 \div n$. It is expressed by Eq. (2).

$$
\begin{aligned}
Y_1 &= \beta_1 + \beta_2 X_{21} + \beta_3 X_{31} + \ldots + \beta_k X_{k1} + u_1 \\
Y_2 &= \beta_1 + \beta_2 X_{22} + \beta_3 X_{32} + \ldots + \beta_k X_{k2} + u_2 \\
&\cdots\cdots\cdots\cdots\cdots\cdots\cdots\cdots\cdots\cdots\cdots\cdots \\
Y_n &= \beta_1 + \beta_2 X_{2n} + \beta_3 X_{3n} + \ldots + \beta_k X_{kn} + u_n
\end{aligned}
\tag{2}
$$

$$
Y = \begin{pmatrix} Y_1 \\ Y_2 \\ Y_3 \\ . \\ . \\ Y_n \end{pmatrix}
\beta = \begin{pmatrix} \beta_1 \\ \beta_2 \\ \beta_3 \\ . \\ . \\ \beta_n \end{pmatrix}
U = \begin{pmatrix} U_1 \\ U_2 \\ U_3 \\ . \\ . \\ U_n \end{pmatrix}
X = \begin{bmatrix} 1 & X_{21} & X_{31} & X_{41} & \ldots & X_{k1} \\ 1 & X_{22} & X_{32} & X_{42} & \ldots & X_{k2} \\ 1 & X_{23} & X_{33} & X_{43} & \ldots & X_{k3} \\ & & & . & & \\ & & & . & & \\ 1 & X_{2n} & X_{3n} & X_{4n} & \ldots & X_{kn} \end{bmatrix}
\tag{3}
$$

Hence, Eq. (1) is expressed by:

$$Y = X\beta + U \tag{4}$$

2.2 Hypothesis of Multivariate Regression Models for Labor Forecast

Hypothesis of multivariate regression models for Labor forecast is expressed by

$$E(U) = E\begin{pmatrix} U_1 \\ U_2 \\ U_3 \\ . \\ . \\ . \\ U_n \end{pmatrix} = \begin{pmatrix} E(U_1) \\ E(U_2) \\ E(U_3) \\ . \\ . \\ . \\ E(U_n) \end{pmatrix} \quad UU^t = \begin{bmatrix} U_1^2 \ U_1U_2 \ U_1U_3 \ \ldots \ U_1U_n \\ U_2U_1 \ U_2^2 \ U_2U_3 \ \ldots \ U_2U_n \\ U_3U_1 \ U_3U_2 \ U_3^2 \ \ldots \ U_3U_n \\ . \\ . \\ U_nU_1 \ U_nU_2 \ U_nU_3 \ \ldots \ U_n^2 \end{bmatrix} \quad (5)$$

+Hypothesis 1: $E(U_1) = 0 \forall i \ or \ E(U) = 0$

+Hypothesis 2: $E(U_{i,} U_j) = \begin{cases} 0 \ \forall i \neq j \\ 1 \ i = j \end{cases} or \ E(UU') = \partial^2 I$

Where I is the n levels matrix

+Hypothesis 3: $(X_1, X_2, \ldots, X_3)$ is determined
+Hypothesis 4: There is no multi-collinearity phenomenon among the explanatory variables or the rank of the matrix X by k: $R(X) = k$

2.3 Applied the Proposed Model for Labor Forecasting

Step 1- Find the best regression function
Calculate a regression function corresponding to each regression function which is expressed by

$$AIC = \left(\frac{RSS}{n}\right)\hat{u}^{\frac{2k}{n}} \qquad (6)$$

Optimize Min $\{AIC_i\}$, which regression function has the smallest AIC is the selected function.

Step 2- Calculate coefficients
Vector $\hat{\beta}_1, \hat{\beta}_2, \ldots, \hat{\beta}_k$ is expressed by

$$\hat{\beta} = (X^T X)^{-1} X^T Y \qquad (7)$$

Step 3-Constructe a regression function for its estimation
$$\hat{Y}_i = \hat{\beta}_1 + \hat{\beta}_2 X_{2i} + \hat{\beta}_3 X_{3i} + \ldots + \hat{\beta}_k X_{ki} \qquad (8)$$

Step 4- Forecast labor demand
Calculate $\hat{Y}_0$ and Var $\left(\hat{Y}_0\right)$.

Calculate standard errors of $\hat{Y}_0$ and SE $\left(\hat{Y}_0\right)$

Forecast the average confidence interval of $\hat{Y}_0$ is expressed by

$$E\left(\frac{Y}{X_0}\right) \in (\hat{Y}_0 - \varepsilon_0; \hat{Y}_0 + \varepsilon_0 \tag{9}$$

$$\varepsilon_0 = SE\left(\hat{Y}_0\right)t_{(n-2,\frac{\alpha}{2})}$$

2.4 Parameter estimation

The most common method used in order to estimate regression coefficients which is the smallest method of normal squares (MLS). The regression functions is ex-pressed by

$$\begin{aligned}
\hat{Y}_i &= \hat{\beta}_1 + \hat{\beta}_2 X_{2i} + \hat{\beta}_3 X_{3i} + \ldots + \hat{\beta}_k X_{ki} \\
Y_i &= \beta_1 + \beta_2 X_{2i} + \beta_3 X_{3i} + \ldots + \beta_k X_{ki}
\end{aligned} \tag{10}$$

$$Y = X\beta + e.\, e = \begin{bmatrix} e_1 \\ e_2 \\ \cdot \\ \cdot \\ \cdot \\ e_n \end{bmatrix} \tag{11}$$

MLS estimates which can be expressed by

$$\sum_{i=1}^{n} e_i^2 = \sum_{i=1}^{n} (Y_i - \hat{\beta}_1 - \hat{\beta}_2 X_{2i} - \ldots - \hat{\beta}_k X_{ki} \rightarrow min \tag{12}$$

$\sum_{i=1}^{n} e_i^2$ is the sum of square of remainder
The symbol $X^T, Y^T, \hat{\beta}^T, e^T$ is the matrix transposition of $X, Y, \hat{\beta}, e$
Hence, it can be expressed by

$$\begin{aligned}
e^T e = \sum_{i=1}^{n} e_i^2 &= \left(Y - X\hat{\beta}\right)^T \left(Y - X\hat{\beta}\right) = \left(Y^T - \hat{\beta}^T X^T\right)\left(Y - X\hat{\beta}\right) \\
&= Y^T Y - \hat{\beta}^T X^T Y - Y^T X\hat{\beta} + \hat{\beta}^T X^T X\hat{\beta} \\
&= Y^T - 2\hat{\beta}^T X^T Y + \hat{\beta}^T X^T X\hat{\beta}
\end{aligned} \tag{13}$$

$$\frac{\partial\left(e^T e\right)}{\partial \hat{\beta}} = -2X^T Y + 2X^T X\hat{\beta} \rightarrow X^T Y = X^T X\hat{\beta}$$

$$\rightarrow \hat{\beta} = \left(X^T X\right)^{-1} X^T Y \tag{14}$$

As referred from Eq. (9) we calculate the matrix of parameter estimates of $\hat{\beta}$.

3   Experimental Results

In this paper, the proposed approach to has been designed to enable the creation of an effective model which provides a basis upon which the prediction of labor, as shown in Fig. 1.

In research labor demand it is influenced by factors of human working jobs as follows [2, 10]:

S_1 Agriculture, fishery and forestry

S_2 Extractive

S_3 Processing, Making industry

S_4 Producing and distributing electricity, gas, hot water, steam and air conditioning

S_5 Water supply; waste and wastewater management and treatment activities

S_6 Construction

S_7 Wholesale and retail, car, motorcycle and motorbikes and other motor vehicles repair

S_8 Transportation, Warehouse

S_9 Accommodation and food services

S_{10} Information and communication

S_{11} Financial, bank and insurance activities

S_{12} Estate business activities

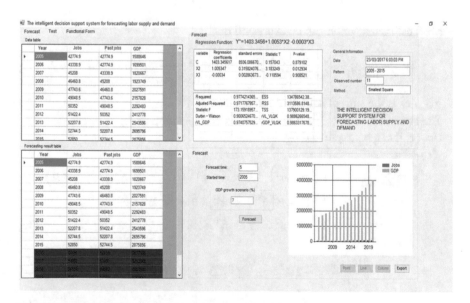

Fig. 1   Results of labor forecast on the screen

S_{13} Professional science and technology activities

S_{14} Administrative activities and support services

S_{15} Activities of the Communist Party, Social and Political organizations; State management. national security; Compulsory social assurance

S_{16} Education and training

S_{17} Health and social assistance activities

S_{18} Art, fun and entertainment

S_{19} Other service activities

S_{20} Activities of hiring jobs in households, production of material products and self-consumption services of households

S_{21} Activities of international organizations and agencies

The proposed model has been tested using data sets from Statistic Government [10]. Experimental results show that labor forecast, as shown in Fig. 2.

Figure 2 shows the annual data on percentage in the employed structure of 5 sectors. It is also reached at 6,34% in 2020. S_7 of wholesale and retail, car, motor-cycle and motorbikes and other motor vehicles repair also accounted for a significant share of the employed structure and experienced little changes, reached at 12,66% in 2015 and expected at 11,31% in 2020. S_8 of labors in transportation and warehouse fluctuated around the mark of 3% to achieve 3,02 and 3,07% at the start and end of the 10 years. By the end of our forecast period, it was expected to fall back to 2,89%. S_9 of accommodation and food services rose sharply from 1,93% in 2005 to 4,62%

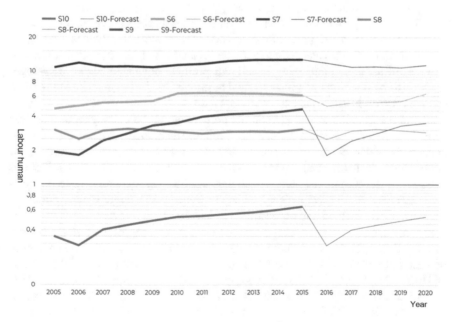

Fig. 2 Results of factors for labor forecast (S6-S10)

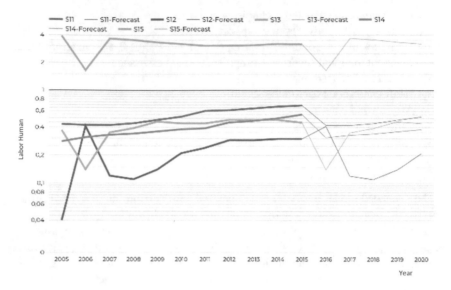

Fig. 3 Results of labor sectors' forecast (S11–S15)

in 10 years later, however it was expected to reach 3,49% at the end of 2020. S_10 of Information and communication is rather insignificant share around 0,5% by 2020.

Figure 3 shows the annual data on the percentage in the employers of 5 sectors from S_{11} to S_{15}. S_{11} of labors of financial, bank and insurance activities, S_{12} of estate business activities, S_{13} of professional science and technology activities, and S_{14} of administrative activities and support services overall rose from 2005 to 2015. S_{15} of activities of the communist party, social and political organizations; state management. national security; compulsory social assurance held a larger share around 3 to 4%, with 3,19% in 2015 and expectedly 3,2% in 2020.

Figure 3 shows the annual data on the percentage in the employers of 5 sectors from S_{11} to S_{15}. S_{11} of labors of financial, bank and insurance activities, S_{12} of estate business activities, S_{13} of professional science and technology activities, and S_{14} of administrative activities and support services overall rose from 2005 to 2015. S_{15} of activities of the communist party, social and political organizations; state management. national security; compulsory social assurance held a larger share around 3 to 4%, with 3,19% in 2015 and expectedly 3,2% in 2020.

Figure 4 shows the annual data on percentage in the employed structure of 6 sectors from S_16 to S_21. S_16 of labors in education and training experienced an overall increase in the data collecting period, from 2,94 to 3% and was expected to rise to 3,41% in 2020. S_19 of other service activities slightly fluctuated, finished at 1,63% in 2015 and was projected to reach 1,4% in 2020. S_17 of health and social assistance activities, S_18 of Art, fun and entertainment, and S_20 of activities of hiring jobs in households, production of material products and self-consumption services of households all accounted for approximately less than 1% of the employed structure.

Fig. 4 Results of labor
sectors' forecast (S16–S21)

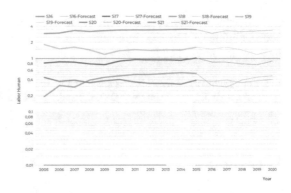

Table 3 Comparisons of the
proposed model and MLRM
performance

	R%	AA%
Multiple linear regression method	94,2	85
The proposed model	97	94,3

Their shares of the total employed in 2015 were 1,02%, 0,52% and 0,39% respectively, while these figures forecasted for 2020 were 0,89, 0,47 and 0,4%. Notably, from 2005 to 2015, S21 of Activities of international organizations and agencies always maintained at 0,01% with an exception of going to 0% in 2014. This labor line pattern was expected to continue going to 2020.

In evaluation, the proposed model has been tested with data sets [10], under the same conditions with the comparisons for the basic method of regression models. Forecast the average confidence interval of $\hat{Y}_0$ is given by Eq. (9). The results show that an average of the proposed model as labor rate through the strong correlation coefficient (R)% for 97,15%, respectively, as well as the percentage of accuracy (AA)% for 94,3%, respectively.

As shown in Table 3, the experimental results also show that the proposed approach performs better than MLRM method in term of correlation coefficient and accuracy.

4 Conclusion

In this paper, we have presented a novel method which targets improvements in labor forecast using real data sets. Our proposed approach uses multivariate regression approach that estimates a single regression model with outcome variable by dealing time series data. Experimental results demonstrate that the proposed model can provide enhanced forecast accuracy with respect to time series in data sets. Furthermore, the results derived from the testing [using real data sets] have confirmed that overall performance of the proposed method shows an improvement over the MLRM method. In summary, our reported results demonstrate the potential of our proposed model.

Further study for improvement in labor forecast accuracy, future research will address influenced factors, indicators using Fuzzy Knowledge Graph [11] to overall index of labor demand forecast in Big data.

Acknowledgements This work was supported by the Vietnam National Foundation for Science and Technology Development (NAFOSTED) under grant number 102.05-2019.316.

References

1. Yas AA (2013) Labor market forecasting by using data mining. Procedia Comput Sci 18:1700–1709
2. Ross G et al (2020) Forecasting extreme labor displacement: a survey of AI practitioners. Technol Forecast Social Change 161:120323
3. Luz A et al (2020) Local labour demand and immigrant employment. Labour Econ 63:101808
4. Yalcinkaya A et al (2021) A new approach using the genetic algorithm for parameter estimation in multiple linear regression with long-tailed symmetric distributed error terms: an application to the Covid-19 data. Chemometr Intell Lab Syst 216:104372
5. Pan Y et al (2021) Cost-effectiveness evaluation of pavement maintenance treatments using multiple regression and life-cycle cost analysis. Constr Build Mater 292:123461
6. Hu X et al (2021) Expanding the scope of multivariate regression approaches in cross-omics research. Engineering 7:1725–1731
7. Consonni V et al (2021) A MATLAB toolbox for multivariate regression coupled with variable selection. Chemometr Intell Lab Syst 213:104313
8. Lucy L, Julie Z (2020) Minimax D-optimal designs for multivariate regression models with multi-factors. J Stat Plan Inference 209:160–173
9. Yihe W, Dave ZS (2021) A nonparametric empirical Bayes approach to large-scale multivariate regression. Comput Stat Data Anal 156:107130
10. General Statistics Office. https://www.gso.gov.vn/en/statistical-data/
11. Lan LTH et al (2020) A new complex fuzzy inference system with fuzzy knowledge graph and extensions in decision making. IEEE Access 8:164899–164921
12. Son LH, Ciaramella A, Thu DT, Stanio A, Tuan TM, Hai PV (2020) Predictive reliability and validity of hospital cost analysis with dynamic neural network and genetic algorithm. Neural Comput Appl 32:15237–15248
13. Hai PV, Dong NT (2021) Hybrid louvain-clustering model using knowledge graph for improvement of clustering user's behavior on social networks. In: The international conference on intelligent systems & networks. Springer, Hanoi, pp 126–133
14. Hai PV, Truong DX (2021) Social network analysis based on combining probabilistic models with graph deep learning. In: Communication and intelligent systems. Springer, Delhi, pp 975–986
15. Hai PV, Dong NT (2020) Graph neural network combined nowledge graph for recommendation system. In: International conference on computational data and social networks. Springer, Montreal, pp 59–70

Twitter Sentiment Analysis on Oxygen Supply During Covid 19 Outbreak

Akash Kashyap⊙, Kunal Yadav⊙, and Sweta Srivastava⊙

1 Introduction

Humans are social beings who interact and, relate to each other by understanding one's thoughts and emotions. To increase social interactions, several platforms were made to interact throughout the world and to get information on the current happenings. One such microblogging site is Twitter, which people use to express their views over any issue and get global information. As of Q2 2021, there are 199 million active users on Twitter posting 500 million tweets per day. The current scenario of the covid second wave has isolated people due to the imposition of lockdown. This has led to an increase in the usage of social media platforms such as Twitter, Facebook, WhatsApp etc. for expressing emotions and exchange worldly information. The covid second wave started in April, reached its peak average value of 4.09 lakhs during 5–8th May [20]. A steep decline in covid cases was observed in September 2020 which prompted the Central government to presume—that the war against the Coronavirus had been won. This early celebration of covid-19 resulted in the second wave of the pandemic which proved more fatal and disastrous to the country. With the increase in cases, deaths also reached their peak, due to a lack of beds, concentrators, and other medical equipment's. The shortage of medical oxygen proved to be a major reason behind increasing deaths.

Thus, the paper uses Twitter data for sentiment analysis as it provides brief, real-time content availability with access to networks of similar discussions through hashtags [8]. The paper aims to analyse public sentiments across India during the Covid second wave using NLP techniques for text processing, TextBlob library to analyse and label the tweets based on polarity score as Positive, Negative and Neutral sentiments and finally, ML techniques for classification, to provide a depth on the impact of the pandemic on lives of people. The following sections advance as follows:

A. Kashyap · K. Yadav (✉) · S. Srivastava
Department of Computer Science and Engineering, Amity University, Noida, India
e-mail: mrkunalyadav7@gmail.com

© The Author(s), under exclusive license to Springer Nature Singapore Pte Ltd. 2022   655
B. Unhelker et al. (eds.), *Applications of Artificial Intelligence and Machine Learning*,
Lecture Notes in Electrical Engineering 925,
https://doi.org/10.1007/978-981-19-4831-2_53

Sect. 2 gives a discussion on Related work in sentiment analysis; Sect. 3 explains the methodology; Sect. 4 explains the implementation of work on how data was collected, pre-processed and its analysis; Sect. 5 displays experimental results and discussion followed by concluding remarks.

2　Literature Review

Trupthi, M., et al. [1] used Twitter API to collect data. A Uni-word Naive Bayes classification model was developed to get a polarity score for sentiment analysis. Entity level-based analysis on user's sentiments was done. 20,00,000 Tweets were transferred in the ML model and preprocessed using map-reduce. NLP and ML algorithms were used for classification. The supervised machine learning approach used here is Naïve-Bayes. By using n-gram classification rather than limiting to uni-gram, the accuracy of the model could be made more efficient.

Research presented by Kausar, M. A., et al. [3] about COVID-19 sentiment analysis using Twitter to analyse public emotions during the pandemic. They proposed a model using the Syuzhet algorithm to generate a sentiment score using fractions for each word between −1 (Negative) and 1 (Positive). Tweets from 11 infected countries were collected from 21st June 2020 to 20th July 2020. The countries included are the USA, Brazil, India, Russia, South Africa, Peru, Mexico, Chile, Spain, UK and Oman. According to the results, the majority of 11 countries had more than 50% positive tweets compared to neutral and negative tweets.

Behl, S., et al. [5] proposed an MLP model. Two public datasets (Nepal and Italy Earthquake) were considered for training and one original Twitter dataset (COVID-19) for testing. The reusability of the model trained on previous disasters was explored. The analysis of the original COVID-19 dataset resulted in 83% accuracy. The problem with re-usability was observed when the Italy dataset was used to train the model. The efficiency of the model was dropped because of the lesser number of tweets.

Pimprikar, et al. [10], tried to calculate the most reliable algorithm to foretell the values of the stock market with the help of machine learning and Twitter sentiments analysis. Linear regression and SVM models were used by them in this process, providing an accuracy of 82 and 60% respectively. They concluded that Twitter sentiments analysis is only influential when more than 80% of tweets are showing positive sentiment otherwise, they get overshadowed by negative and neutral sentiments.

Cristian R., et al. [11], did a Twitter sentiments analysis on the coronavirus dataset using machine learning algorithms and NLP techniques. Around 54% of the users showed positive feelings while the rest showed negative feelings. The method they used can be widely used for various data set.

Wang, H., et al. [12], tried a real-time analysis of the sentiments during the US presidential elections. To get higher accuracy and speed, they made a real-time data processing infrastructure with IBM. They assumed that opinions would have a high subjectivity, they used the k-means algorithm and got an accuracy of around

59%, they used naive Bayes, random trees, simple cart and many algorithms for classification.

Gautam and Yadav [17] presented a customer review on twitter dataset using Maximum entropy, SVM, Naive Bayes, along with the Semantic Orientation based WordNet which root outs the similarity for the content feature. According to Neethu and Rajasheree [18], the challenge with analysis of sentiments using Twitter compared to general sentiment analysis is because of presence of slangs as well as misspellings. They used twitter posts about electronic products like mobiles, laptops etc. using Machine Learning approach to extract public opinion about those products.

Gupta et al. [19] presented a review on some research in sentiment analysis on Twitter. They also discussed the methodologies adopted and models applied, along with describing a generalized Python based approach.

Wagh, R. and Punde, P. [4] presented a review on sentiment analysis approaches by various researchers on Twitter datasets. Machine learning techniques like N-grams, Maximum-Entropy, Semantic Analysis, etc. were discussed.

Suchdev, R., et al. [13], used sanders analytic dataset to perform sentiments analysis for tweets that were made on google and apple. Hybrid methods were used which combined knowledge-based approaches and ML capabilities. They used a feature vector to include components like hashtags, emoticons, etc., and the remaining words were included in the knowledge-based approach. When they integrated these two methods, accuracy of almost 100% was achieved.

3   Research Methodology

The present study performs sentiment analysis for tweets retrieved from Twitter API. Tweets for Indian users were collected from June 20th, 2021 to June 26th, 2021; and was processed using in-built functions, NLP and ML techniques. Thereafter, sentiment analysis was performed based on the polarity score, which lies in the range of [−1, + 1]. Tweets were classified based on polarity scores as negative sentiments (polarity score < 0), neutral sentiments (polarity score = 0), and positive sentiments (polarity score > 0). The sentiment analysis result was summarized based on individual counts and percentages; and displayed using plots. In addition, subjectivity for the tweets was also analysed to understand the nature of tweets. Lastly, the Supervised learning approach was adopted for the classification of the model to get precise outcomes. The labelled sentiments were chosen as the target and their values were mapped in such a way that the negative sentiments were mapped to −1, neutral were mapped to 0 and positive sentiments were mapped to +1. The ML techniques such as Naïve Bayes, SVM, and Logistic Regression were used for the classification of the model and a comparative study was done on each of them to understand their effectiveness in analysing the sentiments precisely. Figure 1 shows a block diagram of the research methodology.

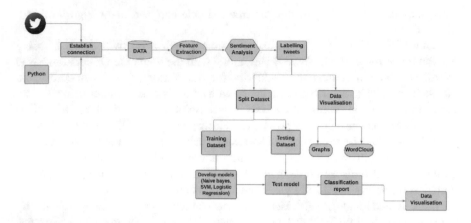

Fig. 1 Proposed block diagram

4 Implementation

The implementation of work is divided into 3 phases:

1. Data collection
2. Pre-processing
3. Sentiment Analysis and classification

4.1 Data Collection

The present study uses the Tweepy library which integrates Twitter API with python.7000, tweets were collected from 20th June 2021 to 26th June 2021, using the Tweepy package in python. For Data collection, Hashtags used are #oxygen, #IndiaFightsCorona, #oxygenconcentrator, #OxygenExpress, #Oxygen-Deaths, #oxygenshortage, #weneedoxygen, #covid, #coronavirus, #covid19, #COVIDEmergency2021, #COVID, #OXYGEN, #MedicalOxygen, #OxygenAudit The collected data was filtered on basis of their location and tweets from India were considered. While retrieving tweets, the retweets and duplicates were dropped. The complete dataset obtained after filtering contain 4638 tweets. Table 1 shows the top 15 most frequent words in the collected tweets.

4.2 Pre-processing

Firstly, Pre-processing is done to analyse tweets and obtain proper sentiment classification. It involves removing noise such as mentions, whitespace, punctuations,

Table 1 Top 15 most frequent words

S. no	Words	Count
1	India	816
2	covid	700
3	oxygen	680
4	report	506
5	Delhi	474
6	vaccination	470
7	vaccine	446
8	doses	444
9	lakh	385
10	people	375
11	administered	373
12	June	369
13	cases	357
14	today	274
15	vaccinated	269

URLs, and stop words from the dataset. For further processing of texts, Natural Language Toolkit (NLTK) a python library is used. Following NLTK libraries were adopted in this advanced pre-processing stage:

Stemming and Lemmatization
The process of reducing or normalizing the words in a text to their root word is called Stemming. The stemmed words are analysis of words morphologically and may not be the same as the root word. It may be Over-stemmed or Under-stemmed which results in distorted words. PorterStemmer, an NLTK module is used for stemming.

Another approach, lemmatization is also used to reduce the words in root form. It deals with the inflected forms of a word so they can be analysed as a single item. After creating an instance of the stemmer, a function is defined such that it takes each sentence of a corpus as input and lemmatized words are returned. The Wordnet module of NLTK is used for lemmatization.

4.3 Sentiment Analysis and Classification

The process of analysing textual data and classifying it based on sentiments as Positive, Negative and Neutral is called Sentiment Analysis. Classification and analysis of tweets are done using NLP and ML techniques to assign sentiment scores for each sentence. A common python library used in this paper is TextBlob for classifying tweets. It is based on NLTK. TextBlob helps in analysing the tweets and provides polarity and subjectivity for them.

Polarity defines the emotion expressed in the analysed text. The range is from [−1, 1]. A polarity score of 1 defines positive sentiment and a score of −1 defines negative sentiment whereas Subjectivity refers to personal opinions and emotions in a sentence. Subjectivity lies in the range of [0, 1].

Polarity scores were labelled as negative, neutral and positive sentiments. Thus, Supervised learning was used on them. These labelled tweets were considered as targets for which model was used to predict the sentiments correctly. To make these labels machine-readable, Label Encoding was done i.e., labels were converted into numeric form to make it machine-readable. The negative tweets were encoded as −1, neutral tweets were encoded as 0, and positive tweets were encoded as 1 [9].

Dataset is split into the training set and test set using scikit-learn modules. Machine learning algorithms of Naïve Bayes classification, Support Vector Machine (SVM) and Logistic regression models were used to evaluate the test set individually. Trigrams were also used on the developed dataset for accurate predictions. The training set is fitted to each technique and model evaluation was done by passing the test dataset, therefore generating the classification report for each technique.

- Multinomial Naïve Bayes: It is a probabilistic classification technique used for textual data analysis. It is mostly used in NLP. Equation 1 shows the Multinomial Naïve Bayes function.

$$Pr Pr(j) = log\,log\,\pi_j \sum_{i=0}^{|V|} log(1 + f_b) \log(Pr(i/j)) \tag{1}$$

- Support Vector Machine (SVM): It is a classification technique which differentiates the classes by plotting a hyperplane between them. Equation 2 represents the function for SVM.

$$L(w) = \sum_{i=1} \left(0, 1 - y_i\left[w^T x_i + b\right]\right) + \lambda\|w\|_2^2 \tag{2}$$

- Logistic Regression: It predicts whether something is true or false. In this paper, multinomial logistic regression is used since the number of variables are more than 2. The Logistic regression function is shown in Eq. 3.

$$log\left[\frac{y}{1-y}\right] = b_0 + b_1x_1 + b_2x_2 + b_3x_3 + \cdots\ldots\ldots + b_nx_n \tag{3}$$

5 Experimental Results and Discussion

Word frequency distribution of tweets was calculated using FreqDist. It is an NLTK library module that tells how many times a word appears in the data frame. It is done to analyse word frequency. Frequency Distribution of most frequent words is shown in Fig. 2.

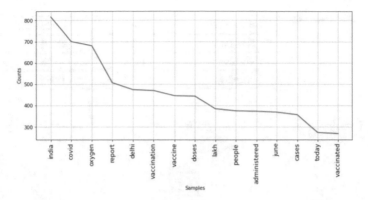

Fig. 2 Frequency distribution of most frequent words

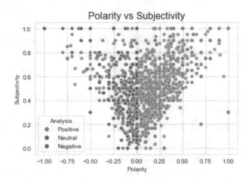

Fig. 3 Polarity vs subjectivity

Polarity lies in the range of [−1, 1] that defines positive and negative statements. Subjective sentences that refers to personal opinion or emotion. Subjectivity also lies in the range of [0, 1]. Figure 3 shows a scatter plot of polarity vs subjectivity, that was created using the Seaborn library with Polarity on X-axis and Subjectivity on Y-axis for understanding of sentiments. The orange dots depict neutral tweets. As their polarity score is '0', it lies in the center. Dots on either side of the Neutral tweets, which are represented by purple and green dots are Negative and Positive respectively.

Further tweets were separated in different data frames based on their sentiments as shown in Table 2. The total number of tweets data in each frame was calculated using the value_counts() method. It's done to get the value and percentage of positive, negative, and neutral tweets.

To identify the label of the tweets, The data set was grouped into training dataset and test dataset, on which machine learning algorithms were applied. Thereby, performance for each model was measured and a comparative study was done of the

Table 2 Value counts of sentiments

Sentiments	Total	Percentage
Neutral	2031	45.58
Positive	1638	36.76
Negative	787	17.66

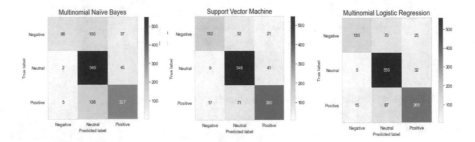

Fig. 4 **a** Confusion matrix for multinomial Naïve Bayes classification, **b** Confusion matrix for SVM, **c** Confusion matrix for multinomial logistic regression

models that are multinomial naïve bayes classifier, multinomial logistic regression, and support vector machines (SVM).

Results obtained from all the three mentioned models were visualized using confusion-matrix. The confusion matrix is N × N matrix that is used to summarize the overall performance of the models i.e., it explains how well the developed classification model is working and what all errors were generated. In this paper, the target has three variables namely negative, neutral, and positive. Thus, a 3 × 3 confusion matrix was generated for each model to give an overall analysis of performance. Here the diagonal shows the correct predicted values. The support tweets for each classification method were 859 comprising of 145 negative tweets, 410 neutral tweets and 304 positive tweets. Figure 4 shows confusion matrix for multinomial naïve bayes classification, SVM and multinomial logistic regression respectively.

The confusion matrix is also used for calculation and analysis of the following parameters of the model:

1. Precision: It is the ratio of correctly predicted positive values to the overall positive predicted values.

$$Precision = \frac{TP}{TP + FP} \tag{4}$$

2. Recall: It gives the ratio of correctly predicted positive values to all the actual positive observations in, the actual class.

$$Recall = \frac{TP}{TP + FN} \tag{5}$$

Table 3 Comparative study on the models (Training set = 80, test set = 20, random state = 123)

Model	Label	Prec.	Recall	F1 Score	Acc. (%)
Naïve Bayes	− 1	0.89	0.41	0.56	77
	0	0.74	0.91	0.81	
	+1	0.79	0.75	0.77	
SVM	− 1	0.84	0.70	0.77	85
	0	0.84	0.90	0.87	
	+1	0.86	0.84	0.85	
Logistic Regression	− 1	0.88	0.64	0.74	85
	0	0.82	0.94	0.88	
	+1	0.88	0.83	0.86	

3. F1-Score: It is the weighted average of recall and precision.

$$F1 - Score = \frac{2 * Recall * Precision}{Recall + Precision} \tag{6}$$

4. Accuracy: It is the ratio of all correctly predicted values to the overall predicted values.

$$Accuracy = \frac{TP + TN}{TP + TN + FP + FN} \tag{7}$$

A comparative analysis based on the mentioned parameters is shown in Table 3, where −1 = Negative, 0 = Neutral, + 1 = Positive, Prec. = Precision, Acc. = Accuracy.

It was observed that both Logistic regression and SVM gave an accuracy of 85%, while Multinomial Naïve Bayes classification gave an accuracy of 77%. The ratio of training and testing set was decided on multiple runs.

For Training set = 70, test set = 30, the accuracy was 75, 84 and 82%. In this case SVM outperformed all the other algorithms. For Training set = 90, test set = 10 was 79, 86 and 86% respectively for Naïve Bayes, SVM and Logistic Regression. Here also both SVM and Logistic regression performed better than Naïve Bayes. In all the 3 algorithms, some improvement was seen in terms of accuracy when the training set was increased but this might lead to overfitting.

Lastly three separate WordCloud were made. A word cloud is a cluster of words arranged in different sizes. The words appearing more frequently in a text the bigger and bolder size it takes in the word cloud. WordCloud for all words in the tweets, only positive tweets, and only negative tweets. The words were masked in Indian map. In Fig. 5 shows word cloud for all frequent words in the tweets, positive words in the tweets and negative words in the tweets.

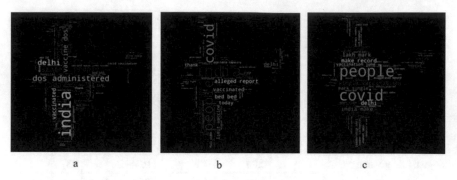

a   b   c

Fig. 5 **a** Word Cloud for all words, **b** Word Cloud for the words in positive tweets, **c** Word Cloud for the words in negative tweets

6 Conclusion

The present study has successfully performed sentiment analysis for Indian tweeter users in the COVID19 pandemic. The study primarily focused on oxygen supply and deficiency. The model can be carried out in a way to perform overall sentiment analysis on entire COVID19 pandemic. The sentiment analysis was carried out using machine learning techniques like Naïve Bayes, SVM, and logistic regression.

Based on the results generated from the available twitter database, tweets were categorized in positive, neutral, and negative sentiments based on subjectivity and polarity scores. Neutral tweets comprise about 45%, positive tweets about 37% and negative tweets about 18%. Later, The Supervised machine learning approach helped in generating classification based on the labelled sentiments and it was observed that SVM and Logistic Regression models proved to be more accurate, providing a better overall score compared to the Naïve Bayes approach. Both SVM and Logistic Regression gave an accuracy score of 85%. Keeping aside the neutral tweets, the positive tweets are almost double the negative tweets. It can be observed that, in the ongoing pandemic, people are becoming more optimistic.

Further studies may be performed with a larger database say tweets for about a couple of months as it will help in training the dataset with larger data thus, it will improve the performance of the model.

References

1. Trupthi M, Pabboju S, Narasimha G (2017) Sentiment analysis on twitter using streaming API. In: 2017 IEEE 7th international advance computing conference (IACC), pp 915–919. IEEE
2. Giachanou A, Crestani F (2016) Like it or not: a survey of twitter sentiment analysis methods. ACM Comput Surv (CSUR) 49(2):1–41
3. Kausar MA, Soosaimanickam A, Nasar M (2021) Public sentiment analysis on twitter data during COVID-19 outbreak. Int J Adv Comput Sci Appl 12(2)

4. Wagh R, Punde P (2018) Survey on sentiment analysis using twitter dataset. In: 2018 second international conference on electronics, communication and aerospace technology (ICECA), pp 208–211. IEEE
5. Behl S, Rao A, Aggarwal S, Chadha S, Pannu HS (2021) Twitter for disaster relief through sentiment analysis for COVID-19 and natural hazard crises. Int J Disaster Risk Reduction 55:102101
6. Garcia K, Berton L (2021) Topic detection and sentiment analysis in Twitter content related to COVID-19 from Brazil and the USA. Appl Soft Comput 101:107057
7. Barnaghi PG, Breslin J, Ghaffari P (2016) Opinion mining and sentiment polarity on twitter and correlation between events and sentiment. In: IEEE second international conference on big data computing service and applications
8. Su Y, Venkat A, Yadav Y, Puglisi LB, Fodeh SJ (2021) Twitter-based analysis reveals differential COVID-19 concerns across areas with socioeconomic disparities. Comput Biol Med 213(3):243
9. Babu NV, Rawther FA (2021) Multiclass sentiment analysis in text and emoticons of Twitter data: a review. In: Palesi M, Trajkovic L, Jayakumari J, Jose J (eds) Second international conference on networks and advances in computational technologies. TRACOSCI. Springer, Cham. https://doi.org/10.1007/978-3-030-49500-8_6
10. Pimprikar R, Ramachandran S, Senthilkumar K (2017) Use of machine learning algorithms and Twitter sentiment analysis for stock market prediction. Int J Pure Appl Math 115(6):521–526
11. Chritian R, Machuca GC, Toasa R (2021) Twitter sentiment analysis on coronavirus: machine learning approach. J Phys Conf Ser 1828(1):012104
12. Wang H, Can D, Kazemzadeh A, Bar F, Narayanan S (2012) A system for real-time Twitter sentiment analysis of 2012 US presidential election cycle. In: Proceedings of the ACL 2012 system demonstrations
13. Suchdev R, Kotkar P, Ravindran R, Swamy S (2014) Twitter sentiment analysis using machine learning and knowledge-based approach. Int J Comput Appl 103(4):36–40
14. Colianni S, Rosales S, Signorotti M (2015) Algorithmic trading of cryptocurrency based on Twitter sentiment analysis. CS229 Project 1–5
15. Le B, Nguyen H (2015) Twitter sentiment analysis using machine learning techniques. In: Le Thi H, Nguyen N, Do T (eds) Advanced computational methods for knowledge engineering. AISC, vol 358, pp 279–289. Springer, Cham. https://doi.org/10.1007/978-3-319-17996-4_25
16. Song J, Kim KT, Lee B, Kim S, Youn HY (2017) A novel classification approach based on Naïve Bayes for Twitter sentiment analysis. KSII Trans Internet Inf Syst (TIIS) 11(6):2996–3011
17. Gautam G, Yadav D (2014) Sentiment analysis of twitter data using machine learning approaches and semantic analysis. In: 2014 seventh international conference on contemporary computing (IC3), pp 437–442. IEEE
18. Neethu MS, Rajasree R (2013) Sentiment analysis in twitter using machine learning techniques. In: 2013 fourth international conference on computing, communications and networking technologies (ICCCNT), pp 1–5. IEEE
19. Gupta B, Negi M, Vishwakarma K, Rawat G, Badhani P (2017) Study of Twitter sentiment analysis using machine learning algorithms on Python. Int J Comput Appl 165(9):29–34
20. https://www.covid19india.org/

Artificial Eye for the Visually Impaired

Aakansha Gupta, Harshil Panwar, Dhananjay Sharma, and Rahul Katarya

1 Introduction

Technology plays a vital role in every human's life, even when it comes to enabling them to complete many day-to-day tasks effectively and efficiently. In the case of differently abled people, supportive technology is an extremely relevant example of how technology allows people to live their lives very comfortably. Realizing the importance of this domain of technology, this product's concept gives a comprehensive solution for the visually impaired by providing them an independent and viable alternative to their current human/animal aids that will also prove to be cheaper in the long run.

Deep learning is a modern computational field that attempts to imitate the human brain's learning process in the best possible manner using extensive mathematical equations. A successfully designed and implemented model can enable the machine to recognize patterns and make predictions accurately. Although the field is still picking up momentum, certain subdivisions have already attained accuracy levels surpassing a human expert in the field.

Object recognition, in deep learning, is a subpart of the image processing division. It involves parsing through any input frame (image or video), locating potential objects, and labeling them accordingly based on the list of objects on which the model has been trained.

A. Gupta · H. Panwar (✉) · D. Sharma · R. Katarya
Department of Computer Science and Engineering, Delhi Technological University, Delhi, India
e-mail: harshilpanwar_2k18co142@dtu.ac.in

D. Sharma
e-mail: dhanajaysharma_2k18co119@dtu.ac.in

R. Katarya
e-mail: rahulkatarya@dtu.ac.in

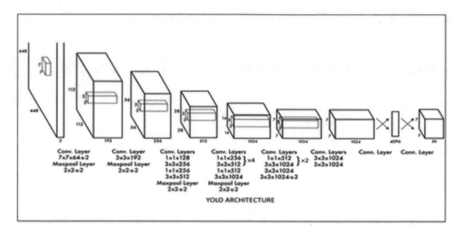

Fig. 1 The architecture of the YOLO object detection network

Most of the detectors employ classifiers to perform detection. To detect an object, classifiers for different objects are tried on the image at various scales. Systems like deformable parts models (DPM) [2] use a sliding window approach where the classifier is run at evenly spaced locations over the entire image.

Object detection using the YOLO [1] algorithm is posed as a regression problem, straight from image pixels to bounding box coordinates and class probabilities.

The YOLO unified model has several benefits over other detection systems (Fig. 1):

1. It is extremely fast. Since it avoids the complex pipeline by posing the problem as regression, also, it avoids the sliding windows technique used by detection systems such as DPM, thus making it faster.
2. YOLO sees the entire image during training and test time, so it encodes contextual information about classes as well as their appearance. Fast Region-Based Convolutional Neural Network (R-CNN) [3], a top detection method, mistakes background patches in an image for objects because it cannot see the larger context.

2 Fundamentals

Image Processing is the division of Deep Learning responsible for converting input images into clean, equal-sized, well-formatted, processable images that any neural network can analyze with ease. This is an essential process, as it allows the algorithms to identify features and patterns efficiently and then use them to solve many real-world problems.

Object Detection, as mentioned earlier, is the subdivision of image processing responsible for scanning images and identifying certain objects present in them. The general approach used by more algorithms is to identify the most promising areas, identify the probability of that area being a particular object, to process and confirm it if the probability value is higher than a particular threshold [5].

Convolutional Neural Networks are neural networks comprising mainly convolutional layers combined with various pooling and dense layers. These networks are the most effective in image feature extraction as they are designed to identify the most relevant pixels accurately, with ease [4].

The YOLO or the You only look once algorithm, in particular, is one of the most efficient object detection algorithms that converts the complex problem of object detection into a simple regression and bounding box probability problem. It is faster and more reliable than its competitors as it processes the entire image in one go.

3 Related Work

Today, most detectors use two-stage object detection algorithms to classify objects such as R-CNN [3] and FPN [12]; this technique is slow in real-time and requires extensive computation. Although the additional step of proposing regions makes the detectors accurate, it increases computation time significantly.

The paper that functioned as a guiding light throughout the bulk of this project is known as "YOLOV3: An Incremental Improvement," written by Joseph Redmon and Ali Farhadi.[1] This paper implements the YOLO network with darknet-53 as the feature extractor, further reducing the time taken by the algorithm to detect objects in real-time. YOLOv3 soon gained popularity in the real-time detection domain because of its high detection speed and exceptional speed-accuracy balance.

Today, most assistive solutions for the visually impaired make use of external hardware for object detection, depth perception, and providing feedback [6]. Hardware-based solutions are not only costly but are also vulnerable to external damage and varying environmental conditions. This work aims to provide a software-based cost-effective solution by making the above-mentioned hardware obsolete.

4 Methodology

4.1 Input

The first task for the project is to obtain the input on which the model will perform. For this, the OpenCV library in python was used. OpenCV can tap into the device's cameras and obtain a photo or a video feed. This will serve as the input for object detection.

4.2 Object Detection

For object detection, the YOLO, or the You Only Look Once algorithm, was implemented. The YOLO algorithm is one of the fastest and simplest (in terms of complexity of the neural network) for performing object detection. Vigorous research over the years has led to a rapid decrease in not only the complexity of the network structure but also the time taken by it to perform object detection.

The way YOLO excels is by converting complex object detection problems into much simpler problems involving regression and class probabilities. First, a frame is divided into a grid of a predetermined size. Now, a sliding window is used to parse through the grid, mini box by mini box, determining the probability of each mini box containing an object.

After a single iteration, all the boxes with a higher probability (say ≥ 0.5) of containing an object class are taken into consideration and bounding boxes are drawn accordingly. These bounding boxes will contain the detected objects, respectively.

To ensure that the same object is not detected more than once, the concept of non-maximum suppression or NMS [1] was introduced. In NMS, the boxes with the highest confidence value are considered and their IOU (Intersection over Union) with other boxes is calculated [1]. The IOU is the ratio of the area common between two bounding boxes and the total area occupied by both boxes, hence the name, Intersection over Union. Now, all the boxes having a high IOU with the arbitrarily chosen box are eliminated as they are most likely to be representing the same object. The same process is repeated for the bounding box with the next highest confidence value until the final bounding box, representing a unique object, is obtained.

As an additional contribution to this step, the YOLO model was trained on additional classes to identify certain harmful objects such as guns. These classes were not a part of the multiple classes already present in the COCO [7] Dataset.

4.3 Estimating Position and Depth and Output

Using the boundary box values for both static images and video files position of each object is estimated, along with depth estimation using the safety index. Using the google text to speech API, the output is given in audio format, which is overlaid and synced with the video using FFMPEG.

As an additional contribution to this step, a custom function that divides the image into a 3*3 matrix and identifies every object's location relative to the frame of the image was built.

Furthermore, a simple yet efficient method for depth perception was employed rather than the state-of-the-art but computationally heavy techniques such as monocular depth perception. Each frame is assumed to be at 2*LDDV, where LDDV is the least distance for a distinct vision for the Human eye. Now for the safe distance, if the index S (Safety index) > 0.6, then the warning symbol is generated for that object.

$$Safety\ Index(S) = I \times C \tag{1}$$

where C is Confidence Score of Bounding Box and,

$$I = Area(Bounding\ Box)/Area(Frame) \tag{2}$$

5 Experimentation

Libraries, frameworks used:

TensorFlow. One of the most popular libraries for constructing and implementing neural networks has been used.

Darknet Framework. A 53 Layered, fully convolutional neural network has been used as the feature extractor.

Pydub. A simple, well-designed python module for audio manipulation.

GTTS. Google Text To Speech conversion API has been used to obtain the final output.

FFMPEG. One of the best audio/video processing libraries has been used.

OpenCV. Python's most popular library for image processing has been used.

Hardware Requirements.

- 2 Core CPU clocked $\geq$ 2.4 GHz
- 8 GB RAM
- 120 GB Disk Space

6 Dataset Description

The YOLO model used to develop the tool was trained on the COCO (common objects in context) dataset [7]. This dataset contains over 80 classes and is one of the most popular datasets used for object detection.

Furthermore, two additional datasets containing about 300 images of harmful objects such as guns [8, 9], were utilized. As these datasets were not labeled, bounding boxes were constructed for every image manually, and then the model was trained on these datasets as well. Therefore, a total of two datasets were utilized.

7 Computation

A Graphics Processing Unit (GPU) is an integrated circuit configured to quickly manipulate and change memory to provide acceleration to create images in a frame buffer. These have various applications in mobile phones, personal computers, etc. They can optimally manipulate images with ease. The parallel structure of the GPUs makes them much more efficient than the CPUs for the purpose of processing large amounts of data simultaneously. Hence in order to increase computational speed and save time when performing computationally heavy tasks such as training networks, GPUs are used.

A CPU usually works on a few cores, averaging between 4–8, while CPUs with 64 and 128 cores are also commonly used in supercomputers, these usually work with one or two threads per core. There are usually a few hundred cores in a GPU, and each of these has tens or hundreds of threads, thus bringing the total to thousands of threads parallelly computing and performing tasks. A GPU works parallelly on tasks that, if performed on a CPU, are done sequentially, using a for loop while in GPU vector addition and vector operations.

CPUs have more powerful cores than GPUs; thus, they can perform better for computationally complex tasks if per-core performance is considered GPUs have a greater number of weaker cores that can outperform CPU when tasks can be parallelly processed, such as in big data analysis or 3D rendering.

7.1 Tensor Processing Unit

TPU is also an alternative to the GPU, an ML-specific ASIC, designed to speed up Linear Algebra operations, specifically heavy matrix multiplications. TPU is one of the most advanced DL platforms. It gives up to 30 times better performance than conventional CPUs and GPUs. It provides very high performance with an effective bandwidth of 12.5 Gbps (Fig. 2).

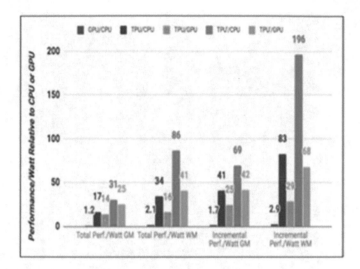

Fig. 2 Relative performance of different processing units

8 Training

8.1 Training the Model

The model was trained on the famous COCO dataset [7] containing a total of 80 classes, all common objects that one might come across in their everyday lives. Apart from this, two additional databases containing images of harmful objects such as guns, etc. [8, 9], were also labeled and used to train the model. The following images and graphs show the loss function and how it changed throughout the process of training along with other frames captured during the live testing of the tool (Figs. 3 and 4).

For feature extraction, Darknet-53 was used. It stood out in terms of fast calculation speed and fewer floating-point operations. Below is the image of the real-time deployment of the tool (Figs. 5 and 6).

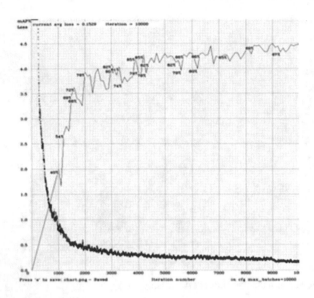

Fig. 3 Yolo training loss plot for COCO dataset

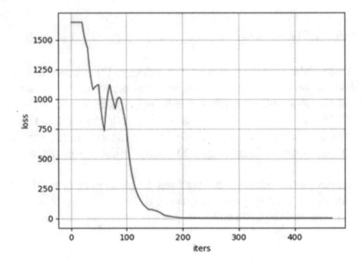

Fig. 4 Yolo training loss plot for custom classes

Fig. 5 Output from static frame

Fig. 6 Frame capture from real-time

9 Experimental Result and Discussion

Through this research work, it can be observed how object detection can be viewed as a regression problem and the advantages of this methodology. The model was trained with new classes with a resultant total loss of 0.0004 over 400 iterations, which resulted in a satisfactory mAP. It was found that with a threshold of 0.6 over the safety index, Position determination and depth estimation were determined satisfactorily, leveraging the expensive computation and delayed responses.

10 Future Work

Apart from being of great use in the support and aid domain, this technology also has the potential to form the base of many other potential applications. When combined with OCR technology, the device can be used to identify objects, recognize their brands, and locate them on popular online sellers such as amazon. This technology can even be used to recognize license plates and scan documents and identity cards.

When combined with pose detection algorithms, this technology can be used in many fields such as athletics, sports, yoga, etc., to identify and analyze the various physical activities being performed by athletes and help them train.

As far as the current tool is considered, it may still be improved in terms of its accuracy and performance in real-time, making it even more viable as a probable replacement to old and traditional methods.

References

1. Redmon J, Divvala S, Girshick R, Farhadi A: You only look once: unified, real-time object detection. arXiv:1506.02640
2. Girshick R, Iandola F, Darrell T, Malik J (2015) Deformable part models are convolutional neural networks. arXiv:1409.5403
3. Girshick R (2015) Fast R-CNN. arXiv:1504.08083
4. Krizhevsky A, Sutskever I, Hinton GE (2012) ImageNet classification with deep convolutional neural networks. Neural Inf Process Syst 25. https://doi.org/10.1145/3065386
5. Zhao ZQ, Zheng P, Xu ST, Wu X (2019) Object detection with deep learning: a review. IEEE Trans Neural Netw Learn Syst 1–21. https://doi.org/10.1109/TNNLS.2018.2876865
6. Rahman S, Debnath C, Trisha TA (2019) Design and implementation of a smart assistive system for visually impaired people using Arduino. Int J Adv Comput Electron Eng 4(11):1–5
7. Lin TY et al (2014) Microsoft COCO: common objects in context. In: Fleet D, Pajdla T, Schiele B, Tuytelaars T (eds) Computer Vision–ECCV 2014. ECCV 2014. LNCS, vol 8693. Springer, Cham. https://doi.org/10.1007/978-3-319-10602-1_48
8. Sasank S (2019) Guns Object Detection, Version 1. https://www.kaggle.com/issaisasank/guns-object-detection
9. Shekhar S (2020) Knife Dataset, Version 1. https://www.kaggle.com/shank885/knife-datasetM. Of, An effective implementation of face recognition using. J Southwest Jiaotong Univ 54(5):1–9 2019. https://doi.org/10.35741/issn.0258-2724.54.5.29
10. Redmon J (2013) Darknet: open-source neural networks in c. 2018
11. Lin T-Y et al (2017) Feature pyramid networks for object detection. In: Proceedings of the IEEE Conference on Computer Vision and Pattern Recognition
12. Girshick R et al (2014) Rich feature hierarchies for accurate object detection and semantic segmentation. In: Proceedings of the IEEE Conference on Computer Vision and Pattern Recognition

Blockchain Network: Performance Optimization

Om Pal, Surendra Singh, and Vinod Kumar

1 Introduction

In Blockchain, records are stored in decentralized manner and stored in form of blocks. To maintain the integrity of data, each block contains the hash of previous block. In Blockchain, decentralized storage provides higher cryptographic security compared to centralized storage. Using PKI, Blockchain provides transparency among the nodes, authenticity of data, integrity, non-repudiation, immutability of data, higher trust over the network when applied in diverse areas.

In Blockchain, transactions are approved with consensus of peer members. There is no central authority for approval of transactions. Data is stored in decentralized manner and transactions are approved with consensus of majority. Copy of the previous transactions is available with each peer member. Hence, there is no possibility of central node attack and also, integrity of the data is quick verifiable. Alteration of records is only possible with consensus of majority.

In Blockchain, data is stored in form of blocks and these blocks are connected with each other in form of chain. Diagram of chain of block is given in Fig. 1.

Each header of block of chain contains many fields like address of previous block, hash of previous block, time stamp, block number, nonce etc. Second part of block is called body and it contains the detail of transactions. Data structure of block is given in Fig. 2.

O. Pal (✉) · S. Singh
Ministry of Electronics and Information Technology, Govt. of India, New Delhi, India
e-mail: ompal.cdac@gmail.com

V. Kumar
Department of Electronics and Communication (Computer Science and Engineering), University of Allahabad, Prayagraj, U.P., India

© The Author(s), under exclusive license to Springer Nature Singapore Pte Ltd. 2022   677
B. Unhelker et al. (eds.), *Applications of Artificial Intelligence and Machine Learning*,
Lecture Notes in Electrical Engineering 925,
https://doi.org/10.1007/978-981-19-4831-2_55

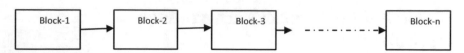

Fig. 1 Blockchain

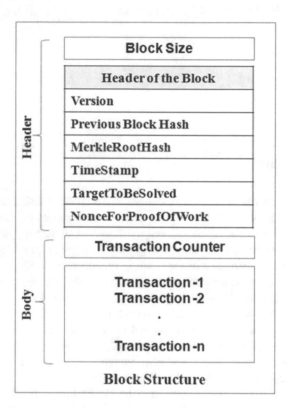

Fig. 2 Block structure

1.1 Types of Blockchain

Blockchain can be either permissionless or permissioned.

Public Blockchain. In Public Blockchain, no permission is required for joining the network. There is no restriction on joining the Blockchain network. Public Blockchain is also called permissionless Blockchain. In this type of network, transactions are validated and approved by the consensus of public. Anyone can become the part of the network and can contribute without any restriction. Bitcoin (BTC) and Ethereum (ETH) are examples of Permissionless Blockchain [1–4].

Private Blockchain. In Private Blockchain, restrictions are imposed on joining the network and contribution. This Blockchain is controlled by few core members or by a single owner. With the permission of core members, anyone can join the Blockchain. For joining such network, permission is required. Therefore, it is also called the permissioned Blockchain. After becoming approved member of the Blockchain, only approved members validate the transactions. Core members of the Blockchain have higher privileges and these core members are responsible for policy level decisions like who can join the network, to whom transactions detail will be visible etc. However, in approval of transactions, all approved members of the Blockchain take the part. Private Blockchain is more centralized in nature.

Quorum, Hyperladger Fabric and R3 Corda are example of Permissioned Blockchain. The private Blockchain framework is more suitable for shipping firms, supermarkets, banks telecommunication companies, hospital management system and even in financial sector [5–8].

A permissioned Blockchain is particularly appropriate for enterprise applications that require authenticated node. Each node in a permissioned Blockchain may be owned by different organizations.

2 Performance of Blockchain

The performance of Blockchain network is the speed at which the Blockchain get the requested service. The different Blockchain networks have their own performance and scalability [9, 10].

2.1 Performance Evaluation

Performance evaluation of the Blockchain network is the process of measuring and testing various network parameters such as throughput, latency time etc. The performance of the Blockchain is based on the steps involved to complete the transaction. The life cycle of the Blockchain includes the following steps:

- Request/Create a Transaction (Creation of Smart Contract in the Ethereum).
- The transaction (Smart Contract in the Ethereum) is send (broadcast) to all participation of Blockchain network.
- Every/specific Node in Blockchain network does the validation of the transaction under validation points.
- Validated transactions are stored in a block.
- This block becomes part of the Blockchain.
- Now the transaction is the part of the Blockchain and cannot be altered in any way.

The above are the common steps involved in life cycle of transaction in any Blockchain network the same is depicted in the given below diagram (Fig. 3):

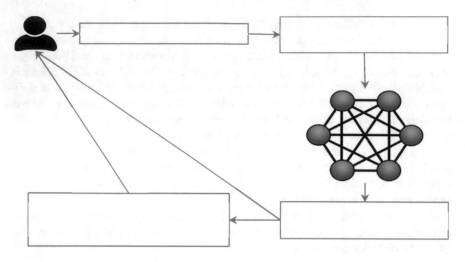

Fig. 3 Transaction life cycle

2.2 Benchmarking

Under Benchmarking, one system is compared with another under standard measurements or same system is compared with own's old performance. In order to evaluate the performance of Blockchain network, data are collected for each transaction. The parameters chosen to be measured the performance of Blockchain network are execution time, latency and throughput.

2.3 Transaction Latency Time

Latency time of Blockchain network is the time between the request is submitted to network and the response is received from network.

$$\text{Transaction Latency Time} = \text{Time of response received/confirmation time} \\ - \text{Time of the request submitted}$$

Transaction latency also referred to as "block time", is the time required to create the next block of transactions in the Blockchain. In other words, Transaction latency is amount of time a user has to wait, after submitting the request to network, to block created in the Blockchain.

2.4 Transaction Throughput

Transaction throughput is the rate at which valid transactions of Blockchain network are committed in a specified time period. This rate is defined as total number transactions committed in one second.

Transaction Throughput = Total committed transactions/Total time in seconds.

3 Blockchain Performance Analysis Tool: Hyperledger Caliper

Hyperledger Caliper is network performance analysis tools which evaluate the Blockchain network under the predefined use cases. It figure out the existing loopholes in the implemented Blockchain network. With help of Caliper tool, user can create the performance report of the tested Blockchain network. Test report contains the performance parameters such as Transactions Per Second (TPS), Transaction latency, Resource Utilization etc. The performance report can be used for enhancing the performance and optimizing the certain parameters of the network.

Hyperledger Caliper is a unified Blockchain benchmark framework, it integrates various Blockchain platforms viz. Hyperledger Burrow, Hyperledger Besu, Hyperledger Fabric, Ethereum, Hyperledger Iroha, FISCO BCOS and Hyperledger Sawtooth. Hyperledger Caliper is available in NPM Packages, Docker Imagse and source code on GitHub. Hyperledger Caliper is operated in local or distributed mode.

3.1 Hyperledger Fabric Blockchain Architecture

Blockchain network primarily contains peer nodes and it is a fundamental element of bockchain. Eeach peer nodes contains a copy of ledger and smart contracts. Hyperledger fabric Blockchain network node contains chaincode in place of smart contracts. In bockchain Network peer nodes are the host of the ledger, smart contract and application. The peer nodes are the contact point to access the Blockchain resources. A peer node can hold more than one ledger and smart contracts. A peer node can interact with more than one application. An application must contact with the peer node to access the ledger. An application may request to the node for simple query or for an update in ledger. If application request for simple query to peer node, the query execute through smart contract (chaincode) with few steps whereas ledger updates event involves a more complex interaction between applications, peers, orderers and Ledgers.

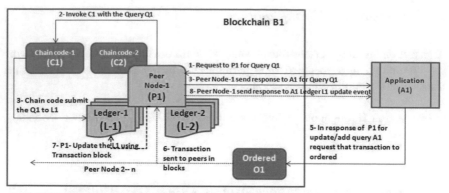

1- Peer Node-1 (P1) host two ledger (L1 and L2) and two chain code (C1 and C2)
2- Application (A1) request to peer node (P1) to access the Bock chain L-1 for the Query Q1
3- Query Q1 may be Select Query or Update/Add request
4- Ordered sent the transactions to peers in blocks
5- Peer Node-1 (P1) add the block in the Ledger L1

Fig. 4 Hyperledger fabric blockchain architecture

In Fig. 4, the connectivity of various elements of Blockchain such as peer nodes, chaincodes, ledgers, ordered nodes and application is given. Application A1 requests to peer P1for query q1. After receiving request, P1connects to chaincode C. After receiving request from A1, C1 update ledger L1 by submitting query Q1. In this case, P1 invoked to C1. C1 generates the desired proposal response for Q1and submit to L1. However, P1 can also update L1 directly if query response is already available with P1. After updation L1, P1 provides the response of Q1 to application A1. After receiving response from P1, A1 prepares transaction using received response and sends the transaction to ordered node (O1) for ordering. O1 receives various transactions across the network. After combining many transactions into one block, O1distribute these blocks to all peers of the network including P1. P1 validates the transaction and commit into L1. After updation of L1, P1 informed to A1 regarding completion of transaction. All Ordered nodes and Peers ensure that all ledger hosted at all peers are updated.

3.2 Performance Analysis on Hyperladger Fabric Blockchain Network

In this section, we study the key parameter which impact the performance of Hyperladger Fabric Blockchain and how the throughput, latency varied on the different parameters [11–15].

Hyperledger Fabric is used for many use cases such as Everledger, SecureKey, Global Trade Digitization. Hyperladeger Febric Blockchain has many phases while processing the Blockchain transactions. These phases are commit phase, validation phase, endorsement phase and ordering phase. Endorsers endorse the Blockchain transactions followed by ordering, validation and finally commit. Hyperladeger Fabric provides the flexibility in selection of various configurable parameters like size of the block, endorsement policies, state database, channels. Hyperledger Fabric uses Go/JAVA/Node.js for running the smart contracts. A Fabric network contains many entities, pluggable consensus protocol and it supports trust model for validation of transactions.

Peer Node (endorse). In Hyperladger fabrics peer node works as endorser and executes the chaincode (smart contracts implemented in Go/JAVA/Nodejs language) and maintains the ledger. The chaincode is executed in un-trusted peer nodes in the network. Chaincode has an endorsement policy which specify a set of peers on a channel that needs to execute chaincode and this policy also endorses the execution results for the valid transaction. The endorsement policy defines the peers who need to "endorse" the execution of a proposal.

Ordering Service Nodes. The ordering service is the second phase of transaction in Hyperladger Fabrics Blockchain network. The ordering service node endorsed transactions from the endorses and arrange these endorsed transaction in a sequence in which these will be verified and committed to the ledger.

Validation and Commit. The validation and commit is third phase of transaction in Hyperladger Fabrics Blockchain network. This phase involves distribution of transactions followed by validation and finally these valid transactions are committed to the ledger.

4   Blockchain Performance Impacting Parameters

4.1   Block Size

Block size is one of the parameter on which performance of Blockchain is heavily dependent. Blockchain throughput is indirectly proportional to the waiting time of a block after submission of block in Blockchain network for addition in Blockchain. Throughput is also dependent on the size of the block. A set of transactions is batched first at orderer and subsequently it is delivered to peers using gossip protocol. Each peer is responsible for processing of one block. Endorsement signature verification is done for each transaction separately but orderer signature is done block -wise. To enhance the performance of the signature verification phase, it is advisable to use light weight cryptographic signature algorithms.

4.2   Endorsement Policy

Throughput of any Blockchain network is directly dependent on the endorsement policy. There are policies for ordering, verification, validation and commitment of the transactions and also final addition of the block in the Blockchain. Endorsement Policy specifies that how many signing and transactions are needed before a transaction request can be submitted to orderers so that transaction can be passed to peers for validation purpose and final commitment of block which contains validated transaction. Validation of a transaction is the process of evaluating whether transaction fulfils the endorsement policy or not. Validation phase also checks whether identify and signature of endorsers are genuine or not. Endorsement process requires coordination and synchronization among the Blockchain members and it directly affects the performance of the Blockchain.

4.3   Channel

In Blockchain channels are used for embedding the parallelism in various phases like execution of smart contracts, oredering phase, verification phase etc. Using channels, transactions are executed in parallel and due to this, it enhances the throughput of the Blockchain system. How many channels should be used and what channels should be used, it is dependent on various aspects including type of applications, size of application etc. Channels has notable performance implications on scalability and performance of Blockchain network. If number of channels exceeds from a limit then it degrades the performance of the network. Therefore, it is advisable to select the number of channels appropriately.

4.4   Resource Allocation

As part of system chaincodes, peers run signature computation and signature verification algorithms on CPUs. Therefore, speed of verification and computation phase is directly dependent on the CPU computing power as well as number of deployed CPUs for this purpose. By varying number of CPUs, performance of Blockchain system can be enhanced. Resource utilization is also a parameter which needs to be looked with increase of computing power.

4.5 Ledger Database

Optimization of Ledger database also enhances the performance of the Blockchain system at a great instant. For maintain the current state in Fabric, it supports GoLevelDB and CouchDB as two alternate for key-value store. GoLevelDB uses client–server model. Therefore, suitable client–server protocols should be used. Database may be optimize in terms of storage, accessibility etc.

4.6 Other Performance Impacting Parameters

Blockchain Performance is also based on the following.

- Consensus protocol (i.e. RAFT, Practical Byzantine Fault Tolerant (PBFT), etc.)
- Geographic Position of Nodes (Nodes Geographically Location)
- Network model (The complexity of the network between Nodes)
- Number of network nodes
- Software component dependencies.

5 Conclusion

Blockchain Technology is one of the emerging technologies which has applications in various fields including healthcare system, financial sector, e-governance, land record system etc. Due to decentralized nature of Blockchain network, coordination among the network nodes is a challenge. Another major concern is the long delay in block addition in the Blockchain. To achieve the benefits of Blockchain, it is necessary to optimize the Blockchain network in the best possible way. In this paper, we discussed the Blockchain network and its various applications. Next, we analyzed various parameters which affect the performance of Blockchain network in terms of throughput, latency etc. Further, we suggested many key points for enhancing the performance of the Blockchain network. By considering performance key parameters and suggested key points, performance of the Blockchain network can be enhanced. In future, our plan is to analyze the performance of the Blockchain network using various tools.

References

1. Lei A et al (2017) Blockchain-based dynamic key management for heterogeous intelligent transportation systems. J IEEE Internet Things 4(6):1832–1843
2. Lin I-C, Liao T-C (2017) A survey of blockchain security issues and challenges. Int J Netw Secur 19(5):653–659

3. Aras ST, Kulkarni V (2017) Blockchain and its applications–a detailed survey. Int J Comput Appl 180(3):29–35
4. Pal O, Alam B, Thakur V, Singh S (2021) Key management for blockchain technology. ICT Express 7(1):76–80
5. Huckle S, Bhattacharya R, White M, Beloff N (2016) Internet of things, blockchain and shared economy applications. In: International Workshop on Data Mining in IoT Systems, Procedia Computer Science, vol 98
6. Park JH, Park JH (2017) Blockchain security in cloud computing: use cases, challenges, and solutions. Symmetry 9(8):164
7. Aza A, Ekblaw A, Vieira T, Lippman A (2016) MedRec: using blockchain for medical data access and permission management. In: 2nd International Conference on Open and Big Data. https://doi.org/10.1109/OBD.2016.11
8. Survey on Blockchain Technologies and Related Services (2016) Nomura Research Institute, Japan's Ministry of Economy, Trade and Industry
9. Thakkar P, Nathan S, Viswanathan B (2018) Performance benchmarking and optimizing hyperledger fabric blockchain platform. In: IEEE 26th International Symposium on Modeling, Analysis, and Simulation of Computer and Telecommunication Systems (MASCOTS), Milwaukee
10. Sukhwani H, Wang Trivedi N, Rindos AK (2018) Performance modeling of hyperledger fabric (permissioned blockchain network). In: IEEE 17th International Symposium on Network Computing and Applications
11. https://hyperledger-fabric.readthedocs.io/en/release-2.2
12. Lincoln NK (2021) Hyperledger fabric 1.4.0 performance information report. IBM Blockchain Developer Tools Ver 1.0
13. Liu M, Richard F, Teng Y, Leung VCM, Song M (2019) Performance optimization for blockchain-enabled industrial internet of things (IIoT) systems: a deep reinforcement learning approach. IEEE Trans Ind Inf 15(6):3559–3570
14. Zhang J, Zhong S, Wang J, Yu X, Alfarraj O (2021) A storage optimization scheme for blockchain transaction databases. Comput Syst Sci Eng (CSSE) 36(3):521–535
15. Fu J, Qiao S, Huang Y, Si X, Li B, Yuan C (2020) A study on the optimization of blockchain hashing algorithm based on PRCA. Secur Commun Netw. https://doi.org/10.1155/2020/8876317

Abstractive Text Summarization Using Attentive GRU Based Encoder-Decoder

Tohida Rehman, Suchandan Das, Debarshi Kumar Sanyal,
and Samiran Chattopadhyay

1 Introduction

The quantity of data around us is increasing at such a high velocity that we all need a mechanism to access correct and quick information that cuts through the noise and is brief enough to be assimilated, yet not lacking in crucial content. We need a method to obtain a correct summary from an outsized volume of data. Automatic text summarization is such a technique through which a large chunk of information can be condensed into a meaningful summary. Extractive and abstractive summarization are two types of text summarization methods. A technique for *extracting* essential sentences or paragraphs from the source text and condensing them into a shorter text is known as extractive summarization. The statistical and linguistic properties of sentences, as well as their extraction and placement in the output text, are used to determine the relevance of sentences. An abstractive summarization technique tries to present the text's primary idea in natural language without the verbatim use of terms from the text. The original text is transformed into a more comprehensible conceptual form in the abstractive summary approach, resulting in a shorter summary of the original text content.

In this paper, we present an encoder-decoder based model to summarize documents. A gated recurrent unit (GRU) has been used to boost a recurrent neural network's memory capacity as well as to make training a model easier. It also helps us to overcome the vanishing gradient problem. In attention mechanism, the context vector concatenated with the previous decoder output. That are fed along

T. Rehman (✉) · S. Das
Jadavpur University, Kolkata, India
e-mail: tohida.rehman@gmail.com

D. K. Sanyal
Indian Association for the Cultivation of Science, Kolkata, India

S. Chattopadhyay
TCG Crest, Jadavpur University, Kolkata, India

© The Author(s), under exclusive license to Springer Nature Singapore Pte Ltd. 2022   687
B. Unhelker et al. (eds.), *Applications of Artificial Intelligence and Machine Learning*,
Lecture Notes in Electrical Engineering 925,
https://doi.org/10.1007/978-981-19-4831-2_56

with previous decoder hidden state into the Decoder GRU component for each time step to generate the output [1]. We have used the CNN/Daily Mail dataset [2, 3]. We obtained higher F1 scores using ROUGH-1 and ROUGH-L compared to some other competitive baselines in the literature.

2   Related Work

Nallapati et al. [2] have proposed baseline encoder and decoder architecture where LSTM has been used. Bidirectional as well as unidirectional LSTM was used at encoder and decoder, respectively. Word level and sentence level bidirectional GRU was used. Performance of basic encoder and decoder model has been improved in Bahdanau et al. [1]. See et al. [3] offered a detailed study of numerous abstractive text summarization models for pointer-generator and RNN seq2seq models that are based on sequence-to-sequence encoder-decoder architecture. Sutskever et al. [4] proposed a multilayer LSTM based end-to-end solution to sequence learning. The input for the encoder was a fixed length of text, and the output for the decoder was the same. Lin et al. [5] proposed global encoding mechanism for abstractive text summarization. In this paper, we have designed GRU based encoder and decoder with one extra attention layer. Shi et al. [6] proposed to "improve seq2seq models, making them capable of handling different challenges, such as saliency, fluency and human readability, and generate high-quality summaries". Generally speaking, most of these techniques differ in one of the three categories: network structure, parameter inference, and decoding/generation. Luong et al. [7] examines two simple and effective classes of attention mechanism: a global approach which always attends to all source words and a local one that only looks at a subset of source words at a time. Ksenov et al. [8] proposed "the encoder and decoder of a Transformer-based neural model on the BERT language model". Recently, a model proposed as "BART: Denoising Sequence-to-Sequence Pre-training for Natural Language Generation, Translation, and Comprehension" [9] captures the simplicity of BERT (Devlin et al.) [10] and GPT (Radford et al.) [11] and other pre-training schemes. BART opens many ways to thinking about fine-tuning models for text summarization applications.

3   Methodology

In this section, we describe the methodology that we have used to design our abstractive text summarizer. Generic work flow of our model is shown in Fig. 1. Here we have used GRU [12] in the seq2seq model. The GRU has gating units to manage flow of information inside the unit.

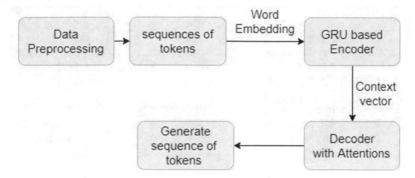

Fig. 1 Generic work flow of our model

Several crucial steps were followed such as data collection and pre-processing, tokenization, encoder and decoder model design, training the model, evaluation of the model and so on to solve the text generation problem for the prediction of semantically meaningful summary.

Let us consider the input sequence

$$X = (X_1, X_2, X_3, \ldots, X_{T_x}) \tag{1}$$

where, T_x is the input sequence size.

The output sequence is

$$Y = Y_1, Y_2, Y_3 \ldots \ldots \ldots \ldots Y_{T_y} \tag{2}$$

where T_y is the output sequence size. Here, $T_y < T_x$, which means the length of output sequence is less than the length of the input sequence.

3.1 Data Collection and Pre-processing

Dataset plays a key role in each and every deep learning process. To get satisfactory results, it is very important to use a good dataset. Various types of data sets are available for different problems. We have used the CNN/Daily Mail dataset [2, 3]. There are different columns present in the data set but we have taken news and summary description to fulfill our purpose. Due to constraints of our system, only 10,000 examples from CNN/Daily Mail dataset have been used.

Before we begin creating the model, we must first complete some basic pre-processing tasks. A decision based on messy and filthy text input could be disastrous. Therefore, we have removed all unneeded symbols, letters, and other elements from the text in the preprocessing stage. We have also removed HTML tags, parentheses, and special characters.

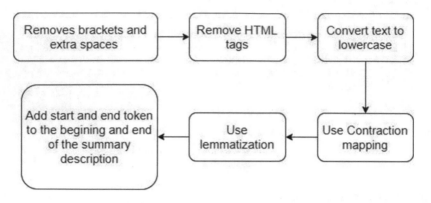

Fig. 2 Steps used in data pre-processing

[['<start> ctress deepika padukone has said that she will not be walking the red carpet at the cannes film festival deepika added right now all my energies are focused on padmavati earlier it was reported that deepika had been ap-pointed the brand ambassador of oral and would represent the brand at the film festival<end>'
'<start> deepika padukone will not be walking the red carpet at the cannes film festival. <end>'],
['<start> beverage giant pepsico ceo indra nooyi received million over crore in compensation for marking increase in her pay this was the fourth consecutive pay raise for nooyi who has been the ceo since the rise in compensation came as efforts to steer the companys port-folio away from sugary products helped earnings <end>'
'<start> pepsico ceo indra nooyi received million over crore<end>']]

Fig. 3 Cleaned news article and summary data

We changed the entirety of the content to lower case, and afterward we split it up with sent_tokenize and word_tokenize [13]. There are different constrictions in the English language, for example, *doesn't*, *aren't*, etc. We have added contraction mapping in the pre-processing phase. Then, at that point we lemmatized the words that have various types of a similar terms. At the beginning and end of the news and summary description, we have included START and END tokens, respectively. Figure 2 represents steps that we have used to clean the data set and prepare the news article and summary pair. Figure 3 shows some cleaned data.

3.2 GRU Based Encoder-Decoder with Attention

Cho et al. [14] introduced the RNN based encoder-decoder where the RNN in the encoder helps to encode a sequence of words into a fixed length vector representation and the RNN in the decoder helps to decode the incoming representation into a sequence of words. We used a bidirectional GRU encoder, and a unidirectional GRU decoder with attention mechanism [2].

Here, the seq2seq model *with Bahdanau attention mechanism* [1] builds a context vector using all the hidden states present in the encoder. It aids in focusing on the most important information in the source sequence. The decoder uses the context vectors associated with the source position and the previously created target words to predict the target word at each time stamp. Below are the steps which describe how the Bahdanau attention mechanism [1] works.

1. The encoder produces the annotation (h_i) for each word x_i, for an input sentence of length T_x words at each time step i. The encoder has a bidirectional GRU, which reads the input sentence in forward as well as in backward direction to generate the (h_i) for each time steps.

$$h_i = [\overrightarrow{h_i^T}, \overleftarrow{h_i^T}]^{\mathrm{T}} \tag{3}$$

2. At each time step, the decoder takes the annotations (h_i) and the previous hidden states s_{i-1} to calculate attention score (e_{ij}). It can be written as follows.

$$e_{ij} = att\left(s_{i-1}, h_j\right) \tag{4}$$

Bahdanau attention is defined as follows:

$$att\left(s_{i-1}, h_j\right) = V^{\mathrm{T}} \tanh(W[s_{i-1}, h_j]) \tag{5}$$

Where W, V are the trainable weights.

3. The attention weights (α_{ij}) are computed as follows:

$$\alpha_{ij} = \frac{\exp(e_{ij})}{\sum_{k=1}^{T_x} \exp(e_{ik})} \tag{6}$$

where T_x is the number of words in the input sequence.

4. Linear sum is computed using attention weight (α_{ij}) and hidden state of encoder to generate the context vector. The context vector c_i depends on a sequence of annotations $(h_1, h_2, \ldots . h_{T_x})$ to which an encoder maps the input sentence. This context vector is calculated as follows:

$$c_i = \sum_{j=1}^{T_x} \alpha_{ij} h_j \tag{7}$$

5. At time step i, the decoder produces the hidden state (s_i) depending upon s_{i-1}, which is the previous hidden state, y_{i-1}, which is the target word at time step $i - 1$, c_i, which is the context vector. Here, f is a non-linear function.

$$s_i = f(s_{i-1}, y_{i-1}, c_i) \tag{8}$$

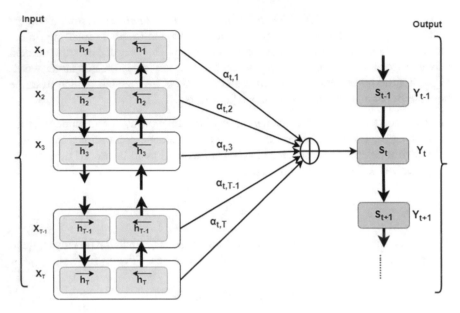

Fig. 4 How attention works in seq2seq encoder-decoder model

6. Steps 2 to 5 are repeated until the end of the sentence or the maximum length of generated tokens is reached. Each word is predicted based on the following rule:

$$P(y_i | y_{i-1}, y_{i-2} \ldots \ldots y_2, y_1, X) = g(y_{i-1}, s_i, c_i) \tag{9}$$

where g is a non-linear function.

Figure 4 shows how attention works in sequence-to-sequence or encoder-decoder model based on GRU.

4 Experiment and Result Analysis

As the computational power of our machines was low, a small dataset has been used. Here, we have used 10,000 examples from CNN/Daily Mail dataset [2, 3], Adam optimizer, a Sparse Categorical Cross-entropy loss function with batch size = 128, embedding dimension = 256, hidden units = 1024. We used 80% of the data for training purposes and 20% for testing purposes. We have trained the model for 100 epochs. Loss has been reduced to 0.0480. Table 1 shows F1 of ROUGH-1 and ROUGH-L score on the basis of the output from the model. We now provide some illustrative examples of the output of our model.

Table 1 ROUGH score (F1) on the basis of output from model

ROUGH-1	ROUGH-L
F1	F1
35.29	35.25

4.1 Sample Output

Input: *"actress deepika padukone has said that she will not be walking the red carpet at the cannes film festival deepika added right now all my energies are focused on padmavati earlier it was reported that deepika had been ap-pointed the brand ambassador of oral and would represent the brand at the film festival".*

Actual Summary: *deepika padukone will not be walking the red carpet at the cannes film festival.*

Predicted Summary: *not walking red carpet at cannes film festival says deepika.*

Input: *"beverage giant pepsico ceo indra nooyi received million over crore in compensation for marking increase in her pay this was the fourth consecutive pay raise for nooyi who has been the ceo since the rise in compensation came as efforts to steer the companys port-folio away from sugary products helped earnings".*

Actual Summary: *pepsico ceo indra nooyi received million over crore.*

Predicted Summary: *pepsico ceo indra nooyi pay rises to crore in year.*

Heatmap We now show heatmaps for predictive outputs. In the attention heatmap plot in Fig. 5, the X axis denotes the actual input words, Y axis denotes the output summary words and the cells indicate the attention weights. Main goal of using attention mechanism is to emphasize on the important information. Figure 5 shows which parts of the input sentence has the model's attention while generating the summary.

In our proposed solution, we have used daily news dataset get 35.29 ROUGH-1 F1 score and 35.25 ROUGH-L F1 score, which are slightly better than those produced by some competitive models in the literature. It generates more semantically meaningful single sentence summary. Table 2 shows the comparison of ROUGH 1 and ROUGH L scores with some existing models. Here, k refers to the size of the beam for generation.

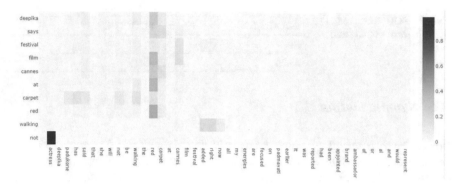

Fig. 5 Attention heatmap

Table 2 Comparison of the ROUGH score (F1) with some existing models

Model	ROUGH-1	ROUGH-L
	F1	F1
Words-lvt5k-1sent [2]	28.61	25.423
Words-lvt2k-temp-att [2]	35.46	32.65
ABS + (Rush et al.) [15]	28.18	23.81
RAS-Elman (k = 10) (Chopra et al.) [16]	33.78	31.15
Our model	**35.29**	**35.25**

5   Conclusion and Future Work

A GRU-based encoder and decoder model with Bahdanau attention mechanism has been used to design an automatic text summarizer. The proposed method provides better result than several other approaches in the literature. A meaningful summary with single sentence has been generated, which can be used for news headline generation. However, we also observed that our model is not always producing the best result. In future, we will use BERT-based pre-trained models to enhance the performance and to generate more meaningful summary. We will try to create summary of Covid-19 related scientific articles which can help the medical community by providing a clean high-quality knowledge base of the pandemic.

References

1. Bahdanau D, Cho K, Bengio Y (2014) Neural machine translation by jointly learning to align and translate, arXiv preprint arXiv:1409.0473
2. Nallapati R, Zhou B, Gulcehre C, Xiang B et al. (2016) Abstractive text summarization using sequence-to-sequence rnns and beyond, arXiv preprint arXiv:1602.06023

3. See A, Liu PJ, Manning CD (2017) Get to the point: summarization with pointer-generator networks, arXiv preprint arXiv:1704.04368
4. Sutskever I, Vinyals O, Le QV (2014) Sequence to sequence learning with neural networks. In: Advances in neural information processing systems, pp 3104–3112
5. Lin J, Sun X, Ma S, Su Q (2018) Global encoding for abstractive summarization, arXiv preprint arXiv:1805.03989
6. Shi T, Keneshloo Y, Ramakrishnan N, Reddy CK (2021) Neural abstractive text summarization with sequence-to-sequence models. ACM Trans Data Sci 2(1):1–37
7. Luong M-T, Pham H, Manning CD (2015) Effective approaches to attention based neural machine translation, arXiv preprint arXiv:1508.04025
8. Aksenov D, Moreno-Schneider J, Bourgonje P, Schwarzenberg R, Hennig L, Rehm G (2020) Abstractive text summarization based on language model conditioning and locality modeling, arXiv preprint arXiv:2003.13027
9. Lewis M, Liu Y, Goyal N, Ghazvininejad M, Mohamed A, Levy O, Stoyanov V, Zettlemoyer L (2019) Bart: denoising sequence-to-sequence pre-training for natural language generation, translation, and comprehension, arXiv preprint arXiv:1910.13461
10. Devlin J, Chang M-W, Lee K, Toutanova K (2018) BERT: pre-training of deep bidirectional transformers for language understanding, arXiv preprint arXiv:1810.04805
11. Radford A, Narasimhan K, Salimans T, Sutskever I (2018) Improving language understanding by generative pre-training. https://s3-us-west-2.amazonaws.com/openai-assets/research-cov ers/language-unsupervised/language_understanding_paper.pdf
12. Cho K, van Merrienboer B, Bahdanau D, Bengio Y (2014) On the properties of neural machine translation: encoder-decoder approaches, arXiv preprint arXiv:1409.1259
13. Masum AKM, Abujar S, Talukder MAI, Rabby ASA, Hossain SA (2019) Abstractive method of text summarization with sequence to sequence RNNs. In: 2019 10th international conference on computing, communication and networking technologies (ICCCNT). IEEE, pp 1–5
14. Cho K, Van Merriënboer B, Gulcehre C, Bahdanau D, Bougares F, Schwenk H, Bengio Y (2014) Learning phrase representations using RNN encoder-decoder for statistical machine translation, arXiv preprint arXiv:1406.1078
15. Rush AM, Chopra S, Weston J (2015) A neural attention model for abstractive sentence summarization, arXiv preprint arXiv:1509.00685
16. Chopra S, Auli M, Rush AM (2016) Abstractive sentence summarization with attentive recurrent neural networks. In: Proceedings of the 2016 conference of the north American chapter of the association for computational linguistics: human language technologies, pp 93–98

Object Detection and Foreground Extraction in Thermal Images

P. Srihari and Harikiran Jonnadula

1 Introduction

Thermal imaging is slightly different from visible imaging as the visible image is formed by the reflection of light from the object whereas thermal image is formed when the infrared rays that are released from the thermal camera focused on the particular area, Thermal image of that particular area is formed as colour maps of image which are different from thermal camera to thermal camera with varying intensities of different colors. The working principle of thermal camera can be shown as Fig. 1. When infrared radiation is incident on object, the reflected infrared radiation [7, 13] is captured through the thermal sensors in thermal camera. The difference in the temperature values is formed as images with colour maps in thermal cameras. Thermal radiation can penetrate through smoke, dust and mist. Visible image cameras suffer from illumination problems but thermal cameras have no illumination problems like if there is too much light like head lights or there is no light in such cases the quality of the image formed is not so good. Thermal cameras can see what humans cannot see.

Applications of thermal images [8] include they are used in building inspections to find cracks in building, Heat losses in buildings, detecting pedestrians, medical, industry, fire monitoring, security surveillance, search, and rescue operations etc. Figure 2a–c presents the advantages of thermal camera over night vision camera. Challenges in object detection include object classification and localization, speed at which object is being detected, object detection under illumination conditions, detection at different scales of image size. The historical approaches used in object detection in thermal images [10, 11, 13] is thresholding. In thermal Thresholding based detection cannot work in situations like Objects or parts of objects may have approximately the same temperature as the background, reflections (Ex. Glass reflects

P. Srihari (✉) · H. Jonnadula
School of Computer Science and Engineering, VIT-AP University, Amravati, India
e-mail: srihari.19phd7018@vitap.ac.in

© The Author(s), under exclusive license to Springer Nature Singapore Pte Ltd. 2022   697
B. Unhelker et al. (eds.), *Applications of Artificial Intelligence and Machine Learning*,
Lecture Notes in Electrical Engineering 925,
https://doi.org/10.1007/978-981-19-4831-2_57

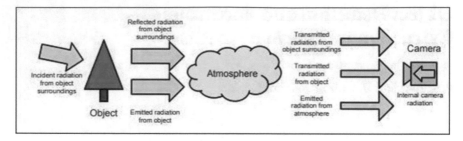

Fig. 1 Thermal camera taking Transmitted radiation as input

more thermal radiation) and versatile background (if the temperature of other objects in the image frame is same). So here instead of using Threshold based object detection in thermal images HOG+SVM, RCNN, Mask-RCNN based detections are considered on CSIR-CSIO moving object Thermal dataset by dividing videos of the dataset into frames and analyzing those frames.

Dataset

Here the benchmark dataset we used is CSIR-CSIO moving object Thermal dataset which consists of videos of objects like persons walking, birds, cars and the video data collected using Micro bolometer thermal detector sensor which is fixed at a height of 4 ft. with Uniform Intensity moving targets like Pedestrians, dog, bird, and non-uniform intensity targets like Ambassador car, innova car, motorcycle and auto-rickshaw etc.

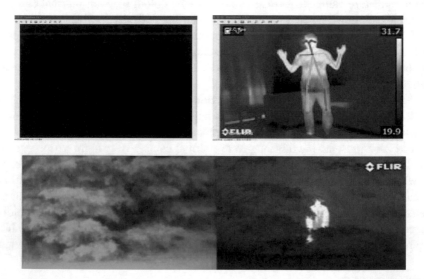

Fig. 2 **a** denotes image of an environment captured by a typical RGB camera and **b** denotes image of an environment captured by a thermal camera and **c** denotes image taken using night vision camera at the left and thermal camera at right

Literature Survey

In [1] proposed by Navneet Dalal SVM based object detection using Histogram of Oriented Gradient (HOG) features was proposed for the images any significant degree of smoothing before calculating gradients damages the HOG results emphasizes that much of the available image information is from abrupt edges at fine scales, and that blurring this in the hope of reducing the sensitivity to spatial position is a mistake and still optimizations are needed to speed up object detection. Boykov, Y., And Jolly in [3] proposed graphcut algorithm for segmentation which takes labelled object in image as input and perform segmentation. Justin F. Talbot and Xiaoqian Xu in [4] explains the working of grabcut algorithm which internally uses graphcut with rough initial Trimap. Although GrabCut is much faster than previous methods, the delay due to the Graph Cut step is still noticeable. The minimization algorithm presented in the GrabCut paper only guarantees convergence to a local minimum. The border matting in [16] discussed in the paper needs to be implemented and compared to other matting techniques, such as Bayesian or Poisson Matting. Tsung-Yi Lin, Piotr Dollár in [5] proposed Feature Pyramid Networks which is backbone for extracting region of proposals, detecting object, segmentation in Mask-RCNN. Chinthakindi Balaram Murthy and Mohammad Farukh Hashmi In [6] explained the general approach for object detection and various approaches were discussed in this paper. M. Narendran and Manonmani Lakshmanan has given various approaches of object detection in Thermal images. Nicola Altini, Giacomo Donato Cascarano in [9] proposed segmentation using Mask-RCNN, Faster-RCNN and various parameter metrics for object detection are proposed. A. Chayeb and N. Ouadah, in [10] proposed multi-object detection system based on detectors in cascade, using histograms of oriented gradients (HOG). [11, 12] focuses on development of a learning based approach to the problem that makes use of a sparse, part-based representation. [12] focuses on Object Class Recognition by Unsupervised Scale-Invariant Learning. [13–15, 18] focuses on neural network based feature extractor approach used in object detection. [17] focuses on FPN based object detection which is used mainly for feature extraction in Mask-RCNN. [17] focuses on implementation of graphcut.

2 Proposed Methodology

Here we considered the video of pedestrian's by dividing video into frames and tried to detect the person in the Thermal frame in two ways. The first way is Extracting HOG features of Person in a Thermal frame [7, 13] and foreground extraction using Grabcut algorithm, and the second is Using Mask-RCNN [12]. As Mask-RCNN is used for the instance segmentation no further algorithm is proposed for foreground and background extraction and the results were analysed. Figure 3 shows the flow process of proposed methodology in general.

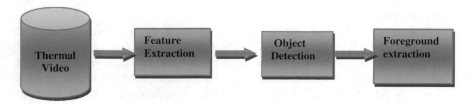

Fig. 3 Procedure for the proposed methodology

Histogram of Oriented Gradients (HOG)

HOG is a texture and shape feature that can be extracted from images. HOG feature is more robust to noise and scale invariant [13]. This feature is widely used in standstill images for object detection. HOG feature extraction can be done using the following steps

1. Pre-processing a given image
2. Compute the gradient images
3. Evaluate the histogram of gradients in cells
4. Block Normalization
5. Computation of HOG feature vector.

Figure 4 show the flow chart for object detection using HOG and SVM. The pre-processing step involves two tasks one is normalizing the pixel values and the second task is reshaping the image in the aspect ratio of **1:2**. The task1 in pre-processing is completely optional but Dalal [1] states that normalizing the pixel values improves the accuracy of object detection. Gamma or power law Normalization or Square Root normalization on each pixel of image can be done. In Gamma or power law Normalization, logarithmic value of each pixel value of image is taken while in Square Root normalization, the logarithmic value of the pixel gets replaced by the square root value of pixel here in the dataset. We do not change the parameters of height and width and during reshaping of an image and it is maintained as mentioned by Dalal [1]. Gradient images can be calculated by applying a convolution operation to obtain the gradient images x-gradient $G_x = I * D_x$ and y-gradient $G_y = I * D_y$ where I is the input image, D_x, D_y is our filters (Sobel) in the x-direction and y-direction. The Sobel filter can compute gradient with smoothing. Sobel operator is used to detect edges whenever there is sharp change in intensity of the region.

$$D_x = [+10 - 1 + 20 - 2 + 10 - 1] \text{ and } D_y = [+1 + 2 + 1000 - 1 - 2 - 1]$$

We can compute the final gradient magnitude of the image using $G = \sqrt{(G_x^2 + G_y^2)}$ and the orientation of the gradient for each pixel in the input image can then be computed using $\theta = tan^{-1}(Gx/Gy)$. Calculating Gradient Images is useful in eliminating the region having same intensity for example constant coloured background which is not useful for feature extraction. After computing the Gradient image using the Sobel filter the image is divided into 8×8 cells (patch size) and a

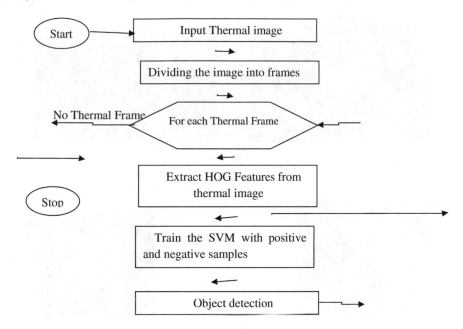

Fig. 4 Flow chart for object detection based on HOG+SVM

histogram of gradients is calculated for each 8×8 cells. One reason for maintaining patch size as 8×8 cells is computing HOG features over a patch is all the necessary features of man like face, forehead, eyes can be extracted using this patch size [2]. HOG features computed from patches of an image are more robust to noise compared to computing HOG feature on entire image at once. As Histogram of the patch is calculated, the final Hog feature is concatenation of all histograms. The histogram of patch is a vector of 9 buckets (numbers) corresponding to angles from corresponding to angles 0, 20, 40 … 160. These are called **unsigned gradients**. It has been shown that unsigned gradients worked better than signed gradients for pedestrian detection. We must vote the corresponding magnitude in one of the nine bins. One drawback of gradients of an image is sensitive to overall lighting illuminations. To make robust to lighting illuminations, we normalize the histogram. Normalization of the histogram makes the feature scale invariant. A 16×16 block is formed by 4 patches where each patch is a 8×8 cell has 4 histograms which can be concatenated to form a 36×1 element vector and it can be normalized. The final feature vector is obtained by the concatenation of all 36×1 vectors. After detecting features using HOG (Histogram of Oriented Gradients) a linear SVM **(Support Vector Machine)** is used which is trained on positive and negative samples? Visualization of HOG features were shown in Fig. 5 taking FLIR sample image as Example. The no of trained positive and negative samples also effects the accuracy of detection.

Figure 6a shows the sample input image and Fig. 6b shows the output of HOG+SVM. Here we used the SVM that is already trained on pedestrian dataset

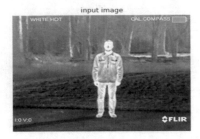
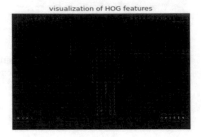

Fig. 5 Illustrating the HOG features

Fig. 6 **a** Input thermal image **b** Output of HOG+SVM

to detect a person for the thermal dataset. After the bounding box of detection is drawn, the resultant image is passed as input to grabcut algorithm.

3 GrabCut Algorithm

Grabcut algorithm can be used for Foreground extraction where Foreground and Background separation in an image is difficult. Grabcut can be extended to a N-Dimensional images Grabcut takes RGB image as input and performs hard segmentation by constructing Foreground Mask. In this algorithm, the region that is present outside the Bounding box can be treated as Background and it is turned black. The region inside the Bounding box is unknown that is we cannot estimate the region blindly as Foreground or Background as it consists of foreground region as well as Background region also initially, and we assume the region inside the Bounding box as Foreground Matte and only Foreground region is extracted during the iterations. For this purpose Gaussian Mixture Model (GMM) is used to decide the pixels present inside the bounding box as Foreground or Background. The GMM learns how to create labels for the pixels inside the bounding box and where every pixel inside the bounding box is clustered in terms of colour statistics.

Grabcut algorithm segments image by considering the texture and shape-based feature (ex. HOG). A graph is generated from this pixel distribution where the pixels inside the bounding box are assumed as vertices or nodes of the graph and two additional nodes are added that is the Source node is Considered as Foreground and Sink node as Background as shown in Fig. 7. Grabcut internally uses Graphcut algorithm. The weights of edges that have edges for source node and sink node from a pixel is determined by the probability of a pixel belonging to the foreground or the background. From ROI (Region of Interest) we passed as a bounding box a graph is constructed as shown in the above Fig. 7. Now we have to find weight of the edges of adjacent pixels. Let N(m, n) [4] denote the weight of the edge where m and n are adjacent pixels is given by

$$N(m, n) = \frac{50}{dist(m, n)} * e^{-\beta ||z_m - z_n||^2} \tag{1}$$

z_m = colour of pixel m. And $dist(m, n) = distance\ between\ the\ pixels$

$D_{Back}(m)$, $D_{Fore}(m)$ gives the likelihoods of the pixel ϵ foreground and background GMMs respectively. These likelihoods are computed as follows for pixel m:

Pixel type	Background	Foreground
m ∈ Foreground	0	K
m ∈ Background	K	0
m ∈ Unknown	$D_{Back}(m)$	$D_{Fore}(m)$

$$D(m) = -log \sum_{i=1}^{k} - \pi(\alpha_m, i) \frac{1}{det \sum(\alpha_m, i)} \times exp\left(\frac{1}{2}[z_m - \mu(\alpha_m, i)]^T \sum inverse(\alpha_m, i) * [z_m - \mu(\alpha_m, i)]\right) \tag{2}$$

$N(m, n)$ does not vary in the algorithm but D_{Fore} and D_{Back} varies from iteration to iteration. We repeat the iterations until the classification converges. The algorithm bipartites the graph into two, separating the source node and the sink node. With the help of a cost function which is minimum sum of the weights of the edges removed during biparting the graph. To partition a given graph, we implement Max Flow Min cut theorem so that the graph is divided as shown in Fig. 7. Each pixel in Grabcut

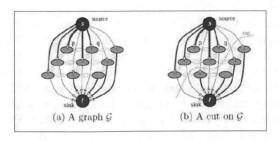

(a) A graph $\mathcal{G}$ (b) A cut on $\mathcal{G}$

Fig. 7 Illustrating GRAPHCUT

 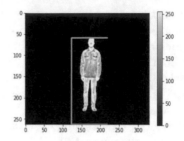

Fig. 8 **a** Input to GRABCUT **b** Output to GRABCUT

requires colour (Z), Trimap (whether the pixel belongs to Foreground or Background or Unknown), Matte information (α). Initially the region inside the bounding box is considered as Matte Foreground, here hard segmentation is performed initially with reference to matte. A Component Index which is a number with in the set$\{1, k\}$ and k is number of Gaussian components in Gaussian Mixture Model (GMM (k)) for each of the foreground and background GMM's. For every Gaussian component, we store the following, mean (μ) of the RGB triple as the input to the grabcut is the RGB image, Σ^{-1} which denotes the inverse of covariance matrix of size 3×3, $\det\Sigma$ which indicates the determinant of the covariance matrix, weight of the component denoted by π. Boykov, Y [3] suggests using k = 5 to 8 in his paper for foreground and background GMMs. Figure 8a shows the Input to GRABCUT and Fig. 8b shows the Output of GRABCUT algorithm which is performed on sample FLIR Thermal image.

4   Mask RCNN

Mask-RCNN is used for multiple object detection and segmentation (instance segmentation) purposes. Mask RCNN is an improvised version of Fast-RCNN which is improvised over RCNN. Fast RCNN works faster than RCNN as we know that though RCNN (Region based CNN) and CNN (Convolutional nueral networks) can be used in object detection, CNN is limited to single object detection where as RCNN can detect multiple objects in a single image in addition to classification. An assumption in R-CNN is that a selective region given is dominated by a single object. Selective search algorithm [18] for object detection is used in R-CNN to generate region of proposals which generate initial sub segmentation of input image which uses graph based segmentation [3, 4, 9] during the iterations the most similar regions gets combined into larger regions until no regions can be combined. Identification of Regions in an image can be done by varying colors, scales, Textures, enclosures and those regions are recursively combined form a large region. R-CNN selects very few windows, approximately uses 2000 regions for an image and each region is warped for removing distortion in shape and is fed to CNN in-order to extract features. The extracted features are flattened when there are supplied as input to fully connected network which is a part of CNN. Support Vector Machines (SVM) based classification is used for the detection of object in the region. Here regression is performed on region of proposals followed by classification. Figure 9 shows the architecture of RCNN. For Feature extraction Fast R-CNN uses single Deep ConvNet [13, 14] and softmax in the place of SVM used in R-CNN for classification of objects. Fast R-CNN [9] uses multi task loss [11, 14, 19] which can be defined as sum of all loss, to fulfill end to end training in Deep ConvNets which improves the algorithm's detection accuracy. In Mask-RCNN FPN (Feature Pyramid Networks) is the backbone which is mainly used for detecting objects in Mask RCNN [7–9]. FPN [17] consists of Features of pyramids of same image at different scales for object detection As FPN consists of multiple Feature of same image that are differently scaled Features extracted at lower layer of the pyramid (bottom to top pathway) can be considered as low-resolution features that are not good enough for object detection. Low resolution strong semantic features get combined with High Resolution Weak features using top-down pathway and lateral connections. The size of the feature map generated by the Top-down Pathway is similar to that of Bottom-up Pathway. The bottom-up pathway is used to extract features from raw images and it can be ConvNet or RESNET or VGG. Lateral connections are Convolution and adding operations of both top-down pathway and bottom-up pathway. In Mask RCNN [9, 16, 17] the first stage is object detection which is followed by Region Proposal identification. While scanning feature map, we need a procedure of binding such feature in raw image for such purpose anchor can be used. Anchors are a set of boxes which are used for multiple detection of objects where each object is having different scale with respect to images. (RPN) Regional Proposal Network using anchors find out at which region the object is and what size the bounding box should be Two types of outputs are being generated from RPN, one is Anchor class and the other output is bounding box refinement. If Anchor class is either Foreground class or Background class. If anchor class is foreground class, indicates the presence of object in that bounding box. Bounding box refinement helps in estimating how well the object gets fit to the anchor. Using RPN only anchors with highest foreground score gets selected and passed as input to ROI (region of Interest). ROI is also having two types of outputs as like RPN one is class another is Bounding box refinement. Class gives the information about to which class does the ROI belongs to for example the class may be objects like Pedestrians, car, etc. Second output is bounding box refinement tells how well the object gets fitted to class. ROI classifiers cannot handle arbitrary sized inputs well. For this we use ROI pooling. Convolving (convolution), down sampling and up sampling operations on an image would keep features staying, the same relative locations does not get disturbed as the objects in original image. The feature maps {C2, C3, C4, C5} corresponds to up sampling and {P2, P3, P4, P5} corresponds to Down sampling. The feature maps (C_i, P_i), where i varies from 2 to 5 are of same spatial size Finally, a 3×3 convolution is appended on each merged map to generate the final feature map, which is to reduce the aliasing effect of up sampling. Figure 9 shows the structure of Mask-RCNN. Figure 10 shows the Input and Output of Mask-RCNN.

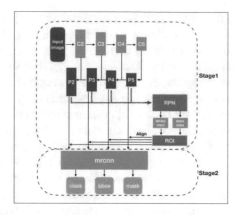

Fig. 9 Structure of mask-RCNN

Input Image

Output Image

Fig. 10 Output of mask-RCNN

5 Experimental Results

Here, we considered Object detectors as single stage object detector and two stage object Detector on CSIR-CSIO moving object Thermal dataset and the results has been analysed by considering Mean average precision. True Positive case occurs when the proposed algorithm correctly identifies the required object by drawing the bounding box around object. similarly True negative case occurs when the proposed algorithm correctly does not identify the unrequired object, For Example we must identify only Humans but given a bird image, algorithm should not draw bounding box to bird. False positive case occurs when the algorithm draws bounding box to unrequired object (For Example we have to only identify Humans but given a bird image, algorithm draw bounding box to bird). False negative case occurs if the proposed algorithm does not identify the required object by drawing the bounding box around object and the formula for precision and recall is given in Eqs. 3 and 4. Average precision is the area under the Precision recall curve. Using the above parameters, the performance of single stage detectors and two stage detectors are

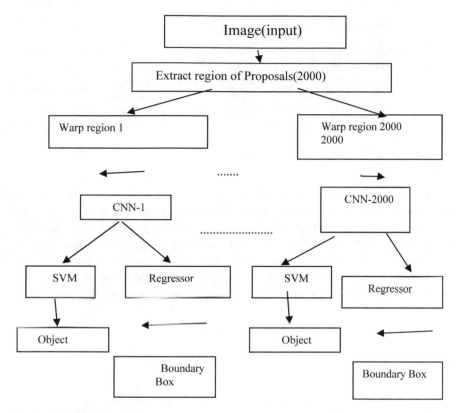

Fig. 11 RCNN architecture

analyzed in the form of a table.

$$precision = \frac{TP}{TP + TN} \tag{3}$$

$$recall = \frac{TP}{TP + FP} \tag{4}$$

where TP = True Positive, FP = False Positive, TN = True Negative, FN = False Negative.

Table 1 Mean average precision of different algorithms on the CSIR-CSIO moving object thermal dataset

S. No	Architecture (object detection)	Mean average precision (map) (%)
1	HOG+SVM	63.10
2	RCNN	66
3	FRCNN	70.00
4	Mask RCNN	78.20

6 Conclusion

In this paper, an alternative to threshold-based object detection in Thermal images was proposed and the accuracy of single stage detectors and two stage detectors are examined. Though two stage object detectors are fast in object detection relatively compared to single stage object detectors which are still good for object detection and easy to implement where time taken for detecting objects is not an issue. Two stage object detectors which uses CNN based approach performed well in terms of mean average precision. Mask-RCNN performs better among the two stage object detectors.

References

1. Dalal N, Triggs B (2005) Histograms of oriented gradients for human detection. In: 2005 IEEE computer society conference on computer vision and pattern recognition (CVPR 2005), vol 1, pp 886–893. https://doi.org/10.1109/CVPR.2005.177
2. Yussiff AL, Yong SP, Baharudin BB (2014) Detecting people using histogram of oriented gradients: a step towards abnormal human activity detection. In: Jeong H, Obaidat SM, Yen N, Park J (eds) Advances in computer science and its applications. Lecture notes in electrical engineering, vol 279. Springer, Heidelberg. https://doi.org/10.1007/978-3
3. Boykov YY, Jolly M (2001) Interactive graph cuts for optimal boundary & region segmentation of objects in N-D images. In: Proceedings eighth IEEE international conference on computer vision, ICCV 2001, vol 1, pp 105–112. https://doi.org/10.1109/ICCV.2001.937505
4. Talbot J, Xu X (2006) Implementing grabcut. Brigham Young University
5. Lin T, Dollár P, Girshick R, He K, Hariharan B, Belongie S (2017) Feature pyramid networks for object detection. In: 2017 IEEE conference on computer vision and pattern recognition (CVPR), pp 936–944. https://doi.org/10.1109/CVPR.2017.106
6. Murthy CB, Hashmi MF, Bokde ND, Geem ZW (2020) Investigations of object detection in images/videos using various deep learning techniques and embedded platforms – a comprehensive review. Appl Sci 10:3280. https://doi.org/10.3390/app10093280
7. Sharma P (2018) A review on object detection in thermal imaging and analysing object and target parameters, November 2017
8. https://liu.diva-portal.org/smash/get/diva2:918038/FULLTEXT01.pdf
9. Altini N, Cascarano GD, Brunetti A et al (2020) A deep learning instance segmentation approach for global glomerulosclerosis assessment in donor kidney biopsies. Electronics 9:1768. https://doi.org/10.3390/electronics9111768
10. Chayeb A, Ouadah N, Tobal Z, Lakrouf M, Azouaoui O (2014) HOG based multi object detection for urban navigation. In: 17th international IEEE conference on intelligent transportation systems (ITSC), pp 2962–2967. https://doi.org/10.1109/ITSC.2014.6958165

11. Agarwal S, Awan A, Roth D (2004) Learning to detect objects in images via a sparse, part-based representation. IEEE Trans Pattern Anal Mach Intell 26(11):1475–1490. https://doi.org/10.1109/TPAMI.2004.108

12. Fergus R, Perona P, Zisserman A (2003) Object class recognition by unsupervised scale-invariant learning. In: 2003 IEEE computer society conference on computer vision and pattern recognition. Proceedings, p II. https://doi.org/10.1109/CVPR.2003.1211479

13. Akula A, Sardana HK (2019) Deep CNN-based feature extractor for target recognition in thermal images. In: TENCON 2019–2019 IEEE region 10 conference (TENCON), pp 2370–2375. https://doi.org/10.1109/TENCON.2019.8929697

14. Chen X, Zhang Q, Han J, Han X, Liu Y, Fang Y (2019) Object detection of optical remote sensing image based on improved faster RCNN. In: 2019 IEEE 5th international conference on computer and communications (ICCC), pp 1787–1791. https://doi.org/10.1109/ICCC47050.2019.9064409

15. Resita E, Sigit R, Harsono T, Rahmawati R (2020) Color RGB and structure GLCM method to feature extraction system in endoscope image for the diagnosis support of otitis media disease. In: 2020 international electronics symposium (IES), pp 599–605. https://doi.org/10.1109/IES50839.2020.9231532

16. Abhilash R (2007) Natural image and video matting. In: International conference on computational intelligence and multimedia applications (ICCIMA 2007), pp 469, 475. https://doi.org/10.1109/ICCIMA.2007.11

17. Lin T, Dollár P, Girshick R, He K, Hariharan B, Belongie S (2017) Feature pyramid networks for object detection. In: 2017 IEEE conference on computer vision and pattern recognition (CVPR), pp 936–944. https://doi.org/10.1109/CVPR.2017.106

18. Felzenszwalb PF, Huttenlocher DP (2004) Efficient graph-based image segmentation. Int J Comput Vis 59:167–181. https://doi.org/10.1023/B:VISI.0000022288.19776.77

19. Kisilev P, Sason E, Barkan E, Hashoul S (2016) Medical image description using multi-task loss CNN. In: Carneiro G et al (eds) Deep learning and data labeling for medical applications. DLMIA 2016, LABELS 2016. Lecture notes in computer science, vol 10008. Springer, Cham. https://doi.org/10.1007/978-3-319-46976-8_13

STABA: Secure Trust Based Approach for Black-Hole Attack Detection

Virendra Dani⊚, Priyanka Kokate, and Jayesh Umre

1 Introduction

Ad hoc [1] networks require the structure seen in managed distant networks. In this network, nodes should act as a server, client, and router in order for the system to function correctly. In general, it is understood that all nodes follow the relevant and protocol requirements. This hypothesis may be incorrect in any scenario due to asset restraints (e.g., low battery control) or implacable behaviour. Accepting flawless conduct may lead to unintended consequences, such as a lack of network competence, excessive resource consumption, and vulnerability to attacks. As a result, a method is needed that enables node to infer the constancy of other nodes [2].

Giving a trust metric to every router is valuable when hubs get out of hand, as well as when hubs trade data. As per the world view of autonomic systems [3], a hub might to be equipped for self-arranging, self-overseeing, and self-learning by method for gathering nearby data and trading data with its neighbors. Along these lines, it is essential to discuss just with dependable neighbors, since speaking with acting up hubs can trade off the self-rule of specially appointed systems [4].

One of the key problems in ad-hoc in the presence of hostile nodes is to build a solid security solution that can defend MANET against various routing attacks. Black-hole attacks are one of the most common and severe attacks in wireless ad hoc networks, and most suggested methods to protect against them employed positioning devices, synchronized clocks, or directional antennas etc. [5] In this paper we are going to proposed trust-based approach to protect the network and increase performance.

V. Dani (✉) · J. Umre
Computer Science and Engineering Department, Shivajirao Kadam Institute of Technology and Management, Indore, India
e-mail: virendradani.cs@gmail.com

P. Kokate
Computer Science and Engineering Department, Shri Govindram Seksaria Institute of Technology and Science, Indore, India

2   Background Study

A study's background is a crucial component of our article. It establishes the study's context and goals. As a result, background research is required to aid in the preparation of various parts of the attack and trust management.

2.1   Black-Hole Attack

As depict in Fig. 1, a Black-hole attack in MANET is a kind of denial-of-service attack wherein a network node eliminates packets rather than transmitting them. Because it happens when a node is hacked for several reasons, the packet dropping attack [6, 7] is especially difficult to recognize and prevent. Black-hole attacks in MANETs may be classified into many types depending on the method used by the malevolent node to initiate the attack.

- It has ability to drop packets originating from or intended for specific nodes that it dislikes.
- The gray-hole attack is utilised, which is a variant of the black hole attack. The rogue node preserves a part of the packets while the remainder is routinely transmitted in this attack.

The compromised node will transmit a message claiming to have the quickest route to a target to facilitate behavior a Black-hole attack. As a result, all packet transfers will be routed through the hacked node, which will be able to drop packets if necessary.

The attack can be detected using conventional networking tools if the attacker node tries to discard all packets [8, 9].

3   Literature Review

In this section, we have listed some of recent development in trust management for MANETs:

Fig. 1 Packet drop attack

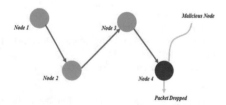

Ningrinla Marchang et al. [10] presented a lightweight trust-based routing mechanism. It's lightweight in the sense that the intrusion detection system (IDS) that determines how much trust one node has in another only consumes a tiny number of computational resources. Additionally, it solely uses local data, ensuring scalability. This technique used AODV protocol.

George Theodorakopoulos et al. [11] focused on the Ad Hoc Networks trust proof evaluation procedure. Due to the complexity of Ad Hoc networks, trust confirmation might be confusing and unfinished. Furthermore, there can be no assumption of pre-existing infrastructure.

To enhance the performance of AODV, G. Arulkumaran et al. [12] presented a fuzzy logic method based on certificate authority, energy auditing, packet veracity check, and trust node to identify black hole attacks. This method used Fuzzy schema which is a type of mathematical logic for wide range of facts.

Mohammed Baqer M. Kamel et al. [13] proposed a stable and trust-based method based on ad hoc on demand distance vectors to increase the defense of the AODV routing protocol (STAODV). Based on their past data, the technique separates hostile nodes attempting to attack the network.

Vishvas Haridas Kshirsagar et al. [14] suggested a routing algorithm that is compatible with current protocols. For a safe communication route, the idea of a trusted list is used. The number of times each node participated in the discussion is shown by the trusted list and trust values.

To ensure secure routing, S. Naveena et al. [15] suggested a trust-based routing scheme. Data retrieval (DR), which detects and retains each node's data transfer mechanism in a routing context, and route design, which predicts a safe way for delivering a data packet to the goal node, are two steps of this routing method.

4 Proposed System

This section explains how to use the proposed method for calculating trust in secure communication measured.

4.1   Concept

The suggested trust computation black-hole attack detection method covers the three
major stages for network security.

- **Assign Parameter:** For node trust computation, we assign several QoS parameters.
- **Trust Computation:** In this phase, we compute average threshold value for estimating trust of the entire selected node.
- **Proposed Algorithm:** Finally, STABA Algorithm is prepared based on mutual trust calculation of node and now we able to detect black-hole attack and try to secure network by malicious nodes.

4.1.1 Methodology
Description: Here, we are describing essential steps the proposed routing algorithm
and flow of work is depicted in Fig. 2:

Firstly, we configured the network in an idealized mode, which is the network
system initiation. Different nodes are created for communication in this condition,
so those nodes have been created using Route Reply and Route Request to route
discovery from source to destination. In this point, the routing protocol starts a
communication for the route discovery process, so initially the source router sends
an RREQ message and waits for response. RREP packets respond as the source
router obtains the path.

To establish routes between source and destination, we follow the process to setup communication before the procedure applied to black-hole detection. For this,

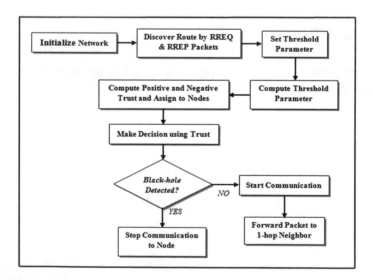

Fig. 2 STABA working flow

set the threshold parameter an important role to find out malicious activity in the network.

The For enabling trust computation of the entire network that detects the black-hole attack firstly, we need we need to assume a limit that makes the decision, consider some percentages amount of Packet Drop Ratio, Buffer Length and Energy Utilization with respect to weight. In this consideration take 25, 50 and 25% of W1, W2, and W3 consequently. Now, assign trust to each node by calculating total amount of percentages of threshold parameters. This criterion denotes the trust factor of nodes. Finally, find average trust value of all nodes by threshold computation and apply different checks that make decision for attacker availability. If there is Black-hole availability, is shows that last hop is malicious node and stop communication to the node. Otherwise, no black-hole is detected then forwards the packets to the 1-hop neighbor and start the communication in entire network.

4.1.2 Parameter Selection

The In order to achieve the required goals the following three main parameters is selected.

- **Packet Drop Ratio (α):** The total number of packets lost by each network node. In this case, it's written as α.
- **Buffer Length (β):** Any node will fill up the queue at any time during communication. This is denoted by the symbol β.
- **Energy Utilization (γ):** The quantity of energy consumed during a period of time. It is given here by the latter γ.

4.1.3 Trust Computation

The network node properties can be used to assess the network nodes' trustworthiness. Therefore, in first a network with 20 nodes is created and configured with the help of AODV routing protocol. After configuring the network, the selected parameters are computed for all the participating nodes. Therefore, the mean values of nodes are computed as:

$$\alpha_n = \frac{1}{N} \sum_{i=1}^{N} \alpha_i \tag{1}$$

$$\beta_n = \frac{1}{N} \sum_{i=1}^{N} \beta_i \tag{2}$$

$$\gamma_n = \frac{1}{N} \sum_{i=1}^{N} \gamma_i \tag{3}$$

These computed values demonstrate the mean values which are required by a normal node to communicate in the network more effectively. Now, using Eq. (1), (2) and (3), we calculate trust values using these node's values such that

$$W = (W_1 * \alpha_1) + (W_2 * \beta_1) + (W_3 * \gamma_1) \tag{4}$$

These coefficients can be chosen at random based on design considerations, but they must be between 0 and 1 and adhere to the specified conditions. We have assumed weights for the selected parameters are 25, 50 and 25% of W_1, W_2, and W_3 consequently.

$$W_1 + W_2 + W_3 = 1 \tag{5}$$

The calculated weight is used to calculate the positive trust for the nodes, and the following equation can be used to compute the negative trust: To represent the trust here we use as π symbol.

$$\pi_{-ve} = 1 - W \tag{6}$$

Now, we have two values positive trust and negative trust for evaluating the normal networks. Thus, these two values are usages to compute the node's trust and responsible for secure communication. The process of secure communication is demonstrated using Table 1. In this context a new network is created with the 20, 40, 60, 80 and 100 nodes and black hole is deployed in network. Using the following process, the malicious nodes are detected.

Table 1 Proposed STABA algorithm

Input: Network Nodes NN // communication entities in the system
Output: Detected Nodes DN // node seeking to deny service to other nodes in the network

Process:

1: RREQ packet send by source node to destination node.

2: Wait for the construction of the reverse direction.

3: For every next hop in route do

    a: Calculate the node's Negative trust π_{-ve} and Positive trust W

    b: If threshold π_{-ve} < node's π_{-ve}

        i: Labeled Node is malicious

    c: If threshold W < node's W

        i: Node is Genuine

        ii: Go to next hop

    d: End if

4: End for

4.2   Algorithm Design

The algorithmic structure in Table 1 can be used to manage the internal process of secure communication to identify black-hole attacks using AODV.

5   Simulation

This section provides the simulation depiction of the black-hole attack detection where packets are forward to the legitimate nodes using different number of nodes.

5.1   Simulation Arrangement

The projected work is carried out with the aid of the NS2 [16] network simulator version 2. In addition, the following setup is ready for performance assessment and simulation (Table 2).

A small network is shaped with 20, 40, 60, 80, and 100 nodes. User Datagram Protocol is a connectionless protocol that makes black hole attack analysis easy, the constant bit rate (CBR) produces packets via a UDP connection. The CBR packet chosen has a size of 512 bytes. In the TCL script, the node locations are manually specified. During this exercise, on a flat, two-dimensional (2D) geometric plane, an Omni antenna is a wireless transmitting or receiving antenna that broadcasts or captures radiofrequency (RF) electromagnetic waves uniformly well in all horizontal directions. Because there is no cable connection, we used the wireless channel to setup the node. To begin, the parameters of the AODV protocol are evaluated while it is in operation.

Table 2 Simulation setup definition

Simulation properties	Values
Antenna model	Omni directional
Topography area	500 × 500
Radio-propagation model	Two ray ground
Channel type	Wireless channel
No of mobile nodes	20, 40, 60, 80,100
Routing protocol	AODV
Traffic model	CBR

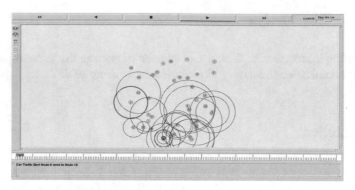

Fig. 3 Attack simulation

5.2 *Simulation Scenario*

To simulate the working and the security of the network the single different experiments are performed under the following network scenarios.

- **Simulation of black hole network:** In this scenario, the network is setup, and its performance is monitored using the recommended trust-based routing protocol. A network simulation screen during a packet drop attack is shown in Fig. 3. The attacker nodes in the network are represented by red nodes, while the normal nodes are represented by green nodes.

6 Result Analysis

6.1 *E to E Delay*

The amount of time it takes for a packet to arrive from a source to a destination system on a network is known as end to end delay. The following formula is used to calculate the delay:

$$\text{E2E Delay} = \text{Packet Receiving Time} - \text{Packet Sending Time} \tag{7}$$

Figure 4 compares the conventional AODV routing under attack with the proposed secure routing methods in terms of end-to-end network delay. The STABA, according to the findings, causes less end-to-end delay under attack situations than traditional routing strategies with black-hole attack. As a result, the suggested method is both efficient and effective. The projected work is carried out with the aid of the NS2 [16] network simulator version 2. End to End Delay is clearly represented in Table 3 by numerical scenario for both approaches which is measured in milliseconds.

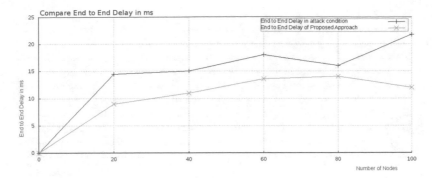

Fig. 4 Comparison of E to E delay

Table 3 Numerical values of E to E delay

Number of nodes	Proposed STABA under black-hole attack	Traditional AODV under black-hole attack
20	8	14
40	12	15
60	13.5	17.5
80	14	16
100	12	22

6.2 PDR

The packet delivery ratio, also known as the PDR ratio, is an efficiency parameter that indicates how efficient a routing protocol is based on the number of successfully transmitted packets to the destination. The following formula can be used to measure PDR:

$$PDR = \frac{\text{Total No.of Delivered Packets}}{\text{Total No.of Sent Packets}} \times 100 \tag{8}$$

The network's comparable packet delivery ratio is depicted in Fig. 5. According to the results,

STABA method sends more packets than the previous strategy although the network includes an attacker node, enabling the proposed technique to avoid the attack consequence and enhance performance of network. We found the numerical result of the PDR in Table 4 for proposed and traditional AODV under attack.

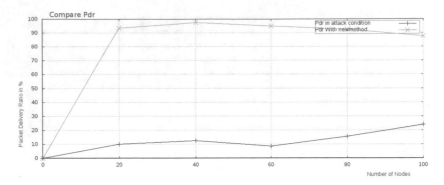

Fig. 5 Comparison of network PDR

Table 4 Numerical values of packet delivery ratio

Number of nodes	Proposed STABA under black-hole attack	Traditional AODV under black-hole attack
20	93	10
40	98	12
60	95	10
80	93	17
100	89	23.5

6.3 Network Throughput

Network throughput is the standard rate of successful message delivery across a communication connection. This data may be sent directly to a network node without passing via a trusted node through a physical or logical connection. The most common unit of measurement is bits per second, although data packets per second or data packets per time slot are also used. The throughput of the network is depicted in Fig. 6. The X-axis depicts the network's nodes, while the Y axis depicts the network's throughput in KBPS.

The suggested methodology, according to the findings, increases network throughput during attack situations and, as a result, avoids the attack impact when compared to standard routing techniques. Below Table 5 is depicts numerical values of throughput for the proposed STABA and traditional AODV.

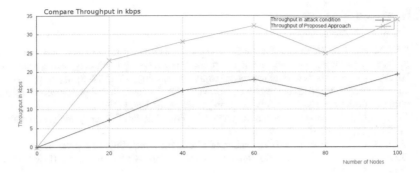

Fig. 6 Comparison of throughput

Table 5 Numerical values of network throughput	Number of nodes	Proposed STABA under black-hole attack	Traditional AODV under black-hole attack
	20	23.5	7.5
	40	27.5	15
	60	23.5	18
	80	25	14
	100	34	19

7 Conclusion

MANET is made up of a variety of mobile devices with varying output capabilities. Any ad-hoc network model proposed does not place unreasonably high communication and computation requirements. It's best if it's as light as possible. Because of the numerous attacks that affect the efficiency of ad hoc networks, protection becomes a critical necessity during deployment. The importance of network trust in detecting node misbehavior cannot be overstated. The idea of trust relevance is multifaceted, complicated, and context dependent. MANETs make trust formation and administration difficult due to significant resource restrictions. The focus of this study is on trust concerns in wireless ad-hoc networks. In addition, a novel trust model for enhancing ad hoc network security is given. In this paper, proposed trust-based concept for black-hole attack detection deployed which is significantly configured using weight computation by finding various threshold parameter e.g., buffer, energy and packet drop of network nodes. This given model is tested against the black hole attacker nodes. STABA implemented using wide reactive based routing protocol AODV. During examination of these attacks the proposed trust-based technique is found efficient and adoptive for improving the network performance and security as well.

References

1. Dani V, Bhati N, Bhati D (2021) EECT: energy efficient clustering technique using node probability in ad-hoc network. In: Innovations in bio-inspired computing and applications, IBICA 2020. Advances in intelligent systems and computing, vol 1372. Springer, pp 187–195
2. Dani V, Birchha V (2015) An improved wormhole attack detection and prevention method for wireless mesh networks. Int J Adv Res Comput Commun Eng 4(12)
3. Jøsang A (2007) Trust and reputation systems. In: Foundations of security analysis and design IV. Springer, Heidelberg, pp 209–245
4. Kephart JO, Chess DM (2003) The vision of autonomic computing. Computer 36(1):41–50
5. Deng H, Li W, Agrawal DP (2002) Routing security in wireless ad hoc networks. IEEE Commun Mag 40(10):70–75
6. Djahel S, Nait-Abdesselam F, Zhang Z (2010) Mitigating packet dropping problem in mobile ad hoc networks: proposals and challenges. IEEE Commun Surv Tutor 13(4):658–672
7. Dani V (2022) iBADS: an improved black-hole attack detection system using trust based weighted method. J Inf Assur Secur 17(3)
8. Pirzada A, Datta A, McDonald C (2004) Trust-based routing for ad-hoc wireless networks. In: Proceedings of the international conference on networks, vol 1, pp 326–330
9. Dani V, Bhonde R, Mandloi A (2022) iWAD: an improved wormhole attack detection system for wireless sensor network. In: International conference on intelligent systems design and applications. Springer, Cham, pp 1002–1012
10. Marchang N, Datta R (2012) Light-weight trust-based routing protocol for mobile ad hoc networks. IET Inf Secur 6(2):77–83
11. Patel NJK, Tripathi K (2018) Trust value based algorithm to identify and defense gray-hole and black-hole attack present in MANET using clustering method. Int J Sci Res Sci Eng Technol 4:281–287
12. Arulkumaran G, Gnanamurthy RK (2019) Fuzzy trust approach for detecting black hole attack in mobile adhoc network. Mobile Netw Appl 24(2):386–393
13. Kamel MBM, Alameri I, Onaizah AN (2017) STAODV: a secure and trust based approach to mitigate blackhole attack on AODV based MANET. In: IEEE 2nd advanced information technology, electronic and automation control conference (IAEAC), pp 1278–1282
14. Kshirsagar VH, Kanthe AM, Simunic D (2018) Trust based detection and elimination of packet drop attack in the mobile ad-hoc networks. Wireless Pers Commun 100:311–320
15. Naveena S, Senthilkumar C, Manikandan T (2020) Analysis and countermeasures of black-hole attack in MANET by employing trust-based routing. In: 2020 6th international conference on advanced computing and communication systems (ICACCS), pp 1222–1227
16. The Network Simulator. NS-2 Available Online at: http://www.isi.edu/nsnam/ns/

Wind Speed Prediction in the Region of India Using Artificial Intelligence

Eeshita Deepta, Neha Juyal, and Shilpi Sharma

1 Introduction

Most nations have put a larger emphasis on increasing renewable energy now, spurred on because of rising levels of greenhouse gases and to maintain energy security. Wind energy has become the most commonly utilized green energy source, accounting for 432,883 MW of worldwide energy consumption [1]. By 2022, India's government hopes to have enhanced renewable resources capacity to above 175 GW. Wind power will contribute 60 GW to this total [2]. At a hub height of 100 m, India's wind resource potential ranges from 49,130 to 302,000 MW, according to CWET, with the wind resource potentially being greater at higher hub heights.

Wind Energy, while a good source of green, renewable and clean energy, can pose some problems. It is arbitrary and random, making its prediction hard [3]. This is why wind speed forecasting (WSF) is important. Longer predictions, ranging from weeks to months, are important to understand where wind farms can be built, and for optimizing and managing energy distribution [3]. Shorter predictions help to control and monitor wind turbines, and to schedule and dispatch the power optimally [1]. This can range from minutes to hours. An accurate short-term prediction can decrease fluctuations and make the grid stable [3]. WSF enables the power grid to make proper energy plans so that it doesn't suffer from excessive wastage, or starvation of power.

Wind power should be predicted at least a day in advance so that the efficiency and penetration of wind power can be increased. This is why precise wind speed predictions is critical. This equation shows the relationship between them:

E. Deepta (✉) · N. Juyal · S. Sharma
Amity University, Noida, Uttar Pradesh, India
e-mail: edeepta1713@gmail.com

N. Juyal
e-mail: juyalneha2@gmail.com

S. Sharma
e-mail: ssharma22@amity.edu

B. Unhelker et al. (eds.), *Applications of Artificial Intelligence and Machine Learning*,
Lecture Notes in Electrical Engineering 925,
https://doi.org/10.1007/978-981-19-4831-2_59

Available power in wind P_w,

$$P_w = \frac{1}{2}\delta A V^3 \tag{1}$$

Extracted power from wind P_b,

$$P_b = \frac{1}{2}\delta A V^3 . C_p \tag{2}$$

Here, δ is density of air, A is swept area of the wind turbine's blade, C_p is coefficient of power and V is the wind speed. As a result, the power generated follows a cubic relationship that is directly proportional to the wind speed. A small change in speed can cause a sizable change in the power generated.

Dynamical models only provide macro-level information, and predicting the wind speed is difficult at the station level. One of the most popular statistical models is the Autoregressive Moving Average (ARMA) model. However, with the recent rapid growth of Artificial Intelligence (AI) applications, AI models based on data-driven technologies have become viable.

AI can learn and improve, and is widely preferred though as they don't need to use any mathematical model which can prove to be beneficial [4]. Support Vector Machines (SVM), Fuzzy Logic, Artificial Neural Networks (ANN), Support Vector Machines (SVM), Fuzzy Logic, and other AI methods are commonly utilized in WSF. They are proven to be more accurate than standard approaches that are statistically reliant [5].

To the authors' knowledge, several research have been done employing various algorithms and comparing their mistakes and predictions, but no study has attempted a thorough assessment of such investigations. This study will look at and evaluate Artificial Intelligence systems based on how far ahead of time they can predict wind speeds. This study focuses solely on research conducted in India.

2 Literature Review

A comparative has been done of three models namely ANN, ARIMA and hybrid (fusion of ANN and ARIMA) for wind speed forecasting in the regions of Dharapuram, Kayathar, and (Tamil Nadu, India). The hybrid model had the least inaccuracy. For Kayathar sites the MAPE values are 14.2%, 15.2% and 25.3% for 1 h, 3 h, 24 h respectively as was noted in [1].

NARX, Linear Regression, and Persistence models were used to assess short-term forecasts (24 h) in the city of Jaisalmer, Rajasthan, according to [2]. Their MAPE was an average of 11% showing much more precision than the other models.

For forecasting WS 10, 20, 30 min, and 1 h ahead, ANN models for WS have been developed and the regression values were 92.04%, 87.10%, 83.83%, and 76.25%, respectively, indicating the usefulness of the proposed approach in [3].

In reference to [4], SVM based on the Neuro-Fuzzy reduces prediction errors. The data from a wind farm in Coimbatore is used for daily (30 h ahead) and weekly forecasting and compared to BPN, BPN + SVM, WT + BPN + SVM and WT + BPN + SVM + QR models. Their model improved on the other models by 15.2% and 13.6% MAPE for daily and weekly forecasts respectively.

Reference [5] suggests that the ANN model in Tamil Nadu for 7 wind farms, using data from 2002 to 2005, was helpful for wind energy producers.

Discussed in reference [6], the writers discuss the significance of green technology, as well as the various methods currently in use for green energy forecasting. They discuss the different time-frames for forecasting and the contribution of different renewable energy sectors in India.

Reference [7] concludes that in comparison to the FFNN model, the AWNN model offers greater approximation and training abilities. A two-stage forecasting system for predicting wind speed output up to 30 look-ahead hours has been devised. Over the course of a year, the forecasting models were tested with a normalized RMSE of 10.22%.

As concluded by [8] the RNN model outperforms the ARIMA model for short-term WSF. A comparison has been done between two models RNN and ARIMA for short term WSF. The MAE for RNN is 18.18% lower than the ARIMA model for both datasets. In addition, for both datasets, MSE and RMSE are around 20 to 25% and 11 to 14% lesser in the RNN model as compared to the ARIMA model.

There has been a comparison between ANN and PSO-based ANN models for their effectiveness in predicting average monthly wind speed from weather and location data. When the two were compared, it was discovered that the PSO-based ANN approach performed more consistently than the ANN-only technique as concluded in [9].

In reference to [10] for long term WSF, ANN model has been considered. They have collected the online dataset of 26 cities of India from NASA. Taking the parameters such as air temperature, earth temperature, relative humidity as its input variables. Forecasting wind speed across 39 cities in Maharashtra, India, validates the performance of the ANN model. For 39 locations in Maharashtra, MSE and MAPE were determined to be 0.48 and 22.8% respectively.

As suggested by [11] GRNN model has been used for monthly WSF. An online dataset of 31 cities was used to assess the performance of the proposed model. For long-term wind, the proposed model is extremely efficient as evidenced by its low MSE and great correlation coefficient.

The goal of reference [12] is to give a broad overview of the link between AI and legal thinking. The nature of AI life-forms is also explored, as well as the philosophy of legal reasoning.

The paper in reference [13] provides a model based on the Fast Correlation Based Filter (FCBF) algorithm, the optimized Radial Basis Function (RBF) model, and the

Fourier distribution for wind speed, which blends artificial intelligence techniques with statistics.

Using machine learning-based text mining and network analytics, reference [14] experimentally explores the evolution of AI algorithms in wind power technology.

Short-term prediction and worldwide model output are combined with static structure high-resolution WRF RTFDDA modelling using numerous artificial intelligence algorithms in a holistic wind power forecasting system developed in conjunction with Xcel Energy in [15].

3 Methodology

The major goal was to evaluate and review the various models employed in India, hence studies outside India are not considered. WSF methods are divided into short, medium and long-term, sometimes including ultra-short-term. We could not find a fixed definition for these terms, but the generally agreed classification we have divided the predictions into are ultra-short-term, short-term, medium-term, and long-term which can be found in Table 1 [6]. This categorization is done as when the time is shorter, the accuracy of the model is higher. Therefore, to achieve a fair analysis, models are compared only if they have the same classification.

To establish accuracy, Root Mean Square Error (RMSE) was used in most models. However, to compare models using different scales, we needed to normalize the errors to make them scale-free. Thus, Normalized Root Mean Square Error (NRMSE) was calculated by us.

$$NRMSE = \frac{RMSE}{\bar{y}} \tag{3}$$

Here, $\bar{y}$ is the average (or mean) of the determined data, which in our instance is the measured wind speed in the specific research. For calculations, data, if directly supplied was used or it was extracted from graphs present.

These were then plotted as a bar graph to see which model fared better in which categorization.

Table 1 The time-bound classifications of WSF methods

Classification	Time
Ultra-short-term	From a few minutes to one hour
Short-term	From more than one hour to less than 24 h
Medium-term	From a day to a week
Long-term	From weeks to months

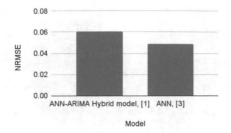

Fig. 1 NRMSE of ANN-ARIMA hybrid model and ANN for 1 h time frame

4 Result

4.1 Ultra-Short-Term WSF

As concluded by [1], there were three different models that were put to the test., out of which the ANN-ARIMA hybrid model revealed the least error with the Kayathar site. This was accurate throughout all the time-frames. In this model, the data was first given to ARIMA for linear trend of the time span, and then to ANN for the non-linear model of the time series. ARIMA combines auto regressive (AR) and the moving average (MA) models, with I standing for integrated. The data that we are considering is for 1 h.

Reference [3] suggests that the ANN model should be aware of the link between the input and the output before it can give accurate results. It is adaptable with any amount of inputs, but needs enough to be able to comprehend the input–output relationship. The data we are considering is also 1 h. Figure 1 depicts the calculated NRMSE of both the models.

4.2 Short-Term WSF

There were only 2 studies which had run their models on the time-frame between one hour to 23 h.

Reference [1] had the ANN-ARIMA model, which was tested for a 3-h forecast for the Kayathar site. This was chosen instead of the 8-h prediction because the second model's time-frame was also nearer to 3 h. In reference to [8], RNN model was used which is one of the finest models in the neural network category owing to it having a short-term memory. The concept of internal memory helps the neurons to better analyses the data.

Figure 2 depicts the estimated NRMSE of the models.

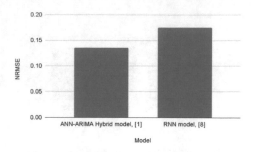

Fig. 2 NRMSE of ANN-ARIMA hybrid model and RNN model for 3 h time frame

4.3 Medium-Term WSF

Since the 24-h time-frame is what is usually needed for the prediction of optimization of wind farms, it was unsurprising to find the most models that concentrated on this window. Not all of them used the same exact time-frame, as the time ranges from 24 h to more than 30 h. However, all of them can help to make WSF a day ahead.

The first model was for the 24 h time-frame in reference [1] for the Kayathar site. The second model, in reference [2], uses the NARX network (Nonlinear Autoregressive Neural Network with Exogenous Inputs) with a dataset from Coimbatore. This network has a feedback connection with several layers, and has a better learning rate than other neural networks, which also makes it faster. Their data was divided into 70% for training, 15% for validation, and 15% for testing.

SVM (Support Vector Machines) based Neuro fuzzy model was used as the third model which was tested in Coimbatore, from reference [4]. This paper determines wind power forecasts instead of wind speed, however they did determine wind speed before wind power, which is what this paper is about. Their model combines wavelet transform data filtering with a computing model which is based on ANFIS network. SVM classifiers are also applied to decrease errors. Instead of deterministic forecasting, they use Quantile regression for probabilistic forecasting.

The fourth model, from reference [7], was of the AWNN model (Artificial Wavelet Neural Network) for. Unlike other models, this is not popularly used and Zhang et al. were the first to suggest it as an alternative to feedforward neural network (FFNN). This paper also focused on wind power forecasting, but their first stage involved a WSF using AWNN, which has a better generalization property. This model was used to estimate wind speed 30 h ahead.

As it has been concluded by reference [8], the fifth model of RNN was proposed predicts wind speed up to 30–35 h ahead of time. The NRMSE of all five models are depicted in Fig. 3.

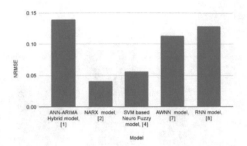

Fig. 3 NRMSE of ANN-ARIMA hybrid model (24 h), NARX model (24 h), SVM based Neuro Fuzzy model (30 h), AWNN model (30 h), and RNN model (30+ hours)

4.4 Long-Term WSF

Instead of predicting wind speeds on a day-to-day or hourly basis, such forecasts are done months or even a year ahead to analyze the best way to build and install a wind farm. It is essential to determine the right place as well as for controlling and scheduling wind power generation. There were three papers which compared models for this ranging from a month to a year.

The first model is from reference [9], which uses PSO-NN (Particle Swarm Optimization Neural Network) on 67 cities in India with 19 inputs to predict the wind speed in cities determined monthly. The PSO-NN model eliminates a disadvantage of the ANN model, where it can become confined to the local minimum. Instead of using arbitrary values, the weights and biases of the network are tailored. It was inspired by how bird flocks act when searching for food, with each bird changing speed and location in response to their prior position and that of its neighbors.

The ANN's use is suggested by reference [10], and is the second model. Datasets obtained from 26 cities in India are used. Data from 22 cities was utilized for training purposes, while data from the remaining four cities was used for evaluation purposes. Monthly wind speeds are predicted for 39 cities in Maharashtra as well. This research makes use of data from Maharashtra's cities. This ANN model uses a two-layer FFNN with varying numbers of hidden neurons.

The third model is GRNN (Generalized Regression Neural Network) based on a four-layer feed neural network, i.e., Layers for input, pattern, and summing and output layer. GRNN's training speed is fast as well. 31 cities of India were used in this model, 26 were used for training and 5 being used for testing as stated by [3]. Figure 4 shows the NRMSE values including all three models.

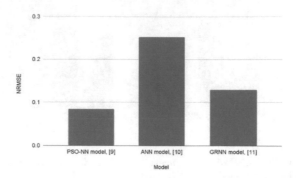

Fig. 4 NRMSE values of PSO-NN model, ANN model and GRNN model for monthly time-frame

5 Conclusion

Forecasting wind speeds is crucial for wind power generation. India's generating potential for wind power is vast, but it has yet to be completely exploited. Artificial Intelligence are self-adaptable and know the tasks they need to perform. They closely mimic the characteristics of natural living system. Using AI will only enhance the efficiency of wind power generation while also stabilizing the country's power grid. AI algorithms have been tested in various places across India to optimize WSF. This paper divides the algorithms in accordance to the time-frame of their forecasting, each having their own uses. The algorithms were reviewed and their errors were normalized to make them scale-free. These were plotted to get a better understanding about the efficiency of each of these methods in different time frames. It was found that for ultra-short-term WSF, ANN model was the best fit. For short-term WSF, ANN-ARIMA Hybrid Model was the better fir while for medium-term WSF, the better fit was NARX model. The best fit for long-term WSF was PSO-NN model. It can therefore be concluded that for different time frames, different models proved to be the better fit, and there was no one algorithm that fit all the different time frames.

6 Future Work

In India, the majority of the effort has been concentrated on medium-term WSF, and so more studies need be carried out to compare the efficiency of the AI models for ultra-short-term and short-term WSF.

Some algorithms such as Genetic Algorithm (GA) have yet not been tested to the knowledge of the authors. Studies based on Hybrid models have also not been tested enough to determine their efficiency.

A large-scale study into the accuracy and efficiency of the common AI methods for WSF will also shed light in this area. More comparative studies need to be carried out separately in order to determine the best AI algorithm for WSF.

References

1. Nair KR, Vanitha V, Jisma M (2017) Forecasting of wind speed using ANN, ARIMA and Hybrid models. In: 2017 international conference on intelligent computing, instrumentation and control technologies (ICICICT). IEEE, pp 170–175
2. Baby CM, Verma K, Kumar R (2017) Short term wind speed forecasting and wind energy estimation: a case study of Rajasthan. In: 2017 international conference on computer, communications and electronics (Comptelix). IEEE, pp 275–280
3. Yadav AK, Malik H (2019) Short-term wind speed forecasting for power generation in Hamirpur, Himachal Pradesh, India, using artificial neural networks. In: Applications of artificial intelligence techniques in engineering. Springer, Singapore, pp 263–271
4. Ranganayaki V, Deepa SN (2017) SVM based neuro fuzzy model for short term wind power forecasting. Natl Acad Sci Lett 40(2):131–134
5. Mabel MC, Fernandez E (2008) Analysis of wind power generation and prediction using ANN: a case study. Renew Energy 33(5):986–992
6. Verma A, Tripathi MM, Upadhyay KG, Kim HA (2017) A review article on green energy forecasting. Asia-Pacific J Adv Res Electr Electron Eng 1:1–8
7. Bhaskar K, Singh SN (2012) AWNN-assisted wind power forecasting using feed-forward neural network. IEEE Trans Sustain Energy 3(2):306–315
8. Sandhu KS, Nair AR (2019) A comparative study of ARIMA and RNN for short term wind speed forecasting. In: 2019 10th international conference on computing, communication and networking technologies (ICCCNT). IEEE, pp 1–7
9. Malik H, Padmanabhan V, Sharma R (2019) PSO-NN-based hybrid model for long-term wind speed prediction: a study on 67 cities of India. In: Applications of artificial intelligence techniques in engineering. Springer, Singapore, pp 319–327
10. Malik H (2016) Application of artificial neural network for long term wind speed prediction. In: 2016 conference on advances in signal processing (CASP). IEEE, pp 217–222
11. Ansari MA, Pal NS, Malik H (2016) Wind speed and power prediction of prominent wind power potential states in India using GRNN. In: 2016 IEEE 1st international conference on power electronics, intelligent control and energy systems (ICPEICES). IEEE, pp 1–6
12. Gupta D, Sharma S (2020) Artificial intelligence approach to legal reasoning evolving 3D morphology and behavior by computational artificial intelligence. In: Proceedings of the third international conference on computational intelligence and informatics. Springer, Singapore, pp 869–875
13. Zhang Y, Pan G, Zhao Y, Li Q, Wang F (2020) Short-term wind speed interval prediction based on artificial intelligence methods and error probability distribution. Energy Convers Manage 224:113346
14. Lee M, He G (2021) An empirical analysis of applications of artificial intelligence algorithms in wind power technology innovation during 1980–2017. J Clean Prod 297:126536
15. Kosovic B, Haupt SE, Adriaansen D, Alessandrini S, Wiener G, Delle Monache L, Liu Y, Linden S, Jensen T, Cheng W, Politovich M, Prestopnik P (n.d.) A comprehensive wind power forecasting system integrating artificial intelligence and numerical weather prediction

Lung Cancer Detection Using Modified Fuzzy C-Means Clustering and Adaptive Neuro-Fuzzy Network

Sajeev Ram Arumugam⬤, Bharath Bhushan⬤, Monika Arya⬤, Oswalt Manoj⬤, and Syed Muzamil Basha⬤

1 Introduction

Air contamination is a combination of diverse particulates and biotic molecules that get instigated into the Earth's atmosphere. There are different causes of air pollution, and they are human-made or natural. Breathe in polluted air are the reason for diseases, allergies, and it may cause the death of humans—people suffer from a respiratory problem, breathing problems, lung disease. There are many chemicals and environmental reasons for lung cancer. Occupationally induced lung cancer is indistinct from lung cancer due to cigarette smoking, and it is one of the reasons for lung cancer and other carcinogens [1]. Research shows that there is a strong bond between fine particles exposure and early death. Fine particles can lead to diseases such as asthma, heart attack, bronchitis. According to the Journal of the American Medical Association, continuous disclosure to $PM_{2.5}$ can cause heart attack and stroke. Scientists have estimated that for every ten micrograms per cubic meter rise in air pollution, there is 4, 6 and 8% more risk of lung cancer mortality. Figure 1 [2] shows a comparison of the relative size of the particulate matter [3].

S. R. Arumugam (✉)
Department of AI&DS, Sri Krishna College of Engineering and Technology, Coimbatore, India
e-mail: imsajeev@gmail.com

B. Bhushan
School of Computer Science and Engineering, VIT, Bhopal, India
e-mail: Bharath.bhushan@vitbhopal.ac.in

M. Arya
Department of CSE, Bhilai Institute of Technology, Durg (C.G), Chhattisgarh, India

O. Manoj
Department of CSBS, Sri Krishna College of Engineering and Technology, Coimbatore, India

S. M. Basha
School of Computer Science and Engineering, REVA University, Bangalore, India

© The Author(s), under exclusive license to Springer Nature Singapore Pte Ltd. 2022   733
B. Unhelker et al. (eds.), *Applications of Artificial Intelligence and Machine Learning*,
Lecture Notes in Electrical Engineering 925,
https://doi.org/10.1007/978-981-19-4831-2_60

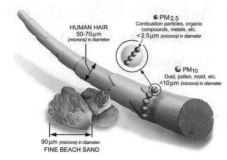

Fig. 1 Size of particulate matter

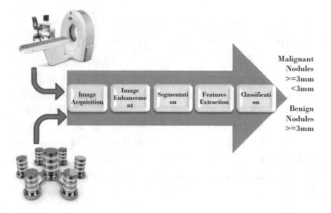

Fig. 2 The architecture of a CAD system

In Asia, contamination caused by particulate matter is trickling to new areas. In Asian cities, air contains smog, and it is a hazard to human health [4]. In numerous cities, the particulate matter levels are more than the dangerous limit, explicitly in high populated, fast-growing countries. The cancer patient persistence rate has improved from the mid-1970s because of changes in the diagnosis process. Progress of the patient is fast for hematopoietic and lymphoid malignancies due to progress in treatment protocols. The relative survival rate for chronic myeloid leukaemia improved from 22% in the mid1970s to 69%. In the US, 81% of lung cancer is due to smoking. In India, lung malignancy is a leading public health problem. Awareness of risks due to smoking and early prediction of nodules in the lungs made a decrease in the death rate of Lung cancer patients, but the cases are still at an inclined rate. To be more precise in the time period between 2007 till 2020 the death rate of lung cancer patients is declined by 15.20% among men and 8.10% among women. The new cancer cases are expected to increase by 24% among men and 21% among women in the same time period [5]. The new cancer patients and the expected number of cancer deaths are graphically represented in Fig. 2.

2 Literature Survey

The medical field combines with computer applications is used in x-ray, Computerized Tomography Scan imaging. The medical organization uses a specific format. CT images will be mostly available in Digital Imaging and Communication in Medicine (DICOM) format. It also combines two-dimensional (2D) and three-dimensional (3D) [6]. Computer-aided detection (CAD) helps to assist in the understanding of the medical images and to diagnose. Computer output is used as a second opinion by the radiologists. In some situations, radiologists are sure about their final decisions about identification. In some cases, radiologists are not sure about their results; a computer output makes the final decision. When the computer performance is higher, the investigative result is better. The precise investigative outcome is produced by merging the experience of the radiologists and the capability of a computer [7]. CAD is used to identify tumours by placing a mark on the doubtful area. CAD systems also produce false-positive results slightly higher than the human being, when comparing sensitivity levels to an experienced radiologist [8].

R. Bellotti et al. [9] projected a tumour identification method to guide the radiologists in predicting lung nodules among lung cancer patients. The CAD system comprises of three steps: (1) segmenting the lung image into several segments which includes pleural nodules, (2) candidate detection and (3) False-positive reduction. False-positive reduction is effectively reduced using thresholding technique and neural network classifier. The system uses 15 CT scans, and the system achieves an accuracy of 88.5% with 6.6 FPs/CT. Jeong Won Lee et al. [10] proposed a method for detecting malignant cells for 15 patients. Classification of malignant cells depends on the size and how they are connected to the adjacent structures. The sensitivities for identifying tumours higher than 5 mm were 83% and for the radiologist 75%.

Lilla et al. [11] projected a system that automatically finds out the best possible set of features and selects the most appropriate features from a database. The support vector machine classifier produces 100% sensitivity and 56.4% specificity using a validation set. Lee et al. [11] proposed a Lung Malignancy Detection System (LMDS) which uses template-based segmentation methods and classification processes. Different classifiers were tried with different input variables and the best performing classifier was implemented. the researchers used CT images taken from 32 patients and the performance of the system was determined using sensitivity and specificity which was found to be 98.33% and 97.11% respectively.

Emre et al. [12] proposed a CAD system capable of separating the malignant nodules by implementing a neural network model and using ANN for classification. The performance of the system was able to have 90.63% accuracy, 92.30% sensitivity and specificity of the system were found to be 89.47%. Mohsen et al. [13] proposed 2D stochastic and 3D anatomical features extract the features and use SVM classifier. The identified malignancy nodules are used in the active contour modelling to extract nodule borders and attain the dice coefficient value. The accuracy of the system is 89%, and the number of false-positive is 7.3/scan.

Kumar et al. [14] developed a classification system for detecting lung nodules. the system uses an autoencoder to extract features from the CT images which were collected from 157 patients. the performance of the system was later calculated using metrics such as accuracy, sensitivity and false positive, which are recorded as 75.01, 83.35% and the false positive was identified to be 0.39/patient. Muzzamil et al. [15] developed a morphological closing technique to detect juxtapleural nodules from the lung region. K-means clustering technique was used to classify malignant nodules. 2D and 3D features are extracted from the nodules to classify them as malignant and benign, which also reduces the False Positives. the sensitivity of the system was reported to be 83.33%

3 Proposed System

The CAD system is used for detecting cancer nodules, and it contains five phases namely image acquisition, pre-processing of the image, segmentation of the pre-processed image, nodule detection or candidate detection and elimination of false positives (FP). CAD aids to improve the performance of radiologists and image understandings. The classification might be performed in two phases as training and testing phases. The classifier is trained to consider the parameters of the framework and the testing phase is to assess the classifier achievement. The radiologist evaluates the analysis and the CAD system performance on the outcomes before waiving the results. It has a significant effect on the clinical analysis in the exposure of toxic diseases. Figure 2 shows the design of the CAD framework.

3.1 Image Acquisition

In the proposed system, the CT images are used, and it is taken from Lung Image Database Consortium (LIDC) which is one of the largest open-sourced datasets and widely used among researchers. The dataset contains both malignant images and benign images. The CT images are generally available in DICOM format, and acquiring CT images from the datasets will be the initial step in the projected method. In the proposed system 200 images were used both for training and testing the CAD System. To make the classes to be even, both images having malignant and benign nodules were kept in equal proportion. A benign tumour does not overrun its adjacent tissue around the body. A malignant tumour spread through its surrounding tissue and around the body. The benign tumour grows slowly, and the malignant tumour proliferates [16].

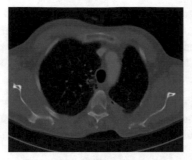

Fig. 3 Wiener filtered image

3.2 Wiener Filter

Wiener Filter can be used for reducing the mean square error, and it is based on a stochastic framework. The noise present in the spectrum is equivalent to the noise variance. Wiener filter is used in the frequency domain. It reduces noise and blurring of images. It is used in the steganography process. Figure 3. shows the wiener filtered image. The estimation of the power spectrum is computed as shown in (1)

$$Power\,Spectrum = \frac{1}{N^2}[y(k,l)y(k,l)] \tag{1}$$

Y(k, l) is the DFT of the observation.

The noise smoothing is performed by the following expression shown in (2)

$$s = \frac{Sa - Sb}{|II^2|} \tag{2}$$

Which is a result of $s_a = s_b + s\,|H^2|$

The power spectrum sa can be calculated using the periodogram estimate. It is implemented as shown in (3)

$$w = \frac{1}{H}\frac{s_{yy}^{per} - sb}{s_{yy}^{per}} \tag{3}$$

3.3 Modified Fuzzy C-Means Clustering

In the FCM algorithm, the number of clusters has to be initialised first. There are no rules for calculating the number of clusters. FCM finds the optimal value of the objective function iteratively. It makes the FCM algorithm computationally complex. The usage of histogram information in the FCM algorithm makes it works more efficient and is it is called a Modified FCM algorithm. The modified FCM algorithm is optimized by calculating the covariance matrix of each cluster. The modified FCM algorithm works as follows:

1. The histogram of the source image is created and the number of clusters was defined using the function $1 < m < \alpha$ where $u_i == 1$ and $i = 1, 2 \ldots c$.
2. The centre of the cluster is calculated as shown in (4):

$$V_i = \frac{\sum_{k=1}^{n} u_{ik}^m P(r)r}{\sum_{k=1}^{n} u_{ik}^m} \qquad (4)$$

3. The covariance matrix is calculated using the formula shown in (5):

$$G = \frac{\sum_{k=1}^{n} \mu_{ik=1}^{n}(xk - vi)(xk - vi)^t}{\sum_{k=1}^{n} \mu_{ik}^m} \qquad (5)$$

4. The new membership function is calculated using the formula shown in (6)

$$u_{ik} = \frac{1}{\sum_{j=1}^{c} \left(\frac{\left| (xk-vi)^T Gi(xk-vi) - \ln|Gi| \right|}{\left| (xk-vj)^T Gj(xk-vj) - \ln|Gj| \right|} \right)^{\frac{2}{m-1}}} \qquad (6)$$

5. Compare U_{k+1} and U_k and if the difference is less than or equal to ϵ, stop the process. Where ϵ is the acceptable changes in U. Figure 4 shows the fuzzy clustering-based segmentation technique applied to the wiener filtered image.

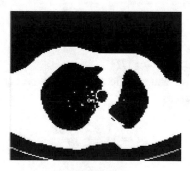

Fig. 4 Modified FCM clustering-based segmentation

3.4 Feature Extraction

Feature extraction reduces the image dimension by which an initial set of raw data reduces to convenient groups for processing. A representative of the extensive database is a collection of features that require much calculation for the process. Feature extraction is the process of selecting or combining features by reducing the amount of data to be processed. The Wiener filter is used for filtering the image. Then modified FCM is used for segmenting the image. The GLCM technique is used to extract image features.

3.5 Random Forest Adaptive Neuro-Fuzzy Classifier

Random Forest Adaptive Neuro-Fuzzy Classifier is a combination of random forest classifier and artificial neuro-fuzzy inference system. The first step is to classify the cancerous and non-cancerous nodules using a random forest classifier. The next step is to classify data using neural networks and fuzzy logic. It is a combination of feedback mechanisms and fuzzy rules. The advantage of a random forest classifier is it will not overfit the model. The random forest takes the features of the testing data and uses the rules of the decision tree to predict the output. It calculates the votes of each predicted output. The predicted output, which has a higher value, is given as input to the adaptive neuro-fuzzy classifier. The adaptive neuro-fuzzy inference system follows a set of fuzzy if–then rules to approximate the functions. In the proposed system, the Gaussian member function is used, and the number of epochs is set to 10 for training the model. For every model, the output with the minimum estimation error was calculated using the Root Mean Square Error (RMSE).

4　Results and Discussions

The entire work has been performed in Matlab 2018 version, and necessary files for performing feature selection and classification process. The front end design is also made using Matlab GUI.

4.1 Training Phase

LIDC dataset is used in training the proposed CAD system. In the proposed CAD system, we used 200 images for training the system. Of them, 100 benign images and 100 malignant images. The images are in DICOM format. The information about the nodules is available in the LIDC database. the proposed system uses 11 features for training namely (1) contrast, (2) correlation, (3) cluster prominence, (4) cluster shade, (5) Dissimilarity, (6) energy, (7) entropy, (8) homogeneity, (9) maximum probability, (10) sum of squares, (11) autocorrelation.

4.2 Testing Phase

To test the proposed system, we used the LIDC database. The system is tested with benign nodule image as well as malignant nodule image. The performance and implementation of statistical features are measured using accuracy, sensitivity and specificity. The formula for accuracy (7), sensitivity (8) and specificity (9) are as follows

$$Accuracy = \frac{TP + TN}{Tp + TN + FP + FN} \tag{7}$$

$$Sensitivity = \frac{TP}{TP + FN} \tag{8}$$

$$Specificity = \frac{TN}{FP + TN} \tag{9}$$

where

TP: Nodule presents in the image are diagnosed correctly.
FP: Nodule is not found in the image, and it is misdiagnosed as nodule is found in the image.
TN: Nodule does not present in the image is predicted correctly.
FN: Nodule found in the image, and it is misdiagnosed as nodule is not found in the image.

Table 1 shows the comparative analysis of performance measures of various segmentation techniques. The Modified Fuzzy C Means Clustering shows high performance rate compared with other segmentation algorithms.

Table 2 provides a comparative analysis of the proposed system with the existing system, and we found that the proposed system shows improved performance than the current system. The proposed system achieves 98.7% accuracy, 95.8% sensitivity and 90% specificity.

Table 1 Performance measures of segmentation technique

Segmentation techniques	Accuracy (%)	Sensitivity (%)	Specificity (%)
Adaptive thresholding	95.9	93	84
Canny edge detection	97.2	96.6	87
Modified Fuzzy C Means Clustering	98.7	95.8	90

Table 2 Comparison between proposed system and existing system

Author name	Techniques used	Accuracy (%)	Sensitivity (%)	Specificity (%)
Al Mohammad et al. [16]	Local Contrast Enhancement + Adaptive Distance Threshold + Fisher Linear Discriminant	78.1		
Albert Balachandar et al. [17]	Isotrophic Resampling + MorpholgicalOpertions + ANN		87.5	
Silva et al. [18]	Histogram Equalisation + SOM + ANN	90.63	92.30	89.47
Javaid et al. [19]	Gray level threshold + 3D shape features + 3DRA		81	
ArulMurugan et al. [20]	Wavelet feature descriptor + ANN	92.6	91.2	100
Proposed system	Wiener Filter + Modified FCM clustering + Random Forest Adaptive NeuroFuzzy classifier	98.7	95.8	90

5   Conclusion

A CAD system for lung cancer malignancy detection using an adaptive neuro-fuzzy network is proposed. In the proposed CAD system, the LIDC database is used for training the database which contains details about the nodule size, type of nodules. The Wiener filter enhances the input image quality by eradicating the noise present in the image. The edges of the image are also to be preserved for better segmentation. The Modified Fuzzy C Means clustering technique is an iterative process, and it segments the image by clustering the pixels in an image. The GLCM is used to attain the testing image features. The system is trained with a set of features, and the testing image features are to be extracted and want to be compared in the database. This step is done by a random forest adaptive neuro-fuzzy classifier. The performance of the proposed system is measured in each step by particular metrics, and the best processing step is chosen. The system accuracy is 98.7%, sensitivity is 95.8%, and specificity is 90%. It is predicted that lung vessels are responsible for the bulk false positive values and small nodules are responsible for the bulk false negative values. The work can be extended by sub classifying the stages of cancer and if it could be done the appropriate treatment could also be suggested with the help of the developed CAD system.

References

1. How air pollution can increase the risk of developing lung cancer? Best rated air purifier (2018)
2. Particulate matter (PM) basics I US EPA. https://www.epa.gov/pm-pollution/particulate-mat ter-pm-basics. Accessed 29 Sept 2021
3. Miettinen V (2019) What is PM2.5 and why you should care. Bliss Air
4. Paul AK Air pollution measures for Asia and the Pacific I Climate & Clean Air Coalition. UNO Environment
5. Corrales L, Rosell R, Cardona AF, Martin C, Zatarain-Barrón ZL, Arrieta O (2020) Lung cancer in never smokers: the role of different risk factors other than tobacco smoking. Crit Rev Oncol Hematol 148:1040–8428
6. Fujita H (2020) AI-based computer-aided diagnosis (AI-CAD): the latest review to read first. Radiol Phys Technol 13(1):6–19
7. Mangrulkar A, Rane S, Sunnapwar V (2020) Image-based bio-cad modeling: overview, scope, and challenges. J Phys Conf Ser 1706(1):12189
8. Shylaja CS, Anandan R, Ram AS (2020) Evolution of lung CT image dataset and detection of disease. In: Intelligent computing and innovation on data science. Springer, pp 439–446
9. Bellotti R et al (2007) A CAD system for nodule detection in low-dose lung CTs based on region growing and a new active contour model. Med Phys 34(12):4901–4910
10. Lee JW, Goo JM, Lee HJ, Kim JH, Kim S, Kim YT (2004) The potential contribution of a computer-aided detection system for lung nodule detection in multidetector row computed tomography. Invest Radiol 39(11):649–655
11. Böröczky L, Zhao L, Lee KP (2006) Feature subset selection for improving the performance of false positive reduction in lung nodule CAD. IEEE Trans Inf Technol Biomed 10(3):504–511. https://doi.org/10.1109/TITB.2006.872063
12. Keshani M, Azimifar Z, Tajeripour F, Boostani R (2013) Lung nodule segmentation and recognition using SVM classifier and active contour modeling: a complete intelligent system. Comput Biol Med 43(4):287–300. https://doi.org/10.1016/j.compbiomed.2012.12.004
13. Benign vs. malignant: definition, characteristics & differences. Study.com (2019)
14. Yusoh M, Phon-On A, Khongkraphan K (2018) Estimating motion blur parameters with gradient descent method. In: 2018 22nd international computer science engineering conference, ICSEC 2018, pp 1–4. https://doi.org/10.1109/ICSEC.2018.8712808
15. Chora RS (2007) Image feature extraction techniques and their applications for CBIR and biometrics systems, vol 1, no 1
16. Al Mohammad B, Brennan PC, Mello-Thoms C (2017) A review of lung cancer screening and the role of computer-aided detection. Clin Radiol 72(6):433–442
17. Chon A, Balachandar N, Lu P (2017) Deep convolutional neural networks for lung cancer detection. Standford Univ.
18. Silva M et al (2018) Detection of subsolid nodules in lung cancer screening: complementary sensitivity of visual reading and computer-aided diagnosis. Invest Radiol 53(8):441–449
19. Shah SIA, Javaid M, Javid M, Rehman MZU (2016) A novel approach to CAD system for the detection of lung nodules in CT images. Comput Methods Programs Biomed 135:125–139. http://dx.doi.org/10.1016/j.cmpb.2016.07.031
20. Arulmurugan R, Anandakumar H (2018) Early detection of lung cancer using wavelet feature descriptor and feed forward back propagation neural networks classifier. Lect Notes Comput Vis Biomech 28:103–110. https://doi.org/10.1007/978-3-319-71767-8_9

Significance of Preprocessing Techniques on Text Classification Over Hindi and English Short Texts

Sandhya Avasthi, Ritu Chauhan, and Debi Prasanna Acharjya

1 Introduction

Hindi is the most spoken language in the Indian subcontinent, and English is known to be the second language of educated people living in India. The census data from 2011 reveals that more than 26% population in India can converse in Hindi [1]. In recent times as the popularity of social media platforms grew, people are expressing their opinion, positing comments, reviews on the latest events and trends. Since they are comfortable in Hindi and English language, the posted text comprises features of both the language. The massive influx of text data from such digital platforms is pushing the situation of digital information explosion and prompted the need for automatic text classification for maintaining, organizing, and discovering knowledge. The unstructured nature of text documents is the fundamental impediment to achieving good classification accuracy. The use of preprocessing steps and strategies transforms unstructured data into structured data, which can alleviate various difficulties [26–28]. Text classification is the process of assigning texts in the documents to some predefined categories based on their content. In the current scenario, organizing and categorizing electronic documents has become a complex task due to the growing number of text documents in different formats. Text classification is successfully applied to various domains such as topic detection, spam e-mail filtering, SMS spam filtering, author identification, web page classification, and sentiment analysis. The raw text documents contain many irregularities which are not good for text classification tasks.

S. Avasthi
Amity Institute of Information Technology, Amity University, Noida, India

R. Chauhan (✉)
Center for Compuational Biology and Bioinformatics, Amity University, Noida, India
e-mail: rituchauha@gmail.com

D. P. Acharjya
VIT, Vellore, India

The commonly used preprocessing task include conversion to lower case, removing punctuation marks, stop-word removal, lemmatization, and stemming. In this context, pre-processing refers to the process of cleaning and preparing texts for classification. It's a reality that unstructured messages on the Internet — especially, in our case, on Twitter — have a lot of noise in them. The term "noise" refers to data that contains no valuable information for the analysis at hand, in this example, sentiment analysis. As described in [2], the percentage of noise is 40% in general that causes problems in text classification and other machine learning algorithms. Users on various social media platforms such as Twitter and blog sites are prone to typographical, spelling errors, and use of local slang. To emphasize their emotions, users may also use a lot of punctuation symbols and emoticons. While performing ML, usually many terms in the texts can be replaced, merged, or just ignored for better interpretation of results. Therefore, we need proper cleaning and normalization of the text data as their quality is the primary factor for the accuracy of machine learning [3].

In this paper, we conversed about the importance of preprocessing techniques in the text classification process pipeline and propose a text classification framework based on preprocessing pipeline. Hindi document sentence segmentation, tokenization (word boundary identification), Hindi stop words deletion, and stemming are all part of the pre-processing phase. The root words for inflected words are found using the Hindi language stemmer. Stemming can be used to increase sentence retrieval efficiency. Any Hindi language text can be used as input in the pre-processing step, and the result is structured text. It also enables the removal of duplicate sentences before the processing stage [4, 25].

The objective of this paper is to gather many common pre-processing techniques from previous studies, as well as a few novel ones such as replacing contractions and applying text normalization to classification accuracy in sentiment classification and the number of features on chosen Hindi and English datasets of tweets and reviews. Many research papers are available which focus on English language text data, here an attempt has been made to provide a deeper understanding of the Hindi language with Devanagari script using supervised machine learning methods. Hindi language is morphologically very rich and considered to be a free order language but there is a lack of annotated data needed for learning models.

Section 1 in the research paper generalizes the introduction, Sect. 2 describes the related works, methodology, and framework are discussed in Sect. 3, text classification algorithms and evaluation metrics described in Sect. 4, results and conclusion are given in the 5 and 6 sections respectively.

2 Related Works

In the paper [5, 6], preprocessing techniques are discussed and evaluated on two different languages datasets of e-mails and news. The methods like stop word removal, lowercase conversion, and stemming for text classification purposes.

Further, they concluded that no fixed combination of pre-processing techniques that can be used for improving accuracy in classification problem. The text preprocessing methods plays important role in the text classification pipeline. The text classification results can be improved significantly if appropriate methods of preprocessing are used. Many research publications make a distinction between these strategies and how they affect classification results. The advantages of pre-processing on Twitter data for the improvement of sentiment analysis are studied in this research [7, 8]. The paper [9] provides an evaluation study of several preprocessing tools for English text classification.

Stop word removal, word stemming, indexing with term frequency (TF), weighting with inverse document frequency (IDF), and normalizing each document feature vector to unit length were all addressed in the paper [10]. Using a linear SVM and varied lengths of a BOW representation, these combinations were applied to two benchmark datasets: Reuters-21578 and 20 Newsgroups. The results of their tests revealed that normalization text classifiers to unit length can always maximize their accuracy.

The authors [11] analyzed the performance of three preprocessing methods on TC for an in-house corpus containing 32,620 news documents divided into ten categories downloaded from various Arabic news websites (stop word removal, word stemming, and normalization of certain Arabic letters that have different forms in the same word to one form). Further, the paper applies kNN, SVM, and Naïve-bayes to the Arabic language to prove that preprocessing significantly improves classification accuracy The stemming technique is described in paper [12] to minimize various grammatical forms or word forms such as noun, adjective, verb, adverb, and so on. Stemming is the process to reduce a word's inflectional form to its most common base form that is occasionally related. This paper compares various stemming strategies in terms of their use, benefits, and downsides.

3 Best Preprocessing Strategies for Hindi and English Text

Both Hindi language text and English language text require different strategies in preprocessing. Few steps such as tokenization, stemming, foreign word removal, removal of numbers, or punctuation marks are common in both languages. For example, conversion to lower case is not required in the Hindi language since Hindi language letters have only one type of letter. The bilingual text data needs various steps of preprocessing and then clean corpus is used in various classification or clustering tasks. The proposed text classification framework that incorporates various text normalization steps is illustrated in Fig. 1.

Remove Numbers. Many classification models do not consider numbers because numbers do not contain any sentiment, hence it's a common strategy to remove them in preprocessing pipeline. However, in some specific application number provides useful insights so numbers are not removed.

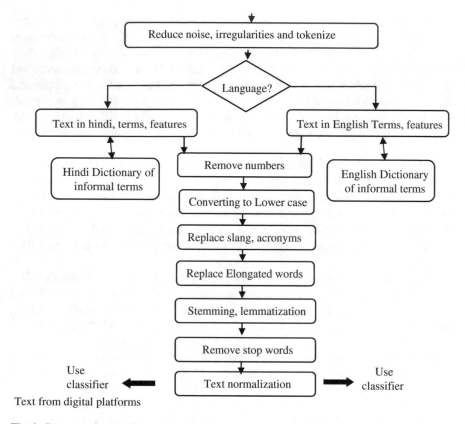

Fig. 1 Preprocessing pipeline

Conversion to Lowercase. It is the most popular preprocessing method; numerous words are merged in this procedure, and the problem's dimensionality is minimized.

Text Imputation. In case of text data, imputation methods are not used. Imputing make sense where regression models and numerical values are to be handled. For example in sentiment analysis or opinion mining, imputing text to fill missing reviews will be frowned upon. The strategies useful in case of missing review is discard that record from dataset table.

Replacing Acronyms and Slang. On social media and business sites, people type informally, thus their writings are full of slang and abbreviations. To be properly comprehend the meaning, such words must be replaced. Each language has its own set of local slang or abbreviations. For example, in the English language 'ty' and 'omg' mean 'thank you' and 'oh my god' respectively [13].

Replace Elongated Words. The word "sweeeet" is elongated because it contains a character that repeats more than two times. It's crucial to replace these terms with their source words so that they can be combined.

Stemming. It is the process of removing the endings of the words to detect their root form. After this step many similar words having the same root can be merged [14]. Stemming is a technique for improving retrieval efficiency and reducing indexing file size. A stemmer will merge all the occurrences of 'user', 'users','used','using','use' to 'use' only.

Lemmatization. In this method, the endings of the words are removed to find their lemmas, i.e., their root forms in the dictionary. It is a very important step in inflected language. With a prefix, suffix, or infix, or another internal modification such as a vowel change, an inflection expresses one or more grammatical categories. Both Hindi and English have different degree of inflection [15].

Remove Stop Words. All the function words with the high frequency of occurrences in sentences are called Stop words. Such words contain no useful information, so it is needless to analyze them. The set of these words are not completely predefined and can be modified as per the application requirements. Each language has their own set of stop words so for Hindi and English, a different set of stop words set were used using NLTK library in python.

Text Normalizations. Some useful text normalizations strategies include replacement of URLS, username, converting transliterated words in Hindi and English, spelling corrections, replacing of emoji's with words, emoticons, removing punctuation, and converting capitalized words [16, 17]. For example, converting transliterated sentence 'yeh chalta nahi hai ab' is converted to Devnagri letters 'यह चलता नहीं है अब'or 'कंप्यूटर साइंस'is converted to 'computer science' in English letters.

4   Text Classification Experiments

Three classifier model for text classification is described briefly here that are used to see the improvement after applying text preprocessing strategies on chosen datasets.

4.1   Datasets and Preprocessing

We used two datasets in the Hindi language, the first one is a dataset of tweets and the other one is movies reviews available on IIT Mumbai webpage [18, 19]. For the English language, the first one is a dataset of tweets [20] and the other one is a dataset of amazon product reviews available at Kaggle [21]. The summary of all the datasets is given in Table 1.

Table 1 Datasets used, Document size, Total count of words

Language	Dataset name	Doc Size D	Word count	Unique words
Hindi	Tweets (hi_3500)	4999	147,138	21,952
Hindi	Movie review (IITB)	718	489,573	22,842
English	Tweets (TC19)	295,665	4,867,733	601,398
English	Product reviews	98,256	53,396,130	845,820

4.2 Classification Techniques

Three classifiers are employed in studies to see how successful preprocessing processes are: Nave-Bayes (NB), Logistic regression (LR), and Support Vector Machine (SVM). In the classification task, both reviews and tweets are classified into three classes positive, negative, and neutral [22].

Naïve-Bayes: The Bayes Theorem is basis for classification algorithm known as Naive Bayes classifier. It is a family of algorithms that share a similar idea, namely that each pair of features being classified is independent of the others. The model can apply to text data based on feature matrix and can classify text.

SVM: In a high- or infinite-dimensional space, a support-vector machine develops a hyperplane or set of hyperplanes that can be used for classification, regression, or other tasks like identifying outliers. Given a set of training examples, each labelled as belonging to one of two categories, an SVM training algorithm provides a model that assigns new examples to one of two categories, making it a non-probabilistic binary linear classifier [23, 24].

Logistic Regression: A supervised learning approach, it is based on regression equation. The equation is used to calculate value of dependent variable, which is based on one or more than one independent variable. In Eq. 1, b_0 and b_1 are regression coefficients.

$$y = b_0 + b_1 * x \tag{1}$$

5   Results and Discussions

Sample text from Hindi dataset1 and 2 are given in Table 2. Sample texts from English datasets 1 and 2 are given in Table 4.

Table 2 Sample tweets from the Hindi datasets

Text Type	Polarity	Dataset
यूनूस सेजवाल का स्क्रीनप्ले बेहद घटिया है और सैकड़ों बार देखे हुए दृश्यों को उन्होंने फिर परोसा है।	Negative	Movie review
डायरेक्टर की तारीफ करनी होगी कि उन्होंने बॉक्स ऑफिस का मोह त्याग कर कहानी और किरदारों को पूरी तरह से माहौल में समेटा है।	Positive	Movie review
लोग वतन तक खा जाते हैं इसका इसे यकीन नहींमान जाएगा तू ले जाकर दिल्ली इसे दिखा ला दोस्त	Negative	Tweets
RT @aajtak: लव जेहाद के नाम पर क्या उत्तर प्रदेश में हो रही है गुंडागर्दी? देखिए #Hallabol @anjanaomkashyap	Neutral	Tweets

Both tweets and reviews datasets were used for Hindi and English languages. For the movie dataset ratings '0', '1', '2' were used as a label to classify both Hindi and English reviews datasets. The polarity score and top 20 phrases connected to people's emotions, such as negativity and positivity, were used to examine the emotion of people's tweets. Table 3 shows some examples of tweets in the English language with their polarity scores. We can see results in Table 4 that describes those preprocessing strategies improve text classification accuracy significantly.

Table 3 Sample tweets from the English datasets

Text type	Polarity	Dataset
I enjoy vintage books and movies so I enjoyed reading this book	Positive	Product review
Love this one! The author did a great job giving substance to her characters	Positive	Product review
My neighbors say the #COVID19 #vaccine is made from dead babies and has microchips in it that makes you not like guns	Negative	Tweets (TC19)
#COVID19 Transmission News: #AstraZeneca's #vaccine appears to reduce transmission of the #virus and offers strong https://t.co/F26c1w2gea	Positive	Tweets (TC19)

Table 4 Classification accuracy for three classifiers

Model	Classifier	Classification accuracy			
		Dataset 1	Dataset 2	Dataset 3	Dataset 4
No preprocessing	NB	0.70	0.74	0.88	0.88
	LR	0.92	0.94	0.90	0.91
	SVM	0.89	0.90	0.89	0.88
After applying preprocessing	NB	0.70	0.71	0.88	0.90
	LR	0.92	0.93	0.92	0.94
	SVM	0.95	0.95	0.93	0.94

6 Conclusion

Real-world text data comes with lots of irregularities in it, and performing advanced machine learning to get better results, we need different preprocessing techniques to make such text corpus tidy. In-text classification, preprocessing techniques can improve classification accuracy significantly. We investigated using three classifiers SVM, Naïve-Bayes, and Logistic regression compared them on two datasets both in Hindi and English languages. The main preprocessing methods are replacement, removal of stop words stemming, text normalization. In comparison to the base scenario with no preprocessing, the modelling results show that utilizing preprocessing enhanced the results. In future work, we can explore further the effect of applying preprocessing techniques using other machine learning algorithms and taking text datasets of few other languages spoken in the world.

References

1. Census report. https://censusindia.gov.in/2011Census/C16_25062018_NEW.pdf
2. Fayyad UM, Piatetsky-Shapiro G, Uthurusamy R (2003) Summary from the KDD-03 panel: data mining: the next 10 years. ACM SIGKDD Explorations Newsl 5(2):191–196
3. Effrosynidis D, Symeonidis S, Arampatzis A (2017).A comparison of pre-processing techniques for Twitter sentiment analysis. In: Kamps J, Tsakonas G, Manolopoulos Y, Iliadis L, Karydis I (eds) Research and Advanced Technology for Digital Libraries. TPDL 2017. LNCS, vol 10450, pp 394–406. Springer, Cham. https://doi.org/10.1007/978-3-319-67008-9_31
4. Desai NP, Dabhi VK (2021) Taxonomic survey of Hindi Language NLP systems. arXiv preprint arXiv:2102.00214
5. Uysal AK, Gunal S (2014) The impact of preprocessing on text classification. Inf Process Manag 50(1):104–112
6. Avasthi S, Chauhan R, Acharjya DP (2021) Techniques, applications, and issues in mining large-scale text databases. In: Goar V, Kuri M, Kumar R, Senjyu T (eds) Advances in Information Communication Technology and Computing. LNNS, vol 135, pp 385–396. Springer, Singapore. https://doi.org/10.1007/978-981-15-5421-6_39
7. Singh T, Kumari M (2016) Role of text pre-processing in twitter sentiment analysis. Procedia Comput Sci 89:549–554
8. Avasthi S, Chauhan R, Acharjya DP (2021) Processing large text corpus using n-gram language modeling and smoothing. In: Goyal D, Gupta AK, Piuri V, Ganzha M, Paprzycki M (eds) Proceedings of the Second International Conference on Information Management and Machine Intelligence. LNNS, vol 166, pp 21–32. Springer, Singapore. https://doi.org/10.1007/978-981-15-9689-6_3
9. Kadhim AI (2018) An evaluation of preprocessing techniques for text classification. Int J Comput Sci Inf Secur (IJCSIS) 16(6):22–32
10. Song F, Liu S, Yang J (2005) A comparative study on text representation schemes in text categorization. Pattern Anal Appl 8(1–2):199–209
11. Ayedh A, Tan G, Alwesabi K, Rajeh H (2016) The effect of preprocessing on Arabic document categorization. Algorithms 9(2):27
12. Jivani AG (2011) A comparative study of stemming algorithms. Int J Comput Tech Appl 2(6):1930–1938
13. Miller GA (1995) WordNet: a lexical database for English. Commun ACM 38(11):39–41

14. Jabbar A, Iqbal S, Tamimy MI, Hussain S, Akhunzada A (2020) Empirical evaluation and study of text stemming algorithms. Artif Intell Rev 53(8):5559–5588

15. Babhulgaonkar A, Shirsath M, Kurdukar A, Khandare H, Tekale A, Musale M (2021) Empirical laws of natural language processing for Hindi language. In: Gunjan VK, Zurada JM (eds) Proceedings of International Conference on Recent Trends in Machine Learning, IoT, Smart Cities and Applications. Advances in Intelligent Systems and Computing, vol 1245, pp 217–223. Springer, Singapore. https://doi.org/10.1007/978-981-15-7234-0_18

16. Makhija P, Kumar A, Gupta A (2020) hinglishNorm–A Corpus of Hindi-English Code Mixed Sentences for Text Normalization. arXiv preprint arXiv:2010.08974

17. Ali MA, Kulkarni SB (2021) Preprocessing of text for emotion detection and sentiment analysis of Hindi movie reviews. SSRN 3769237

18. HaCohen-Kerner Y, Miller D, Yigal Y (2020) The influence of preprocessing on text classification using a bag-of-words representation. PLoS ONE 15(5):e0232525

19. Setiabudi R, Iswari NMS, Rusli A (2021) Enhancing text classification performance by preprocessing misspelled words in Indonesian language. TELKOMNIKA 19(4):1234–1241

20. Covid19 Tweets. https://www.kaggle.com/sandhyaavasthi/covid19-tweetsjuly2020decembe r2020. Accessed 18 Aug 2021

21. Amazon Review: Kindle. https://www.kaggle.com/bharadwaj6/kindle-reviews

22. Kumar A, Dabas V, Hooda P (2020) Text classification algorithms for mining unstructured data: a SWOT analysis. Int J Inf Technol 12(4):1159–1169

23. Pisner DA, Schnyer DM (2020) Support vector machine. In: Machine Learning, pp 101–121 Academic Press

24. Avasthi S, Chauhan R, Acharjya, D.P. (2022). Information extraction and sentiment analysis to gain insight into the COVID-19 crisis. In: Khanna A, Gupta D, Bhattacharyya S, Hassanien AE, Anand S, Jaiswal A (eds) International Conference on Innovative Computing and Communications. AISC, vol 1387, pp 343–353. Springer, Singapore. https://doi.org/10.1007/978-981-16-2594-7_28

25. Kaity M, Balakrishnan V (2020) Sentiment lexicons and non-English languages: a survey. Knowl Inf Syst 62(12):1–36

26. Chauhan R, Kaur H, Chang V (2020) An optimized integrated framework of big data analytics managing security and privacy in healthcare data. Wirel Pers Commun. https://doi.org/10.1007/s11277-020-07040-8

27. Chauhan R, Kaur H, Chang V (2017) Advancement and applicability of classifiers for variant exponential model to optimize the accuracy for deep learning. J Ambient Intell Humaniz Comput. https://doi.org/10.1007/s12652-017-0561-x

28. Chauhan R, Kaur H (2015) SPAM: an effective and efficient spatial algorithm for mining grid data. In: Geo-Intelligence and Visualization through Big Data Trends, pp 245–263. IGI Global, 2015, Web, 9 September 2015. https://doi.org/10.4018/978-1-4666-8465-2.ch010, https://www.igi-global.com/chapter/spam/136107

CD-KNN: A Modified K-Nearest Neighbor Classifier with Dynamic K Value

Khumukcham Robindro, Yambem Ranjan Singh, Urikhimbam Boby Clinton, Linthoingambi Takhellambam, and Nazrul Hoque

1 Introduction

K-nearest Neighbor (KNN) classifier first proposed by T. M. Cover and P. E. Hart has been used in Machine Learning from last five decades [2]. Due to its simplicity and effectiveness, KNN classifier gains popularity in the field of pattern recognition, text classification, and network intrusion detection [5]. The KNN classifier uses two important parameters during classification, viz., (i) a distance measure to find the nearest neighbors and (ii) the number of nearest neighbors, i.e., the value of K. For the first case, we can use different distance measures in KNN but Euclidean distance is one of the widely used measures in KNN classifiers. The second one is an important parameter that plays a vital role in prediction of an unknown test instance. As claimed by Anil k. Ghose [3], there is no specific theoretical measure to compute an optimal K value for small or moderately large sample sizes. The optimal value of K depends on the specific data set and it is to be estimated using the available training sample observations. In this context, people use various measures to compute an optimal value for K from the training samples. But, a static value of K may lead to poor accuracy if the training dataset is unbalanced in terms of classes. To overcome this problem, many researchers and machine learning practitioners developed methods to compute the K value dynamically. In this paper, we discussed a method called CD-KNN that computes K dynamically for each test instance. In our method, we create clusters or group of instances based on their classes and for each cluster, we estimate the density of instances to find a value of K. KNN is a widely used classification method that unlike other classifiers, performs training and testing at the same time. Upon receipt of a testing object, it analyses proximity of the testing object. Major class label of the objects in its proximity is predicted as the

K. Robindro (✉) · Y. R. Singh · U. B. Clinton · L. Takhellambam · N. Hoque
Department of Computer Science, Manipur University, Canchipur, Imphal West, Manipur 795003, India
e-mail: rkbh@manipuruniv.ac.in

© The Author(s), under exclusive license to Springer Nature Singapore Pte Ltd. 2022 753
B. Unhelker et al. (eds.), *Applications of Artificial Intelligence and Machine Learning*, Lecture Notes in Electrical Engineering 925, https://doi.org/10.1007/978-981-19-4831-2_62

class label of the testing object. The liberty to employ different proximity measures such as Euclidean distance, cosine distance and Pearson correlation in this phase of proximity analysis makes KNN one of the most flexible classifiers to meet diverged demands of various application domains. Another feature of KNN lies in tuning the input parameter K, which allows the user to adjust proximity analysis according to the availability of noisy objects or outliers.

In literature, we found a significant number of works on KNN as well as on dynamic KNN [6, 8–10]. Although, KNN classifier is very simple to implement but it has some drawbacks too, viz., (i) computational complexity of KNN is high (ii) selection of K value efficiently, (ii) handling tie condition for multiclass classification. Some of the issues can be handled by reducing the feature space of the dataset such that the feature space consists of only relevant, informative and non-redundant features [4]. Many machine learning practitioners try to improve the performance of KNN classifier using less number of features selected from the original feature sets. On the other hand, classification accuracy can also be improved by choosing the K value dynamically. An improved KNN classifier with dynamic K value is developed by Zhong et al. [11]. The method first considers an interval of K as K_{min}, K_{max} and then computes an optimal value for K based on the distribution of instances in a particular class. A dynamic set self-join method using dynamic nearest neighbor is proposed by Amagata et al. [1]. An adaptive KNN classifier using dynamic K is developed by Ougiaroglou et al. [7]. The authors use three different heuristics in search of K value. The performance of the method is found effective on various datasets. Onyezewe, Anozie, et al. [12] proposed a meta heuristic search algorithm using Simulated Annealing, to select optimal K, in K-Nearest Neighbor algorithm. Their proposed method shows the efficiency over K-Nearest Neighbor in terms of computational. Zheng, Xin, et al. [13] proposed a method called KNN-MT to improve the translation accuracy. Their method determines the number of K for each target token dynamically by introducing a light-weight Meta-k Network, that can be efficiently trained with only a few training samples. Diamantaras et al. [14] proposed a new algorithm with five heuristics for dynamic K determination. Their algorithm is based on a fast clustering pre-processing procedure that builds an auxiliary data structure. The heuristics exploit the information and dynamically determine how many neighbors will be examined.

1.1 Motivation

Traditional KNN with fixed value of K suffers low classification accuracy in unbalanced datasets. So, instead of making the dataset balanced, if we could choose an optimal value for each test object dynamically, then it is expected that the performance of dynamic KNN will be better as compared to fixed KNN classifier. This motivates us to develop a modified KNN classifier known as CD-KNN to classify unknown instances using dynamic K nearest neighbors on unbalanced datasets.

1.2 Problem Definition

Let's consider a dataset D having N numbers of instances. The dataset is divided into two parts say, D_{train} and D_{test}. Now, to classify an unknown instance $T_i \in D_{test}$, compute the nearest neighbors, i.e., K for each instance T_i.

1.3 Contributions

The main contributions of this paper are twofold.

1. A cluster based method is used to find the similar objects of a test instance in each cluster based on Euclidean distance.
2. For each test instance, an optimal value of K is computed by considering the number of similar objects distributed over each cluster.

In the organization of the paper, related work is discussed in Sect. 1. The proposed method and the algorithm are elaborated in Sect. 2. The experimental results are discussed in Sect. 3. Final conclusion and future work is mentioned in Sect. 4.

2 Proposed Method

The proposed method called Cluster-based Dynamic K-Nearest Neighbor (CD-KNN) computes the K value dynamically for each test instance to be classified. First, the proposed method divides the training instances into clusters based on the class labels of the instances. Instances having similar class label belong to the same cluster and the number of cluster is equal to the number of classes of the dataset. The symbols used to explain the steps of the CD-KNN algorithm are defined in Table 1.

Let us consider a dataset D having p number of classes as $C_1, C_2, \cdots, C_p$ and the dataset is divided into two parts, say D_{train} and D_{test}. Suppose, in D_{train} there are total m number of instances and in D_{test} there are n number of instances. Now,

Table 1 Symbols and their meanings

Symbols	Meanings	Symbols	Meanings
D	Dataset	F	Original feature set
D_{test}	Testing set	D_{train}	Training set
p	Total number of classes in D	C_l	Cluster with index l
T_i	A test instance	O_i	A training instance
P	Precision	R	Recall
F_1	$F1$ Score	Acc	Accuracy

the D_{train} data part is again divided into p number of clusters where instances with same class label belong to one cluster. Next, for a test instance $T_i \in D_{test}$, compute the distance of T_i with the instances $O_j \in C_l$ where $1 \leq l \leq p$. From each cluster instances and their corresponding distance to the test instance T_i, compute the mean distance and find out the number of instances (ω_l) whose distance to T_i is less than or equal to the mean distance in that particular cluster. We compute the K value of the test instance w.r.t cluster C_l as shown in the following Eq. 1.

$$K_l = \frac{\omega_l}{N_l} \times \alpha \tag{1}$$

where, ω_l is the number of instances in cluster C_p whose distance to the test instance T_i is less than or equal to the mean distance. N_l is the total number of instances in a cluster C_l and α is a constant whose value lies between 5 to 10. This way we compute p numbers of K values using Eq. 1 and the maximum one is considered as the final K value for the test instance T_i.

2.1 Conceptual Framework of CD-KNN Method

The conceptual framework of the propose method is depicted in the following Fig. 1. The proposed method works based on the framework.

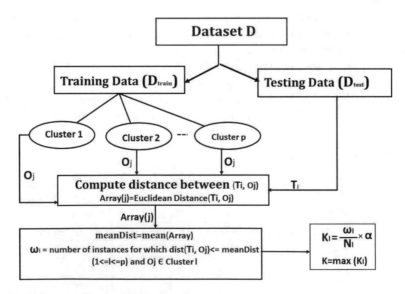

Fig. 1 Framework of the CD-KNN Method

The steps of the CD-KNN algorithm are shown in the Algorithm 1.

Data: Dataset: D having p number of class
Result: Prediction accuracy
Step 1:
Divide the dataset D into D_{train} and D_{test}
Step 2:
Divide the training dataset D_{train} in p number of clusters C_l such that $1 \leq l \leq p$
Step 3:
for each test instance $T_i \in D_{test}$ **do**
   for $l = 1$ to p do
      for each training instance $O_j \in C_l$ **do**
         | array (j) = store Euclidean Distance value between T_i, O_j
      end
      compute mean of array, i.e., meanDist
      ω_l = the number of instance O_j in cluster C_l for which the distance between O_j
         and T_i is less than or equal to meanDist
      N_l =Total number of objects in cluster C_l
      $K_l = \frac{\omega_l}{N_l} \times \alpha$
   end
   K= $max(\prod_l^p k_l)$
   find the K nearest neighbors from the training data D_{train} and assign the class label
   that has maximum occurrences to the test instance T_i
end

Algorithm 1. CD-KNN algorithm

2.2 How Does CD-KNN Differ from KNN?

The proposed CD-KNN method is different from the traditional KNN classifier as pointed out below.

1. Unlike traditional KNN, CD-KNN does not take the value of K as an input parameter.
2. Unlike traditional KNN, the proposed CD-KNN method computes the value of K dynamically.
3. Unlike traditional KNN, CD-KNN considers the distribution of instances in each cluster of the training data.
4. Like traditional KNN, the proposed method predicts the class label of an unknown instance from the K nearest neighbors.

Table 2 Dataset description

Sl. no	Dataset	No. of instance	No. of features	Data type	No. of class
1	Sonar	207	61	Real	2
2	Parkinson disease	195	24	Real	2
3	Tictactoe endgame	958	10	Categorical	2
4	Labor	57	17	Categorical	2
5	Ionosphere	351	35	Integer, real	2
6	Unbalance	856	33	Real, categorical	2
7	Breast cancer	286	10	Real	2
8	Wine	1599	12	Integer, real	6
9	Vote	435	17	Categorical	2

2.3 Complexity Analysis

The complexity of the proposed method depends on the number of training instances as well as test instances. Suppose, the training part has m number of instances and testing part has n number of instances. Since, we make p number of clusters from the training data and the cost of making cluster is say $O(m)$. As compared to m and n, value of p is very negligible. So, in step 3 of the Algorithm 1, although there are three for loop but the overall complexity of the CD-KNN will be $O(m \times n)$. Since p is very negligible, so $O(m \times p \times n) \equiv O(m \times n)$.

3 Experimental Analysis

The proposed CD-KNN method is developed in Windows 10 operating systems having 2.6 GHz processor, 8 GB main memory and 1 TB secondary storage. The method is implemented using Python programming language. We use various python packages like Pandas, Numpy, Scikit-learn, etc., in the implementation of CD-KNN method.

3.1 Dataset Description

To evaluate the performance of the proposed CD-KNN method, we use 9 UCI machine learning datasets available in.[1] The parameters of the datasets such as number of instances, number of features, number of classes, and attributes types are mentioned in the Table 2.

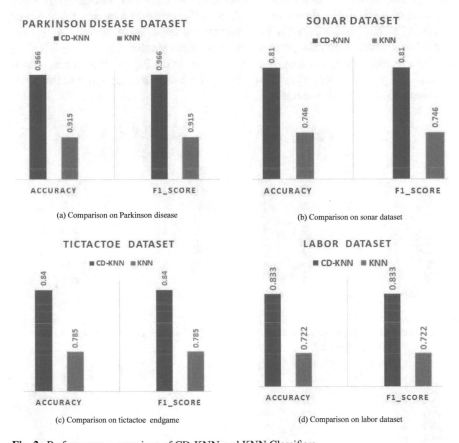

(a) Comparison on Parkinson disease

(b) Comparison on sonar dataset

(c) Comparison on tictactoe endgame

(d) Comparison on labor dataset

Fig. 2 Performance comparison of CD-KNN and KNN Classifiers

[1] https://archive.ics.uci.edu/ml/index/php.

3.2 Result Analysis on UCI Datasets

The proposed CD-KNN method is validated on 9 UCI datasets and the method is compared with traditional KNN classifier in terms of accuracy and F1-Score as shown in Figs. 2, 3 and 4. The CD-KNN method is compared to the traditional KNN classifier in terms of accuracy, precision, recall, and F1-Score as shown in the Table 3. In most of the datasets, our CD-KNN yields better result as compared to traditional KNN classifier. Especially, as shown in Fig. 2, classification accuracy and F1-Score of CD-KNN is much better than KNN on Parkinson disease, sonar, tictactoe endgame, and labor datasets. Also, the proposed method gives similar result to the KNN on ionosphere, unbalance, and vote datasets as shown in the Fig. 3. However, as shown in the Fig. 4 performance of the CD-KNN is a bit lower in comparison to the KNN on wine and breast cancer datasets.

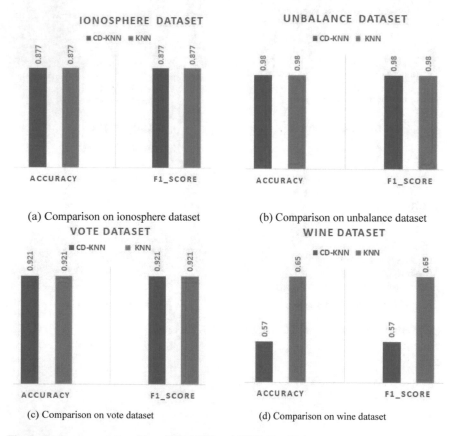

(a) Comparison on ionosphere dataset

(b) Comparison on unbalance dataset

(c) Comparison on vote dataset

(d) Comparison on wine dataset

Fig. 3 Performance comparison of CD-KNN and KNN Classifiers

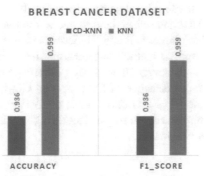

(a) Comparison on breast cancer dataset

Fig. 4 Performance comparison of CD-KNN and KNN Classifiers

Table 3 Comparison of KNN and CD-KNN in terms of accuracy, precision, recall and F1-score measures

Dataset	Type of KNN	Accuracy	F1-score	Precision	Recall
1. Sonar	(CD-KNN)	**0.81**	0.81	0.81	0.81
	(Standard KNN)	0.746	0.746	0.746	0.746
2. Parkinson disease	(CD-KNN)	**0.966**	0.966	0.966	0.966
	(Standard KNN)	0.915	0.915	0.915	0.915
3. Tictactoe endgame	(CD-KNN)	**0.833**	0.833	0.833	0.833
	(Standard KNN)	0.785	0.785	0.785	0.785
4. Labor	(CD-KNN)	**0.833**	0.833	0.833	0.833
	(Standard KNN)	0.722	0.722	0.722	0.722
5. Vote	(CD-KNN)	**0.924**	**0.924**	0.924	0.924
	(Standard KNN)	0.924	0.924	0.924	0.924
6. Ionosphere	(CD-KNN)	**0.877**	0.877	0.877	0.877
	(Standard KNN)	0.877	0.877	0.877	0.877
7. Unbalanced	(CD-KNN)	**0.984**	0.981	0.981	0.981
	(Standard KNN)	0.984	0.984	0.984	0.984
8. Wine	(CD-KNN)	0.571	0.571	0.571	0.571
	(Standard KNN)	0.683	0.683	0.683	0.683
9. Breast cancer	(CD-KNN)	0.936	0.936	0.936	0.936
	(Standard KNN)	0.959	0.959	0.959	0.959

4   Conclusion and Future Work

In this paper, we try to address the issue of choosing an appropriate value for K in KNN classifier. To overcome the issue, we developed a modified KNN classifier known as

CD-KNN that computes the K value dynamically for each test instance. The method is evaluated on various UCI datasets and compared to the traditional KNN classifier in terms of accuracy, precision, recall and F1-Score. From the experimental results we observed that the proposed method outperforms the traditional KNN classifier on most of the datasets. However, on very high dimensional datasets having less number of instances the performance of the proposed CD-KNN may be less. The computational cost of the CD-KNN classifier is equivalent to the KNN classifier. As a future work, we are planning to implement the CD-KNN method in CUDA C parallel programming environment.

References

1. Amagata D, Hara T, Xiao C (2019) Dynamic set KNN self-join. In: 2019 IEEE 35th International Conference on Data Engineering (ICDE), pp 818–829. IEEE
2. Cover T, Hart P (1967) Nearest neighbor pattern classification. IEEE Trans Inf Theory 13(1):21–27
3. Ghosh AK (2006) On optimum choice of k in nearest neighbor classification. Comput Stat Data Anal 50(11):3113–3123
4. Hoque N, Bhattacharyya DK, Kalita JK (2014) MIFS-ND: a mutual information based feature selection method. Expert Syst Appl 41(14):6371–6385
5. Hoque N, Bhuyan MH, Baishya RC, Bhattacharyya DK, Kalita JK (2014) Network attacks: taxonomy, tools and systems. J Netw Comput Appl 40:307–324
6. Kramer O (2013) K-Nearest Neighbors. In: Dimensionality Reduction with Unsupervised Nearest Neighbors, pp. 13–23. Springer, Berlin, Heidelberg. https://doi.org/10.1007/978-3-642-38652-7_2
7. Ougiaroglou S, Nanopoulos A, Papadopoulos AN, Manolopoulos Y, Welzer-Druzovec T (2007) Adaptive k-nearest-neighbor classification using a dynamic number of nearest neighbors. In: Ioannidis Y, Novikov B, Rachev B (eds) Advances in Databases and Information Systems. ADBIS 2007. LNCS, vol 4690, pp 66–82. Springer, Berlin, Heidelberg. https://doi.org/10.1007/978-3-540-75185-4_7
8. Xia S, Xiong Z, Luo Y, Dong L, Zhang G (2015) Location difference of multiple distances based k-nearest neighbors algorithm. Knowl-Based Syst 90:99–110
9. Yunus R, Ulfa U, Safitri MD (2021) Application of the k-nearest neighbors (K-NN) algorithm for classification of heart failure. J Appl Intell Syst 6(1):1–9
10. Zhang Z (2016) Introduction to machine learning: k-nearest neighbors. Ann Transl Med 4(11)
11. Zhong XF, Guo SZ, Gao L, Shan H, Zheng JH (2017) An improved K-NN classification with dynamic k. In: Proceedings of the 9th International Conference on Machine Learning and Computing, pp 211–216
12. Onyezewe A et al (2021) An enhanced adaptive k-nearest neighbor classifier using simulated annealing. Int J Intell Syst Appl 13(1):34–44
13. Zheng X et al (2021) Adaptive nearest neighbor machine translation. arXiv preprint arXiv:2105.13022
14. Ougiaroglou S, Evangelidis G, Diamantaras KI (2020) Dynamic k-NN classification based on region homogeneity. In: Darmont J, Novikov B, Wrembel R (eds) New Trends in Databases and Information Systems. ADBIS 2020. Communications in Computer and Information Science, vol 1259, p 27. Springer, Cham. https://doi.org/10.1007/978-3-030-54623-6_3

Automated Classification of Hyper Spectral Image Using Supervised Machine Learning Approach

Rajashree Gadhave and **R. R. Sedamkar**

1 Introduction

A hyperspectral image is an assortment of a hundred to thousand numbers of highly correlated and informative spectral bands. Various colour tones imitate various bands or frequencies and same hyperspectral bands are apprehended by exceptionally designed hyperspectral sensors from the electromagnetic range of light from remote satellites or committed flying sensors [1]. Hyperspectral sensors offer better adequacy in deciding spectral data and thus more productive when contrasted with multispectral sensors [4, 6, 7]. The main objective of segmentation of hyperspectral image is to allot every pixel of the picture a proper spectral class. Hyperspectral imaging has many wide varieties of applications in horticulture, astronomy, mineralogy and so on. Most of the researchers try to decrease the errors in the characterization of HSI towards a little value by using different features like spectral, spatial and spectral-spatial strategies. For achieving better precision, it is important to select appropriate features. This paper investigates the aggregate impact of features on the order precision just as on the exactness patterns. Clustering methods for hyperspectral imaging extensively characterized into three primary classes: spectral, spatial and spectral-spatial. Numerous effective actions for characterization of hyperspectral imaging utilizing these investigation techniques are acknowledged [13, 14, 17].

For various shaded items, various bands mirrored that creates hyperspectral image gigantic if there should arise an occurrence of spectral dimensionality. But since of the closeness between objects the bands captured are exceptionally associated. These profoundly associated bands can be abandoned for classification and consequently

R. Gadhave (✉)
University of Mumbai, Mumbai, India
e-mail: rajashree86@gmail.com

R. R. Sedamkar
Thakur College of Engineering, Mumbai, India
e-mail: rr.sedamkar@thakureducation.org

© The Author(s), under exclusive license to Springer Nature Singapore Pte Ltd. 2022
B. Unhelker et al. (eds.), *Applications of Artificial Intelligence and Machine Learning*,
Lecture Notes in Electrical Engineering 925,
https://doi.org/10.1007/978-981-19-4831-2_63

the component of hyperspectral images decreased significantly. The other technique which is considered to decrease dimensionality is extraction where associated bands are joined together to produce another arrangement of features. From resulting hyperspectral features, the most instructive arrangement of features is chosen which brings about a decrease in dimensionality. Principal component analysis (PCA) produces a low-dimensional portrayal of the information made for the most part converging on variance of the information [8]. A linear basis of low dimensionality for an information is produced considering the maximal measure of variance of the information [3]. This paper represents a HSI classification with spatial and spectral feature extraction and feature selection using the supervised machine learning techniques, SVM and MLP for the betterment of classification accuracy of two HSI dataset.

2 Related Work

Hyperspectral picture order performed by different specialists is summed up and talked about in this segment. A. A. Happiness et al. [2] assumed the approach of feature extraction particularly, Principal Component Analysis (PCA) and Linear Discriminant Analysis (LDA) and the two of them consolidated and afterward characterized the dataset utilizing Support Vector Machine classifier. The exploratory outcomes show that LDA approach gives the best exactness of 86.53% among the three different feature reductions. L. Younus et al. [3] examine the impact of reduction of dimension in hyperspectral information classification frameworks. A non-linear dimensionality optimization method, isometric component planning (ISOMAP) is executed. The Support Vector Machine (SVM) and K-Nearest Neighbours (KNN) classifiers used on both reduce and original dataset to demonstrate the viability of the executed dimensionality reduction strategy. A. Farooq et al. [4] explore diverse texture and shape-based feature extraction techniques to separate three distinctive grass weed classes utilizing hyperspectral pictures. Extraction of features strategies including Histogram of Oriented Gradients (HoG), Local Binary Pattern (LBP) and Gabor features are assessed. Hyperspectral imaging is used by D. C. Liyanage et al. [5] to explain RGB images and semantic division for independent driving on unstructured terrain applications. Utilizing semantic segmentation network ResNet18, physically commented on learning information will be contrasted and hyperspectral strategy helps explain information by grouping territory situations. K. Djerriri et al. [6] assesses the ability of proposed descriptors named multiband minimal surface units. Using two HSI datasets, the suggested assessment is carried out with reference to a fix-based arrangement viewpoint. Objects were created using superpixel division for this. In the item feature space, a random forest method is used to classify the items. T. Miftahushudur et al. [9] look at how radiance control in hyperspectral images may be improved by using Correlated Color Temperature (CCT) as the DA. Finally, an ensemble method and a switching technique were used to improve the characterisation findings. P. Burai et al. [10] implemented AI techniques utilizing crown portions for picture classification. High spatial resolution hyperspectral pictures and LiDAR information applied to segregate among tree types of mixed forest. A. V. Miclea et al. [11] suggests a network characterization structure dependent on multiresolution local binary pattern strategies and convolutional neural network (CNN) models. The proposed strategy has comparative outcomes with other LBP-1D-CNN characterization organizations, however, has a more modest number of boundaries and decreased time for preparing and testing. Using the dimensionally reduced spectral component vectors of available named tests, the proposed approach first trains a set of individually specific one-class classifiers. M. Dowlatshah et al. [13] extracted spatial characteristics from hyperspectral images using property channels and fractional remaking. Furthermore, named tests, the proposed approach first trains a set of individually specific one-class classifiers. M. Dowlatshah et al. [13] extracted spatial characteristics from hyperspectral images using property channels and fractional remaking. Furthermore, named tests, the proposed approach first trains a set of individually specific one-class classifiers. M. Dowlatshah et al. [13] extracted spatial characteristics from hyperspectral images using property channels and fractional remaking. Furthermore, the 3-D Gabor channel bank is used to extract spectral-spatial characteristics at the same time. A. M. Ahmed et al. [15] presented another idea of utilizing Logical Analysis of Data (LAD) as another classifier for hyperspectral information. Spectral patterns produced by LAD to recognize hyperspectral subclasses and classes; these spectral pattern (s) are exceptional for specific materials inside the equivalent dataset, while it is diverse for a similar material in another dataset. D. K. Pathak et al. [16] presented a classification method using SVM which extracts features utilizing both spatial and spectral information. The developed method features SVM to encode spatial-spectral information of pixels with classification tasks. K. Bhardwaj et al. [17] discussed a spatial-spectral active learning model based on super-pixel profile for hyperspectral image classification having samples of scarce labels. The model proposed is efficient and pretty promising in spatial-spectral classification of HSI having samples with limited labels. A. I. Champa et al. [18] presented a hybrid system for reduction in feature by combining feature selection and extraction. The method developed to detect subspace that delivers better accuracy than existing approaches. M. Ihsan et al. [19] want to lower computing costs and reduce overfitting by removing characteristics that aren't linked to the goal.

3 Materials and Methods

The proposed model for automated HSI classification on hyperspectral image dataset, University of Pavia and Indian Pines consists of mainly two phases, training and testing as shown in Fig. 1. The proposed method comprises following stages: (i) dataset splitting (ii) pre-processing (iii) feature extraction and selection (iv) model training, hyper-parameter tuning and classification.

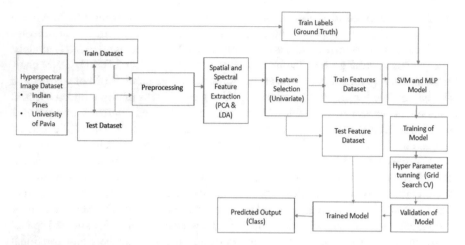

Fig. 1 Proposed block diagram for HSI classification

A. *Dataset Splitting*

In machine learning approach, to train a model based on HSI dataset, need to fragment the data into train and test data. Here in our case, we used the train to test split ratio to fragment data in 80:20 ratio i.e. 80% of data used to train model while 20% used to test the model that is built out of it.

B. *Pre-processing.*

The raw data for the Hyperspectral Image (HSI) dataset is available in *.mat format and may be accessed using the Python programming language. It has a spectral-spatial frequency band that is distinct. We retrieve data pixel by pixel, with a spectral orientation in mind. The extraction of pixel information from the HSI dataset requires a lot of pre-processing. This procedure aids in the collection of data and the use of machine learning techniques such as classification, clustering, and others.

C. *Feature Extraction and Selection.*

Feature Extraction is a mechanism of searching novel features by choosing or/and merging existing features to generate reduced feature subset, while precisely and entirely relating to a dataset deprived of loss of information. Linear Discernment Analysis (LDA) and Principal component analysis (PCA) is used for experimentation. Spectral bands extracted from HSI dataset utilizing methodologies LDA and PCA both features combined trailed by their classification using supervised machine learning algorithms. Employing the LDA and PCA methods on the original features $xi1, xi2, ..., xin$, got two different sets of extracted features. After applying PCA method, we retrieved a set of features that can be exemplified as $xk1, xk2, ..., xkn$. The LDA method created a feature set that represented as $xj1, xj2, ..., xjn$. As a combined feature set, amalgamated these features altogether and created a new feature dataset frame represented as $xk1, xk2, ..., xkn, xj1, xj2, ..., xjn$.

Feature Selection is a procedure of choosing dimensions of features of dataset which subsidizes mode to machine learning techniques like clustering, classification, etc. It can be attained by utilizing various methods such as univariate analysis, correlation analysis, etc. In our case, Univariate feature selection is used, which works by selecting the best features based on univariate statistical tests.

D. *Model Training, Hyper-parameter Tuning and Classification.*

Train and test features are extracted from the train and test datasets, respectively, during the feature extraction procedure. After that, two supervised machine learning algorithms, Support Vector Machine (SVM) and Multi-layer Perceptron Neural Network, are employed to train the model (MLP). So, utilising hyper-parameter tuning, effective parameters must be tweaked or set for the best classification result of the test dataset when training.

Support Vector Machines (SVM) is a method for supervised classification. The fundamental concept of SVM assists in mapping non-linear data to a new space in which linearly separable data may be mapped by employing a hyper-plane that appropriately separates that data by obeying two essential conditions: Because different classes of vectors will be on opposite sides, the distance between the hyperplane and the vectors must be utilised. The hypothesis function h has the following definition:

$$h(x_i) = \begin{Bmatrix} +1 \ if \ w.x \ + \ b \geq 0 \\ -1 \ if \ w.x \ + b \ < 0 \end{Bmatrix} \tag{1}$$

Class +1 will be assigned to points above or on the hyperplane, whereas class -1 will be assigned to points below the hyperplane.

Multi-Layer Perceptron is a kind of perceptron that has many layers. An Artificial Neural Network (ANN) with one or more hidden layers is known as a neural network. A perceptron is a type of neural network that consists of a single neural model. It is used to represent high non-linear functions, which is the foundation for deep learning neural networks. The degree of inaccuracy in an output node j in the nth data point (training example) can be represented by

$$e_j(n) = d_j(n) - y_j(n) \tag{2}$$

where, 'd' is the goal value and 'y' is the perceptron's output value. The node weights can then be changed depending on adjustments that reduce the overall output error, as determined by

$$\in (n) = \frac{1}{2} \sum_j e_j^2 (n) \tag{3}$$

The challenge of selecting a set of appropriate hyper-parameters for a model learning algorithm is known as hyper-parameter tuning or optimization. A hyper-parameter is a value for a parameter that is used to regulate the learning process. The grid search technique, perhaps the most fundamental hyper-parameter tuning approach, is utilised. We simply create a model for each conceivable combination of all of the hyper-parameter values provided, evaluate each model, and choose the architecture that provides the best results using this method. The procedure for hyper-parameter tweaking the model is explained further down.

step 1: Defining a machine learning model.

step 2: For selected approach, define the range of feasible values for all hyper-parameters.

step 3: Define a sample technique for hyper-parameter values.

step 4: Define an evaluative criterion for the model's evaluation.

step 5: Define a cross-validation technique for determining system efficiency.

4   Results and Discussion

A. *Experimental Setup*

There are two datasets considered for hyper-spectral image classification, University of Pavia and Indian Pines dataset. The classification was carried using Pycharm IDE with Anaconda distribution and Scikit Learn Machine Learning Toolkit on a system configured with Intel i5 CPU @ 2.80 GHz, 16 GB RAM and Windows 10 (64 bit) operating system.

B. *Dataset Description*

The University of Pavia has a total of 103 spectral bands. Pavia University has a resolution of 610 × 610 pixels, 1.3 m is the geometric resolution. The image ground truths distinguish nine types. The image was captured by the AVIRIS sensor above the Indian Pines test site in North-western Indiana and consists of 145 × 145 pixels and 224 spectral reflectance bands in the wavelength range of 0.4–2.5 10^ (–6) metres. Two-thirds of the scene in Indian Pines is agricultural, with the remaining one-third being forest or other natural permanent flora. Ground truth is divided into sixteen groups. The dataset has been downloaded from HRSS. Sample band and its respective ground truth is shown in Fig. 2 for both datasets.

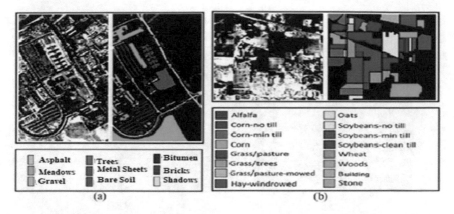

Fig. 2 Spectral band information and its equivalent ground truth label of **a** University of Pavia HSI Dataset and **b** Indian Pines HSI Dataset

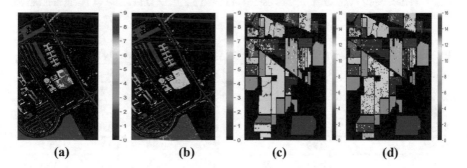

Fig. 3 Predicted output using **a** SVM for UoP **b** MLP for UoP **c** SVM IP **d** MLP for IP

C. *Experimental Results and Performance Analysis*

Proposed method is applied on both dataset and found the predicted output image against ground truth, for both algorithms used SVM and MLP as shown in Fig. 3 and Fig. 4. For evaluation of parameters of proposed model, classification report from Scikit Learn Toolkit, in which Precision, Recall, F-score, is evaluated and shown in Tables 1 and 2 for both datasets. Tables 3, 4, 5 and 6 shows the confusion matrix Table 7 shows the overall Accuracy, Matthews's Correlation Coefficient (MCC) and Kappa Score. Finally overall performance is shown graphically in Fig. 4, it is indicated that MLP gives best performance measure results than SVM. Hence, performance of MLP is compared in Table 8, shows the proposed system efficiency with respect to existing approach for different dataset used.

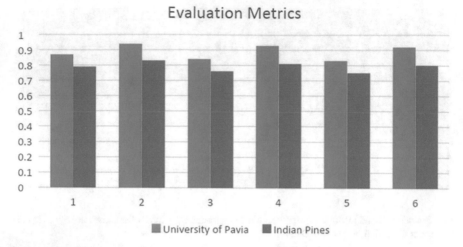

Fig. 4 Performance evaluation metrics using SVM and MLP for University of Pavia and Indian Pines dataset

Table 1 Classification report parameters using SVM and MLP for University of Pavia

Classes	Precision		Recall		F-score	
	SVM	MLP	SVM	MLP	SVM	MLP
Asphalt	0.93	0.94	0.94	0.96	0.94	0.95
Meadows	0.83	0.98	1.00	0.96	0.90	0.97
Gravel	0.89	0.78	0.76	0.86	0.82	0.82
Trees	0.97	0.93	0.92	0.98	0.95	0.95
Painted metal sheets	0.99	1.00	1.00	1.00	1.00	1.00
Bare Soil	0.96	0.90	0.30	0.92	0.46	0.91
Bitumen	0.94	0.86	0.82	0.90	0.88	0.88
Self-Blocking Bricks	0.84	0.90	0.91	0.82	0.87	0.85
Shadows	1.00	1.00	0.99	0.99	1.00	1.00

In the instance of the University of Pavia (UoP), precision, recall, and f-score parameter values are more likely to be calculated using the MLP method rather than the SVM algorithm as given Table 1. In the case of the Indian Pine (IP) Dataset, the same kind of findings can be seen in Table 2. In terms of overall system performance, the University of Pavia dataset produces more accurate results than the Indian Pine dataset for both machine learning methods. Also, comparative results are also shown in Table 8 with respect to existing literature.

Table 2 Classification report parameters using SVM and MLP for Indian Pines

Classes	Precision		Recall		F-score	
	SVM	MLP	SVM	MLP	SVM	MLP
Alfalfa	0	0.89	0	0.89	0	0.89
Corn-notill	0.77	0.77	0.65	0.76	0.7	0.76
Corn-mintill	0.84	0.67	0.51	0.8	0.64	0.73
Corn	0.77	0.72	0.49	0.7	0.6	0.71
Grass-pasture	0.75	0.94	0.96	0.93	0.84	0.93
Grass-trees	0.86	0.95	0.98	0.95	0.91	0.95
Grass-pasture-mowed	0	0.75	0	0.6	0	0.67
Hay-windrowed	0.91	0.97	1	1	0.96	0.98
Oats	0	0.5	0	0.25	0	0.33
Soybean-notill	0.77	0.79	0.67	0.86	0.72	0.82
Soybean-mintill	0.67	0.89	0.87	0.71	0.75	0.79
Soybean-clean	0.79	0.58	0.74	0.86	0.76	0.69
Wheat	0.95	0.91	0.93	1	0.94	0.95
Woods	0.95	0.96	0.97	0.92	0.96	0.94
Buildings-Grass-Trees	0.8	0.71	0.51	0.74	0.62	0.73
Stone-Steel-Towers	1	1	0.95	0.95	0.97	0.97

Table 3 Confusion matrix table using SVM for University of Pavia

Classes		Predicted Class								
		1	2	3	4	5	6	7	8	9
Actual Class	1	1244	11	9	0	1	0	15	46	0
	2	0	3712	0	16	0	2	0	0	0
	3	20	4	320	0	0	0	0	76	0
	4	0	47	0	565	0	1	0	0	0
	5	0	0	0	0	269	0	0	0	0
	6	1	702	0	1	1	300	0	1	0
	7	47	0	0	0	0	0	219	0	0
	8	18	13	30	0	0	8	0	668	0
	9	1	0	0	0	0	0	0	0	188

Table 4 Confusion matrix table using MLP for University of Pavia

Classes		Predicted Class								
		1	2	3	4	5	6	7	8	9
Actual Class	1	1269	0	7	0	0	1	34	15	0
	2	0	3587	4	35	0	100	0	4	0
	3	8	1	363	0	0	0	1	47	0
	4	0	10	0	599	0	3	0	1	0
	5	0	0	0	0	269	0	0	0	0
	6	0	69	0	12	1	922	0	2	0
	7	26	0	0	0	0	0	240	0	0
	8	44	0	89	0	0	0	3	601	0
	9	1	0	0	0	0	0	0	0	188

Table 5 Confusion matrix table using SVM for Indian Pines

Classes		Predicted class															
		1	2	3	4	5	6	7	8	9	10	11	12	13	14	15	16
ACTUAL CLASS	1	0	0	0	0	4	0	0	4	0	0	1	0	0	0	0	0
	2	0	186	1	3	1	1	0	0	0	19	71	4	0	0	0	0
	3	0	8	85	2	0	0	0	0	0	4	66	1	0	0	0	0
	4	0	2	4	23	1	3	0	1	0	0	13	0	0	0	0	0
	5	0	0	0	0	93	0	0	1	0	0	1	0	0	2	0	0
	6	0	0	0	0	1	143	0	0	0	0	1	0	0	0	1	0
	7	0	0	0	0	2	0	0	3	0	0	0	0	0	0	0	0
	8	0	0	0	0	0	0	0	96	0	0	0	0	0	0	0	0
	9	0	0	0	0	0	4	0	0	0	0	0	0	0	0	0	0
	10	0	6	3	0	3	1	0	0	0	130	37	14	0	0	0	0
	11	0	38	5	0	7	0	0	0	0	9	425	4	0	0	3	0
	12	0	1	3	2	0	0	0	0	0	6	19	88	0	0	0	0
	13	0	0	0	0	0	1	0	0	0	0	0	0	38	0	2	0
	14	0	0	0	0	2	1	0	0	0	0	0	0	0	246	4	0
	15	0	0	0	0	10	13	0	0	0	0	1	1	2	11	39	0
	16	0	1	0	0	0	0	0	0	0	0	0	0	0	0	0	18

Table 6 Confusion matrix table using MLP for Indian Pines

Classes		Predicted class														
		1	2	3	4	5	6	7	8	9	10	11	12	13	14	15	16
A C T U A L C L A S S	1	8	0	0	0	0	0	1	0	0	0	0	0	0	0	0	0
	2	0	216	15	5	0	0	0	0	0	13	17	20	0	0	0	0
	3	0	6	133	3	0	0	0	0	0	3	12	9	0	0	0	0
	4	0	3	6	33	0	1	0	1	0	1	1	1	0	0	0	0
	5	0	0	0	1	90	0	0	0	0	1	2	0	1	0	2	0
	6	0	0	0	0	1	138	0	0	0	0	1	0	0	2	4	0
	7	0	0	0	0	1	0	3	1	0	0	0	0	0	0	0	0
	8	0	0	0	0	0	0	0	96	0	0	0	0	0	0	0	0
	9	0	0	0	0	0	2	0	0	1	0	0	0	1	0	0	0
	10	0	3	4	1	0	0	0	0	0	166	9	11	0	0	0	0
	11	0	48	39	1	1	0	0	0	0	19	351	31	0	0	1	0
	12	0	2	3	2	0	0	0	0	0	7	3	102	0	0	0	0
	13	0	0	0	0	0	0	0	0	0	0	0	0	41	0	0	0
	14	0	0	0	0	3	0	0	0	0	0	0	0	0	234	16	0
	15	1	0	0	0	0	5	0	1	1	0	0	3	2	7	57	0
	16	0	1	0	0	0	0	0	0	0	0	0	0	0	0	0	18

Table 7 Evaluation metric performance for different datasets

Dataset	Accuracy		MCC		Kappa score	
	SVM	MLP	SVM	MLP	SVM	MLP
University of Pavia	0.87	0.94	0.84	0.93	0.83	0.92
Indian Pines	0.79	0.83	0.76	0.81	0.75	0.80

Table 8 Comparative analysis for different datasets

Dataset	References	Accuracy
University of Pavia	[12]	0.8697
	[16]	0.9102
	Our Work	0.94
Indian Pines	[2]	0.8251
	[11]	0.8241
	[12]	0.7984
	Our Work	0.83

5 Conclusion

Proposed method for automated classification of hyperspectral image (HSI) using spectral-spatial features with grid search hyper-tuning optimization is presented. From the performance analysis it is quite clear that MLP provide better result than SVM. Multi-layer perceptron is preferred as classifier as it delivers the most promising and consistent results. The best results achieved for University of Pavia data set, due to a lower number of classes. It has been proven that the propounded technique performs better than existing techniques. Additionally, the performance of the system also studied for SVM and MLP classifiers using Accuracy, Precision, Recall, F-score, MCC and Kappa Score. In future, the classification using Deep Learning method could be used to improve further performance of system. Also, hybrid features could be selected and optimized as per dataset requirements.

References

1. Lv W, Wang X (2020) Overview of hyperspectral image classification. J Sens 1–13. https://doi.org/10.1155/2020/4817234
2. Joy AA, Hasan MAM, Hossain MA (2019) A comparison of supervised and unsupervised dimension reduction methods for hyperspectral image classification. In: 2019 International Conference on Electrical, Computer and Communication Engineering (ECCE), Cox'sBazar, Bangladesh, pp 1–6. https://doi.org/10.1109/ECACE.2019.8679360
3. Younus L, Kasapoglu NG (2019) Dimension reduction and its effects in hyperspectral data classification. In: 2019 6th International Conference on Electrical and Electronics Engineering (ICEEE), Istanbul, Turkey, pp 359–366. https://doi.org/10.1109/ICEEE2019.2019.00076
4. Farooq A, Jia X, Zhou J (2019) Texture and shape features for grass weed classification using hyperspectral remote sensing images. In: IGARSS 2019–2019 IEEE International Geoscience and Remote Sensing Symposium, Yokohama, Japan, pp 7208–7211. https://doi.org/10.1109/IGARSS.2019.8900132
5. Liyanage DC, Hudjakov R, Tamre M (2020) Hyperspectral imaging methods improve RGB image semantic segmentation of unstructured terrains. In: 2020 International Conference Mechatronic Systems and Materials (MSM) Bialystok, Poland, pp 1–5. https://doi.org/10.1109/MSM49833.2020.9201738
6. Djerriri K, Safia A, Adjoudj R, Karoui MS (2019) Improving hyperspectral image classification by combining spectral and multiband compact texture features. In: IGARSS 2019–2019 IEEE International Geoscience and Remote Sensing Symposium, Yokohama, Japan, pp 465–468. https://doi.org/10.1109/IGARSS.2019.8900211
7. Shinde SR, Bhavsar K, Kimbahune S, Khandelwal S, Ghose A, Pal A (2020) Detection of counterfeit medicines using hyperspectral sensing. In: 2020 42nd Annual International Conference of the IEEE Engineering in Medicine & Biology Society (EMBC) Montreal, QC, Canada, pp 6155–6158. https://doi.org/10.1109/EMBC44109.2020.9176419
8. Asghari Beirami B, Mokhtarzade M (2020) Band grouping SuperPCA for feature extraction and extended morphological profile production from hyperspectral images. IEEE Geosci Remote Sens Lett 17(11):1953–1957. https://doi.org/10.1109/LGRS.2019.2958833
9. Miftahushudur T, Heriana O, Prini SU (2019) Improving hyperspectral image classification using data augmentation of correlated color temperature. In: 2019 International Conference on Radar, Antenna, Microwave, Electronics, and Telecommunications (ICRAMET), Tangerang, Indonesia, pp 126–130. https://doi.org/10.1109/ICRAMET47453.2019.8980420

10. Burai P, Beko L, Lenart C, Tomor T, Kovacs Z (2019) Individual tree species classification using airborne hyperspectral imagery and Lidar data. In: 2019 10th Workshop on Hyperspectral Imaging and Signal Processing: Evolution in Remote Sensing (WHISPERS) Amsterdam, Netherlands, pp 1–4. https://doi.org/10.1109/WHISPERS.2019.8921016
11. Miclea AV, Terebes R, Meza S (2020) One dimensional convolutional neural networks and local binary patterns for hyperspectral image classification. In: 2020 IEEE International Conference on Automation, Quality and Testing, Robotics (AQTR), Cluj-Napoca, Romania, pp 1–6. https://doi.org/10.1109/AQTR49680.2020.9129920
12. Singh PS, Singh VP, Pandey MK, Karthikeyan S (2020) One-class classifier ensemble based enhanced semisupervised classification of hyperspectral remote sensing images. In: 2020 International Conference on Emerging Smart Computing and Informatics (ESCI), Pune, India, pp 22–27. https://doi.org/10.1109/ESCI48226.2020.9167650
13. Dowlatshah M, Ghassemian H, Imani M (2019) Spatial-Spectral feature extraction of hyperspectral images using attribute profile with partial reconstruction and 3-D Gabor filter bank. In: 2019 5th Iranian Conference on Signal Processing and Intelligent Systems (ICSPIS), Shahrood, Iran, pp 1–6. https://doi.org/10.1109/ICSPIS48872.2019.9066038
14. Wei J, Su W, Fan Y, Li J (2016) Development and application of hyperspectral image classification technology, pp 143–149. https://doi.org/10.14257/astl.2016.123.28
15. Ahmed AM, Ibrahim SK, Yacout S (2019) Hyperspectral image classification based on logical analysis of data. In: 2019 IEEE Aerospace Conference Big Sky, MT, USA, pp 1–9. https://doi.org/10.1109/AERO.2019.8742023
16. Pathak DK, Kalita SK (2019) Spectral spatial feature based classification of hyperspectral image using support vector machine. In: 2019 6th International Conference on Signal Processing and Integrated Networks (SPIN), Noida, India, pp 430–435. https://doi.org/10.1109/SPIN.2019.8711731
17. Bhardwaj K, Das A, Patra S (2020) Spectral-Spatial active learning with superpixel profile for classification of hyperspectral images. In: 2020 6th International Conference on Signal Processing and Communication (ICSC), Noida, India, pp 149–155. https://doi.org/10.1109/ICSC48311.2020.9182764
18. Champa AI, Rabbi MF, Banik N (2019) Improvement in hyperspectral image classification by using hybrid subspace detection technique. In: 2019 International Conference on Sustainable Technologies for Industry 4.0 (STI), Dhaka, Bangladesh, pp 1–5. https://doi.org/10.1109/STI47673.2019.9067973
19. Hyperspectral Remote Sensing Scenes Dataset. http://www.ehu.eus/ccwintco/index.php/Hyperspectral_Remote_Sensing_Scenes

An Ensemble Model for Network Intrusion Detection Using AdaBoost, Random Forest and Logistic Regression

Nitesh Singh Bhati and Manju Khari

1 Introduction

Network activity has become a crucial part of almost every person or every organization. Concurrently, various threats and attacks are consistently increasing to harm network activities and to steal confidential information from the network environment [1]. Valuable and confidential information is always pleasing for the attackers and therefore, vulnerable to various cyber-attacks [2]. Hence, the development of an adequate model of the intrusion detection system becomes a necessity for securing the network environment. Intrusion Detection System (IDS) is a novel security technique for preventing, sensing and detecting suspicious activity insecure network environment. IDS play an essential role in performing safe and secure network activities [3]. Intrusion is a process when an attacker crawls into the system server, forwards malicious packet or interrupt the network for stealing, modifying and corrupting the confidential information.

IDS tackle the network attacks by observing and controlling the network environment. It performs various steps, to gather intrusion-related knowledge, that occurred at the time of monitoring and analyzing them as a sign of intrusion [4]. IDS is used to identify all types of attack in a network environment by sensors and prevent from misuse the information, unwanted access, network corruption and penetration of malicious user inside the private network. The existing intrusion detection system has various manual definitions for normal and abnormal behaviour detection, but the latest research shows that it is possible to identify abnormality automatically in a network through machine learning techniques [5, 6]. It has stated that ensemble techniques as bagging, boosting, voting and stacking provide a better result than

N. S. Bhati (✉)
Ph.D Scholar, USICT, Guru Gobind Indraprastha University, Delhi, India
e-mail: niteshbhati07@gmail.com

M. Khari
JawahrLal Nehru University, Delhi, India

© The Author(s), under exclusive license to Springer Nature Singapore Pte Ltd. 2022   777
B. Unhelker et al. (eds.), *Applications of Artificial Intelligence and Machine Learning*,
Lecture Notes in Electrical Engineering 925,
https://doi.org/10.1007/978-981-19-4831-2_64

single learner in the terms classification accuracy and lesser false alarms [7]. Even so, attackers are continuously trying new patterns to generate novel attacks to smash the security of signature-based technique [8–10]. Figure 1 depicts the working of an IDS. It senses the network traffic communication.

IDS has categorized into two parts based on audit data: Host-based and Network-based. Host-based works by storing logs into the database for analysis, and the Network-based IDS works by investigating the packets of network traffic [11]. An attacker moves between several nodes to find the origin of attack or search the vulnerable area of the system to steal or modify the confidential information.

The permanent desire of this work is to improve the classification algorithms of Machine-Learning algorithms for the IDS by developing an ensemble model. In this research work, we propose, an ensemble framework of an IDS for attack detection. The main novel idea of this work has to combine AdaBoost, Random Forest and Logistic Regression ensemble technique to build one new ensemble approach for the implementation of the intrusion detection system. KDDCUP99 open-source labelled dataset has been used to test the experiments. The dataset has classified into five types of attacks: DoS attack, Probe attack, r2l attack, u2r attack and normal. Our ensemble model integrates three ensemble learners to generate better results for identifying the attacks and produced promising results by the measurement unit's accuracy, recall, precision and f1-score [12–14]. The confusion matrix is also drawn by the tested consequence to find the average conclusive results. Further, portion of this work has organized into various sections as Sect. 2. Presents the literature work of machine learning related intrusion detection system, Sect. 3. Depicts the proposed methodology of the research, Sect. 4. Presents the experimental results, Sect. 5. Presents the comparative analysis of result and Sect. 6. Shows the conclusion and future work.

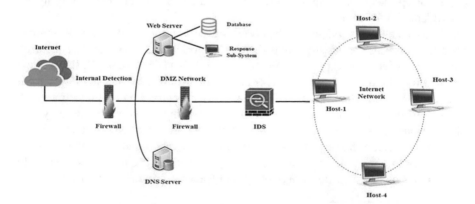

Fig. 1 Intrusion detection system

2 Literature Review

This section presents the previously proposed work of the machine learning technique for the IDS. There are very few benchmark datasets available for performing the experimental setup for detecting the attacks by the IDS. Few of them are NSL-KDD Dataset, UNSWNB-15, CAIDA-2007 and KDD-CUP 99.

Haijun Xiao et al. [15] proposed various ensemble classifiers which are combined by weight voting rule for an intrusion detection system, and they have stated that the combined approach provides improved results than a single classifier. The results of their experiment show that the Ensemble model had scored 0.9900 detection rate and 0.0018 false-positive rates.

Giorgio Giacinto et al. [16] proposed a Multiple Classifier System (MCS) based unlabeled Network Anomaly IDS proposed and experiments were performed on KDD-cup 1999 dataset. The results of the experiment show that the proposed model outperformed with 94.31% detection rate and 9.49% false alarm rate.

Deepak Rathore et al. [8] proposed the ensemble cluster classification technique using the SOM network for mixed variable malicious attacks that were generated by the software. The KDD-99 dataset had been used for empirical evaluation. The best results that were generated by ECC-SOM model was Accuracy 97.13%, Precision 94.52% and Recall 93.67%.

Shalinee Chaurasia et al. [17] proposed an Ensemble model with the combination of K-nearest neighbour and neural network by using KDD-99 dataset for intrusion detection system and stated that the model produced better results than the other individual learner. Performance of the model was measured in terms of Accuracy = 96.84%, Precision = 88.88%, Recall = 91.52%, True Positive Rate (TPR) = 88.52% and False Positive Rate (FPR) = 0.21%.

Zhuo Chen et al. [18] proposed an Extreme Gradient Boosting (XGB) method has used for the Software-Defined Network (SDN) cloud detection, and topology of SDN has built by Mininet. For the empirical evaluations, flow packet dataset has collected from TcpDumps. They stated that XGBoost method achieved more significant results than Random Forest and SVM.The results that were achieved by XGBoost was Accuracy 98.52%, False Positive Rate 0.008% and Training Time (sec) 11.07.

Sukhpreet Singh Dhaliwal et al. [19] proposed an XGBoost technique that had been employed on NSL-KDD dataset for getting the higher accuracy and robust Intrusion Detection System. The performance of the model was measured by accuracy = 98.70%, Precision = 98.41%, Recall = 99.11% and F1-Score = 98.76%.

Sweta Bhattacharya et al. [20] proposed a novel XGBoost classification based PCA-Firefly hybrid model for the intrusion detection system. Dataset for experiments had extracted from Kaggle that had 43 attributes holding categorical and numerical data. The proposed model outperformed in terms of Accuracy 99.09%, Sensitivity 93.01% and Specificity 99.09%.

Table 1 Chronological summary of machine learning-based intrusion detection system

Serial No	Year	Author	Proposed-Work
[1]	2007	Haijun Xiao et al. [15]	The various ensemble classifiers have combined by weight voting rule for an intrusion detection system, and it has stated that the combined approach provides improved results than a single classifier
[2]	2008	Giorgio Giacinto et al. [16]	Multiple Classifier System (MCS) based unlabeled Network Anomaly IDS proposed and experiments were performed on KDD-cup 1999 dataset
[3]	2012	Deepak Rathore et al. [8]	The ensemble cluster classification technique using the SOM network has proposed for mixed variable malicious attacks that were generated by the software. The KDD-99 dataset had been used for empirical evaluation
[4]	2014	Shalinee Chaurasia et al. [17]	The Ensemble model with the combination of K-nearest neighbour and neural network has proposed by using KDD-99 dataset for intrusion detection system and stated that the model produced better results than the other individual learner
[5]	2018	Zhuo Chen et al. [18]	Extreme Gradient Boosting (XGB) method has used for the Software-Defined Network (SDN) cloud detection, and topology of SDN has built by Mininet. For the empirical evaluations, flow packet dataset has collected from TcpDumps. It has stated that XGBoost method achieved more significant results than Random Forest and SVM
[6]	2018	Sukhpreet Singh Dhaliwal et al. [19]	XGBoost technique had employed on NSL-KDD dataset for getting the higher accuracy and robust Intrusion Detection System
[7]	2020	Swcta Bhattacharya et al. [20]	A novel XGBoost classification based PCA-Firefly hybrid model had proposed for the intrusion detection system. Dataset for experiments had extracted from Kaggle that had 43 attributes holding categorical and numerical data
[8]	2020	Preethi Devan et al. [21]	XGBoost-DNN (Deep Neural Network) model had proposed and applied on NSL-KDD dataset for intrusion detection

(continued)

Table 1 (continued)

Serial No	Year	Author	Proposed-Work
[9]	2021	Moualla et al. [22]	A dynamically scalable multiclass machine learning based IDS has been proposed, which is tested on the UNSW-NB15 Dataset

Preethi Devan et al. [21] proposed an XGBoost-DNN (Deep Neural Network) model and applied on NSL-KDD dataset for intrusion detection. The results of their experiment show that XGBoost-DNN outperformed with Accuracy 97.6%, Precision 97%, Recall 97% and F1-Score 97%.

Moualla et al. [22] proposed a dynamically scalable multiclass machine learning based IDS, which is tested on the UNSW-NB15 Dataset. Results show that the proposed system performs better in terms of accuracy, false alarm rate, Receiver Operating Characteristic (ROC), and Precision-Recall Curves (PRCs).

The various machine learning techniques like Support Vector Machine, Naïve Bayes, K-Nearest Neighbors, and Decision Tree are prevalent for attack detection. However, in the latest research, many ensemble techniques as XGBoost, Ada-Boost, Gradient-Boost, Random-Forest and Extra-Trees came into the picture for the implementation of the intrusion detection system. Table 1 presents the chronological summary of machine learning-based intrusion detection research.

3 Proposed Methodology

As described earlier, the major goal of this research has to develop an ensemble model to improve the performance of the intrusion detection system and decreasing false alarms. The objective of the proposed model is to monitor and analyze the network activities in a computer network environment and collects the network logs. After that, collected logs has analyzed for feature selection through data mining techniques. To accomplish this task, the training and testing of three individual classifiers was done and integrated them for ensemble development. A majority voting approach has been used for getting the opinion of every classifier for better results. In majority voting, decisions are made based on various individual classifiers and remove the least-performed classifier. The KDDCUP99 dataset has been used for empirical evaluation of the proposed model. Before the implementation of machine learning classification techniques, preprocessing steps as data cleaning, feature extraction and vectorization has applied on a dataset to convert into machine-understandable form. AdaBoost, Random Forest and Logistic Regression were applied as base learners and combined them by majority voting as to produce an ensemble model that is superior to any individual learner. Proposed model calculates

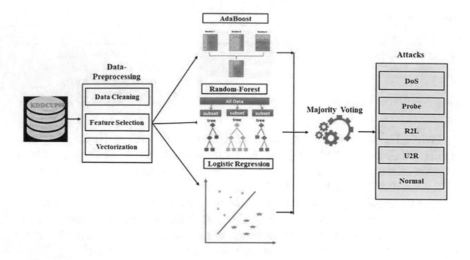

Fig. 2 Proposed framework

the results as: Majority voting = mode {AdaBoost(X), Random-Forest(X), Logistic-Regression(X)}. Figure 2 presents the whole structure of the proposed ensemble model for the intrusion detection system.

3.1 Dataset for Empirical Evaluation

To evaluate the proposed model, the Knowledge Discovery and Data Mining 1999 (KDDCUP99) dataset [23] has been selected. It is a standard benchmark dataset for intrusion detection related experiments. This dataset holds thousands of records related to the network connection. Total 41 qualitative and quantitative features have been selected for each TCP/IP connection, and every element of a single connection holds one observation either it is a standard or malicious state [24]. This dataset has been classified into four attacks (Denial of Service, Probe, Remote to Local, User to Local) and one normal state. For correctly analyzing the performance of each classifier, the dataset has been divided into two parts: one as a training set and one as a testing set. This dataset contains total 4,94,020 samples of network activities, and it has divided into the ratio of 80:20.

I. Training Set: This set contains 80% sample of total record and used to train every expert in the ensemble.
II. Testing Set: This set contains 20% sample of total record and used for evaluating the performance of each base learner of the model, as well as the performance of the proposed ensemble model.

3.2　Data Preprocessing

Data preprocessing is a process which cleans the dataset and converts into machine understandable form using various steps. Different methods are required to perform preprocessing according to the need and variation of datasets. For the implementation of the proposed work, the following three necessary steps are being followed:

I.　Data Cleaning: Initial process like filling missing values and removing outliers has accomplished in this step, and dataset has been prepared for further feature selection and vectorization process.
II.　Feature Selection: This process filtered the relevant features of the data. It reduces the number of input variables for reducing the computational time and cost. Also, speed up the execution process with accurate results.
III.　Vectorization: In this step, nominal categorical data has converted into vectors. For applying the vectorization, label encoding technique has been used.

3.3　Base Learning Techniques

As discussed above, the multi classifier-based ensemble technique has proposed in this research work to get better performance of the intrusion detection system. Mainly three classifiers have been used as a base learner to develop final ensemble model.

AdaBoost Classifier. AdaBoost is one of the most comprehensive machine learning technique in the current years. It was the first boosting algorithm that successfully implemented for binary classification [25]. Classification is a predictive technique that is used to precisely forecast the target class [26]. It is more capable than Self-Organizing Maps (SOM), Artificial Neural Network (ANN) and SVM techniques for handling the voluminous data in less time. For this reason, it has been used as a base learner to build our ensemble model of intrusion detection.

Random Forest Classifier. Random forest is an ensemble technique, which is superlative in performance and accuracy among the latest machine learning techniques [3]. 1 The random forest creates various trees, and every tree has contracted by a different bootstrap sample from the original dataset. After forming the forest new object is put down for classification in the forest. Decisions are made based on the vote given by each tree, and the maximum voted tree has selected [27].

Logistic Regression Classifier. Logistic Regression is a statistical technique that is applied in machine learning frequently in recent years. It has used to handle the binary classification problem. This technique has based on the logistic function that is also called a sigmoid function for describing the properties of population growth [28].

4   Experimental Setup and Results

4.1   Dataset

The dataset of our empirical evaluation had generated by the 1999 DARPA IDS evaluation program in Lincoln Laboratory by MIT [29]. KDDCUP99 dataset, which has used for our evaluation, is a part of the DARPA program. This dataset contains a total 24 attacks that are further categorized into four major categories as DoS, Prob, U2R and R2L. This dataset has 41 attributes for every connection record with one class label.

4.2   Experiment

All experiments have been performed on Intel Core(TM) PC, 1.60 GHz CPU, 64 BIT O/S with 4 GB Ram using ANACONDA Jupyter Notebook. Majority voting ensemble method has applied to develop a new ensemble approach for intrusion detection by combining three base ensemble learners. Classical learner use only single learner on the training dataset, but ensemble learners combine various learners to get best results. The evaluation has done based on majorly three features as Protocol_type1, protocol_type2 and flag with dependent feature response_class. Figure 3 presents the majority voting scheme and Table 2. Presents the proposed algorithm for describing the experimental procedure.

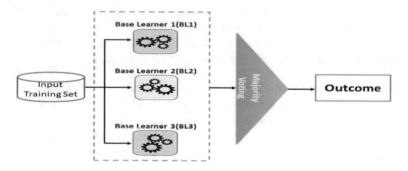

Fig. 3 Majority voting technique

Table 2 Proposed ensemble algorithm of intrusion detection [31, 32]

	Description
Step1:	Input: Preprocessed KDDCUP99 dataset. Split data: Training Set (TrS) Testing Set (TeS)
Step2:	Base Learner 1: AdaBoost a) $H(x) = sign\left\{\sum_{t=1}^{T} \alpha_t h_t(x)\right\}$ b) $BL_t(x) = H(x)$
Step3:	Base Learner 2: Random-Forest a) $\frac{RFf_i = \sum_{j \in alltrees} normf_{ij}}{T}$ b) $BL_2 = RFf_i$
Step4:	Base Learner 3: Logistic Regression a) $p = \frac{1}{1+e^{-(\beta_0 + \beta_1 x_1 + \beta_2 x_2 + \cdots + \beta_n x_n)}}$ b) $BL_3 = p$
Step5:	Combine Base Learners: Majority Voting a) $Result(x) = mode\{BL_1(x), BL_2(x), BL_3(X)\}$

Table 3 Performance metric

		Actual Results	
		Intrusion	Normal
Predicted	Intrusion	True Positive (TP)	False Positive (FP)
Results	Normal	False Negative (FN)	True Negative (TN)

4.3 Results

The experimental results of our proposed model have been presented in terms of classification accuracy, and Confusion matrix has chosen for more averaged forms of the result. Table 3 shows the criteria of evaluation matrix [30].

I. **True Positive (TP):** Presents to the situation when an attack identifies correctly by IDS.
II. **True Negative (TN):** Presents to the situation when no attack is identified by IDS.
III. **False Positive (FP):** It is a situation when IDS ring the alarm for the indication of an attack. When there was no attack performed actually.
IV. **False Negative (FN):** It shows the situation when IDS fail to identify the attack.

Whereas, the accuracy rate and false positive rate have calculated by these formulas:

$$AccuracyRate = \frac{TP + FN}{TP + TN + FP + FN} \tag{1}$$

Table 4 Classification report of proposed ensemble model

Attack	Precision	Recall	F1-score	Support
DoS	1.00	1.00	1.00	78,157
Normal	0.99	1.00	1.00	19,579
Probe	1.00	0.85	0.92	843
R_2L	0.95	0.91	0.95	214
U_2R	1.00	0.36	0.53	11

$$Falsepositive = \frac{FP}{TP + FP} \tag{2}$$

Further, the confusion matrix has drawn for measuring the performance of the proposed model. For generating the confusion matrix Precision, Recall, and f1-Score have calculated as [33, 34]:

Precision: It is a factor that presents the correct identification ability of a classifier.

$$Precision = \frac{TP}{TP + FP} \tag{3}$$

Recall: It shows the ability of the classifier to identify all the certain instances correctly.

$$Recall = \frac{TP}{TP + FN} \tag{4}$$

F1-Score: It calculates the harmonic mean of precision and recall. Where 1 shows to the best score and 0 shows to the worst score.

$$F1 - Score = 2 * \frac{Precision * Recall}{Precision + Recall} \tag{5}$$

Support: It shows the occurrence of a particular class in a specific dataset.

Table 4 presents the classification report of our ensemble model and Table 5. Presents the confusion matrix of the model. The proposed ensemble model outperformed and provided 99.86% accuracy on KDDCUP99 dataset.

5 Comparative Analysis

The work presented in the Sect. 2, includes various machine learning based models, many of which are ensemble based as well, however their results have not been up

Table 5 Confusion matrix report of the proposed ensemble model

Attacks	DoS	Normal	Probe	R$_2$L	U$_2$R
DoS	78,156	1	0	0	0
Normal	2	19,575	1	1	0
Probe	15	114	714	0	0
R$_2$L	1	18	1	194	0
U$_2$R	0	6	0	1	4

Table 6 Comparative report of various machine learning-based IDS models based on accuracy

Author	Model	Score (%)
Nitesh Singh Bhati	Proposed voting ensemble	99.86
Dhaliwal et al. [19]	Extreme Gradient Boosting (XGB)	98.70
Chen et al. [21]	XGBoost with SDN	98.52
Boro et al. [15]	Meta ensemble	98.42
Rathore et al. [16]	Cluster classification	97.13
Chaurasia et al. [17]	Ensemble model	96.84
Giorgio Giacinto et al. [14]	Multiple classifier system	94.31
Zhang et al. [13]	Random forest based hybrid model	94.7
Govindarajan et al. [8]	Hybrid model	85.19

to the mark as per the current industry standards, the technology has improved and so has the types of attacks seen in the current network based IDSs. Therefore, the permanent desire of this work is to provide an improvement in the classification algorithms of Machine-Learning algorithms for the IDS by developing an ensemble model of an IDS for attack detection using the combination of AdaBoost, Random Forest and Logistic Regression ensemble technique.

The KDDCUP datasets are used by various researchers, for the implementation of IDS based proposed schemes. Table 6 presents the different results acquired by various researchers regarding classification models. The performance of these models has been shown in Table 6 to establish why our IDS based proposed ensemble model looked at as a powerful model of machine learning for the intrusion detection system. After analyzing the work of various researchers, as referenced in Table 6, our proposed ensemble scheme outperformed.

The performance and accuracy achieved by our ensemble model is 99.86%, which is better as compared with other models.

6 Conclusion and Future Work

Through empirical evaluations, it has illustrated that the effectiveness of a specific machine learning technique can be improved by combining the votes of various experts. The majority voting approach has applied to combine votes of various experts AdaBoost, Random-Forest and Logistic Regression. This study aims to present a comparative analysis that gives the evidence that our new ensemble model outperforms over previous ensemble models. The best results that achieved by our new model are Accuracy 99.86% on KDDCUP99 dataset for an IDS. It has proven that our model outperformed. The results show that the proposed voting model has capable of solving the attack detection problem of the intrusion detection system efficiently.

For future work, it has recommended that the researcher should work on optimization techniques to enhance the performance and decrease the false positive rate of an intrusion detection system. A researcher can also develop a new approach for detecting the unknown attacks in a network environment. We will further expand this area in our future work by creating new ensemble techniques for the intrusion detection system.

References

1. Bukhtoyarov V, Zhukov V (2014) Ensemble-distributed approach in classification problem solution for intrusion detection systems. In: Corchado E, Lozano JA, Quintián H, Yin H (eds) Intelligent Data Engineering and Automated Learning–IDEAL 2014. IDEAL 2014. LNCS, vol 8669, pp 255–265. Springer, Cham. https://doi.org/10.1007/978-3-319-10840-7_32
2. Aljawarneh S, Aldwairi M, Yassein MB (2018) Anomaly-based intrusion detection system through feature selection analysis and building hybrid efficient model. J Comput Sci 25:152–160
3. Aburomman AA, Reaz MB (2016) A novel SVM-kNN-PSO ensemble method for intrusion detection system. Appl Soft Comput 38:360–372
4. Gudadhe M, Prasad P, Wankhade LK (2010) A new data mining based network intrusion detection model. In: 2010 International Conference on Computer and Communication Technology (ICCCT), pp 731–735. IEEE
5. Harish BS, Aruna Kumar SV (2017) Anomaly based intrusion detection using modified fuzzy clustering. IJIMAI 4(6):54–59
6. Bhati NS, Khari M (2021) A survey on hybrid intrusion detection techniques. In: Kumar R, Quang NH, Kumar Solanki V, Cardona M, Pattnaik PK (eds.) Research in Intelligent and Computing in Engineering. AISC, vol 1254, pp 815–825. Springer, Singapore. https://doi.org/10.1007/978-981-15-7527-3_77
7. Gaikwad DP, Thool RC (2015) Intrusion detection system using bagging with partial decision treebase classifier. Procedia Comput Sci 49:92–98
8. Giacinto G, Perdisci R, Del Rio M, Roli F (2008) Intrusion detection in computer networks by a modular ensemble of one-class classifiers. Inf Fusion 9(1):69–82
9. Lee W, Stolfo SJ, Mok KW (1998) A data mining framework for adaptive intrusion detection. In: Proceedings of the 7th USENIX Security Symposium
10. Dongre SS, Wankhade KK (2012) Intrusion detection system using new ensembleboosting approach. Int J Model Optim 2(4):488
11. Alsafi HMA, Basamh SS (2013) A review of intrusion detection system schemes in wireless sensor network. J Emerg Trends Comput Inf Sci 4(9):688–697.Chicago

12. Khonde S, Ulagamuthalvi V (2018) A machine learning approach for intrusion detection using ensemble technique-a survey. Int J Sci Res Comput Sci Eng Inf Technol 3(1):328–338
13. Chebrolu S, Abraham A, Thomas JP (2005) Feature deduction and ensemble design of intrusion detection systems. Comput Secur 24(4):295–307
14. Bhati NS, Khari M, García-Díaz V, Verdú E (2020) A review on intrusion detection systems and techniques. Int J Uncertain Fuzziness Knowl Based Syst 28(Supp02):65–91
15. Saidi A, Bendriss E, Kartit A, El Marraki M (2017) Techniques to detect DoS and DDoS attacks and an introduction of a mobile agent system to enhance it in cloud computing. IJIMAI 4(3):75–78
16. Xiao H, Hong F, Zhang Z, Liao J (2007) Intrusion detection using ensemble of SVM classifiers. In: Fourth International Conference on Fuzzy Systems and Knowledge Discovery (FSKD 2007), vol 4, pp 45–49. IEEE
17. Rathore D, Jain A (2012) Design Hybrid method for intrusion detection using ensemble cluster classification and SOM network. Int J Adv Comput Res 2(3):181
18. Chaurasia S, Jain A (2014) Ensemble neural network and k-NN classifiers for intrusion detection. Int J Comput Sci Inf Technol 5:2481–2485
19. Chen Z, Jiang F, Cheng Y, Gu X, Liu W, Peng J (2018) XGBoost classifier for DDoS attack detection and analysis in SDN-based cloud. In: 2018 IEEE international conference on big data and smart computing (bigcomp), pp 251–256. IEEE
20. Dhaliwal SS, Nahid AA, Abbas R (2018) Effective intrusion detection system using XGBoost. Information 9(7):149
21. Bhattacharya S, Kaluri R, Singh S, Alazab M, Tariq U (2020) A novel PCA-firefly based XGBoost classification model for intrusion detection in networks using GPU. Electronics 9(2):219
22. Devan P, Khare N (2020) An efficient XGBoost–DNN-based classification model for network intrusion detection system. Neural Comput Appl 32(16):12499–12514
23. Moualla S, Khorzom K, Jafar A (2021) Improving the performance of machine learning-based network intrusion detection systems on the UNSW-NB15 dataset. Comput Intell Neurosci 2021
24. http://kdd.ics.uci.edu/databases/kddcup99
25. Tiwari D, Singh N (2019) Ensemble approach for twitter sentiment analysis
26. Tiwari D, Kumar M (2020) Social media data mining techniques: a survey. In: Tuba M, Akashe S, Joshi A (eds) Information and Communication Technology for Sustainable Development. AISC, vol 933, pp 183–194. Springer, Singapore. https://doi.org/10.1007/978-981-13-7166-0_18
27. Hu W, Hu W (2005) Network-based intrusion detection using Adaboost algorithm. In: The 2005 IEEE/WIC/ACM International Conference on Web Intelligence (WI 2005), pp 712–717. IEEE
28. Breiman L (2001) Random forests. Mach Learn 45(1):5–32
29. Jiong Z, Zulkernine M (2005) Network intrusion detection using random forests. In: PST 2005
30. Random Forest. https://towardsdatascience.com/the-mathematics-of-decision-trees-random-forest-and-feature-importance-in-scikit-learn-and-spark-f2861df67e3. Accessed 16 Aug 2020
31. MIT Lincoln Laboratory. http://www.ll.mit.edu/IST/ideval
32. Tiwari D, Nagpal B (2020) Ensemble methods of sentiment analysis: a survey. In: 2020 7th International Conference on Computing for Sustainable Global Development (INDIACom), pp 150–155. IEEE
33. Syarif I, Zaluska E, Prugel-Bennett A, Wills G (2012) Application of bagging, boosting and stacking to intrusion detection. In: Perner P (eds) Machine Learning and Data Mining in Pattern Recognition. MLDM 2012. LNCS, vol 7376, pp 593–602. Springer, Berlin, Heidelberg. https://doi.org/10.1007/978-3-642-31537-4_46
34. Bhati BS, Chugh G, Al-Turjman F, Bhati NS An improved ensemble based intrusion detection technique using XGBoost. Trans Emerg Telecommun Technol e4076

Real Time Location Tracking for Performance Enhancement and Services

Gaurav Dubey, Anant Kumar Jayaswal, Akhilesh Srivastava, and Anurag Mishra

1 Introduction

At present time, numerous conveyance organizations endless supply of their administrations in assignment of workers as indicated by request. These plans of action use GPS and Web Services interminably [1]. Through our undertaking, Thesis based on an application that works in an ongoing climate and can be utilized to create some productive information that can be utilized for building up a Machine Learning Model [2].

A similar model can be utilized for the improvement of a specific business to encourage the utilization and coordination of the operationalized model and the code they contain. In this way, discussing the current situation in a considerable lot of the unmistakable conveyance frameworks, numerous representatives who are explicitly planned for the conveyance of the results of their associations, that the customer has requested for, are utilized for the entire term of the day, in any event, when there are fewer demands for the conveyance of the item, prompting the wastage of Labor just as the time. This is fundamental because of the absence of a legitimate examination of the constant information.

G. Dubey · A. Mishra
Department of Computer Science, KIET Group of Institutions, Delhi-NCR, Ghaziabad, Uttar Pradesh, India
e-mail: gaurav.dubey@kiet.edu

A. Mishra
e-mail: Anurag.mishra@kiet.edu

A. Srivastava
ABES Engineering College, Ghaziabad, Uttar Pradesh, India
e-mail: Akhilesh.srivastava@abes.ac.in

A. K. Jayaswal (✉)
Amity University, Noida, Uttar Pradesh, India
e-mail: akjayaswal@amity.edu

© The Author(s), under exclusive license to Springer Nature Singapore Pte Ltd. 2022   791
B. Unhelker et al. (eds.), *Applications of Artificial Intelligence and Machine Learning*,
Lecture Notes in Electrical Engineering 925,
https://doi.org/10.1007/978-981-19-4831-2_65

The framework comprises of the following module which gather information from GPS empowered gadgets on continuous basis.

i.   The framework gives ongoing observing of conveyance vehicles.
ii. This can advance the assignment of conveyance representatives by breaking down the information gathered on constant.
iii. Using AI apparatuses like relapse it can anticipate how much conveyance representatives are needed in a specific time allotment.

2 Background Details and Related Work

K. Subha and Dr. S. Sujatha et al. [3] proposed a system which presents GPS and GSM based global positioning framework has different applications in the present time in this world [2]. For instance vehicle and auto-following, youngsters following areas of interest, and any gear following and so forth A productive Real-Time Tracking System is utilized to follow different things like following focal points for different purposes, following of representatives, following clinical frameworks and so on With the assistance of GPS, GSM different equipment like modem and the microcontroller are implanted alongside the product and web administrations with the point of empowering clients to find their things with and that too in an advantageous way. Yet, this incorporates the contribution of equipment alongside programming [3]. This framework is equipped for giving the client the office to follow their things distantly through the versatile organization and web administrations. This paper presents the advancement of the global positioning framework with equipment models and incorporation of programming. The directions are changed over into intelligible tends that can be perused in a helpful way and consequently can be followed.

Younes Charfaoui et al. [4] proposed some points the which were as follows:

i.   Choosing a model that would not expect you to standardize such highlights. You can utilize non-scaled highlights in certain models like Decision Tree, however it is likely for a similar model to improve its exactness with scaled/standardized highlights. So, we should perform include change.
ii. Performing Reverse Geo-coding. You can utilize libraries, for example, geopy and reverse geo-coder to recuperate address from topographical area. Nonetheless, Arseniy composes on his medium post that the con of accepting data through HTTP, that it isn't in every case quick and that you may arrive at API limits in the event that you are working with enormous information [5]. Moreover, addresses (city, country names) may contain grammatical mistakes which require further cleaning of your changed information.
iii. Converting geo-location information into zones [7]. You can utilize bunching calculation like k-Nearest Neighbor calculation to aggregate your geo-area information (utilizing few possible groups) and relegate each group or a gathering a one-of-a-kind id. These exceptional ids would then be able to supplant your scope and longitude segment.

3   Proposed Approach

Here, Machine Learning is used as the key technology for the collection of data, analysis of data and prediction of the outcomes for the delivery services. The major steps included in Machine Learning process are:

i.   **Data Collection:** The quality and quantity of the data represents the efficiency and accuracy of the model. The data is called training data set, which will then be given to the algorithm [6].

ii. **Data Preparation:** Rectify the data, and prepare it for training. It may include removing of the duplicates, dealing with missing values, correction of errors, normalization, conversion of various data types, etc.

iii. **Choose a Model:** The model that is used here is the "Time Series Model". A **time series** is a collection of observations in chronological order. These could be daily stock closing prices, weekly inventory figures, annual sales, or countless other things. Time Series model is used to predict the outcomes based on previous data. Time series analysis, then, is nothing more than analyzing plotting, identifying patterns, etc.

Time Series model has further as follows:

1. Moving Average Model
2. LSTM model (Long-Short-Term Memory)
3. RNN model (Recurrent Neural Network)

The type of time series model is underlying criteria for the accuracy of the prediction [9].

iv. **Train the Model:** The main motive of the training process is to answer a question or make a prediction correctly as much as possible [2].

v.   **Evaluate the model for classification:**

1. Use of some of the performance metric or combination of metrics to measure expected performance of the mode.
2. Test the classification model against the previously unseen data and check the outcomes.

vi. **Parameter Tuning:** Parameter Tuning is done to improve the performance of classification model. It includes improvement in learning rate, distribution and initialization values.

vii. **Make Predictions:** Using further test set data which has been obtained after parameter tuning, are used to test the model, which is a better approximation of how the model will actually perform in the real world. It is the final step in the process of Machine Learning, which is to predict the data and also classification of the data [8] (Fig. 1).

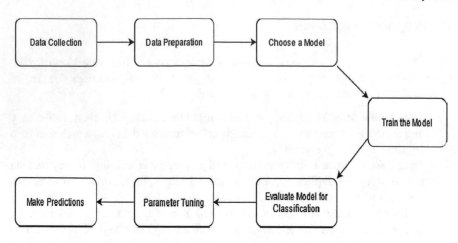

Fig. 1 The process flow of the application

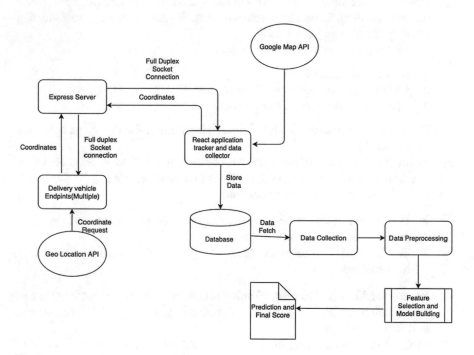

Fig. 2 Data flow diagram of the application

3.1 Data Flow Diagram for the Application

In the diagram given below, the components required, the technology used and the internal working of the application can be seen. It also describes the work flow of the application (Fig. 2).

3.2 Deep Learning Using LSTM Model

Deep learning is the subset of machine learning which uses the machine learning concepts based on human neural network. In a neural network, there are two fixed layers input layer and output layer and one is the hidden layer that may be there in some of the networks. LSTM model is an artificial recurrent neural network architecture [10]. LSTM model is mainly used for classifying, processing and making predictions based on the time series data. It has feedback connections instead of having feedforward connections. Apart from processing single data points, it is also capable of processing the entire sequence of data. It processes the data passing on the information after propagating it forward. It takes the input from input layer and process it in hidden layers and then give it to the output layer, by providing the continuous feedback from output layer to the input layer [11] (Fig. 3).

Equations Involved in the Process

$$f_t = \text{ag}(M_f \times \text{st} + U_f \times h_{t-1} + b_f) \tag{1}$$

Input Rate is calculated using Eq. (2),

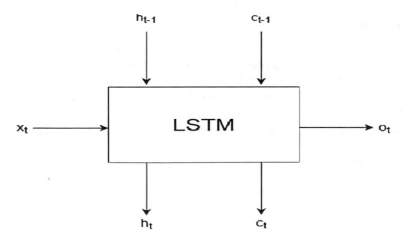

Fig. 3 LSTM inputs, outputs and corresponding equations for a single timestamp

$$i_t = ag(M_i \times st + U_i \times h_{t-1} + b_i) \tag{2}$$

Output is called as sigmoid, and can be calculated using Eq. (3)

$$o_t = ag(M_o \times st + U_o \times h_{t-1} + b_o) \tag{3}$$

Cell states are calculated using Eqs. (4) and (5),

$$c_t' = ac(M_c \times st + U_c \times h_{t-1} + b_c) \tag{4}$$

$$c_t = (ft + ct - 1 + it.cut) \tag{5}$$

The Eq. (6) calculates the value of hidden state.

$$h_t = (ot.ac(ct)) \tag{6}$$

σ_g = sigmoid σ_c = tanh f_t = function rate i_t = input rate c_t = cell state h_t = hidden state.

3.3 Advantages of Time Series Model

i.   Time Series Analysis: Helps You Identify Patterns.
ii. Time Series Analysis: Creates the Opportunity to Clean Your Data.
iii. Time Series Forecasting: Can Predict the Future.

4 Results

The data set is collected and plotted here for a month for each week and is shown here where on vertical axis there is Number of employees and on horizontal axis there is Time for 24 h. The data is collected for four weeks and predicted for fifth week (Fig. 4).

Then a graph is plotted to show the variation between Real time data and the predicted data. This is basically the prediction for first day. The model that is used is Moving Average Model (Fig. 5).

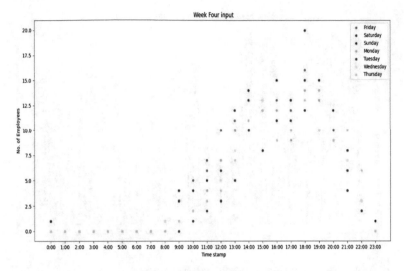

Fig. 4 Data set for each day of a particular week

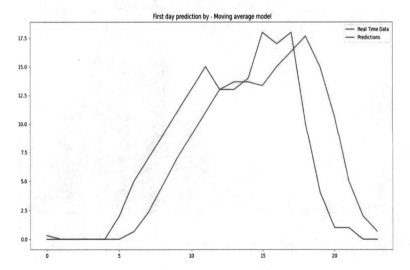

Fig. 5 First day prediction by moving average model

After plotting the data for one day, the analysis for full month is done by plotting the curves. The model used is Moving Average Model. Here the variations between real time data and the predicted data can be seen (Fig. 6).

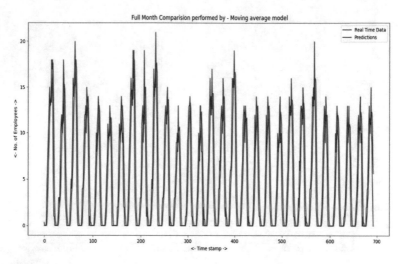

Fig. 6 First full month comparison by moving average model

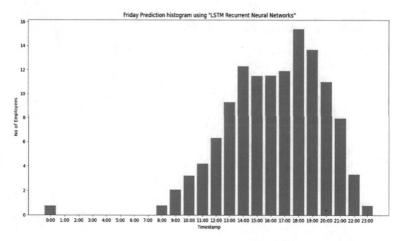

Fig. 7 First friday long short term memory prediction

To check the accuracy of the prediction on a specific day, another model under time series model is also used which is Long Short Term Memory Model, which is representing the prediction that, between which time duration, there should be lesser number of employees and between which particular duration, the number of employees needs to be more (Fig. 7).

To check the variation between the real time data and predicted data, the curves are plotted based upon Long Short Term Memory Model, and the results can be seen, that without using the application, how the organizations used to hire the employees, and after using this application, how it is going to benefit them (Fig. 8).

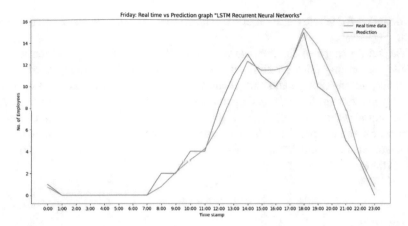

Fig. 8 Long short term memory model for real time data and predicted data on friday

5 Challenges, Conclusion and Future Scope

i. **Impact due to traffic**

The application fetches the co-ordinates of the delivery vehicle or delivery person. It is recording the co-ordinates at different span of time. If there is no change in the co-ordinates that is constant co-ordinates are there, then the inference that has been drawn is that there is no movement of the vehicle during that particular period of time. But it cannot be inferred that either the delivery has ended or the vehicle is stuck in traffic or traffic light is red.

ii. **Proper Internet connectivity or bandwidth**

As the application completely relies on the latitude and longitude and timestamps of the employees (delivery person), so a proper internet connection or proper bandwidth is required, otherwise the location and timestamps would be inaccurate.

iii. **National or Festival Holidays**

So far this application is concerned; it is not taking National and Festival holidays into consideration. So, the proper data for analysis and prediction would not be there, so this may compromise the utility of the application.

Suppose, there are many demands of a particular product during the evening time than morning time, but the employees are hired for the whole duration of the day irrespective of the number of demands at that time. This results in consumption of time as well as resources like remuneration given to those employees. This application can be the resource and time saver for the organizations in which delivery of the products is a major part. With the help of this application, those organizations can decide, when to hire more number of employees and when to hire relatively lesser number of employees who are intended for the delivery of products. So, using this application, the organization can earn a considerable amount of profit. The employees also are not bounded for whole duration of the day, so they can also do some other work as well.

The future scope of the application is equally good. In today's scenario itself, people are relying on the online sites for their needs like ordering food, buying groceries, buying clothes and gadgets and what not. The deliveries are done by the delivery persons. In future, this demand is only going to increase due to expansion of the online businesses. To cater all those demands there will be need of delivery persons. So the organizations have to use strategy to hire the employees in order to ensure customer satisfaction as well as to earn more profit, and this application fulfills that.

References

1. MDN Developers: Geolocation API Documentation. https://developer.mozilla.org/en-US/docs/Web/API/Geolocation_API
2. Tan CW, Bergmeir C, Petitjean F, Webb GI (2020) Time series regression
3. Subha K, Sujatha S (2016) Ongoing global positioning framework dependent on ARM7 GPS and GSM innovation, vol 6, Issue 5, May 2016. ISSN: 2277 128X
4. Charfaoui Y (2019) Latitude and Longitude information in my AI and ML
5. Developers: Location and Map. http://developer.android.com/guide/topics/location/index.html
6. Hamid FS (2020) Analyzing real-time objects on mobile telephone tower stations, April 2020
7. Charfaoui Y (2019) Working with Geospatial Data in Machine Learning, Google certified ML Engineer and Android Developer, 13 November 2019. https://heartbeat.fritz.ai/working-with-geospatial-data-in-machine-learning-ad4097c7228d
8. pyAF: Automatic Time Series Forecasting (2020). https://github.com/antoinecarme/pya
9. AutoTS: Model Selection for Multiple Time Series (2020). https://github.com/winedarksea/AutoTS
10. Bellec G, Salaj D, Subramoney A, Legenstein R, Maass W (2018) Long short-term memory and learning-to-learn in networks of spiking neurons. In: Advances in Neural Information Processing Systems, pp 787–797
11. Cen Z, Wang J (2019) Crude oil price prediction model with long short term memory deep learning based on prior knowledge data transfer. Energy 169:160–171

Enhanced Contrast Pattern Based Classifier for Handling Class Imbalance in Heterogeneous Multidomain Datasets of Alzheimer Disease Detection

C. Dhanusha, A. V. Senthil Kumar, and Lolit Villanueva

1 Introduction

A kind of chronic condition which leads to memory enervation because of brain cells deterioration is known as Alzheimer Disease (AD). It affects mainly the elderly persons and there is no treatment to stop or reverse its progression. In recent years there is a steady growth of people affected due to AD, as it is incurable, detection of alzheimer at their earlier stages with appropriate treatment can regulate the neurons degeneration [1]. Data mining is widely used in different medical areas. Advancement in technologics highly benefits the medical applications for improved data accessing and discovering symptoms for different disease in their earlier stages. cognitive mental issue like forgetfulness, anxiety, confusion are also other symptoms of Alzheimer's.

One of the most vital issue in both machine learning and data mining is for classification and pattern mining. The patterns can be used for either finding the potential features or it is used for generation rules. The classification model utilizes these features or else rules for produce accurate results in prediction process. Additionally, pattern mining in structured domains is considered as propositionalizing model which helps to conduct propositional machine learning and data mining methods. One of the desirable properties of classification is understandability, these kinds of classification model is used for solving real time applications [2]. Among the understandable classification methods, contrast pattern related classification is the best suitable model for decision making because it provides more accurate results while comparing other start of the art classifiers used for Alzheimer Disease detection.

C. Dhanusha (✉) · A. V. S. Kumar
Department of MCA, Hindusthan College of Arts and Science, Coimbatore, India
e-mail: kunjukannan020817@gmail.com

L. Villanueva
ECE Department, Xavier University, Cagayan de Oro City, Philippines

© The Author(s), under exclusive license to Springer Nature Singapore Pte Ltd. 2022
B. Unhelker et al. (eds.), *Applications of Artificial Intelligence and Machine Learning*,
Lecture Notes in Electrical Engineering 925,
https://doi.org/10.1007/978-981-19-4831-2_66

Contrast Pattern classifier is a patter which appears in potentially in a specific class corresponding to the other classes. There are three main reasons of low performance of the standard pattern-based classifier are listed below.

- Contrast patterns with minority class will have less support value while compared to majority class. Because these classifiers are related to the support of the patterns.
- Discriminative power of contrast patterns relies on large collection of patterns which reduce the classification efficiency.
- Each class score is identified by computing its normalization by finding its median and thus it is biased due to its score distribution.

2   Related Work

Jack Albright [3] developed a machine learning model to detect the alzheimer disease with multiclass labels by processing the dataset with 1737 patients by applying all pairs model. This work compares all the possible pairs of temporal data and to predict the alzheimer it uses artificial neural network to predict the progression of alzheimer.

Jyoti Islam and Yanqing [4] in their work deep learning model is designed to predict the presence of alzheimer in Open Access Series of Imaging Studies dataset. The multiclass detection is done using Brain MRI data to understand the depth knowledge using convolutional neural network.

Ramesh Paudel et al. [5] stated that using mining approaches and machine learning models for handling aging population related health issues can predict its symptoms at their earlier stages. The dataset collected from smart homes sensors to predict whether an elderly person suffers from cognitive impairment without distributing them by analyzing their daily tasks.

Robben and Krose [6] designed a prediction model to discover the behavior of the elderly persons to improve the safety and quality of the residents in their home environment. The functional health assessment is done on the elderly persons by collecting information about their longitudinal ambient monitored by sensors connected to the appliances in the home. Their motors skill and indoor activities which is known as high level features.

Suzuki et al. [7] performed the cognitive disability detection by analyzing the mini mental state examination and indoor activities. Both the daily activities examination and the clinical information about the persons are analyzed for detecting the alzheimer disease more prominently.

Dodge et al. [8] explored the relationship among gait parameters and reasoning parameter for performing alzheimer disease detection. They used latent trajectory approach for discovering the cognitive disabilities at their early stages.

3 Methodology: Handling Class Imbalance in Alzheimer Detection Using Enhanced Contrast Pattern Based Classifier

The main objective of this proposed work is to handle the problem of class imbalance for effective alzheimer disease detection. While handling the disease dataset it is very challenging to work with the imbalanced class labels, because all the standard classifiers is highly influenced by most frequently occurring classes. To overcome this issue, this research work developed an enhanced contrast pattern-based mining model which highly balances both the most frequently occurring classes and least occurring classes to improve the accuracy of alzheimer detection. The overall architecture of the proposed model is shown in the Fig. 1.

3.1 Dataset Description

This work used two different datasets namely CASAS and OASIS dataset. The CASAS dataset [9] comprised of information collected from the smart home sensors in order to analyse the day to day activities of elderly persons without interrupting their works. This dataset used single resident apartment equipped with 1BHK, sensors are fixed to ceilings, doors, light and cabinets as shown in the Fig. 2 and 3. The continuous data collected from the sensors is used in the data server. Second dataset is collected from Open Access Series of Imaging Studies (OASIS) [10], which comprised of MRI data information with 373 instances and 12 attributes with one class label.

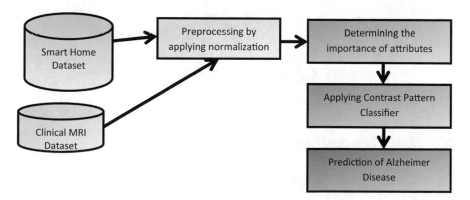

Fig. 1 Architecture of proposed enhanced contrast pattern-based classifier for Alzheimer disease detection

Fig. 2 Smart home equipped with sensor

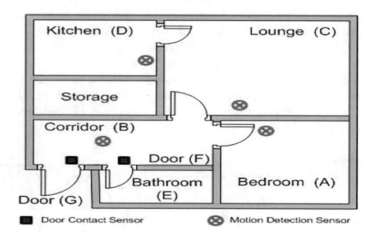

Fig. 3 CASAS-single resident apartment

3.2 Data Preprocessing

The two datasets CASAS and OASIS are preprocessed by applying min–max normalization to make the attributes values fall under same range, and it is done by applying the formula

$$M - M(X) = \frac{x - min}{max - min} \tag{1}$$

where min and max are the minimum value and maximum value of each attributes range, which is normalized between 0 to 1.

3.3 Mutual Information for Feature Elimination

The features which is not influencing or contributing in the process of classification are eliminated by applying mutual information-based feature elimination. This feature selection is used to calculate the expected mutual information of term t and class c. It reveals how much information the presence or absence of an attribute contributes to making the correct classification decision on the class label of alzheimer dataset. The mutual information of an attribute with the class label is formulated as [11]

$$I(U, C) =$$
$$\sum_{e_{att}\in\{1,0\}} \sum_{e_{cls}\in\{1,0\}} prob(U = e_{att}, C = e_{cls})log_2\frac{prob(U = e_{att}, C = e_{cls})}{prob(U = e_{att})prob(C = e_{cls})}$$

$$(2)$$

where U is a random variable, whose value will be $e_{att} = 1$ if the instance contains the attribute and $e_{att} = 0$ if the instance doesn't have the attribute. C is a random variable whose $e_{cls} = 1$ when the instance belongs to class c, else $e_{cls} = 0$ when it is not present. The pattern is a kind of expressing specific language which describes a collect of objects. A contrast pattern is a pattern which seems to appear more frequently in a class and infrequently in the remaining classes. It is essential to choose contrast patterns for pattern classifiers to overcome the problem of class imbalance. While using contrast patterns in extracts patterns with high support of majority class and low support for minority class. This results in a bias classification output towards majority class and if the classification model produces a greater number of patterns will increase the computational complexity. Contrast classification has three stages they are extracting patterns using mining, obtaining high quality patterns based on filtering and combining patterns to classify the contrast patterns [12].

To discover the possible combination of features many searching algorithms are available. While using only the categorical variables, it is easy to compute the support value in a straight manner. While a combination of different attributes is involved then quantitative contrast mining for for finding the contrast pattern classifier for alzheimer disease detection is developed in this work as shown in the Fig. 4.

This model discovers the itemsets that are contrasts among the groups and it must have the interest measure i.e. their support difference must be larger than the present minimum. The main idea starts with dividing discretize the continuous attribute to compute the interest measures based on top down approach and find out whether to further explore or to stop searching. Next, it merges identical continuous spaces in a bottom-up manner and refines the range.

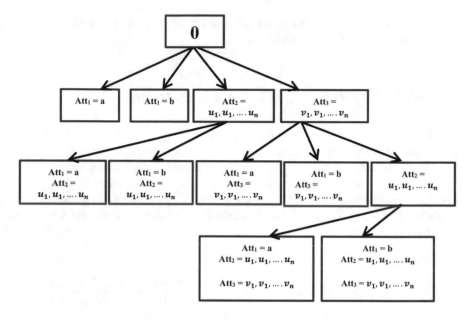

Fig. 4 Dataset with different data types

Handling Continuous Attributes Using Top Down Approach

A stepwise refinement approach which is known as Top- Down approach is used for handling the attributes with continuous values for contrast pattern-based classifier to detect the presence of alzheimer. This model breaks down the system to reveal the insight of the base element by breaking the whole dataset into smaller segments.

Assume the Alzheimer dataset AD which contains group of interest, ct refers to categorical itemset and cnt refers to continuous attributes to be explored. The two input parameters ⌊ and © is considered as the user inputs. Initially the interest measure of parents is assigned to 0. The Min-Sup is set to the present minimum support into top – k contrasts of the current list. If the list is null it means it doesn't have k contrasts then the Min-Sup is assigned as ©. The procedure starts by partition the continuous attribute in a top-down fashion by applying median. Next, it identifies the space combinations among continuous attributes. The number of spaces is computed by 2^{cnt}, these spaces defines the primary boundaries of bin.

The procedure continues while each space created by verifying the lookup table or calculation some information and storing it in lookup table to decide whether to prune space or not. If the space is found in the lookup table then it will be pruned. The support of the itemset in each group is computed and the interest measure is updated by finding the support difference. To decide whether to explore the space further for optimistic estimates for the child space. The optimistic is calculated as

$$Sup_1(ca_t) = \frac{count_1(ca_t)}{|i_1|} \tag{3}$$

where t is the current space, ca_t is the itemset found at t space and $count_k$ (ca_t) is the number of instances belongs to group k in space t. $|i_1|, |i_2| \dots |i_n|$ and the computation of group 1 is shown in the Eq. (1). Likewise, same formula is used to define the same itemset in for all n groups. The maximum instances child space is computed as follows

$$max\_insts\_child = \frac{|AD|}{2^{lvl+1} * |cnt|} \tag{4}$$

where max_insts_child refers to maximum child spaces instances for the current level lvl, and cnt refers to number of continuous attributes. Continuous attributes are divided at the median and thus it distributes child spaces in equal proportion to the data points. For itemset ca_t in group 1 its maximum support is formulated as

$$max\_sup\_grp1 = min\left(\frac{max\_insts\_child}{|grp_1|}, sup_1(ca_t)\right) \tag{5}$$

Here, maximum support conceivable in child space for group 1 is found, the median is computed corresponding to all the given instances and there is a possibility of the imbalance among groups. Dividing the space alone doesn't results in support reduction proportionally in all groups, if the possible instances in child space is larger than the number of instances in group 1, then first value inside the max function will be greater than 1. As this is not possible, in the second part the support is monotonically reduced as the space reduced and if the current space support is less than possible maximum support of child space, then current support will be the child space with maximum support. The same can be applied for other groups.

$$ot\_isnts\_grp1 = |AD| - cnt_1(ca_t) \tag{6}$$

where ot_isnts_grp1 refers to number of instances of the other groups other than group 1 in present space t.

$$min\_insts\_grp1 = max\_isnts\_child - ot\_insts\_grp1 \tag{7}$$

here the min_insts_grp1 value will be negative if the majority of the elements does not belong to grp1. The minimum support of group 1 is computed as follows

$$min\_sup\_grp1 = max\left(0, \frac{min\_insts\_grp1}{|grp_1|}\right) \qquad (8)$$

The optimistic child space estimate is formulated as shown below

$$\text{Opt - est} = max(\forall_i \forall_j, \, m\bar{n}, \, max\_sup\_grp_i - min\_sup\_grp_j) \quad i, j = 1, 2, \ldots n \qquad (9)$$

The child space will be recursively explored when the optimized estimate computed is greater than minimum support. If the child space has better contrast pattern, then it is added to the current list, else the present contrast pattern is huge and prominent then it is included in the current list of contrast CD. The present itemset is attached to CD if the interest measure is better than its parents. Else, until the spaces are discovered the algorithm will wait and adds it if at least the interest portion in one space is larger than its parents.

Once contrast spaces are discovered, the similar and continuous spaces are merged to get more common and details contrasts in the bottom up approach. To merge the partitions, the spaces are sorted in increasing order of size. The proposed model discovers fewer and more significant itemsets meanwhile there is more chance for merging smaller size itemsets.

The interest measure is computed by finding the different among support, the purity ratio is used to define the interest measure which explains the homogenous current region corresponding to the group. For finding the purity ration a value closer to 1 signifies that the current space comprised most instances from same group. Let I and j are the groups are contrasting, dz is the discretized quantitative attributes itemset, the support of group (i) which is denoted as sp_{idz} in dz itemset space and its purity ratio is defined as

$$PUR(c) = 1 - \frac{min(sp_{idz}, sp_{jdz})}{max(sp_{idz}, sp_{jdz})} \qquad (10)$$

The purity ratio doesn't consider size of the itemsets so to overcome this problem surprising measure is included in this contrast pattern classifier

$$SRM(dz) = PUR(dz) * Diff(dz) \qquad (11)$$

By finding the surprising measure the size of the contrast is also considered as a significant measure so that it gives equal importance to both groups.

3.4 Pruning Redundant Itemsets

The itemset is considered to be redundant when its support is equal to one of its subsets supports. The support difference of the redundant itemset is same as its ancestors this is because of missing value or typing mistakes. The attributes with high correlation are also supposed to have redundant contrasts. The testing dataset of alzheimer are considered as the population sample and statistical tests are essential to make the decision based on it. If any two itemsets have same difference in support then the central limit theorem is used. This is because the normal distribution is done for multiple population samples, the best rough calculation of the means of the support difference for the population is the variance in support in the present example.

To find the bounds of the difference dif_{bnd} among each subset of two groups x and y with the size of $|grp_x|$, $|grp_y|$ and their corresponding wsupport $sup_x(c)$ and $sup_y(c)$ long with current difference dif_{cur} and dif_{sbst} is computed as

$$h(c) = \frac{sup_x(c) * (1 - sup_x(c))}{|grp_x|} \quad l(c) = \frac{sup_y(c) * (1 - sup_y(c))}{|grp_y|} \tag{12}$$

$$dif_{bnd} = dif_{sbst} \pm \beta * \sqrt{h + l} \tag{13}$$

If dif_{cur} is within the boundary range dif_{bnd} then the current itemset and its subsets have same support and they won't produce any interesting pattern so it is pruned.

The detailed procedure of Enhanced Contrast Pattern Classifier for Alzheimer Disease Detection with Class imbalance.

Algorithm 1: ENHANCED CONTRAST PATTERN CLASSIFIER FOR ALZHEMIER DISEASE DETECTION

Input: **AD** dataset with group attribute, categorical items ct in
itemset, continuous attributes cnt, ™, ®, min_sup, measure of parent mp
Output: Alzheimer contrast patterns of set

Begin

$AD \leftarrow$ *Contrast itemset list (initially = 0)*

$AD_{temp} \leftarrow$ *itemsets list which may be contrast ($AD_{temp}=0$)*

%continous attribtues in alzheimer dataset is partitioned using median
pcnt = partition_continuous attributes(cnt)
% Discover possible all ranges of combination identified by pcnt
t = discover _comb(pcnt)
*for each space in t **do***

*if prune (ca_t) **then***

Itemset is added to the prune_list
Compute sp(ca_t) for each group
Compute user defined interest measure it_ms (ca_t)
*if Opt-est (t) > min_sup **then***
AD_{child} = *call ECPCAD ($AD_t, ct,,, min\_sup, it\_s$ (ca_t))*

*if AD_{child} > not empty **then***
Attach (AD, AD_{child})
else

*if ca_t is significant and large **then***
*if ca_t > mp **then***
Attach (AD,ca_t)
else

Attach (AD_{temp}, ca_t)

*if length (AD) > 0 **then***
Attach (AD, AD_{temp})
else

return []
*if level == 1 **then** (spaces are sorted in ascending order*
***FRIS** = SORT*
While** No other space left to combine in FRIS **do
//Verify two contiguous spaces if combination is Conceivable
*if Cmb_psble **then***
itemsets are combined;
contrast set is updated;
Return AD

4   Results and Discussions

This section discusses in detail about the performance analysis of the proposed Enhanced Contrast Pattern Based classifier (ECPB) for Alzheimer disease detection by handling the class imbalance problem very effectively. The proposed model ECPB is developed using Matlab software. The two different datasets used for Alzheimer disease detection is collected from CASAS which collects sensor data of daily activities about elderly persons in smart homes. The OASIS dataset consists of MRI information Alzheimer patients. The performance of ECPB is compared with Decision Tree (DT) and support vector machine [13].

This Fig. 5 depicts the performance of the three different classification models for detecting alzheimer disease based on two different datasets. The Activities of Daily living collected from the sensors of the smart home it is collected from CASAS dataset. The clinical information about alzheimer patients are used from OASIS dataset. The proposed enhanced contrast pattern-based classifier produces high precision rate compared to standard decision tree and support vector machine. Because ECPB handles imbalance of class in alzheimer dataset by focusing equally both instances with highest occurrence of class label and low occurrence of class label.

Detection of alzheimer at its earlier stages avoids further progression and serious cause to other cognitive disabilities, thus this proposed work highly concentrates on contrast pattern based alzheimer disease. By using three different sub process namely itemset with high and low support finding, performing interest measure and pruning redundant itemsets greatly influences recall rate of proposed ECPB while comparing with DT and SVM. The class imbalance problem which mainly occurs during testing phase is prominently tackled by ECPB and thus it produced highest recall rate as depicted in the Fig. 6.

The Fig. 7 explores that the accuracy obtained by the proposed model ECPB using two different datasets CASAS and OASIS. ECPB produced highest accuracy rate by finding contrast patterns in datasets with continuous as well as categorical attributes. By applying contrast pattern-based classifier avoids redundant itemsets and thus its works independently to maintain class imbalance in alzheimer dataset. The decision tree and support vector machine fail to handle instances with alzheimer symptom as

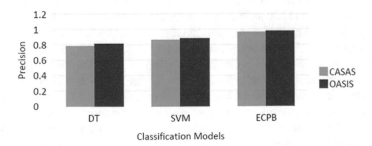

Fig. 5 Performance comparation based on Precision

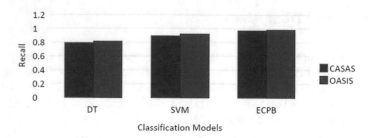

Fig. 6 Performance comparation based on recall

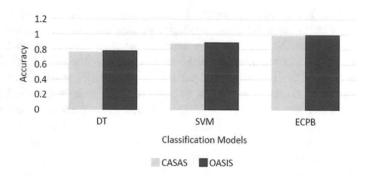

Fig. 7 Performance comparation based on accuracy

its proportion is very less and it doesn't influence accurate classification when there is very less instances with presence of alzheimer. Thus, the accuracy of both DT and SVM is less compared to the ECPB.

5 Conclusion

The main problem in accurate detection of Alzheimer disease detection is the class imbalance while performing classification. The class imbalance occurs when one of the two classes (i.e. in alzheimer disease presence of absence) having more instances than other classes. This problem makes the classifier to produce bias results towards the majority class and the accuracy will be greatly affected during prediction process. Hence, this work introduced a significant feature elimination model and contrast pattern based classifier to handle the presence of class imbalance in alzheimer data. The mutual information approach is used to discover the importance of each attribute and it eliminates the irrelevant attributes. With the significant attributes, the contrast pattern-based classifier is applied to detect the presence or absence of alzheimer. This works used two datasets which is collected from sensor data of smart home and MRI data both with continuous and categorical attributes. The contrast among the instances are discovered and redundant instances are removed to handle the class imbalance issue. The efficacy of the proposed ECPBC achieves better accuracy while comparing with the standard classifiers namely decision tree and support vector for early detection of alzheimer's in two different datasets.

References

1. Ouchi Y, Akanuma K, Meguro M, Kasai M, Ishii H, Meguro K (2012) Impaired instrumental activities of daily living affect conversion from mild cognitive impairment to dementia: the Osaki-Tajiri Project. Psychogeriatrics 12(1):34–42
2. Chaytor N, Schmitter-Edgecombe M, Burr R (2006) Improving the ecological validity of executive functioning assessment. Arch Clin Neuropsychol 21(3):217–27
3. Albright J (2019) Forecasting the progression of Alzheimer's disease using neural networks and a novel preprocessing algorithm Alzheimer's & Dementia. Transl Res Clin Interv 5:483–491
4. Islam J, Zhang Y (2017) A novel deep learning based multi-class classification method for Alzheimer's disease detection using brain MRI data. In: International Conference on Brain Informatics, pp 1–11
5. Paudel R, Dunn K, Eberle W, Chaung D: Cognitive health prediction on the elderly using sensor data in smart homes. In: The Thirty-First International Florida, Artificial Intelligence Research Society Conference (FLAIRS-31), pp 317–322
6. Robben S, Pol M, Krose B (2014) Longitudinal ambient sensor monitoring for functional health assessments. In: Proceedings of the 2014 ACM International Joint Conference on Pervasive and Ubiquitous Computing Adjunct Publication–UbiComp 2014 Adjunct. ACM Press, New York, New York, USA, Sep. 2014, pp 1209–1216
7. Suzuki T, Murase S (2010) Influence of outdoor activity and indoor activity on cognition decline: use of an infrared sensor to measure activity. Telemed J e-health J Am Telemed Assoc 16(6):686–690
8. Dodge HH, Mattek NC, Austin D, Hayes TL, Kaye JA (2012) In home walking speeds and variability trajectories associated with mild cognitive impairment. Neurology 78(24):1946–1952
9. Cook DJ, Crandall AS, Thomas BL, Krishnan NC. CASAS (2013) A smart home in a box. IEEE Computer
10. https://www.oasis-brains.org/
11. Hoque N, Bhattacharyya DK, Kalita JK (2014) MIFS-ND: a mutual information-based feature selection method. Expert Syst Appl 41(14):6371–6385
12. Zhu S, Ju M, Yu J, Cai B, Wang A (2015) A review of contrast pattern based data mining. In: Proceedings of the SPIE 9631, Seventh International Conference on Digital Image Processing (ICDIP 2015), p 96311U, 6 July 2015
13. Salas-Gonzalez D et al (2009) Computer aided diagnosis of Alzheimer disease using support vector machines and classification trees. In: Köppen M, Kasabov N, Coghill G (eds) Advances in Neuro-Information Processing. ICONIP 2008. LNCS, vol 5507. Springer, Berlin, Heidelberg. https://doi.org/10.1007/978-3-642-03040-6_51
14. Dhanusha C, Senthil Kumar AV (2019) Intelligent intuitionistic fuzzy with elephant swarm behaviour based rule pruning for early detection of Alzheimer in heterogeneous multidomain datasets. Int J Recent Technol Eng (IJRTE) 8(4):9291–9298. ISSN: 2277-3878
15. Dhanusha C, Senthil Kumar AV (2020) Enriched neutrosophic clustering with knowledge of chaotic crow search algorithm for Alzheimer detection in diverse multidomain environment. Int J Sci Technol Res (IJSTR) 9(4):474–481. Scopus Indexed, April 2020 Edition. ISSN:2277-8616
16. Dhanusha C, Senthil Kumar AV, Musirin IB (2020) Boosted model of LSTM-RNN for Alzheimer disease prediction at their early stages. Int J Adv Sci Technol 29(3):14097–14108

17. Dhanusha C, Senthil Kumar AV: Deep recurrent Q reinforcement learning model to predict the Alzheimer disease using smart home sensor data. In: International Conference on Computer Vision, High Performance Computing, Smart Devices and Network, Jawaharlal Nehru Technological University, Kakinada, Andhra Pradesh

18. Dhanusha C, Kumar AV, Musirin IB, Abdullah HM (2021) Chaotic chicken swarm optimization based deep adaptive clustering for Alzheimer disease detection. In: International Conference on Pervasive Computing and Social Networking ICPCSN 2021